Soil Fertility
and Fertilizers

FOURTH EDITION

Soil Fertility and Fertilizers

Samuel L. Tisdale
President (Retired), The Sulphur Institute
Formerly Professor of Soils
North Carolina State University

Werner L. Nelson
Senior Vice President, Potash & Phosphate Institute
Formerly Professor of Agronomy
North Carolina State University

James D. Beaton
Northwest Director, Potash & Phosphate Institute
Formerly Instructor of Soils
University of British Columbia

Macmillan Publishing Company   NEW YORK

Collier Macmillan Publishers   LONDON

Copyright © 1985, Macmillan Publishing Company, a division of Macmillan, Inc.

PRINTED IN THE UNITED STATES OF AMERICA

Earlier editions copyright © 1956, 1966, and 1975 by Macmillan Publishing Company

Macmillan Publishing Company
866 Third Avenue, New York, New York 10022

Collier Macmillan Canada, Inc.

Library of Congress Cataloging in Publication Data

Tisdale, Samuel L.
  Soil fertility and fertilizers.

  Includes bibliographies and index.
  1. Fertilizers. 2. Soil fertility. 3. Plants—
Nutrition. I. Nelson, Werner L. II. Beaton, James D.,
1930-   . III. Title.
S633.T66   1984   631.4'2   83-9902

ISBN 0-02-420830-2 (Hardcover Edition)
ISBN 0-02-946760-8 (International Edition)

Printing: 1 2 3 4 5 6 7 8   Year: 5 6 7 8 9 0 1 2 3

ISBN 0-02-420830-2

Preface

The purpose of the fourth edition of *Soil Fertility and Fertilizers* is the same as that of the first three: to present the fundamental principles of soil fertility and fertilizer manufacture and use in a manner suitable for students of agriculture at the junior and senior levels in college and graduate students. As with the first edition, this text will be of greatest value to those who have first completed a fairly comprehensive beginning course in soils. Adequate training in the fields of inorganic chemistry and crop science is also desirable.

Minor changes have been made in the sequence of chapters, and the subject matter in several chapters has been reorganized. There has been extensive updating of the material covered, with numerous references being made to the literature, particularly since the mid-1970s. Information has been drawn more extensively from many regions of the United States and Canada and elsewhere in the world.

More emphasis has been placed on describing for each nutrient its occurrence, forms, behavior in soils, and the factors influencing availability and uptake. Plant factors affecting nutrient needs and responses have received greater attention.

Chapter 7 is now devoted entirely to coverage of potassium. Sulfur is discussed in greater depth, together with calcium and magnesium, in Chapter 8. Treatment of the micronutrients has been expanded, as has that of the beneficial elements chlorine, sodium, silicon, selenium, and vanadium. Material related to commercial fertilizers has been reorganized and condensed into a single chapter, Chapter 10.

The role of aluminum in the development of soil acidity is stressed in Chapter 11, "Soil Acidity and Liming." This chapter also includes details on the latest ideas and developments in the liming of soils in the warm, humid regions of the United States, as well as in cool, humid areas.

Chapters 12 through 16 describe the latest ideas related to soil fertility evaluation, fertilizer application, the economics of plant nutrient use, and good soil management. Discussions on critical nutrient ranges, nutrient ratios, and the Diagnosis and Recommendation Integrated System (DRIS) have been added to Chapter 12, "Soil Fertility Evaluation."

Recent developments in conservation tillage and other aspects of soil management are covered in Chapter 14, "Cropping Systems and Soil Management." The resultant modifications in soil fertility management are also discussed in this chapter. In addition, the sections on the nature and use of animal wastes and sewage sludges and effluents have been updated and revised. A discussion on remote sensing in crop management has been added to this chapter.

A new Chapter 17, "Interactions of Plant Nutrients in a High-Yield Agriculture," focuses on the numerous interactions of plant nutrients with other crop management inputs. The impact of these interactions in modern high-yield agricultural systems is emphasized.

A summary and list of questions are included at the end of each chapter as was the case with the third edition.

The cost/return figures in this text are used largely for illustrative purposes. The world agricultural picture is in a state of flux brought about by a number of factors, and prices have fluctuated dramatically. The physical relations between plants and their environment, however, remain the same regardless of prices, but to determine whether a given practice is economically sound, current prices must be applied to the physical crop response data.

World demand for food and fiber has increased tremendously since 1965, and it remained strong throughout the decade of the 1970s. In spite of the slackening in this demand in the early 1980s, it is still expected to increase substantially to at least the end of this century. Accompanying this expanding need for food and fiber there has been a corresponding increase in the production and consumption of fertilizers.

Between 1972 and 1984 there were wide swings in the availability and cost of energy for the production of ammonia, the basis of almost all chemical nitrogen fertilizer. During the late 1970s it appeared that energy costs for this purpose would soon be so expensive that alternative methods of supplying nitrogen for plant growth would be needed. Among the alternatives were the greater use of sewage sludge, legume crops, and animal wastes, practices that were largely unattractive just a few years ago.

Although in the early 1980s energy supplies increased and fertilizer costs declined markedly in many areas, improved efficiency of all phases of crop production, including fertilization, has become of deep concern due to the combination of low crop prices and diminished world demand for agricultural products. Consequently, there is now an urgent need to achieve high yields through efficient crop production systems. Much improvement in efficiency can be gained by utilizing existing technology that costs little or nothing extra. Taking advantage of the numerous positive interactions that exist among the use of plant nutrients and other crop production inputs is one of the most promising ways for survival of farming enterprises in unfavorable economic times.

A discussion of the factors affecting agricultural prices and energy supplies is beyond the scope of a text such as this, but an understanding of the principles of soil fertility and plant growth set forth in this book can be of considerable help in meeting the production problems that confront the farmer in periods of either high production costs or low crop prices and restricted world demand. To manipulate input/output economics successfully in a fluid agricultural price situation, an understanding of the relation between plant growth and the factors affecting it, including soil fertility, is a necessity. It is that central theme around which this text has been built.

We have drawn liberally from the published work of our colleagues in North America and in other parts of the world. Many of the illustrations contained in previously published reports were supplied by the authors of these papers. To these people, too numerous to list here, we are especially grateful. Dr. T. L. Jackson and many of our other colleagues have made valuable suggestions for changes, and to them we extend our thanks.

We are grateful for the efforts of Dr. V. V. Rendig, Dr. M. E. Sumner, and Dr. G. E. Wilcox in reviewing this revised edition. The assistance of Dr. A. S. R. Juo and Dr. G. Uehara in the review of certain sections is also sincerely appreciated.

Special acknowledgment is made to the following publishers for their generosity in permitting the reproduction of tables and figures: The American Society of Agronomy, The American Chemical Society, Academic Press, Inc., The British

Society of Soil Science, The Canadian Societies of Plant and Soil Science, The David McKay Publishing Co., The Williams & Wilkins Co., John Wiley & Sons, The Soil Science Society of America, The Potash & Phosphate Institute, The Sulphur Institute, Rheinhold Publishing Co., Marcel Dekker, Inc., and Martinus Nijhoff.

Last but by no means least, our thanks to Mrs. Doris Beaton for her invaluable assistance in the typing and preparation of the manuscript. Our gratitude is also expressed to Mr. J. C. Blyth and Mrs. Jeanette Nelson for their work on the manuscript, to Mrs. Allyne Tisdale for assistance in the preparation of the index, and to Don and Andrea Hare for their help in obtaining reference material.

S. L. T.

W. L. N.

J. D. B.

Contents

Chapter 7

Soil and Fertilizer Potassium   249

Chapter 8

Soil and Fertilizer Sulfur, Calcium, and Magnesium   292

Chapter 9

Micronutrients and Other Beneficial Elements in Soils and Fertilizers   350

Chapter 16

Fertilizers and Efficient Use of Water   699

Chapter 17

Interactions of Plant Nutrients in a High-Yield Agriculture   720

Appendix

Common Conversions and Constants   733

Introduction: Fertilizers in a Changing World

Now more than ever the importance of an adequate supply of plant nutrients to ensure efficient crop production is being recognized. Growers are continually striving to overcome nutrient deficiencies as well as use improved management practices in order that yields may more nearly approach the genetic limit of crop plants. As a result of this effort, great progress in fertilizer technology and in the use of plant nutrients has been made in the United States, Canada, and in many other countries, especially since 1960. A wider understanding of plant and soil chemistry has led to improved fertilization and cultural practices. Improved technology has led to the production of more efficient forms of fertilizer.

Until about 1900 the demands for greater agricultural yields in the United States were met primarily by bringing new land into cultivation. Although some new land can still be made arable by irrigation, drainage, or clearing of forested areas, the relative increases in acreage of cropland by these means will be quite small. Actually, the annual loss of agricultural land to urban expansion, roads, and industry is about 700,000 acres per year. Similar conversion of farmland in Canada between 1961 and 1971 amounted to nearly 3 million acres, primarily from the most climatically favored areas. It is certain that any substantial improvement in agricultural production must come from larger yields on land already in cultivation. Higher yields per acre mean a higher net profit per acre and lower unit production costs.

The tremendous increase in plant-nutrient consumption from 1950 to 1982 indicates that the importance of fertilizers to crop production in the United States is being accepted. During this period the respective tonnage increases in the use of nitrogen, phosphorus, and potassium were as follows:

Years	Tons Used (USDA)		
	N	P_2O_5	K_2O
1949–1950	956,000	1,930,000	1,070,000
1964–1965	4,605,442	3,785,230	2,828,458
1972–1973	8,339,000	5,072,000	4,412,000
1981–1982	11,079,000	4,818,000	5,614,000

Canada, with a much smaller agricultural land base, has also shown a dramatic growth in usage of plant nutrients during the past 23 years:

| Years | Tons Used (Agriculture Canada) | | |
	N	P_2O_5	K_2O
1958–1959	62,100	145,000	88,000
1964–1965	170,700	266,400	122,800
1972–1973	409,600	415,200	190,700
1981–1982	949,800	583,800	355,000

In spite of the progress made, the rates of application in the United States and Canada, as shown in Table I-1, are much less than those of many countries. However, the use per capita in the United States is among the top five countries listed, while Canada ranks seventh. This high use of fertilizer in relation to population in North America helps to increase opportunities for export of agricultural products. In Europe the land is old in terms of agricultural use, and it must be maintained at a high level of production because of the high population density. In comparison with European soils, the soils of the United States are relatively new agriculturally and dependence is still placed on native fertility in some areas. Then, too, many millions of acres in this country are in dryland or desert areas, where water rather than plant nutrients is the limiting factor. The information given in Table I-1, however, is an indication of the importance of and the need for fertilizers in various parts of the world.

TABLE I-1. Fertilizer Consumption in Relation to Arable Land and Per Capita in Selected Countries

| | Kilograms of Fertilizer Consumed per Hectare of Arable Land | | | Per Capita Total |
	N	P_2O_5	K_2O	
United States	56.1	25.8	29.7	93.4
USSR	35.6	24.1	21.1	70.4
Belgium	220.6	116.6	161.7	42.8
Denmark	141.0	41.8	53.6	122.5
France	115.1	95.1	90.6	104.5
East Germany	149.3	77.2	98.6	97.8
West Germany	206.9	111.8	152.7	57.4
Greece	84.9	40.1	9.1	54.9
The Netherlands	560.7	96.2	131.8	48.0
United Kingdom	177.2	57.7	58.6	36.5
Canada	20.3	14.3	8.3	80.0
Brazil	14.6	32.1	21.1	34.3
Peru	24.2	5.0	3.3	6.3
Mexico	37.7	10.8	3.2	17.3
China	122.1	27.7	4.8	15.4
India	20.8	6.5	3.7	7.6
Japan	125.8	141.4	104.9	15.4
Turkey	22.2	17.3	1.7	25.9
Algeria	11.8	15.4	4.8	12.7
South Africa	34.4	33.3	10.3	36.1
Egypt	194.0	35.7	2.6	15.8
Zaire	0.6	0.3	0.3	0.3
Australia	5.6	19.2	2.9	84.8
New Zealand	46.4	761.6	209.7	148.7
World	41.5	21.7	16.7	26.2

Source: Fertilizer Yearbook, Vol. 31, Rome: Food and Agriculture Organization of the United Nations, 1981

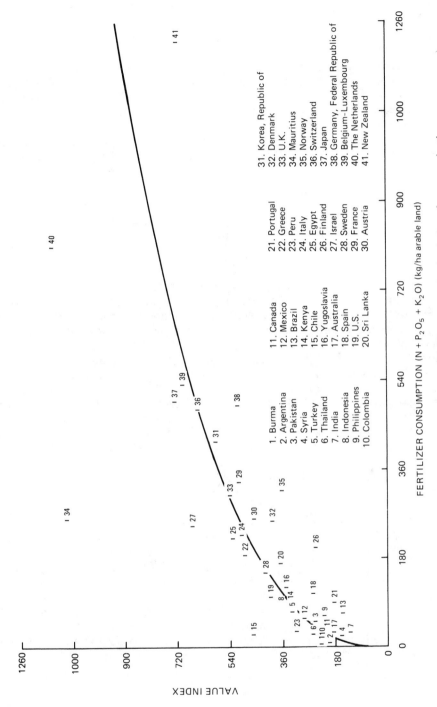

FIGURE I-1. Average relationship between fertilizer use and value index of crop production per arable hectare in 41 countries, 1979–81 (Courtesy of John Couston, Food and Agricultural Organization of the United Nations, Rome, 1982).

World use of plant nutrients is spreading rapidly, but future requirements, as estimated by the Food and Agricultural Organization of the United Nations, Tennessee Valley Authority, and other organizations, will be about 50% greater in 1990 than in 1982.

The relation between fertilizer use and the value index of crop production per arable acre for 41 countries is shown in Figure I-1. Fertilizers are more than just an index to modern agricultural methods; they are also a powerful factor that motivates the improvement of other cultural practices. The whole of the increase in yields shown in Figure I-1 is, of course, not the result of the use of fertilizers, but these increases do indicate what can be achieved when fertilizers are adapted to improved cropping practices.

It is of interest that the lower-use countries have made considerable strides in fertilizer use and value index. FAO reports the following $N/P_2O_5/K_2O$ ratios in fertilizer use:

	N	P_2O_5	K_2O
Developed countries	100	62	57
Developing countries	100	38	16

A significant development in the 1940s in the United States was the increase in use of liming materials. The application of limestone reached a peak during 1946 and 1947, but since that time usage has declined. This is unfortunate, for greater use of nitrogen and more intensive cropping add to the need for lime. It has been estimated that in the United States more than 94 million tons of limestone are required annually, but only 34 million tons are being applied. The effect of limestone is both direct and indirect, and its use on acid soils is essential if maximum returns are to be obtained from fertilization.

The gradually increasing yields of the principal crops in the United States (see the table) are the result of improved varieties, cultural practices, and pest control as well as the use of fertilizers. However, the actual fertility of some of our soils is decreasing, for greater quantities of plant nutrients are being removed than are being added.

Year	Corn (bu.)	Wheat (bu.)	Soybeans (bu.)	Cotton (lb. lint)	Alfalfa (tons)
1950	37.6	14.3	21.7	269	2.1
1964	62.1	26.2	22.8	524	2.4
1972	96.9	32.7	28.0	495	2.9
1982	114.4	35.6	32.4	613	3.4

It is estimated that two-thirds of the 4.8 billion people in the world exist on an inadequate diet. This condition, coupled with the prediction that the population will reach 6 to 7 billion by the year 2000, presents an imposing challenge to the world's producers of food and fiber.

Chapter 1

Soil Fertility—Past and Present

$\mathbf{F}$OOD is a basic necessity for human beings, and during most of our existence on this planet it has been procured by hunting and gathering. The various systems of agriculture that have provided food in the past and the estimated global population each system was capable of sustaining are summarized in Table 1-1. All of these systems exist in various parts of the globe today since the entire world has not advanced in unison from one system to the next. Even at present there is starvation in large areas where the systems for producing and transporting food are inadequate. On the other hand, developed countries utilizing fertilizers and pesticides are, when taken as a whole, self-sufficient.

By 1975 the world's population had reached 4 billion, close to the value of 4.2 billion appearing in Table 1-1. With global population anticipated to reach 6.2 billion by the year 2000 and to stabilize at 11 billion in 2075, it is obvious that large and sustained increases in food production will be needed to nourish adequately populations of this magnitude. Since cultivated land areas are expected to expand by only about 20%, there will be an absolute need to intensify agricultural production, in which fertilizers will play a vital role.

Because of the importance of soil fertility in food production it is fitting to trace the highlights in our understanding and development of sound soil fertility practices. The period in the development of the human race during which we began the cultivation of plants marks the dawn of agriculture. The exact time, of course, is not known, but it was certainly several thousand years before the birth of Christ. Until then man hunted almost exclusively for his food and was nomadic in his habits.

Ancient Records

As the years went by man became less of a wanderer and more of a settler. Families, clans, and villages developed, and with them came the development of the skill we call agriculture. It is generally agreed that one area in the world that shows evidence of a very early civilization is Mesopotamia, situated between the Tigris and Euphrates rivers in what is now Iraq. Writings dating back to 2500 B.C. mention the fertility of the land. It is recorded that the yield of barley was 86-fold and even 300-fold in some areas, which means, of course, that for every unit of seed planted 86 to 300 units were harvested.

Herodotus, the Greek historian, reporting on his travels through Mesopotamia

TABLE 1-1. Capability of Agricultural Systems to Produce Food and Support Population

Agricultural System	Cultural Stage or Time	Cereal Yield (t/ha)	World Population (millions)	Hectares per Person
Hunting and gathering	Paleolithic		7	
Shifting agriculture	Neolithic (10,000 years ago)	1	35	40.0
Medieval rotation	500–1450 A.D.	1	900	1.5
Livestock farming	Late 1700s	2	1800	0.7
Fertilizers/pesticides	Twentieth century	4	4200	0.3

Source: McCloud, *Agron. J.,* **67:**1 (1975).

some 2,000 years later, mentions the phenomenal yields obtained by the inhabitants of this land. The high production was probably the result of a well-developed irrigation system and soil of high fertility, attributable in part to annual flooding by the river. Theophrastus, writing around 300 B.C., referred to the richness of the Tigris alluvium and stated that the water was allowed to remain on the land as long as possible so that a large amount of silt might be deposited.

In time man learned that certain soils would fail to produce satisfactory yields when cropped continuously. The practice of adding animal and vegetable manures to the soil to restore fertility probably developed from such observations, but how or when fertilization actually began is not known. Greek mythology, however, offers one picturesque explanation: Augeas, a legendary king of Elis, was famous for his stable, which contained 3000 oxen. This stable had not been cleaned for 30 years. King Augeas contracted with Hercules to clean the stable out and agreed to give him 10% of the cattle in return. Hercules is said to have accomplished this task by turning the River Alpheus through the stable, thus carrying away the accumulated filth and presumably depositing it on the adjacent land. Augeas then refused payment for this service, whereupon a war ensued, and Augeas was put to death by Hercules.

In the Greek epic poem *The Odyssey*, attributed to the blind poet Homer, who is thought to have lived between 900 and 700 B.C., the manuring of vineyards by the father of Odysseus is mentioned. The manure heap, which would suggest its systematic collection and storage, is also referred to. Argos, the faithful hound of Odysseus, was described as lying on such a heap when his master returned after an absence of 20 years. Having recognized his master, Argos wagged his tail feebly and "went down into the blackness of death." These writings suggest that manuring was an agricultural practice in Greece nine centuries before the birth of Christ.

Theophrastus (372–287 B.C.) recommended the abundant manuring of thin soils but suggested that rich soils be manured sparingly. He also endorsed a practice considered good today—the use of bedding in the stall. He pointed out that this would conserve the urine and bulk and that the humus value of the manure would be increased. It is interesting to note that Theophrastus suggested that plants with high nutrient requirements also had a high water requirement.

The truck gardens and olive groves around Athens were enriched by sewage from the city. A canal system was used, and there is evidence of a device for regulating the flow. It is believed that the sewage was sold to farmers. The ancients also fertilized their vineyards and groves with water that contained dissolved manure.

Manures were classified according to their richness or concentration. Theophrastus, for example, listed them in the following order of decreasing value: human, swine, goat, sheep, cow, oxen, and horse. Later, Varro, an early writer on Roman agriculture, developed a similar list but rated bird and fowl manure as superior to human excrement. Columella recommended the feeding of snail clover to cattle because he felt that it enriched the excrement.

Not only did the ancients recognize the merits of manure, but they also observed the effect that dead bodies had on increasing the growth of crops. Archilochus made such an observation around 700 B.C., and the Old Testament records are even earlier. In Deuteronomy it is directed that the blood of animals should be poured on the ground. The increase in the fertility of land that has received the bodies of the dead has been acknowledged down through the years, but probably most poetically by Omar Khayyam, the astronomer-poet of Persia, who around the end of the eleventh century wrote

> I sometime think that never blows so red
> The rose as where some buried Caesar bled;
> That every hyacinth the garden wears
> Dropt in her lap from some once lovely head.
>
> And this delightful herb whose tender green
> Fledges the rivers lip on which we lean—
> Ah, lean upon it lightly! for who knows
> From what once lovely lip it springs unseen.

The value of green-manure crops, particularly legumes, was also soon recognized. Theophrastus noted that a bean crop (*Vicia faba*) was plowed under by the farmers of Thessaly and Macedonia. He observed that even when thickly sown and large amounts of seed were produced the crop enriched the soil.

Virgil (70–19 B.C.) advocated the application of legumes, as indicated in the following passage:

> Or, changing the season, you will sow there yellow wheat, whence before you have taken up the joyful pulse, with rustling pods, or the vetch's slender offspring and the bitter lupine's brittle stalks, and rustling grove.

The use of what might now be called mineral fertilizers or soil amendments was not entirely unknown to the ancients. Theophrastus suggested the mixing of different soils as a means of "remedying defects and adding heart to the soil." This practice may have been beneficial from several standpoints. The addition of fertile soil to infertile soil could lead to increased fertility, and the practice of mixing one soil with another may have provided better inoculation of legume seed on some fields. Again, the mixing of coarse-textured soils with those of fine texture, or vice versa, may have caused an improvement in the water and air relations in the soils of the fields so treated.

The value of marl was also recognized. The early dwellers of Aegina dug up marl and applied it to their land. The Romans, who learned this practice from the Greeks and Gauls, even classified the various liming materials and recommended that one type be applied to grain and another to meadow. Pliny (A.D. 62–113) stated that lime should be spread thinly on the ground and that one treatment was "sufficient for many years, though not 50." Columella also recommended the spreading of marl on a gravelly soil and the mixing of gravel with a dense calcareous soil.

The Bible records the value of wood ashes in its reference to the burning of briars and bushes by the Jews, and Xenophon and Virgil both report the burning of stubble to clear fields and destroy weeds. Cato advised the vine keeper to burn prunings on the spot and to plow in the ashes to enrich the soil. Pliny states that the use of lime from lime kilns was excellent for olive groves, and some farmers burned manure and applied the ashes to their fields. Columella also suggested the spreading of ashes or lime on lowland soils to destroy acidity.

Saltpeter, or potassium nitrate, was mentioned both by Theophrastus and Pliny as useful for fertilizing plants and is referred to in the Bible in the book of Luke. Brine was mentioned by Theophrastus. Apparently recognizing that palm trees required large quantities of salt, early farmers poured brine around the roots of their trees.

Virgil wrote on the soil characteristic known today as bulk density. His advice on determining this property was

> ... first, you shall mark out a place with your eye, and order a pit to be sunk deep in solid ground, and again return all the mold into its place, and level with your feet the sands at top. If they prove deficient, the soil is loose and more fit for cattle and bounteous vines; but if they deny the possibility of returning to their places, and there be an over plus of mold after the pit is filled up, it is a dense soil; expect reluctant clods and stiff ridges, and give the first plowing to the land with sturdy bullocks.

Virgil describes another means which might today be considered the prototype of a chemical soil test.

> But saltish ground, and that which is accounted bitter, where corn can never thrive, will give proof of this effect. Snatch from the smoky roofs bushels of close woven twigs and the strainers of the wine press. Hither let some of the vicious mold, and sweet water from the spring, be pressed brimful; be sure that all the water will strain out and bid drops pass through the twigs. But the taste will clearly make discovery; and in its bitterness will distort the wry faces of the tasters with the sensation.

Columella also suggested a taste test to measure the degree of acidity and salinity of soils, and Pliny stated that the bitterness of soils might be detected by the presence of black and underground herbs.

Pliny wrote that "among the proofs of the goodness of soil is the comparative thickness of the stem in corn," and Columella stated simply that the best test for the suitability of land for a given crop was whether it would grow.

Many of the early writers (and, for that matter, many people today) believed that the color of the soil was a criterion of its fertility. The general idea was that black soils were fertile and light or gray soils infertile. Columella disagreed with this viewpoint, pointing to the infertility of the black marshland soils and the high fertility of the light-colored soils of Libya. He felt that such factors as structure, texture, and acidity were far better guides to an estimation of soil fertility.

The age of the Greeks from perhaps 800 to 200 B.C. was indeed a Golden Age. Many of the people of this period reflected genius that was unequaled, or at least not permitted expression, for centuries to come. Their writings, their culture, their agriculture were copied by the Romans, and the philosophy of many of the Greeks of this period dominated human thinking for more than 2000 years.

Soil Fertility During the First Eighteen Centuries A.D.

After the decline of Rome there were few contributions to the development of agriculture until the publication of *Opus ruralium commodorum*, a collection of local agricultural practices, by Pietro de Crescenzi (1230–1307). De Crescenzi is referred to by some as the founder of modern agronomy, but his manuscript seems to be confined to the work of writers from the time of Homer. His contribution was largely that of summarizing the material. He did, however, suggest an increase in the rate of manuring over that in use at the time.

After the appearance of de Crescenzi's work little was added to agricultural knowledge for many years, although Palissy in 1563 is credited with the observation that the ash content of plants represented the material they had removed from the soil.

Around the beginning of the seventeenth century Francis Bacon (1561–1624) suggested that the principal nourishment of plants was water. He believed that the main purpose of the soil was to keep the plants erect and to protect them from heat and cold and that each plant drew from the soil a substance unique for its own particular nourishment. Bacon maintained further that continued production of the same type of plant on a soil would impoverish it for that particular species.

During this same period Jan Baptiste van Helmont (1577–1644), a Flemish physician and chemist, reported the results of an experiment which he believed proved that water was the sole nutrient of plants. He placed 200 lb of soil into an earthen container, moistened the soil, and planted a willow shoot weighing 5 lb. He carefully shielded the soil in the crock from dust, and only rain or distilled water was added. After a period of five years, van Helmont terminated the experiment. The tree weighed 169 lb and about 3 oz. He could account for all but about 2 oz of the 200 lb of soil originally used. Because he had added only water, his conclusion was that water was the sole nutrient of the plant, for he attributed the loss of the 2 oz of soil to experimental error.

Van Helmont's work and his erroneous conclusions were actually valuable contributions to our knowledge, for even though they were wrong, his conclusions stimulated later investigations, the results of which have led to a much better understanding of plant nutrition.

The work of van Helmont was repeated several years later by no less a figure that Robert Boyle (1627–1691) of England. Boyle is probably best known for expressing the relation of the volume of a gas to its pressure. He was also interested in biology and a great believer in the experimental approach to the solution of problems dealing with science. He believed that observation was the only road to truth. Boyle confirmed the findings of van Helmont, but he went one step further. As a result of the chemical analyses he performed on plant samples, he stated that plants contained salts, spirits, earth, and oil, all of which were formed from water.

About this same time J. R. Glauber (1604–1668), a German chemist, suggested that saltpeter (KNO_3) and not water was the "principle of vegetation." He collected the salt from soil under the pens of cattle and argued that it must have come from the droppings of these animals. He further stated that because the animals ate forage the saltpeter must have come originally from the plants. When he applied this salt to plants and observed the large increases in growth it produced, he was convinced that soil fertility and the value of manure were due entirely to saltpeter.

John Mayow (1643–1679), an English chemist, supported the viewpoint of Glauber. Mayow estimated the quantities of niter in the soil at various times during the year and found it in its greatest concentration in the spring. Failing to find any during the summer, he concluded that the saltpeter had been absorbed, or sucked up, as he put it, by the plant during its period of rapid growth.

About 1700, however, a study was made which was outstanding and which represented a considerable advance in the progress of agricultural science. An Englishman by the name of John Woodward, who was acquainted with the work of Boyle and van Helmont, grew spearmint in samples of water he had obtained from various sources: rainwater, river water, sewage water, and sewage water plus garden mold. He carefully measured the quantity of water transpired by the plants and recorded the weight of the plants at the beginning and end of the experiment. He found that the growth of the spearmint was proportional to the amount of impurities in the water and concluded that terrestrial matter, or earth, rather than water, was the principle of vegetation. Although his conclusion is not correct in its entirety, it represents an advance in knowledge, and his experimental technique was considerably better than any that had been used before.

There was much understandable ignorance concerning the nutrition of plants during this period. Many quaint ideas came into being and "bode their hour or two and went their way." Not the least of these ideas were introduced by another enterprising Englishman, Jethro Tull (1674–1741). Tull was educated at Oxford, not a common event for a person of agricultural leanings. He appears to have been interested in politics, but ill health forced his retirement to the farm. There he carried out numerous experiments, most of which dealt with cultural practices. He believed that the soil should be finely pulverized to provide the "proper pabulum" for the growing plant. According to Tull, the soil particles were actually ingested through openings in the plant roots. The pressure caused by the swelling of the growing roots was thought to force this finely divided soil into "the lacteal mouths of the roots," after which it entered the "circulatory system" of the plant.

Tull's ideas about plant nutrition were, to say the least, a bit odd. However, his experiments led to the development of two valuable pieces of farm equipment, the drill and the horse-drawn cultivator. His book *Horse Hoeing Husbandry*, was long considered an authoritative text in English agricultural circles.

Around 1762, John Wynn Baker, a Tull adherent, established an experimental farm in England, the purpose of which was the public exhibition of the results of experiments in agriculture. Baker's work was praised later by Arthur Young, who, however, admonished his readers to beware of giving too much credit to calculations based on the results of only a few years' work, an admonition that is as timely today as it was when originally made.

One of the more famous of the eighteenth-century English agriculturists was Arthur Young (1741–1820). Young conducted pot tests to find those substances that would improve the yield of crops. He grew barley in sand to which he added such materials as charcoal, train oil, poultry dung, spirits of wine, niter, gunpowder, pitch, oyster shells, and numerous other materials. Some of the materials produced plant growth, others did not. Young, a prolific writer, published a work entitled *Annals of Agriculture*, in 46 volumes, which was highly regarded and made a considerable impact on English agriculture.

Many of the agricultural writings of the seventeenth and eighteenth centuries reflected the idea that plants were composed of one substance, and most of the workers during this period were searching for this *principle of vegetation*. Around 1775, however, Francis Home stated that there was not only one principle but probably many, among which he included air, water, earth, salts, oil, and fire in a fixed state. Home felt that the problems of agriculture were essentially those of the nutrition of plants. He carried out pot experiments to measure the effects of different substances on plant growth and made chemical analyses of plant materials. His work was considered to be a valuable stepping-stone in the progress of scientific agriculture.

The discovery of oxygen by Priestley was the keystone to a number of other discoveries that went far toward unlocking the mystery of plant life. Jan Ingenhousz (1730–1799) showed that the purification of air took place in the presence of light, but in the dark the air was not purified. Coupled with this discovery was the statement by Jean Senebier (1742–1809), a Swiss natural philosopher and historian, that the increase in the weight of van Helmont's willow tree was the result of air!

Progress During the Nineteenth Century

These discoveries stimulated the thinking of Theodore de Saussure, whose father was acquainted with the work of Senebier. He attacked two of the problems on which Senebier had worked—the effect of air on plants and the origin of salts in plants. As a result, de Saussure was able to demonstrate that plants absorbed oxygen and liberated carbon dioxide, the central theme of respiration. In addition, he found that plants would absorb carbon dioxide with the release of oxygen in the presence of light. If, however, plants were kept in an environment free of carbon dioxide, they died.

DeSaussure concluded that the soil furnishes only a small fraction of the nutrients needed by plants, but he demonstrated that it does supply both ash and nitrogen. He effectively dispelled the idea that plants spontaneously generate potash and stated further that the plant root does not behave as a mere filter. Rather, the membranes are selectively permeable, allowing for a more rapid entrance of water than of salts. He also showed the differential absorption of salts and the inconstancy of plant composition, which varies with the nature of the soil and the age of the plant.

De Saussure's conclusion that the carbon contained by plants was derived from the air did not meet with immediate acceptance by his colleagues. No less a figure than Sir Humphry Davy, who published his work *The Elements of Agricultural Chemistry* about 1813, stated that although some plants may have received their carbon from the air, the major portion was taken in through the roots. Davy was so enthusiastic in this belief that he recommended the use of oil as a fertilizer because of its carbon and hydrogen content.

The middle of the nineteenth to the beginning of the twentieth century was a time during which much progress was made in the understanding of plant nutrition and crop fertilization. Among the men of this period whose contributions loom large was Jean Baptiste Boussingault (1802–1882), a widely traveled French chemist who established a farm in Alsace on which he carried out field-plot experiments. Boussingault employed the careful techniques of de Saussure in weighing and analyzing the manures he added to his plots and the crops he harvested. He maintained a balance sheet which showed how much of the various plant-nutrient elements came from rain, soil, and air, analyzed the composition of his crops during various stages of growth, and determined that the best rotation was that which produced the largest amount of organic matter in addition to that added in the manure. Boussingault is called by some the father of the field-plot method of experimentation.

Justus von Liebig (1803–1873), a German chemist, very effectively deposed the humus myth. The presentation of his paper at a prominent scientific meeting jarred the conservative thinkers of the day to such an extent that only a few scientists since that time have dared to suggest that the carbon contained in plants comes from any source other than carbon dioxide. Liebig made the following statements:

1. Most of the carbon in plants comes from the carbon dioxide of the atmosphere.
2. Hydrogen and oxygen come from water.
3. The alkaline metals are needed for the neutralization of acids formed by plants as a result of their metabolic activities.
4. Phosphates are necessary for seed formation.
5. Plants absorb everything indiscriminately from the soil but excrete from their roots those materials that are nonessential.

Not all of Liebig's ideas, of course, were correct. He thought that acetic acid was excreted by the roots. He also believed that the NH_4^+ form of nitrogen was the one absorbed and that plants might obtain this compound from soil, manure, or air.

Liebig firmly believed that by analyzing the plant and studying the elements it contained, one could formulate a set of fertilizer recommendations based on these analyses. It was also his opinion that the growth of plants was proportional to the amount of mineral substances available in the fertilizer.

The *law of the minimum* stated by Liebig in 1862 is a simple but logical guide for predicting crop response to fertilization. This law states that

> every field contains a maximum of one or more and a minimum of one or more nutrients. With this minimum, be it lime, potash, nitrogen, phosphoric acid, magnesia or any other nutrient, the yields stand in direct relation. It is the

> factor that governs and controls . . . yields. Should this minimum be lime. . .
> yield . . . will remain the same and be no greater even though the amount of
> potash, silica, phosphoric acid, etc. . . . be increased a hundred fold.

Liebig's law of the minimum dominated the thinking of agricultural workers for
a long time thereafter and it has been of universal importance in soil fertility
management.

Liebig manufactured a fertilizer based on his ideas of plant nutrition. The formulation of the mixture was perfectly sound, but he made the mistake of fusing
the phosphate and potash salts with lime. As a result, the fertilizer was a complete
failure. Nonetheless, the contributions that Liebig made to the advancement of
agriculture were monumental, and he is perhaps quite rightly recognized as the
father of agricultural chemistry.

Following on the heels of Liebig's now famous paper was the establishment in
1843 of an agricultural experiment station at Rothamsted, England. The founders
of this institution were J. B. Lawes and J. H. Gilbert. Work here was conducted
along the same lines as that carried out earlier by Boussingault in France.

Lawes and Gilbert did not believe that all of the maxims set down by Liebig
were correct. Twelve years after the founding of the station they settled the following points:

1. Crops require both phosphorus and potash, but the composition of the plant
   ash is no measurement of the amounts of these constituents required by the
   plant.
2. Nonlegume crops require a supply of nitrogen. Without this element, no growth
   will be obtained, regardless of the quantities of phosphorus and potassium
   present. The amount of ammonia nitrogen contributed by the atmosphere is
   insufficient for the needs of crops.
3. Soil fertility could be maintained for some years by means of chemical fertilizers.
4. The beneficial effect of fallow lies in the increase in the availability of nitrogen
   compounds in the soil.

The problem of soil and plant nitrogen remained unsolved. Several workers had
observed the unusual behavior of legumes. In some instances they grew well in
the absence of added nitrogen, whereas in others no growth was obtained. Nonlegumes, on the other hand, always failed to grow when there was insufficient
nitrogen in the soil.

In 1878 some light was thrown on the confusion by two French bacteriologists,
Theodore Schloessing and Alfred Müntz. These scientists purified sewage water
by passing it through a filter made of sand and limestone. They analyzed the
filtrate periodically, and for 28 days only ammonia nitrogen was detected. At the
end of this time nitrates began to appear in the filtrate. Schloessing and Müntz
found that the production of nitrates could be stopped by adding chloroform and
that it could be started again by adding a little fresh sewage water. They concluded
that nitrification was the result of bacterial action.

The results of these experiments were applied to the soils by Robert Warrington
of England. He showed that nitrification could be stopped by carbon disulfide and
chloroform and that it could be started again by adding a small amount of un-
sterilized soil. He also demonstrated that the reaction was a two-step phenomenon,
the ammonia first being converted to nitrites and the nitrites subsequently to
nitrates.

Warrington was unable to isolate the organisms responsible for nitrification.
This task remained for S. Winogradsky, who effected the isolation by the use of a
silica-gel plate, rather than the usual agar medium, because the organisms are
autotrophic and obtain their carbon from the carbon dioxide of the atmosphere.

As to the erratic behavior of legume plants with respect to nitrogen, two Germans, Hellriegel and Wilfarth, in 1886 concluded that bacteria must be present in the nodules attached to legume roots. Further, these organisms were believed to assimilate gaseous nitrogen from the atmosphere and to convert it to a form that could be used by higher plants. This was the first specific information regarding nitrogen fixation by legumes. Hellriegel and Wilfarth based their arguments on observations made in certain of their experiments. They did not, however, isolate the responsible organisms. This was later done by M. W. Beijerinck, who called the organism *Bacillus radicicola*.

The Development of Soil Fertility in the United States

Although most of the advances made in agriculture during the eighteenth century were accomplished on the Continent, a few early American contributions were sufficiently significant to mention. In 1733, James E. Oglethorpe established an experimental garden on the bluffs of the Savannah River, which is the present site of the city of Savannah, Georgia. This garden was devoted to the production of exotic food crops and is said to have been a place of beauty while it was maintained. Interest in it was lost, however, and it soon ceased to exist, but because it was largely the result of British interests it probably cannot be truly considered an "American" undertaking.

Benjamin Franklin demonstrated the value of gypsum. On a prominent hillside he applied gypsum to the land in a pattern which outlined the words, "This land has been plastered." The increased growth of pasture in the area to which the gypsum had been applied served as an effective demonstration of its fertilizer value.

In 1785 a society was formed in South Carolina which had among its objectives the setting up of an experimental farm. Eleven years later President Washington, in his annual message to the Congress, pleaded for the establishment of a national board of agriculture. Some of the most important contributions to early American agriculture were made by Edmond Ruffin of Virginia from about 1825 to 1845. He is believed to have been one of the first to use lime on humid-region soils to replace fertility elements lost by crop removal and leaching. Ruffin was a careful observer, a studious reader, and possessed of a keen and inquiring mind. Although his use of lime to bolster the yields of crops was known to the ancients, it was apparently a new experience in a new America.

It was not until 1862 that the Department of Agriculture was established, and, in the same year, the Morrill Act provided for state colleges of agriculture and mechanical arts. The first organized agricultural experiment station, set up in 1875 at Middletown, Connecticut, was supported by state funds. In 1877 North Carolina established a similar station, followed closely by New Jersey, New York, Ohio, and Massachusetts. In 1888 the Hatch Act called for the setting up of state experiment stations to be operated in conjunction with the land-grant colleges, and an annual grant of $15,000 was made available to each state for their support. Although much of the early experimental work was largely demonstration, a "scientific" approach to agricultural problems was gradually developed in this country.

The idea of extracting soils with acids to determine their fertility status persisted, and E. W. Hilgard (1833–1916) found that the maximum solubility of soil minerals in hydrochloric acid (HCl) was obtained when the acid was at a specific gravity of 1.115, which happens to correspond to the concentration of the acid obtained on prolonged boiling. Hilgard attached particular significance to this fact. The strong acid digestion became quite popular, and numerous analyses of soils were made by this method. It was later shown that there was little foundation for assuming that this technique would give data of predictive value, and its use was discontinued.

Two early workers who contributed much to the development of interest in soil fertility in the United States were Milton Whitney and C. G. Hopkins. Around the beginning of the twentieth century these two men engaged in a controversy which attracted nationwide attention and which, in fact, became quite bitter. Whitney maintained that the total supply of nutrients in soils was inexhaustible and that the important factor from the standpoint of plant nutrition was the rate at which these nutrients entered into the soil solution. Hopkins, on the other hand, felt that this philosophy would lead to soil depletion and a serious decline in crop production. He made a survey of the soils of Illinois and reduced soil fertility to a system of bookkeeping. As a result of these exhaustive studies, he concluded that Illinois soils required the addition of only lime and phosphate. So effectively did he preach this doctrine that the use of lime and rock phosphate in a corn, oats, and clover rotation was a continuous practice in this state for many years.

The controversy between Whitney and Hopkins finally waned. Whitney's ideas were shown to be at least partly incorrect, but the argument, although bitter, did much to stimulate the thinking of agricultural scientists of the period.

Just after the turn of the twentieth century most experiment stations had established field plots which showed the remarkable benefits of fertilization. As a result of these experiments, the major problem of soil fertility could be delimited in a broad way. It was shown, for example, that there was a widespread need for phosphatic fertilizers, that potassium was generally lacking in the coastal plains regions, and that nitrogen was particularly deficient in the soils of the South. The soils east of the Mississippi River were generally acid and needed lime, whereas those west of the river were as a rule fairly well supplied with calcium. Even though the broad outline of the fertility status of soils in the United States had been fairly well defined, it was soon apparent that blanket fertilizer recommendations based on such knowledge could not be made. Each farm required individual attention, as did each field within that farm. The interest in soil tests sprang up anew.

During the last 30 years much headway has been made toward an understanding of the problems of soil fertility. To enumerate the men whose contributions have advanced our knowledge would require far more space than is available here. These advances have not been the work of scientists of any one country. The English, who began around 1600, have continued to make great strides. The agricultural workers of France, Germany, Scandinavia, the USSR, Canada, Australia, New Zealand, as well as the United States and other countries, have unraveled many problems which have held back the progress of the science. The fruits of these studies are everywhere apparent, for agricultural production in advanced countries is higher today than it has ever been, and the free world as a whole is generally better fed, clothed, and housed than at any time in the past. This could not be so if the production of crops today were at the level of Europe in the Dark Ages, when the average yield of grain was 6 to 10 bu/A.

Some appreciation of the progress made in increasing long-term average yields of major U.S. crops can be obtained from Figure 1-1. Although corn yields changed very little from 1900 to the early 1940s, there have been significant improvements during the last 30 to 40 years. Some of the principal factors contributing to higher corn yields include the commercial introduction of hybrids in the 1940s, the availability of relatively inexpensive fertilizers, especially nitrogen, in the early 1950s, and the use of higher rates of nitrogen and more effective pesticides and improved mechanization in the 1950s and early 1960s.

Corn yields exceeding 200 bushels or more now occur frequently in the United States. Some of the management practices used by growers to achieve this high level of production are shown in Table 1-2. It is apparent that adequate supply of plant nutrients is an important component of their programs.

Yields of soybeans have not shown the marked increases that corn has (Figure 1-1). This crop did not gain prominence until after the 1940s and it is only recently that the benefits of agronomic research in breeding, fertilization, and management

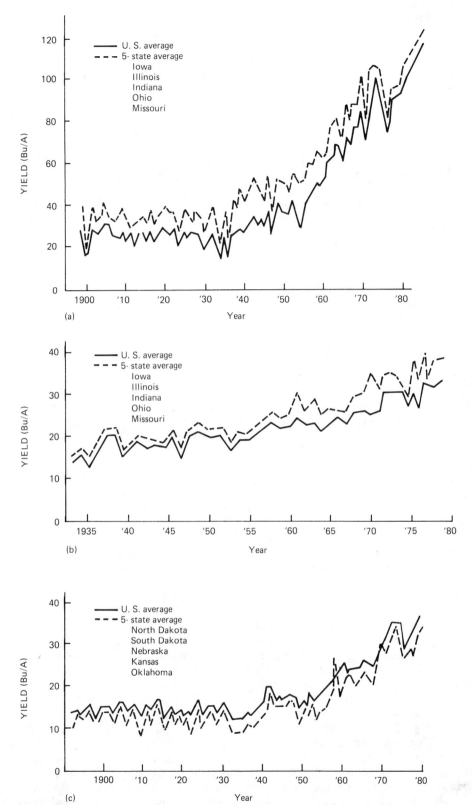

FIGURE 1-1. Long-term trends in average annual yield of (a) corn, (b) soybeans, and (c) wheat. [Hueg, *ASA Spec. Publ. 30*; pp. 78–79 (1977).]

TABLE 1-2. Average Management Practices by Region Associated with 200+ bu/A Corn Yields, Throughout the United States, 1975–1976

	Overall	West	Great Plains	Midwest	Northeast	Southeast
			*Region**			
Yield (bu/A)	218	220	219	216	217	221
Planting date	Apr. 22	Apr. 17	Apr. 27	Apr. 29	May 6	Mar. 28
Harvest population	25,500	27,350	26,300	24,400	23,950	25,550
Fertilizer rate (N-P_2O_5-K_2O)	220-95-109	227-65-61	234-59-29	204-120-144	184-90-93	258-113-172
Number of growers	549	99	102	275	15	58

*West: CA, OR, WA, UT; Great Plains: CO, KS, OK, NE, NM, TX; Midwest: IL, IN, IA, KY, MI, MN, MO, OH, WI; Northeast: DE, MD, NJ, NY, PA, VA; Southeast: AL, FL, GA, LA, NC, SC, TN, MS.
Source: Dibb and Walker, *Better Crops Plant Food,* **62**:17 (Winter 1978–1979).

have begun to be realized, as can be seen in the higher yields of the late 1970s. The steady improvement in wheat yields since the 1940s (Figure 1-1) is attributed to the introduction of new types, such as the short-strawed varieties, and increased use of fertilizers and pesticides.

Looking to the Twenty-first Century

As civilization moves nearer to the twenty-first century and as world population continues to increase, the importance of a continuing increase in food production is obvious. Continuing research on all phases of agricultural production is necessary if this increasing population is to be fed and clothed. Research advances are being made, at agricultural and other types of research centers, which may contribute to increased agricultural production in the future.

Some of the advancements looming on the horizon, or, in fact, already being put into practice, show great promise of pushing crop yields higher and increasing the efficiency of agricultural production. These advances will put additional pressure on the land and will increase even further the importance that soil fertility plays in crop production.

Remote sensing, which is infrared photography from high altitudes, is used to determine crop conditions. Problems arising from soil, irrigation, or pest conditions can often be detected and corrected in time to prevent serious depressions in yield.

Research is showing that conservation tillage can increase water-use efficiency and decrease soil erosion while increasing crop yields. This kind of management can have considerable effect on crop requirements for certain nutrients, especially nitrogen, phosphorus, potassium, and sulfur, and further research is needed to develop soil fertility practices suitable for use with conservation tillage.

In many areas of the United States and Canada, vast acreages that were heretofore unsuitable for crop production because of a lack of water have been brought into a higher state of productivity with the development of center-pivot irrigation systems. Wells are drilled in the center of the fields, the water is transferred to the field through spray nozzles attached to a conduit that is self-propelled and moves in a circle around the wellhead. Several hundred acres can be irrigated with the larger systems. Fluid fertilizers and pesticides may also be distributed through these systems. Because of the elimination of moisture as a limiting factor, greater fertilizer-use efficiency can be obtained and unit production costs lowered. These systems have become widespread in the Plains states in the United States. They are also finding their way into other areas, including the central and western provinces of Canada, and the southeastern, southwestern, and western areas of the United States. For this expensive source of water to be most efficiently utilized by the crop, the supply of nutrients available must be optimized.

In these same semiarid regions of the United States and Canada, there are promising developments in water capture and more efficient use of moisture for crop production. Crop production systems involving these modern methods of moisture management, in conjunction with other high-yield factors, such as fertilization, hybrid or variety, and seeding date, need to be examined.

Irrigation efficiency is of concern in many areas of the world where supplies of water for agricultural uses are limited. Drip or trickle irrigation may reduce by 50% the water now used in conventional irrigation systems. Fertilizer needs and systems for their application under trickle irrigation warrant further study.

Soil and plant analysis as a means of determining fertilizer and lime requirements for crops have been used for many years. This use continues and is increasing. However, considerable additional information is needed before these tests can become more than a somewhat refined general guide to crop fertilization and liming. Foliar application of elements other than just the micronutrients shows promise of becoming a general agricultural practice in some areas. However, because of inconsistent results, more research is needed to determine those conditions conducive for response from this method of application.

Improvements have been and will continue to be made in the development of more effective fertilizer materials. Some of the materials that have been or are being developed include slow-release nitrogen fertilizers, high-analysis polyphosphates, magnesium compounds suitable for use in complete fluid fertilizers, chelated or sequestered micronutrients, and high-analysis sulfur materials for use in both liquid and solid fertilizers.

Needed together with the development of these new techniques and products is the continued evaluation of their effectiveness through short- and long-term experimentation. Such field experimentation is a needed requirement for continued increases in crop production efficiency. Higher crop yields impose different nutrient requirements. Fertilizer rates that gave satisfactory responses with corn yields of 150 bu/A will not be adequate at yield levels of 200 or more bu/A. Further, with soil tests often calibrated years ago, rates of applied nutrients needed for different soil test values may well be too low for the higher crop yields of today and the even higher crop yields of the future.

One new development looming large on the horizon and which may have a profound impact on agricultural production is the recently developed science of molecular genetics. Through the technique of gene transplants, desirable qualities of one genus or species may be transferred to another. If and when this science becomes perfected, it is conceivable that greater photosynthetic efficiency, higher protein and vitamin content, better disease and insect resistance, and other factors can be introduced into otherwise desirable crop species. Such genetic alterations could have a marked impact on nutrient requirements, and hence on fertilization practices.

Progress in agriculture depends on research of a high caliber. For every problem solved by the scientist today, many more are raised. Agricultural scientists must delve into questions of a fundamental nature, questions that deal more with the *why* of things than with the *what*.

It is not the purpose of this chapter to cover all of the significant events in the development of the science of soil fertility. Much has been omitted, and much more could have been written. Certainly, the advances made toward the end of the nineteenth century and in the twentieth have been largely responsible for the present state of our learning. These advances were covered only superficially in this chapter, but the remaining chapters, in the information they contain, confirm the importance of these events to progress in soil fertility. Although brief, it is hoped that this sketch will give the student some idea of the time, effort, and thought that has been devoted in the last 4500 years to accumulating what is still insufficient knowledge.

Selected References

Diagnostic Techniques for Soils and Crops. Washington, D.C.: Potash Institute of North America, 1948.

Fussell, G. E., "John Wynne Baker: an improver in eighteenth-century agriculture." *Agr. Hist.*, 5:151–161 (1931).

Gruber, J. W., "Irrigation and land use in ancient Mesopotamia." *Agr. Hist.*, 22:69–77 (1948).

Holland, J. W., "The beginnings of public agricultural experimentation in America: the trustee's garden in Georgia." *Agr. Hist.*, 12:271–298 (1938).

Hueg, W. F., Jr., "Focus on the future with an eye to the past." *ASA Spec. Publ. 30*, pp. 73–85. Madison, Wis., American Society of Agronomy, 1977.

Jacob, K. D., "Fertilizers in retrospect and prospect." *Proc. Fert. Soc. (London)* (April 1963).

Lucretius Carus Titus, *De Rerum Naturum.* Translated by William Ellery Leonard. New York: Dutton, 1921.

McCloud, D. E., "Presidential address: Man and his food." *Agron. J.*, 67:1 (1975).

Miller, A. C., Jr., "Jefferson as an agriculturist." *Agr. Hist.*, 16:65–78 (1942).

Olson, L., "Columella and the beginnings of soil science." *Agr. Hist.*, 17:65–72 (1943).

Olson, L., "Pietro de Crescenzi: the founder of modern agronomy." *Agr. Hist.*, 18:35–40 (1944).

Olson, L., "Cato's views on the farmer's obligation to the land." *Agr. Hist.*, 19:129–133 (1945).

Plinius Secundus Caius, *Natural History*, Book XVII. Translated by John Bostock and H. T. Riley. London: Henry G. Bohn, 1857.

Russell, E. W., *Soil Conditions and Plant Growth*, 10th ed. London: Longmans, Green, 1973.

Semple, E. C., "Ancient Mediterranean agriculture." *Agr. Hist.*, 2:61 (1928).

Virgil (*Publius Vergilius Maro*), *Georgicus.* Translated by Theodore Williams. Cambridge, Mass.: Harvard University Press, 1915.

Xenophon, *Oeconomicus.* Translated by E. C. Marchant. New York: Putnam.

Chapter 2

Growth and the Factors Affecting It

STATED in its most elemental terms, the success of the farming operation depends largely on the growth of the crop. If plant growth and yield of the harvest are good, farmers will have been successful; and barring an unfavorable economy, they will receive a return on their investment in labor and capital in keeping with the quality and quantity of the crop. On the other hand, if plant growth has been poor and the yield of the harvest is low, farmers will receive a correspondingly lower return, if any at all.

From the practical standpoint of profitable agriculture, growth and the factors affecting it occupy a place of importance second to none in the farming enterprise. Because of this, some of these factors, and the effect they may have on limiting plant growth responses to an adequate supply of plant nutrients, are discussed briefly.

Growth

Growth is defined as the *progressive development of an organism*. There are, however, several ways in which this development can be expressed. It may refer to the development of some specific organ or organs or to the plant as a whole. Growth may be expressed in terms of dry weight, length, height, or diameter. The various ways in which growth has been expressed will be treated later in this chapter. From an agricultural point, however, the factors that influence plant growth are of the greatest importance, for a knowledge of these factors and how they affect crop production can make it possible for farm operators to manipulate them to advantage in maximizing the return on farming operations.

Factors Affecting Plant Growth

Some specialists have identified 52 factors that affect crop growth and have assigned mathematical values to each. A partial list of these factors appears in Table 2-1. Growers are said to have the capability of controlling up to 45 of the factors. For high yields the 45 factors must operate in unison, as many of them are interrelated. Growers are unable to manage seven of these factors: (1) temperature, (2) sunlight, (3) violent storms, (4) flooding, (5) rainfall, (6) carbon dioxide, and (7) altitude. Growers can, however, modify the rainfall factor by irrigation and increase carbon dioxide supply by applying fresh decomposable animal manures and crop residues.

TABLE 2-1. Some Factors Affecting the Potential Yield of a Crop

Organic matter of the soil	Water percolation rate
Clay concentration	Rainfall amount and distribution
Cation exchange capacity	Altitude
Quantity and intensity of light	Latitude
Slope and topography	Specific crop growth characteristics
Percent sunshine	Aeration factor
Evaporation/transpiration	Carbon dioxide
ratio	Wind velocity
Relative humidity	Available soil water
Temperature	Depth of root zone
Days over 90°F	Manure
Irrigation	
Tiles	

Source: Strauss, *Fert. Solut.,* **22:**(1):68 (1978).

The 18 factors designated by Kunkel and his co-workers in Washington as being essential for maximum yield of potatoes are listed in Table 2-2. Seven of the factors, all related to climate and environment, are considered to be beyond grower control.

The two major factors often acknowledged as establishing the upper limit of the potential yield of crops are (1) the amount of moisture available during the growing season and (2) the length of the growing season. Potential potato production is said to be determined by the amount of radiant energy available and the number of frost-free days. To produce the maximum crop possible it is generally agreed that crop plants must utilize a high percentage of the available solar energy.

The potential maximum yield of most crops far exceeds current yield levels. For example, it has been estimated by R. R. Johnson and D. B. Peters of the University of Illinois and by others that maximum yields of corn and soybeans are of the order of 490 to 580 bu/A and 140 to 225 bu/A, respectively. Wheat genotypes presently available have the potential for producing 12 to 15 tons/ha of grain or 178 to 223 bu/A.

In 1983, R. J. Flannery of the New Jersey Agricultural Experiment Station grew 338 bu/A of corn and 118 bu/A of soybeans. These are recognized as world records

TABLE 2-2. Factors Affecting Yield and Quality of Potatoes

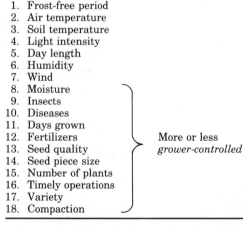

1. Frost-free period
2. Air temperature
3. Soil temperature
4. Light intensity
5. Day length
6. Humidity
7. Wind
8. Moisture
9. Insects
10. Diseases
11. Days grown
12. Fertilizers
13. Seed quality
14. Seed piece size
15. Number of plants
16. Timely operations
17. Variety
18. Compaction

More or less *grower-controlled*

Source: Kunkel et al., *Proc. 11th Annu. Wash. Potato Conf. Trade Fair,* p. 87 (1972).

TABLE 2-3.   Yield of Corn Hybrids of Different Eras Grown from 1971 to 1973

Period Hybrid Developed	Poor Conditions		Good Conditions	
	kg/ha	% Increase over 1930	kg/ha	% Increase over 1930
1930	3709	—	6538	—
1940	4464	20	7544	15
1950	4778	29	7670	17
1960	4902	32	8550	31
1970	5972	61	8990	38

Source: Hueg, *ASA Spec. Publ.* 30, p. 81 (1977).

in replicated research plots. Corn yields greater than 300 bu/A have been recorded nine times since 1951 by U.S. farmers.

Those factors known to be involved in crop growth can be classified as genetic or environmental. A brief review of the major factors follows.

Genetics. The importance of genetics in the growth of agricultural crops is illustrated by the large increases in yield resulting from the introduction of new corn hybrids (Table 2-3) and improved wheat varieties (Table 2-4). Yields obtained with corn hybrids of the recent past were 38 to 61% greater under good and poor conditions, respectively, than they were with hybrids developed in the 1930s. Variety improvement led to a 79% increase in yield of hard red spring wheat.

Variety and Plant Nutrient Needs. It is obvious that the high crop yields produced with modern hybrids, varieties, lines, and so on, will require more plant nutrients than was necessary for the lower yields of the past. This important fact, which has often been overlooked in shifts to higher yielding varieties, is evident in Table 2-5 for several varieties and lines of soybeans in Iowa. Under low-fertility conditions a new high-yielding variety cannot develop its full yield potential. In fertile soils the same new variety will deplete the soil more rapidly, and eventually yields will decline if supplemental nutrients are not provided.

An example of the importance of genetics as a yield-limiting factor is shown in Figure 2-1. Notice especially the inability of the Arksoy variety to produce the high yields obtained with the other varieties on the moderate and high-fertility soils. One of the first steps in a successful crop farming enterprise is the selection of hybrids or varieties that are genetically capable of producing high crop yields and of utilizing to the fullest extent the supply of plant nutrients that will be made available to them.

In research leading to the development of new hybrids or varieties, as well as in the use of these new lines in the farming enterprise, pest control measures must

TABLE 2-4.   Spring Wheat Variety Performance

Variety	Year Released	Yield* (kg/ha)	Percent Greater Than Marquis
Marquis	1926	2028	—
Thatcher	1935	2230	10
Lee	1958	2425	16
Chris	1967	2735	35
Era	1971	3623	79

* All tested in 1974 at three locations.
Source: Hueg, *ASA Spec. Publ. 30,* p. 79 (1977).

TABLE 2-5.   Nutrient Removal in Seed and Nutrient Concentrations in Leaves of Various Varieties and Lines of Soybeans

Soybean Variety or PI Line	Yield of Seed (kg/ha)	Nutrient Removal in Seed (kg/ha)			Nutrient Concentration in Leaves (%)		
		N	P	K	N	P	K
PI92561	2914	196	20	58	4.9	0.38	1.1
Ford	2903	195	17	57	4.7	0.34	0.8
Adams	2882	191	17	59	4.9	0.39	1.0
Seneca	2840	183	16	56	4.3	0.38	1.4
PI80536	1082	75	8	21	4.8	0.39	1.4
PI200479	1024	73	6	19	3.6	0.29	0.8
PI200482	924	67	6	17	3.4	0.34	0.9
PI227212	830	56	5	16	4.6	0.38	1.3

Source: deMooy et al., in B.E. Caldwell, ed., *Soybeans: Improvement, Production, and Uses*, pp. 267–352. *ACS Agron. Monog. 16.* Madison, Wis.: American Society of Agronomy, 1973.

be adopted whenever necessary. Frequently, greater vegetative growth results from the use of these newer materials under higher fertility conditions, and insects, disease, and weeds may be encouraged. If such pests are not controlled, many otherwise promising high-yielding materials will be discarded in a research program or fail to perform adequately on the farm. This is another example of the limiting factor concept.

Variety–Fertility Interactions.   The differential response of crop varieties to applied plant nutrients has been demonstrated on numerous occasions. In general, varieties that have a small range of adaptation tend to show significant variety–fertilizer interactions, whereas those with a wide range of adaptation do not. As early as 1922 the Tennessee Experiment Station was recommending varieties of corn on the basis of the fertility level of the soil. The varieties selected for poor soils were entirely different from those suggested for the more fertile soils.

In most developed agricultural countries of the world it is customary to supply adequate plant nutrients on the low-fertility soils. Thus recommendations for varieties or hybrids generally do not need to be made on the basis of the fertility

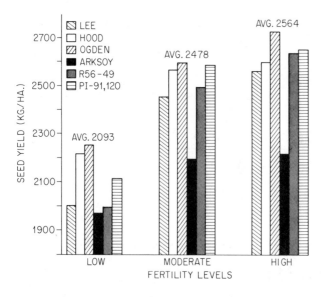

FIGURE 2-1.   Average yields of six genetic lines of soybeans when grown at three levels of soil fertility for six years, Stuttgart, Arkansas. [Caviness and Hardy, *Agron. J.*, **62**:236 (1970).]

level of the soil but rather on their ability to withstand insects, diseases, or unfavorable conditions of moisture or temperature. It should be noted, however, that one strategy for managing land areas infertile because of low iron availability or an excess of chemical substances such as aluminum or soluble salts is to select a cultivar or species capable of tolerating the specific condition. Although soil fertility need no longer be a limiting factor in most instances, knowledge of nutrient uptake mechanisms and their cultivar specificity is needed.

      Importance of Progressive Research in Plant Genetics. The genetic constitution of a given plant species limits the extent to which that plant may develop. No environmental conditions, no matter how favorable, can further extend these limits. It is imperative that there be a forward-moving plant-breeding program that will produce new varieties or hybrids capable of achieving maximum yields under specified conditions. Crop-breeding studies under nonlimiting environments were started by Lambert and Cooper at the University of Illinois in 1974 and by Cooper at Wooster, Ohio, in 1977.

What has been achieved in plant breeding and genetics is clearly evident in the crop production records kept by the better growers in leading agricultural countries of the world. What can be done will be determined by tomorrow's needs and the imagination and skill of geneticists and plant breeders the world over. There are promising indications that it will be possible to breed widely divergent species and create new crop varieties by using innovative methods which go beyond those of conventional plant breeding. Some of the promising techniques for genetic manipulation or molecular genetics include in vitro techniques for asexual approaches, broad crosses between crop species, single-cell culture, anther and pollen culture, and somatic hybridization. Also, genetic fortification offers a great challenge and opportunity for improving the nutritive value of the world's feed grains by increasing protein content and amino acid distribution.

      Environmental Factors. Environment is defined as the *aggregate of all the external conditions and influences affecting the life and development of an organism.* Among the environmental factors known to influence plant growth, the following are probably the most important:

1. Temperature.
2. Moisture supply.
3. Radiant energy.
4. Composition of the atmosphere.
5. Soil structure and composition of soil air.
6. Soil reaction.
7. Biotic factors.
8. Supply of mineral nutrient elements.
9. Absence of growth-restricting substances.

Many environmental factors do not behave independently. An example is the inverse relation that exists between soil air and soil moisture or between the content of oxygen and carbon dioxide in the soil atmosphere. As the soil moisture increases, the soil air decreases, and as the carbon dioxide content of the soil air increases, the oxygen content decreases.

Another example is the relation between the diffusion rate of oxygen in the soil and soil temperature. It will be shown in a subsequent section of this chapter that the partial pressure of oxygen in the root environment is extremely important to the growth of plants. The maintenance of this pressure is related to the diffusion rate of oxygen to the root surface, which in turn is influenced by the soil temperature.

The effects of a number of these environmental factors acting individually and

TABLE 2-6.   Relationship of Climatic Factors to Potential Yields of Corn in Three of the Great Plains States (bu/a)

	Nonirrigated			Irrigated
	S. Dakota	Nebraska	Kansas	Kansas
Western	42	54	59	270
Central	89	132	178	319
Eastern	188	209	217	327

Source: Strauss, *Fert. Solut.,* **22**(1):68 (1978).

collectively were used to estimate the potential corn yields shown in Table 2-6. A higher yield potential exists in the eastern parts of these selected states because of more favorable growing conditions, including cooler temperatures, higher rainfall, lower altitude, less wind velocity, and better soil physical conditions. An appreciation of the impact of frost-free days or the length of growing season can be gained by comparing yields by state in a north-to-south direction.

These examples are offered simply to illustrate the dependence on other factors of so-called independent variables of plant environment. Even though these factors are not independent of one another, they will be separately treated for simplicity of discussion.

Temperature.   Temperature is a measure of the intensity of heat. Physicists consider that the temperature of our universe ranges from a low of $-273°C$ to a high of several million degrees near the center of the sun. In terms of biological life as we know it, this is an almost unbelievably wide range. The limit of survival of those living organisms on this planet has generally been reported to be between -35 and $+75°C$. The range of growth for most agricultural plants, however, is usually much narrower—perhaps between 15 and 40°C. At temperatures much below or above these limits growth decreases rapidly. Consequently, the range of temperature in which earth life may continue is startlingly small when contrasted with the known range of temperatures.

Optimum temperatures for plant growth are dynamic since they change with the species and varieties, duration of exposure, age of the plant, stage of development, and the particular growth criterion used to evaluate performance. Temperature directly affects the plant functions of photosynthesis, respiration, cell-wall permeability, absorption of water and nutrients, transpiration, enzyme activity, and protein coagulation. This influence is reflected in the growth of the plant. A plant's capacity for growth of new photosynthetic area can substantially influence total photosynthesis and plant productivity. Therefore, the initiation and expansion rate of new leaves and duration of the various phases of plant development contribute greatly to crop productivity. Figure 2-2 shows the effect of temperature on encouraging development of corn leaves.

The effect of temperature on photosynthesis is complex and differs with plants of various species as well as with the carbon dioxide content of the atmosphere, the intensity of light, and the duration of light of a given intensity. The consensus among physiologists is that if light is limiting, temperature has little effect on photosynthesis rate. If, however, carbon dioxide is limiting and light intensity is not, photosynthesis is increased by an increase in temperature. The complexity of these relationships is illustrated by the data arranged graphically in Figure 2-3.

Respiration is also affected by changes in temperature. It takes place more slowly at low temperatures and increases as the temperature rises. At very high temperatures the rate of respiration is initially great but is not maintained. After a few hours at elevated temperatures, respiration rates for at least some plants drop off rather rapidly.

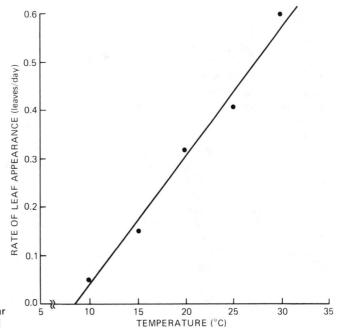

FIGURE 2-2. Influence of temperature on the rate of corn leaf appearance. [Tollenaar et al., *Crop Sci.*, **19**:363 (1979).]

For may crop plants of the temperate zone the temperature optimum for photosynthesis is lower than that for respiration. This has been suggested as one reason for the higher yield of starchy crops, such as corn and potatoes, in cool climates as contrasted with the yield of these crops in warmer regions. It is possible that under conditions of prolonged temperatures above the optimum, a plant may literally suffer from starvation simply because respiration is taking place more rapidly than photosynthesis.

Transpiration, or the loss of water vapor from the stomata of leaves, is influenced by temperature. Transpiration rates as a rule are low at low temperatures and

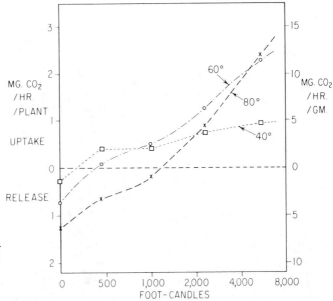

FIGURE 2-3. Net carbon dioxide exchange rates of five-week-old rice plants as related to temperature and light intensity. [Ormrod, *Agron. J.*, **53**:94 (1961).]

increase with rising temperatures. Under conditions of excessive transpiration, water losses may exceed the water intake by the plant and wilting soon follows.

The absorption of water by plant roots is affected by temperature. Again, the influence of temperature is modified by species. However, with a number of plants adapted to conditions of the temperate zone, absorption increases with a rise in temperature of the rooting medium from 0°C to about 60 or 70°C. Above this point there is a leveling off of the rate of absorption.

Low soil temperature may adversely affect the growth of plants by its effect on the absorption of water. If soil temperatures are low, yet excessive transpiration is taking place, the plant may be injured because of tissue dehydration. The moisture supply of the soil may also be influenced to some extent by temperature, for unusually warm weather produces more rapid evaporation of water from the soil surface.

Temperature also affects mineral element absorption. Results from numerous experiments have indicated that in a number of plant species the absorption of solutes by roots is retarded at low soil temperatures. This may be caused by lower respiratory activity or by reduced cell membrane permeability, both of which could affect uptake itself as well as the rate and extent of root permeation in the soil. Nutrient availability and movement to roots are also influenced by temperature.

The effect of temperature on the uptake of nutrients by the potato plant is illustrated by the data shown graphically in Figure 2-4. Notice the effect of temperature on the content of the various nutrients in both tops and roots. For example, the content of phosphorus in both tops and roots was increased with an increase in temperature. On the other hand, an increase in temperature resulted in a decrease in the root content of potassium.

Temperature exerts its influence on plant growth indirectly by its effect on the microbial population of the soil. The activity of the nitrobacteria as well as of most heterotrophic organisms increases with a rise in temperature. Soil pH may change with temperature, which, in turn, may affect plant growth. It has been observed that the soil pH increases in winter and decreases in summer. This is generally considered to be related to the activities of microorganisms, since microbial activity

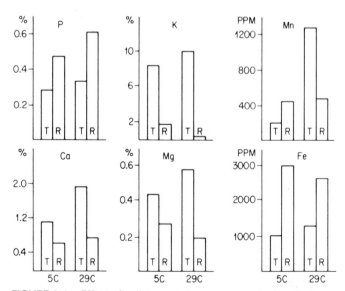

FIGURE 2-4. Effect of soil temperature on several minerals in potato tops (T) and roots (R). [Epstein, *Agron. J.*, **63**:664 (1971).]

is accompanied by the release of carbon dioxide which combines with water to form carbonic and other acids. In soils that are only slightly acid this small change in pH may influence the availability of such micronutrient elements as manganese, zinc, or iron.

Numerous studies on the direct relationship between yield or dry-matter production and temperature have been made. This relation varies for different crop species and varieties. The impact of soil temperature may also be modified by a number of soil conditions, including amount of clay, moisture content, and drainage. Some examples of the influence of root-zone temperature on the growth of corn tops and roots under greenhouse conditions are summarized in Table 2-7. A 5°C rise within the temperature ranges investigated frequently increased top growth 100% or more. Figure 2-5 shows how heating of field soil in northern Ohio improved dry-matter production and grain yield of corn. These data illustrate the importance of the limiting factor concept stated earlier in this chapter. The temperature–plant function relations illustrated here point up the need for a working knowledge of this concept, for the planting of a crop or variety not adapted to prevailing temperatures will result in lowered yields and a lower return from inputs of labor and capital.

Temperature may also alter the composition of the soil air; this again is the result of increases or decreases in the activity of microorganisms. When the activity of the micropopulation is great, there will be higher partial pressure of the carbon dioxide of the soil atmosphere as the oxygen content decreases. Under conditions restricting the diffusion of gases into and out of the soil, a decrease in the O_2 pressure resulting from such activity might influence the rate of respiration of the plant roots, hence their ability to absorb nutrients.

Practical application has been made of the plant growth–temperature relationship to develop what is termed the heat unit concept. Heat units have been described in various ways: degree-days, optimum days, and effective degrees. All of the terms express the amount of heat energy that has been absorbed by the soil over a given period of time. The number of units required to bring a crop to maturity (or some particular stage of growth) has been determined for various crop plants. Some commercial farmers, particularly the growers of canning and fresh-frozen crops, make use of this technique for determining planting and harvesting dates with a degree of precision never before obtainable. Use of this technique has considerably improved the efficiency of canning and freezing operations, for it permits planting dates to be staggered so that harvests will provide a steady flow of high-quality produce to the canner or freezer. For a more complete discussion of this phase of temperature–plant growth relationship, the reader is referred to Katz (1952) and Gilmore and Rogers (1958).

Moisture Supply. The growth of many plants is proportional to the amounts of water present, for growth is restricted both at very low and very high levels of soil moisture. Water is required by plants for the manufacture of carbohydrates, to maintain hydration of protoplasm, and as a vehicle for the translocation of foods and mineral elements. Internal moisture stress causes reduction both in cell division and cell elongation, hence in growth.

Plant water stress results when the extractable water in the root zone is insufficient to meet the plant's transpirational demands. If the latter is excessive, symptoms of moisture stress may appear even though the extractable water supply is present at a level otherwise considered adequate. Variations in soil water deficits are mainly responsible for year-to-year fluctuations in crop yields. The various physiological processes in plants are affected differently by water stress. For example, leaf elongation is more sensitive to soil water deficits than the other processes. Also, elongation will cease before all the extractable soil water is consumed.

TABLE 2-7. Effect of Root-Zone Temperature on the Growth of Corn

Crop Part	Growth Medium	Air Temperature (°C)	Relative Yields	Reference
Tops	Soil	20°C average in greenhouse	13°C 20°C 25°C 100 325 (with and without P combined)	Ketcheson, *Can. J. Soil Sci.*, **37**:41 (1957)
Tops	Mardin silt loam	Greenhouse conditions	15°C 20°C 25°C 100 240 361 (three rates of P combined)	Knoll et al, *Agron. J*, **56**:146 (1964)
Tops	Honeoye silt loam	"	15°C 20°C 25°C 100 132 263 (three rates of P combined)	"
Tops	Sand culture	Varied from about 20 to 30°C in greenhouse	15°C 20°C 25°C 100 238 531 (four rates of P combined)	Knoll et al., *Soil Sci. Soc. Am. J.*, **28**:401 (1964)
Roots	Sand culture	"	15°C 20°C 25°C 100 230 368 (four rates of P combined)	"
Tops	Sand culture	Greenhouse Daytime: 31°C ± 4°C Nighttime: 20°C ± 2°C	15°C 20°C 25°C 100 218 291 (three rates of P combined)	Patterson et al., *Crop Sci.*, **12**:227 (1972)

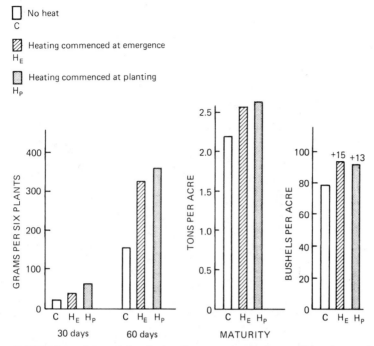

FIGURE 2-5.   Dry weight of corn plants at several stages of development and corn yield as affected by soil heating. [Mederski and Jones, *SSSA J.*, **27**:188 (1963).]

Roots grow best when soils are well supplied with moisture, but as demonstrated in Figure 2-6, growth can occur in even relatively dry soils. When water deficits limit root development the uptake of nutrients and water will be curtailed, especially if there is severe water stress.

The effect of irrigated and dryland moisture regimes and increasing rates of

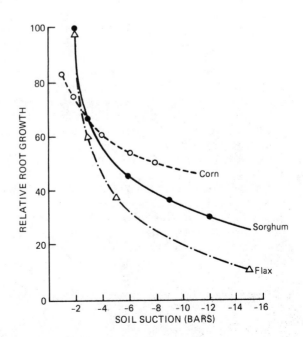

FIGURE 2-6.   The relative growth of roots of maize, sorghum, and flax as influenced by soil water potential (suction in bars). [Hurd and Spratt, *Physiological Aspects of Dryland Farming,* p. 183. New Delhi: Oxford & IBH, (1975).]

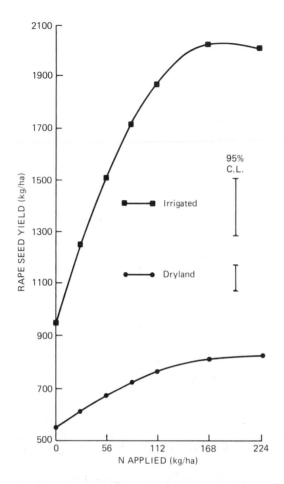

FIGURE 2-7. Effects of nitrogen fertilization on yield of rapeseed under irrigated and dryland conditions (mean data of eight site-years for each moisture regime). [Henry and MacDonald, *Can. J. Soil Sci.*, **58**:305 (1978).]

fertilizer nitrogen on the yield of rapeseed is shown in Figure 2-7. At any given level of applied nitrogen, the addition of irrigation water raises the yield of this oilseed crop. Conversely, increasing nitrogen rates up to 168 kg/ha under both sets of moisture conditions increases the rapeseed yields. High rates of nitrogen in combination with irrigation resulted in nearly fourfold increases in yield. This illustrates very clearly the limiting effect that too little moisture can have on plant responses to applied fertilizer. It should also be noted that it is necessary to provide sufficient fertilizer to make the greatest use of available water.

Yield is not the only plant property affected by soil moisture. Protein content of grain is frequently influenced by the degree of available water. Higher percentages of protein are generally associated with low levels of extractable soil moisture, as demonstrated for barley in Figure 2-8. A strong moisture–fertilizer nitrogen interaction is also seen in this figure.

Soil moisture level also has a pronounced effect on the uptake of plant nutrients. Low levels of extractable water in the root zone retard nutrient availability by impairing each of the three major processes involved in nutrient uptake. These processes are (1) diffusion, (2) mass flow, and (3) root interception and contact exchange. As a general rule, there is an increase in nutrient uptake when extractable water is high rather than low. The effect of decreasing soil moisture stress on phosphorus uptake by soybean tops is apparent in Table 2-8. Flooding of soil pores by excessive amounts of moisture is detrimental since the resultant lack of oxygen restricts root respiration and ion absorption.

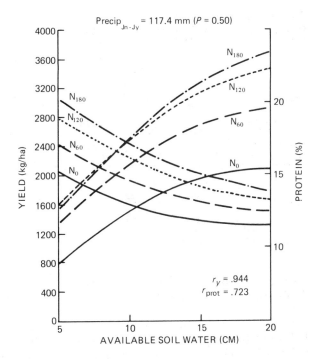

FIGURE 2-8. Increase in yield and decrease in protein content of barley as a function of available soil water at four fertilizer nitrogen levels (kg/ha), with an average (117 mm) June–July precipitation. Lines descending left to right are protein; those ascending left to right are yield. [Bole and Pittman, *Proc. 1978 Sask. Soil Crops Workshop* (February 8–9, 1978).]

The tendency for plant nutrients to be taken up more readily as supplies of extractable moisture increase has a favorable effect on water-use efficiency of plants. Water-use efficiency (WUE) is the amount of dry matter that can be produced from a given quantity of water. It is also described as the weight of dry matter derived from an acre-inch of water (or a hectare-centimeter of water). Table 2-9 summarizes the effects of two fertilizer application programs and three water management systems on the WUE of corn grain production. Repeated small applications of fertilizer throughout the growing season, coupled with the tensiom-

TABLE 2-8. Influence of Soil Moisture Stress in the Phosphorus-Fertilized Zone on Phosphorus Uptake by Soybeans

Soil Moisture Status* (tension in bars)	P Content of Tops† (mg)			P (%)
	Total	Derived from:		
		Fertilizer	Soil	
Phosphorus, 24 ppm				
Continuously < 2	15.3	3.6	11.7	0.254
Depletion from 0.03 to 15	13.0	3.1	9.9	0.230
Maintained near 15	9.5	2.0	7.5	0.164
Phosphorus, 243 ppm				
Continuously < 2	30.3	20.9	9.4	0.214
Depletion from 0.03 to 15	24.5	15.9	8.6	0.170
Maintained near 15	10.1	6.7	3.4	0.105
Phosphorus, 1,456 ppm				
Continuously < 2	67.1	54.3	12.8	0.407
Depletion from 0.03 to 15	65.6	53.5	12.1	0.242
Maintained near 15	45.6	35.7	9.9	0.266

* For the 10 days preceding harvest.
† Mean of the six replications; LSD 5% for total P content, 10.8 mg; SEM, 3.77.
Source: Marais and Wiersma, *Agron. J.,* **67:**778 (1975).

TABLE 2-9.   Yield of Corn and Water Use Efficiency as Affected by Water Management and Fertilizer Application Method*

Fertilizer Treatment	Natural Rainfall: 49 cm		Daily Irrigation (0.64 cm): 49 + 50 cm		Tensiometer Scheduled: 49 + 26 cm	
	Yield (kg/ha)	WUE	Yield (kg/ha)	WUE	Yield (kg/ha)	WUE
Conventional†	2860	59	3380	34	4760	64
Improved program**	2710	56	4950	50	6600	80

* WUE = kg/(ha · cm).
† 112, 98, and 280 kg/ha of N, P, K, broadcast preplant plus two later topdressings each of 112 kg/ha of N.
** 336, 98, and 280 kg/ha of N, P, K, broadcast in increments of 5%, 5%, 10%, 20%, 20%, 20%, and 20% during the growing season.
Source: Rhoads et al., *Agron J.*, **70**:306 (1978).

eter-scheduled irrigation system, produced the highest WUE on this sandy soil, which is subject to leaching of plant nutrients and rapid depletion of extractable soil moisture.

Placement of fertilizer nutrients is an important consideration in cropping situations where upper portions of the root zone are subject to rapid and prolonged drying. Fertilizers placed at depths in the root zone where soil is moist will be more effective. In arid and semiarid regions, where leaching is not a problem, improved distribution of fertilizer nutrients in the root zone can result from occasional heavy surface dressings. Fertilizer placement is discussed at length in subsequent chapters, particularly in Chapter 13.

Soil moisture level also influences plant growth indirectly by its effect on the behavior of soil microorganisms. At extremely low or extremely high moisture levels the activity of organisms responsible for the transformation of nutrients into plant-available forms is inhibited, with the result that plants may not be adequately supplied with the essential plant food elements.

Activity of plant pathogens in the soil can also be affected by soil moisture. Take-all, caused by the soilborne fungus *Gaeumannomyces graminis*, can be a serious disease of recropped wheat grown under conditions of plentiful soil moisture. On the other hand, growth of dryland crown and foot rot caused by *Fusarium* is favored by droughty soil conditions.

Radiant Energy. Radiant energy is a significant factor in plant growth and development. The quality, intensity, and duration of light are all important. Clear-day radiation is a useful indicator of the amount of solar energy available for physiological processes within plants. The clear-day radiation values for the periods between tassel initiation (anthesis) and maturity in corn suggest that the highest production potential for this crop in the United States exists near the 40° latitude, a zone running east to west from the southern border of Pennsylvania, through the middle of Ohio, Indiana, and Illinois, across northern Missouri and astride the Nebraska and Kansas borders. Much of the Corn Belt's top production occurs between 38 and 43° latitude. Many of the records of extremely high corn yields of the past few years have occurred at latitudes between 42 and 43°.

Studies have been made of the effect of light quality on plant growth, but such experiments are difficult to conduct because of the necessity for controlling simultaneously the wavelength and the intensity of the radiation. Even though the results of these studies have suggested that the full spectrum of sunlight is gen-

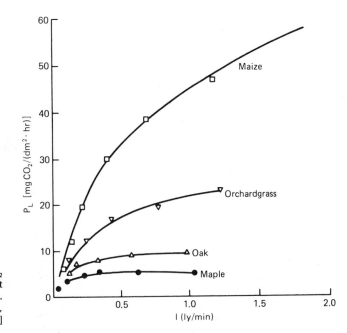

FIGURE 2-9. Relation of CO_2 assimilation (P_L) to light intensity (I) among species. [Hesketh and Baker, *Crop Sci.*, 7:286 (1967).]

erally most satisfactory for plant growth, light quality is also an influence. The impact of this variable has not been studied to the same extent as have the effects of light intensity and photoperiod. Even though light quality is known to affect plant growth, it is not likely in the foreseeable future that this factor can be controlled on a large-scale field basis. It may, however, be quite feasible for small acreages and for high-value specialty crops.

The intensity of light as a factor in plant development has also been investigated. It has been shown that most plants are generally able to make good growth at light intensities of less than full daylight. However, plants do differ in their response to light of varying intensity, as illustrated by the curves in Figure 2-9. Note that the two forest species, oak and maple, were the least responsive to increasing radiation intensities. Light intensity in their usual forest habitats is considerably less than that of full sunlight, especially where maple is present in the understory.

For a crop such as corn, which continues to respond to increasing insolation, the interception of solar radiation by its canopy and associated efficiency of photosynthesis is a principal determinant of growth and yield. Net photosynthesis rate in corn is almost linearly proportional to radiation interception, provided that soil moisture is adequate (Figure 2-10). The ability of a crop to intercept radiation thus becomes an important consideration.

Changes in light intensity caused by shading can exert considerable influence on crop growth. With high plant populations, light penetration to lower positions in the plant canopy may be inadequate for bottom leaves to carry on photosynthesis. Table 2-10 shows how corn yields were increased by providing artificial light low in the row. The importance of light distribution within crop canopies is further displayed in Figure 2-11, where corn grain yields are compared for hybrids with diverse leaf orientation grown at different plant densities. An advantage for leaf erectness is evident in the yield response obtained in both years.

Shading of crops can also occur when two different species are grown in a mixture, such as a grass–clover pasture. This problem has been studied by several workers in Australia. Balanced growth between the grass and clover is an important problem in good pasture management. It is known, for example, that excessive

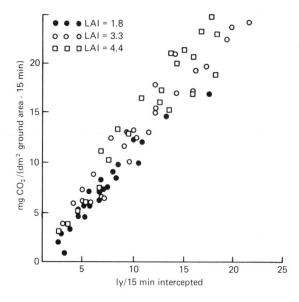

FIGURE 2-10. Apparent photosynthesis in corn [mg CO_2/dm^2 ground area · 15 min] versus intercepted light (ly/15 min) at three leaf area indexes. [Hesketh and Baker, *Crop Sci.,* **7**:291 (1967).]

nitrogen fertilization of grass–clover pastures will drive the clover out of the mixture. C. M. Donald and his associates in Australia studied this problem and reached the following conclusions:

1. Increased rates of nitrogen application give increased yields of grass.
2. Increased yields of grass give higher leaf areas of grass disposed above the clover–leaf canopy.
3. Higher leaf areas above the clover reduce the light density at the clover-leaf canopy.
4. Reduced light density at the clover-leaf canopy causes reduced growth of clover.

The loss of clover from clover-grass pastures heavily fertilized with nitrogen is frequently observed in many areas of the United States. This has been attributed largely to competition for moisture and nutrients, particularly potassium, but Donald's work suggests the possible limiting effect of too low a light density.

Studies by Japanese workers, who used wheat as the test plant, indicate that the absorption of ammonium, sulfate, and water was increased with increasing light intensity but that absorption of calcium and magnesium was little affected. Light intensity had marked effects on the uptake of phosphate and potassium. It

TABLE 2-10. Yield per Acre As Related to Amount of Artificial Light, 1971 Crop Year

Population (plants/A)	Artificial Light*	Ears per Plant	Yield† (bu/A)
40,672	None	0.77	196
41,404	300	0.94	240
39,228	600	0.94	245
34,848	900	0.93	266
41,394	1800	1.00	321

* Expressed in watts per 16 ft of row.
† Adjusted to 15.5 percent H_2O.
Source: Graham et al., *Trans. ASAE,* **15**(3):578 (1972).

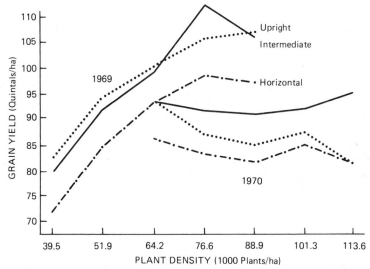

FIGURE 2-11. Mean yield response of hybrids grouped by leaf angle. [Hicks and Stucker, *Agron. J.*, **64**:486 (1972).]

was also observed that oxygen uptake by the roots increased with increasing light intensity.

Photoperiodism. Even though light quality and intensity may be of limited significance from the standpoint of field-grown crops, the duration of the light period is important. The behavior of the plant in relation to day length is termed photoperiodism. On the basis of their reaction to the photoperiod, plants have been classed as short-day, long-day, or indeterminate. Short-day plants are those that will flower only when the photoperiod is as short or shorter than some critical period of time. If the time of exposure to light is longer than this critical period, the plants will develop vegetatively without completing their reproductive cycle. Mammoth tobacco and coleus are examples.

Long-day plants are those that will bloom only if the period of time during which they are exposed to light is as long or longer than some critical period. If the plants are exposed to light for periods shorter than this critical time, they develop only vegetatively. Grains and clovers are members of this group.

Plants that flower and complete their reproductive cycle over a wide range of day lengths are classed as indeterminate. Cotton and buckwheat are representative of this group.

The phenomenon of photoperiodism was first described by Garner and Allard. They observed that a variety of burley tobacco known as Maryland Mammoth failed to bloom in the field during the summer months. When one of these plants was transferred to the greenhouse for the winter, however, it flowered profusely. Subsequent work showed this behavior to be related to the length of time that the plant was exposed to light. Since this discovery numerous other plants have been found to exhibit photoperiodism. From the standpoint of agriculture this property is of obvious importance, and the failure of crops to bloom or to set seed may frequently be related to day length.

Flowering and maturity of soybeans are largely determined by the length of days or photoperiod. Thus the commonly grown varieties are limited to local areas rather than to broad regions. Most varieties adapted to northern areas have an indeterminant growth habit and flower under long days, whereas varieties developed for southern areas are determinant and flower under short days. Semideterminant varieties are, however, being developed for northern areas.

Production of chrysanthemums is also based on knowledge of the growth–photoperiod relationship. This flower crop can be made to bloom at any predetermined time under greenhouse conditions simply by controlling illumination or the photoperiod.

Composition of the Atmosphere. Carbon is required for plant growth and, except for water, is the most abundant material within plants and other living things. The principal source of carbon for plants is carbon dioxide gas in the atmosphere. It is taken into their leaves and through photosynthetic activity is chemically bound in organic molecules.

Plants do not always obtain sufficient carbon dioxide since its concentration in the atmosphere is usually only about 0.03% by volume (300 ppm). Consequently, concentration of atmospheric carbon dioxide is a dominant factor as a rate determinant in photosynthesis. Many economic plants respond to elevated levels of carbon dioxide by increased growth and productivity (Table 2-11). Plants that fix carbon dioxide through the C_3 enzyme route (ribulose 1,5-diphosphate carboxylase) have a greater potential for response to increased atmospheric levels of carbon dioxide than do C_4 plants, which utilize this source of carbon via phosphoenolpyruvate carboxylase.

Carbon dioxide is continually being returned to the atmosphere as a product of respiration of animals and plants. Microbial decomposition of organic residues is an important source of this gas. One of the beneficial effects of farmyard manure and crop residues on crop yields may be the carbon dioxide that is released into the air trapped in and below dense plant canopies.

Although the normal value of carbon dioxide in the atmosphere is approximately 0.03%, the concentration may range from one-half to several times this figure. Within a thick plant canopy on a still day the carbon dioxide concentration may become measurably less during daylight hours when there is a high rate of photosynthesis. Similarly, in a dense forest the content may drop considerably. Under greenhouse conditions the carbon dioxide level may drop appreciably below the average concentration in outside air.

For more than 60 years it has been realized that crops will grow more rapidly if they are supplied with supplementary carbon dioxide. About 0.1% (1000 ppm) is generally thought to be ideal. At lower concentrations a full response to enrichment is not obtained, and higher concentrations usually do not produce any further increase in yield. Depressed productivity has been observed with concentrations in the range 0.3 to 0.5%.

From the standpoint of present commercial application to food production, use

TABLE 2-11. **Differences in Leaf Photosynthesis Among Plants and Percentage Increases in Growth or Yield from Elevated Atmospheric Levels of Carbon Dioxide**

Plant	$CO_2 \ [mg/(dm^2 \cdot hr)]$ (Normal CO_2 Levels)	Increase in Growth or Yield at Elevated CO_2 Levels (%)
Corn, grain sorghum, and sugarcane	60–75	100
Rice	40–75	135
Sunflower	50–65	130
Cotton	40–50	100
Soybean, sugar beet	30–40	56
Oats, wheat, barley	30–35	66
Tobacco	20–25	67
Tomato, cucumber, lettuce	20–25	50
Tree species, grapes and ornamentals, citrus	10–20	40

Source: Wittwer, *Proc. Agr. Res. Inst. 21st Annu. Meet.*, pp. 69–86 (1973).

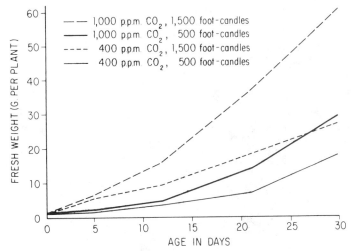

FIGURE 2-12. Yield of Bibb lettuce influenced by light intensity and carbon dioxide level of the atmosphere in controlled environment growth chambers. [Wittwer et al., *Econ. bot.,* **18**:334 (1964).]

of supplementary carbon dioxide seems to have its greatest potential in green-houses. It is more practical to maintain higher-than-normal carbon dioxide concentrations within the confines of a greenhouse than in open-field situations, where wind action usually results in natural mixing of air. However, in densely planted crops which restrict air movement it may be possible to raise carbon dioxide concentration through microbial breakdown of fresh, active sources of organic matter.

The effect of increasing light intensity and carbon dioxide content on the growth of lettuce in growth chambers is shown in Figure 2–12. Yield of lettuce in terms of 10 head weights was improved by raising only carbon dioxide concentration (Table 2-12). As demonstrated in Figure 2-13, supplementary carbon dioxide raised

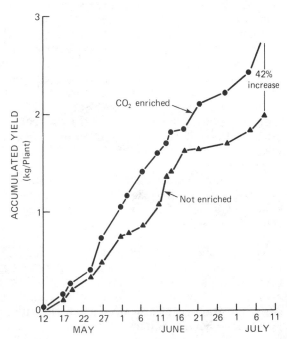

FIGURE 2-13. Effect of CO_2 enrichment on tomato yield. Yields from each harvest were added to the yields from all previous harvests to provide the accumulated yield values. [Hicklenton and Jolliffe, *Can. J. Plant Sci.,* **58**:808 (1978).]

TABLE 2-12. Yields of Lettuce Varieties Influenced by Carbon Dioxide Added to the Greenhouse Atmosphere

Lettuce Variety	Yields* (lb/10 heads)	
	$-CO_2$	$+CO_2$†
Bibb	1.2	1.9
Cheshunt No. 5B	1.6	2.8
Grand Rapids H-54	1.3	2.6
Mean (all varieties)	1.4	2.4

* Yields of each variety, and the mean of all varieties, significantly greater at $+CO_2$ (odds 99:1).

† Levels of carbon dioxide ranged from 125 to 500 ppm, in $-CO_2$ and from 800 to 2000 ppm in $+CO_2$ during the daylight hours, except when ventilators were open.

Source: Wittwer et al., *Econ. Bot.,* **18**:334 (1964).

greenhouse tomato yields by as much as 42%. The quality of tomatoes has also been reported to be favorably influenced by carbon dioxide treatments.

Successful carbon dioxide supplementation studies have been conducted in the field using transparent chambers (1 to 2 m²) to isolate crops in the study area. The chambers are ventilated with air containing predetermined levels of carbon dioxide. The results in Table 2-13 from such a study on dwarf spring wheat reveal that carbon dioxide enrichment increased grain yield by up to 23%. The largest yield increases generally occurred when the crop was exposed to extra carbon dioxide in the second and third months after seeding.

Other plants, among which are cucumbers, flower crops, foliage crops (greens), peas, beans, and potatoes, have also responded to increased carbon dioxide concentrations.

When the level of carbon dioxide is increased, the light requirement may be also. Maximum carbon dioxide fixation at normal atmospheric levels occurs at relatively low light intensities of 1500 to 2000 footcandles for some plants. Some recent studies, however, have shown that higher light intensities are required for certain crops. As Wittwer and Robb point out, the greatest potential benefits to be derived from carbon dioxide enrichment of a greenhouse atmosphere come when normal light intensities are in excess of the saturation value for a normal (0.03%) carbon dioxide atmosphere. In this connection it should also be noted that as carbon dioxide levels of the atmosphere are increased, photosynthesis becomes more sensitive to temperature.

The foregoing discussion illustrates that plant growth can be increased by supplemental carbon dioxide. Its use under greenhouse conditions appears quite feasible, but whether its use under field conditions can be developed commercially remains to be seen.

Toxic Atmospheric Substances. The quality of the atmosphere surrounding aboveground parts of plants may under certain conditions influence growth. Certain gases, such as sulfur dioxide (SO_2), carbon monoxide (CO), and hydrofluoric acid (HF), when released into the air in sufficient quantities, are toxic to plants. Although the exception rather than the rule, isolated cases of injury from these gases have been reported.

TABLE 2-13. Effect of Carbon Dioxide Supplementation on the Yield of Dwarf Spring Wheat

Days from Seeding When Crop Received Extra Carbon Dioxide	Increase in Grain Dry Weight (%) Due to Exposure to Extra Carbon Dioxide (750 ppm) in:		
	1971	1972	1973
12–39	12	− 1	7
40–73	23	− 3	15
68–98	7	21	14
96–124	11	13	1

Source: Fischer and Aguilar, *Agron. J.*, **68**:750 (1976).

Strong acids such as sulfuric, nitric, and hydrochloric have lowered the pH of rain and snow falling on much of northern Europe and the eastern sections of the United States and Canada to between 4 and 5. Values between pH 2.1 and 3.0 have been observed for individual storms at various locations. Acid rain is often due mainly to relatively high concentrations of sulfur dioxide and sulfates. Some of the effects that acid rain can have on plants and soil include increased leaching of inorganic nutrients and organic substances from foliage; accelerated cuticular erosion of leaves; leaf damage when pH values fall below 3.5; altered response to associated pathogens, symbiants, and saprophytes; lowered germination and establishment of conifers; reduced availability of soil nitrogen; decreased respiration; and increased leaching of nutrient ions from soils.

Injury to vegetation by fluorine released during the manufacture of metallic aluminum and the production of phosphatic fertilizers has been reported. Damage to crops, however, may not be so important as the toxicity to grazing livestock.

Soil Structure and Composition of Soil Air. The structure of soils, particularly those containing appreciable quantities of silt and clay, has a pronounced influence on both the root and top growth of plants. Soil structure to a great extent determines the bulk density of a soil. As a rule, the higher the bulk density, the more compact the soil, the more poorly defined the structure, and the smaller the amount of pore space. This is quite frequently reflected in restricted plant growth. The effect of fertilizer, moisture, and soil compaction on the dry weight of corn plants is given by the data in Table 2-14. The marked reduction in both top and

TABLE 2-14. Effect of Moisture, Fertility Level, and Degree of Soil Compaction on the Growth of Corn Plants

Treatment	Weight of Tops (g)	Weight of Roots (g)	Top/Root Ratio	Weight of Total Plant (g)
Loose, wet, fertilized	39.4	14.8	1:0.38	54.2
Loose, wet, unfertilized	23.5	10.1	1:0.43	33.7
Loose, dry, fertilized	27.5	9.3	1:0.34	36.8
Loose, dry, unfertilized	20.3	9.3	1:0.46	29.6
Compact, wet, fertilized	16.0	6.5	1:0.40	22.5
Compact, wet, unfertilized	17.0	7.7	1:0.45	24.7
Compact, dry, fertilized	20.1	11.3	1:0.56	31.4
Compact, dry, unfertilized	19.3	9.9	1:0.51	29.2

Source: Bertrand and Kohnke, *SSSA Proc.*, **21**:137 (1957).

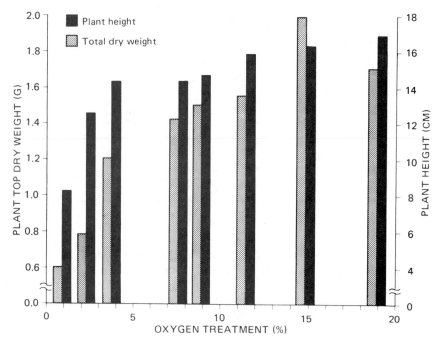

FIGURE 2-14. Effect of oxygen concentration of soil air on the growth of snapdragon plants. [Letey et al., *SSSA Proc.,* **25**:184 (1961).]

root growth as a result of soil compaction is well illustrated in this table. Also described is the limiting effect that soil compaction can have on plant response to applied fertilizer. Compare treatments 1, 2, and 5.

Bulk density is really a measure of pore space in the soil; the higher the bulk density for a given textural class, the smaller the amount of pore space present. Pore space, of course, is occupied by air and water, the amount of one being inversely related to the amount of other.

High bulk densities inhibit the emergence of seedlings. Anthocyanin tends to accumulate in tomato plants grown on soils with bulk densities of 1.4 to 1.7, and these same plants tend to be high in protein and low in sugar. High bulk densities offer increased mechanical resistance to root penetration. They almost certainly influence the rate of diffusion of oxygen into the soil pores, and root respiration is directly related to a continuing and adequate supply of this gas. The effect of increasing the percentage of oxygen on growth of snapdragon plants is illustrated in Figure 2-14. Although bulk density has an important influence on plant growth, the effect of the amount of oxygen in the soil air is of equal or greater importance, as shown in Figure 2-15.

Air with the oxygen percentages indicated in Figure 2-15 was maintained over the surface of each of the containers throughout the growth period of the snapdragons, which were the test plants. The soil in the containers had a bulk density of 1.2. The effect on root growth of the increasing oxygen percentage is obvious.

Under field conditions oxygen diffusion into the soil is determined largely by the moisture level of the soil if bulk density is not a limiting factor. On well-drained soils with good structure, oxygen content is not likely to retard plant growth except for possible infrequent periods of flooding, during which it may become a serious consideration because of the impact that oxygen supply has on ion uptake. The increasing importance of oxygen with decreasing moisture tension is illustrated in Figure 2-16. Note that raising the oxygen supply at low moisture tensions results in a continued increase in Rb ion uptake by the roots to an oxygen percentage of

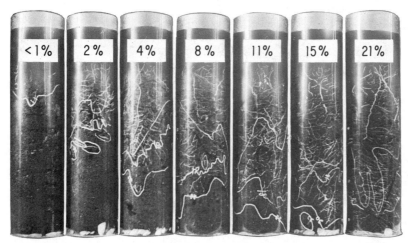

FIGURE 2-15. Effect of oxygen concentration maintained at the soil surface on root development. [Stolzy et al. *SSSA Proc.,* **25**:464 (1961).]

8 to 10. At higher moisture tensions ion uptake is reduced and, quite possibly because of the limiting effect of the high moisture tension, much less oxygen is required for the maximum.

The oxygen supply at the root absorbing surface is critical. Hence not only is the gross oxygen level of the soil air important, but also the rate at which oxygen diffuses through the soil to maintain an adequate partial pressure at the root surface. The influence of various rates of oxygen diffusion on the growth of pea plants on soils of three different fertility levels is illustrated in Figure 2-17. At low rates of oxygen diffusion small increases in diffusion rate have a much greater impact on corn grown on soils of medium or high fertility levels than on soils of low fertility level.

Agricultural plants differ widely in their sensitivity to soil oxygen supply. Paddy rice is grown under conditions of complete soil submergence. Tobacco is so sensitive to poor soil aeration that flooding a field for only a few hours causes serious damage or complete loss of the crop. Some pasture species are more tolerant of poorly aerated conditions than others, and species adapted to the conditions peculiar to the area should be selected. Good soil structure and aeration are imperative for

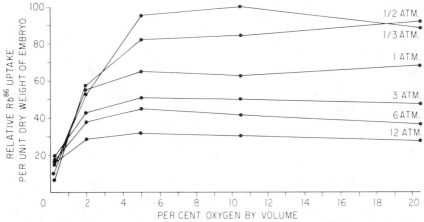

FIGURE 2-16. Effect of oxygen level and soil moisture tension on Rb[86] uptake by corn seedlings. [Danielson et al., *SSSA Proc.,* **21**:5 (1957).]

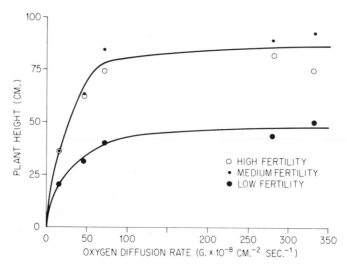

FIGURE 2-17. Effect of oxygen diffusion rate and fertility level on the growth of pea plants. [Cline and Erickson, *SSSA Proc.*, **23**:334 (1959).]

maximum yields of most agricultural crops, and the limiting effect that inadequate root oxygen supply can have on growth must be considered in sound crop production programs.

Soil Reaction. Soil reaction (soil acidity, pH) may affect plant development by its influence on the availability of certain elements required for growth. Examples are the increased sorption and precipitation of phosphates in acid soils high in iron and aluminum and of manganese in high organic matter soils with high pH values. A decline in the availability of molybdenum results from a decrease in soil pH. Acid mineral soils are frequently high in soluble aluminum and manganese, and excessive amounts of these elements are toxic to plants. Although the solubility of Fe^{3+} is low in well-aerated acid soils, iron oxides can still have a major role in the reversion of fertilizer phosphorus. Concentrated bands or particles of phosphorus fertilizer are powerful dissolving agents of iron oxide.

When ammoniacal nitrogen fertilizers are left on the surface of soil with pH values greater than 7, ammonia may be lost by volatilization. If losses are large, the expected responses to the applied nitrogen will not be obtained. High soil pH values of about 7.5 to 8.0 and above will favor the conversion of water-soluble fertilizer phosphorus into less soluble forms of lower availability to crops. Certain soilborne diseases are influenced by soil pH. Scab of Irish potatoes, pox of sweet potatoes, and black root rot of tobacco are favored by neutral-to-alkaline conditions. These diseases can be almost completely controlled by lowering the soil pH to 5.5 or less. The preference of many pathogenic fungi for ammonium nitrogen is apparently related to the acidification produced in the rhizosphere as a result of crop uptake of this form of nitrogen.

The importance of soil acidity to crop growth and the availability of plant nutrients are treated in detail in Chapter 11. Soil acidity is a property of the greatest importance to the grower and one that is easily and economically altered.

Biotic Factors. The many biotic factors that can limit plant growth present a constant hazard to farming operations and pose a potential threat of reduced crop yields, if not of crop failure. Heavier fertilization may encourage greater vegetative growth and better environmental conditions for certain disease organisms. The imbalance of nutrients available to plants may also be a reason for

TABLE 2-15.   Stem Rot Incidence in Rice and Grain Yield Index As Affected by Fertilizer

Treatment (lb/A)				
N	P_2O_5	K_2O	Percent Diseased	Yield Index
0	0	0	47.0	100
120	0	60	7.2	163
120	60	60	4.4	187
120	60	0	69.2	66

Source: Ismunadji, *Proc. 12th Colloq. Int. Potash Inst.,* pp. 47–60 (1976).

increased incidence of disease. The data in Table 2-15 show how proper balance of nutrients, especially phosphorus and potassium in combination with nitrogen, reduced stem rot of rice and increased yields of this crop. Controlling disease is essential for maximum crop yields and the greatest responses to applied fertilizer.

Certain pests may impose an added fertilizer requirement. Viruses and nematodes, for example, attack the roots of certain crops and reduce absorption. It is then necessary to supply a greater concentration of nutrient elements in the soil to provide reasonable growth. These pests attack many species of crop plants and cause serious reductions in yield. Fortunately, they can be controlled by practices such as the production of virus-free seed, crop rotation, and by chemical treatment of the soil. One of numerous examples of the beneficial effects on crop yield of controlling nematodes is given in Figure 2-18. Note the beneficial effect on soybean yield of using both potassium and a nematicide.

Closely allied with disease is the problem of insects. Any infestation may seriously limit plant growth, and uncontrolled it may be responsible for the failure of a farming enterprise. Numerous examples can be cited. Heavier fertilization may encourage certain insects, such as the cotton boll weevil, by greater vegetative growth. Definite advances have been made in breeding insect-resistant strains of certain crops, and great strides have been made in the development of insecticides over the last two decades.

Weeds are another serious deterrent to efficient crop production, for they compete for moisture, nutrients, and in many instances light. In addition to these competitive effects of weeds, crop growth may also be suppressed by biochemical inter-

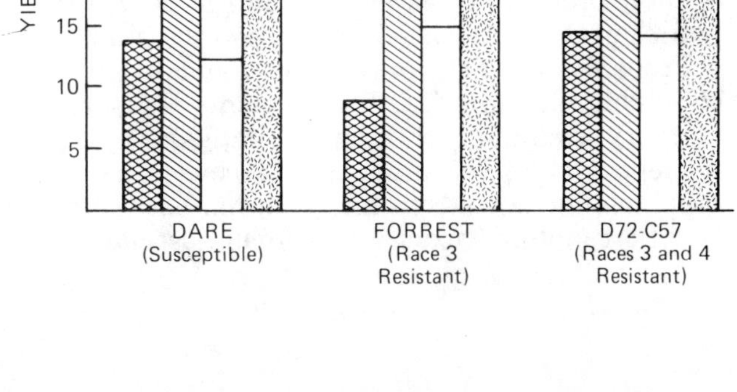

FIGURE 2-18. Yield of soybean varieties as affected by nematicide treatment and potassium fertility (East Prairie, Missouri, 1975–1976). [Shannon et al., *Better Crops Plant Food,***61** (1) :14 (1977).]

TABLE 2-16.   Effect of Wild Oats on the Yield of Principal Field Crops in Western Canada

Crop	Number of Wild Oats in Crop	Yield (bu/A)
Wheat	$191/m^2$	44
	None	82
Rapeseed	$187/m^2$	15
	None	37
Barley	$161/m^2$	71
	None	114
Flax	$220/m^2$	6
	None	33

Source: Dew, Agriculture Canada Research Station, Lacombe, Alberta (1978).

ference or allelopathy. Weeds are known to produce and release harmful substances into the root environment.

Chemical weed control is an established practice with most economic crops. The use of chemicals reduces the chance of root injury, which frequently results from mechanical cultivation, and lessens the structural damage to soils caused by repeated use of heavy tillage equipment. Preemergent herbicides have the advantage of destroying weeds before they begin to impose competitive and allelopathic effects on crops. Serious impairment of yields can occur in heavily weed-infested fields, as shown in Table 2-16. The importance of adequate weed control for less tolerant crops such as flax is obvious.

The subjects discussed in preceding sections of this chapter are more appropriately treated in other texts. They could not be passed over without mention, however, because of the limiting effect they may have on the efficient use of plant nutrients.

Supply of Mineral Nutrient Elements.   About 5 to 10% of the dry weight of plants is composed of the nutrient elements nitrogen, phosphorus, potassium, calcium, magnesium, sulfur, boron, chlorine, copper, iron, manganese, molybdenum, and zinc. Soil is the prime source of these essential nutrients, as well as other elements that are beneficial for plant growth. Various aspects of supply, availability, and uptake of these elements will be discussed in subsequent chapters.

Absence of Growth-Restricting Substances.   Normal development of plants can be restricted or stopped completely by toxic substances. Almost all soil elements, even those essential and beneficial for plant growth, will become toxic to plants when they are present in abnormally high concentrations in the root zone. It is not definitely known if the toxicity of these elements is a direct result of their presence in excessive quantities or if it is due to their interference with plant uptake of essential nutrients with similar chemical properties.

Probably one of the most widespread toxicity problems involves excessive levels of soluble aluminum. Other elements that are potentially toxic are nickel, lead, mercury, cadmium, chromium, manganese, copper, zinc, selenium, arsenic, molybdenum, chloride, boron, and fluorine. Other elements that pose potential toxicity problems include strontium, lithium, beryllium, vanadium, bismuth, iodine, and tin. Fortunately, toxicity problems associated with these various elements are

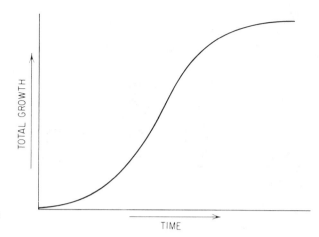

FIGURE 2-19. Generalized curve illustrating the growth pattern of an annual plant.

infrequent in most agricultural soils. They are most likely to occur in situations involving disposal of waste materials such as those from mines and metallurgical operations, sewage systems, animal and poultry enterprises, food-processing industries, pulp and paper mills, garbage collection, and others.

Organic compounds, including phenol, cresols, hydrocarbons, substituted ureas, and chlorinated hydrocarbon insecticides, can be toxic if they are present in high concentrations. Such chemicals are not usually a problem at low concentrations since soil microorganisms can acquire the ability to decompose them.

Growth Expressions

The manner in which plant growth responds to various inputs of plant nutrients has been the special study of a group of scientists who might be termed biometricians. These scientists have used various mathematical models to describe or define plant growth, and several of these concepts will be briefly presented in the following sections. If such mathematical models can be developed, they can be useful in predicting what crop yields can be expected if the supply of elements available to the crop is known.

Growth as Related to Time. Whether growth is expressed as the increase in dry weight or in height of the plant, there is a fairly constant relationship between the measure of growth employed and time. The general pattern is one of initially small increases in size, followed by large increases, and then by a period during which the size of the plant increases slowly or not at all. This pattern is illustrated by the generalized growth curve shown in Figure 2-19.

Growth Related to the Factors Affecting It. Growth curves are helpful to an understanding of the general pattern of plant development. They indicate nothing, however, of the factors affecting growth, such as the supply of mineral nutrient elements, light, carbon dioxide, and water. The plant is a product of both its genetic constitution and its environment. The genetic pattern is a fixed quantity for a given plant and determines, so to speak, its potential for maximum growth under an environment favorable to its development. From the standpoint of that particular plant, attainment of this maximum development is dependent on a favorable environment. In other words, plant growth is a function of various environmental or growth factors, which may be considered as variables, the magnitude and combination of which determine the amount of growth that will be made. Symbolically, this may be expressed as

$$G = f(x_1, x_2, x_3, \ldots, x_n) \tag{1}$$

where

G = some measure of plant growth

and

$x_1, x_2, x_3, \ldots, x_n$ = the various growth factors

Further, if all but one of the growth factors are present in adequate amounts, an increase in the quantity of this limiting factor will generally result in increases in plant growth. Symbolically,

$$G = f(x_1)x_2, x_3, \ldots, x_n$$

This, however, is not a simple linear relationship. Although the results of experiments, such as those of Liebig, show linear responses over a portion of the yield response curve, more often the addition of each successive increment of a growth factor results in progressively smaller increases in growth.

Liebig's law of the minimum and his other contributions to agricultural chemistry were discussed in Chapter 1. This law can be simply stated as follows: "Even if all but one of the essential elements be present, the absence of that one constituent will render the crop barren."

Mitscherlich's Laws. Mitscherlich was among the first to quantify the relationship between plant growth response and the addition of a growth factor. Two laws that resulted from his work are defined next.

Mitscherlich's Law of Physiological Relationships. "Yield can be increased by each single growth factor even when it is not present in the minimum as long as it is not present in the optimum."

Mitscherlich's Growth Law. "Increase in yield of a crop as a result of increasing a single growth factor is proportional to the decrement from the maximum yield obtainable by increasing the particular growth factor."

Mitscherlich's Equation. The growth equation discussed in the preceding section is a generalized expression relating growth to all the factors involved. In 1909, E. A. Mitscherlich of Germany developed an equation that related growth to the supply of plant nutrients. He observed that when plants were supplied with adequate amounts of all but one nutrient, their growth was proportional to the amount of this one limiting element that was supplied to the soil. Plant growth increased as more of this element was added, but not in direct proportion to the amount of the growth factor added. The increase in growth with each successive addition of the element in question was progressively smaller. Mitscherlich expressed this mathematically as

$$dy/dx = (A - y)C \tag{2}$$

where dy is the increase in yield resulting from an increment dx of the growth factor x, A is the maximum possible yield obtained by supplying all growth factors in optimum amounts, y is the yield obtained after any given quantity of the factor x has been applied, and C is a proportionality constant which might be considered as an efficiency coefficient or factor.

Spillman's Equation. As is so often the case in scientific investigations, this same principle was independently developed several years later by W. J.

Spillman. Spillman expressed the relation as

$$y = M(1 - R^x) \tag{3}$$

where y is the amount of growth produced by a given quantity of the growth factor x, x is the quantity of the growth factor, M is the maximum yield possible when all growth factors are present in optimum amounts, and R is a constant.

Further work by Spillman showed that, though of a different form, his equation and that of Mitscherlich could both be reduced to

$$y = A(1 - 10^{-cx}) \tag{4}$$

where y is the yield produced by a given quantity of the growth factor x, A is the maximum yield possible, and c is a constant dependent on the nature of the growth factor. None of these expressions is conveniently handled as written, but they may also be stated as the integral, in common log form, of equation (2):

$$\log (A - y) = \log A - c(x) \tag{5}$$

The symbols used are the same as those in equation (4). If the function is graphed, the curve obtained appears as shown in Figure 2-20.

The Baule Unit. Reference has been made so far only to units of a growth factor. Certainly this is a rather meaningless term, and it is necessary to understand its relation to familiar terms before proceeding with the discussion. This unit has been designated as a "baule," after the German mathematician who collaborated with Mitscherlich. Baule suggested that the unit of fertilizer, or any other growth factor, be taken as that amount necessary to produce a yield that is 50% of the difference between the maximum possible yield and the yield before that unit was added. As almost everyone knows, plants require different absolute amounts of nitrogen, phosphorus, and potassium, but the amount in pounds of each required to produce a yield that is 50% of the maximum possible is termed 1 baule unit. Obviously, according to this concept, 1 baule of a growth factor is equivalent to 1 baule of any other growth factor in terms of growth-promoting ability. The values of the baule unit in pounds per acre of N, P_2O_5, and K_2O are 223, 45, and 76, respectively, as calculated from the results of Mitscherlich's work.

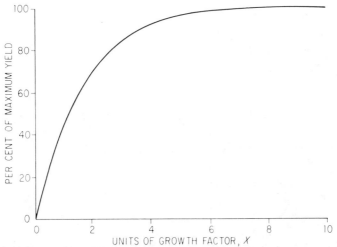

FIGURE 2-20.   Percentage of maximum yield as a function of increasing additions of a growth factor, x.

Calculation of the Value of the Proportionality Factor C. The constant c in equation (5) becomes 0.301 when yields are expressed on a relative basis of $A = 100$ and x is stated in baule units. This is shown by first rewriting equation (5) as follows:

$$\log A - \log (A - y) = cx$$

or

$$\log \frac{A}{A - y} = cx$$

When the nutrient supply is increased by 1 baule unit,

$$\frac{A}{A - y} = \frac{100}{50} = 2$$

Thus $\log 2 = c(1)$ and $c = 0.301$.

When conventional units of yield are used, the value of c varies with the particular growth factor. Mitscherlich found that the value of c was 0.122 for N, 0.60 for P_2O_5, and 0.40 for K_2O. He claimed that it was constant for each fertilizer nutrient, independent of the crop, the soil, or other conditions. The average value for c in British experiments conducted before 1940 was 1.1 for N, 0.80 for P_2O_5, and 0.80 for K_2O. It has been observed in numerous other investigations that c is not a constant term and that it varies rather widely for different crops grown under different conditions.

The significance of the c term is that it gives an indication of whether the maximum yield level can be achieved by a relatively low or high quantity of the specific growth factor. When the value of c is small, a large quantity is needed, and vice versa.

Calculation of Relative Yields from Addition of Increasing Amounts of a Growth Factor. If A, the maximum yield, is considered to be 100%, equation (5) reduces to

$$\log (100 - y) = \log 100 - 0.301(x) \tag{6}$$

It is possible to determine the relative yield expected from the addition of a given number of units of x. It will be helpful if the student observes how these calculations are made.

If none of the growth factor is available, that is, $x = 0$, then $y = 0$; but suppose that 1 unit of x is present. Then

$$\log (100 - y) = \log 100 - 0.301(1)$$
$$\log (100 - y) = 2 - 0.301$$
$$\log (100 - y) = 1.699$$
$$100 - y = 50$$
$$y = 50$$

the addition of 1 unit of the growth factor x results in a yield that is 50% of the maximum.

Assume, however, that 2 units of the growth factor were present. In this instance

$$\log (100 - y) = \log 100 - 0.301(2)$$
$$\log (100 - y) = 2.000 - 0.602$$
$$\log (100 - y) = 1.398$$
$$100 - y = 25$$
$$y = 75$$

The same operation may be repeated until 10 units of the growth factor have been added. The result of such a series of calculations is given in tabular form.

Units of Growth Factor, x	Yield (%)	Increase in Yield (%)
0	0	—
1	50	50
2	75	25
3	87.5	12.5
4	93.75	6.25
5	96.88	3.125
6	98.44	1.562
7	99.22	0.781
8	99.61	0.390
9	99.80	0.195
10	99.90	0.098

It is obvious that the successive increases of a growth factor result in a yield increase that is 50% of that resulting from addition of the preceding unit until a point is reached at which further increases, for all intents and purposes, are of no consequence.

Yield Increases in Response to More Than One Growth Factor. An extension of the baule unit concept has been developed for cases in which two or more growth factors are limiting. It has been shown that when all growth factors but one, x, were present in optimum amounts, the addition of 1 unit of this factor x would produce a yield that was 50% of the maximum possible. Suppose, however, that all except two growth factors, x_1 and x_2, were present in optimum amounts. If 1 baule of each is added, the yield obtained will not be 50% but 50% times 50%, or 25% of the maximum. If all growth factors but three were present in optimum amounts, the simultaneous addition of 1 unit of each of the three would result in a yield that was $50 \times 50 \times 50$, or 12.5% of the maximum. This relationship is expressed by the general equation

$$y = A(1 - 10^{-0.301x_1})(1 - 10^{-0.301x_2})(1 - 10^{-0.301x_3}) \tag{7}$$

in which x_1, x_2, and x_3 are quantities of the growth factors to be added. The equation is very simply handled by determining the percentage yields that correspond to given values for x_1, x_2, and x_3 and then multiplying these figures together.

It is of interest to note that the rule of decreasing increments applies even though all growth factors are not present in optimum amounts. This makes possible a calculation of the maximum possible yield under a given set of climatic and genetic conditions when fertilizers are added in increasing amounts.

It also makes possible the calculation of the greatest possible crop production from a limited amount of fertilizer, for the greatest increases in yield result from the addition of the first unit increment. The German people made extensive use of this concept during World War II. It is probably fair to say that the practical application of Mitscherlich's ideas was largely responsible for preventing a greater degree of starvation in Germany than actually occurred during the war years.

The Agrobiology of O. W. Wilcox. O. W. Wilcox has employed the Mitscherlich equation to extend a branch of science that he calls quantitative agrobiology. This science, he states, "comprises the general and specific *quantitative* relations between plants and the other factors of their growth and yield." Wilcox assumes that the Mitscherlich equation is valid, an opinion that is by no means shared by many agricultural scientists. He further develops from the Mitscherlich equation the inverse nitrogen–yield concept, which says in effect that the yield of a crop is inversely proportional to its nitrogen content. Symbolically,

$$Y = \frac{k}{n} \tag{8}$$

where Y is the yield, n is the percentage of nitrogen in the crop, and k is a constant. To support this contention, Wilcox cites a number of examples of crop yields and nitrogen percentages that follow this general rule.

From the Mitscherlich equation, Wilcox further extended the inverse yield–nitrogen concept and evaluated the constant in equation (8). He finds that the value of k is 318 lb/A and states that this is the maximum amount of nitrogen that can be absorbed in one season by an annual crop growing on an acre of land. This presumably would occur with the production of a maximum yield, which could happen only if none of the various growth factors was limiting. It is obvious that crop yields would increase ad infinitum if only the nitrogen content of the crop on a percentage basis could be made to decrease.

Although there are numerous examples reported in which the Mitscherlich–Baule–Wilcox equations fail to describe plant growth adequately as a function of plant nutrient supply, the concepts, at least in many situations, provide useful expressions for quantitatively relating nutrient levels and plant growth responses. The ideas of Mitscherlich, Baule, and Wilcox are not completely without foundation. Plant growth as a function of nutrient inputs *is* logarithmic and generally follows a pattern of diminishing increases, as expressed in the Mitscherlich equation. The growth of annual plants *does* tend to reach a maximum with increasing inputs of nutrients under a particular set of environmental conditions, and often the plants that produce the highest yield of dry matter have the lowest percentage of nitrogen in their tissues. However, it remains for posterity to determine whether a single expression can be developed that will universally predict the amount of growth that can be produced from the input of a given quantity of plant nutrients when environmental and genetic growth factors are adequately described.

An example of the use to which the Mitscherlich concept can be put in soil fertility research was developed by workers at the Tennessee Valley Authority at Muscle Shoals, Alabama. The effectiveness of various fertilizer materials as sources of nitrogen for corn was determined. The yield data obtained followed the Mitscherlich diminishing yield response pattern. The model so developed provided for a common limiting yield and a common intercept. It can be fitted by standard regression techniques, and the coefficients obtained can be readily compared by statistical techniques. A comparison of the observed and values predicted by this technique is shown in Figure 2-21.

Factorial Experiments and Regression Equations. Much fertilizer research has made use of what is known as the factorial experiment. In such studies the effect on crop yield of several levels of different input factors is evaluated simultaneously in one experiment. For example, the effect of several levels of nitrogen and phosphorus on the yield of a crop can be studied simultaneously in one experimental layout. If three levels of nitrogen and three levels of phosphorus are examined, the experiment is said to be a 3^2 factorial and requires nine treatments to evaluate the effect on yield of all combinations of nitrogen and phosphorus. The results are usually subjected to various statistical treatments. An equation, normally termed a regression equation, is developed in which the yield is functionally related to the inputs of the fertilizer variables.

Research workers at the Iowa and North Carolina Experiment Stations were among the first to make extensive use of this type of experiment, which is now popular in the United States. When climate, soil type, plant population, fertilizer placement, and other factors are uniform and constant, such studies help to predict

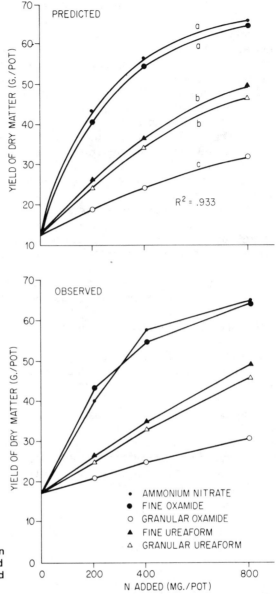

FIGURE 2-21. Observed and predicted corn forage yield as affected by rate, source, and granule size of fertilizer nitrogen. [Engelstad and Khasawneh, *Agron. J.,* **61**:473 (1969).]

fertilizer requirements, but the equations developed from them are not universally applicable. The models generated by the data from these studies are commonly of the diminishing-return type, similar to those of the Mitscherlich and related expressions.

Waggoner and Norvell at the Connecticut Agricultural Experiment Station in 1979 reported a statistical method for fitting Liebig's law of the minimum by least-squares regression to yield responses to two essential factors. Their method provided a simpler, more logical, and better-fitting model than did multiple linear regression.

It should be noted that regression techniques can only organize data already collected under a particular set of conditions and have little predictive value for other situations. The latter is of interest and significance in making fertilizer recommendations.

Nutrient Interactions. Interations between plant nutrients are often overlooked even though they can have considerable influence on plant growth. The interplay of plant nutrients is best studied in multifactorial experiments which test each nutrient at three or more rates. The main effects of nutrients are often unrelated to their interactions, and interactive influence may not decline with increasing rates of addition, as primary effects do.

Two or more growth factors are said to interact when their influence individually is modified by the presence of one or more of the others. An interaction takes place when the response of two or more inputs used in combination is unequal to the sum of their individual responses. There can be both positive and negative interactions in soil fertility studies (Figure 2-22). In addition, there can be circumstances where there is no interaction, with the action of factors being only additive.

In negative interactions, the two nutrients combined increase yields less than when they are applied separately. This kind of interaction can be the result of substitution for and/or interference of one treatment on the other. Lime × P, lime × Mo, Mo × P, and Na × K are common negative interactions involving apparent substitution effects. Changes in soil pH will result in numerous interactions where one ion or nutrient interferes with or competes with the uptake and utilization of other nutrients by plants.

Positive interactions are in accordance with Liebig's law of the minimum. If two factors are limiting, or nearly so, addition of one will have little effect on growth, whereas provision of both together will have a much greater influence. In severe deficiencies of two or more nutrients, all fertilizer responses will result in strong positive interactions.

Yield increases from an application of one nutrient can reduce the concentration of a second nutrient, but the higher yields result in greater uptake of the second nutrient. This is a dilution effect, which should be distinguished from an antagonistic effect.

In addition to interactions between two or more nutrients, there are numerous opportunities for other kinds of interactions: for example, nutrients and disease, nutrients and cultural practice, nutrients and crop species, nutrients and hybrid or variety, nutrients and seeding date, nutrients and plant population or spacing, and nutrients and environmental conditions. Many of these types of interactions are discussed in Chapter 17.

Bray's Nutrient Mobility Concept. A modification of the Mitscherlich–Baule–Spillman concept was proposed by R. Bray and his co-workers at the University of Illinois. In brief, it is claimed that crop yields obey the percentage

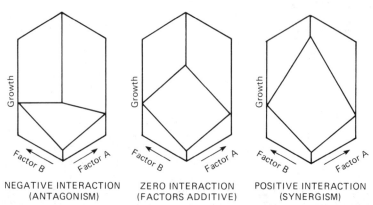

NEGATIVE INTERACTION   ZERO INTERACTION   POSITIVE INTERACTION
(ANTAGONISM)   (FACTORS ADDITIVE)   (SYNERGISM)

FIGURE 2-22. Influence of interactions between two nutrient factors on the growth of a crop. (Sumner and Farina, in J. K. Syers, Ed., *Phosphorus in Agricultural Systems.* Elsevier, New York, 1983.)

sufficiency concept of Mitscherlich for such elements as phosphorus and potassium, which are relatively immobile in the soil. This concept, in turn, is based on Bray's nutrient mobility concept, which states that

> as the mobility of a nutrient in the soil decreases, the amount of that nutrient needed in the soil to produce a maximum yield (the soil nutrient requirement) increases from a variable net value, determined principally by the magnitude of the yield and the optimum percentage composition of the crop, to an amount whose value tends to be a constant.

The magnitude of this constant is independent of the amount of crop yield, provided that the kind of plant, planting pattern and rate, and fertility pattern remain constant and that similar soil and seasonal conditions prevail. He further states that for a mobile element such as nitrate nitrogen, Liebig's law of the minimum best expresses the growth of a crop. Bray has modified the Mitscherlich equation to

$$\log (A - Y) = \log A - C_1 b - Cx \qquad (9)$$

where A, Y, and x have the connotation already given; C_1 is a constant representing the efficiency of b for yields in which b represents the amount of an immobile but available form of nutrient, such as phosphorus or potassium, measurable by some suitable soil test; and C represents the efficiency factor for x, which is the added fertilizer form of the nutrient b.

On the basis of work done in Illinois, Bray has shown that the values for C_1 and C are specific and fairly constant over a wide area in the state, regardless of yield and season for each of the following crops: corn, wheat, and soybeans. The factors that will alter the values, however, are wide differences in soil type, plant population and planting patterns, and the form and distribution in the soil of the immobile nutrient element under study. Hence, as management practices and fertilizer placement methods are changed to obtain higher yields, the values change and must be reexamined.

Limited Applications of Growth Expressions. It is obvious that there is a lack of agreement among the various concepts developed to describe the relation between plant growth and nutrient element input. Steenbjerg and Jakobsen of Denmark, in commenting on the variability among growth response curves, point out that "the constants in formulas are not constants because the variables in the formulas are not independent variables." Factors other than nutrient interactions obviously affect the shape of yield curves. They include other environmental factors, which were discussed in the preceding sections. The change in shape and position of yield–plant nutrient input curves with changes in environmental conditions is of the greatest importance to the practical agriculturalist. The understanding of the interplay among these factors and their successful manipulation by the farm operator to bring about maximum crop production at a minimum production cost determines in the final analysis the degree of success of the farm operation.

The term *limiting growth factor* or more simply *limiting factor*, used frequently throughout this book, has been clearly illustrated by the variable nature of the response curves and surfaces previously discussed. If, for example, a crop has inadequate moisture, the application of a given amount of fertilizer will provide a lower yield than if moisture were adequate. Another example, and an important one, is the application of fertilizer to a crop growing on a soil that is too acid for maximum growth, regardless of the amount of fertilizer added. If lime is not applied, acidity becomes the limiting factor that keeps yield responses to the added fertilizer low and reduces the farmer's return on investment. The importance to practical farm operations of the concept of a growth pattern and how it may be altered by various "limiting factors" cannot be overstated.

An adequate supply of mineral nutrient elements is required for maximum agricultural production, but alone they are no guarantee of maximum yield because of the possible limiting effect of the numerous items that influence plant growth. The remaining chapters in this book deal largely with the role that mineral elements play in the production of agricultural crops.

Summary

1. Plant growth as a function of time, genetic makeup of the plant, and environmental factors was discussed. The importance of selecting crops that are genetically capable of making maximum utilization of the supply of available plant nutrients was pointed out.

2. The environmental factors were considered in relation to their effect on plant growth as well as their impact on limiting crop response to applied plant nutrients. The concept of limiting growth factors was discussed and the need for recognizing the importance of this concept in practical farming operations pointed out.

3. Growth of annual plants follows a well-defined pattern. Plant responses to environmental conditions, including the supply of plant nutrients, also follow a set pattern. When growth is plotted as a function of increasing amounts of applied nutrients, successive increments of fertilizer give successively smaller increases in plant growth. Such curves, known as response curves, have been studied by numerous investigators and various mathematical formulas have been developed to describe them.

Questions

1. Equations that express plant growth as a function of inputs of a plant nutrient differ from one another to varying degrees. To what do you attribute these differences?

2. Despite the differences among the various equations referred to in Question 1, they all seem to have one thing in common. What is it?

3. Crop yields have been increasing over the years because of improvements in tillage, varieties, pest control, fertilization, and so on. From a *theoretical* standpoint, what do you consider to be the factor that will *ultimately* limit further increases in plant growth? From a *practical* standpoint, what do you think this factor will be?

4. Roger Bray of Illinois had unusually good results in predicting crop responses, especially by corn, to applications of potassium and phosphate fertilizers from the equations and soil tests that he has developed. What are some of the factors that may have contributed to this success?

5. Plant yields in terms of forage, grain, or fruit are the criteria of the effectiveness of various fertilizer inputs. If one is interested in the effect of the imposed treatment on the plants' ability to convert radiant energy to a usable form, are these yield figures in your opinion the best criteria of treatment effect? Why? (*Hint:* Compare the highest yields of soybeans with the highest yields of corn.) What criterion would you use?

6. What growth factor in the past has been frequently overlooked by plant breeders in developing new crop varieties?

7. According to the Mitscherlich–Baule concept, what percentage of maximum growth would be produced if only three plant nutrients were the factors limiting growth and they were supplied to the extent of 6, 4, and 2 baules, respectively?

8. Among environmental factors limiting crop response to applied nutrients, which is probably the most easily and inexpensively overcome?

9. What are some of the environmental factors affecting plant response to applied nutrients that are more easily overcome or controlled in the greenhouse than in the field? If you were planning to control these factors in a greenhouse operation set up for the commercial production of crops, what items would you consider before instituting control?

10. In your opinion, could the limited or partial control of the carbon dioxide content of the aboveground atmosphere under field conditions be undertaken successfully? Under what conditions would you expect such control to be successful?

11. Of *all* the growth factors limiting crop production, which is most easily corrected in commercial farming operations?

12. Study the various growth curves discussed in this chapter. Should a commercial grower always attempt to produce maximum yields? Why? At what point along the yield–fertilizer input curve do you feel the grower should operate? Elaborate.

13. Why is soil structure so important in influencing crop responses to applied plant nutrients? Be specific. What can be done about soil structure? On which type of soil is structure the more important— loamy sands or silty clay loams? Why?

14. Light or radiant energy was listed as an environmental factor affecting growth and response to applied plant nutrients. In what ways does light influence growth under field conditions? Under controlled conditions? What can be done about the effect of light on growth?

15. Although not listed as such, human beings comprise the biotic factor that influences crop growth to the greatest extent. Why is this statement made, especially in relation to commercial farming?

16. In the strictest sense, is it correct to refer to the various environmental factors as independent variables? Why?

17. Which plant physiological process (or processes) is (or are) most affected by soil water deficits?

18. List the principal atmospheric pollutants that might affect plant growth. Are any of them beneficial for plants?

19. Describe the various ways that weeds compete with crops.

20. Are there differences in the ability of crop species to compete with a given weed species?

21. Identify one major growth factor that may strongly influence successful corn production in the U.S. Corn Belt.

22. Why is temperature an important factor affecting crop growth?

23. Are current yields of corn and soybeans approaching their potential maximum yields?

24. High-yield potential is an important plant characteristic. What are other important properties?

25. What plant growth factors can be largely controlled by man? Which ones are beyond our control?

26. Is fertilizer response related to soil moisture supply?

27. Does soil moisture supply influence plant nutrient availability and uptake? How does it affect nutrient availability?

28. Does photoperiod affect crop growth? How?
29. What is an interaction?

Selected References

Adams, J. E., "Effect of soil temperature on grain sorghum growth and yield." *Agron. J.*, **54**:257 (1962).

Black, C. A., and O. Kempthorne, "Willcox's agrobiology: I. Theory of the nitrogen constant 318." *Agron. J.*, **46**:303 (1954).

Black, C. A., and O. Kempthorne, "Willcox's agrobiology: II. Application of the nitrogen constant 318." *Agron. J.*, **46**:307 (1954).

Bole, J. B., and U. J. Pittman, "The effect of fertilizer N, spring moisture, and rainfall on the yield and protein content of barley in Alberta." *Proc. 1978 Soils Crop Workshop*, Publ. 390, Univ. of Saskatchewan Ext. Div. 114 (1978).

Bray, R. H., "Confirmation of the nutrient mobility concept of soil–plant relationships." *Soil Sci.*, **95**:124 (1963).

Caviness, C. E., and G. W. Hardy, "Response of six diverse genetic lines of soybeans to different levels of soil fertility," *Agron. J.*, **62**:236 (1970).

Ching, P. C., and S. A. Barber, "Evaluation of temperature effects on K uptake by corn." *Agron. J.*, **71**:1040 (1979).

Cline, R. H., and A. E. Erickson, "The effect of oxygen diffusion rate and applied fertilizer on the growth, yield, and chemical composition of peas." *SSSA Proc.*, **23**:333 (1959).

Cook, R. J., and E. Reis, "Cultural control of soil borne pathogens of wheat in the Pacific Northwest of the U.S.A." In J. F. Jenkyn and R. T. Plumb, Eds., *Strategies for the Control of Cereal Diseases*. London: Blackwell Scientific Publications, 1981, pp. 167–177.

Cooke, G. W., "Value of 'blueprints' in research and advisory work." *IPI Symp.*, Warsaw (June 23, 1981).

Danielson, R. E., and M. B. Russell, "Ion absorption by corn roots as influenced by moisture and aeration." *SSSA Proc.*, **21**:3 (1957).

de Geus, J. G., *Fertilizer Guide for the Tropics and Subtropics*, 2nd ed. Zurich: Centre d'Étude de l'Azote, 1973.

Dew, D. A., "An index of competition for estimating crop loss due to weeds." *Can. J. Plant Sci.*, **52**:921 (1972).

Dew, D. A., and C. H. Keys, "An index of competition for estimating loss of rape due to wild oats." *Can. J. Plant Sci.*, **56**:1005 (1976).

Donald, C. M., "The interaction of competition for light and for nutrients." *Australian J. Agr. Res.*, **9**:421 (1958).

Engelstad, O. P., and F. E. Khasawneh, "Use of a concurrent Mitscherlich model in fertilizer evaluation." *Agron. J.*, **61**:473 (1969).

Evans, S. A., "The place of fertilizers in 'blueprints' for the production of potatoes and cereals." *Proc. 13th Colloq. IPI*, p. 231 (1977).

Fischer, R. A., and I. Aguilar M., "Yield potential in a dwarf spring wheat and the effect of carbon dioxide fertilization." *Agron. J.*, **68**:749 (1976).

Flocker, W. J., and D. R. Nielson, "The absorption of nutrient elements by tomatoes associated with levels of bulk density." *SSSA Proc.*, **26**:183 (1962).

Flocker, W. J., J. A. Vomocil, and F. D. Howard, "Some growth responses of tomatoes to soil compaction." *SSSA Proc.*, **23**:188 (1959).

Gilmore, E. C., and J. S. Rogers, "Heat units as a method of measuring maturity in corn." *Agron. J.*, **50**:611 (1958).

Graham, E. R., P. L. Lopez, and T. M. Dean, "Artifical light as a factor influencing yields of high-population corn." *Trans. ASAE*, **15**(3):576 (1972).

Harper, L. A., D. N. Baker, J. E. Box, Jr., and J. D. Hesketh, "Carbon dioxide and the photosynthesis of field crops: a metered carbon dioxide release in cotton under field conditions." *Agron. J.*, **65**:7 (1973).

Hicklenton, P. R., and P. A. Jolliffe, "Effects of greenhouse CO_2 enrichment on the yield and photosynthetic physiology of tomato plants." *Can. J. Plant Sci.*, **58**:801 (1978).

Ismunadji, M., "Rice diseases and physiological disorders related to K deficiency," in *Fertilizer Use and Plant Health*, Proc. 12th Colloq. IPI, pp. 47–60. Bern: International Potash Institute, 1976.

Jackson, T. L., "Interactions in soil fertility," in *Résumé of Papers, Irrigated Agr. Fert. Conf.*, Pasco, Wash. (January 5, 1983).

Johnson, R. R., "How high can yields go?" *Crops Soils*, **32**(9):9 (1980).

Jones, M. B., C. M. McKell, and S. S. Winans, "Effect of soil temperature and nitrogen fertilization on the growth of soft chess (*Bromus mollis*) at two elevations." *Agron. J.*, **55**:44 (1963).

Kilmer, V. J., O. L. Bennett, J. F. Stahly, and D. R. Timmons, "Yield and mineral composition of eight forage species grown at four levels of soil moisture." *Agron. J.*, **52**:282 (1960).

Knoll, H. A., D. J. Lathwell, and N. C. Brady, "The influence of root zone temperature on the growth and contents of phosphorus and anthocyanin of corn." *Soil Sci. Soc. Am. J.*, **28**:400 (1964).

Kramer, P. J., "Water stress and plant growth." *Agron. J.*, **55**:31 (1963).

Kunkel, R., M. Campbell, and R. Thornton, "Potato fertilizer and moisture management—What's needed?" *Proc. 11th Annu. Wash. Potato Conf. Trade Fair*, p. 87 (1972).

Letey, J., O. R. Lunt, L. N. Stolzy, and T. E. Szuszkiewicz, "Plant growth, water use, and nutritional response to rhizosphere differentials of oxygen concentration." *SSSA Proc.*, **25**:183 (1961).

Letey, J., L. H. Stolzy, N. Valoras, and T. E. Szuszkiewicz, "Influence of soil oxygen on growth and mineral content of barley." *Agron. J.*, **54**:538 (1962).

Nielsen, K. F., R. L. Halstead, A. J. McLean, R. M. Holmes, and S. J. Bourget, "The influence of soil temperature on the growth and mineral composition of oats." *Can. J. Soil Sci.*, **40**:255 (1960).

Ormrod, D. P., "Photosynthesis rates of young rice plants as affected by light intensity and temperature." *Agron. J.*, **53**:93 (1961).

Parks, W. L., and J. L. Knetsch, "Utilizing drought days in evaluating irrigation and fertility response studies." *SSSA Proc.*, **24**:289 (1960).

Phillips, R. E., and D. Kirkham, "Soil compaction in the field and corn growth." *Agron. J.*, **54**:29 (1962).

Rhykerd, C. L., R. Langston, and G. O. Mott, "Influence of light on the foliar growth of alfalfa, red clover, and bird's-foot trefoil." *Agron. J.*, **51**:199 (1959).

Scarsbrook, C. E., O. L. Bennett, and R. W. Pearson, "The interaction of nitrogen and moisture on cotton yields and other characteristics." *Agron. J.*, **51**:718 (1959).

Shannon, J. G., C. H. Baldwin, Jr., G. W. Colliver, and E. E. Hartwig, "Potash fertilization helps fight soybean cyst nematode." *Better Crops Plant Food*, **61**(1):12 (1977).

Steenbjerg, F., and S. T. Jakobsen, "Plant nutrition and yield curves." *Soil Sci.*, **95**:69 (1963).

Sumner, M. E., and P. M. W. Farina, "Phosphorus interactions with other nutrient and lime in field cropping systems," in J. K. Syers, Ed., *Phosphorus in Agricultural Systems*. New York: Elsevier, 1983.

Van der Paauw, F., "Critical remarks concerning the validity of the Mitscherlich effect law." *Plant Soil*, **4**:97 (1952).

Viets, F. G., Jr., "Fertilizers and the efficient use of water." *Adv. Agron.*, **14**:223 (1962).

Waggoner, P. E., and W. A. Norvell, "Fitting the law of the minimum to fertilizer applications and crops yields." *Agron. J.*, **71**:352 (1979).

Waggoner, P. E., D. N. Moss, and J. D. Hesketh, "Radiation in the plant environment and photosynthesis." *Agron. J.*, **55**:36 (1963).

White, W. C., and C. A. Black, "Willcox's agrobiology: **III**. The inverse yield-nitrogen law." *Agron. J.*, **46**:310 (1954).

Willcox, O. W., *The ABC of Agrobiology*. New York: Norton, 1937.

Wittwer, S. H., "Maximum production capacity of food crops." *Bioscience*, **24**(4):216 (1974).

Wittwer, S. H., "Environmental and societal consequences of a possible CO_2-induced climate change on agriculture." *Annu. Meet. Am. Assoc. Adv. Sci.*, (January 5, 1980). (*Michigan Agr. Exp. Sta. Publ. 9270.*)

Wittwer, S. H., and W. Robb, "Carbon dioxide enrichment of greenhouse atmospheres for vegetable crop production." *Econ. Bot.*, **18**:343 (1964).

Chapter 3

Elements Required in Plant Nutrition

Essentiality of Elements in Plant Nutrition

ALTHOUGH Arnon's criteria of essentiality of elements for plant growth may be more exact, the definition of essentiality proposed by D. J. Nicholas of the Long Ashton Research Station is more practical and thus is employed in this text. Nicholas advanced the term "functional or metabolic nutrient" to include any mineral element that functions in plant metabolism, whether or not its action is specific. This definition avoids the confusion that sometimes occurs when the more rigid criteria of essentiality are imposed, requiring for example, the necessity of establishing that symptoms of a deficiency can be prevented or corrected by supplying that particular element. Elements such as chlorine, silicon, sodium, and vanadium are classed as essential when the less restrictive definition of essentiality is used.

Several terms commonly used to describe levels of nutrient elements in plants are: deficient, insufficient, toxic, and excessive. Their definitions follow:

Deficient: when an essential element is at a low concentration that severely limits yield and produces more or less distinct deficiency symptoms. Extreme deficiencies will lead to death of the plant.

Insufficient: when the level of an essential plant nutrient is below that required for optimum yields or when there is an imbalance with another nutrient. Symptoms of this condition are seldom evident.

Toxic: when the concentration of either essential or other elements is sufficiently high to reduce plant growth severely. Severe toxicity will result in death of plants.

Excessive: when the concentration of an essential plant nutrient is sufficiently high to result in a corresponding shortage of another nutrient.

Elements Required in Plant Nutrition

Carbon, hydrogen, oxygen, nitrogen, phosphorus, and sulfur are the elements of which proteins, hence protoplasm, are composed. In addition to these six, there are 14 other elements which are essential to the growth of some plant or plants: calcium, magnesium, potassium, iron, manganese, molybdenum, copper, boron, zinc, chlorine, sodium, cobalt, vanadium, and silicon. Not *all* are required for *all* plants, but *all* have been found to be essential to *some*. These mineral elements, in addition to phosphorus and sulfur, usually constitute what is known as the *plant ash*, or the mineral remaining after the "burning off" of carbon, hydrogen, oxygen, and nitrogen. Each of the 20 plays a role in the growth and development of plants, and when present in insufficient quantities can reduce growth and yields.

The carbon, hydrogen, and oxygen contained in plants are obtained from carbon dioxide and water. They are converted to simple carbohydrates by photosynthesis and ultimately elaborated into amino acids, proteins, and protoplasm. These elements are not considered to be mineral nutrients, and, with the exception of the control man exerts over water and, to a lesser extent, carbon dioxide, there is little of practical importance that can be done to alter the supply to plants.

Plant content of mineral elements is affected by a host of factors. Their percentage composition in crops therefore varies considerably and should be kept in mind when consulting tables of data showing plant composition of the various elements. Some of the older data still in use may list figures that are too low because of the inadequacy of earlier analytical methods. A case in point is sulfur. Recent studies have shown that some of the earlier percentage figures for the plant content of this element are too low. As pointed out by Venema, values of 2- to a 100-fold higher for sulfur are found today with modern analytical methods. Also, earlier chemical analyses may be low because they were determined on crops yielding much below the levels required in modern agriculture.

Plant composition data have in some cases been mistakenly used as the sole basis for formulating a fertilizer program, the idea being that the quantities of the elements removed by the crop should be the quantities replaced by the fertilizer. This approach ignores such important factors as losses by leaching, fixation by the soil in an unavailable form of certain elements, efficiency of various plants in absorbing certain elements, and so on. When considered with these factors, however, such data can be a helpful guide to the formulation of a sound fertility program.

Shown in Table 13-1 later in the text are the contents of some of the mineral elements in a selection of common crop plants. Although the latest available data have been used, the figures should be regarded only as averages. Soil, climate, crop variety, and management factors exert considerable influence on plant composition and in individual cases may cause appreciable variation from the values in the table. The roles of the various elements in plant growth are covered briefly in the following sections.

Nitrogen. Nitrogen is a vitally important plant nutrient, the supply of which can be controlled by man. Plants normally contain between 1 and 5% by weight of this nutrient. It is absorbed by plants in the form of nitrate and ammonium ions and as urea. In moist, warm, well-aerated soils the NO_3^- form is dominant.

Once inside the plant, nitrate is reduced to $NH_4 - N$ using energy provided by photosynthesis. Based on dark metabolism, glucose consumption for protein production is about 50% higher when N is provided as NO_3^- rather than as NH_4^+. Some reduction occurs in roots, the amount depending on species, but a larger fraction is reduced in the shoot. The NH_4-N so produced combines with various organic compounds, including glutamate, with the amide glutamine as a product. Other likely products of the initial reaction, again differing with species, are asparagine, and the amino acids aspartic and glutamic acids.

Nitrogen in these initial products of assimilation can be transferred as NH_2 to other substrates, and thus each of the 100 or so kinds of amino acids in plants are synthesized. Only a fraction of them, about 20, are used in protein synthesis. The sequence in which they are linked in proteins is under genetic control.

The proteins formed in plant cells are largely functional rather than structural; that is, they are enzymes. As such, they control the metabolic processes that take place in plants, including those involved in the reduction of NO_3^- and the synthesis of protein. These functional proteins are not stable entities, of course, for they are continually being degraded and resynthesized.

A complex group of proteins known as nucleoproteins is involved in the control of developmental and hereditary processes. One of these proteins, deoxyribonucleic acid (DNA), is present in the nucleus and the mitochondria of the cell. During meristematic growth DNA duplicates all the genetic information that the cell possesses and passes this information via chromosomes to each daughter cell. Ribonucleic acid (RNA) is also present in the nucleus, but more of it is found in the surrounding cell contents, the cytoplasm. The essential role of RNA is to execute the instructions coded within the DNA molecules.

In addition to its role in the formation of proteins, nitrogen is an integral part of chlorophyll, which is the primary absorber of light energy needed for photosynthesis. The basic unit of chlorophyll's structure is the porphyrin ring system, composed of four pyrrole rings, each containing one nitrogen and four carbon atoms. A single magnesium atom is bonded in the center of each porphyrin ring.

An adequate supply of nitrogen is associated with vigorous vegetative growth and a dark green color. An imbalance of nitrogen or an excess of this nutrient in relation to other nutrients, such as phosphorus, potassium, and sulfur, can prolong the growing period and delay crop maturity. Stimulation of heavy vegetative growth early in the growing season can be a serious disadvantage in regions where soil moisture supplies are often low. Early-season depletion of soil moisture without adequate replenishment prior to the grain-filling period can depress yields.

The effect of nitrogen on delaying maturity is not as important as it was once thought to be. In fact, if nitrogen is used properly in conjunction with other needed soil fertility inputs, it can, as demonstrated in Table 3-1, speed the maturity of crops such as corn and small grains. Applications of up to 300 lb of nitrogen per acre lowered the percentage of water in corn grain at harvest. This favorable influence saves energy required to dry grain to 15.5% and/or permits earlier harvest.

The supply of nitrogen is related to carbohydrate utilization. When nitrogen supplies are insufficient, carbohydrates will be deposited in vegetative cells, which will cause them to thicken. When nitrogen supplies are adequate, and conditions are favorable for growth, proteins are formed from the manufactured carbohydrates. Less carbohydrate is thus deposited in the vegetative portion, more protoplasm is formed, and, because protoplasm is highly hydrated, a more succulent plant results.

Excessive succulence in some crops may have a harmful effect. With crops such as cotton, a weakening of the fiber may result. With grain crops, lodging may occur,

TABLE 3-1.   Effect of Nitrogen on the Moisture Content and Yield of Corn Grain (Averages for the Years 1967–1977)

N (lb/A)	Yield (bu/A)	Moisture in Grain at Harvest (%)
0	66	36.1
60	101	30.0
120	135	27.9
180	158	26.9
240	167	28.2
300	168	27.2

Source: Ohio State Univ., *17th Annu. Agron. Demonstration, Farm Sci. Rev.* (1979).

particularly when potassium supplies are inadequate or when varieties not adapted to high levels of nitrogen fertilization are used. Excessive nitrogen fertilization will also reduce the sugar content of sugar beets. In some cases excessive succulence may make a plant more subsceptible to disease or insect attack.

It is not intended to imply that nitrogen fertilization is detrimental. Quite the contrary. Supplemental nitrogen has been shown to be beneficial for growth of crops such as rice, rapeseed, and barley. Crude protein concentration in small grains, grass forages, and other crops is frequently raised substantially by adequate nitrogen nutrition.

Protein quality is important and the influence of nitrogen fertilization on the amino acid composition of wheat grain has been investigated by many researchers. Dubetz and co-workers at the Lethbridge Research Station in Canada reported that nitrogen fertilization increased the proportion of glutamic acid, proline, phenylalanine, cystine, methionine, and tyrosine in the grain, whereas lysine, histidine, arginine, aspartic acid, threonine, glycine, valine, and leucine decreased. This comes about because different nitrogen levels are reflected in the proportions of different kinds of protein being synthesized, not because of a change in the amino acid makeup of individual proteins. Rendig and Broadbent at the University of California have found similar responses of corn to nitrogen fertilization.

As pointed out earlier in this section, it is only under the unusual conditions of excessive applications of nitrogen combined with inadequate supplies of the other elements or use of unadapted varieties that harmful effects are observed. One of the greatest boons to the development of a strong and efficient agriculture has been the production and use of commercial nitrogen fertilizers. When used in conjunction with other plant nutrients in a sound crop-management program, nitrogen fertilizers greatly increase crop yields and growers' net incomes.

When plants are deficient in nitrogen, they become stunted and yellow in appearance. This yellowing, or chlorosis, usually appears first on the lower leaves; the upper leaves remain green. In cases of severe nitrogen shortage the leaves will turn brown and die. In grasses the lower leaves usually "fire," or turn brown, beginning at the leaf tip and progressing along the midrib until the entire leaf is dead. The appearance of normal and nitrogen-deficient corn plants is shown in Figure 3-1.

The tendency of the young upper leaves to remain green as the lower leaves yellow or die is an indication of the mobility of nitrogen in the plant. When the roots are unable to absorb sufficient amounts of this element to meet the growing requirement, nitrogen compounds in the older plant parts will undergo lysis. The protein nitrogen thus is converted to a soluble form, translocated to the active meristematic regions, and reused in the synthesis of new protoplasm.

Phosphorus. Phosphorus, like nitrogen, potassium, calcium, magnesium, and sulfur, is classed as a macronutrient. It occurs in most plants in concentrations between 0.1 and 0.4%, a range considerably lower than that typically found for nitrogen and potassium. Plants can absorb phosphorus as either the primary $H_2PO_4^-$ ion or smaller amounts of the secondary HPO_4^{2-} orthophosphate ion. Since the former is most abundant over the range in soil pH prevailing for most crops, it is usually the principal form absorbed. Studies with some plants have shown that there are about 10 times as many absorption sites on plant roots for $H_2PO_4^-$ as there are HPO_4^{2-}. Absorption of the $H_2PO_4^-$ species is greatest at low pH values, whereas uptake of the HPO_4^{2-} ion is greatest at higher values of soil pH.

Other forms of phosphorus, including the pyrophosphates and metaphosphates, which are components of certain commercial fertilizers, are suitable for crops. Since in aqueous solutions both of these forms hydrolyze to orthophosphate, their absorption probably occurs mainly after conversion to orthophosphate.

FIGURE 3-1. Corn plants receiving adequate and inadequate nitrogen. Notice that the nitrogen-deficient plants in the center rows have a thin, spindly appearance and light color compared to the vigorous growth and dark color of the rows to the right and far left.

Plants may also absorb certain soluble organic phosphates. Nucleic acid and phytin are taken in by plants from sterile sand or solution cultures. Both compounds may occur as degradation products of the decomposition of soil organic matter and as such could be utilized directly by growing plants. The nucleic acid DNA and monopotassium phosphate were found to be equally effective phosphate sources for barley grown on a phosphorus-deficient soil. Phosphatidylcholine, a soil phospholipid, was inferior to these two substances. Because of the instability of many of these organic phosphorus compounds in the presence of an active microbial population, their importance as sources of phosphorus as such for higher plants under field conditions is limited.

Of the many essential functions that phosphorus has in plant life, its role in energy storage and transfer is singly the most important. Phosphate compounds act as "energy currency" within plants. Energy obtained from photosynthesis and metabolism of carbohydrates is stored in phosphate compounds for subsequent use in growth and reproductive processes.

The most common phosphorus energy currency is that found in adenosine di- and triphosphates (ADP and ATP). It occurs in the form of high-energy pyrophosphate bonds between phosphate molecules located at a terminal position on ADP and ATP structures, as indicated in Figure 3-2. ATP has two pyrophosphate bonds between its three phosphate molecules, and hence contains the most energy. When the terminal phosphate molecule from either ATP or ADP is split off, a relatively large amount of energy is liberated, 12,000 cal/mol.

Adenosine triphosphate is the source of energy that powers practically every energy-requiring biological process in plants. Almost every metabolic reaction of any significance proceeds via phosphate derivatives. Table 3-2 lists some of the more important metabolic processes which involve ATP or its equivalent.

Donation or transfer of the energy-rich phosphate molecules from ATP to energy-requiring substances in the plant is known as phosphorylation. In this reaction

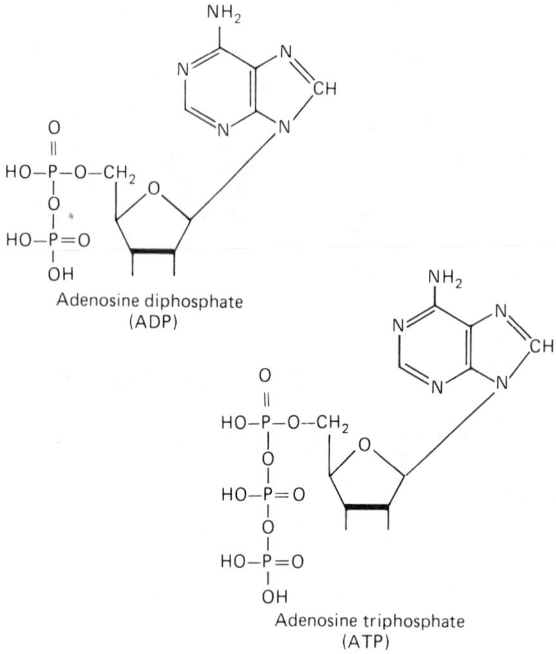

Figure 3-2. Structure of adenosine diphosphate (ADP) and adenosine triphosphate (ATP). (Wallingford, in *Phosphorus for Agriculture: A Situation Analysis,* p. 7. Atlanta, Ga.: Potash & Phosphate Institute, 1978.)

ATP is converted back to ADP, or ADP back to adenylic acid, with a phosphate molecule being left attached to the phosphorylated compound. The compounds ADP and ATP are formed and regenerated in the presence of sufficient phosphorus at sites of energy production, such as oxidative reactions of respiration.

In addition to this vital metabolic role, phosphorus is an important structural component of a wide variety of biochemicals, including nucleic acids, coenzymes, nucleotides, phosphoproteins, phospholipids, and sugar phosphates. The significance of the two nucleic acids, DNA and RNA, in the control of developmental and hereditary processes was reviewed briefly in the preceding discussion on nitrogen. An adequate supply of phosphorus early in the life of a plant is important in laying down the primordia for its reproductive parts. Large quantities of phosphorus are found in seed and fruit and it is considered essential for seed formation. Phytin, composed of calcium and magnesium salts of phytic acid, is the principal storage form of phosphorus in seeds.

Phospholipids such as lecithin and cephalin appear to be involved in the structural framework of the protoplasm. Thus phospholipids occur as a part of the chloroplast structure.

TABLE 3-2. Processes or Pathways Involving ATP

Membrane transport	Generation of membrane electrical potentials
Cytoplasmic streaming	Respiration
Photosynthesis	Biosynthesis of cellulose,
Protein biosynthesis	pectins, hemicellulose, and lignin
Phospholipid biosynthesis	Lipid biosynthesis
Nucleic acid synthesis	Isoprenoid biosynthesis → steroids and gibberellins

Source: Glass et al., *Proc. Western Canada Phosphate Symp.,* p. 358 (1980).

A good supply of phosphorus has been associated historically with increased root growth. Ohlrogge and his associates at Purdue University have shown that when soluble phosphate compounds and ammonium nitrogen are applied together in a band, plant roots proliferate extensively in that area of treated soil. There is also a greatly increased uptake of phosphorus, which is not observed if nitrate nitrogen is used instead of the ammonium form. This influence of ammonium is not fully understood, but it is possible that high concentrations of this cation are instrumental in formation of more soluble and available phosphorus reaction products. Also, the net exudation of hydrogen ions that occurs when plants absorb ammonium is expected to increase phosphorus solubility in calcareous and high pH soils.

Drew of the Letcombe Laboratory in England has reported on the effect of a localized supply of phosphate, nitrate, ammonium, and potassium on root form. Nutrient solutions were used in sand culture, which thus avoided the complications arising from soil chemical reactions. Figure 3-3 demonstrates how the exposure of parts of the main seminal roots of barley to zones of high phosphorus concentration caused initiation and prolific development of both first- and second-order laterals.

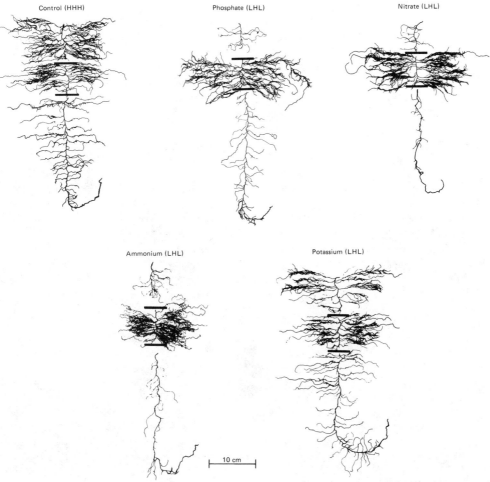

FIGURE 3-3.   Effect of a localized supply, of phosphate, nitrate, ammonium, and potassium on root form. Control plants (HHH) received the complete nutrient solution to all parts of the root system. The other roots (LHL) received the complete nutrient solution only in the middle zone, the top and bottom being supplied with a solution deficient in the specified nutrient. [Drew, *New Phytol.*, **75**:486 (1975).]

This resulted in considerable modification to root form but had little influence on the extension of the main seminal roots. Similar responses to those shown for phosphorus also occurred with variations in concentration of nitrate and ammonium forms of nitrogen. These results help to explain some of the effects that banding nitrogen and phosphate fertilizers together have on root development. The greatly increased root proliferation should encourage extensive exploitation of the treated soil areas for nutrients and moisture.

Several other gross quantitative effects on plant growth are attributed to phosphate fertilization. Phosphorus has long been associated with early maturity of crops, particularly in grain crops. The effect of ample phosphorus nutrition on reducing the time required for grain ripening is illustrated in Figure 3-4.

An adequate supply of phosphorus is associated with greater strength of cereal straw. The quality of certain fruit, forage, vegetable, and grain crops is said to be improved and disease resistance increased when these crops have satisfactory phosphorus nutrition. The effect that phosphorus has on raising the tolerance of small grains to root-rot diseases is particularly noteworthy. Also, the risk of winter damage with resultant poor yields of small grains can be substantially lowered by applications of phosphorus, particularly on phosphorus-deficient soils and when growing conditions are unfavorable.

Phosphorus is readily mobilized in plants, and when a deficiency occurs, the element contained in the older tissues is transferred to the active meristematic regions. However, because of the marked effect that a deficiency of this element has on retarding overall growth, the striking foliar symptoms that are evidence of a deficiency in certain other nutrients, such as nitrogen or potassium, are seldom observed.

FIGURE 3-4. Effect of phosphate fertilization on the maturity of small grains. Notice the more advanced maturity of the small grains receiving the phosphorus *(left)* in contrast to those that have received no phosphorus *(right)*. (Courtesy of O. H. Long, Univ. of Tennessee.)

Potassium. The third macronutrient required for plant growth is potassium. Concentration of this nutrient in plants typically ranges between 1 and 4 to 5%, but it can be somewhat higher. It is absorbed from the soil solution as the potassium ion K^+. Potassium exists in several forms in soils, and the fraction that is readily available to plants usually represents a very small proportion of the total soil potassium. A discussion of potassium equilibria in soils appears in a later chapter.

Plant requirements for this element are quite high. When potassium is present in short supply, characteristic deficiency symptoms appear in the plant. Typical potassium deficiency symptoms of alfalfa are shown in Figure 3-5.

Potassium is a mobile element which is translocated to the younger, meristematic tissues if a shortage occurs. As a result, the deficiency symptoms usually appear first on the lower leaves of annual plants, progressing toward the top as the severity of the deficiency increases. Unlike nitrogen, sulfur, phosphorus, and several other plant nutrients, potassium does not combine with other elements to form such plant components as protoplasm, fats, and cellulose. Instead, it exists in mobile ionic form and its function appears to be primarily catalytic in nature. The functions of potassium in plants fall into the following six areas:

1. *Enzyme activation.* Over 60 enzymes have been identified that require potassium for their activation. These enzymes are involved in so many important plant physiological processes that enzyme activation is regarded as potassium's single most important function. Activation occurs when one or more potassium ions attach to the surface of the enzyme molecule, changing the molecule's shape and exposing the enzymes's active site. These enzymes tend to be most abundant in meristematic tissue at the growing points, both above and below ground level, where cell division takes place rapidly and where primary tissues are formed.

2. *Water relations.* The predominance of potassium over other cations in plants makes its role in osmotic regulation particularly important. Potassium provides

FIGURE 3-5.   Potassium-deficient alfalfa. [*Western Potash News Lett.,* **W-29** (April 1963).]

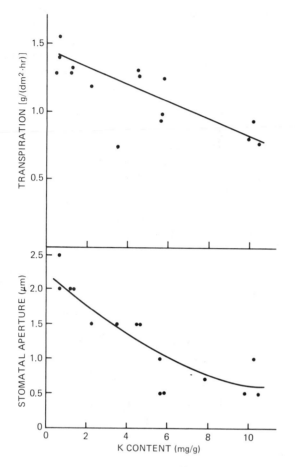

FIGURE 3-6. Improved potassium nutrition reduced the transpiration rate of peas due to smaller stomatal opertures. [Brag, *Physiol. Plant.,* **26**:254 (1972).]

much of the osmotic "pull" that draws water into plant roots. Plants that are potassium deficient are less able to withstand water stress, mostly because of their inability to make full use of available water.

Maintenance of plant turgor is essential to the proper functioning of photosynthetic and metabolic processes. The opening of stomata occurs when there is an increase of turgor pressure of the guard cells surrounding each stoma, which is brought about by an influx of potassium. Malfunctioning of stomata due to a deficiency of this nutrient has been related to lower rates of photosynthesis and less efficient use of water.

Transpiration is the loss of water through stomata, and it accounts for the major portion of a plant's water use. Potassium can affect the rate of transpiration and water uptake through regulation of stomatal opening. An example of how improved potassium nutrition reduced transpiration rate of peas by more complete closing of the stomata is given in Figure 3-6.

3. *Energy relations.* Plants require potassium for the production of high-energy phosphate molecules (ATP), which are produced in both photosynthesis and respiration. The amount of carbon dioxide that is assimilated into sugars during photosynthesis drops off sharply with insufficient potassium. In corn, the photosynthetic rate of the lower leaves is the first to be affected.

Potassium plays an important role in photosynthesis, and up to 50% of the total quantities of this element in leaves is concentrated in the chloroplasts. The favorable effect of increasing levels of potassium in corn leaves on rate of photosyn-

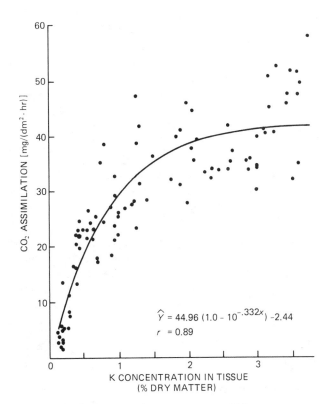

$$\hat{Y} = 44.96 \,(1.0 - 10^{-.332x}) -2.44$$
$$r = 0.89$$

K CONCENTRATION IN TISSUE
(% DRY MATTER)

FIGURE 3-7. Adequate potassium in corn leaves increases photosynthesis as measured by carbon dioxide fixation. [Smid and Peaslee, *Agron. J.*, **68**:907 (1976).]

thesis, as measured by the rate of carbon dioxide fixation, is readily apparent in Figure 3-7.

4. *Translocation of assimilates.* Once carbon dioxide is assimilated into sugars during photosynthesis, the sugars must be transported to organs of the plant where they will be stored or used for growth. The plant's transportation system uses energy in the form of ATP—which requires potassium for its synthesis.

The translocation of sugar from leaves of sugarcane is greatly reduced in potassium-deficient plants. It was found, for example, that in normal cane plants the downward movement of sugars from the leaf proceeded at a rate of approximately 2.5 cm/min. In potassium-deficient plants the rate of translocation was reduced to less than half that in plants receiving sufficient potassium.

5. *Nitrogen uptake and protein synthesis.* Potassium is required for nitrogen uptake and protein synthesis in plants. The total nitrogen taken up is usually lower and protein synthesis reduced in potassium-deficient plants, as indicated by a buildup of amino acids. Again, the involvement of potassium is through the need for ATP for both processes.

Studies with sugarcane have shown that nonprotein nitrogen accumulates in the leaves of potassium-deficient plants. In other investigations, free amino acids were found to accumulate in the leaves of potassium-deficient barley plants, and in extremely deficient plants the concentration of these free acids decreased with an increase in the concentration of amides.

A buildup of amide nitrogen and a reduction in conversion to protein have been observed in certain grasses grown with inadequate potassium. When potassium nutrition was improved, however, there was an increase in protein nitrogen and a corresponding decrease in amide nitrogen.

6. *Starch synthesis.* Starch synthetase is the key enzyme controlling the rate of

TABLE 3-3.   Effect of K Nutrition on Growth Factors of Winter Wheat (Average of Five Varieties)

Growth Factor	Relative Responses to Increasing K Nutrition			
	Deficient	Mildly Deficient	Adequate	High
Grain yield	100	143	199	225
Single kernel wt.	100	132	181	192
Kernels per head	100	109	111	112
Heads per plant	100	101	99	105

Source: Forster, *Proc. 8th Inter. Fert. Congr.*, Vol. 1; p. 41 (1976).

incorporation of glucose into long-chain starch molecules. Potassium is required for the activation of starch synthetase. Conversion of soluble sugars into starch is a vital step in the grain-filling process. Heavy test weight of kernels is a significant factor in determining grain yields.

A high level of potassium nutrition has been shown by Forster in West Germany to be especially important for grain filling in spring and winter wheat. As demonstrated in Table 3-3, the beneficial effect of potassium on increased yields of winter wheat was due mainly to higher kernel weights. This striking effect of adequate potassium on elevating both yield of winter wheat grain and kernel test weight resulted from prolongation of the physiologically active period of flag leaves during the late stages of grain filling. Increasing either photosynthetic capacity or the productive life of flag leaves is of extreme importance because they often account for up to 80% of grain filling.

Potassium is reported to have a beneficial effect on symbiotic N_2 fixation by legumes such as alfalfa, fababeans, and hairy vetch. High potassium supply has increased nodule mass, N_2 fixation rate, nitrogenase activity, and plant growth. Nitrogenase is the enzyme responsible for reducing atmospheric N_2 to NH_3 in the cells of *Rhizobium* bacteria. The NH_3 thus formed is released into cells of the host legume plant, where it is used for synthesis of amino acids. The intensity of this N_2 reduction process depends on carbohydrate supply. It seems that potassium enhances carbohydrate assimilate transport to nodules and utilization for the synthesis of amino acids.

The gross impact of these numerous functions of potassium on crop production is reflected in several ways. Perhaps the first visible indication of a potassium deficiency is the development of characteristic symptoms in leaves, such as those shown in Figure 3-5 for alfalfa.

Another symptom of insufficient potassium is weakening of straw in grain crops, which produces lodging of small grains and stalk breakage in crops such as corn (Figure 3-8) and sorghum. The results in Table 3-4 are indicative of how seriously stalk breakage can affect production through impaired yields and harvesting losses.

Potassium deficiencies greatly reduce crop yields. In fact, serious yield reductions may occur without the appearance of deficiency symptoms. This phenomenon has been termed "hidden hunger" and is not necessarily restricted to the element potassium. Crops may exhibit hidden hunger for other elements as well, a point that is discussed more fully in Chapter 12.

Decreased resistance to certain plant diseases is associated with potassium deficiency. An in-depth review by S. Perrenoud of the literature on the relationship between potassium fertilizer use and plant disease lead to a number of important conclusions, including:

1. Potassium improved plant health in 65% of the studies and was deleterious 23% of the time.

FIGURE 3-8. Response of corn to potassium on a low-potassium soil. Note the poor growth and lodged condition of the crop on the right. (Courtesy of the Potash & Phosphate Institute, Atlanta, Ga.)

2. Potassium reduced bacterial and fungal diseases 70% of the time, insects and mites 60% of the time, and nematodes and virus influences in a majority of the cases.

As discussed earlier, inadequate potassium nutrition often results in the buildup of soluble sugars and amino acids. These low-molecular-weight compounds encourage activity of plant pathogens.

Soybeans are highly susceptible to pod and stem blight caused by the fungus *Diaporthe sojae* L. The gray, moldy seed resulting from infection by this pathogen

TABLE 3-4.   **Effect of Nitrogen and Potassium on Yields and Stalk Breakage of Corn**

K_2O Applied (lb/A)	Nitrogen Applied (lb/A)		
	0	80	160
	Yield (bu/A)		
0	48	33	38
80	73	116	119
160	59	122	129
	Stalk Breakage (%)		
0	9	57	59
80	4	3	8
160	4	4	4

Source: Schulte, *Proc. Wisconsin Fert. and Aglime Conf.*, p. 58 (1975).

TABLE 3-5.   Effect of Potassium on Soybean Seed Yield and Disease

KCl or K_2SO_4 (g/cylinder)	Seeds per Plant		Diseased Seed (%)*	
	Var. A	Var. B	Var. A	Var. B
Control	254	200	87	62
2	262	207	65	58
10	275	209	21	33
30 + 10 sidedress	264	200	13	14
	NS†		LSD = 6.0	

* Percent gray, moldy seed (*D. sojae* infected).
† Not significant.
Source: Crittenden and Svec, *Agron. J.,* **66:**697 (1974).

means not only lower yield, but also lower seed quality. The relationship between percent of soybean seed infected by *D. sojae* L. and potassium treatment is illustrated in Table 3-5. Higher rates of potassium as either potassium chloride or potassium sulfate markedly decreased the incidence of disease in each variety.

The quality of some crops, particularly fruits and vegetables, is decreased with insufficient potassium. Figure 3-9 shows the striking influence of adequate potassium on quality of soybeans.

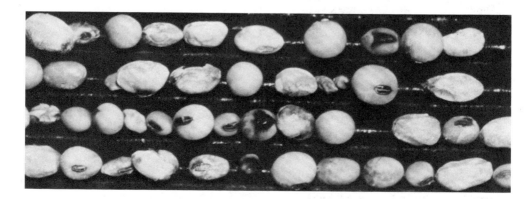

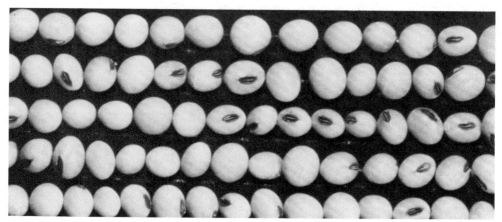

FIGURE 3-9.   Low potassium causes moldy, diseased, low-quality soybeans. (Courtesy of the Potash & Phosphate Institute, Atlanta, Ga.)

Calcium. Calcium is another macronutrient required by all higher plants; absorbed as the ion Ca^{2+}, it is abundant in leaves. Its normal concentration ranges from 0.2 to 1.0%. While calcium frequently occurs as free Ca^{2+} in cell sap, it may also be associated with such immobile, organic ions as carboxyl, phosphoryl, and phenolic hydroxyl. It may also exist as deposits of calcium oxalate, carbonate, and phosphate in cell vacuoles. The phytin in many seeds may be an acid calcium magnesium salt of phytic acid. Calcium pectate is one of the components of the middle lamella of cells.

A deficiency of calcium manifests itself in the failure of the terminal buds of plants to develop. The same applies to the apical tips of roots. Because of this cessation in meristematic activity at the growing points, plant growth ceases in the absence of an adequate supply of this element. In corn, a shortage of calcium prevents the emergence and unfolding of new leaves, the tips of which are almost colorless and are covered with a sticky gelatinous material which causes them to adhere to one another.

The most frequent indicator of insufficient calcium supply seems to take the form of disorders in the storage tissues of fruit and vegetables. Examples of calcium-related disorders are blossom-end rot in tomato and bitter pit of apples. The unsatisfactory condition of apples affected by bitter pit is evident in Figure 3-10.

As indicated earlier, calcium has an essential role in cell elongation and division. However, its exact function in these processes is not fully understood. Calcium accumulates during respiration by mitochondria and it increases their protein content. In view of the role played by mitochondria in aerobic respiration, hence salt uptake, there is apparently a direct relationship between calcium and ion uptake in general.

Calcium has long been known to have an important role in the structure and permeability of cell membranes. Lack of calcium produces a general breakdown of membrane structures, with resultant loss in retention of cellular diffusible compounds. Calcium enhances uptake of nitrate nitrogen and therefore is interrelated with nitrogen metabolism.

Unlike other cations, calcium does not have a major role in enzyme activation. Small amounts of calcium are required to produce normal mitosis, and it has been suggested that this element has a specific function in the organization of chromatin or of the mitotic spindle. There are indications that calcium is directly involved in chromosome stability and that it is a constituent of chromosome structure. Calcium has been shown to have an effect on carbohydrate translocation in plants.

Finally, calcium is generally considered to be an immobile element. There is very little translocation of calcium in the phloem, and for this reason there is often a poor supply of calcium to fruits and storage organs. Downward translocation of Ca^{2+} is also limited in roots, which usually prevents them from entering soils not well supplied with calcium.

Magnesium. Magnesium is the fifth element in the group of six macronutrients. It is absorbed in the form of the ion, Mg^{2+} and its usual concentration in crops varies between 0.1 and 0.4%. Magnesium is the only mineral constituent of the chlorophyll molecule, and is located at its center, as described in our earlier discussion on nitrogen. The importance of magnesium is obvious, for without chlorophyll the autotrophic green plant would fail to carry on photosynthesis. Chlorophyll formation usually accounts for about 15 to 20% of the total magnesium content of plants.

Magnesium also serves as a structural component in ribosomes. It appears to stabilize the ribosomal particles in the configuration necessary for protein synthesis. Also, magnesium probably activates the formation of polypeptide chains from amino acids. As a consequence of magnesium deficiency, the proportion of

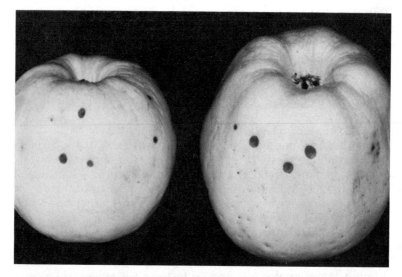

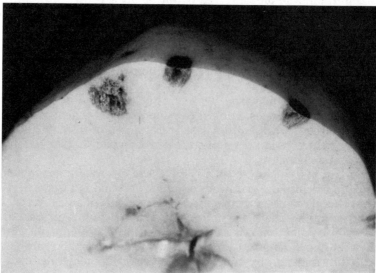

FIGURE 3-10. *Top:* Bitter pit development in a Golden Delicious apple. *Bottom:* Cross-sectional view of bitter pit development in a Golden Delicious apple. (Courtesy of B. Fleming and Dr. G. Neilsen, Agriculture Canada, Summerland, B.C.)

protein nitrogen decreases and that of nonprotein nitrogen generally increases in plants.

Magnesium is a mobile element and is readily translocated from older to younger plant parts in the event of a deficiency. As much as 70% of the total magnesium in plant tissues is diffusible and associated with inorganic anions and organic anions such as malate and citrate. However, it also combines with non-diffusible anions, including oxalate and pectate. Seeds of such crops as the small grains contain magnesium as a salt of phytic acid.

Magnesium is involved in a number of physiological and biochemical functions. It is associated with transfer reactions involving phosphate-reactive groups. Magnesium is required for maximal activity of almost every phosphorylating enzyme in carbohydrate metabolism. Most reactions involving phosphate transfer from adenosine triphosphate (ATP) require this element. It has been suggested that magnesium forms a chelated structure with the phosphate groups, establishing the configuration that allows maximal activity in the transfer reactions. Since the fundamental process of energy transfer occurs in photosynthesis, glycolysis, the tricarboxylic acid cycle (citric acid or Krebs cycle), and respiration, magnesium has an important involvement throughout plant metabolism.

Several enzymes other than those involved in phosphate transfer require Mg^{2+} as a cofactor. Insufficient magnesium for these enzymes can restrict carbon dioxide assimilation and, in turn, photosynthesis. It has a vital role in the activation of enzyme RuDP carboxylase, which is found in chloroplasts. Affinity of the enzyme for carbon dioxide is enhanced by magnesium.

Magnesium is related to the synthesis of oil in plants. With sulfur, it brings about significant increases in the oil content of several crops.

Because of the mobility of a substantial portion of plant magnesium and its ready translocation from older to younger plant parts, deficiency symptoms often appear first on the lower leaves. In many species shortage of magnesium results in an interveinal chlorosis of the leaf, in which only the veins remain green. In more advanced stages the leaf tissue becomes uniformly pale yellow, then brown and necrotic. In other species, notably cotton, the lower leaves may develop a reddish-purple cast, which gradually turns brown and finally necrotic. Magnesium deficiency of corn is illustrated in Figure 3-11.

Sulfur. Sulfur, the sixth element in the macronutrient category, is absorbed by plant roots almost exclusively as the sulfate ion, SO_4^{2-}. Low levels of SO_2 can be absorbed through plant leaves and utilized within plants, while high concentrations of this gaseous form of sulfur are toxic. Elemental sulfur dusted on fruit tree leaves finds its way in small amounts to the internal plant system relatively soon after application. The mechanism by which this water-insoluble source of sulfur penetrates the plant is not known.

FIGURE 3-11.   Magnesium-deficiency symptoms in corn. (Courtesy of Dr. G.R. Hagstrom, Duval Sales Corp., Houston, Tx.)

FIGURE 3-12.   Normal (*right*) and sulfur-deficient (*left*) rapeseed plants. (Courtesy of K. D. McLachlan, CSIRO, Canberra, Australia, and The Sulphur Institute, Washington, D.C.)

Typical concentrations of sulfur in plants range between 0.1 and 0.4%. Like nitrogen, much of the SO_4^{2-} is reduced in the plant, and the sulfur is found in the —S—S and —SH forms. Sulfate sulfur in large amounts may also occur in plant tissues and cell sap. Sulfur is present in equal or lesser amounts than phosphorus in such plants as wheat, corn, beans, and potatoes but in larger amounts in alfalfa, cabbage, and turnips.

A deficiency of sulfur, which has a pronounced retarding effect on plant growth, is characterized by uniformly chlorotic plants—stunted, thin-stemmed, and spindly. In many plants these symptoms resemble those of nitrogen and have undoubtedly led to many incorrect conclusions as to the cause of the trouble. Unlike nitrogen, however, sulfur does not appear to be easily translocated from older to younger plant parts under stress caused by its deficiency.

Sulfur-deficient cruciferous crops such as cabbage and canola/rapeseed will initially develop a reddish color on the underside of the leaves. In canola/rapeseed the leaves are also cupped inward. As the deficiency progresses in cabbage, there is a reddening and purpling of both upper and lower leaf surfaces; the cupped leaves turn back on themselves, presenting flattened-to-concave surfaces on the upper side. The characteristic appearance of sulfur-deficient rapeseed is shown in Figure 3-12.

Crop deficiencies of sulfur have long been known in localized areas throughout the world. Reports of sulfur deficiency have become much more frequent and extensive because of the following factors:

1. Increased use of high-analysis fertilizers containing little or no sulfur.
2. Higher-yielding crops and multiple cropping remove greater amounts of sulfur from the soil.
3. Greater control of industrial emissions of sulfur and decreased use of high-sulfur fuels.
4. Decreased use of sulfur in insecticides and herbicides.
5. Declining reserves of sulfur in surface soils caused by losses in organic matter through erosion and mineralization.

The functions of sulfur in plant growth and metabolism are numerous and important. A summary of some of these follows:

1. It is required for the synthesis of the sulfur-containing amino acids cystine, cysteine, and methionine, which are essential components of protein. Approximately 90% of the sulfur in plants is found in these amino acids. For the majority of proteins, cysteine and methionine account for 3 to 7% of the amino acids present, but the proportions can vary greatly.

One of the main functions of sulfur in proteins or polypeptides is the formation of disulfide bonds between polypeptide chains. This bridging is achieved through the reaction of two cysteine molecules forming cystine. Linking of two cysteine units within a protein by a disulfide bond (—S—S—) will cause the protein to fold. Disulfide linkages are therefore important in stabilizing and determining the configuration of proteins. This, in turn, is directly relevant to the catalytic or structural properties of the protein.

The sulfhydryl group (—SH) of cysteine is involved in the catalytic function of enzymes. In some cases the quantity of this form of sulfur in plants has been associated with increased cold resistance.

2. It is needed for the synthesis of other metabolites, including coenzyme A, biotin, thiamin or vitamin B_1, and glutathione. Coenzyme A is probably the most important of these substances since it is involved in many basic processes, such as the oxidation and synthesis of fatty acids, the synthesis of amino acids, and the oxidation of certain intermediates of the tricarboxylic acid or citric acid cycle.

Biotin serves as a cofactor for various carboxylation reactions, and thiamine pyrophosphate is a cofactor for the decarboxylation of oxoacids and transketolase reactions. Glutathione is a reducing agent which has a very important role in the detoxification of certain metabolites injurious to cells.

3. It is a component of other sulfur-containing substances, including S-adenosylmethionine, formylmethionine, lipoic acid, and sulfolipid. Methylation reactions are dependent on S-adenosylmethionine, which is a methyl group donor. Formylmethionine is the amino acid at the N-terminal end of proteins during synthesis and is often cleaved subsequently. Lipoic acid is a cofactor necessary for the oxidation of oxoacids in the tricarboxylic acid cycle. Sulfolipid is an important structural material of chloroplast membranes.

4. It is a vital part of the ferredoxins, a type of nonheme iron sulfur protein occurring in the chloroplasts. Ferredoxin participates in oxidoreduction processes by transferring electrons. The oxidized form of ferredoxin is the recipient of electrons expelled from chlorophyll during photooxidation of chlorophyll in the light reactions of photosynthesis. Reduced ferredoxin is probably the eventual source of reducing power for the reduction of carbon dioxide in the dark reaction of photosynthesis.

In addition to photosynthetic processes, ferredoxin has a significant role in nitrite reduction, sulfate reduction, the assimilation of N_2 by root nodule bacteria and free-living nitrogen-fixing soil bacteria, and in mitochondrial electron transport.

5. Although not a constituent, sulfur is required for the synthesis of chlorophyll. Table 3-6 shows the importance of adequate sulfur nutrition in the occurrence of chlorophyll in red clover.

6. It occurs in volatile compounds responsible for the characteristic taste and smell of plants in the mustard and onion families. Sulfur enhances oil formation in crops such as flax and soybeans. A number of proteolytic enzymes, such as the papainases, are activated by sulfur.

The data in Table 3-7 show that, within limits, the sulfur-containing amino acid content of plants can be altered by sulfur fertilization. Increasing levels of sulfur nutrition in the culture medium raised the methionine, cystine, and total sulfur

TABLE 3-6. Effect of a High Level of Sulfur Nutrition on the Chlorophyll Content of Kenland Red Clover

Applied Sulfate (ppm S)	Chlorophyll Content (% dry weight)
0	0.49
5	0.54
10	0.50
20	1.02
40	1.18

Source: Rendig et al., Agron. Abstr. Annu. Meet. Am. Soc. Agron., p. 109 (1968).

content in both of the experimental strains of alfalfa. There were large differences, however, in the ability of the two strains to synthesize methionine and cystine.

From the standpoint of animal and human nutrition, the ability of different cultivars to utilize effectively the supply of mineral nutrients in synthesizing nutritionally important compounds is a very desirable property. The potential improvement of nutritive value of crops by increasing protein levels and modifying amino acid distribution through genetic fortification was referred to in Chapter 2.

While nitrogen-deficient plants usually contain small amounts of soluble forms of nitrogen and normal quantities of sulfate, plants suffering sulfur deficiency accumulate nonprotein nitrogen in the form of amide and nitrate, in contrast to their very low levels of sulfate. Accumulations of nitrate can be readily explained by the known effect of sulfur deprivation on decreased nitrate reductase activity. Working with sulfur-deprived corn seedlings, researchers in Wisconsin found a considerable decline in soluble protein concentration coupled with a rise in nitrate concentration before deficiency symptoms appeared.

The results in Table 3-8 from a study conducted in Washington State clearly show the influence of adequate sulfur nutrition on controlling concentrations of nonprotein and nitrate in orchardgrass. It is also apparent that sulfur fertilization improved the quality of this forage by narrowing the nitrogen-to-sulfur ratio. A nitrogen to sulfur (N/S) ratio of 15:1 has in the past been considered adequate for ruminants. It is now known, however, that a ratio of between 9:1 and 12:1 is needed for effective use of recycled nitrogen by rumen microorganisms. This beneficial effect of sulfur fertilization on improving crop quality through reductions in the N/S ratio is often overlooked.

TABLE 3-7. Effect of Sulfur on the Content of Sulfur-Containing Amino Acids in Alfalfa

SO_4^{2-} Ion Concentration (ppm)	Methionine (mg/g N)		Cystine (mg/g N)		Sulfur (%)	
	C_3	C_{10}	C_3	C_{10}	C_3	C_{10}
0	10.6	17.6	21.5	24.4	0.100	0.089
1	20.8	27.6	28.6	35.2	0.103	0.098
3	33.6	34.9	37.0	43.6	0.129	0.121
9	38.0	40.3	38.9	42.9	0.186	0.200
27	41.4	43.9	42.9	45.0	0.229	0.227
81	43.4	44.3	43.6	46.0	0.244	0.242

Source: Tisdale et al., Agron. J., **42:**221 (1950).

TABLE 3-8. Effects of Elemental Sulfur Application on the Yield and Quality of Orchardgrass (113 kg/ha of Nitrogen Applied After Each Cutting)

Sulfur* (kg/ha)	Yield† (metric tons/ha) of Cutting		Nonprotein Nitrogen (%) in Cutting:		Nitrate Nitrogen (%) in Cutting:		Nitrogen/ Sulfur Ratio in Cutting:	
	1	3	1	3	1	3	1	3
0	3.74	1.77	1.05	1.22	0.064	0.211	21.3	21.4
23	3.72	2.55	0.64	0.85	0.037	0.184	15.3	18.7
45	3.63	2.62	0.59	0.49	0.051	0.144	14.3	14.8
90	3.40	2.89	0.51	0.44	0.037	0.137	12.2	13.4
113	3.40	2.76	0.49	0.37	0.033	0.106	10.8	10.0

* Applied in 1965 and 1967.
† Harvests taken in 1968.
Source: Baker et al., *Sulphur Inst. J.,* **9**(1):15 (1973).

Sulfur is also involved in the formation of oil in crops such as flax, soybeans, peanuts, and rapeseed. This effect is not widely recognized. It has been found that additions of magnesium to the sulfur further enhances the production of oil. The influence of sulfur and magnesium in combination with potassium on the oil content of soybeans is illustrated by the data in Table 3-9.

The effects of sulfur nutrition on the oil content of crops have received more attention in Europe than they have in the United States, where scientists have focused on increasing the oil content of crops by breeding and selection rather than by upgrading of nutrition.

Boron. Boron is the first of the micronutrients that will be covered briefly in this chapter. Its concentration in monocotyledons and dicotyledons generally varies between 6 and 18 ppm and 20 and 60 ppm, respectively. Levels of boron in mature leaf tissue of most crops are usually adequate if over 20 ppm.

Most of the boron in soils occurs as undissociated boric acid (H_3BO_3, pK 9.2) and as such is absorbed by plants. Much smaller amounts of other forms, such as $B_4O_7^{2-}$, $H_2BO_3^-$, HBO_3^{2-}, and BO_3^{3-}, may be present, but they generally do not contribute significantly to plant needs for this element. The mechanisms involved in boron uptake are not fully understood, but it does appear that the uptake process is mainly passive through water flow to roots.

Boron deficiency is the most widespread of all the micronutrients. It has been found lacking in at least 43 states in the United States and in nearly all of Canada's 10 provinces. A large number of field, fruit, and vegetable crops suffer from a lack of boron, with this nutritional problem occurring frequently in alfalfa. Other examples of crops with a high boron requirement include asparagus, carrots, celery, lettuce, onions, sugar beets, sunflowers, and various brassicas. Boron fertilizer

TABLE 3-9. Effect of Chlorine, Sulfur, and Magnesium on the Oil Content of Soybeans

Treatment	Seed Yield (kg/ha)	Oil Content (%)	Oil Yield (kg/ha)
N + P	566	22.9	129.6
N + P + K + Cl	645	24.0	154.8
N + P + K + SO$_4$	720	24.8	178.6
N + P + K + SO$_4$ + Mg	784	25.4	199.1

Source: Venema, *Potash Trop. Agric.,* **5** (October 1962).

FIGURE 3-13. Boron-deficient sugar beet roots. Note blackened heart tissue. (Courtesy of Professor R. L. Cook, Michigan State Univ., and reprinted from *Hunger Signs in Crops,* 3rd ed., 1964, with permission of the David McKay Co., New York.)

rates used for these high-requiring crops may cause injury to low-requiring ones, such as the small grains, peas, and beans.

Boron plays an essential role in the development and growth of new cells in the plant meristem. Since it is not readily translocated from older to actively growing tissues, the first visual deficiency symptom is cessation of the growth of the terminal bud, followed shortly thereafter by death of the young leaves. Flowering and fruit development are also restricted by a shortage of this nutrient.

In boron-deficient plants the youngest leaves become pale green, losing more color at the base than at the tip. The basal tissues break down, and, if growth continues, the leaves have a one-sided or twisted appearance. Usually, the leaves die and terminal growth ceases.

Boron-deficiency symptons will often appear in the form of thickened, wilted, or curled leaves; a thickened, cracked, or water-soaked condition of petioles and stems; and a discoloration, cracking, or rotting of fruit, tubers, or roots. Internal cork of apple is caused by a deficiency of this element and a lack of boron in citrus results in uneven thickness of the peel, lumpy fruit, and gummy deposits in the albedo of the fruit. The breakdown of internal tissues in root crops gives rise to darkened areas referred to as brown or black heart. Boron-deficient sugar beets are shown in Figure 3-13.

Plants require boron for a number of growth processes, especially the following:

New cell development in meristematic tissue.
Proper pollination and fruit or seed set.
Translocation of sugars, starches, nitrogen, and phosphorus.
Synthesis of amino acids and proteins.

Nodule formation in legumes.
Regulation of carbohydrate metabolism.

Boron is not incorporated in enzymes or structural macromolecules and, unlike Mn^{2+}, Zn^{2+}, and Mg^{2+}, it does not bring about reactions by chelation of enzyme and substrate. It does not undergo valency changes in its physiological role as do iron, manganese, and molybdenum. The borate ion tends to resemble phosphate in its reactions with OH^- groups from sugars, alcohols, and organic acids to form esters of boric acid.

A multiplicity of functions for boron in plants has been postulated, many of which are related to the unusual affinity of borate ion for polyhydroxyl compounds with ortho configurations. Some of these are:

1. Stability and fine structure of cell walls is increased by the polyhydroxyl compounds formed by the reaction of boron with cell-wall components. In this role boron influences fast-growing meristematic tissues, the stability of pollen tubes, and the germination and growth of pollen.
2. The polyhydroxyl complexes that boric acid forms with alcohols and sugars facilitate the translocation of sugar through plants.
3. Creation of a stable positive charge or "hole" in plasma membranes, thereby accelerating the movement of negatively charged growth substances and metabolites into receptor cells.
4. Carbohydrate metabolism with a tendency, where boron nutrition is inadequate for the pentose phosphate pathway, to supplant normal glycolysis. Such a shift leads to an elevated level of phenolic compounds, which is compounded by a reduction in the capacity of the detoxification mechanism. It has been proposed that adequate boron nutrition tends to inhibit starch synthesis, thereby maintaining sugars in easily translocated soluble forms, and increasing synthesis of sucrose, which is the major compound translocated by plants.
5. RNA metabolism is affected through the roles that boron has in nitrogen base metabolism and in controlling the rate of phosporus incorporation into nucleotides.
6. Phytohormonal activity is altered by boron, with an accumulation of auxin in plants deficient in this nutrient. Boron is apparently involved in protecting the indoleacetic acid oxidase systems.

Regardless of our inability to explain the precise role of boron in plants, it is essential in varying, but usually small quantities for the growth of many important agricultural crops. Although it is required for higher plants and some algae and diatoms, boron is not needed by animals, fungi, and microorganisms.

Iron. The sufficiency range of iron in crops is normally between 50 and 250 ppm. In general, when iron values are 50 ppm or less in the dry matter, deficiency is likely to occur. It functions both as a structural component and as a cofactor for enzymatic reactions. Iron can reach plant roots as Fe^{2+}, Fe^{3+}, and as organically complexed or chelated iron. The ion required in metabolism is Fe^{2+}, and it is absorbed in this form by plants. In the Fe^{2+} form, iron is more mobile and available for incorporation into biomolecular structure. Some plant tissues containing large quantities of Fe^{3+} may exhibit iron-deficiency symptoms.

Iron deficiency has been observed in many species. It is most frequently seen in crops growing on calcareous or alkaline soils, although the presence of high phosphate levels may also induce this condition in acid soils in some species. Citrus and deciduous fruit often exhibit iron chlorosis. It is also fairly common in blueberries (which are grown on very acid soils) and in sorghum, grown in neutral-to-alkaline soils. Other crops known to display deficiencies of this element are soybeans, bush

FIGURE 3-14. Iron-deficient sorghum. Note chlorosis and other deficiency symptoms in iron-deficient plot on left. (Courtesy of M. Marsolek and Dr. G. R. Hagstrom, Duval Sales Corp., Houston, TX.)

beans, corn, strawberries, avocado, vegetable crops, and many ornamentals. Iron deficiency of soybeans has become quite common in the western Corn Belt in the United States.

A deficiency of iron shows up first in the young leaves of plants. It does not appear to be translocated from older tissues to the tip meristem, and as a result, growth ceases. The young leaves develop an interveinal chlorosis, which progresses rapidly over the entire leaf. In severe cases the leaves turn entirely white. Iron-deficiency symptoms in sorghum are illustrated in Figure 3-14.

The chemical properties of iron make it an important part of oxidation–reduction reactions. It is a transition metal capable of existing in more than one oxidation state, whereby it can accept or donate electrons according to the oxidation potential of the reactants. Iron can also combine with electron donors or ligands to form complexes. Chelation results when the ligand can donate more than one electron. Iron will form stable chelates with molecules containing oxygen, sulfur, or nitrogen. The movement of electrons between the organic molecule and iron provides the potential for many of the enzymatic transformations in which iron is essential. Major functional roles of iron in plant nutrition are listed below. There is a very close association between structural and enzymatic functions.

1. Structural component of porphyrin molecules: cytochromes, hemes, hematin, ferrichrome, and leghemoglobin. These substances are involved in oxidation–reduction reactions in respiration and photosynthesis. As much as 75% of the total cell iron is associated with the chloroplasts and up to 90% of the iron in leaves occurs with lipoprotein of the chloroplast and mitochondria membranes.

The localization of iron in chloroplasts reflects the presence of cytochromes for performing various photosynthetic reduction processes and of ferredoxin as an initial electron acceptor. Reduction of oxygen to water during respiration is the most commonly recognized function of iron-containing compounds.

2. Structural component of nonheme molecules: ferredoxins that are stable Fe-S proteins. Ferredoxin is the first stable redox compound of the photosynthetic electron transport chain.

3. Enzyme systems.

   a. Cytochrome oxidase.
   b. Catalase.
   c. Peroxidase.
   d. Aconitase.
   e. Chlorophyll synthesis, γ-aminolevulinate dehydratase, γ-aminolevulinate synthetase, and ferrochelatase.
   f. Peptidylproline hydrolase.
   g. Nitrogenase. This enzyme is at the center of the N_2 fixation process in nitrogen-fixing microorganisms.

Iron may also be capable of partial substitution for molybdenum as the metal cofactor necessary for the functioning of nitrate reductase in soybeans.

Manganese. Manganese is another of the transition metals required as a micronutrient. Normal concentrations of this element in plants are typically from 20 to 500 ppm. Deficiencies of manganese are usually associated with levels of between 15 and 25 ppm in upper plant parts.

It is absorbed by plants as the manganous ion Mn^{2+}, and in molecular combinations with certain natural and synthetic complexing agents. Foliar applications of either form can be taken in directly through the leaves.

Like iron, manganese is a relatively immobile element. The symptoms of deficiency usually show up first in the younger leaves. In broadleaved plants the condition appears as an interveinal chlorosis, which also occurs in members of the grass family, but it is less conspicuous. Leaves from manganese-deficient soybean plants are shown in Figure 3–15. Manganese deficiency of several crops has been described by such terms as gray speck of oats, marsh spot of peas, and speckled yellows of sugar beets.

The involvement of manganese in photosynthesis, particularly in the evolution of oxygen, is well known. It also takes part in oxidation–reduction processes and decarboxylation and hydrolysis reactions. Manganese can substitute for magnesium in many of the phosphorylating and group-transfer reactions.

Although it is not necessarily specifically required, manganese is needed for maximal activity of many enzyme reactions in the citric acid cycle. In the majority of enzyme systems, magnesium is as effective as manganese in promoting enzyme transformations. Manganese influences auxin levels in plants and it seems that high concentrations of this micronutrient favor the breakdown of indoleacetic acid.

A summary of the functional roles of manganese follows:

1. Electron transport in photosystem II.
2. Maintenance of chloroplast membrane structure.
3. Manganin.
4. Enzyme systems, including
   a. Chromatin-bound RNA polymerase.
   b. Synthesis of tRNA-primed oligoadenylate.
   c. Synthesis of phosphatidylinositol.
   d. Inactivation of indoleacetic acid protectors.
   e. NAD malic enzyme of aspartate-type C_4 plants.

Plants are injured by excessive amounts of manganese. Crinkle leaf of cotton is a manganese toxicity that is sometimes observed on highly acid red and yellow

FIGURE 3-15.   Different stages of manganese deficiency of soybean leaves. (Courtesy of Dept. of Agronomy, Purdue Univ., and reprinted from *Hunger Signs in Crops,* 3rd ed., 1964, with permission of the David McKay Co., New York.)

soils of the old Cotton Belt in the southern United States. Manganese toxicity has also been found in tobacco, soybeans, tree fruits, and canola/rapeseed growing on extremely acid soils. Upward adjustment in soil pH by liming will readily correct this problem.

Copper.   The properties of copper that make it essential to plant nutrition are somewhat similar to those of iron. Its normal concentration in plant tissue ranges from 5 to 20 ppm. Deficiencies are probable when copper levels in plants fall below 4 ppm in the dry matter.

Copper is absorbed by plants as the cupric ion, Cu^{2+}, and may be absorbed as a component of either natural or synthetic organic complexes. Copper salts and complexes are also absorbed through the leaves.

Deficiencies of copper have been reported in numerous plants, although it is more prevalent among crops growing on peat and muck soils. Crops most susceptible to copper deficiency include alfalfa, wheat, barley, oats, lettuce, onions, carrots, spinach, and table beets. Other crops responding to copper fertilization include clover, corn, and fruit trees.

Symptoms of copper deficiency vary with the crop. In corn the youngest leaves become yellow and stunted, and as the deficiency becomes more severe, the young leaves pale and the older leaves die back. In advanced stages dead tissue appears along the tips and edges of the leaves in a pattern similar to that of potassium deficiency. Copper-deficient small-grain plants lose color in the younger leaves, which eventually break and the tips die. In many vegetable crops the leaves lack turgor. They develop a bluish-green cast, become chlorotic, and curl, and flower production fails to take place. Lack of copper causes iron to accumulate in corn plants, especially in the nodes.

Copper provides a plant with a metal which in its reduced form readily binds and reduces O_2. In the oxidized form the metal is readily reduced and in protein complexes copper has a high redox potential. These properties of copper are exploited by enzymes that can hydroxylate monophenols, oxidize them to create complex polymers such as lignin and melanin, terminate electron transfer chains, detoxify superoxides, oxidize amines, and act generally as cytoplasmic oxidases. Copper is unique in its involvement in enzymes and it cannot be replaced by any other metal ion.

The recognized functional roles of copper in plant nutrition follow:

1. Oxidase enzymes, including tyrosinase, laccase, and ascorbic acid oxidase.
2. Terminal oxidation by cytochrome oxidase.
3. Photosynthetic electron transport mediated by plastocyanin.
4. Indirect effect on nodule formation.

Zinc. Zinc is the last element in the first transition series to be required by plants. The normal concentration range is 25 to 150 ppm of zinc in the dry matter of plants. Deficiencies of this element are usually associated with concentrations of less than 20 ppm, and toxicities will occur when the zinc leaf levels exceed 400 ppm.

Plant roots absorb zinc as the ion Zn^{2+} and as a component of synthetic and natural molecular complexes. Soluble zinc salts and zinc complexes can also enter the plant system directly through leaves.

Corn and beans are particularly sensitive to zinc deficiency. Citrus and deciduous fruit trees such as peach are also very sensitive to inadequate zinc nutrition. Other crops classified as being very sensitive to insufficient zinc are flax, grapes, hops, onions, pecans, pine, and soybeans. Some of the mildly sensitive crops are alfalfa, clovers, cotton, potatoes, sorghum, Sudan grass, sugar beets, and tomatoes.

Zinc deficiency can often be identified by distinctive visual symptoms which occur most frequently in the leaves. Sometimes the deficiency symptoms will also appear in the fruit or branches or are evident in the overall development of the plant. Symptoms common to many crops include:

Occurrence of light green, yellow, or white areas between the veins of leaves, particularly the older, lower leaves.
Death of tissue in these discolored, chlorotic leaf areas.
Shortening of the stem or stalk internodes, resulting in a bushy, rosetted appearance of the leaves.
Small, narrow, thickened leaves. Often the leaves are malformed by continued growth of only part of the leaf tissue.
Early loss of foliage.
Stunted growth.
Malformation of the fruit, often with little or no yield.

Zinc deficiency causes the characteristic little leaf and rosetting or clustering of leaves at the top of fruit tree branches which have become mainly bare. In corn and sorghum, zinc deficiency is called "white bud" and in cotton it is known as "little leaf." The deficiency is referred to as "mottle leaf" or "frenching" in citrus crops and is described as "fern leaf" in Russet Burbank potato. Zinc-deficient potato is shown in Figure 3-16.
Zinc fills a need in plant cells for stable metalloenzyme complexes in which the coordination is basically tetrahedral. It can form stable bonds of covalent character with nitrogen and sulfur ligands, and zinc is not subject to oxidation—reduction in biological systems. Zinc is involved in a diversity of enzymatic activities, but it is not definitely known whether zinc acts as a functional, structural, or regulatory cofactor.

FIGURE 3-16. Zinc-deficient (right) and normal (left) shoots of potato. (Courtesy of Louis C. Boawn [Retired], Agricultural Research Service, USDA.)

The functional roles attributed to zinc include:

1. Auxin metabolism.
   a. Tryptophan synthetase.
   b. Tryptamine metabolism.
2. Dehydrogenase enzymes: pyridine nucleotide, alcohol, glucose-6-phosphate, and triose phosphate.
3. Phosphodiesterase.
4. Carbonic anhydrase (localized in chloroplasts).
5. Superoxide dismutase.
6. Promotes synthesis of cytochrome c.
7. Stabilizes ribosomal fractions.

Molybdenum. Molybdenum, with the properties of a nonmetal, is the only period 5 transition element required by plants. It is available to plants in the MoO_4^{2-} form. Molybdate is a weak acid and can form complex polyanions such as phosphomolybdate by six-coordination. Sequestering of molybdenum in this form may explain why it can be taken up in relatively large amounts without any apparent toxicity.

Normally, the molybdenum content of plant material is less than 1 ppm in the dry matter. Plants deficient in this nutrient usually contain less than 0.2 ppm in the dry matter. Molybdenum concentrations in plants are frequently low because of the extremely small amounts of MoO_4^{2-} in the soil solution. In some cases however, molybdenum levels in crops may exceed the range 1000 to 2000 ppm.

Molybdenum is an essential component of the major enzyme nitrate reductase in plants. Most of the molybdenum in plants is concentrated in this enzyme, which is a soluble molybdoflavoprotein occurring in the envelope of chloroplasts in leaves. It has also been isolated from roots. The molybdenum requirement of plants is influenced by the form of inorganic nitrogen supplied to plants, with either nitrite or ammonium effectively lowering its need. It is also a structural component of nitrogenase, the enzyme actively involved in nitrogen fixation by root-nodule bacteria of leguminous crops, by some algae and actinomyctes, and by free-living nitrogen fixing organisms such as *Azotobacter*. Molybdenum concentrations in the nodules of legume crops of up to 10 times higher than those in leaves have been observed. Molybdenum is also reported to have an essential role in iron absorption and translocation in plants.

Chlorine. It was not until the mid-1950s that this element was shown to be essential for plant growth. Plant requirements for chlorine are generally assumed to be almost as great as for sulfur. Chlorine is absorbed by plants as the Cl^- ion through both roots and aerial parts.

Although the normal range in chlorine accumulation in plants is from about 0.2 to 2.0% in the dry matter, levels as high as 10% are not uncommon. All of these values are much greater than the physiological requirement of most plants. Concentrations of 0.5 to 2.0% in the tissues of sensitive crops can lower yields and quality. Similar reductions in yield and quality can occur when chlorine levels approach 4% in tolerant crops like sugar beets, barley, corn, spinach, and tomatoes.

Chlorine has not been found in any true metabolite in higher plants. The essential role of chlorine seems to lie in its biochemical inertness, which enables it to fill osmotic and cation neutralization roles which may have biochemical and/or biophysical consequences of importance. Chlorine can be readily transported in plant tissues. A useful function for chlorine is as the counterion during rapid potassium fluxes, thus contributing to turgor of leaves and other plant parts.

The observations that partial wilting and loss of leaf turgor are symptoms of chloride deficiency lend support to the concept that chloride is an active osmotic agent. Investigations currently under way at Oregon State University indicate that a high level of chloride nutrition will increase total leaf water potential and cell sap osmotic potential in wheat plants. Accompanying these favorable changes in internal moisture conditions were tendencies for the flag leaves to be more erect and to be retained longer.

Chlorine does appear to have a definite role in the evolution of oxygen in photosystem II in photosynthesis. Extremely high chloride concentrations of nearly 11% have been detected in chloroplasts.

Uptake of both NO_3^- and SO_4^{2-} can be reduced by the competitive effects of chlorine. Lower protein concentrations in winter wheat resulting from high levels of chloride nutrition are attributed by V. Haby of the Montana State University to the strong competitive relationship between chloride and nitrate.

Disease Control. It has been found in Oregon and Washington that chloride-containing fertilizers greatly depress take-all root-rot infections in winter wheat. Some of the inhibitory effect of chloride is apparently related to restricted uptake of nitrate and to a less favorable rhizosphere pH for activity of the pathogen when the crop obtains much of its N requirement in the form of NH_4^+. The influence of Cl^- on water potential components within plants may also be a factor in controlling take-all since the pathogenic fungus responsible for this disease grows best at high water potentials or under moist conditions. Increased osmotic potential of cell sap, resulting in a lower energy status of water, seems to be involved in greater resistance to take-all.

A further disease-suppressive effect of chloride treatments was observed in Oregon. Winter wheat plants receiving ammonium chloride were less subject to stripe rust infection than they were where ammonium sulfate was used. Jackson and Christensen and their co-workers at Oregon State University have also found that *Septoria*, a leaf disease that infects wheat heads, is suppressed by the addition of chloride-containing materials.

Fusarium dryland root rot infection in spring barley was found by V. Haby of the Montana State University station at Huntley to be reduced more by potassium chloride than by comparable rates of K^+ supplied as potassium sulfate.

The severity of stalk rot in corn has been reduced by applications of potassium chloride. The suppressive effect could be the result of Cl^- rather than K^+, since additions of potassium sulfate had no influence on the disease. According to Von Uexkull of the Potash and Phosphate Institute, chloride curbs certain leaf and root rot diseases of oil palms and coconuts.

Osmotic Effects. Investigators at Oregon State University observed that leaves of winter wheat plants fertilized with ammonium chloride were more erect and were retained longer than those of plants receiving the same amount of nitrogen as ammonium sulfate. Differences in leaf habit were most pronounced at midday during bright sunny weather. The Cl^- ion was found to affect plant water potentials through osmotic adjustments, such as turgor pressure, leaf water potential, and osmotic potential. A less well known influence of chlorine on crop growth is the hastening of maturity of small grains, which has been observed in Montana and Washington.

Chlorosis in the younger leaves and an overall wilting of the plants are the two most common symptoms of chlorine deficiency. Necrosis in some plant parts and leaf bronzing may also be seen in chlorine-deficient plants. Marked reduction in root growth has also been observed in chlorine-deficient plants. Tissue concentrations below 70 to 700 ppm are usually indicative of chlorine deficiency.

Excesses of chlorine can be harmful and crops vary widely in their tolerance to this condition. Tobacco, peach, avocado, and some legumes are among the most sensitive crops. Leaves of tobacco and potatoes become thickened and tend to roll when these crops accumulate excessive amounts of chlorine. The storage quality of potato tubers and the smoking quality of tobacco are adversely affected by surplus uptake of chlorine.

Responses to chlorine have been observed in the field for a large number of crops, including tobacco, tomatoes, buckwheat, peas, lettuce, cabbage, carrots, sugar beets, barley, wheat, corn, potatoes, cotton, coconut, and oil palms. The latter two tropical crops are especially responsive to chlorine. The chlorine needs of high yields of most temperate region crops are usually satisfied by only 4 to 10 kg/ha.

Cobalt. It has been shown that cobalt is essential for microorganisms fixing atmospheric nitrogen. Cobalt is thus needed in the nodules of both legumes and alder, as well as in nitrogen-fixing algae. The normal cobalt concentration in the dry matter of plants ranges from 0.02 to 0.5 ppm. Only 10 ppb of cobalt in nutrient solution was found to be adequate for nitrogen fixation by alfalfa.

The essentiality of cobalt for the growth of symbiotic microorganisms such as the rhizobia, free-living nitrogen-fixing bacteria, and blue-green algae is its role in the formation of vitamin B_{12}. Cobalt forms a complex with nitrogen atoms in a porphyrin ring structure which provides a prosthetic group for association with a nucleotide in the B_{12} coenzyme. This cobalt complex is termed the cobamide coenzyme.

Other functions attributed to cobalt include leghemoglobin metabolism and ribonucleotide reductase in *Rhizobium*. It has been reported to be one of several metals that activate enolase and succinic kinase.

Soviet workers have reported responses by cotton, beans, and mustard to applications of cobalt. It was said that it improved growth, transpiration, and photosynthesis and that in beans and mustard the reducing activity and chlorophyll content of the leaves were raised. In cotton an increase in boll number and a decrease in the number of fallen squares were attributed to the addition of cobalt. It was further stated that water content and catalase activity in leaves were increased and that a decrease in the concentration of the cell sap resulted from cobalt treatments. An increase of 9 to 21% in the yield of cotton resulted from the cobalt addition.

Vanadium. Low concentrations of vanadium are reported to be beneficial for the growth of microorganisms, animals, and higher plants. Although it is definitely considered to be essential for the green alga *Scenedesmus*, there is no decisive evidence as yet that vanadium is essential for higher plants. Some workers suggest that vanadium may partially substitute for molybdenum in fixation of atmospheric nitrogen by microorganisms such as the rhizobia. It has also been speculated that it may function in biological oxidation–reduction reactions. Increases in growth attributable to vanadium have been reported for asparagus, rice, lettuce, barley, and corn. The vanadium requirement of plants is said to be less than 2 ppb dry weight, whereas normal concentration in plant material will average about 1 ppm.

Sodium. This element is essential for halophytic plant species which accumulate sufficient of its salts in vacuoles to maintain turgor and growth. The succulence of such plants is sometimes increased by sodium. Crops that require sodium for optimum growth include celery, mangold, spinach, sugar beet, Swiss chard, table beet, and turnip. Favorable effects of sodium are also reported to occur with cabbage, kale, kohlrabi, mustard, radish, and rapeseed. The increased growth produced by salt in halophytes is believed to be due to increased turgor.

Sodium is absorbed by plants as the ion Na^+. Its concentration varies widely from 0.01 to approximately 10% in leaf tissue dry matter. Sugar beet petioles frequently contain levels at the upper end of this range. Many plants that possess the C_4 dicarboxylic photosynthetic pathway require sodium as an essential nutrient. It also has a role in inducing crassulacean acid metabolism, which is considered part of a general response to water stress. It is not definitely known how sodium affects C_4 and crassulacean acid metabolism. There is some evidence, however, that sodium increases the activity of phosphoenolpyruvate carboxylase, the primary carboxylating enzyme in C_4 photosynthesis. Lack of sodium will cause certain plant species to shift their carbon dioxide–fixation pathway from C_4 to C_3. Provision of adequate sodium can restore these plants to their normal C_4 carbon fixation.

Water economy in plants seems to be related to the C_4 dicarboxylic photosynthetic pathway of plants. Many plant species that have the extremely efficient C_4 carbon dioxide–fixing system occur naturally in arid, semiarid, and tropical conditions, where the closure of stomata to prevent wasteful water loss is essential for growth and survival. Carbon dioxide entry must also be restricted when stomata tend to remain closed. The ratio of weights of carbon dioxide assimilated to water transpired by C_4 plants is often double that of C_3 plants. It is also noteworthy that C_4 plants are often found in saline habitats.

Other functions of sodium in plant growth include:

1. Oxalic acid accumulation.
2. Potassium-sparing action.
3. Stomatal opening.
4. Regulation of nitrate reductase.

Sugar beets appear to be particularly responsive to sodium. It influences water relations in this crop and increases the resistance of sugar beets to drought. In low-sodium soils the beet leaves are dark green, thin, and dull in hue. The plants wilt more rapidly and may grow horizontally from the crown. There may also be an interveinal necrosis similar to that observed in potassium deficiency in sugar beets. Some of the effects ascribed to sodium may also be due to chlorine since the usual source of sodium has been sodium chloride. The reverse is also true for chlorine.

Silicon. Silicon is one of the most abundant elements in the lithosphere and is present in many plant species. This element is known to accumulate in the roots of plants and it has been associated in the past with drought resistance and the mechanical strengthening of plants. Cereals and grasses contain 0.2 to 2.0% silicon in their dry matter, while dicotyledons may accumulate only one-tenth of these concentrations or less. Concentrations of up to 10% occur in silicon-rich plants. The involvement of silica in root functions is believed to contribute to the drought tolerance of crops such as sorghum.

Silicon seems to be essential for plants such as rice, grasses, sugarcane, and horsetail, which contain large amounts of it. Growth of other crops, including barley, cucumbers, desmodium clover, gherkins, and lettuce, has been improved by silicon.

Soluble silica exists mainly as monosilicic acid, $Si(OH)_4$, and plants are believed to absorb it in this form from the soil solution. Entry of silica into plants appears to require the expenditure of metabolic energy, since the process is sensitive to both metabolic inhibitors and variations in temperature. Monosilicic acid can form complexes with polyhydroxyl compounds just as boric acid does.

Silicon apparently contributes to the structure of cell walls. Grasses, sedges, nettles, and horsetails accumulate 2 to 20% of the foliage dry weight as hydrated polymer or silica gel. This silica primarily impregnates the walls of epidermal and vascular tissues, where it appears to strengthen the tissues, reduce water loss, and retard fungal infection. Where large amounts of silica are accumulated, intracellular deposits known as plant opals can occur.

Although no biochemical role for silicon in the development of plants has been positively identified, it has been proposed that enzyme–silicon complexes form in sugarcane which act as protectors or regulators of photosynthesis and enzyme activity. Silicon can suppress the activity of invertase, peroxidase, polyphenol oxidase, phosphatase, and adenosinetriphosphatase in sugarcane. Suppression of invertase activity results in greater sucrose production. A reduction in phosphatase activity is believed to provide a greater supply of essential high-energy precursors needed for optimum cane growth and sugar production.

Silicon additions have improved the growth of sugarcane in Florida, Hawaii, Mauritius, Puerto Rico, and Saipan. The beneficial effects of silicon have been attributed to correction of soil toxicities arising from high levels of available manganese, ferrous iron, and active aluminum; prevention of localized accumulations of manganese in sugarcane leaves; plant disease resistance; greater stalk strength and resistance to lodging; increased availability of phosphorus; reduced transpiration; and unknown physiological functions.

Freckling, a necrotic leaf spot condition, is a symptom of low silicon in sugarcane receiving direct sunlight. Ultraviolet radiation seems to be the causative agent in sunlight since plants kept under plexiglass or glass do not freckle. There are suggestions that silica in the sugarcane plant filters out harmful ultraviolet radiation.

A further role of silica in light relationships within leaf cells of sugarcane is indicated by research at the University of Michigan. Silica cells, filled with silica gel, apparently provide "windows" in the epidermal system, allowing more light to be transmitted to photosynthetic mesophyll and cortical tissues below the epidermis of leaves and stems, respectively, than would occur if silica cells were absent. It has not been proven yet if the presence of an abundance of silica cells in the internode or leaf surface results in greater CO_2 fixation, more growth, and increased sugar production.

This element has also had a favorable influence on rice production in Ceylon, India, Japan, and Taiwan. Japanese workers have observed that silicon tends to maintain erectness of rice leaves, increases photosynthesis because of better light interception, and results in greater resistance to diseases and insect pests. The oxidizing power of rice roots and accompanying tolerance to high levels of iron and manganese were found to be very dependent on silicon nutrition. Supplemental silicon was beneficial when the silica concentration in rice straw fell below 11%.

From previous discussions in this chapter, readers will perceive that many of the favorable effects of silicon on plant growth, such as disease resistance, stalk strength, and reductions in lodging, have also been attributed to potassium. Since potassium chloride is the most commonly used potassium fertilizer, the question arises as to the possibility that one or both components of this fertilizer salt enhance silicon uptake and metabolism.

Summary

1. All elements absorbed by plants are not necessarily essential to plant growth. The term *functional or metabolism nutrient* was introduced to include any element that functions in plant nutrition, regardless of whether its action is specific. It was suggested that this term might avoid the confusion that sometimes occurs in a definition of *essential* plant nutrients.

2. Twenty elements have been found to be essential to the growth of plants. Not all are required by all plants, but all are necessary to some plants.

3. The elements required by plants are carbon, hydrogen, oxygen, nitrogen, phosphorus, potassium, calcium, magnesium, sulfur, boron, iron, manganese, copper, zinc, molybdenum, chlorine, cobalt, vanadium, sodium, and silicon. The first three, with nitrogen, phosphorus, and sulfur, constitute their living matter or protoplasm. Elements other than carbon, hydrogen, and oxygen are termed *mineral nutrients*. The elements nitrogen, phosphorus, potassium, calcium, magnesium, and sulfur are classed as macronutrients, and the remaining mineral elements as micronutrients.

4. Nitrogen is used largely in the synthesis of proteins, but structurally it is also a part of the chlorophyll molecule. Many proteins are enzymes, and the role of nitrogen can be considered as both structural and metabolic.

5. Phosphorus is essential in supplying phosphate, which acts as a linkage unit or binding site. The stability of phosphate enables it to participate in the numerous energy capture, transfer, and recovery reactions which are vital for plant growth.

6. Potassium is necessary to many plant functions, including carbohydrate metabolism, enzyme activation, osmotic regulation and efficient use of water, nitrogen uptake and protein synthesis, and translocation of assimilates. It also has a role in decreasing certain plant diseases and in improving quality.

7. Sulfur plays an important part in protein synthesis and the functioning of several enzyme systems. The synthesis of chlorophyll and the activity of nitrate reductase are very dependent on sulfur.

8. The remaining mineral elements are generally involved in the activation of various enzyme systems. Magnesium, in addition, is an essential component of the chlorophyll molecule.

9. Sodium and chlorine are electrolytes necessary for osmotic pressure and acid–base balance. Chlorine activates the oxygen producing enzyme of photosynthesis.

10. Silicon contributes to the structure of cell walls, thereby imparting greater disease resistance, stalk strength, and resistance to lodging.

Questions

1. Can you as a commercial grower do anything to supply plants with carbon, hydrogen, and oxygen? What, specifically?
2. In what ways does nitrogen function in plant growth?
3. Visually, how would you differentiate between nitrogen and potassium deficiencies of corn?
4. Is nitrogen a mobile element in plants? What visual proof is there?
5. Phosphorus is important in many plant functions. What, however, is probably its most important overall function?
6. Nitrogen, phosphorus, and potassium are arbitrarily classed as macro- or major elements. In terms of their importance in plant nutrition, is this terminology justified? Why? What justification is there for such a classification?
7. Fruit growers in a certain region of the Pacific Northwest decided to change their nitrogen fertilizer and use ammonium nitrate. A year or two later their trees turned a uniform light yellow-green. Tissue tests showed that the leaves were high in nitrogen. The trees were irrigated with water from a glacier-fed river into which no industrial wastes had been emptied. The area was far from industrial activities. Phosphorus, potassium, and magnesium levels were adequate, and soil pH was satisfactory. None of the microelements was in short supply. What element was most likely to be deficient? How would you correct this deficiency?
8. A deficiency of calcium is sometimes observed under very dry soil conditions. Can you explain this?
9. In what ways do the symptoms of magnesium and potassium deficiencies resemble each other? In what ways are they dissimilar?
10. What function of magnesium is unique?
11. Which of the essential or metabolic elements are structurally a part of protoplasm?
12. What element is specifically involved in nitrate reduction in plants?
13. What element is required specifically by rhizobia in the fixation of elemental nitrogen?
14. Sulfur is an integral part of certain amino acids. Name the amino acids. Can ruminants synthesize these sulfur-containing amino acids from inorganic sulfur and nitrogen compounds?
15. What crops have responded to applications of sodium?
16. What precaution must be observed in applying elements such as copper, zinc, boron, cobalt, molybdenum, and magnesium to crops?
17. Name several elements, deficiencies of which are exhibited first in the apical region of the growing plant. What does this imply?
18. If you saw a field of dwarfed corn, reddish purple in color, and the corn plant tissue tested high in nitrate nitrogen, you might suspect that these plants were deficient in what element?
19. List the essential mineral elements required in plant nutrition, and give the principal functions of each.

Selected References

Adams, S. N., "The effect of sodium and potassium fertilizer on the mineral composition of sugar beet." *J. Agr. Sci.*, **56**:383 [*Soils Fert.*, **24**:375 (1961)].

Ahmed, S., and H. J. Evans, "Cobalt: a micronutrient element for the growth of soybean plants under symbiotic conditions." *Soil Sci.*, **90**:205 (1960).

Amberger, A., "The effect of boron nutrition on respiration intensity and quality of crops." *Landwirtsch. Forsch. Sonderh.*, **14**:107 [*Soils Fert.*, **23**:347 (1960).]

Baker, A. S., W. P. Mortensen, and P. Dermanis, "The effect of N and S fertilization on the yield and quality of orchardgrass." *Sulphur Inst. J.*, **9**(1):14 (1973).

Barta, A. L., "Response of symbiotic N_2 fixation and assimilate partitioning to K supply in alfalfa." *Crop. Sci.*, **22**:89 (1982).

Benson, N. R., E. S. Degman, I. C. Chmelir, et al. "Sulfur deficiency in deciduous tree fruit." *Proc. Am. Soc. Hort. Sci.*, **83**:55 (1963).

Brag, H., "The influence of K on the transpiration rate and stomatal opening in *Triticum aestivum* and *Pisum sativum*." *Physiol. Plant.*, **26**:250 (1972).

Brownell, P. F., and C. J. Crossland, "The requirement for sodium as a micronutrient by species having the C_4 dicarboxylic photosynthetic pathway." *Plant Physiol.*, **49**:794 (1972).

Cairns, R. R., and R. B. Carson, "Effect of sulfur treatments on yield and nitrogen and sulfur content of alfalfa grown on sulfur-deficient and sulfur-sufficient grey wooded soils." *Can. J. Plant Sci.*, **41**:715 (1961).

Clark, K. W., "Fababeans require K for growth and nitrogen fixation in Manitoba." *Better Crops Plant Food*, **64**:14 (Fall 1980).

Cock, J. H., and Shouichi Y., "An assessment of the effects of silicate application on rice by a simulation method." *Soil Sci. Plant Nutr. (Tokyo)*, **16**(5):212 (1970).

Delwiche, C. C., C. M. Johnson, and H. M. Reisenauer, "Influence of cobalt on nitrogen fixation by *Medicago*." *Plant Physiol.*, **36**:73 (1961).

Dubetz, S., and E. E. Gardiner, "Effect of nitrogen fertilizer treatments on the amino acid composition of Neepawa wheat." *Cereal Chem.*, **55**(3):166 (1979).

Fox, R. L., J. A. Silva, O. R. Younge, D. L. Plucknett, and G. D. Sherman, "Soil and plant silicon and silicate response by sugar cane." *SSSA Proc.*, **31**:775 (1967).

Fox, R. L., J. A. Silva, D. L. Plucknett, and D. Y. Teranishi, "Soluble and total silicon in sugarcane." *Plant Soil*, **30**(1):81 (1969).

Friedrich, J. W., and L. S. Schrader, "Sulfur deprivation and nitrogen metabolism in maize seedlings." *Plant Physiol.*, **61**:900 (1978).

Fujiwara, A., *Ammonium Chloride Fertilizer Produced in Japan*, rev. ed. Tokyo: Japan Ammonium Chloride Fertilizer Association, 1976.

Gascho, G. J., and H. J. Andreis, "Sugarcane response to calcium silicate slag applied to organic and sand soils." *Proc. Int. Soc. Sugar Cane Technol.*, **15**:543 (1974).

Glass, A. D. M., J. D. Beaton, and A. Bomke, "The role of P in plant nutrition." *Proc. Western Canada Phosphate Symp.* (Alberta Soil Sci. Workshop, March 1980), pp. 357–368 (1980).

Jackson, T. L., R. L. Powelson, and N. W. Christensen, "Combating take-all root rot of winter wheat in western Oregon." *Oregon Ext. Serv. FS.250* (1980)

Jackson, W. A., and H. J. Evans. "Effect of Ca supply on the development and composition of soybean seedlings." *Soil Sci.*, **94**:180 (1962).

Johnson, C. M., P. R. Stout, T. C. Broyer, and A. B. Carlton. "Comparative chlorine requirements of different plant species." *Plant Soil*, **8**:337 (1957).

Jung, G. A., and D. Smith. "Influence of soil potassium and phosphorus content on the cold resistance to alfalfa." *Agron. J.*, **51**:585 (1959).

Kilmer, V. J., S. E. Younts, and N. C. Brady. Eds., *The Role of Potassium in Agriculture*. Madison, Wis.: American Society of Agronomy, Crop Science Society of America, Soil Science Society of America, 1968.

Levitt, J., C. Y. Sullivan, N. O. Johansson, and R. M. Pettit, "Sulfhydryls—a new factor in frost resistance: I. Changes in SH content during frost hardening." *Plant Physiol.*, **6**:611 (1961).

Lewin, J., and B. E. F. Reimann, "Silicon and plant growth." *Annu. Rev. Plant Physiol.*, **20**:289 (1969).

Lynd, J. Q., E. A. Hanlon, Jr., and G. V. Odell, Jr., "Potassium effects on improved growth, nodulation, and nitrogen fixation of hairy vetch." *Soil Sci. Soc. Am. J.*, **45**:302 (1981).

McKercher, R. B., and T. S. Tollefson, "Barley response to phosphorus from phospholipids and nucleic acids." *Can. J. Soil Sci.*, **58**:103 (1978).

Mengel, K., and E. A. Kirby, *Principles of Plant Nutrition*, 3rd. ed. Bern: International Potash Institute, 1982.

Odelien, M., "The effect of sulfur supply on the quality of plant products." *Tidsskr. Nor. Landbruk*, **70:**35 (1963). Translated by J. Platou, The Sulphur Institute, Washington, D.C.

Ollagnier, M., and R. Ochs, "The chlorine nutrition of oil palm and coconut." *Oléagineux*, **26**(6):367 (1971).

Possingham, J. V., and D. Spencer, "Manganese as a functional component of chloroplasts." *Australian J. Biol. Sci.*, **15:**58 (1962).

Reisenauer, H. M., "Cobalt in nitrogen fixation by a legume." *Nature (London)*, **186:**375 (1960).

Reisenauer, H. M. "The effect of sulfur on the absorption and utilization of molybdenum by peas." *SSSA Proc.*, **27:**553 (1963).

Schulte, E. E., "Fertility–disease–tillage interaction in corn." *Proc. Wisconsin Fert. Aglime Conf.*, p. 58, Soil Science Dept., Univ. of Wisconsin, Madison (1975).

Sherman, G. D., "Crop growth response to application of calcium silicate to tropical soils in Hawaiian Islands." "Agr. Dig., **18:**11 (1969).

Smid, A. E., and D. E. Peaslee, "Growth and CO_2 assimilation by corn as related to K nutrition and simulated canopy shading." *Agron. J.*, **68:**904 (1976).

Spencer. K., "Growth and chemical composition of white clover as affected by sulfur supply." *Australian J. Agr. Res.*, **10:**500 (1959).

Sprague, H. B., Ed., *Hunger Signs in Crops: A symposium*, 3rd ed., New York: David McKay, 1964.

Tisdale, S. L., R. L. Davis, A. F. Kingsley, and E. T. Mertz, "Methionine and cystine content of two strains of alfalfa as influenced by different concentrations of the sulfate." *Agron. J.*, **42:**221 (1950).

Turner, J. R., "Boron for crop production." *Fert. Prog.*, **11**(3):24 (1980).

Venema, K. C. W., "Some notes regarding the function of the sulfate-anion in the metabolism of oil producing plants, especially oil palms." *Potash Trop. Agr.*, **5**(3) (July 1962).

von Vexkull, H. R., "Response of coconuts to (potassium) chloride in the Philippines." *Oléagineux*, **27**(1):13 (1972).

Wallace, T., *The Diagnosis of Mineral Deficiencies in Plants by Visual Symptoms*, 2nd ed. New York: Chemical, 1961.

Wallingford, W., "Phosphorus functions in plants," in *Phosphorus for Agriculture: A Situation Analysis*, pp. 6–12. Atlanta, Ga.: Potash & Phosphate Institute, 1978.

Wallingford, W., "Functions of potassium in plants," in *Potassium for Agriculture: A Situation Analysis*, pp. 10–27. Atlanta, Ga.: Potash & Phosphate Institute, 1980.

Zinc in Crop Nutrition. New York: International Lead Zinc Research Organization, Inc., and Zinc Institute, Inc., 1974.

Chapter 4

Basic Soil–Plant Relationships

THE purpose of this chapter is to review briefly the phenomenon of ion exchange in soils and to consider some of the suggested mechanisms for the movement of ions in the soil solution and into the cells of the absorbing roots.

Ion Exchange in Soils

Ion exchange is a reversible process by which one type of cation or anion held on the solid phase is exchanged with another kind of cation or anion in the liquid phase. If two solid phases are in contact, exchange of ions may also take place between their surfaces. Of the two phenomena, cation and anion exchange, the first is generally considered to be more important, since the anion and molecular retention capacity of most agricultural soils is much smaller than the cation retention capacity. This property of cation exchange is one of the major distinguishing features between soils and other plant-rooting media.

Cation Exchange

Soils are composed of the three forms of matter: solids, liquids, and gases. The solid phase is made up of organic and inorganic materials. The organic fraction consists of the residues of plants and animals in all stages of decomposition, and the stable part is usually termed humus.

The inorganic fraction of soil solids is composed of primary and secondary minerals with different particle sizes. The soil fractions that are seats of ion exchange are the organic and the mineral components with effective particle diameters of less than 20 μm. This includes a portion of the silt and all of the clay fraction (< 2 μm) as well as colloidal organic matter.

Because cations are positively charged, they are attracted to surfaces that are negatively charged. In the organic fraction these arise from the dissociation of H^+ from certain functional groups, particularly from carboxylic ($-COOH$) and phenolic ($-C_6H_4OH$) groups. Many carboxylic groups will dissociate at pH values below 7, leaving a negative charge at the site of the functional group, as shown in the following equation.

$$-COOH \rightleftharpoons -COO^- + H^+$$

It is estimated that 85 to 90% of the negative charge of humus is due to these two functional groups alone. Two other groups, enol ($-$ COH $=$ CH) and imide ($=$ NH), also contribute to the negative charge of organic matter.

Negative charges in the inorganic clay fraction generally arise from two sources. The first is isomorphous substitution in layer silicate minerals such as smectite, and the second is caused by the deprotonation of both (1) hydroxyl $-$ OH groups attached to the silicon atoms at the broken edges of the tetrahedral planes, and (2) exposed AlOH groups in layer silicates. The charge resulting from isomorphous substitution arises from the replacement of a silicon or aluminum atom by an atom of similar geometry but of lower charge (e.g., Mg^{2+} for Al^{3+} or Al^{3+} for Si^{4+}). This produces a net negative charge which is fairly uniformly distributed over the plate-shaped clay particles. Isomorphic substitution occurs mainly during crystallization of layer silicate minerals, and once the charge is created it is largely unaffected by future changes in the environment. The charge resulting from isomorphous substitution is responsible for the permanent charge of soils.

With increasing pH, negative charges form at the edge of clay plates by reactions such as the following:

$$- \text{SiOH} + \text{OH}^- \rightleftharpoons - \text{SiO}^- + H_2O$$

$$- \text{AlOH} + \text{OH}^- \rightleftharpoons - \text{AlO}^- + H_2O$$

Layer silicate clay minerals in soils are of three general classes: 2:1, 2:1:1, and 1:1. The 2:1 clays are composed of layers, each of which consists of two silica sheets between which is a sheet of alumina. Examples of the 2:1 clays are smectites (montmorillonite), illite, and vermiculite. Muscovite and biotite are examples of 2:1 primary minerals which are often abundant in silt and sand fractions.

Chlorites are examples of 2:1:1 layer silicates commonly found in soils. This clay mineral consists of an interlayer hydroxide sheet in addition to the 2:1 structure referred to above.

The 1:1 clays are composed of a series of layers each of which contains one silica sheet and one alumina sheet. Kaolinite and halloysite are two of the most important clay minerals in this group.

Isomorphic substitution is the principal source of negative charge for the 2:1 and 2:1:1 groups of clay minerals but is of minor consequence for the 1:1 group of clays. Deprotonation or dissociation of H^+ from OH groups at the broken edges of clay particles is the prime source of negative charge in the 1:1 clay minerals. High pH values favor this deprotonation of exposed hydroxyl groups.

The oxides and hydrous oxides that are abundant in highly weathered soils have pH-dependent charges. These materials occur as coatings and interlayers of crystalline clay minerals. On exposure to moisture their surfaces become hydroxylated. Charges develop on these hydroxylated surfaces either through amphoteric dissociation of the surface hydroxyl groups or by adsorption of H^+ or OH^- ions. The total charge of soil particles usually varies with the pH at which the charge is measured. The positive charge developed at low pH and the excess negative charge formed at high pH are termed pH-dependent charge. Only about 5 to 10% of the negative charge on 2:1 clays is pH dependent, while 50% or more of the charge developed on 1:1 clay minerals can be pH dependent.

The negative charge that develops on organic and mineral colloids is neutralized by cations attracted to the surfaces of these colloids. The quantity of cations expressed in milliequivalents per 100 g of oven-dry soil is termed the *cation exchange capacity* (CEC) of the soil. It is one of the important chemical properties of soils and is usually closely related to soil fertility. A thorough understanding of cation exchange is necessary to an understanding of soil fertility and acidity. Therefore, the following brief review is given of the way in which this quantity is determined.

Procedures differ for measuring CEC of soils, but the following simplified description illustrates the basic features.

Cation exchange, as pointed out previously, means the exchange of one cation for another in a solution phase. Soil colloids have adsorbed to their exchange sites numerous cations, including calcium, magnesium, potassium, sodium, ammonium, aluminum, iron, and hydrogen. These ions are held with varying degrees of tenacity, depending on their charges and their hydrated and unhydrated radii. As a rule, ions with a valence of 2 or 3 are held more tightly than monovalent cations. Also, the greater the degree to which the ion is hydrated, the less tightly it will be held.

A conventional method of CEC measurement is to extract a soil sample with neutral 1 N ammonium acetate. All of the exchangeable cations are replaced by ammonium ions and the CEC becomes saturated with ammonium. If this ammonium-saturated soil is extracted with a solution of a different salt, say 1.0 N KCl, the potassium ions will replace the ammonium ions. If the soil–potassium chloride suspension is filtered, the filtrate will contain the ammonium ions that were previously adsorbed by the soil. The quantity of ammonium ions in the leachate is a measure of the CEC of the soil in question and can easily be determined.

To illustrate, suppose that 20 g of oven-dry soil was extracted with 200 ml of 1.0 N NH$_4$Ac (ammonium acetate). The extraction is accomplished by intermittent shaking over a period of 30 minutes. The soil-ammonium acetate solution is filtered and the soil is washed with alcohol to remove the *excess* solution. The soil containing the adsorbed ammonium ions is next extracted with 200 ml of a solution of 1.0 N KCl. The soil–potassium chloride solution is filtered and the ammonium contained in the filtrate is determined. Suppose that 0.054 g of NH$_4^+$ were found. This was, of course, retained by the 20 g of soil extracted (0.054 g is 3 meq—i.e., 0.054/0.018 = 3, as 0.018 g is the milliequivalent weight of 1 meq). Because 3 meq were present in 20 g of soil, the CEC of the soil is 15 meq/100 g.

The CEC of a soil will obviously be affected by the nature and amount of mineral and organic colloid present. As a rule, soils with large amounts of clay and organic matter will have higher exchange capacities than sandy soils low in organic matter. Also, soils with predominately 2:1 colloids will have higher exchange capacities than soils with predominately 1:1 mineral colloids.

Generally, 1:1 mineral colloids have CEC values of 1 to 10 meq per 100 g; 2:1 mineral colloids such as montmorillonite and vermiculite, 80 to 150 meq per 100 g; 2:1:1 chlorites and 2:1 micas, 20 to 40 meq per 100 g; and organic colloids, 100 to 300 meq per 100 g.

Effective CEC. The use of neutral ammonium acetate displacement for determining the CEC of soils has been and is still used by many laboratories in the United States. However, some workers believe that CEC can better be estimated by extraction with an unbuffered salt which would give a measure of the CEC at the soils normal pH. Use of neutral N ammonium acetate will result in a high CEC value if the soil is acid simply because of the adsorption of NH$_4^+$ ions to the so-called pH-dependent exchange sites.

Coleman, Kamprath, Thomas, and other workers in North Carolina defined exchangeable cations in acid soils as those cations extracted with a neutral unbuffered salt, with the sum of these cations being termed the effective cation exchange capacity. Such an unbuffered salt solution (1.0 N KCl in this case) will extract only the cations held at active exchange sites at the particular pH of the soil. The exchangeable acidity thus extracted is due to aluminum and hydrogen. To measure the effective CEC, one sample of soil is extracted with neutral normal NH$_4$Ac to determine the exchangeable basic cations, such as K, Ca, Mg, and Na. Another sample of the same soil is extracted with 1.0 N KCl to determine the exchangeable aluminum and hydrogen. The sum of the milliequivalents of calcium, magnesium, potassium, and sodium, plus aluminum and hydrogen, is the effective CEC.

Base Saturation. One of the important properties of a soil is its degree of base saturation, which generally reflects the extent of leaching and weathering of the soil. It is defined as the percentage of total CEC occupied by such basic cations as calcium, magnesium, sodium, and potassium. To illustrate how this quantity is calculated, suppose that in the example given the following ion quantities were found in the ammonium acetate extract from the leaching of the 20 g of soil:

Ca	0.02 g
Mg	0.006 g
Na	0.0115 g
K	0.0195 g

The milliequivalent weights of calcium, magnesium, sodium, and potassium are, respectively, 0.02, 0.012, 0.023, and 0.039. The milliequivalents of each of these ions present is

Ca	= 0.02/0.02	= 1	meq
Mg	= 0.006/0.012	= 0.5	meq
Na	= 0.0115/0.023	= 0.5	meq
K	= 0.0195/0.039	= 0.5	meq
Total		2.5 meq per 20 g of soil	

(2.5 meq of bases per 20 g of soil is 12.5 meq per 100 g of soil). The total CEC of this soil was 15 meq per 100 g, so the percentage of base saturation is (12.5/15) × 100, or 83.3.

As a general rule, the degree of base saturation of normal uncultivated soils is higher for arid than for humid region soils. Although not always true, especially in humid regions, the degree of base saturation of soils formed from limestones or basic igneous rocks is greater than that of soils formed from sandstones or acid igneous rocks. Base saturation is related to soil pH and to the level of soil fertility. For a soil of any given organic and mineral composition, the pH and fertility level increase with an increase in the degree of base saturation.

The ease with which cations are absorbed by plants is related to the degree of base saturation. For any given soil the availability of the nutrient cations such as calcium, magnesium, and potassium to plants increases with the degree of base saturation. For example, a soil with a base saturation of 80% would provide cations to growing plants far more easily than the same soil with a base saturation of only 40%. The relation between percent base saturation and cation availability is modified by the nature of the soil colloid. As a rule, soils with large amounts of organic or 1:1 colloids can supply nutrient cations to plants at a much lower degree of base saturation than soils high in 2:1 colloids.

As will be seen in the following section, the CEC, hence the percentage of base saturation, can be rather arbitrary figures unless the method by which they are measured is clearly defined. As a general rule, however, the statements made concerning base saturation and plant availability of cations are true.

Nature of Charge and CEC. The foregoing discussion of cation exchange and degree of base saturation was deliberately oversimplified to set forth clearly the fundamental reaction in this basic soil phenomenon. Actually, studies have shown that the cation exchange capacity of a soil is not a fixed quantity but is dependent on the pH and concentration of the extracting solution used for its determination. The total negative charge on soil colloids which gives rise to cation exchange is caused by isomorphous substitution of ions in the lattice structure of clay minerals, the ionization of hydroxyl groups from hydrated iron and aluminum oxides, and organic matter. Numerous studies have shown that the cation exchange capacity of soils is a continuous function of pH, with this value being lowest in the acid range, pH 3 to 4, and increasing continuously as the pH increases up to the alkaline range, pH 8 to 9. This increase in CEC with increasing pH is caused by the ionization of the OH groups at the edges of the clay lattice and on the hydrous Al and Fe oxides and from the carboxyl and phenolic groups present in soil organic matter.

When the CEC of a soil is determined using an unbuffered neutral salt solution, the value obtained will be lower than would be the case if it were measured using a highly buffered solution at a pH of 7, 8, or 9. The effective CEC discussed in the preceding section therefore is probably a more meaningful value as far as plant growth, fertilizer additions, and liming are concerned than the CEC determined with the buffered solutions at high pH values. The CEC values found by using neutral N ammonium acetate are somewhere between the values found by using the unbuffered salt solution and barium chloride-triethanolamine. For routine CEC determinations, the ammonium acetate method is rapid and convenient and is still used in many laboratories in the United States. If one keeps in mind that the value obtained will be greater than that obtained using the unbuffered salt solution and accordingly makes the necessary allowances, it is a perfectly satisfactory method for measuring this important soil property.

Anion Exchange

It has been known for a long time that phosphates do not leach from soils but are retained in forms that may be removed only by extraction with various salt, acid, and alkaline solutions. A fraction of the phosphorus appears to be held in forms that are quite insoluble. The topic of phosphate retention is covered more fully in Chapter 6. More recently it has been found that much larger amounts of sulfate can be extracted from soils high in 1:1 clays and the hydrous oxides of iron and aluminum with a solution of potassium phosphate than can be extracted with water.

These findings have led to the realization that soils do indeed possess anion exchange properties, and subsequent studies have shown that anions such as chlorides and nitrates may be adsorbed, although not to the extent of phosphates and sulfates.

Contrary to cation exchange, the capacity for retaining anions increases with a decrease in soil pH. Further, anion exchange is much greater in soils high in 1:1 clays and those containing hydrous oxides of iron and aluminum than it is in soils with predominately 2:1 clays.

The mechanisms responsible for anion retention in soils are much more complex than the simple electrostatic attractions involved in most cation exchange reactions. Anions may be retained by soil particles through a number of reactions, some of which are simply electrostatic and are described as being nonspecific. Specific adsorption or chemisorption reactions of a nonelectrostatic nature are also possible.

The positive charge sites responsible for electrostatic adsorption and exchange of anions originate in the broken bonds, primarily in the alumina octahedral sheet, exposing OH groups on the edges of clay minerals. Anion exchange may also occur with OH groups on the hydroxyl surface of kaolinite. Displacement of OH ions from hydrous iron and aluminium oxides is also considered to be an important mechanism for anion exchange, particularly in highly leached soils of the tropics and subtropics, and it is in such soils that anion exchange is greatest. Clay minerals in the montmorillonite group of expansible layer silicates usually have anion exchange capacities of less than 5 meq per 100 g. On the other hand, kaolinites can have an anion exchange capacity as high as 43 meq per 100 g at an acidic equilibrium pH of 4.7.

The pH of most productive soils in the United States and Canada is usually too high for full development of anion exchange capacities. As a result, anions, with the exception of phosphate and to a lesser degree sulfate, will not be retained. There are some soils, however, in the southeastern United States and others formed under high rainfall conditions in Hawaii, Washington, Oregon and in the Canadian province of British Columbia, that have acid subsoils, high in iron and aluminum hydrous oxides, capable of adsorbing anions, especially sulfate. The retention of chlorides and nitrates by anion adsorption is generally not considered to be of any great practical significance in most agricultural soils in North America. A more extensive treatment of phosphate and sulfate retention in soils will be found in Chapters 6 and 8.

Contact Exchange

The discussion of ion exchange so far has dealt only with the exchange of ions between liquid and solid phases. It is believed by some that this exchange can take place between ions held on the surfaces of solid-phase particles and that it does not have to occur via the liquid phase. The extension of this theory leads to the conclusion that ions attached to the surface of root hairs (such as H^+ ions) may exchange with those held on the surface of clays and organic matter in soils because of the intimate contact that exists between roots and soil particles.

The mechanism that permits such an exchange could be described in this way. Clays and plant roots both have CEC properties. Ions are believed to be held at certain spots or sites on both roots and colloidal soil surfaces. The ions held by electrostatic or van der Waals forces at these sites tend to oscillate within a certain volume. When the oscillation volumes of two ions overlap, the ions exchange places. In this way a calcium ion on a clay surface could then presumably be absorbed by the root and utilized by the plant.

There is uncertainty about the importance of contact exchange in the nutrition of plants. It has been estimated that roots may grow to only about 3% of the available nutrients in soils. In addition, there is some question about soil and roots coming into close enough contact for the exchange of ions to occur. There is a strong possibility, however, that the root–soil gap is eliminated by the mucilaginous gel that can occur around root surfaces. This mucigel could serve as a contact complex since it is known to be penetrated by soil particles.

The presence of ectotrophic mycorrhiza, a symbiotic association between fungi and the roots of plants, enhances the uptake of several plant nutrients, particularly phosphorus. This beneficial effect of mycorrhiza is greatest when plants are growing in impoverished soils. The hyphal threads of mycorrhizal fungi act as an extension of plant root systems, resulting in greater exploitation of soil in the root zone. It is possible that contact exchange becomes more significant when mycorrhiza are active.

Root Cation Exchange Capacities

Soil colloids are not the only component of the soil–plant system to exhibit cation exchange properties. It has been observed that plant roots themselves may also possess this property. Capacities ranging from less than 10 to almost 100 meq per 100 g have been measured.

The exchange properties of roots appear to be attributable mainly to carboxyl groups (— COOH) present in pectic substances. Such sites may account for from

70 to 90% of the exchange properties of roots. Uptake of exchangeable ions by roots is considered to be a passive process that is distinct from intake into the interior, living portions of cells.

Plants differ considerably in the magnitude of their measured root CEC values. Legumes and other dicotyledons generally have values at least double the capacities reported for monocotyledons, including the grasses. Legumes and other plant species with high CEC values tend to absorb divalent cations such as calcium preferentially over monovalent cations, whereas the reverse occurs with grasses. These cation exchange properties of roots help to explain why, in grass–legume pastures on soils containing less than adequate K^+, the grass survives but the legume disappears. The grasses are considered to be more effective absorbers of potassium than are the legumes.

Movement of Ions from Soils to Roots

For ions to be absorbed by plant roots, they must come in contact with the root surface. There are generally three ways in which nutrient ions in soil may reach the root surface: (1) root interception, with the possibility of contact exchange, mentioned previously; (2) diffusion of ions in the soil solution; and (3) movement of ions by mass movement with the soil solution. Making certain assumptions about the occurrence and behavior of ions in soil, Barber at Purdue University has estimated the relative importance of these mechanisms in providing nutrients to corn. The values he obtained are shown in Table 4-1. The contribution of diffusion was estimated by the difference between total nutrient needs and the amounts supplied by interception and mass flow.

Root Interception. The importance of root interception and contact exchange as a mechanism for ion absorption is enhanced by the growth of new roots throughout the soil mass and perhaps also by mycorrhizal infections. As the root system develops and exploits the soil more completely, soil solution and soil surfaces retaining adsorbed ions are exposed to the root mass and absorption of these ions by the contact exchange mechanism is accomplished. The quantity of nutrients that can come in direct contact with the plant roots is the amount in a volume of soil equal to the volume of roots. It can be assumed that roots usually occupy 1% or less of the soil. Because roots grow through soil pores which may have higher

TABLE 4-1.   Relative Significance of the Principal Ways in Which Plant Nutrient Ions Move from Soil to the Roots of Corn

Nutrient	Amount of Nutrient Required for 150 bu/A of Corn (lb/A)	Percentage Supplied by		
		Root Interception	Mass Flow	Diffusion
Nitrogen	170	1	99	0
Phosphorus	35	3	6	94
Potassium	175	2	20	78
Calcium	35	171	429	0
Magnesium	40	38	250	0
Sulfur	20	5	95	0
Copper	0.1	10	400	0
Zinc	0.3	33	33	33
Boron	0.2	10	350	0
Iron	1.9	11	53	37
Manganese	0.3	33	133	0
Molybdenum	0.01	10	200	0

Source: Barber and Olson, in L. B. Nelson et al., Eds., *Changing Patterns in Fertilizer Use,* p. 169. Madison, Wis.: Soil Science Society of America, 1968.

than average nutrient content, it is estimated that roots would contact a maximum of 3% of the available nutrients in the soil.

Mass Flow. Movement of ions in the soil solution to the surfaces of roots is an important factor in satisfying the nutrient requirement of plants. This movement is accomplished largely by mass flow and diffusion. Mass flow, a convective process, occurs when plant nutrient ions and other dissolved substances are transported in the flow of water to the root that results from transpirational water uptake by the plant. Some mass flow can also take place in response to evaporation and percolation of soil water.

Amounts of nutrients reaching roots by mass flow are determined by the rate of water flow or the water consumption of plants and the average nutrient concentrations in the soil water. The level of a particular nutrient around the root will fluctuate depending on the balance between the rate at which it reaches this zone by mass flow and the rate of uptake by the root. Mass flow supplies an overabundance of calcium and magnesium in many soils and most of the mobile nutrients, such as nitrogen and sulfur, if concentrations in the soil are sufficient (Table 4-1).

Diffusion. It is apparent in Table 4-1 that most of the phosphorus and potassium moves to the root by diffusion. Diffusion occurs when an ion moves from an area of high concentration to one of low concentration by random thermal motion. As plant roots absorb nutrients from the surrounding soil solution, a diffusion gradient is set up. Plant roots absorbing nutrients in this manner thus create a sink to which nutrients diffuse. A high plant requirement or a high root "absorbing power" results in a strong sink or a high diffusion gradient, favoring ion transport.

The three principal soil factors influencing the movement of nutrients into the root are the diffusion coefficient, concentration of the nutrient in the soil solution, and the buffering capacity of the solid phase of the soil for the nutrient in the soil solution phase. Of these factors, the diffusion coefficient is the most important since it controls how far nutrients can diffuse to the root. For a given spacing of roots in soil, it determines the fraction of the nutrients in the soil that can reach the root during a specific period of plant growth.

Using the general equation from Fick's law of diffusion it can be shown that the effective diffusion coefficient, D_e, for the diffusion of an ion in soil, is influenced by three principal factors: volumetric water percentage, θ; tortuosity factor, f, which expresses the irregular and indirect pathway of diffusion in the pores of the soil; and the buffering capacity, b. The relationship of these parameters to the diffusion coefficient is

$$D_e = D_w \theta f \frac{1}{b} \tag{1}$$

where D_w is the diffusion coefficient for the particular nutrient in water. From equation (1) it can be predicted that diffusion can be increased by high soil moisture contents or volumetric water percentage. Raising θ also reduces tortuosity, which in turn increases diffusion. High soil moisture levels thus have a substantial influence on diffusion.

Buffering capacity usually decreases as nutrient levels in the soil are raised. Reduction in buffering capacity is associated with a rise in the rate of diffusion. Fertilization of part of the soil or localized placement is a practical means of increasing the effective diffusion coefficient.

Although not indicated in the foregoing equation, the effective diffusion coefficient varies directly with the square of the temperature (absolute), thus making it very sensitive to soil temperature. Movement of nutrient ions by diffusion is slow under most soil conditions and occurs over very short distances in the vicinity of the root surface. Although values vary greatly with soils and soil moisture conditions, typical average distances for diffusion to the root are nitrogen, 1 cm; phosphorus, 0.02 cm; and potassium, 0.2 cm. The mean distance between corn roots in the top 15 cm of soil is about 0.7 cm, indicating that some nutrients would need to diffuse half this distance, or 0.35 cm, before they would be in position for absorption by the plant root.

Under some conditions the concentration of certain ions may build up at the root surface because the root is unable to absorb them at a sufficiently rapid rate. This results in a phenomenon known as "back diffusion," in which the concentration gradient, and hence the movement of certain ions, will be away from the root surface and back toward the soil solution. Normally such a condition will not occur, but as roots do not absorb all nutrient ions at the same rate, there may on occasion be a buildup of those ions that are less rapidly absorbed, particularly during periods when the plant is absorbing moisture rapidly. It should be noted that elevated levels of one or more nutrients in the rhizosphere can have important effects on the uptake of other nutrients.

The importance of both diffusion and mass flow in supplying the root surface with ions for absorption depends on the ability of the solid phase of the soil to supply the liquid phase with these ions. Solution concentrations of ions will be influenced by the nature of the colloidal fraction of the soil and the degree to which these colloids are saturated with basic cations. The nature of the adsorbed cations is also important. Results of several studies have shown, for example, that the ease of replacement of calcium from colloids, either by dilute hydrochloric acid or by plant uptake, varies in this order: peat > kaolinite > illite > montmorillonite. Mehlich in North Carolina, for example, showed that an 80% calcium-saturated beidellite clay (a 2:1 clay) gave the same percentage release of this ion as a 35% calcium-saturated kaolinite or a 25% calcium-saturated peat. Other workers have observed similar relationships.

Complementary Ion Effect. There is another phenomenon affecting the nature of ions in the soil solution. It is known as the complementary ion effect and is defined as the influence of one adsorbed ion on the release of another from the surface of a colloid. When only a portion of the cations held by soil colloids is being exchanged, the release of a given cation from exchangeable form is easier as the retention strength or the strength of bonding of the complementary exchangeable cations increases.

An appreciation of the complementary ion effect can be gained from an example where ammonium in the soil solution is exchanging with calcium on soil colloids. This exchange will take place more readily when the complementary cation on the exchange complex is aluminum rather than sodium. Ammonium will replace much more sodium than it will aluminum. By ammonium becoming involved in exchange for sodium, there is less of it available for replacement of calcium. The more strongly held trivalent aluminum tends to satisfy a greater part of the cation exchange capacity and permits the exchange of ammonium for calcium to proceed more completely.

The usual ease of cation replacement in montmorillonite- and organic matter-dominated soil systems follows the order $Na > K = NH_4 > Mg > Ca > Al$, where sodium is released most easily and aluminum least readily. Since the divalent cations are held more tightly than monovalent cations, excessively high levels of exchangeable ions such as ammonium or potassium can induce deficiencies of divalent cations such as calcium and magnesium. The occurrence of hypomagnesemia of ruminants is sometimes caused by the presence of high levels of exchangeable potassium, aluminum, and/or ammonium and the resultant depression in absorption of magnesium by plants. Ruminants consuming large amounts of low-magnesium forage may develop the nutritional disorder known as grass tetany or hypomagnesemia.

Fertilizer Additions. Application of fertilizers to soils helps, at least temporarily, to maintain a high concentration of nutrient ions in the soil solution. Addition of nutrients such as phosphorus and potassium with restricted mobility in the soil because of their reactions with soil components will produce enriched soil pockets which can persist for long periods. As plant roots absorb ions, the presence of added fertilizer will increase the likelihood of a favorable diffusion gradient, thereby enhancing ion movement to root surfaces. Fertilizer applications can also have a positive influence on nutrient uptake by raising the concentrations of nutrient ions carried to the root in the transpirational flow of water.

Fertilization helps to offset soil conditions that retard the movement of nutrient ions to roots. Some of the principal soil factors influencing ion transport are reviewed in the next section.

Soil Factors Influencing Ion Transport. A number of soil factors, the most important of which are soil texture, moisture content, and temperature, modify delivery of ions to the plant root surface by diffusion and mass flow. The finer the texture of the soil, the less rapid will be the movement of soil moisture and the diffusion of ions through the water. Also, ions diffusing through soil moisture in clay soils are much more likely to be attracted to adsorption sites on the clay than in a sandy soil. The tortuosity parameter in the equation given earlier for the effective diffusion coefficient accounts for these effects of soil texture.

A reduction in soil moisture has a similar effect on both water movement and ion diffusion. As soil moisture is reduced (an increase in soil moisture tension), water movement slows down. Thus the movement of moisture to the root surface is slowed. Similarly, as the moisture content of the soil is lowered, the moisture films around the soil particles become thinner and the diffusion of ions through these films becomes more tortuous. Transport of nutrients to the root surface is probably most effective at a soil moisture content corresponding to field capacity.

Low temperatures will slow down the transport of ions by both diffusion and mass flow. As explained previously in this chapter, diffusion rates are proportional to temperature squared. The movement of nutrients ions by mass flow will be reduced at low temperatures because the transpirational demands of plants will be substantially less at low temperatures than under warmer conditions. In addition, the transport of ions in the flow of water evaporated at the soil surface will diminish at low soil temperatures.

Uptake of ions at the root surface, which is responsible for creating and maintaining diffusion gradients, is strongly influenced by temperature. Within the range of about 10 to 30°C, an increase of 10°C usually causes the rate of ion absorption to go up by a factor of 2 or more. The stream of water moving from the soil into the root is also very sensitive to temperature, since this water must move through endodermal cytoplasm and is therefore subject to factors such as temperature which influence metabolism.

Ion Absorption by Plants

Ion absorption by plants is a topic of great complexity and one that can be given only passing treatment in this book. The present consensus concerning the mechanics of ion absorption is that ions enter roots by exchange, by diffusion, and by the action of carriers or metabolic ion-binding compounds. These three mechanisms are associated with two components of the root system. One is termed *outer space*, or apparent free space, and the other is termed *inner space*. Absorption of ions into the outer space is believed to be governed by the processes of simple

TABLE 4-2. Characteristics of Ion Movement into Inner and Outer Cell Spaces

Outer	Inner
Diffusion and exchange adsorption	Ion-binding compounds or carriers
Nonlinear with time, equilibrium approached in short times	Linear with time for periods up to several hours
Ions stoichiometrically exchangeable for other ions	Ions essentially nonexchangeable and dialyzable
Not highly selective	Specific with regard to site and entry
Nonmetabolic	Dependent on aerobic metabolism
Ions in solution or adsorbed in outer space	Ions in vacuoles and partly in cytoplasm

Source: Gauch, *Annu. Rev. Plant Physiol.,* **8**:31 (1957). Reprinted with permission of the author and the publisher, Annual Reviews, Inc., Palo Alto, Calif.

diffusion and exchange adsorption. Absorption of ions into the inner space is metabolic; that is, an expenditure of energy by the root cell is required for this type of absorption. In contrast to absorption in the outer space, it is largely irreversible.

The characteristics of the modes of entry of ions into the two types of root space have been summarized in Table 4-2.

Inner Space. The concept of an ion-binding compound or carrier is generally accepted, and active accumulation (accumulation against a concentration gradient) apparently involves a combination of the ions with a protoplasmic component, thus accounting for the selectivity of ions in this type of absorption. The various carrier theories all envision a metabolically produced substance that combines with free ions. This carrier–ion complex can then cross membranes and other barriers not permeable to free ions. After the transfer is accomplished, the ion–carrier complex is broken, the ion is released into the inner space of the cell, and the carrier is believed in some cases to be restored.

Two different mechanisms are involved in transport of ions into the inner space. For some ions a "mechanism 1" operates at very low concentrations, while at high concentrations above about 1 mM a "mechanism 2" with different properties comes into play. The plasmalemma, or external cytoplasmic membrane, is implicated as the locale of these dual mechanisms.

As indicated in Table 4-2, transfer into the inner space of cells is a highly selective process. Although potassium, rubidium, and cesium compete for the same carrier, they do not compete with elements such as calcium, strontium, and barium. The last three elements do, however, compete among themselves for another carrier. Selenium will compete with sulfate but not with phosphate or with monovalent anions. Interestingly, $H_2PO_4^-$ and HPO_4^{2-} apparently have separate carriers and do not compete with one another for entry into the inner space.

The nature of the ion-binding compounds is not known exactly, but it is likely that they are directly connected with proteins or are themselves proteins. A number of ion-binding polypeptides have been found. It has been suggested that adenosine triphosphate may react with specific metabolic intermediates in the Krebs cycle to form or destroy ion carriers. Another hypothesis suggests that the carriers are ribonucleoproteins in which nucleic acid binds the cations and the protein moiety binds the anions. Still other workers suggest that the carriers are phosphorylated nitrogen-containing intermediates in protein synthesis and that the carrier releases the ions on incorporation into the protein at the site of synthesis.

The cell mitochondria have also been suggested as ion carriers, and some work has indicated that cations and anions accumulate in these bodies. Considerable work has been done in Sweden which indicates that the cytochromes are implicated in the active transport of ions. It has been pointed out by other workers, however, that the observed high correlation between cytochrome activity and ion uptake does not constitute proof that ion transport is achieved by the operation of the cytochrome system.

Outer Space. A considerable fraction of the total volume of the root is accessible for the passive absorption of ions. The "outer" or "apparent free space," where the diffusion and exchange of ions occurs, is located in the walls of the epidermal and cortical cells of the root and in the film of moisture lining the intercellular spaces. Walls of the cells of the cortex are apparently the principal locale of the outer space. This extracellular space is outside the outermost membrane, the plasmalemma, which is a barrier to diffusion and exchange of ions.

Extracellular spaces exist in the mesophyll cells of leaves where ions are able to diffuse and exchange. Most of the nutrient ions reach the "outer" space of leaves via the xylem, from the roots. Mineral ions in rain, irrigation water, and in foliar applications penetrate leaves through the stomata and cuticle to reach the interior of leaves, where they become available for absorption by mesophyll cells.

The possible relation between outer space and translocation of ions to the tops of plants is of considerable interest. Some workers maintain that ions actively accumulated by roots are not exchangeable, hence are not free to move to the tops of plants. The ions in the outer space, however, apparently are transported quite freely to plant tops. Other workers have observed that the movement of ions from roots to shoots is determined by the rates of water absorption and transpiration, suggesting that mass flow may be important in the movement of ions. They also observed that the relative contributions of the active and passive components to the total accumulation of ions in shoots depended both on water absorption and the concentration of the medium. The passive component is more important at high than at low solute concentrations.

Because active absorption is believed to result in fixation in the tissue of the ions so absorbed, the concepts of free space and passive absorption are useful in explaining the movement of salts through root systems. Many workers believe that the cells of roots and plant tops may accumulate ions from the transpiration stream, for obviously any ions that enter the roots and are carried upward by mass flow come in contact with many adsorption sites on numerous actively accumulating cells. This, in turn, would lead to a considerable gradient in the concentration of ions in the transpiration stream and may support the observation that ions from dilute external solutions do not appear to reach plant tops through passive absorption.

Mass flow as a means of explaining ion transport is not without its flaws. Selectivity of ions in mass flow is most difficult to explain. An excellent example is found in sodium, which is virtually excluded from the tops of some plant species. Despite the incompleteness of the picture in regard to ion absorption by plants, the present concepts of outer and inner space and their relation to active and passive absorption of ions are useful in explaining many of the observed facts. As in many other incompletely understood biological phenomena, future research will undoubtedly unveil more and more that is presently hidden from view.

Little has been said to this point about the differences between plants in selectivity of nutrient uptake. Many aspects of absorption, transport, and utilization of mineral nutrients in plants are under genetic control. Considerable evidence has been assembled which shows that genotypes within a species may differ greatly in various features of mineral nutrition, including rates of absorption and translocation of specific elements, efficiency of metabolic utilization, tolerance to high concentrations of elements in the growth medium, and other factors. The differential tolerance of soybean varieties to iron stress is exhibited in Table 4-3. The Bragg cultivar was able to absorb sufficient iron to grow satisfactorily, whereas the Forrest cultivar developed severe deficiencies of this nutrient. In addition to genetically controlled differences in mechanisms of mineral nutrition, the mor-

TABLE 4-3.   Differential Response of Soybean Varieties to Iron Stress

Cultivar	Yield* (g dry weight)			Iron Concentration in Tops (ppm)		
	Quinlan	Tripp	Millville	Quinlan	Tripp	Millville
North						
Amsoy 71	1.17	2.83	1.50	32	55	38
Corsoy	1.16	2.68	2.22	33	39	43
Hodgson	1.55	3.03	1.96	43	40	43
Bonus	1.32	2.57	1.99	42	51	48
Williams	1.40	3.21	2.32	29	39	45
South						
Forrest	1.07	2.70	1.60	20	28	22
Davis	0.98	2.12	1.41	27	49	43
Lee	1.15	3.06	1.61	26	37	39
Bragg	1.46	2.87	1.87	39	49	45
Tracy	1.36	2.99	2.05	37	36	47

* Soils from Oklahoma, Kansas, and Utah, respectively, which have produced iron stress in other crops.

Source: Brown and Jones, *Agron. J.,* **69:**401 (1977).

phology of roots can also significantly influence the uptake of nutrient ions. Some varieties are better able to exploit soil for nutrients and moisture because of larger or more finely branched root systems.

Summary

1. Ion exchange, which is defined as the reversible process by which cations and anions are exchanged between solid and liquid phases, was reviewed with emphasis on the phenomenon as it occurs in soils.

2. The determination of cation exchange capacity (CEC) in soils was reviewed, and the factors affecting this important property were discussed. The CEC of a soil is related to the nature and amounts of the mineral and organic colloids present. It increases as soil pH rises.

3. Base saturation, which is the degree to which the exchange capacity of a soil is saturated with basic cations (i.e., Ca^{2+}, Mg^{2+}, K^+, and Na^+), was covered in relation to the nature of the charge on the exchange complex. The dependence of the CEC on the way in which this property is measured was discussed.

4. Anion exchange can take place in soils, but for all practical purposes it is confined to the phosphate and sulfate ions. Unlike cation exchange, anion exchange increases with a decrease in soil pH. The reason that other anions do not undergo adsorption in most agricultural soils was explained.

5. Contact exchange, which is the exchange of ions between the surfaces of two solids without movement through a liquid phase, was described. Its possible importance in soils was considered because plant roots themselves exhibit the property of cation exchange. The latter phenomenon was also discussed briefly.

6. Plant nutrient ions are brought into contact with the absorbing surfaces of roots by (a) root interception and contact exchange, which is enhanced by the growth of roots through the soil mass; (b) diffusion of ions in the soil solution; and (c) mass flow of soil water, which brings the nutrient ions into contact with the plant roots. Absorption of moisture by the roots is one of the principal causes of the mass flow of soil water.

7. Mass flow supplies large amounts of calcium and magnesium and a large proportion of the nitrogen and sulfur requirements of plants. Most of the phosphorus and potassium moves from the soil to the root surface by diffusion. Mass flow and diffusion are influenced by soil factors such as clay content, moisture levels, and temperature. Nutrient concentrations in the soil solution have an important effect on the transport of ions to root surfaces.

8. Fertilizer applications can improve the mass flow and diffusion of ions to the root by increasing nutrient concentrations in the soil solution. Diffusion is slowed by increasing clay content, low soil moisture, and low temperatures. The addition of phosphorus and potassium helps to offset these retarding effects by increasing the diffusion gradient.

9. Some of the suggested mechanisms by which ions are actually absorbed by plants were considered. Ions are absorbed by both active and passive mechanisms. Active absorption is thought to take place by metabolically produced carriers which transfer the ions across otherwise impassable barriers. Passive absorption is governed largely by exchange adsorption and diffusion phenomena. Actively absorbed ions are believed to be taken into what is termed the *inner space* in roots, whereas ions that are passively absorbed move into the *outer space* of roots. It was pointed out that much is still to be learned about the mechanics of ion absorption by roots.

Questions

1. Define ion exchange.
2. From what sources does the charge on soil colloids arise?
3. Why does anion adsorption appear to be of no importance in most agricultural soils?
4. Potassium acetate solution was used to determine the cation exchange capacity of a soil. Ten grams of oven-dry soil was extracted with 200 ml of 1.0 N KAc. It was found that 0.078 g of potassium was retained by the soil. What is its CEC?
5. Can one determine by the method described in Question 4 the percent base saturation of this soil? Why?
6. In the example listed on page 98, what assumption must be made if the percentage of base saturation measured is to be considered valid? Under what conditions would you assume that this assumption is valid?
7. Cation exchange in soils appears to increase as the pH _____ Anion exchange in soils appears to increase as the pH _____
8. A soil was found to have a CEC of 24 meq per 100 g. If the exchange capacity were saturated with sodium, to how many grams of sodium chloride would this be equivalent?
9. The CEC of a soil as measured using 1.0 N neutral ammonium acetate gives lower values than those obtained when a solution of BaCl$_2$-triethanolamine buffered at pH 8.3 is used. Why?
10. What is the origin of the effective CEC in mineral soil colloids?
11. What is contact exchange?
12. A soil has a CEC of 25 meq per 100 g on an oven-dry basis. Suppose that this soil was 5% saturated with potassium and you wished to increase the saturation to 9%. Assuming that all of the added potassium would be adsorbed, how much 100% potassium chloride would have to be added to an acre furrow slice of this soil to raise

the potassium saturation to the desired level? Assume that an acre furrow slice of oven-dry soil weighs 2 million pounds.

13. What mechanisms are believed to be the cause of anion adsorption?

14. Under what soil conditions will the SO_4^{2-} ion be adsorbed to the greatest extent?

15. Based on methods presently used to measure the CEC of roots, what type of plant root appears to have the greater CEC—grasses or legumes? Has the CEC been related to the behavior of these plants growing in the field? Explain.

16. What is your opinion as to the relative importance of root interception and contact exchange, simple diffusion, and mass flow in bringing nutrient ions into contact with the absorbing surfaces of plant roots? Would the importance of these three mechanisms be altered by soil texture? By plant species? Why?

17. What soil factors influence diffusion of nutrient ions to roots? Describe and explain a practical way of improving the diffusion of nutrient ions.

18. Assume that 50 lb of sulfur as sulfate was lost per acre per year by leaching and runoff. Assume that another 30 lb was lost through crop removal. How many milliequivalents of cations per 100 g of soil would have to accompany this sulfur to maintain electrical neutrality? If calcium were the only ion involved, to how many pounds per acre of calcium would this removal amount?

19. What is active and passive absorption of elements by plant root cells? In what way are these types of absorption related to the inner and outer space in roots?

20. What, generally, is the mechanism that has been proposed to account for the active absorption of ions by roots?

21. What is the complementary ion effect? In what way does it influence plant uptake of ions? Is it of any practical consequence to commercial crop production? Explain. Give some specific examples to support your answer.

22. Why is cation exchange such an important item in a study of soil fertility and in commercial crop production as well, for that matter?

23. Assume that you are addressing a group of farmers and business managers who are well versed in crop production but who are not so conversant with the technical aspects of plant nutrition and soil fertility. Your mission is to explain to this group the nature of cation exchange and why it is important to crop production. How would you proceed?

24. During the discussion after your speech, one member of the audience asks why chlorides and nitrates will leach from soils but phosphates, which also have a negative charge, will not. What is your answer?

25. Can genotypes within a species of plant differ in various features of mineral nutrition? How will differences in root morphology influence the ability of plants to obtain moisture and nutrients?

Selected References

Adams, F., "Ionic concentrations and activities in soil solutions." *SSSA Proc.*, **35:**420 (1971).

Asher, C. J., and P. G. Ozanne, "The cation exchange capacity of plant roots and its relation to the uptake of insoluble nutrients." *Australian J. Agr. Res.*, **12:**755 (1961).

Baliger, V. C., and S. A. Barber, "Genotypic differences of corn for ion uptake." *Agron. J.*, **71**:870 (1979).

Barber, S. A., "A diffusion and mass flow concept of soil nutrient availability." *Soil Sci.*, **93**:39 (1962).

Barber, S. A., "Efficient fertilizer use," in *Agronomic Research for Food*, ASA Spec. Publ. 26. Madison, Wis.: American Society of Agronomy, 1976.

Barber, S. A., and R. A. Olson, "Fertilizer use on corn," in L. B. Nelson, M. H. McVickar, R. D. Munson, L. F. Seatz, S. L. Tisdale, and W. C. White, Eds., *Changing Patterns in Fertilizer Use*. Madison, Wis.: Soil Science Society of America, 1968.

Barber, S. A., J. M. Walker, and E. H. Vasey, "Mechanisms for the movement of plant nutrients from the soil and fertilizer to the plant root." *Agr. Food Chem.*, **11**:204 (1963).

Baver, L. D., W. H. Gardner, and W. R. Gardner, *Soil Physics*, 4th ed. New York: Wiley, 1972.

Black, C. A., *Soil–Plant Relationships*, 2nd ed. New York: Wiley, 1968.

Bohn, H. L., B. L. McNeal, and G. A. O'Connor, *Soil Chemistry*. New York: Wiley, 1979.

Bradfield, R., "A quarter century in soil fertility research and a glimpse into the future." *SSSA Proc.*, **25**:439 (1961).

Brown, J. C., "Genetic improvement and nutrient uptake in plants." *Bioscience*, **29**:289 (1979).

Brown, J. C., and W. E. Jones, "Fitting plants nutritionally to soils: I. Soybeans." *Agron. J.*, **69**:399 (1977).

Chao, T. T., and M. E. Harward, "Nature of acid clays and relationships to ion activities and ion ratios in equilibrium solutions." *Soil Sci.*, **93**:246 (1962).

Clark, R. B., "Differential response of maize inbreds to Zn." *Agron. J.*, **70**:1057 (1978).

Clark, R. B., and J. C. Brown, "Differential mineral uptake by maize inbreds." *Commun. Soil Sci. Plant Anal.*, **5**(3):213 (1974).

Clarkson, D. T., and J. B. Hanson, "The mineral nutrition of higher plants." *Annu. Rev. Plant Physiol.*, **31**:239 (1980).

Coleman, T. R., E. J. Kamprath, and S. B. Weed, "Liming." *Adv. Agron.*, **10**:475 (1958).

Cooke, G. W., "Chemical aspects of soil fertility." *Soils Fert.*, **25**:417 (1962).

Crooke, W. M., and A. H. Knight, "An evaluation of published data on the mineral composition of plants in light of the cation exchange capacity of their roots." *Soil Sci.*, **93**:365 (1962).

Dixon, J. B., and S. B. Weed, Eds., *Minerals in Soil Environments*. Madison, Wis.: Soil Science Society of America, 1977.

Epstein, E., "Mineral nutrition of plants: Mechanisms of uptake and transport." *Annu. Rev. Plant Physiol.*, **7**:1 (1956).

Epstein, E., *Mineral Nutrition of Plants: Principles and Perspectives*. New York: Wiley, 1972.

Evans, L. T., and I. F. Wardlaw, "Aspects of the comparative physiology of grain yield in cereals." *Adv. Agron.*, **28**:301 (1976).

Franklin, R. E., "Exchange and absorption of cations by excised roots." *SSSA Proc.*, **30**:177 (1966).

Fried, M., and H. Broeshart, *The Soil–Plant System*. New York: Academic Press, 1967.

Gauch, H. G., "Mineral nutrition of plants." *Annu. Rev. Plant Physiol.*, **8**:31 (1957).

Haag, W. L., M. W. Adams, and J. V. Wiersma, "Differential responses of dry bean genotypes to N and P fertilization of a central American soil." *Agron. J.*, **70**:565 (1978).

Harley, J. L., and R. S. Russell, *The Soil–Root Interface*. London: Academic Press (under the aegis of the *New Phytologist*), 1979.

Heintze, S. G., "Studies on cation-exchange capacities of roots." *Plant Soil*, **13**:365 (1961).

Huffaker, R. C., and A. Wallace, "Possible relationships of cation exchange capacity of plant roots to cation uptake." *SSSA Proc.*, **22**:392 (1958).

Huffaker, R. C., and A. Wallace, "Variation in root cation exchange capacity within plant species." *Agron. J.*, **51:**120 (1959).

Jackson, M. L., "Aluminum bonding in soils: a unifying principle in soil science." *SSSA Proc.*, **27:**1 (1963).

Khasawneh, F. E., "Solution ion activity and plant growth." *SSSA Proc.*, **35:**426 (1971).

Lagerwerff, J. V., "The contact exchange theory amended." *Plant Soil*, **13:**253 (1960).

Mengel, K., and E. A. Kirby, *Principles of Plant Nutrition*, 3rd ed. Bern: International Potash Institute, 1982.

Mouat, M. C. N., and T. W. Walker, "Competition for nutrients between grasses and white clover: I. Effect of grass species and nitrogen supply." *Plant Soil*, **11:**30 (1959).

Mouat, M. C. N., and T. W. Walker, "Competition for nutrients between grasses and white clover: II. Effect of root cation-exchange capacity and rate of emergence of associated species. *Plant Soil*, **11:**41 (1959).

Nielsen, N. E., and S. A. Barber, "Differences among genotypes of corn in the kinetics of P uptake." *Agron. J.*, **70:**695 (1978).

Ohlrogge, A. J., "Some soil–root–plant relationships." *Soil Sci.*, **93:**30 (1962).

Oliver, S., and S. A. Barber, "An evaluation of the mechanisms governing the supply of Ca, Mg, K, and Na to soybean roots." *SSSA Proc.*, **30:**82 (1966).

Olsen, R. A., "The driving force on an ion in the absorption process." *SSSA Proc.*, **32:**660 (1968).

Pearson, R. W., and F. Adams, Eds., *Soil Acidity and Liming*, p. 1. Agronomy Series 12. Madison, Wis.: American Society of Agronomy, 1967.

Rovira, A. D., "Plant-root excretions in relation to the rhizosphere effect: I. Nature of root exudate from oats and peas." *Plant Soil*, **7:**178 (1956).

Rovira, A. D., "Plant-root excretions in relation to the rhizosphere effect: II. A study of the properties of root exudate and its effect on the growth of micro-organisms isolated from the rhizosphere and control soil." *Plant Soil*, **7:**195 (1956).

Rovira, A. D., "Plant-root excretions in relation to the rhizosphere effect: III. The effect of root exudate on the numbers and activity of micro-organisms in soil." *Plant Soil*, **7:**209 (1956).

Rovira, A. D., "Plant-root exudates in relation to the rhizosphere microflora." *Soils Fert.*, **25:**167 (1962).

Schenk, M. K., and S. A. Barber, "Phosphate uptake by corn as affected by soil characteristics and root morphology." *Soil Sci. Soc. Am. J.*, **43:**880 (1979).

Schuffelen, A. C., "Growth substance and ion absorption." *Plant Soil*, **1:**121 (1949).

Sommerfeldt, T. D., "Effect of anions in the system on the amount of cations adsorbed by soil materials." *SSSA Proc.*, **26:**141 (1963).

Theng, B. H. K., Ed., *Soil with Variable Charge*. Lower Hutt, New Zealand: New Zealand Soil Science Society, 1981.

Torrey, J. G., and D. T. Clarkson, Eds., *The Development and Function of Roots* (Third Cabot Symposium). New York: Academic Press, 1975.

Uehara, G., and G. Gilman, *Chemistry, Physics, and Mineralogy of Soils with Variable Charge Clays*. Boulder, Colo.: Westview Press, 1981.

Wiersum, L. K., "Utilization of the soil by the plant root system." *Plant Soil*, **15:**189 (1961).

Wiersum, L. K., and K. Bakema, "Competitive adaptation of the CEC of roots." *Plant Soil*, **11:**287 (1959).

Williams, D. E., "Anion-exchange properties of plant root surfaces." *Science*, **138:**153 (1962).

Williams, D. E., and N. T. Coleman, "Cation exchange properties of plant root surfaces." *Plant Soil*, **2:**243 (1950).

Yost, R. S., and R. L. Fox, "Contribution of mycorrhizae to P nutrition of crops growing on an Oxisol." *Agron. J.*, **71:**903 (1979).

Chapter 5

Soil and Fertilizer Nitrogen

THE ultimate source of the nitrogen used by plants is the inert gas N_2, which constitutes about 78% of the earth's atmosphere. In its elemental form, however, it is useless to higher plants. The primary pathways by which nitrogen is converted to forms usable by higher plants are:

1. Fixation by rhizobia and other microorganisms that live symbiotically on the roots of legumes and certain nonleguminous plants.
2. Fixation by free-living soil microorganisms and perhaps by organisms living on the leaves of tropical plants.
3. Fixation as one of the oxides of nitrogen by atmospheric electrical discharges.
4. Fixation as ammonia, NO_3^-, or CN_2^{2-} by any of the various industrial processes for the manufacture of synthetic nitrogen fertilizers.

The supply of elemental nitrogen is inexhaustible. This inert nitrogen is in dynamic equilibrium with the various fixed forms. Even as nitrogen is fixed by the different processes just indicated, so is there a release of elemental nitrogen to the atmosphere from these fixed forms by microbiological and chemical processes which are discussed in this chapter. The nitrogen cycle, which illustrates these transformations, is shown in Figure 5-1.

Nitrogen Fixation by Rhizobia and Other Symbiotic Bacteria

For centuries the use of legumes in crop rotations and the application of animal manures were the principal ways of supplying additional nitrogen to nonleguminous crops. Although they are still important sources of fixed nitrogen for agriculture, the importance of legumes and manure is dwindling with each passing year because of the rapid increase in the production of low-cost synthetic nitrogen compounds. In 1980, 11.4 and 0.8 million tons of nitrogen in the form of various synthetic fertilizers were applied to crops grown in the United States and Canada, respectively. It is estimated that by 1990 U.S. consumption will be over 15 million tons. The growth in consumption of synthetic nitrogen materials is largely the result of the efficiency with which the nitrogen industry is operated and the low cost of these materials in relation to the prices received for crops.

Amounts of Nitrogen Fixed. The quantity of nitrogen fixed by properly nodulated legumes averages about 75% of the total nitrogen used in the growth of the plant. Nitrogen present in soil or additions of fertilizer must make up the difference. Amounts of nitrogen fixed by rhizobia differ with the *Rhizobium* strain,

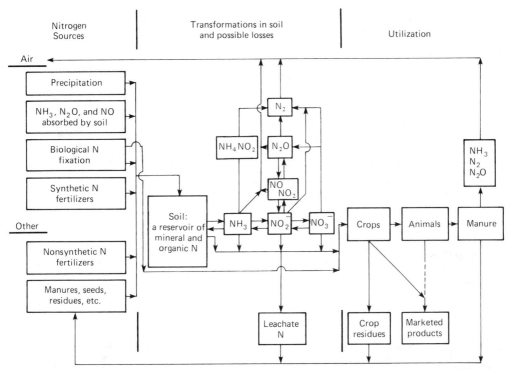

| Nitrogen Sources | Transformations in soil and possible losses | Utilization |

FIGURE 5-1. Nitrogen cycle in soil. [Allison, F. E. 1965 *In* W. V. Bartholomew & F. E. Clark (Editors), "Soil Nitrogen." *Agronomy*, **10**:573–606.]

the host plant, and the environmental conditions under which the two develop. In New Zealand amounts as high as 600 lb of nitrogen have been fixed by clover, growing in mixed stands with grass, and values ranging between 150 and 300 lb/A have been obtained frequently in Australia and New Zealand. Climatic conditions in New Zealand are very favorable year-round for legume growth and nitrogen fixation. Most of the nitrogen required for crop production in New Zealand is still obtained through fixation by rhizobia.

Amounts of nitrogen typically fixed by a selection of legume crops are shown in Table 5-1. Alfalfa, clovers, and lupines generally have greater nitrogen-fixing capacity than peanuts, beans, and peas. Soybeans and peas are inefficient nitrogen fixers in comparison to fababeans. Nitrogen fixation in most legume stands in the temperate zone amounts to about 100 lb/A per year, while in intensively managed pastures, values of 100 to 200 lb/A are common. Short-season annual legumes will often fix between 50 and 100 lb of nitrogen per acre per year, while perennial legumes have the capacity for fixing considerably larger quantities.

The species of the genus *Rhizobium* are numerous and require specific host legume plants. For example, the bacteria that live symbiotically with soybeans will not do so with alfalfa. It is imperative that legume seed be inoculated with the correct inoculum and with one that has been properly prepared and stored prior to use. Inoculation is recommended the first time a field is planted to a new legume species and when the effectiveness of indigenous rhizobia is questionable.

Transfer of Nitrogen Fixed by Legumes. Yields of corn, small grains, and forage grasses are often increased when they are grown in combination with legumes. Some of the benefit appears to be related to improved nitrogen supply for the nonlegume plants. The transfer of nitrogen from the roots of legume crops to the companion crop is not fully understood. Small amounts of nitrogen in the

TABLE 5-1.   Amounts of Nitrogen Fixed by Legumes

	Nitrogen Fixed [lb/(A/yr)]	
Legume	Range in Reported Values	Typical
Alfalfa	50–450	194
Ladino clover	—	179
Sweet clover	–267	119
Red clover	76–169	114
Clovers (general)	50–300	—
Kudzu	—	107
White clover	—	103
Cowpeas	58–116	90
Lespedezas (annual)	—	85
Vetch	80–138	80
Peas	30–140	72
Soybeans	58–160	100
Winter peas	—	50
Peanuts	—	42
Beans	– 71	40
Fababeans	51–148	130
Fababeans (shaded)	–648	—

form of amino acids and other nitrogenous compounds may be excreted by the legume. Microbial decomposition of the sloughed-off root and nodular tissue of the legume may also contribute to the nitrogen supply of crops growing with legumes.

Considerable nitrogen may be released from the roots of legumes when the plants become senescent, are killed, or when the tops are removed by harvesting or grazing. Part of this nitrogen may be transferred to and used by nonlegume crops. This process is important in situations where nonlegume crops follow a legume.

Under some conditions little nitrogen transfer seems to take place and applications of fertilizer nitrogen are necessary to provide most, if not all, of the nitrogen required for satisfactory crop production. There can also be situations where for one reason or other the legumes do not actively fix nitrogen, and associated non-legume crops, as well as the legume, will benefit from nitrogen fertilization.

Supplemental Nitrogen for Legumes. Legume fixation of nitrogen is at a maximum only when the level of available soil nitrogen is at a minimum. It is sometimes advisable to include a small amount of nitrogen in the fertilizer of agricultural legume crops at planting time to ensure that the young seedlings will have an adequate supply until the rhizobia can become established on their roots. Early spring application of nitrogen can be beneficial for legume crops where rhizobial activity is restricted by cold, wet soil conditions. Nitrogen supplements may also be needed for rapid recovery between cuttings on intensively managed legume stands and for very high yields of soybeans. Applications of 200 lb of nitrogen per acre were used by Cooper of the U.S. Department of Agriculture at Wooster, Ohio, in achieving soybean yields of 102 bu/A. Nitrogen, at the rate of 75 lb/A, was also used for the record-setting 118 bu/A of soybeans grown by R. Flannery in replicated research plots in New Jersey.

Soil Reaction and Rhizobial Activity. Soil acidity is a major factor restricting the survival and growth of rhizobia in soil. There are, however, differences in the sensitivity of rhizobial species to soil acidity. The detrimental effect of soil pH levels below 6.0 on *Rhizobium meliloti* numbers in the root zone, nodulation, and alfalfa yields is clearly evident in Figure 5-2. On the other hand, soil pH values between 4.5 and 7.0 had little influence on *R. trifoli*.

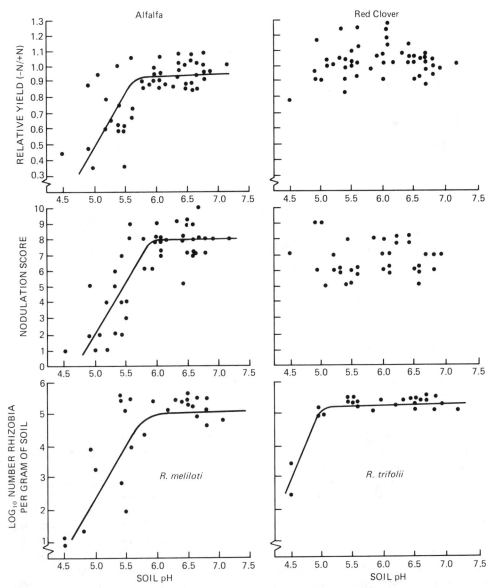

FIGURE 5-2. Effect of soil pH on rhizobia numbers, nodulation, and relative yields (yield without added nitrogen versus yield with 224 kg N/ha per year) of alfalfa and red clover grown at 28 locations in Alberta and northeastern British Columbia. [Rice et al. *Can. J. Soil Sci.*, **57**:197 (1977).]

Application of lime to acid soils is an obvious way of improving conditions for crops such as alfalfa which are dependent on *R. meliloti*. For locations where economical sources of lime may not be available or where transportation costs are prohibitive, alternative means of growing alfalfa must be used. Success in establishing alfalfa under acid soil conditions has been achieved by using special inoculation techniques. These include high levels of inoculum, where excessive amounts of soluble manganese and aluminum are not a problem, and by rolling inoculated seeds in a slurry of pulverized lime. Another approach is to select and use acid-tolerant strains of *Rhizobium*. The effectiveness of some strains of *R. meliloti* at low soil pH is illustrated in Figure 5-3.

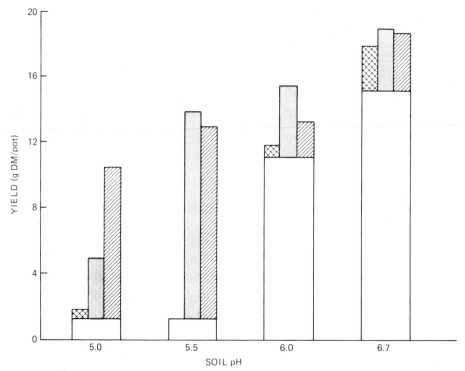

FIGURE 5-3. Yield increase of alfalfa inoculated with three different strains of *R. meliloti* (narrow shaded bars) and grown at various soil pH levels. The wide open bars represent the yield without inoculation. [Rice et al., *Can. J. Soil Sci.,* **57**:197 (1977).]

Effective Nodulation. Where there is reliance on legumes for providing a major portion of the nitrogen requirement for associated and following crops, careful attention should be paid to rhizobial effectiveness. The presence of nodules on the root system of legumes is in itself no guarantee of nitrogen fixation, as it is important that the strain of rhizobia present in the nodules be of high nitrogen-fixing capacity. Mature effective alfalfa nodules tend to be large, elongated (2 to 4 by 4 to 8 mm), often clustered on the primary roots, and have pink-to-red centers. The red color is attributed to the occurrence of leghemoglobin, which is confined to those nodule cells containing rhizobia that are fixing nitrogen. Ineffective nodules are small (< 2 mm in diameter), usually numerous, and are scattered over the entire root system, or in some cases they are very large (> 8 mm in diameter) and few in number. They have white or pale green centers.

Fixation by Leguminous Trees and Shrubs. Although not of great concern to agriculture as practiced in most advanced countries, fixation of nitrogen by leguminous trees is important to the ecology of tropical and subtropical forests. Numerous legume tree species, widely distributed throughout the tropical and temperate zones of the world, fix appreciable amounts of nitrogen. Two well-known examples in the United States are *Mimosa* and *Acacia.* Black locust is another leguminous tree that can produce significant accumulations of nitrogen in its root zone.

Some nonleguminous plants also fix nitrogen by a mechanism similar to that of the symbiotic relationship between legumes and rhizobia. Such plants are widely distributed. Certain members of the following plant families are known to bear root nodules and to fix nitrogen: Betulaceae, Elaeagnaceae, Myricaceae, Coriariaceae, Rhamnaceae, and Casurinaceae. Alder and *Ceanothus* are two species from

TABLE 5-2.   Economically Important Bacteria Involved in Biological Nitrogen Fixation

Organisms	General Properties	Use in Agriculture
Azotobacter	Aerobic, free fixers, live in soil, water, rhizosphere (area surrounding the roots) leaf surfaces	Proposed benefit to crops has not been confirmed; hormonal effect on root and plant growth
Azospirillum	Microaerobic, free fixers, or in association with roots of grasses Inside root symbiosis?	Potential use in increasing yield of grasses; inoculation benefits crops Hormonal effect on roots and plant growth
Rhizobium	Fix nitrogen in legume–*Rhizobium* symbiosis	Legume crops are benefited by inoculation with proper strains
Actinomycetes, Frankia	Fix nitrogen in symbiosis with nonlegume wood trees—alder, Myrica, Casuarina	Potentially important in reforestation, wood production
Blue-green algae, Anabaena	Contain chlorophyll, as in higher plants; aquatic and terrestrial	Enhance crop of rice in paddy soils; *Azolla* (a water fern)–*Anabaena azolla* symbiosis is used as green manure

Source: Okon, *Phosphorus Agr.,* **82**:3 (1982).

this group of plant families which are commonly found in the Douglas-fir forest region of the Pacific Northwest. These two woody nodulated plants can potentially contribute substantial nitrogen to the ecosystem on a continuing basis. *Frankia*, an actinomycete, is the microorganism responsible for nitrogen fixation by these nonleguminous woody plants (Table 5-2).

Free-Living Soil Microorganisms

Nitrogen fixation in soils is also brought about by certain strains of free-living bacteria and blue-green algae (Table 5-2). Alexander at Cornell University has compiled a comprehensive list of the organisms that are able to effect this reaction. A discussion of some of the most important of these organisms follows.

Blue-Green Algae. The blue-green algae occur under a wide range of environmental conditions, including rock surfaces and barren wastelands. They are completely autotropic and require only light, water, free nitrogen (N_2), carbon dioxide (CO_2), and salts containing the essential mineral elements. Their numbers are normally far greater in flooded than in well-drained soils. Because they need light, they probably make only minor contributions to the nitrogen supply in upland agricultural soils after closure of crop canopies. In desert or semiarid regions, blue-green algae or lichens containing them become active following occasional rains and they may fix considerable quantities of nitrogen during the short-lived flushes of activity while moisture conditions are favorable. Nitrogen fixation by blue-green algae is of economic significance in hot climates, particularly in tropical rice soils. The nitrogen made available to other organisms by blue-green algae is probably of considerable importance during the early stages of soil formation.

There is a noteworthy symbiotic relationship between *Anabaena azolla* (a blue-green algae) and *Azolla* (a water fern) in temperate and tropical waters. The blue-green algae located in cavities in leaves of the water fern is protected from external adverse conditions and it is capable of supplying all of the nitrogen needs of the host plant. An important feature of this association is the water fern's very large light-harvesting surface, a property that limits the N_2-fixing capacity of free-living blue-green algae.

Studies in California and in the Philippines clearly indicate that this water fern–blue green algae system is potentially beneficial for rice production. *Azolla* receiving adequate P nutrition can be used both as a green manure during the fallow season and as a cover crop for the rice crop. Thriving growth of this water fern–blue green algae association under Davis, California, conditions has provided 105 kg/ha of nitrogen per growth season or about 75% of the nitrogen requirements of rice. Used as a green manure, the fern, containing 50 to 60 kg/ha of nitrogen, has increased rice yields substantially over those of unfertilized controls in California.

The agricultural importance of nitrogen fixation by free-living bacteria is greater than that of the blue-green algae. These organisms, with the exception of *Rhodospirillum*, require a source of available energy in the form of organic residues. Part of the energy from the oxidation of these residues is used to fix elemental nitrogen. There has been considerable speculation about the amounts of nitrogen actually fixed by these free-living organisms. Some estimates have been as high as 20 to 45 lb/A annually, but a more generally accepted figure based on recent work is about 6 lb/A.

Fixation in Association with Field Crops. Certain nitrogen-fixing bacteria can grow on root surfaces and to some extent within root tissues of a number of plants, including corn, grasses, millet, rice, sorghum, wheat, and many other higher plant species. *Azospirillum* is the nitrogen-fixing bacterium that has been identified and two species are recognized: *Azospirillum brasilense* and *A. lipoferum*. Energy demands of the *Azospirillum* are satisfied by carbonaceous exudates from plants. It has been estimated that nitrogen fixation varies from 0.0018 to 1 lb of nitrogen per acre per day.

Asymbiotic (Free-Living) Nitrogen Fixers. Considerable attention has been given to the rhizosphere of plant roots, the soil area immediately adjacent to the roots, high in energy-rich materials because of their exudation of organic compounds and their sloughing off of tissue. It has been suggested that this zone is the site of nitrogen fixation by *Azotobacter* and *Clostridium*. Soviet agriculturalists have claimed that inoculation of seed with these organisms brought about increased plant growth. The inability of workers in the U.S. Department of Agriculture to confirm these claims tends to support the prevailing assumption that activity of *Azotobacter* and *Clostridium* is of little consequence in soil nitrogen relationships in intensive agriculture. Steyn and Delwiche in their study of non-symbiotic nitrogen fixation at four sites in California found that it amounted to only about 5 kg of nitrogen per hectare per year under the most favorable environment examined, and on a more arid site with native vegetation the quantity fixed was less than half this amount. More nitrogen was fixed in the winter months than at any other time of the year, and it seemed that soil moisture and soluble available energy sources were the principal factors limiting this phenomenon.

The organism *Beijerinckia* inhabits the leaf surfaces of many tropical plants and is thought by some to engage in its nitrogen-fixing activities on these leaves rather than in the soil. *Beijerinckia* is found almost exclusively in the tropics, and it has been suggested that it is a leaf inhabitant rather than a true soil bacterium. For appreciable nitrogen fixation by hetertrophic soil microorganisms, their population must be large, growing, and multiplying rapidly. In addition, most of the nitrogen they contain must come directly from the atmosphere.

Addition from the Atmosphere

Nitrogen compounds are present in the atmosphere and are returned to the earth in rainfall. The nitrogen is in the form of ammonia, NO_3^-, NO_2^-, and nitrous oxide and in organic combinations. The ammonia comes largely from industrial sites where ammonia is used or manufactured. Some undoubtedly is present in the ammonia that escapes from the soil surface because of the reactions taking place. The organic nitrogen can probably be accounted for by the finely divided organic residues which are swept into the atmosphere from the earth's surface.

The soil has a pronounced capacity for adsorbing ammonia gas from the atmosphere. Laboratory studies carried out in New Jersey with six soil types and with atmospheres to which known amounts of ammonia gas had been added indicated that from 50 to 67 lb of NH_3 per acre per year could be adsorbed by these soils. Sorption was positively related to NH_3 concentration and to temperature. In localized areas where atmospheric NH_3 concentrations are above normal, significant quantities of this gas may be adsorbed by soils. This, of course, is independent of that which may be added in rainfall.

Because of the small amount of NO_2^- present in the atmosphere, it is usually lumped in with the figures reported for NO_3^-. The presence of NO_3^- has been attributed to its formation during atmospheric electrical discharges, but recent studies suggest that only about 10 to 20% of the NO_3^- in rainfall and atmosphere arises in this way. The remainder is thought to come from industrial waste gases or possibly from the soil. Atmospheric nitrogen compounds are continually being returned to the soil in rainfall. The total amount of fixed nitrogen thus brought down has been variously estimated to range between 1 and 50 lb/A annually, depending on location. These figures are generally higher around areas of intense industrial activity and as a rule are greater in tropical than in polar or temperate zones. The involvement of NO_3^- as nitric acid in the acid rain problem was referred to in Chapter 2.

Industrial Fixation of Nitrogen

From the standpoint of commercial agriculture, the industrial fixation of nitrogen is by far the most important source of this element as a plant nutrient. Because of the scope of this topic, it is treated in Chapter 10, which deals with the fundamentals of fertilizer manufacture.

Forms of Soil Nitrogen

The total nitrogen content of soils ranges from less than 0.02% in subsoils to more than 2.5% in peats. Nitrogen concentration in the top 1 ft of most cultivated soils in the United States normally varies between 0.03 and 0.4%. The nitrogen present in soil can generally be classed as inorganic or organic. Ninety-five percent or more of the nitrogen in surface soils usually occurs in organic form.

Inorganic Nitrogen Compounds. The inorganic forms of soil nitrogen include ammonium (NH_4^+), nitrite (NO_2^-), nitrate (NO_3^-), nitrous oxide (N_2O), nitric oxide (NO), and elemental nitrogen (N_2). The last form of nitrogen is inert except for its utilization by rhizobia and other nitrogen-fixing microorganisms.

From the standpoint of soil fertility, the NH_4^+, NO_2^-, and NO_3^- forms are of greatest importance; N_2O and NO are also important in a negative way, for they represent forms of nitrogen that are lost through denitrification. The ammonium, nitrite, and nitrate forms arise either from the normal aerobic decomposition of soil organic matter or from the additions to the soil of various commercial fertilizers. These three forms usually represent from 2 to 5% of the total soil nitrogen.

Organic Nitrogen Compounds. The organic forms of soil nitrogen occur as consolidated amino acids or proteins, free amino acids, amino sugars, and other complex, generally unidentified compounds. The latter group is believed to include materials that result from (1) the reaction of ammonium with lignin, (2) polymerization of quinones and nitrogen compounds, and (3) the condensation of sugars and amines. The proportion of total soil nitrogen usually accounted for in these various fractions is as follows: bound amino acids, 20 to 40%; amino sugars such as the hexosamines, 5 to 10%; and purine and pyrimidine derivatives, 1% or less. Very little is known about the chemical nature of the 50% or so of the organic nitrogen not found in these fractions.

Proteins are commonly found in combination with clays, lignin, and perhaps other materials. This has been suggested as one of the reasons for their resistance to decomposition. The existence of these proteins is deduced from the presence of amino acids found in acid soil hydrolyzates. It is assumed that because proteins are formed by a combination of amino acids the presence of these amino acids in the hydrolyzates is proof of the existence of proteins in soils.

Analytical techniques are now available, making it possible to isolate free amino acids from soils which are not in peptide linkages or in combination with high-molecular-weight organic polymers, clays, or lignin. The suitability of these substrates for biological oxidation would imply that they will not build up in large quantities in soils. The ease with which they are decomposed also suggests that they may be a more important source of NH_4^+, the substrate for the nitrifying bacteria, than the nitrogen in the more insoluble consolidated amino acids, the amino sugars, and the lignin and humic complexes. Relative to other forms, the quantities of free amino acids in soils are low.

Nitrogen Transformations in Soils

Plants absorb most of their nitrogen in the NH_4^+ and NO_3^- forms, and uptake of this nutrient is complicated because plants usually have access to both forms. Nitrate is often the dominant source of nitrogen since it generally occurs in higher concentrations than NH_4^+ and it is free to move to the roots by mass flow and diffusion. Some NH_4^+ is always present and will influence plant growth and metabolism in ways that are not completely understood.

Preference of plants for either NH_4^+ or NO_3^- is determined by age, type of plant, environment, and other factors. Cereals, corn, potatoes, sugar beets, pineapple, rice, and ryegrass use either form of nitrogen. Tomatoes, kale, celery, bushbeans, squash, and tobacco grow best when provided with some NO_3^-. Some plants, such as blueberries, *Chenopodium album*, and certain rice cultivars, cannot tolerate NO_3^-. Flue-cured tobacco is a crop that is adversely affected by extended exposure to NH_4^+ in the soil.

Nitrate. The rate of NO_3^- uptake is usually high and it occurs by active absorption. Nitrate uptake is favored by low-pH conditions. Its absorption can be competitively depressed by NH_4^+.

When plants are nourished with high levels of NO_3^-, there is an increase in organic anion synthesis within the plant coupled with a corresponding increase in the accumulation of inorganic cations (Ca, Mg, K). The growth medium will become alkaline and some HCO_3^- can be released from the roots in exchange for excess organic anion formation.

Ammonium. Ideally, NH_4^+ is the preferred nitrogen source since energy will be saved when it is used instead of NO_3^- for synthesis of protein. Nitrate must be reduced before it can be incorporated into protein. This reduction is an energy-requiring process which uses two NADH molecules for each NO_3^- ion reduced. Also, NH_4^+ is less subject to losses from soil by leaching and denitrification.

Plant uptake of NH_4^+ proceeds best at neutral pH values and is depressed by increasing acidity. Absorption of NH_4^+ by roots reduces the concentration of inorganic cations such as calcium, magnesium, and potassium in plant tissues while

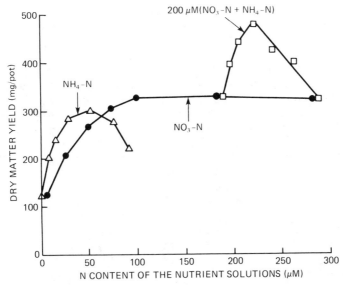

FIGURE 5-4. Effect of nitrogen source, concentration, and the combination of NO_3-N + NH_4-N on the yield of wheat seedlings. [Cox and Reisenauer, *Plant Soil,* **38**:363 (1973).]

raising levels of inorganic anions, including phosphorus, sulfur, and chlorine. Soluble carbohydrates and organic acids may be lower in plants supplied with NH_4^+ compared to those receiving NO_3^-. On the other hand, amide-nitrogen (especially asparagine), amino-nitrogen, total carbohydrate, soluble organic nitrogen, and protein contents may be increased in plants supplied with NH_4^+.

A decline in pH of the rhizosphere occurs when plants are fed NH_4^+. This acidification can have an important effect on both the availability of nutrients and other biological activity in the vicinity of roots. Differences in rhizosphere pH of up to 2.2 units have been observed for NH_4^+- versus NO_3^--fed wheat plants.

Ammonium and Nitrate Combinations. Growth of plants is often improved when the plants are nourished with both NO_3^- and NH_4^+ rather than with either NO_3^- or NH_4^+ singly. Figure 5-4 demonstrates the favorable influence that NH_4^+ in combination with NO_3^- had on the growth of wheat in solution cultures.

Leyshon and others at the Agriculture Canada, Swift Current Research Station observed that when typical nitrogen rates were applied and maintained in the NH_4^+ form, grain yields of barley and wheat were greater than with comparable NO_3^- treatments. The persistence of NH_4^+ in the soil, thereby providing adequate nitrogen nourishment of the crop, may have been partially responsible for its superiority over NO_3^-.

Ammonium tolerance limits are narrow, with excessive levels producing toxic reactions. High levels of NH_4^+ can retard growth and restrict uptake of potassium and produce symptoms of potassium deficiency. In contrast, plants tolerate large excesses of NO_3^- and accumulate it to comparatively high levels in their tissues.

Form of Inorganic Nitrogen and Plant Diseases. The influence of nitrogen nutrition on the occurrence and severity of plant diseases should not be overlooked, particularly the form of inorganic nitrogen available to plants. D. M. Huber of Purdue University and his associates emphasize that the specific form of nitrogen, rather than nitrogen per se, is a major factor influencing disease severity. Some diseases are most severe when NH_4^+ is the primary form of inorganic nitrogen in the root zone; others are most severe when NO_3^- predominates.

The effect that these forms of nitrogen have on rhizosphere soil pH seems to be at least partially responsible for the differences observed in incidence and severity of diseases.

The quantity of NH_4^+ and NO_3^- presented to the roots of agricultural plants depends largely on the amounts supplied as commercial nitrogen fertilizers and released from the reserves of organically bound soil nitrogen. The amount released from these organic reserves (and to a certain extent, those remaining as such in the soil after the addition of ammonium or nitrate fertilizers) depends on the balance that exists among the factors affecting nitrogen mineralization, immobilization, and losses from the soil. By way of definition, nitrogen mineralization is simply the conversion of organic nitrogen to a mineral (NH_4^+, NO_2^-, NO_3^-) form. Nitrogen immobilization is the conversion of inorganic or mineral nitrogen to the organic form. Reactions associated with these phenomena, as well as nitrogen losses from the soil, are discussed in the following sections.

Organic–Mineral Nitrogen Balance in Soil. Soil organic matter is an ill-defined term used to cover organic materials in all stages of decomposition. Broadly speaking, soil organic matter can be placed in two categories. The first is a relatively stable material, termed *humus*, which is somewhat resistant to further rapid decomposition. The second includes those organic materials that are subject to fairly rapid decomposition, materials that range from fresh crop residues to those that, by a chain of decomposition reactions, are approaching a degree of stability.

Nitrogen in some form as well as other nutrients are needed by heterotrophic soil microorganisms that decompose organic matter. If the decomposing organic material has a small amount of nitrogen in relation to the carbon present (wheat straw, mature corn stalks), the microorganisms will utilize any NH_4^+ or NO_3^- present in the soil to further the decomposition. This nitrogen is needed to permit rapid growth of the microbial population which accompanies the addition to the soil of a large supply of carbonaceous material.

If, on the other hand, the material added contains much nitrogen in proportion to the carbon present (alfalfa or clover turned under), there will normally be no decrease in the level of mineral nitrogen in the soil. There may even be a fairly rapid increase in this fraction of soil nitrogen, caused by its release from the decomposing organic material.

Carbon-to-Nitrogen Ratio. The ratio of the percentage of carbon to that of nitrogen, the C/N ratio, defines the relative quantities of these two elements in fresh organic materials, humus, or in the soil as a whole. Nitrogen content of most

TABLE 5-3.   Carbon-to-Nitrogen Ratios in a Selection of Organic Materials

Organic Substances	C/N Ratio	Organic Substances	C/N Ratio
Sweet clover (young)	12:1	Coal liquids and shale oils	124:1
Barnyard manure (rotted)	20:1	Oak	200:1
Clover residues	23:1	Pine	286:1
Green rye	36:1	Crude oil	388:1
Corn stover	60:1	Sawdust (generally)	400:1
Grain straw	80:1	Spruce	1000:1
Timothy	80:1	Fir	1257:1
Bitumens and asphalts	94:1		

Sources: Beaton, "Land Reclamation Short Course," Univ. of British Columbia, pp. Ba–B24 (1974); McGill et al. and Paul and Ladd, Eds., *Soil Biochemistry*, Vol. 5, p. 238. New York: Marcel Dekker, 1980.

TABLE 5-4.   Nitrogen Mineralized from Various Vegetable Residues Following Incubation Under Laboratory Conditions

Plant Residue	C/N Ratio	Total N Mineralized (mg)		
		Constant Airflow	Flask	Uptake by Tomatoes
Check soil	1.8	0.269	0.425	0.294
Tomato stems	45.3	0.029	0.029	0.051
Corn roots	48.1	0.041	0.044	0.007
Corn stalks	33.4	0.128	0.217	0.038
Corn leaves	31.9	0.037	0.123	0.020
Tomato roots	27.2	0.034	0.032	0.029
Collard roots	19.6	0.331	0.184	0.311
Bean stems	17.3	0.769	0.676	0.823
Tomato leaves	15.6	0.665	0.895	0.835
Bean stems	12.1	1.077	1.356	1.209
Collard stems	11.2	1.907	1.788	2.254
Collard leaves	9.7	1.521	1.432	1.781

Source: Iritani and Arnold, *Soil Sci.*, **89**:74 (1960).

forms of humus or stable soil organic matter is about 5.0 to 5.5% and carbon content is 50 to 58%, giving a C/N ratio ranging between 9 and 12. Table 5-3 lists the C/N ratio for a variety of organic substances commonly encountered in soil management.

As shown in Table 5-4, the C/N ratio of organic materials added to soils will have a pronounced effect on positive and negative nitrogen release. In this study with various crop residues, a C/N ratio of approximately 20:1 was the dividing line between immobilization and release of nitrogen. Generally, when organic substances with C/N ratios wider than 30:1 are added to soil, there is immobilization of soil nitrogen during the initial decomposition process. For ratios between 20 and 30, there may be neither immobilization nor release of mineral nitrogen. If the organic materials have a C/N ratio of less than 20, there is usually a release of mineral nitrogen early in the decomposition process. These are general rules of thumb *only*, for many factors other than the C/N ratio influence the decomposition of organic materials and the release or immobilization of nitrogen.

The pattern just discussed is illustrated diagrammatically in Figure 5-5. During the initial stages of the decomposition of fresh organic material there is a rapid increase in the numbers of heterotrophic organisms, accompanied by the evolution of large amounts of carbon dioxide. If the C/N ratio of the fresh material is wide, there will be a net immobilization of nitrogen, as shown in the shaded area under the top curve. As decay proceeds, the C/N ratio narrows and the energy supply (carbon) diminishes. Some of the microbial population dies because of the decreased food supply, and ultimately a new equilibrium is reached. The attainment of this new equilibrium is accompanied by the release of mineral nitrogen (indicated by the crosshatched area under the top curve). The result is that the final soil level of this form of nitrogen may be higher than the original level. There may also be an increase in the level of stable organic matter or humus, depending on the quantity and type of fresh organic material originally added. The time required for this decomposition cycle to run its course depends on the quantity of organic matter added, the supply of utilizable nitrogen, the resistance of the material to microbial attack (a function of the amount of lignins, waxes, and fats present), temperature, and moisture levels in the soil.

Nitrogen Concentration. The total nitrogen content of the organic substance being added to soil is a factor to be considered when predicting effects on nitrogen release. Concentrations of between 1.5 and 1.7% nitrogen are usually sufficient to minimize immobilization of soil nitrogen. In the study referred to in

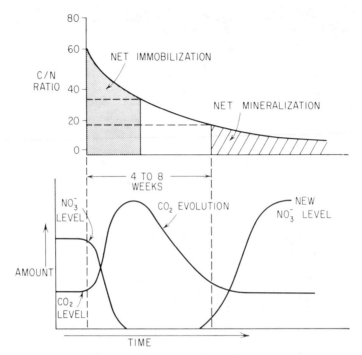

FIGURE 5-5. Changes in nitrate levels of soil during the decomposition of low-nitrogen crop residues. (Courtesy of B. R. Sabey, Univ. of Illinois.)

Table 5-4, mineral nitrogen was released when the incorporated residues contained more than 1.7 to 1.9% nitrogen.

Carbon-to-Nitrogen-to-Sulfur Ratio. There is relative uniformity of the C/N/S ratio within groups of soils, and in a number of Australian soils the ratio was approximately 108:7.7:1. In Saskatchewan this ratio ranged from 58:6.4:1 in arid Chernozemic Brown soils to 129:10.6:1 in leached Grey Wooded soils. The residues produced by decomposition of organic matter added to soils tend to have C/N/S ratios similar to the soil in which they are formed. Low sulfur concentration in the organic substances being added could, just as was pointed out for nitrogen, disturb sulfur mineralization in the soil. An extremely wide N/S ratio (e.g., 20:1 or more) could restrict organic matter breakdown if the soil was already deficient in this element.

In undisturbed (uncultivated) soil the humus content tends to reach a level that is determined by soil texture, topography, and climatic conditions. As a rule, the humus level is higher in cooler than in warmer climates. Further, for any given level of mean annual temperature and type of vegetation the content of stable soil organic matter rises with an increase in effective precipitation. In general, humus contents are greater in fine-textured than in coarse-textured soils. Organic matter contents are higher under grassland vegetation than under forest cover. These relations are generally true for well-drained soil conditions. Under conditions of poor drainage or waterlogging aerobic decomposition is impeded and organic residues build up to high levels, regardless of temperature or soil texture.

Organic Carbon. As a rule, the C/N ratio of the undisturbed *topsoil* in equilibrium with its environment is about 10 or 12 to 1. It narrows in the subsoil in many cases, partly because of the higher content of NH_4^+ nitrogen and the

TABLE 5.5.   Cumulative Effects of Straw Residue at Different Rates During Fallow on Soil Properties in an 8-Year Wheat–Fallow Rotation

Soil Property and Depth	Residue Rate (kg/A)			
	0	*680*	*1360*	*2725*
Organic carbon (%)				
0–7.6 cm	1.79	1.99	2.11	2.20
7.6–15.2 cm	1.33	1.40	1.50	1.71
15.2–30.5 cm	1.11	1.12	1.25	1.32
Total soil nitrogen (%)				
0– 7.6 cm	0.089	0.097	0.096	0.102
7.6–15.2 cm	0.072	0.074	0.083	0.087
15.2–30.5 cm	0.063	0.068	0.069	0.068
N mineralization (ppm) (8 weeks)	18.2	20.9	22.7	24.4
NaHCO$_3$ soluble P (ppm)	7.8	8.4	8.8	9.6
Exchangeable K (meq/100 g)	0.71	0.82	0.91	1.01

Source: Black and Siddoway, *J. Soil Water Conserv.*, **34**:220 (1979).

generally lower amounts of carbon. Under equilibrium conditions the soil microbial population remains about the same, a consistent amount of organic residues is returned to the soil, depending on the vegetative cover, and there is a fairly fixed and usually low rate of nitrogen mineralization. If the soil is disturbed, as in plowing, there is an immediate and rapid increase in mineralization. Continued cultivation without the return of adequate crop residues with sufficient nitrogen will ultimately lead to a decline in the humus content of soils.

The long-term benefits of returning straw residues on organic carbon and total nitrogen concentrations in a Montana soil are shown in Table 5-5. Not only was there a greater buildup of organic matter and nitrogen from the heavier rates of straw, there were also substantial increases in the amounts of mineralizable nitrogen, phosphorus, and potassium.

Continued cultivation with adequate use of commercial fertilizers, coupled with the return of crop residues, not only can maintain the level of soil organic matter but may actually increase it. This is well illustrated in Figure 5-6.

The importance of organic matter is certainly not to be underestimated. It is

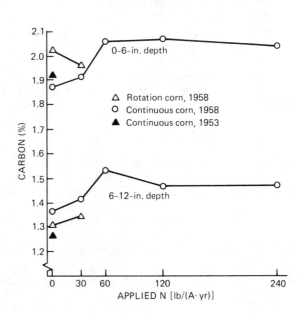

FIGURE 5-6.   Effect of rate of applied fertilizer nitrogen on the carbon content of soil. [Sutherland et al., *Agron. J.,* **53**:339 (1961).]

necessary to maintain good soil structure, especially in fine-textured soils. It increases the cation exchange capacity, thereby reducing leaching losses of elements such as potassium, calcium, and magnesium. It serves as a reservoir for soil nitrogen. It improves water relations, and its mineralization provides a continuous, although limited, supply of nitrogen, phosphorus, and sulfur to the crop. The idea that the maintenance of a high level of soil organic matter should be an end in itself in the farming enterprise is wrong. The ultimate objective of any farming enterprise is *sustained* maximum economical production. The judicious use of lime, fertilizers, and sound management and cultural practices will lead to this objective and, incidentally, will help maintain and even increase the level of soil organic matter.

The mineralization and immobilization of soil nitrogen and the turnover of organic materials in the soil are effected by the heterotrophic soil organisms, including bacteria and fungi. Their requirement for energy is met by their oxidation of the carbonaceous material in the soil. This decomposition increases with a rise in temperature. It is further enhanced by adequate, although not excessive, soil moisture and a good supply of oxygen. Decomposition proceeds under waterlogged conditions, although at a slower rate, and is incomplete. Aerobic, and to a lesser extent anaerobic, respiration releases the contained nitrogen in the form of NH_4^+. This is the first step in the mineralization of nitrogen, the subject to be next considered.

Tillage. Reduced tillage or conservation tillage systems are now being practiced in many areas to reduce wind and water erosion, to make more efficient use of precipitation, and to lower fuel, labor, and equipment costs. Contrasted with conventional tillage, there are fewer operations and less thorough incorporation of crop residues under conservation tillage. With zero tillage, crop residues are allowed to remain on the soil surface rather than being worked into the soil. The insulating and shading effect of crop residues laying on the surface will restrict nitrogen and sulfur mineralization because of lower soil temperatures. The physical nature of loose, coarse accumulations of crop debris is also far from ideal for rapid turnover of organic materials and release of nitrogen and sulfur.

Mineralization of Nitrogen Compounds. The mineralization of organic nitrogen compounds takes place in essentially three step-by-step reactions: aminization, ammonification, and nitrification. The first two are effected by heterotrophic microorganisms and the third largely by autotrophic soil bacteria. The heterotrophs require organic carbon compounds for their source of energy. Autotrophic organisms obtain their energy from the oxidation of inorganic salts and their carbon from the carbon dioxide of the surrounding atmosphere.

During a single growing season in the temperate zones, 1 to 4% of the total humus nitrogen is usually converted by these processes into inorganic forms used by plants.

Aminization. The population of heterotrophic soil microorganisms is composed of numerous groups of bacteria and fungi, each of which is responsible for one or more steps in the numerous reactions in organic-matter decomposition. Bacteria are believed to dominate in the breakdown of proteins in neutral and alkaline environments with some involvement of fungi and possibly also actinomycetes. Under acid conditions the fungi prevail. The end products of the activities of one group furnish the substrate for the next, and so on, down the line until the material is decomposed. One of the final stages in the decomposition of nitrogenous materials is the hydrolytic decomposition of proteins and the release of amines and amino acids. This step is termed *aminization* and is a function of some of the heterotrophic organisms. It is represented schematically by the following:

proteins $\rightarrow$ R — NH_2 + CO_2 + energy + other products

Ammonification. The amines and amino acids so released are further utilized by still other groups of heterotrophs with the release of ammoniacal compounds. This step is termed *ammonification* and is represented as follows:

R — NH_2 + HOH $\rightarrow$ NH_3 + R — OH + energy

+ H_2O

$\searrow NH_4^+$ + OH^-

A very diverse population of bacteria, fungi, and actinomycetes is capable of liberating ammonium. The ammonifying populations include both aerobic and anaerobic microorganisms.

The ammonium released into soil systems is subject to several fates:

1. It may be converted to nitrites and nitrates by the process of nitrification.
2. It may be absorbed directly by higher plants.
3. It may be utilized by heterotrophic organisms in further decomposing organic carbon residues.
4. It may be fixed in a biologically unavailable form in the lattice of certain expanding-type clay minerals.
5. It could, according to the thermodynamic considerations, be slowly released back to the atmosphere as elemental nitrogen.

Nitrification. Some of the NH_4^+ released by the processes of ammonification is converted to nitrate nitrogen. This biological oxidation of ammonia to nitrate is known as nitrification. It is a two-step process in which the ammonia is first converted to nitrite (NO_2^-) and thence to nitrate (NO_3^-). Conversion to nitrite is brought about largely by a group of obligate autotrophic bacteria known as *Nitrosomonas* by a reaction that can be represented by the following equation:

$2NH_4^+$ + $3O_2$ $\rightarrow$ $2NO_2^-$ + $2H_2O$ + $4H^+$

It has also been shown that numerous heterotrophic organisms can convert reduced nitrogen compounds to nitrite (NO_2^-). The organisms include bacteria, actinomycetes, and fungi. The substrates from which the nitrite is produced include not only NH_4^+ but also amines, amides, hydroxylamines, oximes, and a number of other reduced nitrogen compounds. *Nitrosomonas*, however, is considered to be the most important of the soil organisms bringing about the conversion of NH_4^+ to NO_2^-. The conversion from nitrite to nitrate is effected largely by a second group of obligate autotrophic bacteria termed *Nitrobacter*. The equation representing this reaction may be written as follows:

$2NO_2^-$ + O_2 $\rightarrow$ $2NO_3^-$

Although *Nitrobacter* is probably by far the most important organism bringing about the conversion of NO_2^- to NO_3^-, some few heterotrophs, mostly fungi, will also produce nitrates, although a few bacterial strains will also effect this conversion. *Nitrosomonas* and *Nitrobacter* are usually referred to collectively as the *Nitrobacteria*.

Three important and very practical points are brought out by these nitrification equations, an understanding of which will make clearer the reactions taking place when commercial nitrogen fertilizers of the organic or ammoniacal form are applied to the soil. In the first place the reaction requires *molecular* oxygen. This means simply that it will take place most readily in well-aerated soils. A second point is that the reaction releases hydrogen ions (H^+). It is the release of these ions that results in the acidification of the soil when ammoniacal and most organic nitrogen fertilizers are converted to nitrates. Continued use of such forms of nitrogen will lower the soil pH. The judicious use of lime in a farming program, however, will prevent this acid condition from developing. A third point of importance is that because microbial activity is involved, the rapidity and extent of the transformation will be greatly influenced by soil environmental conditions such as moisture supply and temperature. This point is considered in a subsequent section of this chapter.

In well-drained neutral to slightly acid soils the rate of oxidation of NO_2^- to NO_3^- is normally higher than that of NH_4^+ to NO_2^-. The rate of NO_2^- formation is equal to or greater than the rate of formation of NH_4^+. As a consequence, nitrate is the form that tends to accumulate in soils or, if plants are growing thereon, will be the form most used by them.

Factors Affecting Nitrification. Factors influencing the activity of the nitrifying bacteria have a pronounced effect on the amount of nitrates produced and consequently on the utilization of nitrogen by plants. As a general rule of thumb, the environmental factors favoring the growth of most upland agricultural plants are those that also favor the activity of the nitrifying bacteria.

Factors affecting the nitrification pattern in soils are (1) supply of the ammonium ion, (2) population of nitrifying organisms, (3) soil reaction, (4) soil aeration, (5) soil moisture, and (6) temperature.

Supply of the Ammonium Ion. Because the substrate for the nitrifying bacteria is the ammonium ion, a supply of this ion is the first requirement for nitrification. If conditions do not favor the release of ammonia from organic matter (or if ammonium-containing fertilizers are not added to the soils), there will be no nitrification. Temperature and moisture levels favorable to nitrification are also favorable to ammonification. If, however, the ratio of carbon to nitrogen in the soil is too wide, any ammonia released from organic matter will be appropriated by the heterotrophic population that is decomposing the organic material.

This phenomenon is of practical agricultural importance. If large amounts of small grain straw, mature dry corn stalks, or similar materials are plowed into soils with only limited quantities of nitrogen, this nitrogen will be used by the microorganisms in the decomposition of the carbonaceous residues. If crops are planted on such areas immediately after the plowing in of these residues, they may suffer from a shortage of nitrogen. This shortage can be prevented by the addition of sufficient fertilizer nitrogen at the time of disking in the material to supply the needs of the microorganisms as well as those of the growing crop. Such nitrogen deficiencies induced by organic matter are not common but localized examples have been observed in the field.

Population of Nitrifying Organisms. Soils differ in their ability to nitrify added ammonium compounds, even under similar conditions of temperature, moisture, and level of added ammonium. One factor that may be responsible is the variation in the numbers of nitrifying organisms present in the different soils. The impact that this could have on soil nitrification patterns was investigated by workers at Iowa State University.

The presence of different-sized populations of nitrifiers would probably result in differences in the lag time between the addition of the ammonium source and the buildup of nitrate nitrogen in the soil. Because of the tendency of microbial populations to multiply rapidly in the presence of an adequate supply of substrate, the total amount of nitrification taking place in soils would likely not be affected by the number of organisms initially present, provided temperature and moisture conditions were favorable for sustained nitrification.

Others have suggested that differences in the nitrification patterns of soils may be attributed in part to volatile losses of nitrogen resulting from the accumulation of nitrite and its subsequent decomposition.

Soil Reaction. The range of reaction over which nitrification takes place has generally been given as pH 5.5 to about 10.0, with the optimum around 8.5. It is known that nitrates are produced in some soils at pH values of 4.5, and nitrification has been reported in a pasture soil with a pH value of 3.8. Low levels of nitrates have also been detected in acid forest soils, particularly after treatment with nitrogen sources such as urea which temporarily raise soil pH.

The nitrifying bacteria need an adequate supply of calcium and phosphorus, and a proper balance of the elements iron, copper, manganese, and perhaps others. The exact requirement for these mineral elements has not been determined. The influence of both soil pH and available calcium on the activity of the nitrifying organisms suggests the importance of liming in the farming enterprise. Enhancing nitrification during the growing season of the crop is one means of ensuring higher yields.

Soil Aeration. The nitrobacteria, as indicated previously, are obligate autotrophic aerobes. They will not produce nitrates in the absence of molecular oxygen. The relation between oxygen level and nitrification is shown in Figure 5-7. In this study air with the indicated concentration of oxygen was passed through soil to which ammonium sulfate had been added. The soil was then incubated under conditions of adequate moisture and temperature. Maximum nitrification occurred when the percentage of oxygen reached 20, which is about the same as the concentration of this gas in the aboveground atmosphere.

This example illustrates the importance of maintaining conditions that permit rapid diffusion of gases into and out of the soil. Soils that are coarse-textured or possess good structure (by virtue of an adequate supply of humus) facilitate this rapid exchange of gases and ensure an adequate supply of oxygen for the nitro-bacteria.

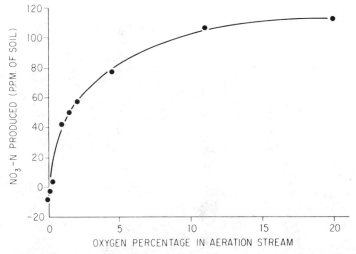

FIGURE 5-7. Production of nitrate nitrogen in a Carrington loam incubated with added ammonium sulfate and aerated with air–nitrogen mixtures with varying oxygen percentages. (Black, *Soil–Plant Relationships,* 1957. Reprinted with permission of John Wiley & Sons, Inc., New York.)

Soil Moisture. Nitrobacterial activity is sensitive to soil moisture. The rates of nitrogen mineralized (NH_4^+ and NO_3^-) are generally highest at soil water contents equivalent to low matric suctions of about $1/3$ bar. Water occupies about 80 to 90% of the total pore space at this matric suction. Nitrogen mineralization tends to be impeded in wet soils with moisture contents exceeding $1/3$ bar or field capacity. Between 15 bars and air dryness, nitrogen mineralization continues to decline gradually.

The data in Figure 5-8 illustrate the effect of high matric suction (low moisture content) on nitrification in a Millville soil incubated for various periods at two matric suctions. Even at the approximate wilting point of 15 bars more than half the ammonium was nitrified in a period of 28 days. At 7 bars, 100% of the ammonium was converted to nitrate at the end of 21 days. Apparently, the *Nitrobacter* are able to function well even in dry soils.

Temperature. Since most biological reactions are influenced by temperature, it is consistent that nitrification is also influenced by this environmental factor. The temperature coefficient, Q_{10}, of nitrogen mineralization is two over the range 5 to 35°C. Thus a twofold change in the mineralization rate is associated with a shift of 10°C within this temperature range. The curve in Figure 5-9 depicts a Q_{10} of two over the temperature range 5 to 35°C in a southern Idaho soil. Below 5 and above 40°C the rate of nitrogen mineralization usually drops off, with the optimum commonly lying between 30 and 35°C. It should be noted, however, that slow formation of nitrate has been detected down almost to the freezing point of water. Significant amounts of nitrate were formed in just two months in certain Iowa soils when the temperature ranged between 0 and 2°C.

Studies in Georgia have shown that some nitrification of added ammonium compounds took place at 37°F. At 42°F, nitrification was appreciable at the end of both three- and six-week incubation periods. Nitrification was essentially complete at the end of nine weeks when the temperature was raised to 52°F. It took 12 weeks, however, for complete conversion of aqua ammonia to nitrate at this higher temperature. The effect of free ammonia on soil microorganisms is discussed in a subsequent section of this chapter.

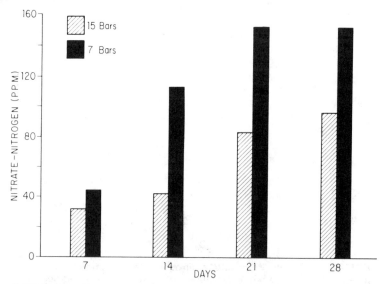

FIGURE 5-8. Effect of moisture levels near the wilting point on the nitrification of 150 ppm of nitrogen applied as ammonium sulfate to a Millville loam and incubated at 25°C. [Justice et al. *SSSA Proc.*, **26**:246 (1962).]

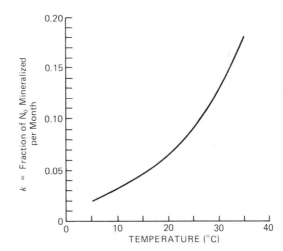

FIGURE 5-9. Fraction of nitrogen
mineralized per month, k, in relation to
temperature (k was estimated graphically
for observed average monthly air
temperatures). [Stanford et al. *Agron. J.,*
69:303 (1977).]

In an Iowa study, the effects on nitrification of temperature over the range 16 to 30°C (61 to 86°F) and varying periods of time were investigated. Nitrification took place at all soil temperatures but was greatest at 30°C. Regardless of temperature, extending the incubation period resulted in greater nitrate production.

Constant temperatures do not persist under most field conditions. Temperature fluctuations will determine the extent of nitrification during the winter months. Thus, if an ammonium fertilizer is added to the soil in the winter in an area in which the mean temperature during the cold months is 37°F, there may be fluctuations in soil temperature that will cause appreciable nitrification. Some Canadian workers addressed themselves to this problem, with the results shown in Figure 5-10. The percentage of nitrification shown on the ordinate of this graph refers to the percentage of added nitrogen (as ammonium sulfate) which was nitrified at the end of 24 days. It is obvious that the occurrence of high temperatures preceding low temperatures results in greater nitrification than if the reverse situation occurred.

A strong interactive effect of soil moisture and temperature on net nitrogen mineralization was observed in a California study. Mineralization increased as temperatures were raised from 15 to 30°C. At suboptimal soil moisture levels there was increased net nitrogen mineralization in the 30°C treatment above that expected from strictly additive effects. The existence of this interaction indicates that these factors should not be considered independently.

In areas with low soil temperatures and/or limited precipitation during the winter, off-season application of ammoniacal fertilizers can mean a saving to the grower of both time and money. It is important that winter temperatures be low enough to retard formation of nitrate, thereby reducing the risk of leaching and denitrification losses of added fertilizer nitrogen before it can be used by the crop in the following spring. Fall applications of ammonium-containing or ammonium-forming fertilizers are expected to be most efficient when daily minimum air temperatures are below 40°F (4.4°C) or when soil temperatures are 50°F (10°C) or lower.

Even if temperatures are occasionally high enough to permit nitrification of fall-applied ammoniacal fertilizers, this itself is not detrimental if leaching does not occur. In many areas of the east and west North Central states, moisture movement through the soil profile during the winter months is insuffecent to remove any nitrates that may accumulate because of temperature fluctuations. For example, ammoniacal nitrogen may be applied in late summer or early fall in the Great Plains to meet the total needs of winter wheat. The same is true farther north for

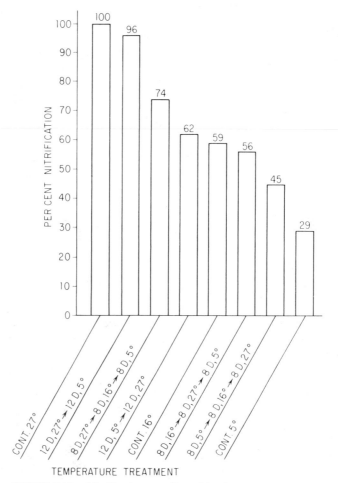

FIGURE 5-10. Nitrification as affected by time–temperature relationships. [Chandra, *Can. J. Soil Sci.,* **42**:314 (1962).]

spring cereal crops in most of the prairie provinces in Canada. Improved positioning and distribution of nitrogen will often result from its over winter movement in dry regions. In other areas of the United States, water movement through the soil profile is excessive, and nitrate losses will occur. Whether or not ammoniacal fertilizers can be fall-applied without significant nitrate loss depends on local soil and weather conditions. Information concerning these patterns is available from local government agencies, universities, and representatives of the fertilizer industry.

Retention of Ionic Nitrogen in Soil

The cationic nature of NH_4^+ permits its adsorption and retention by soil colloidal material. The role of NH_4^+ in the internal nitrogen cycle is shown in Figure 5-11. Nitrate tends to remain outside this cycle.

There is a rapid turnover of the ammonium pool. Uptake of NH_4^+ by heterotrophs and subsequent microbial synthesis produces an active organic phase consisting of the living microbial biomass and its decay products. This active organic phase generally represents 5 to 15% of the total soil nitrogen and it is the main source

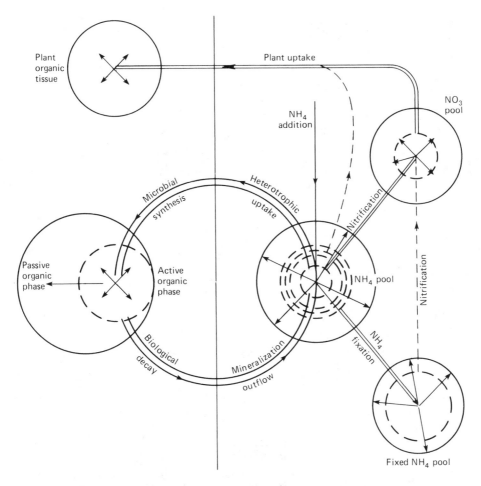

FIGURE 5-11.   Relations between internal nitrogen cycle and addition of mineral nitrogen to soil. [Jansson, K. *Lantbruks-Högskol. Ann.,* **24**:101 (1958).]

of mineral nitrogen during the growing season. A small proportion of the active phase eventually stabilizes as passive organic nitrogen.

As already indicated, it is possible to apply fertilizers containing ammonium nitrogen in the fall in cool and/or dry climates to soils of fine texture without appreciable loss by leaching, provided that temperatures remain below 37 to 40°F. The presence of nitrogen in the cationic form, however, does not ensure its loss against leaching. It is necessary that the soil have a sufficiently high exchange capacity to retain the added ammonium nitrogen or it will be removed in percolating water. Sandy soils with low exchange capacities permit appreciable movement of ammonium nitrogen into the subsoil.

Nitrate Mobility.   Nitrate supplied in commercial fertilizers or produced by nitrification of ammonium is subject to leaching. The NO_3^- form of nitrogen is completely mobile and within limits moves largely with the soil water. Under conditions of excessive precipitation or irrigation it is leached out of the upper horizons of the soil. During extremely dry weather and when capillary movement of water is possible there is an upward movement with the upward movement of

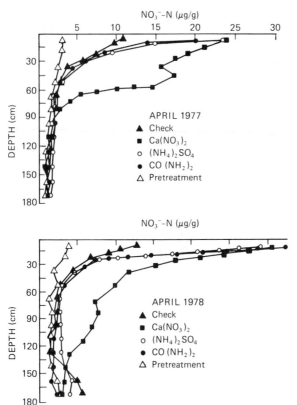

FIGURE 5-12. Overwinter nitrate distribution in the Maddock fine sandy loam following fall application of calcium nitrate, ammonium sulfate, and urea. [Bauder and Montgomery, *Soil Sci. Soc. Am. J.,* **43**:744 (1979).]

water. Under such conditions nitrates will accumulate in the upper horizons of the soil or even on the soil surface.

The pattern of overwinter nitrate distribution in an irrigated fine sandy loam following fall application of several nitrogenous fertilizers is shown in Figure 5-12. Significant nitrate leaching occurred from the calcium nitrate treatment and some nitrate moved to depths of between 75 and 180 cm. Approximately 45 to 55% of the nitrogen applied as ammonium sulfate and urea was recovered as NO_3^- in the 0- to 30-cm depth. This soil receives about 10.7 cm of overwinter precipitation.

Nitrogen Immobilization. Immobilization of nitrogen is the reverse of mineralization and it occurs when large quantities of low-nitrogen crop residues such as cereal crop straw begin decomposing in soil. The high amounts of carbo-hydrate in such residues cause the population of soil microflora to build up quickly. As new cells are formed, nitrogen and other essential elements are used to build protoplasm. Almost invariably this leads to a decrease in the levels of inorganic nitrogen for crops. A shortage of nitrogen can be avoided in such situations by supplying enough fertilizer nitrogen to compensate for immobilization and to meet the crop requirements.

An example of the immobilization of added nitrate nitrogen in the presence of a large amount of rapidly decomposing organic material (wheat straw) is given in Figure 5-13. During the period of active decomposition as indicated by high carbon dioxide evolution, the added nitrogen was rapidly immobilized. As the rate of microbial activity subsided there was a gradual release of the immobilized nitrogen.

Ammonium Fixation. One of the possible fates of NH_4^+ nitrogen in soils is its fixation by clays with an expanding lattice. The mechanism of NH_4^+ fixation

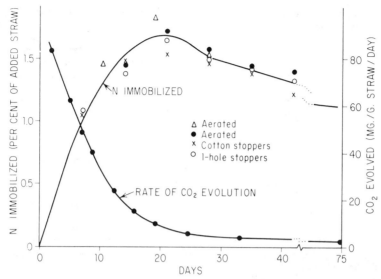

FIGURE 5-13. Immobilization and release of nitrogen and the rate of carbon dioxide formation in a soil receiving wheat straw and nitrate nitrogen. [Allison et al., *Soil Sci.,* **93**:383 (1962). Reprinted with permission of The Williams & Wilkins Co., Baltimore.]

is similar to that of K^+ fixation. It comes about by a replacement of NH_4^+ for interlayer cations in the expanded lattices of clay minerals. The fixed ammonium can be replaced by cations that expand the lattice (Ca^{2+}, Mg^{2+}, Na^+, H^+) but not by those that contract it (K^+, Rb^+, Cs^+).

Certain clay minerals, particularly vermiculite and illite, are chiefly responsible for the fixation of NH_4^+. Fixation of freshly applied NH_4^+ was observed by Kowalenko and Ross of Agriculture Canada to occur in clay-, silt-, and sand-size particles, all of which contained substantial amounts of vermiculite. Coarse clay (0.2 to 2 μm) was found to be quantitatively the most important fraction in fixing added NH_4^+ as well as being the fraction containing the largest portion of native fixed NH_4^+. Fine silt (2 to 5 μm) was the next most important fraction.

It appears that there are small but significant amounts of native fixed ammonium in subsoils. The moisture content and temperature of the soil will affect the fixation of added ammonium compounds. Some of these effects are illustrated in Table 5-6. The data indicate that, at least in the soil types included in this study, appreciable quantities of native fixed ammonium were present and that freezing and drying increased the fixation.

The presence of K^+ will often restrict NH_4^+ fixation since this ion can also fill fixation sites. Consequently, it has been suggested that K fertilization prior to NH_4^+ application is a practical way of reducing NH_4^+ fixation where it is a problem in the field.

Fixation of ammonium by high-organic-matter soils has been investigated by workers at the University of California. The fixation of added ammonium was linearly related to the percentage of carbon in the organic matter, illustrated in Figure 5-14. The effect of oxygen and clay on fixation of the added ammonia is also illustrated. The mechanism of this fixation reaction is not completely understood, although it is suggested that hydroxyl groups present in the organic matter may be the site of the reaction with the added ammonia.

Clay fixation of fertilizer NH_4^+ occurs relatively quickly in eastern Canadian soils while it is released slowly. Under field conditions, a significant proportion (66%) was released in the first 85 days, with the remainder strongly fixed over

TABLE 5-6. **Average Amounts of Native Fixed Ammonium and Added Ammonium Fixed Under Moist, Frozen, and Oven-Dry Conditions in Several Wisconsin Soils**

Horizon Groupings	Average Native Fixed Ammonium (meq/100 g)	Average Fixation of Applied Ammonium (meq/100 g) Under Three Conditions		
		Moist	Frozen	Oven-Dried
Gray-brown podzolic soils				
$Ap + A_1$	0.54	0.08	0.14	0.68
$A_2 + A_3$	0.41	0.06	0.06	0.35
$B_1 + B_2$	0.60	0.15	0.25	0.82
Brunizem soils				
$Ap + A_1$	0.64	0.07	0.10	0.56
A_3	0.65	0.07	0.11	0.72
$B_1 + B_2$	0.60	0.15	0.16	0.67

Source: Walsh et al., *Soil Sci.,* **89**:183 (1960). Reprinted with the permission of The Williams & Wilkins Co., Baltimore.

the next 426 days. Alternate cycles of wetting–drying and freezing–thawing were believed to contribute to the stability of some of the recently fixed NH_4^+ in these soils, as was shown earlier in Table 5-6. In other regions the reported availability of fixed NH_4^+ has ranged from negligible to relatively high. Clay fixation of NH_4^+ will provide some degree of protection against rapid nitrification and subsequent leaching which can be important in management of soil N.

There is evidence that fixed NH_4^+ is in equilibrium with exchangeable NH_4^+. It has been demonstrated by tracer methods that recently fixed NH_4^+ can at least

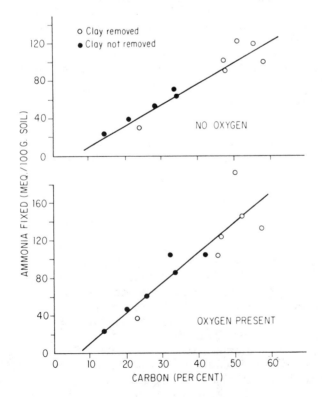

FIGURE 5-14. Relation between the carbon content and ammonia fixation of soil in the presence and absence of oxygen. [Burge et al., *SSSA Proc.,* **25**:199 (1961).]

partially replace native fixed NH_4^+. Furthermore, it has been found that N-Serve nitrification inhibitor reduces the plant availability of recently fixed NH_4^+, which lends support to the theory that some of the fixed NH_4^+ will become exchangeable, thereby becoming susceptible to nitrification.

Significance of Ammonium Fixation. Although the agricultural signif-icance of ammonium fixation is not generally considered to be great, it can be of importance in certain soils, and clay-fixed NH_4^+ has received attention in many parts of the world. It was reported in 1958 that ammonium is fixed by subsoil materials derived from widely scattered locations in central and eastern Wash-ington. A group of 10 soils from Oregon and 7 from Washington were found to fix anhydrous ammonia in a nonexchangeable or difficultly exchangeable form. Of the ammonia retained by these 17 soils, 1 to 8% in the surface and 2 to 31% in the subsurface horizons was fixed by the mineral fraction.

In certain soils of eastern Canada relatively large portions of fertilizer NH_4^+ are clay-fixed, often ranging from 14 to 60% in surface soil and as high as 70% in subsurface soil. Native fixed NH_4^+ is significant in many of these soils and it can amount to about 10 to 31% of the total fixation capacity. The total amount of fixed ammonium in 4-ft profiles of five Saskatchewan soils varied from 2600 to 4600 lb/A and ranged from 7% of the total N in surface soil to as much as 58% in soils at the 4-ft depth. Most of the fixed ammonium occurred in the clay and silt fractions of the three fine-textured soils included in the Saskatchewan study.

Gaseous Losses of Nitrogen

There are losses of nitrogen from soil in ways other than leaching and crop removal. Gaseous losses occur primarily as nitrous oxide (N_2O) and dinitrogen (N_2) during reductive (denitrification) and oxidative (nitrification) processes according to the following reactions:

$$NO_3^- \rightarrow NO_2^- \rightarrow NO \rightarrow N_2O \uparrow \rightarrow N_2 \uparrow$$

$$NH_4^+ \rightarrow NH_2OH \rightarrow (e.g., H_2N_2O_2) \rightarrow NO_2^- \rightarrow NO_3^-$$
$$\downarrow$$
$$N_2O$$

Other mechanisms leading to gaseous losses of nitrogen from soil include:

1. Chemical decomposition of nitrite under aerobic conditions to form N_2, NO plus NO_2, and small quantities of N_2O.
2. Nonbiological volatilization of free ammonia.

Recent unpublished data by J. C. Ryder of the Grasslands Research Institute in the United Kingdom have suggested that denitrification of NH_4^+ does not take place. He obtained significant gaseous losses from ammonium and calcium nitrates but no significant loss from ammonium sulfate.

Denitrification

When soils become waterlogged, oxygen is excluded and anaerobic decomposition takes place. Some anaerobic organisms have the ability to obtain their oxygen from nitrates and nitrites with the accompanying release of nitrogen and nitrous oxide. The most probable biochemical pathway leading to these losses is indicated in the following equation:

$$2HNO_3 \xrightarrow[-2H_2O]{+4H} 2HNO_2 \xrightarrow[-2H_2O]{+2H} 2NO \xrightarrow[-H_2O]{+2H} N_2O \xrightarrow[-H_2O]{+2H} N_2$$

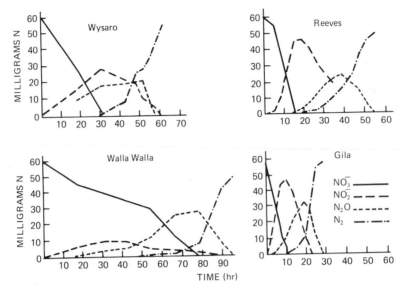

FIGURE 5-15. Sequence and magnitude of nitrogen products formed and utilized during anaerobic denitrification of Reeves loam (pH 7.8), Gila loam (pH 7.9), Wysaro clay (pH 6.1), and Walla Walla silt loam (pH 6.1) at 30°C. [Cooper and Smith, *Soil Sci. Soc. Am. J.,* **27**:659 (1963).]

Examples of the sequence and magnitude of nitrogen products formed and utilized during anaerobic denitrification in four soils of the western United States are given in Figure 5-15.

Only a few particular kinds of facultative aerobic bacteria are responsible for denitrification, and the active species belong to the genera *Pseudomonas, Bacillus,* and *Paracoccus.* A few species of *Chromobacterium, Corynebacterium, Hyphomicrobium,* and *Serratia* are implicated in denitrification. Several autotrophs also capable of bringing about denitrification include *Thiobacillus denitrificans* and *Thiobacillus thioparus.*

There are large populations of these denitrifying organisms in arable soils and they are most numerous in the vicinity of plant roots. Carbonaceous exudates from actively functioning roots are believed to support growth of denitrifying bacteria in the rhizosphere. The potential for denitrification is immense in most field soils, but conditions must arise which cause these organisms to shift from aerobic respiration to a denitrifying type of metabolism involving use of NO_3^- as an electron acceptor in the absence of O_2.

The magnitude and rate of denitrification are strongly influenced by several environmental factors, the most important of which are amount and nature of organic matter present, moisture content, aeration, soil pH, soil temperature, and level and form of inorganic nitrogen at hand.

Decomposable Organic Matter. The amount of readily decomposable organic matter in soil is a critical determinant in the rate of denitrification. A highly significant relationship between denitrification capacity and both water-soluble carbon and mineralizable carbon is evident in Figure 5-16. Stanford and his co-workers in the U.S. Department of Agriculture demonstrated that extractable glucose carbon was a useful index of the quantity of carbon sources associated with loss of nitrate during anaerobic incubation.

The following equations were used by Burford and Bremner at Iowa State University to calculate the amount of available carbon required for microbial reduction

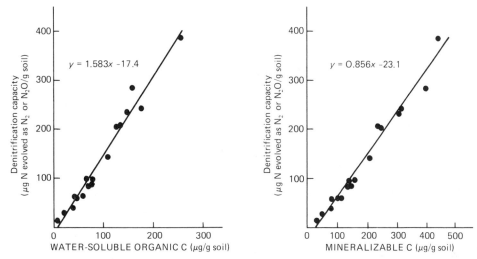

FIGURE 5-16. Relationship between denitrification capacity and (a) water-soluble organic carbon and (b) mineralizable carbon [Burford and Bremner, *Soil Biol. Biochem.*, **7**:389 (1975).]

of nitrate to N_2O or N_2:

$$4(CH_2O) + 4NO_3^- + 4H^+ = 4CO_2 + 2N_2O + 6H_2O$$

$$5(CH_2O) + 4NO_3^- + 4H^+ = 5CO_2 + 2N_2 + 7H_2O$$

According to these equations, 1 ppm of available carbon is required for the production of 1.17 ppm of nitrogen as N_2O or of 0.99 ppm of nitrogen as N_2.

Most of the basic information related to denitrification in soils has been obtained from laboratory investigations with air-dried samples which have been stored for varying lengths of time prior to use. It was shown by Patten and co-workers at Iowa State University that drying and air storage of soils greatly increases their ability to denitrify nitrate under anaerobic conditions. These pretreatments substantially increased the amount of soil organic matter readily utilized by denitrifying microorganisms.

Soil Water Content. Waterlogging of soil results in rapid denitrification by impeding the diffusion of O_2 to sites of microbiological activity. The effect of increasing degree of waterlogging on denitrification is clearly shown in Figure 5-17. Saturation of soil with water during snowmelt in the spring is now suspected of causing major denitrification losses of nitrogen in the Intermountain states such as Utah and in certain soils in the northern parts of Alberta and Saskatchewan. The duration of snow cover on fields and when the melting took place are two factors that seem to affect denitrification associated with spring thawing.

Aeration. Aeration or oxygen availability affects denitrification in two apparently contrasting ways. Formation of NO_3^- and NO_2^-, which are the forms of nitrogen denitrified, is dependent on an ample supply of O_2. Their denitrification, however, proceeds only when the O_2 supply is too low to meet microbiological requirements. The denitrification process can operate in seemingly well-aerated soil, presumably in anaerobic microsites where biological O_2 demand exceeds the supply.

Decreased partial pressure of O_2 will increase denitrification losses. These losses do not become appreciable, however, until the oxygen level is drastically reduced.

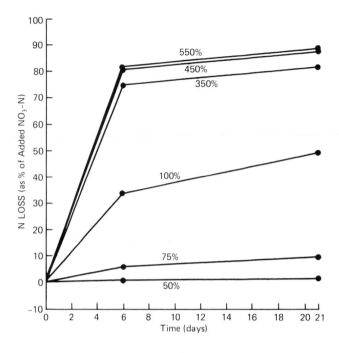

FIGURE 5-17. Effect of moisture, expressed as water-holding capacity, on denitrification in soil receiving glucose. [Bremner and Shaw, *J. Agr. Sci. (Camb.)*, **51**:40 (1958).]

For example, laboratory studies in North Carolina showed that nitrogen losses as a result of denitrification at an atmospheric oxygen content of 7.0 to 8.5% were only 20% of the loss observed when the oxygen content was 4.0 to 5.7%. When the oxygen was reduced to 1.0 to 1.6%, losses of nitrogen at the 7 to 8.5% level of oxygen were only 4% of those sustained at the lowest level (1.0 to 1.6%). In Swedish investigations, the effect of O_2 partial pressures became important when soil moisture contents dropped below 60 to 70% of the water-holding capacity.

Another study conducted by scientists at the U.S. Department of Agriculture showed that reductions in the oxygen content of the atmosphere over incubating soil samples increased the amount of denitrification losses. This is illustrated in Table 5-7. These data show quite clearly the impact of a lowered oxygen content on the increased loss of gaseous nitrogen. Also of importance is the greater loss of nitrogen gas in the presence of a large amount of oxidizable carbonaceous material, in this case, glucose.

TABLE 5-7. Nitrogen Loss from Soils Aerated with 0.46 and 2.27% Oxygen in Nitrogen Gas (All Results per 100 g Soil)

Incubation Period (Days)	No Glucose				With 0.5% Glucose			
	Nitrate-N		Total N		Nitrate-N		Total N	
	Found (mg)	Decrease (%)	Found (mg)	Lost (mg)	Found (mg)	Decrease (%)	Found (mg)	Lost (mg)
0.46% Oxygen								
0	78.3		154.7		78.3		154.7	
5	72.5	7.3	147.9	6.8	23.2	70.4	150.8	3.9
10	63.5	18.8	143.4	11.3	1.4	98.2	114.8	39.9
2.27% Oxygen								
0	78.3		154.7		78.3		154.7	
5	79.0	1.0	149.0	5.7	55.5	29.1	150.5	4.2
10	76.5	2.2	152.4	2.3	58.5	25.2	143.6	11.1

Source: Allison et al., *SSSA Proc.*, **24**:283 (1960).

Soil Reaction. Soil acidity can have a marked influence on the denitrification process since many of the bacteria responsible for this biochemical reduction are sensitive to low pH values. As a result, many acid soils contain small populations of denitrifiers. The results in Figure 5-18 show that the rate of denitrification is greatly influenced by soil pH, being very slow in acid soils (pH 3.6 to 4.8) and very rapid in soils of high pH (8.0 to 8.6). Denitrification losses were of

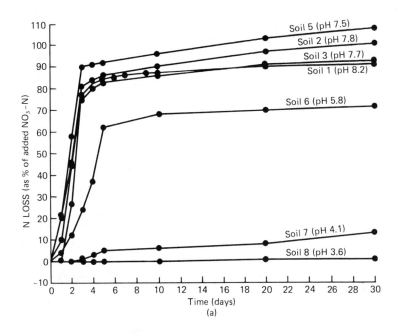

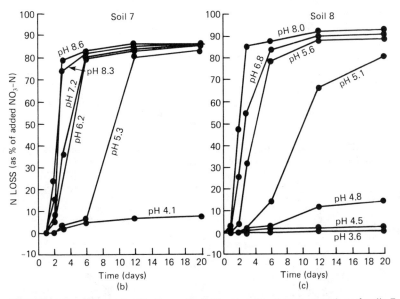

FIGURE 5-18. Effect of soil pH on denitrification. Five-gram samples of soils 7 and 8 previously adjusted to different pH values by addition of calcium hydroxide were incubated at 25° C with 11 ml of water containing 5 mg of NO_3-N (as KNO_3) and 15 mg of carbon (as glucose). [Bremner and Shaw, *J. Agr. Sci. (Camb.)*, **51**:40 (1958).]

little consequence at soil pH values below 5. A Swedish worker, Nommik, also observed that this process was more rapid in neutral soils than in those below 6.0.

Acidity also regulates the sequence and relative abundance of the various nitrogen containing gases formed during denitrification. At pH values below about 6.0 to 6.5, nitrous oxide predominates and it can frequently represent more than half of the nitrogenous gases released in acid environments. Formation of nitric oxide is usually confined to low pH conditions, usually less than about pH 5.5. Nitrous oxide may be the first gas detectable at neutral or slightly acid reaction, but it is reduced microbiologically, so that elemental nitrogen gas tends to be the principal product above pH 6. The occurrence of nitrous oxide under acid conditions is believed to be due to its resistance to further reduction to nitrogen gas.

Temperature. The process of denitrification is very sensitive to soil temperature and its rate increases rapidly in the 2°C to 25°C range. Denitrification will proceed at slightly higher rates when the temperature is increased in the range 25 to 60°C. It is inhibited by temperatures greater than 60°C. The rapid increase in denitrification at elevated soil temperatures suggests that thermophilic microorganisms have a major role in denitrification.

It would seem that the serious denitrification losses coinciding with spring thawing, which were referred to in the section on soil water content, can be related to the greatly accelerated rate of denitrification taking place when soils are quickly warmed from about 2 to 5°C to 12°C or higher.

Nitrate Levels. From the discussions thus far it is obvious that a supply of nitrate and/or nitrite in soil is a prerequisite for denitrification. High nitrate concentrations increase the rate of denitrification and exert a strong influence on the ratio of nitrous oxide (N_2O) to elemental nitrogen (N_2) in the gases released from soil by denitrification. An example of how high levels of nitrate encourage production of nitrous oxide is given in Table 5-8.

Agricultural and Environmental Significance. Both rate and extent of denitrification losses of nitrogen under field conditions are uncertain. Estimates of denitrification losses in the field are approximate because amounts of fertilizer nitrogen are small in relation to the much larger bulk of total soil nitrogen. Also, the ^{15}N tracer technique which has been widely used to study denitrification and other nitrogen transformations in soils is apparently invalid for assessing the contributions of soil and fertilizer nitrogen to nitrous oxide levels in the atmosphere. The discrimination between ^{14}N and ^{15}N that occurs during denitrification of nitrate in soils nullifies the reliability of this tracer technique.

TABLE 5-8. Effect of Nitrate Level on the Proportion of Gaseous Nitrogen as N_2O in the Gases Released Through Denitrification of Nitrate in Soil

Nitrate-N Added* (μg/g soil)	$(N_2 + N_2O)$-N Released (μg/g soil)		Percent of $(N_2 + N_2O)$-N as N_2O	
	2 Days	14 Days	2 Days	14 Days
100	55	97	42	0
200	56	124	48	24
300	61	121	77	42
500	61	119	92	63
1000	57	127	95	86

* 10-g samples of Webster soil were incubated (30°C; 245-ml bottle) under helium after treatment with 6 ml of water containing 1 to 10 mg of nitrate-N as KNO_3.

Source: Bremner, in D. R. Nielsen, Ed., *Nitrogen in the Environment,* Vol. 1, p. 477. New York: Academic Press, 1978.

Fertilizer nitrogen enters a labile "pool" of soil nitrogen which must be subject to continuous denitrification losses to the atmosphere. Since the earth's atmosphere is largely N_2 while its oceans are virtually nitrate-free, denitrification is probably the process responsible for returning nitrogen to the atmosphere, thus offsetting gains from biological N_2 fixation.

The balance-sheet calculations which are possible from the Park Grass experiment at Rothamsted, England, where 96 kg of nitrogen per hectare has been applied every spring since 1856, show that about 30% of the applied nitrogen is lost through denitrification and leaching. Most of the loss has been attributed to denitrification.

Serious reductions in the effectiveness of fall-applied nitrogen for winter wheat have been observed in Utah in years when heavy winter snows persist late into spring. In spite of the fall-applied nitrogen, severe nitrogen deficiencies show up following these unusual snow conditions. The deficient areas may occupy as much as 75% of a field with yields ranging from 5 to 10 bu/A, while adjacent normal areas may yield from 35 to 50 bu. Leaching of nitrogen is not responsible for these shortages; the loss appears to be due to denitrification. Similar results have been obtained in field studies conducted on soils of northern Alberta, where losses of between 25 and 50% in efficiency of nitrogen fertilizers are attributed mainly to denitrification.

Growing season emissions of N_2O from a typical well-managed northern Colorado cornfield amounted to about 2.6 kg of nitrogen per hectare, or about 1.3% of the applied fertilizer nitrogen. Almost 60% of the total N_2O emissions occurred during the week after the field's first irrigation, when restricted oxygen diffusion would favor denitrification. Other researchers have reported that denitrification occurs in periodic bursts, in response to changes in oxygen status, against a background of slow yet continuous denitrification.

There is concern that increased use of nitrogen fertilizers may substantially increase emissions of nitrous oxide (N_2O) from soils and thereby lead to partial destruction of the stratospheric ozone layer protecting the biosphere from biologically harmful ultraviolet radiation from the sun. Although there is evidence that denitrification of fertilizer-derived nitrate is responsible for emission of N_2O, contributions from nitrate produced by the natural transformations of soil organic matter and fresh crop residues have been largely ignored or discounted.

Denitrification can be useful for removal of excessive amounts of NO_3^- from irrigation water and from various wastewaters. For direct treatment of water it may be necessary to inoculate with denitrifying organisms and provide sufficient readily mineralizable carbon in forms such as methanol. Where treatment systems involve disposal of contaminated wastewaters onto soil, measures must also be taken to ensure that levels of mineralizable carbon are adequate in the soil areas being treated.

Chemical Reactions Involving Nitrite

In addition to the volatilization of nitrogen by microbiological denitrification, there is evidence that nitrogen can also be released to the atmosphere by nonenzymatic decomposition of NO_2^- accumulated under apparently well-aerated conditions and from basic (alkaline) soils. There is considerable evidence that these gaseous losses of nitrogen occur while ammonium and ammonium-producing fertilizers are being nitrified. It appears that chemical reactions of NO_2^- are partly responsible for these fertilizer-induced emissions of nitrogen gases such as nitrous oxide and nitric oxide.

Factors Favoring Nitrite Accumulation. Nitrite does not usually accumulate in soil, but when it does occur in significant amounts it can adversely affect plants and microorganisms. Buildup of toxic levels of nitrite is generally

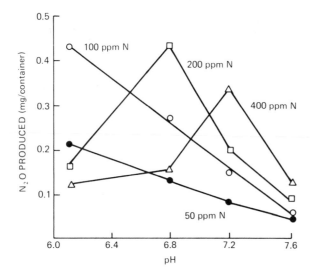

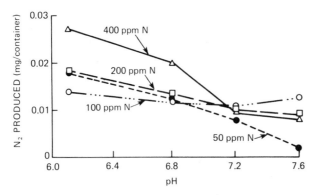

FIGURE 5-19. (Top) Effect of pH on N_2O production as $NaNO_2$ concentration is increased from 50 to 400 ppm N (from 1 to 8 mg per container) t = 3 days. (Bottom) Effect of pH on N_2 production as $NaNO_2$ concentration is increased from 50 to 400 ppm Nt = 3 days. [Christianson et al., *Can. J. Soil Sci.*, **59**:147 (1979).]

caused by two factors, alkalinity and high ammonium levels. Accumulation of nitrite is attributed to the suppressive effect of ammonium salts at alkaline soil pH values on the *Nitrobacter* group of organisms. At pH values of 7.5 to 8.0, the rate of nitrite production exceeds that of nitrate production, but at neutral pH values the potential for converting nitrite to nitrate exceeds that for converting ammonia to nitrite.

Losses of nitrogen from NO_2^- by chemodenitrification have been shown to increase with increasing organic-matter content. It has been proposed that phenolic sites in soil organic matter are responsible for the reduction of nitrous acid to N_2 and N_2O, with nitrosophenols formed as intermediates. The nitrosophenols are believed to tautomerize to quinone oximes, which subsequently reduce some of the nitrous acid to N_2O or N_2. All the N_2O or N_2 evolved is thought to come entirely from the accumulated NO_2^-.

Although buildup of NO_2^- in soil is favored by high pH, its breakdown into gaseous forms of nitrogen is restricted by high soil pH. An example of the tendency for release of N_2O and N_2 to decline with increasing pH of soil, particularly at lower rates of $NaNO_2$ supplementation, is given in Figure 5-19.

Nitrite formed in the fall may undergo chemical denitrification, even if soils freeze. An example of significant rates of chemodenitrification at temperatures below freezing is given in Figure 5-20. The rise in chemodenitrification rate in frozen soil could be the result of forcing dissolved salts, including NO_2^-, into a narrow unfrozen water layer near the surface of soil colloids. This effectively in-

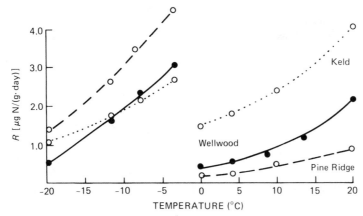

FIGURE 5-20. Chemical denitrification intensity of three soils under frozen and unfrozen conditions; 100 ppm $NO_2 - N$. [Christianson and Cho, *23rd Annu. Manitoba Soil Sci. Meet.*, p. 109, Univ. Manitoba (1979).]

creases NO_2^- concentration, which in turn enhances the chemical reactions involving this form of nitrogen.

Influence of Fertilizers on Nitrite Accumulation. Fertilizer materials such as urea, anhydrous ammonia, aqua ammonia, and diammonium phosphate form alkaline microsites at the point of application in soils. Both NH_4^+ concentration and pH are temporarily elevated in these soil pockets and as a consequence NO_2^- buildup is encouraged regardless of the original bulk soil pH.

When any of the fertilizers named above are used at high rates or when they are applied in bands, creation of these localized zones with basic pH and high NH_4^+ concentration is accentuated. As might be expected, spacing of the bands will also influence the rate of NH_4^+ disappearance and the appearance of NO_2^- and NO_3^-. Extremely large particles of urea or special application techniques that place regular-size urea particles in nests or clusters are expected similarly to intensify transitory rises in pH and NO_2^- concentration.

Restoration of Conditions Suitable for Nitrite Oxidation. Diffusion and/or dilution of the NH_4^+ in the region of the alkaline microsites will help to restore conditions suitable for the normal conversion of NO_2^- to NO_3^-. Diffusion coefficients of NH_4^+ from three nitrogen sources and three soils are listed in Table 5-9. These coefficients varied depending on the types of soil and the fertilizer ma-

TABLE 5-9. Calculated Diffusion Coefficient ($cm^2/sec \times 10^6$) of NH_4^+ from Three Nitrogen Carriers in Three Soils

$(NH_2)_2CO$	$(NH_4)_2SO_4$	NH_4OH
	Keld	
0.64	0.82	0.70
	Wellwood	
1.52	0.73	1.93
	Morton	
1.89	1.24	3.73

Source: Pang et al., *Can. J. Soil Sci.*, **53**:331 (1973).

terials. Small diffusion coefficients were associated with restricted movement of NH_4^+ away from the zone of application and retarded transformation to NO_3^-.

Nitrite can also diffuse beyond the microsite region high in pH and NH_4^+ to reach a soil environment, where the normal functioning of *Nitrobacter* will quickly convert it to NO_3^-.

Pathways of Nitrite Loss. Some of the loss mechanisms involved may include the following:

1. Decomposition of NH_4NO_2.
2. Self-decomposition of HNO_2 at pH values below 5.0 with resultant formation of NO plus NO_2.
3. Dissimilation of NO_2^- by reducing organic compounds.
4. Fixation of NO_2^- by soil organic matter and partial conversion of some NO_2^- to N_2 and N_2O.
5. Catalytic reaction of NO_2^- with reduced transitional metals such as copper, iron, and manganese.

The relative importance of these loss mechanisms will probably vary among soils and the nitrogen management systems practiced.

In the transformation of NO_2^- in soil, part of the nitrogen becomes organically bound or fixed, while some nitrogen is concurrently evolved as N_2, N_2O, and NO + NO_2. This fixation of NO_2^- and the associated release of nitrogen gases is significantly affected by soil pH and organic-matter content. Nitrite fixed by soil organic matter is resistant to mineralization, probably because of the stability of the resultant heterocyclic compounds.

Fertilizer-Induced Emissions That Probably Involve Nitrite. Breitenbeck and co-workers at Iowa State University found that small quantities of N_2O were generated in the course of nitrification of ammonium and ammonium-forming fertilizers. Their results indicate that the fertilizer-induced emissions of N_2O 96 days after treatment with ammonium sulfate and urea amounted to between 0.11 and 0.18% of the fertilizer nitrogen applied. Further research by this Iowa group suggested that N_2O emissions are substantially more when anhydrous ammonia is the nitrogen source.

Thirty percent of the total N_2O emitted during the growing season in a Colorado study occurred from late June to the middle of July, a time when ammonia fertilizer was being oxidized to nitrate by soil bacteria. Apparently, N_2O was produced in the course of autotrophic nitrification of ammonia.

From the practical standpoint it is not known how serious these chemodenitrification losses are under field conditions. If they are significant, it may be possible to modify fertilizer programs so that high concentrations of NH_4^+ in alkaline soil environments do not persist for extended periods.

Volatilization of Ammonia

Ammonium-containing or ammonium-forming fertilizer salts will react with $CaCO_3$ in soil to form $(NH_4)_2CO_3$ and calcium precipitates. The following general equations were developed by Fenn and Kissell of Texas A&M University to represent the reactions that take place:

$$X(NH_4)_zY + NCaCO_3(s) \leftrightharpoons N(NH_4)_2CO_3 + Ca_nY_x \qquad (1)$$

where Y represents the ammonium anion and N, X, and Z are dependent on the valences of the anion and cation. The final reaction product, $(NH_4)_2CO_3$, is unstable and decomposes as follows:

$$(NH_4)_2CO_3 + H_2O \leftrightharpoons 2NH_3 \uparrow + H_2O + CO_2 \uparrow$$
$$\updownarrow$$
$$2NH_4OH \tag{2}$$

The amount of NH_4OH formed in a given time would depend on the solubility of Ca_nY_x and its rate of formation. If Ca_nY_x is insoluble, the reaction would proceed to the right causing more $(NH_4)_2CO_3$ and consequently more NH_4OH to be formed. If no insoluble precipitate is formed, no appreciable quantity of $(NH_4)_2CO_3$ will exist. When $(NH_4)_2CO_3$ decomposes according to equation (2), CO_2 is lost from solution at a faster rate than NH_3 thereby producing additional OH^- ions and an increase in $[OH^-]$. Consequently, more solution NH_4^+ becomes electrically balanced by OH^- which would favor NH_3 loss as represented by the following reaction:

$$NH_4^+ + OH^- \leftrightharpoons NH_4OH \leftrightharpoons NH_3 \uparrow + H_2O \tag{3}$$

If Ca_nY_x is soluble, then the NH_3 loss that occurs will be dependent on the resultant pH of the soil. The NH_3-NH_4^+ equilibrium is pH dependent with lower pH values favoring the NH_4^+ forms.

Factors Influencing Ammonia Volatilization. Many greenhouse and laboratory studies have shown that ammonia volatilization is influenced by soil factors such as pH, calcium carbonate content, cation exchange capacity, exchangeable cations, texture, temperature, moisture content, and species of ammonium-containing or ammonium-forming fertilizer salt. Various aspects of fertilizer management, including rate of NH_4^+ application and depth of incorporation, also have important effects on ammonia losses.

Results from these numerous investigations conducted in the laboratory and under artificial conditions revealed that ammonia losses usually became greater with increases in soil pH, calcium, carbonate content, temperature, and rate of NH_4^+ applied. The percentage of free ammonia increases very rapidly with rising pH, as is evident in Table 5-10. The equilibrium equation for ionized and free ammonia is an integral part of the ammonia volatilization mechanisms advanced by Fenn and Kissel and Feagley and Hossner. Ammonia loss is thus favored by naturally high soil pH or by reactions that raise it, even temporarily.

Some of the major factors that have the opposite effect and retard ammonia volatilization are increased cation exchange capacity, clay content, moisture content, and depth of incorporation of the applied ammonium-bearing or ammonium-forming fertilizer material.

Ammonia loss is greatest from ammonium fertilizer salts, which react with calcium carbonate to form precipitates of low solubility. Figure 5-21 demonstrates the impact that anions of various ammonium salts can have on ammonia volatil-

TABLE 5-10.   Relationship of pH to Percentage of Nonionized Ammonia

pH of Solution	Percentage of Nonionized or Free Ammonia
6	0.1
7	1.0
8	10.0
9	50.0

Source: Parr and Engibous, in *Anhydrous Ammonia Agronomy Workshop*, pp. 4-1 to 4-8. Memphis, Tenn.: Agricultural Ammonia Institute, 1966.

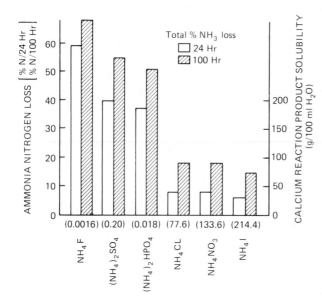

FIGURE 5-21. Total NH_3-N loss at the end of 24 and 100 hr at 22°C, as influenced by the anion of several ammonium salts. The calcium reaction product solubility is shown in parentheses above the chemical formulas. Ammonium nitrogen was applied on soil surface at 550 kg N/ha. [Fenn and Kissel, *Soil Sci. Soc. Am. J.*, **37**:855 (1973).]

ization. It is noteworthy that a rise in soil pH accompanied the formation of insoluble precipitates.

Reliability of Laboratory Measurements of Ammonia Loss. Although substantial losses of ammonia have been measured in laboratory studies, their validity should be closely examined. It should be recognized that experimental systems will impose artificial conditions of air movement, temperature, and relative humidities quite different from those occurring naturally. It is imperative that the laboratory conditions truly represent field conditions because of the far-reaching consequences of conclusions that are drawn.

Gaseous Loss of NO_3^-. It has also been suggested that there may actually be volatile losses of NO_3^- as nitric acid in soils containing appreciable amounts of KCl-exchangeable Al^{3+} plus H^+. Under laboratory conditions greater losses of NO_3^- were observed on those soils with the greater amount of exchangeable acidity, and the losses increased with increases in temperature. The extent of such losses under field conditions is not known.

Ammonia Exchange by Plants

Nitrogen balance sheets frequently ignore ammonia exchange by plants. Porter and his U.S. Department of Agriculture co-workers at Fort Collins, Colorado, demonstrated that corn seedlings were a natural sink for atmospheric ammonia, absorbing up to 43% of the ammonia from air containing 1 ppm of this form of nitrogen. Further studies by Hutchinson of the same organization suggested that field crops exposed to air containing normal atmospheric concentrations of NH_3 might obtain as much as 10% of their total nitrogen requirements by direct absorption of NH_3 from the air. Australian researchers with CSIRO showed that the large amounts of NH_3 produced near the ground surface of grass–clover pasture were almost completely absorbed by the plant cover.

The opposite reaction, one of ammonia volatilization from plant foliage, has been observed from a number of crops, including alfalfa pasture, corn, Rhodes grass, and winter wheat. Several investigators showed that ammonia release was related to stage of plant growth, with losses occurring during ripening and senescence.

Hooker and others at the University of Nebraska suggested that as much as one-third of the nitrogen in a wheat crop was volatilized as ammonia after anthesis. Losses of unspecified forms of gaseous nitrogen have also been reported for rice and soybeans.

Research results indicate that both absorption and loss of NH_3 can occur in field crops. The net ammonia transfer depends on the wetness of the soil surface and the extent of evaporation, which both influence the amount of NH_3 released into the air coming in contact with plant canopies.

Fertilizer Nitrogen Materials

During the last two decades a phenomenal increase in the consumption of commercial fertilizers has been noted in the United States and in other countries as well. The increase in the use of fertilizer nitrogen has been particularly impressive and, as already stated, is due in no small part to the efficiency of the nitrogen producers and the low cost with which these materials have been offered to growers.

Types of Nitrogen Fertilizer

Nitrogen fertilizers may be classified broadly as either *natural organic* or *chemical*. The natural organic materials are of plant or animal origin; the chemical sources are neither plant nor animal.

Organic Forms. Before 1850 virtually all of the fertilizer nitrogen consumed in the United States was in the form of natural organic materials. In the period 1978–1980 these materials accounted for only about 0.1% or less of the total nitrogen usage in the United States. For all intents and purposes these materials are of historical interest only, although small amounts still find their way into specialty fertilizers for lawns, gardens, and shrubs and in some fertilizers applied to flue-cured tobacco. The average nitrogen concentration in natural organics is typically between 1 and 13%.

Natural organic materials at one time were thought to release their nitrogen slowly, thereby supplying the crop with its nitrogen requirement as needed, therein avoiding excessive uptake and reducing potential losses of this element by leaching and denitrification. This was shown not to be the case, however, as most of the nitrogen that became available did so within the first three weeks.

Under conditions optimum for nitrification at best only about half of the total nitrogen was converted to a form available to plants at the end of 15 weeks. In addition, of the nitrogen mineralized during the 15-week period, 80% had been converted to NO_3^- at the end of the first three weeks. It is obvious that under warm, moist conditions slow release of nitrogen from these materials is not effected and that the amount becoming available to the crop is but a fraction of the total amount they contain.

Chemical Sources of Nitrogen. Synthetic or chemical fertilizers are by far the most important sources of fertilizer nitrogen. Anhydrous ammonia is the basic building block for almost all of these chemically derived nitrogen fertilizer materials. Most of the ammonia in the world is produced synthetically by reacting nitrogen and hydrogen gas, although some is still recovered as a by-product of coking coal. The fundamentals of the production of nitrogen materials and other fertilizers are covered in Chapter 10.

From the basic compound, NH_3, many different fertilizer nitrogen compounds are manufactured. A few materials do not originate from synthetic ammonia, but they constitute only a small percentage of the nitrogen fertilizer tonnage used in the world. For convenience the various nitrogen compounds are grouped into four

TABLE 5-11.   Typical Composition of Some Common Chemical Sources of Fertilizer Nitrogen

Source	Percent:						
	N	P_2O_5	K_2O	CaO	MgO	S	Cl
Ammonium sulfate	21.0	—	—	—	—	24.0	—
Anhydrous ammonia	82.0	—	—	—	—	—	—
Ammonium chloride	25.0–26.0	—	—	—	—	—	66
Ammonium nitrate	33.0–34.0	—	—	—	—	—	—
Ammonium nitrate-sulfate	30.0	—	—	—	—	5.0–6.0	
Ammonium nitrate with lime (ANL)	20.5	—	—	10.0	7.0	0.6	—
Ammoniated ordinary superphosphate	4.0	16.0	—	23.0	0.5	10.0	0.3
Monoammonium phosphate	11.0	48.0–55.0	—	2.0	0.5	1.0–3.0	—
Diammonium phosphate	18.0–21.0	46.0–54.0	—	—	—	—	—
Ammonium phosphate-sulfate	13.0–16.0	20.0–39.0	—	—	—	3.0–14.0	—
Ammonium polyphosphate solution	10.0–11.0	34.0–37.0	—	—	—	—	—
Ammonium thiosulfate solution	12.0	—	—	—	—	26.0	—
Calcium nitrate	15.0	—	—	34.0	—	—	—
Calcium cyanamide	22.0	—	—	54.0	—	0.2	—
Potassium nitrate	13.0	—	44.0	0.5	0.5	0.2	1.2
Sodium nitrate	16.0	—	—	—	—	—	0.6
Urea	45.0–46.0	—	—	—	—	—	—
Urea-sulfate	30.0–40.0	—	—	—	—	6.0–11.0	—
Urea-sulfur	30.0–40.0	—	—	—	—	10.0–20.0	—
Urea-ammonium nitrate (solution)	28.0–32.0	—	—	—	—	—	—
Urea-ammonium phosphate	21.0–38.0	13.0–42.0	—	—	—	—	—
Urea-phosphate	17.0	43.0–44.0	—	—	—	—	—

categories: ammoniacal, nitrate, slowly available, and other. The composition of some common chemical sources of nitrogen is shown in Table 5-11.

Ammoniacal Sources.   The properties and behavior in soil of the principal ammoniacal compounds used as sources of fertilizer nitrogen are covered in the following section. Anhydrous ammonia and urea will be emphasized because of their popularity in North America. For example, in the United States in the period 1977–1980, anhydrous ammonia plus a minor contribution from aqua ammonia represented from 38 to 41% of the total nitrogen used. If urea consumption is added to the total for ammonia and aqua ammonia, between 47 and 49% of total nitrogen usage is accounted for.

Anhydrous Ammonia (NH_3).   Some of the properties of this important source of nitrogen are listed in Table 5-12. It contains approximately 82% nitrogen, the highest analysis of any of the commonly used nitrogenous fertilizers. In some respects it resembles water in its behavior since they both have solid, liquid, and gaseous states. The great affinity of anhydrous ammonia for water as a solvent is apparent from the solubility figure given in Table 5-12. This strong attraction of ammonia to water is featured in its behavior in the soil.

Under normal atmospheric conditions, anhydrous ammonia in an open vessel will be constantly boiling and escaping into the atmosphere. To prevent escape and to conserve material, it is stored in pressure vessels designed to withstand

TABLE 5-12.   Properties of Anhydrous Ammonia

Color	Colorless
Odor	Pungent, sharp
Chemical formula	NH_3
Molecular weight	17.03
Weight per gallon of liquid at 60°F	5.15 lb
Specific gravity of the gas (air = 1)	0.588
Specific gravity of the liquid (water = 1)	0.617
Boiling point	− 28°F
Vapor pressure at 0, 68, and 100°F	16, 110, and 198 psig, respectively
One gallon of liquid at 60°F expands to	113 standard ft.^{3} of vapor
One pound of liquid at 60°F expands to	22 standard ft.^{3} of vapor
One cubic foot of liquid at 60°F expands to	850 standard ft.^{3} of vapor
Solubility in water at 60.8°F	0.578 lb/lb of water

	ppm
Slight detectable odor	1
Detectable odor but no adverse effects on unprotected workers for exposure periods of up to 8 hours	25
Noticeable irritation of the eyes and nasal passages within a few minutes	100
Irritation to eyes and throat; no direct adverse effects but exposure should be avoided	400–700
May be fatal after short exposure	2000
Convulsive coughing, respiratory spasms, strangulation, and asphyxiation	5000 +

Sources: Sharp, in *Agricultural Anhydrous Ammonia: Technology and Use.* Madison, Wis.: American Society of Agronomy and Soil Science Society of America, 1966; and *Agricultural Anhydrous Ammonia Operator's Manual.* Washington, D.C.: The Fertilizer Institute, 1973.

pressure up to 250 lb/in.^{2}. Storage in low-pressure tanks at atmospheric pressure is possible under refrigeration (− 28°F), as is often done at large modern bulk storage facilities.

When liquid ammonia is released from a pressure vessel, it expands rapidly, vaporizes, and produces a white cloud of water vapor. This cloud is formed by the condensation of water vapor in the air surrounding the liquid ammonia as it vaporizes.

AMMONIA APPLICATION

The physical characteristics just described provide the basis for handling and applying ammonia. All equipment involved in the handling of ammonia must be designed to withstand the prevailing pressure. Because ammonia vaporizes quickly, it must be injected into the soil and sealed. Typical injection depths vary from 3 to 6 to 6 to 8 in. below the soil surface. Actual injection depths are usually only slightly deeper than those achieved in tillage before or during application.

Equipment utilized for direct application of ammonia consists of a nurse tank, an ammonia applicator, an ammonia transfer system to fill the applicator from the nurse tank, and a tractor to pull the applicator. In North America most of this nitrogen source is applied to agricultural lands through either custom application services provided by fertilizer dealers or by farmers furnishing the tractor to pull dealer-owned applicators. As a consequence, farmers are saved the expense of purchasing their own tanks and equipment for on-the-farm handling and storage.

A typical ammonia applicator consists of a frame, a tank, a tool bar, applicator knives, hoses, flow regulators, valves, and a distribution manifold to divide the flow of ammonia equally to each knife or shank. Applicators may vary in size from those with 6-ft tool bars with 60-gallon tanks to others 55 ft or more wide equipped with tanks of 3000-gallon capacity.

With greater interest in reducing the number of field operations, tillage implements such as cultivators, disks, harrows, plows, and so on, are often fitted for simultaneous tillage and application of anhydrous ammonia. In irrigated areas, either anhydrous ammonia or aqua ammonia is commonly added to surface irrigation systems. Problems related to the addition of these fertilizers to water high in calcium and volatilization losses of nitrogen will be discussed in Chapter 13.

Because it is a gas at atmospheric pressure, some anhydrous ammonia may be lost to the aboveground atmosphere during and after application. Factors associated with this loss are the physical condition of the soil during application, soil texture and moisture content, and depth and spacing of placement. If the soil is hard or full of clods during application, the slit behind the applicator blade will not close or fill, and some of the released ammonia will escape to the atmosphere.

Anhydrous ammonia convertors, which were first introduced in the United States in 1975, have facilitated application by reducing the need for deep injection and preapplication tillage. Power requirements and time spent in application are lowered by shallower application depths. The convertors simply serve as depressurization chambers for hot compressed anhydrous ammonia gas stored in the applicator or nurse tank. Anhydrous ammonia expands in the convertors and in doing so it freezes, separating the liquid ammonia from the vapor and greatly dropping the pressure. Temperature of the liquid ammonia is about $-32°C$ ($-26°F$). Actually, only about 85% of the anhydrous ammonia turns to liquid; the remainder stays in vapor form. The liquid form flows by gravity through regular application equipment into the soil. Vapor collected at the top of the convertor is injected into the soil in the usual manner.

AMMONIA RETENTION ZONES

Soil Conditions Within Injection Zones. Immediately after injection of anhydrous ammonia into soil, a localized zone high in both ammonia and ammonium is formed. The zone will often have a horizontal, roughly circular to oval shape with a diameter ranging from about $1\frac{1}{2}$ to 4 or 5 in. (3 to 13 cm), depending on method and rate of application, spacing, cation exchange capacity, soil texture and tilth, and soil moisture content. Ammonia distribution patterns are commonly oval around the release point, with distance of lateral movement directly proportional to application rate. Vertical movement can be limited to less than 5 cm, with most of it directed upward toward the soil surface.

A number of temporary yet dramatic changes occur in ammonia retention zones, which will markedly influence chemical, biological, and physical conditions of the soil. Some of the conditions that develop include:

1. Increased concentrations of ammonia and ammonium, reaching levels of 1000 to 3000 ppm.
2. Higher pH values of up to 9 to 9.5 or above.
3. Elevated nitrite concentrations in the range 100 to 300 ppm or more.
4. Osmotic suction of soil solution exceeding 10 bar.
5. Lower populations of soil microorganisms.
6. Solubilization of organic matter.

The amount of free or nonionized ammonia and the duration of its existence are of special importance since ammonia in this form is extremely toxic to microorganisms, higher plants, and animals. Free ammonia can readily penetrate cell membranes, while these tissue barriers are relatively impermeable to ammonium. There is a very close relationship between pH and concentration of free or nonionized ammonia and ammonium. Between pH 6.0 and 9.0, there is a 500-fold increase in concentration of nonionized ammonia.

Conditions in the Retention Zone Affecting Nitrite and Nitrate Formation. Figure 5-22 summarizes schematically the effects of pH, osmotic suction, and/or ammonium concentration on the formation of nitrite and nitrate. The influence of

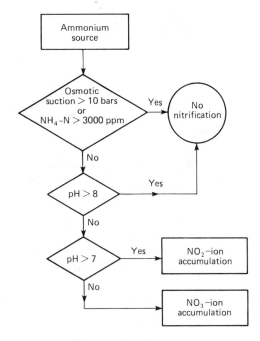

FIGURE 5-22. Diagram to indicate schematically the effects of osmotic suction and pH on nitrification. [Wetselaar et al., *Plant Soil*, **36**:168 (1972).]

high osmotic suction or high ammonium concentration in the soil solution is primarily on the nitrite formers. Activity of the nitrite formers is retarded by pH values above 8.0, especially in the presence of high amounts of ammonia. Nitrite will accumulate at pH values between 7 and 8, while below pH 7 nitrate becomes abundant.

Ammonium Retention Mechanisms. Gaseous ammonia would be lost to the atmosphere if it did not react rapidly with various organic and inorganic soil components. Parr and Papendick, former Tennessee Valley Authority researchers, identified the following possible ammonia-retention mechanisms, which are categorized as either chemical or physical.

1. Chemical.
   a. Reaction of ammonia with a proton to yield ammonium.
   b. Solubilization of ammonia in water.
   c. Reaction of ammonia with hydroxyl groups of clay minerals.
   d. Reaction of ammonia with the tightly bound water of montmorillonitic clay minerals.
   e. Reaction with water of hydration around the exchangeable cations on the exchange complex.
   f. Formation of coordination compounds (ammoniates) and complex ions (amines) of ammonia in clay systems with the exchangeable cations serving as the nuclei around which coordination and condensation occur.
   g. Precipitation of Ca^{2+} and Mg^{2+} as carbonates in the presence of CO_2 and freeing the exchange sites for reaction with NH_4^+.
   h. Reaction with organic matter.
2. Physical.
   a. Entrapment of NH_3 between the lattices of smectite type clays.
   b. The replacement of interlayer cations in an expanded clay lattice by NH_4^+, which in turn causes a contraction of the crystal lattice and an entrapment of the NH_4^+.
   c. Adsorption by clay minerals and organic components through hydrogen bonding.

The relative importance of these mechanisms will vary from soil to soil and they will also be influenced by environmental conditions.

FACTORS INFLUENCING AMMONIA RETENTION IN SOIL.
The factors influencing the size of anhydrous ammonia injection zones were mentioned earlier. The effects that these factors have on ammonia retention will be described more fully in this section.

Soil Moisture. Capacity of soils to retain ammonia increases with soil moisture content. Soil moisture content at or near field capacity provides maximum retainment of ammonia. As soils become either drier or wetter than field capacity, they lose their ability to hold ammonia (Figures 5-23 and 5-24).

The size of the initial ammonia retention zone will decrease with increasing soil moisture. Diffusion of ammonia from the injection zone is impeded by high soil moisture and the strong affinity of ammonia for water may also be a factor.

Texture. The ammonia-holding capacity of soils increases with clay content. On the basis of size of retention zone and volatilization losses, ammonia movement is greater in sandy soils than in clay soils. Gases such as ammonia are able to diffuse more freely in the larger pores found in coarse-textured soils.

Soil textural differences in ammonia retention are probably often obscured by other properties, such as type and amount of soil minerals, organic matter, moisture content, and other chemical characteristics.

Application Depth. As might be expected, retention of ammonia tends to be greater with increasing depth of injection. The depth of release needed to prevent ammonia loss varies considerably depending on soil properties and conditions. Studies have shown that an injection depth of 5 cm was effective for a silt loam soil, but placement at 10 cm was necessary in a fine sandy loam soil. In dry soil, ammonia loss declines with progressive increases in depth of release.

Spacing. The spacing between anhydrous ammonia release points can significantly influence ammonia retention. Close spacings usually result in the most satisfactory retention of ammonia in the treated soil. At a given rate of anhydrous

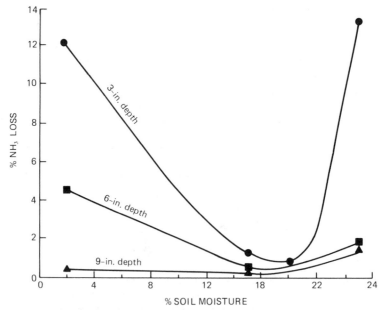

FIGURE 5-23. Losses of ammonia from a Putnam silt loam soil as influenced by depth of application and soil moisture. Anhydrous ammonia applied at the rate of 100 lb of nitrogen per acre in 40-in. spacings. [Stanley and Smith, *Soil Sci. Soc. Am. J.*, **20**:557 (1956).]

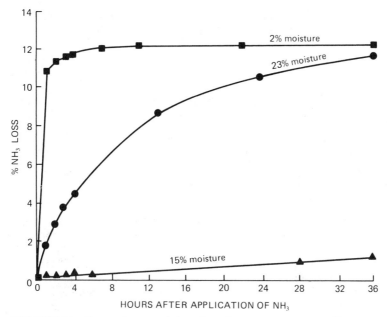

FIGURE 5-24.   Rates of ammonia loss from Putnam silt loam at different soil moistures when applied 3 in. below the surface. Anhydrous ammonia applied at the rate of 100 lb of nitrogen per acre in 40-in. spacings. [Stanley and Smith, *Soil Sci. Soc. Am. J.*, **20**:557 (1956).]

ammonia fertilization, the concentration of ammonia received per unit volume of soil will be reduced as the release-point spacings are brought closer together. With the greater efficiency of retention achieved with narrow spacings, there is less chance of ammonia being volatilized, particularly in sandy soils with limited capacity for holding ammonia.

Injection Pressure.   Orifice size and number of orifices on the application equipment will regulate pressure of the ammonia being injected into soil. Low pressures will produce uneven and inadequate distribution in the soil.

Soil Tilth.   Information on the effects of structure and tilth on the ammonia-holding capacity of soil is limited. There are reports indicating that coarse tilth and cloddy structure are not necessarily detrimental to ammonia retention. In one study ammonia movement and retention were greater under cloddy soil conditions than where the soil had a good granular structure.

Organic Matter.   The organic-matter component of soils contributes significantly to ammonia retention. At least 50% of the ammonia-holding capacity of soils is attributed to organic matter.

Effects of Ammonia on Soil Properties.
The drastic modifications within the soil zone effected by high concentrations of ammonia from anhydrous ammonia, aqua ammonia, urea, and other ammoniacal fertilizers exert considerable influence on various soil properties. The nature and extent of alterations in soil properties can have an important bearing on crop responses to additions of ammonium-containing or ammonium-forming nitrogen fertilizer materials. A brief description of the repercussions resulting from these changes follows.

Effect on Soil Microorganisms and Nitrification.   The high concentration of ammonia and ammonium in combination with basic soil pH and high osmotic potential results in a partial and temporary sterilization of soil within the retention zone, such as is evident in Table 5-13. Bacterial activity is probably affected most by free ammonia while fungi are depressed by high pH. Partially sterilized conditions

TABLE 5-13. **Numbers of Fungi, Bacteria, and Actinomycetes in Arredondo Loamy Fine Sand in the Ammonia Injector Row Compared with Untreated Areas**

Day After Treatment	Bacteria ($\times 10^6$/g)		Actinomycetes ($\times 10^6$/g)		Fungi ($\times 10^3$/g)	
	Check	NH_3	Check	NH_3	Check	NH_3
0	2.3	0.3	1.5	0.4	20.1	5.1
3	1.3	6.3	0.9	1.0	20.2	10.4
10	3.1	9.2	0.9	2.0	15.0	9.3
24	1.3	4.2	0.5	1.3	22.7	9.2
31	4.5	3.4	0.4	1.0	20.0	13.3
38	0.9	0.9	0.3	0.7	24.0	4.0

Source: Eno and Blue, *Soil Sci. Soc. Am. J.*, **18**:178 (1954).

at the center of the retention zone are known to persist for as long as several weeks. A rather rapid recovery in numbers of bacteria and actinomycetes is shown in Table 5-13.

Nitrification is restricted by the conditions that arise in soil zones affected by additions of ammonium-forming or ammonium-containing fertilizers. As a consequence, production of both nitrite and nitrate will be reduced until conditions return to normal.

Effect on Germination and Seedling Growth. The damaging effect of high concentrations of NH_3 and NH_4^+ and of NO_2^- on germinating seedlings is well known. An example of how rates of anhydrous ammonia reduced a stand of corn in field experiments in Illinois is shown in Figure 5-25. Deeper injection offset the harmful action of high rates of ammoniacal nitrogen more than did extending the time for the fertilizer effects to dissipate. Closer spacing of the ammonia release points would probably also temper the injurious effect of large amounts of ammonia. Concentrations in excess of 1000 ppm of NH_3 near the seed were associated with substantial reductions in numbers of corn plants.

Effect on Organic Matter. The ammonium hydroxide created by the reactions of alkaline-hydrolyzing fertilizers in soil will dissolve and hydrolyze certain fractions of soil organic matter. Additions of anhydrous ammonia have been observed to produce dark areas in the affected retention zones, to result in dark-colored water extracts, and to increase the amount of water soluble organic matter.

Most of these effects on organic matter are considered to be only temporary and disappear with time.

Effect on Soil Structure. Long term studies in Kansas have shown no difference among N sources in effects on soil physical properties. Contrasting beneficial and harmful effects have been reported following use of anhydrous ammonia. Increased aggregate stability has been interpreted as being due to both more uniform distribution of dissolved organic matter from the fertilizer-affected retention zone for binding soil particles and the formation of new organic substances which have a favorable effect on soil structure.

On the other hand, harmful effects from ammonia applications which reduced permeability of soils have been reported. Slaking of soil in the center of ammonia retention zones has also been observed. Impairment of soil structure is not expected to be serious or lasting except in situations involving low-organic--matter soils, where any alteration or loss of organic matter would likely be harmful.

Effect on Nutrient Availability. Solubilization of organic matter is expected temporarily to increase the availability of nutrients associated with the organic fraction of soils. There have been reports that anhydrous ammonia applications resulted in substantial but transitory rises in levels of readily extractable phosphorus.

Soluble phosphorus in leachates from forest soil humus was greatly increased by treatment with urea, oxamide, and urea-containing fertilizers. These same

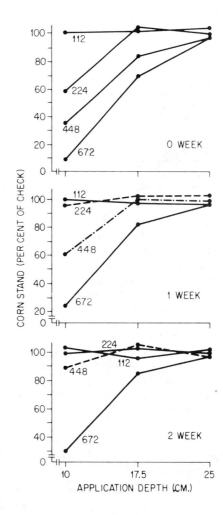

FIGURE 5-25. Effect of time, depth, and rate of NH_3 application on stand 27 days after planting (numbers by lines are kg/ha of nitrogen; Bayes $lsd_{(0.05)}$ = 6%). [Colliver and Welch, *Agron. J.*, **62**:341 (1970).]

fertilizer materials also greatly increased the concentration of soluble organic matter in the leachates.

Aqua Ammonia (20 to 25% N). The simplest nitrogen solution is aqua ammonia, which is made by forcing compressed ammonia gas into a closed container of water. It has a pressure of less than 10 lb/in.² and usually is comprised of 25 to 29% by weight of ammonia.

Transportation and delivery costs limit aqua ammonia production to local nitrogen stations or small fluid fertilizer plants. It is used for direct soil applications within a few miles of its point of manufacture, or it can be used in the production of other liquid fertilizers.

The ammonia in this nitrogen source will volatilize quickly at temperatures above 50°F, and for this reason aqua ammonia is usually injected in soil to depths of 2 to 4 in. At temperatures over 50°F, surface applications of aqua ammonia should be immediately incorporated into the top 1 to 3 in. of soil. Because of its high water content, aqua ammonia can be injected into soils regardless of their moisture content.

Nonpressure Nitrogen Solutions. Of the liquid materials used for direct application in the United States, nonpressure nitrogen solutions are next to anhydrous ammonia in popularity. Consumption of these solutions has been increas-

TABLE 5-14. Physical and Chemical Characteristics of Urea–Ammonium Nitrate Nonpressure Nitrogen Solutions

	Grade (% N)		
	28	30	32
Composition by weight (%)			
Ammonium nitrate	40.1	42.2	43.3
Urea	30.0	32.7	35.4
Water	29.9	25.1	20.3
Specific gravity at 15.6°C (60°F)	1.283	1.303	1.32
Salt-out temperature, °C (°F)	−18 (+1)	−10 (+14)	−2 (+28)

Source: International Fertilizer Development Center and United Nations Industrial Development Organization, *Fertilizer Manual*. Muscle Shoals, Ala.: IFDC, 1979.

ing at a faster rate than anhydrous ammonia. In 1979 and 1980 usage of nonpressure nitrogen solutions was 6.07 and 6.62 million tons, respectively. These tonnages are equivalent to approximately 16.3 and 16.8% of the total nitrogen consumed.

These nitrogen solutions are usually produced from urea and ammonium nitrate solutions plus water. They are often referred to as *UAN solutions*. The composition and properties of three of the principal nonpressure solutions currently used in the United States are compiled in Table 5-14. In addition to the constituents shown here, these nitrogen solutions usually contain one of the corrosion inhibitors listed in Table 5-15 which allows them to be stored and used in mild or carbon steel equipment. Anhydrous ammonia is the most frequently used inhibitor and it is added at the rate of about 10 lb/ton of product.

Nitrogen solutions containing dissolved salts exhibit a phenomenon known as *salting out*, which is simply the precipitation of the dissolved salts when the temperature of the solution reaches a certain level. The salting-out temperature determines the extent to which outside storage may be practiced in the winter and the time of year at which these solutions may be field applied. Salting-out temperatures vary directly with the concentration of plant nutrients in solution.

Some of the main reasons for the rapid growth in use of nitrogen solutions are:

1. Nonpressure nitrogen solutions are easier to handle and to apply than other nitrogen material, such as ammonia or dry products.
2. They can be applied more uniformly and accurately than can solid nitrogen sources.
3. Many pesticides are compatible with them and both fertilizer and chemical can be applied simultaneously, thus eliminating the need for one pass across the field.
4. Nonpressure nitrogen solutions can be applied through various types of irrigation systems and are particularly well suited for use in center pivots and other pressure systems.

TABLE 5-15. Corrosion Inhibitors for Urea–Ammonium Nitrate Nonpressure Solutions

Inhibitor	Concentration
Ammonia	0.5% (pH 7.0–7.5)
10-34-0	0.2% P_2O_5
Ammonium thiocyanate (NH_4CNS)	0.2%
Sodium arsenite ($Na_2HA_5O_3$)	0.1%

Source: International Fertilizer Development Center and United Nations Industrial Development Organization, *Fertilizer Manual*. Muscle Shoals, Ala.: IFDC, 1979.

5. They can be transported easily in pipelines, barges, and railcars. These transportation facilities are less expensive than those required for anhydrous ammonia, and fewer hazards are involved.
6. Low-cost storage facilities such as earthen pits with leakproof liners can be used to store them more economically than those required for most other nitrogen products.
7. The nonpressure nitrogen solutions are excellent sources of nitrogen for use in formulation of NPKS fluid mixtures.
8. Their cost of production is lower than that for most solid sources of nitrogen.
9. They are safer to handle than anhydrous ammonia.

The one feature of nonpressure solutions that particularly stands out is their ease of handling and application. Most of the nitrogen solutions are used for direct application and are applied broadcast using pump and spray nozzle systems.

Application Equipment.

Equipment for application of nonpressure nitrogen solutions alone or in fluid blends basically consists of a tank, pump, hose, gage, bypass valve, boom, and spray nozzles. The applicator wagons used to broadcast the nitrogen solutions vary in size from small field sprayers pulled with a farm tractor to the very large, self-propelled high-flotation applicators with tank capacities ranging from 1200 to 2400 U.S. gallons and spray booms 45 to 85 ft wide. These large flotation-type applicators minimize soil compaction and are particularly useful under wet soil conditions. With high-flotation applicators it is possible to apply normal amounts of fluid fertilizer at rates of 1.75 A/min or faster. Under good operating conditions the modern floater applicators are capable of treating from 600 to 700 acres daily.

Nonpressure nitrogen solutions are often added directly to grasses and small grains. When grasslands are not dormant, spray applications of UAN can cause scorching of foliage. A temporary leaf burn usually lasting less than a week will sometimes occur when broadleaf herbicides and nitrogen combinations are sprayed on small grains. Spray applications of nitrogen solutions have been used to kill weeds in corn and cotton.

For bromegrass in Kansas, dribbling of UAN solution in coarse bands about 6 in. (15 cm) wide and spaced on 18-in. (46-cm) centers tended to be superior to delivering this solution through fan nozzles as relatively small droplets. Results showing this comparison together with those obtained for broadcasting solid ammonium nitrate at two sites in the years 1975 to 1977 are tabulated in Table 5-16. Higher yields of warm-season grasses in Texas have also been obtained by employing the dribble method for applying nitrogen solutions. The apparent greater

TABLE 5-16.   Yield of Bromegrass Influenced by the Method of Nitrogen Application

| | Yield (lb/A) | | | | |
| | Riley Co. | | | Franklin Co. | |
Nitrogen treatment *	1975	1976	1977	1976	1977
None	4707	1257	3664	615	4017
UAN-sprayed	6412	4758	6002	3274	6911
UAN-dribble banded	6925	5787	5900	4064	6872
Ammonium nitrate (solid)—bdct	6301	5389	6144	3893	6658
LSD$_{0.05}$ rate	495	697	476	337	NS†

* Average for the rates 60, 120, and 180 lb/A of nitrogen.
† Not significant.

Source: Lamond et al. Effects of methods of nitrogen solution application on yield of smooth bromegrass. *in* Kansas Fertilizer Research Report of Progress 313. Kansas Agr. Exp. Sta. December 1977, pp. 108–115.

effectiveness of the dribble technique seems to be related to less retention of nitrogen solution on foliage and better penetration of the thatch by flooding a localized zone with a high concentration of nitrogen. It should be noted that scorching will occur in the band where the UAN is dribbled on. Early in the growing season in some years, depending on climatic conditions, there can be a streaking between the dribble bands due to restricted nitrogen movement.

Ammonium Nitrate(NH$_4$NO$_3$) and Ammonium Nitrate with Lime. Fertilizer-grade ammonium nitrate normally contains between 33 and 34% nitrogen. It is the most popular nitrogen fertilizer in many European countries and in some temperate-zone countries. Ammonium nitrate is also in demand in North America but not to the same extent as in Europe. In 1979 and 1980, respectively, it provided 7.84 and 7.76% of the total fertilizer nitrogen consumed in the United States.

The nitrate component of ammonium nitrate is readily available to crops and as a consequence this nitrogen fertilizer is widely used in the United States in cropping situations, where sidedressing and topdressing additions of nitrogen to growing crops is practiced.

Ammonium nitrate has some disadvantages, which include the following:

1. It is quite hygroscopic and care must be taken to prevent caking and physical deterioration in storage and handling.
2. There is some risk of fire or even explosions unless suitable precautions are taken. When in intimate contact with oxidizable carbonaceous substances such as fuel oil it forms an explosive mixture which is widely used as a blasting agent.
3. It is reported to be less effective for flooded rice than urea or ammoniacal fertilizers.
4. It is more prone to leaching and denitrification than ammoniacal products.

Use of straight ammonium nitrate is prohibited in some European countries. Where these restrictions exist, mixtures of ammonium nitrate with calcium carbonate, limestones, or dolomite, called ammonium nitrate-limestone (ANL) or calcium carbonate-nitrate (CAN), are permitted. Initially, CAN was composed of 60% ammonium nitrate and had a grade of 20.5% nitrogen. At present the most common analysis is 26% nitrogen corresponding to about 75% ammonium nitrate. Consumption of ammonium nitrate-limestone in the United States in 1980 amounted to only about 2600 tons of nitrogen.

Ammonium Nitrate-Sulfate. The Tennessee Valley Authority has produced two grades of this fertilizer, with analyses of 30-0-0-5(S) and 27-0-0-11(S). Of the two grades, the former, which contains about 21% ammonium sulfate and 79% ammonium nitrate was the more popular. Both grades were homogeneous, granular products made by neutralizing nitric and sulfuric acids with ammonia. Ammonium nitrate-sulfate is less hygroscopic than either constituent individually.

A 30-0-0-5(S) is manufactured commercially in the western United States for both bulk blending and direct application. It has been used very successfully for direct application to forage, grass seed crops, and small grains.

Ammonium Sulfate [(NH$_4$)$_2$SO$_4$]. This is one of the oldest sources of ammoniacal nitrogen, having been manufactured for many years as a by-product of the coking of coal. It now accounts for approximately 4 million tons of plant nutrient sulfur worldwide. In addition to the ammonium sulfate made available by recovery of coke-oven gas from the steel industry, it is also a by-product from metallurgical and chemical operations. Approximately one-fourth of its world production stems from caprolactam manufacture, a raw material for the production of synthetic fibers. There are about 3 to 4 lb of by-product ammonium sulfate for each pound of caprolactam produced.

The main advantages of ammonium sulfate are low hygroscopicity, chemical stability, and agronomic suitability. It is a good source of both nitrogen and sulfur. The strongly acid-forming reaction of ammonium sulfate in soil can be advantageous in high-pH soils and for acid-requiring crops such as tea. Its use can be undesirable in acidic soils already in need of liming.

The main disadvantage of ammonium sulfate is its relatively low nitrogen concentration which when used solely for this nutrient will add substantially to packaging, storage, and transportation costs. Consequently, its cost delivered to the farm will usually be higher per unit weight of nitrogen than that of urea or ammonium nitrate. It can, however, be an economical source of nitrogen when transportation costs are low, when it is a relatively inexpensive by-product, and when credit is given for its sulfur content.

Ammonium Phosphates. Mono $- (NH_4H_2PO_4)$ and diammonium phosphate $[(NH_4)_2HPO_4]$ and ammonium phosphate-sulfate are generally considered to be more important as sources of phosphorus than of nitrogen. Therefore their properties and reactions in the soil are covered in Chapter 6.

Ammonium Chloride (NH_4Cl). Fertilizer-grade ammonium chloride usually contains 25% nitrogen. About two-thirds of the world capacity for manufacture of this material is located in Japan, with the remaining one-third situated in India. Most of it is produced by the dual-salt process, in which ammonium chloride and sodium carbonate are formed simultaneously. Another production method is the direct neutralization of ammonia with hydrochloric acid.

Some of its advantages include a higher nitrogen concentration than ammonium sulfate and in Japan it has a somewhat lower cost per unit of nitrogen. Ammonium chloride is considered to be superior to ammonium sulfate for rice culture in some countries, such as Japan.

Although ammonium chloride is best known for its use on rice, it is also suitable for a variety of other crops, including barley, wheat, corn, sorghum, fiber crops, and sugarcane. As pointed out in Chapter 3, coconut and oil palms are very responsive to chlorides. Ammonium chloride is an excellent source of both nitrogen and chlorine for these palm crops.

Ammonium chloride is as acid-forming as ammonium sulfate per unit of nitrogen, and this effect will be undesirable in acid soil areas, especially if liming costs are excessive. Other shortcomings are its low nitrogen analysis in comparison to urea or ammonium nitrate, and its high chlorine content will limit its use to tolerant crops.

Urea $[CO(NH_2)_2]$. Although urea holds a historic position in the annals of organic chemistry, having been isolated from urine in 1773, only within the past 40 or so years has it received attention as a fertilizer material. Urea became the first organic compound to be synthesized from inorganic substances when in 1828 the German chemist Wöhler showed that it could be formed by heating ammonium cyanate, a compound made from the inorganic raw materials ammonia and carbon dioxide. Commercial production of urea first began in Germany in 1922, in 1932 in the United States, and in 1935 in England. It was also manufactured as early as 1920 in eastern Canada.

Doubts about the agronomic suitability of urea impeded its adoption. Many agriculturists had reservations about using urea because of potential problems related to (1) harmful effects of biuret, an impurity normally found at low concentrations, on germination and early growth of seedlings; (2) phytotoxicity of urea to seed and seedlings due to locally high concentration of ammonia released during the hydrolysis stage and/or the accumulation of nitrite during nitrification; and (3) nitrogen loss as ammonia from urea exposed on the soil surface, particularly at high soil temperatures. Practical experience with urea during the past 15 to 20 years or so has shown that it is as good as any other nitrogen fertilizer if used properly.

For efficient use of urea a higher degree of understanding of its properties and behavior in soil is necessary than for other nitrogen sources, such as ammonium nitrate and ammonium sulfate. There are, for example, limits to the amount of urea that can be routinely placed with or near crop seeds. In addition, unfavorable climatic conditions, particularly high temperatures, should be avoided for broadcast applications of urea on bare surface soil and on grasslands in humid areas.

Urea is considered a slow-release fertilizer in Europe since it must undergo both hydrolysis and nitrification before it becomes abundantly available to most crops. These transformations, which proceed rapidly in warm, moist soil, will be retarded in cold, wet soils, such as might occur in the spring in northern Europe.

For paddy rice production, urea is preferable to nitrate containing nitrogen fertilizers since nitrates tend to be denitrified and lost to the atmosphere when they are applied to flooded anaerobic soils. Also, the rice plant is fully able to utilize the ammonium form of nitrogen.

The satisfactory properties of the physically improved granular urea plus favorable economics of manufacturing, handling, storage, and transportation have made it a very competitive source of fertilizer nitrogen. In 1978 urea displaced ammonium nitrate as the principal form of dry fertilizer nitrogen in the United States. During 1978–1980 urea consumption in the United States accounted for between 8.4 and 9.2% of total nitrogen usage. Its anticipated rapid rise in popularity throughout the world is reflected in the projections of production capacities for principal dry nitrogen fertilizers, shown graphically in Figure 5-26.

BIURET LEVELS.

Concentration of biuret (NH_2-CO-NH-CO-NH_2) is of special concern because of its phytotoxicity. Biuret levels of up to 2% can be tolerated in most fertilizer programs. Because citrus and other crops, including pineapple, are sensitive to biuret in urea applied as a foliar spray, a product containing less than 0.25% biuret is recom-

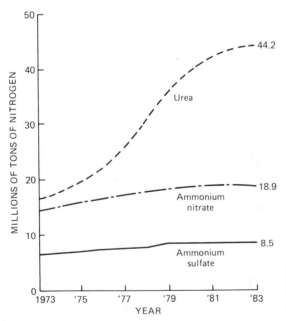

FIGURE 5-26. Trends in world production of dry nitrogen fertilizer materials. [TVA unpublished fertilizer statistics (1978)]

mended. Solutions made from urea containing 1.5% biuret were acceptable for foliar dressings on corn and soybeans. Placement of urea high in biuret near or in the seed row should be avoided.

Formaldehyde Treatment.

Internal conditioners such as formaldehyde are now commonly used as hardeners and anticaking agents. The hardness and durability of urea particles can be increased substantially by the inclusion of such additives. There is little information available on the identity and properties of the substances produced in the formaldehyde treatment of urea fertilizer. It appears that the small amounts of the low-molecular-weight urea-formaldehyde polymers that form do not alter the chemical properties of urea.

Formaldehyde-treated urea seems to be more waterproof and less subject to dissolution by light showers or heavy dew than when urea is conditioned with surface dusts and powders. Protection of urea in this way is deemed desirable since losses of ammonia from urea seem to be the most serious when initially moist surface soil undergoes rapid drying for 4 to 10 days after application or where following the addition of urea there is only slight wetting such as would occur with light showers or heavy dew.

In addition to the marked improvements in size, strength, and density of granular urea, this fertilizer material has a number of other noteworthy characteristics. These include (1) less tendency to stick and cake than ammonium nitrate, (2) lack of sensitivity to fire and explosion, and (3) less corrosive to handling and application equipment.

Substantial savings in handling, storage, transportation, and application costs are possible because of urea's high nitrogen content. A given volume of urea, even after allowance is made for differences in bulk density, contains considerably more nitrogen than either ammonium nitrate or ammonium sulfate.

ADVANCES IN UREA PRODUCTION.

A significant advance in lowering the cost of manufacturing urea occurred in the late 1960s with the development of large single-train plants using modern centrifugal compressors and capable of producing 1000 or more tons daily. The introduction in 1965 of Cominco's process for drum granulation of urea was a major breakthrough in eliminating all of the disadvantages of prilled urea. The granular urea made by this and other processes is larger, stronger, and more dense than prilled urea. Granular urea's crushing strength and resistance to mechanical breakdown is more than twice that of urea prills.

BEHAVIOR OF UREA IN SOIL.

Upon application to soil, urea is acted on by the enzyme urease, which hydrolyzes it to unstable ammonium carbamate. This hydrolytic reaction and the decomposition of the ammonium carbamate intermediate to ammonia and carbon dioxide is represented by the following equation:

$$CO(NH_2)_2 + H_2O \rightarrow H_2NCOONH_4 \rightarrow 2NH_3 + CO_2$$

The ammonia released during the breakdown of ammonium carbamate will react in soil in the same way as ammonia from anhydrous and aqua ammonia. There is also a marked rise in pH of soil in close proximity to particles of urea fertilizer similar to the conditions that develop following the addition of anhydrous ammonia. These changes in soil properties influence nitrogen transformations and crop development just as they do in the vicinity of an anhydrous ammonia retention zone.

In the presence of adequate water or other hydrogen donors, ammonia resulting from the decomposition of urea will be converted to and retained in the soil as ammonium.

Urea hydrolysis proceeds rapidly when soil conditions are also favorable for crop growth. In warm, moist soils much of the added urea nitrogen may be transformed to ammonium in several days.

Urease. This enzyme, which catalyzes the hydrolysis of urea, occurs universally and is abundant in soils. Large numbers of bacteria, fungi, and actinomycetes in soils possess urease. A small group of bacteria, known as urea bacteria, have exceptional ability to decompose urea. Activity of urease tends to increase with the size of the soil microbial population and increases with organic matter content. The presence of relatively fresh plant residues often results in abundant supplies of urease.

Greatest activity of urease is reported to occur in the rhizosphere, where microbial activity is high and where it can be excreted from plant roots. Activity of rhizosphere urease varies depending on plant species and the season of the year.

Although warm temperatures up to 37°C favor urease activity, the hydrolysis of urea occurs at significant rates at temperatures down to 2°C and lower. Bremner at Iowa State University has reported urease-induced decomposition of urea in soils at subzero temperatures as low as -10 to $-20°C$. This evidence of urease functioning at low temperatures, combined with urea's ability to melt ice at temperatures down to 11°F ($-12°C$), suggests that a portion of fall- or early winter-applied urea may be converted to ammonia and ammonium well in advance of spring.

The effects of soil moisture levels on urease activity are generally small in comparison to the influence of temperature and pH. Hydrolysis rates are probably highest at soil moisture contents in the readily available range to plants. It has been observed by Gould and his co-workers at the University of Alberta that moisture contents between 24 and 100% had little effect on the hydrolysis rate of urea.

Free ammonia inhibits the enzymatic action of urease. Since significant concentrations of free ammonia can occur at pH values above 7, it is entirely possible that some temporary inhibition of urease by free ammonia will take place after the addition of urea because soil pH in the immediate vicinity of the urea source may reach values in excess of 8.0 and sometimes up to 9.0. High rates of urea fertilization and its confinement to bands and other methods of localized placement could thus create conditions restrictive to the enzymatic action of urease.

Environmental Factors Influencing the Effectiveness of Urea.
Formation of an alkaline environment in the vicinity of urea particles following hydrolysis is the main condition that will in some instances reduce the effectiveness of this source. Development of alkaline microsites at or near the soil surface can result in ammonia volatilization and if formed close to seed rows, ammonia and ammonium can be harmful to germinating seedlings. The impact of this behavior of urea on crop growth is variable because of the interaction of a number of related soil and environmental factors. A brief description of the main factors follows. Many of them are related to ammonia retention in general, and the information given next will supplement the previous discussions on this subject.

Initial Soil pH. Concentration of free ammonia and volatilization of ammonia increases substantially as the pH rises above 7. Thus reductions in performance of urea related to the formation of ammonia are likely to be more severe in high-pH soils than in acidic soils. However, the effects related to alkalization resulting from ureolysis can also be substantial in acid soils, especially if they are poorly buffered. Recent additions of lime to acid soils may contribute to increases in concentration of free ammonia as a result of higher pH values.

Cation Exchange Capacity (CEC). Retention of ammonia is greatest in soils with the highest cation exchange capacity, that is, soils high in clay and organic matter. The calculated loss of ammonia nitrogen from surface applications of 100 lb of nitrogen per acre as urea is 20% if the CEC is less than 10 meq per 100 g; about 10% for CEC values of 20 meq per 100 g and less than 10% will be lost with CEC values greater than 20 meq per 100 g.

Buffering Capacity. The ability of a soil to resist sharp changes in pH or its buffering capacity is probably as important as initial soil pH in determining the extent to which urea utilization is affected by the customary rise in pH following its hydrolysis. The principal factors influencing buffering capacity are the amount and kind of clay and the quantity of organic matter in soil. Soils containing large amounts of clay and organic matter are usually highly buffered.

Soil Temperature. Urea hydrolysis is relatively slow at soil temperatures below 10°C (50°F) and as a result ammonia forms more slowly at low temperatures. Rates of urea hydrolysis have been observed to be two to six times greater at 25°C (77°F) than at 1°C (34°F). However, as noted earlier, urease activity has been found at subzero temperatures in the range −10 to −20°C.

Temperature also influences the partial pressure of ammonia over ammonium carbonate, with pressures declining as temperature decreases.

Soil Moisture and Weather Patterns. Surface applications of urea are most efficient when they are washed into the soil or applied to soils of low ureolytic activity. Conditions for best performance of surface dressings of urea are soils in either cold or dry states at the time of application and/or the occurrence of significant precipitation, probably more than 0.25 cm (0.1 in.), within the first three to six days following fertilization.

Movement of soil moisture, containing retained ammonia, toward the soil surface during the drying process can probably also contribute to atmospheric losses of ammonia. Further, the diffusion of moisture vapor to the soil surface during the drying process could sweep or carry desorbed ammonia with it and thus assist in the volatilization of ammonia situated at or near the soil surface.

Urea moves in soils about as readily as nitrate. Soils high in organic matter tend to restrict the movement of urea. Loss of urea can occur if there is significant precipitation soon after application and before there has been sufficient time for hydrolysis to be completed. Some downward movement of surface applied or shallow incorporated urea is expected to be beneficial in most arid and semiarid regions since it would result in improved distribution of the added nitrogen deeper within the rooting zone (i.e., a form of deep placement).

Available moisture supply in the soil has an important effect on how urea placed in the vicinity of small grain seeds will influence germinating seedlings. With adequate soil moisture in medium-textured loam soils at seeding time, urea at rates equivalent to 30 lb of nitrogen per acre can be used without reductions in germination and crop emergence. On the other hand, in light-textured (sandy loams) soils of low moisture status the addition of urea at rates exceeding 10 to 20 lb of nitrogen per acre will often reduce both germination and crop yields. Seedbed moisture is less critical in heavy textured (clay and clay loam) soils and urea can usually be drilled-in at rates up to 30 lb of nitrogen per acre.

Improving the Effectiveness of Urea.
Since most of the reductions in effectiveness of urea are related to soil alkalization which takes place following hydrolysis, measures for moderating or compensating for this condition are of interest. Some of the practical safeguards that can be taken are outlined below.

Incorporation and Placement. Alkaline reaction zones, resulting from the hydrolysis of urea at or near soil surfaces, may permit the volatilization of ammonia. Incorporation of top-dressed urea into soil will minimize such losses by providing much greater volumes of soil to retain free ammonia and because any free ammonia not converted in the soil must diffuse much greater distances before reaching the atmosphere. If soil and other environmental conditions appear unfavorable for adequate ammonia retention, deep incorporation is preferred over some form of shallow surface tillage.

There does not appear to be as much information available on conditions within the area of soil affected by bands or other localized placements of urea as there is for anhydrous ammonia. Movement of urea from banded applications equivalent to 70 lb of nitrogen per acre was found to be 2.5 cm (1 in.) within 2 days after its addition, while appreciable amounts of ammonium were observed at distances of 3.8 cm (1½ in.) from the band. After dilution or dispersion of the band by moisture movement, hydrolysis began within three to four days or less under favorable temperature conditions. These findings support suggestions made earlier that band placement of urea will probably result in soil changes comparable to those produced by applications of anhydrous ammonia. It is believed that some of the alterations in properties of soil in anhydrous ammonia bands may result in improved management of fertilizer nitrogen.

Physical Form. Although it has been generally believed that there is little or no difference in effectiveness between surface applications of urea in granular, crystalline, or solution forms, Nommik observed in Sweden that volatilization of ammonia during the first 14 days after application of urea to forest soil was very much less for large than from small pellets.

Nyborg and his co-workers at the University of Alberta have found that the effectiveness of fall-applied solid urea is greatly improved by placing at depths of 5 cm in soil very large particles or, alternatively, nests of regular-size product in amounts ranging between 1.7 and 6.8 g. Spacing of the nests was on a grid 30 cm by 30 cm or 60 cm by 60 cm. Urea placed in this manner nitrified little from fall to spring, whereas there was substantial nitrification of ordinary urea fertilizer incorporated in the soil in the fall.

Considerable improvement in crop yields has been achieved by applying urea as either big particles or nests. An example of the gains that can be realized by this approach to urea fertilization is presented in Table 5-17. Fall-applied urea in nests or as large particles was superior to standard urea incorporated in the fall at all six locations. Fall application of nests or large particles, however, gave both higher and lower yields than urea incorporated in the spring.

Similar to the techniques described above of applying urea in nests or large particles, the International Fertilizer Development Center at Muscle Shoals, Alabama, has found that point placement of urea supergranules will substantially increase the agronomic efficiency of urea for rice production. The usual practice on small rice farms in developing countries is to broadcast regular urea into the floodwater. This method of urea fertilization frequently leads to an efficiency of nitrogen utilization by the plant as low as 20 to 30%. Simply by incorporating the initial dressing of urea into puddled soil, efficiency can be increased to 35 to 44%. Hand placement of large urea particles (supergranules or briquettes) has essentially doubled efficiency to 75 to 85%.

Placement Away from Seed. The harmful effects of urea placed in the seed row can be eliminated or greatly reduced by banding it directly beneath, or below, and at least 2.5 cm (1 in.) to the side of the seed row of most crops. There has been a lack of suitable drills for either of these fertilizer placements during the seeding of small grains. For the most effective use of drilled-in applications of urea, rates

TABLE 5-17. **Effect on Barley Yields of Fall Applications of Urea in Large Particles or Nests**

Urea Applied at Rates of 56 and 84 kgN/ha	Average Yield Increase of Barley Grain at Six Locations During 1975–1978 (kg/ha)
Incorporated in soil in fall	1120
Large particles or nests in fall	1790
Incorporated in soil in spring	1980

Source: Nyborg et al., "Placement of urea in big pellets or nests," In *Effective Use of Nutrient Resources in Crop Production,* Proc. 1979 Alberta Soil Sci. Workshop, pp. 99–112 (1979).

should, as discussed previously, be limited to between 10 and 20 lb of nitrogen per acre, particularly when moisture supplies in seedbeds are low or where soil properties tend to accentuate the detrimental effects of urea on germinating seedlings.

Simultaneous Addition of Other Compounds. Urea is seldom placed alone with seed in the drill-in fertilizer programs practiced in most areas. In some regions it is applied in blends containing varying amounts of monoammonium phosphate (11-55-0, 11-48-0, 16-20-0, etc.). Provided that other fertilizer salts applied in combination with urea do not unduly increase osmotic pressure of the soil solution, they may favorably influence the performance of drill-in urea by serving as buffering agents which help to curtail sudden rises in soil pH and the resultant occurrence of ammonia. Inclusion of monoammonium phosphate with urea in fertilizer blends placed with seed of small grains has improved the effectiveness of urea in several research trials in the prairie provinces of Canada. The contribution of these acidic phosphate materials to the buffering capacity of the soil is believed to be mainly responsible for the reported beneficial interactions with urea. The advantage of acidic effects of other fertilizer materials may largely disappear in calcareous soils because of neutralization taking place before the full effects of urea hydrolysis occur.

Use of calcium or magnesium nitrates or chlorides to stabilize nitrogen supplied in surface dressings of urea has been studied by Fenn of the Texas Agricultural Experiment Station. These calcium and magnesium salts apparently delay urea hydrolysis and restrict the formation of unstable ammonium carbonate, which will decompose and release ammonia. Incomplete retention of free ammonia in soil can lead to the much publicized losses by volatilization. Where such losses are serious, it has been reported that this Texas process will hold nitrogen losses to between 0 and 10%.

The Tennessee Valley Authority reported in 1978 that losses of nitrogen from surface applications of solid urea were essentially nil from mixtures of solid urea with solid calcium nitrate tetrahydrate or calcium chloride dihydrate. Ammonia losses from urea have also been reduced by adding other acidifying substances, such as ammonium chloride, triple superphosphate (monocalcium phosphate monohydrate), orthophosphoric acid, and orthoboric acid.

Modifications to Reduce Rates of Release and to Control Urease Activity. The alkalization of soil as a result of urea hydrolysis is a temporary condition due to the rapid production of ammonia and is less likely to occur if urea becomes available for hydrolysis slowly. Reductions in rate of release of urea can be accomplished by forming sparingly soluble urea-aldehyde compounds (e.g., ureaforms, crotonylidene diurea, isobutylidene diurea, etc.). Another approach is to coat urea particles with materials such as sulfur, which impart slow-release properties. It is difficult to develop products with the desired degree of control of rate of release of urea which are at the same time satisfactory sources of nitrogen for crops during the year of application.

Prevention or retardation of the hydrolytic action of urease following the addition of urea to soil would help avoid difficulties associated with ammonia formation and alkalization by allowing sufficient time for urea to diffuse from the point of application into larger soil volumes. This type of control of even a few days would greatly increase the chance of some precipitation falling and carrying fertilizer urea into the soil before significant hydrolysis occurred. A great many substances are urease inhibitors, but very few meet the rather specific requirements of being (1) effective at low concentrations, (2) relatively nontoxic to higher forms of life, (3) inexpensive, and (4) compatible with urea.

The compound thiourea, which has received some attention by researchers interested in improving efficiency of nitrogen fertilizers, inhibits both urease and nitrification.

The information discussed in this section clearly indicates that the effectiveness

of urea is dependent on many factors. As a result of these factors and their interactions, some variability in crop response to urea will be experienced. In most
instances urea will be about as effective as the other common nitrogen sources.
Occasionally, urea will be either inferior or superior to these other materials, for
reasons that are not readily explained.

Urea-Sulfur. Urea and liquid sulfur are completely miscible and a prilled
40-0-0-10 fertilizer was produced and marketed on a limited scale by one manufacturer in the western United States. Although this product is no longer available,
it should be noted that it had excellent storage and handling properties. Also, the
sulfur component did not oxidize rapidly enough to provide adequate available
sulfur early in the growing season, although it subsequently became available.

An experimental drum-granulated urea-sulfur containing 35 to 40% nitrogen
and 12 to 20% sulfur made in western Canada is now being evaluated at a number
of locations in that region and also in the western United States. It has very
acceptable physical properties and the elemental sulfur constituent is sufficiently
finely divided to provide acceptable rates of SO_4^{2-} formation. There are indications
that nitrification of urea may be delayed in some of the formulations.

A 40-0-0-10 made by coating urea with finely divided sulfur and using oil and
calcium lignosulfonate as a binder is being used on rice and coastal Bermuda grass
in the West South Central states and in Mississippi and New Mexico.

Urea-Ammonium Phosphates. Fertilizers composed mainly of urea and
ammonium phosphate have been produced in Japan for many years. Ammonium
phosphate can be supplied either as a slurry produced by ammoniation of phosphoric acid or in solid form as mono- or diammonium phosphate. Various forms of
urea were used, including crushed prills or crystals, a melt, or a concentrated
solution. Granulation was carried out principally in rotary drums, but pan granulators, pugmills, and other types were also used.

Some of the most recent urea-ammonium phosphate grades made in Japan are
28-28-0, 22-22-11, 18-18-18, and so on. The usual raw materials are crushed urea
prills, spray-dried ammonium phosphate (12-50-0), and potash salts.

The Tennessee Valley Authority (TVA) has actively investigated urea-ammonium phosphate processes with the early work involving incorporation of urea into
its diammonium phosphate slurry granulation process. Urea was added to the
granulator either as concentrated solution, crystals, or prills. Grades ranging from
38-13-0 to 21-42-0 were produced in the pilot plant.

A melt granulation process is now used at the TVA to formulate urea-ammonium
polyphosphates analyzing 28-28-0 and 36-17-0. Ammonium polyphosphate (11-55-
0) melt and molten urea are cogranulated in a pugmill in this process. The TVA
has also successfully prilled urea-ammonium polyphosphate in an oil-prilling pilot
plant.

Norsk Hydro has a process for prilling urea-monoammonium phosphate in air
with or without inclusion of potash. Grades of 29-29-0 and 38-16-0 have been
produced with satisfactory physical properties.

Urea-Phosphate [$CO(NH_2)_2{\cdot}H_3PO_4$]. It is a crystalline adduct that can
be formed readily by the reaction of urea with orthophosphoric acid. Production
of this compound in relatively pure form (17-44-0 grade) from impure wet-process
phosphoric acid has been studied extensively by TVA and others, chiefly as an
intermediate for further processing into clear liquid fertilizers. Solid urea-phosphate analyzing 17-43-0 and with acceptable granule hardness and crushing strength
has been made by TVA. Other grades which have been prepared by adding supplemental urea, during production of 17-43-0, include 27-27-0, 34-18-0, 37-13-0,
and 39-10-0. These products containing additional urea tend to be more hygroscopic

than the original 17-43-0. Urea phosphates of lower purity standards may be adequate for production of suspension fertilizers and for fertigation.

Urea-Sulfate. Granular urea-sulfate with grades ranging from 40-0-0-4 to 30-0-0-13 have been made at the TVA by oil-prilling preparations of ammonium sulfate fines in molten urea. It was also made by coating ammonium sulfate fines with urea in a granulator and by air prilling. The granules have been shown to be more resistant to attrition and less hygroscopic than urea prills and may be further improved by the addition of gypsum, which forms a complex with urea. The N/S ratio in this product may be varied from 3:1 to 7:1, thus providing enough scope to correct nitrogen and sulfur deficiencies in most soils.

A 40-0-0-6 urea-ammonium sulfate was produced commercially during the late 1970s in western Canada. Ammonium sulfate fines were coated with urea in a drum granulator. It was well received in the western United States and Canada for application to grass seed, forage, and small-grain crops.

Mechanically blended urea and ammonium sulfate with a grade of 34-0-0-11 has been sold for many years in western Canada. Handling and storage characteristics of this formulation have not been completely satisfactory.

The TVA has produced a urea-ammonium sulfate suspension with an analysis of 29-0-0-5. It was made by reacting ammonia and sulfuric acid with simultaneous addition of a urea-water solution. After cooling, 1.5% by weight of clay was added. The product has good handling and storage properties and its N/S ratio of roughly 6:1 is favorable.

Nitrate Sources. In addition to the ammonium nitrate materials referred to earlier in this section, several other nitrate-containing fertilizer salts, including sodium nitrate ($NaNO_3$), potassium nitrate (KNO_3), and calcium nitrate [$Ca(NO_3)_2$], should be mentioned because of their importance in certain locales. These nitrate sources are quite soluble and thus very mobile in the soil solution. They are quickly available to crops and because of their mobility they are susceptible to leaching under conditions of high rainfall. They may also be appropriated by soil microorganisms in the decomposition of organic residues. They are subject to denitrification, like all other nitrate sources.

The nitrates of sodium, potassium, and calcium are not acid forming as are ammonium nitrate and the ammonium fertilizers in general. Because the nitrate is often absorbed by crops more rapidly than the accompanying cation, an excess of basic cations can remain in the soil creating a somewhat more alkaline condition. Prolonged use of sodium nitrate, for example, will maintain or even raise the original soil pH.

Sodium Nitrate. At one time sodium nitrate of natural origin was the major source of nitric acid and chemical nitrogen for fertilizer purposes in many countries. Most of it originated in a large ore body on the eastern part of the Chilean coastal range. Nitrate production continues to be a major industry in Chile, and in 1976 production was about 650,000 tons of sodium nitrate annually.

Substantial amounts of synthetic sodium nitrate were made at one time in Europe and the United States. Its manufacture has declined since World War II and now only insignificant amounts are produced from by-product sources. Its main use was for direct surface applications on cotton, tobacco, and some vegetable crops. Use of this fertilizer material in the United States now amounts to only about 11,000 to 13,000 tons of nitrogen annually.

Potassium Nitrate. This fertilizer material containing two essential nutrients is manufactured by a high-temperature distillation process based on the direct attack of concentrated nitric acid on potassium chloride. Potassium nitrate, chlorine, and nitrosyl chloride are formed by this reaction. The intense corrosion problems inherent in this process can be obviated by an alternative low-temperature solvent extraction procedure which yields potassium nitrate and hydrochloric acid. Potassium nitrate may also be formed as a coproduct in certain nitrophosphate processes.

Now commercially available in the United States and certain other countries, KNO_3 finds its greatest use in fertilizers for intensively grown crops such as tomatoes, potatoes, tobacco, leafy vegetables, citrus, peaches, and other crops. The properties of potassium nitrate that make it attractive for these crops include moderate salt index, nitrogen present as nitrate, favorable N/K_2O ratio, negligible chlorine content, and alkaline residual reaction in soil. Its low hygroscopicity allows for considerable flexibility in its use for direct application and in mixtures.

Calcium Nitrate. Most supplies of calcium nitrate originate in Europe, where it is produced in two principal ways. In one method, calcium carbonate is treated with nitric acid and the resulting solution containing about 40% calcium nitrate is clarified and then concentrated in a vacuum evaporator unit followed by prilling. The other source of calcium nitrate is its occurrence as a coproduct in some nitrophosphate processes, where it is separated by crystallization and filtration or centrifugation.

It is extremely hygroscopic and this property detracts from its utility as a fertilizer. Except in very dry climates, calcium nitrate is prone to liquefaction, and storage in moisture-proof bags is usually mandatory. As with other nitrate fertilizer salts, sensitization of calcium nitrate by impregnation with carbonaceous substances should be avoided.

Because of its fast-acting nitrate component, calcium nitrate is a useful fertilizer for winter-season vegetable production. It is sometimes used in foliar sprays for celery, tomatoes, and apples. On sodium-affected soils, calcium nitrate has a special advantage since the calcium constituent can displace unwanted sodium from soil colloids. In the late 1970s and early 1980s, U.S. consumption of calcium nitrate was about 70,000 tons annually.

Slowly Available Nitrogen Compounds. From the preceding discussions in this chapter it is apparent that nitrogen provided by commercial synthetic fertilizers is subject to many different fates in soil. As a consequence, crop recoveries of nitrogen seldom exceed 60 to 70% of that added in the fertilizer. There has been an ongoing search for nitrogen fertilizers with greater efficiency since, as will be explained in a later chapter, their production is energy intensive and also because of environmental concerns over excessive movement of nitrogen into surface and groundwaters, as well as the effects of gaseous nitrogen losses on the upper atmosphere.

Interest in a number of agronomic matters has also served to prompt activity in developing improved nitrogen fertilizers. It would be desirable to have sources capable of releasing nitrogen over an extended period, thus avoiding the need for repeated applications of conventional water-soluble products. Such fertilizer materials would probably also offer other advantages, including less chance of over stimulation due to luxury consumption of nitrogen and disruption of nutrient balance and reduced hazards of injury to germinating crops when used at high rates in or near the seed row.

The ideal product would seem to be one that liberated its nitrogen components in accordance with crop needs throughout the growing season. Most, if not all, of the materials and processes that have been developed or evaluated for controlled nitrogen availability can be grouped as follows:

1. Substances of low water solubility that must undergo chemical and/or microbial decomposition to release plant-available nitrogen.

2. Sparingly soluble minerals.
3. Soluble or relatively water-soluble substances which gradually decompose.
4. Conventional water-soluble products treated to impede dissolution.
5. Ion exchange resins.
6. Nitrification and urease inhibitors.

Substances of Low Solubility Requiring Decomposition. This group is comprised of chemical compounds that are only slightly soluble in water or in the soil solution. Rate of nitrogen liberation from them is related to their water solubility and to the rate of microbiological action and chemical hydrolysis. The rate of microbial and chemical decomposition is related to rate of solution, which is dependent on solubility, particle size, and other factors.

Urea-Formaldehyde.
Urea-aldehyde compounds are the major representatives of this group that are commercially available. It is technically feasible to react urea with a number of aldehydes to form sparingly soluble compounds containing over 30% nitrogen and which decompose slowly in soil by chemical and/or biological action.

One of the best known products in this category is urea-formaldehyde, often referred to as "ureaform." It is produced by about six manufacturers in the United States and in several other countries. They are white, odorless solids analyzing about 38% nitrogen which are made by reacting urea with formaldehyde in the presence of a catalyst. Ureaform instead of being a definite compound consists of methylene urea polymers varying in chain length, and with the larger molecules, in degree of cross-linking. A whole series of compounds, ranging from quite soluble to completely insoluble, is possible, depending on the mole ratio of urea to formaldehyde and on the pH, time, and temperature of reaction.

Trimethyleneurea appears to be the most suitable single component as a slow-release nitrogen material. Unfortunately, it is very difficult to produce commercially a pure trimethyleneurea.

A typical ureaform may contain 30% of its nitrogen in forms that are soluble in cold water (25°C). Nitrogen in the cold-water fraction nitrifies almost as quickly as urea. Solubility in hot boiling water is a measure of the quality of the remaining 70% of its nitrogen. At least 40% of the nitrogen insoluble in cold water should be soluble in hot water for acceptable agronomic response; typical values are 50 to 70%. An activity index is used by the Association of Official Agricultural Chemists to evaluate the suitability of urea-formaldehyde compounds. It is defined as follows.

$$AI = \frac{\% \text{ CWIN} - \% \text{ HWIN}}{\% \text{ CWIN}} \times 100$$

where AI is the activity index, CWIN is the percent nitrogen insoluble in cold water (25°C), and HWIN is the percent nitrogen insoluble in hot water (98 to 100°C).

The suitability of these compounds as fertilizers is dependent on the following:

1. The *quantity* of cold-water insoluble nitrogen, which is the source of the slowly available nitrogen.
2. The *quality* of the cold-water insoluble nitrogen determined by its activity index, which reflects the rate at which the cold-water insoluble nitrogen will become available.

In addition to ureaforms for provision of nitrogen, they can be used in the formulation of compound fertilizers containing other nutrients such as phosphorus and potassium. Urea-formaldehyde solutions originally intended for use in production of mixed fertilizers have been modified for direct application in the lawn-care industry.

Consumption of ureaform in the United States, not including the quantity resulting from inclusion of urea-formaldehyde solutions in production of compound fertilizers, is approximately 50,000 tons annually. Most of it is used in nonfarm markets for turfgrass, landscaping, ornamental, horticulture, greenhouse crops, and as an aid in overcoming planting shock of transplanted coniferous seedlings.

Crotonylidene Diurea (CDU).

This slow-acting nitrogen compound is formed by the reaction of urea with crotonaldehyde or acetaldehyde. Powdered CDU containing 30% nitrogen has been used directly as a fertilizer. A granular product, Floranid, with an nitrogen content of 28% is composed of 90% CDU and 10% NO_3-N. Granular NPK fertilizer with a grade of 20-5-10 has been made utilizing CDU for 75% of the nitrogen content. The microbial transformation of chemically bound nitrogen in CDU is known to be temperature dependent.

Isobutylidene Diurea (IBDU).

It is produced commercially in Japan and is a condensation product of urea and isobutyraldehyde reacted in a 2:1 mole ratio. When pure it contains 32% nitrogen and has a solubility in water of only 0.1 to 0.01%, depending on pH and temperature. The rate-limiting step in conversion of nitrogen in IBDU to plant-available forms appears to be the dissolution of the fertilizer particles, with the smaller ones dissolving fastest. Its mineralization in soil is accelerated when pH, temperature, and moisture conditions are favorable for plant growth.

Fertilizer-grade IBDU contains about 30% nitrogen. It can also be granulated with soluble NPK sources to produce a range of compound fertilizers. IBDU has been marketed in the United States under the trade name Vigoro Once.

Urea-Z (UZ).

Urea-Z is formed by the reaction of urea and acetaldehyde, yielding a mixture of ethylene diurea and diethylene, together with some unreacted urea. The polymers considered for fertilizer use contain about 33 to 38% nitrogen, of which about 15% is initially water-soluble and the remainder is slowly soluble. Its decomposition and release of nitrogen in soil is more rapid than most of the more common slow-release nitrogen materials. This material, originally tested in Germany, is currently not in commercial production, but its manufacture has been considered in Japan.

Other Urea-Aldehyde Condensates.

Glycouril, a reaction product of urea and glyoxal, has only slight water solubility (0.2% at 30°C) and it has been patented for use as a slow-release nitrogen fertilizer. There has been small-scale production in Japan of difurfurylidene triureide by a process utilizing by-product furfural. Some interest in MFU-A, a condensation product of urea and a mixture of formaldehyde and acetaldehyde, has been reported in the USSR.

Oxamide.

Desirable slow nitrogen release characteristics of oxamide, the diamide of oxalic acid, were reported in the late 1950s by Japanese workers. Its formula is

$$NH_2 - \overset{\overset{\displaystyle O}{\|}}{C} - \overset{\overset{\displaystyle O}{\|}}{C} - NH_2$$

and it is a white nonhygroscopic compound with a nitrogen concentration of 31.8%. Oxamide is practically insoluble in cold water and at 25°C its solubility is only 0.4% by weight.

Dissolution rate of oxamide is a direct function of particle size and hardness, with the effect of the former being especially striking. Following dissolution, hydrolysis occurs primarily by means of microbial cleavage of the carbon-to-carbon bond, resulting in the formation of ammonium carbonate. Powdered and fine particles of oxamide nitrify faster than ammonium sulfate and at a rate similar to urea in acid soils.

Oxamide is currently not in commercial production for lack of an economical process. The TVA has resumed studies on methods for synthesizing it and several attractive processes are now being considered. If an economical manufacturing process is developed, considerable tonnages of oxamide will likely be consumed as a nonburning source of nitrogen for turf, and for horticultural and other speciality uses.

Triazines (Urea Pyrolyzates).

Triazines are compounds which are made by heating urea in the presence of ammonia. The nitrogen content of these compounds is 32.4, 43.7, 49.4, and 66.5%, respectively, for cyanuric acid, ammeline, ammelide, and melamine. Although not produced on a commercial scale, they are promising potential slow-release nitrogen fertilizers because of their high nitrogen contents, low dissolution rates, and relative ease of synthesis. Cyanuric acid is reported to be toxic in the early stages of nitrification.

Sparingly Soluble Minerals. Magnesium ammonium phosphate is the most familiar member of a group of compounds with the general formula $MeNH_4PO_4 \cdot XH_2O$ or $MeKPO_4 \cdot XH_2O$, where Me represents a divalent metal such as copper, cobalt, manganese, iron, or zinc. All of these compounds exhibit only slight solubility in water. They are unique in that all of the elements in them may be necessary plant nutrients.

The nitrogen content of these materials is low, usually between 5.7 and 9%, while the P_2O_5 content varies from 29 to 45%. Rates of nitrification and release of plant nutrients can be controlled by use of different granule sizes, with the larger ones delaying these processes. A commercial grade of magnesium ammonium phosphate sold in the United States is primarily in the monohydrate form, and it normally analyzes 8% nitrogen, 40% P_2O_5, and 25% MgO. This material, sold under the trademark MagAmp, has a solubility of 0.014 g per 100 ml of water at 25°C and the water-soluble nitrogen content varies between 1.0 and 2.0%.

Magnesium ammonium phosphate can be applied to the soil surface at relatively high rates without burning plants. It has been widely used for a variety of purposes, including turf, tree seedlings, ornamentals, citrus, vegetable, and field crops. Some of the most impressive results from this product were obtained in treatment of nursery beds for tree seedlings and as basal dressing when the seedlings were outplanted.

Soluble Substances That Gradually Decompose. Guanylurea sulfate (GUS) and guanylurea phosphate (GUP) are unique materials in that they are soluble in water but still have slow-release properties. They are adsorbed on soil colloids and in that state they mineralize very slowly, particularly under aerobic conditions. Mineralization is apparently much more rapid under anaerobic conditions, such as in flooded rice soils.

The nitrogen concentration is 37.0 and 27.8%, respectively, in GUS and GUP. Small amounts of the former, produced by direct acidulation of calcium cyanamide, are marketed for use on rice in Japan.

Conventional Water-Soluble Products Treated to Impede Dissolution. Rather inert water-resistant coatings or membranes, including a wide range of natural or synthetic polymers (e.g., polyurethane, polyethylene, etc.), waxes, paraffins, asphaltic compounds, and elemental sulfur can successfully control the rate of release of nutrients from granules of water-soluble fertilizers. The coatings can be of three types:

1. The semipermeable membrane where water diffuses in through the membrane and dissolves the core of water-soluble nitrogen, creating an increase in osmotic pressure which forces the dissolved nitrogen out through the coating into the soil solution.

2. The perforated impermeable membrane, where pinholes through the membrane coating permit outward passage of water-soluble nitrogen.
3. The solid impermeable membrane, comprised of materials that are degraded by soil microorganisms, thus creating openings through which dissolved nitrogen can be liberated.

Coated Materials, Encapsulations, and Matrix Coatings.
Coatings, encapsulations, and matrixes have been used to impede the dissolution of water-soluble fertilizer materials containing nitrogen and other plant nutrients. Hauck and Koshino in their comprehensive review of slow-release and amended fertilizers characterized the following three main types of coatings which have been used:

1. Impermeable coatings with tiny pores through which the dissolved nitrogen and other constituents diffuse.
2. Semipermeable coating through which water diffuses until the internal solution pressure is sufficient to cause disruption.
3. Continuous impermeable coating that must be broken by chemical, microbial, or abrasive action before the water-soluble contents are released.

Coatings may act as only a barrier or they may also be a source of plant nutrients.

Many substances have been used to coat fertilizers and some of them include gums, oils, waxes, paraffins, tars, asphalts, sulfur, and a variety of polymers. The selected coatings must be thin; otherwise, they occupy too much of the total fertilizer particle volume and significantly lower the nutrient content. The relatively porous, rough, and irregular surface of most fertilizer particles makes it difficult to obtain uniform coatings.

Impermeable Coatings with Tiny Pores Permitting Nutrient Diffusion. This type of controlled-release fertilizer involves enclosing single or multiple particles of water-soluble fertilizer materials, containing nitrogen alone or in multinutrient combinations, with perforated impermeable membranes. Polyethylene film with a limited number of very small punctures has been used successfully for fertilizer encapsulation.

Long-term fertilization of pine trees was shown as early as 1965 to be possible with conventional commercial fertilizers enclosed in polyethylene film packages perforated with holes of various sizes. One example of a commercial product based on this approach is the Rootcontact Pakets developed for fruit trees and other horticultural crops, ornamental plants, and for forest seedlings.

Semipermeable Membranes. Osmocote controlled-release fertilizers are examples of water-soluble nitrogen, as well as phosphorus and potassium, materials coated with a semipermeable membrane, permitting entry of water until the internal solution pressure becomes great enough to disrupt the coating. They are made by applying multiple layers of a resinous polymer, mainly a copolymer of dicyclopentadiene with glycolester. The type and thickness of coating applied largely determines the release characteristics of a given formulation. Thin coatings produce materials with more rapid rates of release and lifetimes of a few months, while heavier coatings provide slower release of longer duration. Formulations with life spans of as little as three and as much as nine months have been commercially produced.

Release of nitrogen and other nutrients from this type of fertilizer is reduced by low temperatures and intermittent moderate drying. Premature release of enclosed nutrients has occurred when previously wetted coated granules were subjected to prolonged periods of excessive drying.

Three of the most widely used grades of Osmocote are 14-14-14, 18-9-9, and 16-4-8. The coating weight ranges from 10 to 15% of the gross weight. This type of controlled-release fertilizer is recommended for turf, floriculture, nursery stock, and high-value row crops.

Solid Impermeable Membranes. One of the most promising approaches to achieving controlled release of nitrogen with a popular dry product such as urea has been to apply a relatively insoluble, inexpensive material such as elemental sulfur. Coatings of this type are solid, impermeable membranes requiring breakdown by soil microorganisms and by chemical and mechanical action.

Sulfur-coated urea (SCU) is a controlled-release nitrogen fertilizer first made in about 1961 by the TVA in laboratory- and bench-scale tests. Since that time, TVA has continued extensive research into methods of making and handling the product and the agronomic benefits of utilizing it under various conditions for a variety of crops. TVA is now producing SCU in a demonstration plant that operates smoothly and which turns out a product of excellent quality.

The basic process involves formation of a sulfur shell around each urea particle, which is accomplished by spraying molten sulfur on a falling curtain of urea particles in a rotating drum. In a second drum, molten sealant (a mixture of polyethylene and brightstock oil) is applied to seal the microscopic pores and cracks in the sulfur coating. Nitrogen concentration in present formulations lies between 36 and 38% and all of it is supplied as urea, with the coatings controlling the time and rate at which urea leaves the granule. Nonuniform coatings among granules result in the release of urea at varying periods of time after SCU is wetted in soil. The release rate of SCU can be adjusted by changing the quantity of sulfur used for coating.

The sulfur coating on TVA's SCU is vulnerable to mechanical impact or abrasion. Excessive impact or abrasion will break down coatings and alter or destroy the slow-release characteristics of the product. Handling should be kept to a minimum, and moving the product by screw- or drag-type conveyors is not recommended because of abrasive effects. Substantial coating damage has been found when SCU in bulk blends is formulated with either paddle-type mixers or mixers fashioned from an inclined screw conveyor. Also, serious physical deterioration can be caused in SCU applied by fan-type spreaders with high-spread fans (800 to 1500 rev/min) and with the fertilizer fed off-center.

Sulfur-coated urea has the greatest potential for use in situations where multiple applications of soluble nitrogen sources are needed during the growing season, particularly on sandy soils under high rainfall or irrigation. It is advantageous for use on sugarcane, pineapple, grass forages, turf, ornamentals, fruits such as cranberries and strawberries, and rice under intermittent or delayed flooding. SCU might also find general use under conditions where decomposition losses are significant.

Another advantage of SCU is its sulfur content. Although elemental sulfur in the coating may not be sufficiently available to correct deficiencies during the first year after application, it can be an important source of plant available sulfur in succeeding years.

Two firms, one each in the United States and Canada, are now producing and marketing SCU. It is also being manufactured in England.

Matrixes. There are numerous patented processes for the dispersal of soluble fertilizer salts containing nitrogen, phosphorus, potassium, and so on, into asphalt, waxes, paraffins, oils, gels, polymers, and resins. Nitrogen compounds of limited water solubility, such as urea-formaldehyde, have also been incorporated into a matrix or have been embodied in expanded vermiculite, perlite, clay, glass frit, and similar materials.

Controlled-release matrix fertilizers produced by mixing finely ground ammonium sulfate and an asphalt-wax binder followed by extrusion into pellets and coating with wax were found to be greatly superior to uncoated ammonium sulfate for rice production. A similar "matrix" fertilizer was more effective than standard fertilizers on pasture and vegetable crops.

Mulches. Synthetic mulches (plastic, paper, or asphalt) are another means for controlling the release of nitrogen from water-soluble sources. These coverings placed above the fertilizer application zone will prevent leaching in sandy soils under conditions of high rainfall or intense irrigation. Plastic mulches are used in vegetable crop production in Florida to maintain desirable nutrient gradients emanating from band applications of nitrogen and other soluble elements. Soluble fertilizer materials supplying nitrogen and other nutrients have been converted to slow-release sources by adding them to the water phase of emulsions with asphalt and wax.

Inhibition of Microbial Activity

NITRIFICATION INHIBITORS.

Certain substances are toxic to the nitrifying bacteria and will, when added to the soil, temporarily inhibit nitrification. Many chemicals have been tested in recent years for their ability selectively to inhibit nitrification in soils and thereby more effectively manage additions of fertilizer nitrogen. A nitrification inhibitor should ideally (1) be nontoxic to plants, other soil organisms, fish, and mammals; (2) block the conversion of ammonium to nitrate by specifically inhibiting *Nitrosomonas* growth or activity; (3) not interfere with the transformation of nitrite by *Nitrobacter*; (4) be able to move with the fertilizer or fertilizer solution so that it will be distributed uniformly throughout the soil zone contacted by nitrogen fertilizer; (5) be stable enough for its inhibitory action to last for an adequate period of time, usually from several weeks to months; and (6) be relatively inexpensive, so that it can be used on a commercial scale. At least two companies in the United States and 10 in Japan hold patents on chemicals developed specifically as nitrification inhibitors.

Probably the two best known and more generally effective compounds are N-Serve or 2-chloro-6(trichloromethyl)pyridine frequently referred to as nitrapyrin and AM, which is a substituted pyrimidine (2-amino-4-chloro-6-methylpyrimidine). By the early 1980s about 2.5 million acres of U.S. cropland were being treated annually with nitrapyrin. Terrazole, or 5-ethoxy-3-trichloromethyl-1,2,4-thiadiazole, has been marketed since the late 1970s in the United States. ATC (4-amino-1,2,4-triazole hydrochloride) is used in Japan and it is now being evaluated at several locations in the United States and Canada. In the late 1960s 2-sulfanilamidothiaszole (sulfathiazole or ST) was introduced in Japan and at about the same time DCS or N-2,5-dichlorophenylsuccinamic acid was identified in that country as being a satisfactory inhibitor. Guanylthiourea (1-amido-2-thiourea or ASU) has also been patented in Japan as a nitrification inhibitor.

Dicyandiamide (NH_2—$\overset{\overset{\displaystyle NH}{\|}}{C}$—NH—C≡N), or DCD, has been tested as both a nitrification inhibitor and as a slow-release nitrogen source. In Japan, it has been added to mixed fertilizers and a product containing ureaform plus 10% by weight of DCD is manufactured there. Dicyandiamide is used in both cases to provide nitrification control and to increase the content of water-soluble nitrogen.

A fertilizer comprised of urea and DCD in a 4:1 ratio is commercially available in West Germany. Ashworth, formerly of the Rothamsted Experimental Station, reported that DCD is readily soluble and stable in anhydrous ammonia. Nitrification of anhydrous ammonia was effectively inhibited for up to three months by the addition of 15 kg/ha of DCD.

Other chemicals that have been studied or developed specifically for use as nitrification inhibitors include 2-benzothiazolesulfane morphdine (KN or KNE), 2-

amino-4-methyl-6-trichloromethyl triazine (MAST), 3-mercapto-1,2,4-triazole (MT), isothiocyanates, halophenols, nitroanilines, and haloanilines.

Thiourea, methionine, cyanoguanidine, sodium chlorate, sodium and potassium azides, dithane, dithiocarbamates, moniodoactic, and so on, may also be effective nitrification inhibitors. Carbon disulfide and related compounds such as the trithiocarbonates, which decompose to release CS_2, have been shown to be strong nitrification inhibitors, and their costs may be substantially less than some of the commercially available inhibitors.

There is interest in England and western Canada in another class of compounds, the xanthates, which also decompose to release CS_2 but at a generally slower rate than the trithiocarbonates. These compounds therefore have a potential for inhibiting nitrification for more extended periods. Although it takes rather high concentrations of xanthates to restrict nitrification significantly, as can be seen in Figure 5-27, they are nevertheless promising because of their low cost.

It must be understood that inhibitors will prevent nitrogen losses only when conditions suitable for unwanted transformations of NO_3^- coincide with the effective period of the particular chemical. If conditions are such that no NO_3^- losses occur, treatment with an inhibitor will not produce any change in nitrogen efficiency. Also, protective action is unlikely when the situations for NO_3^- loss develop after the effects of the inhibitor have dissipated. Although the circumstances conducive to loss of NO_3-N are generally known, it is very difficult to predict accurately when and how much nitrogen will be lost.

UREASE INHIBITORS.
The ability to control the rate of hydrolysis of urea by the enzyme urease has a number of practical applications in situations where the performance of fertilizer urea has proven to be unsatisfactory. A large number of compounds have been evaluated as urease inhibitors. All of the known inhibitors fall into three categories. The first group of substances inhibits urease activity by blocking essential sulfhydryl groups at active sites on the enzyme.

Metal ions including Ag^+, Hg^{2+}, and Cu^{2+} belong in this first category, and inhibition is inversely proportional to the solubility product of the metal-sulfide complex. Benzoquinones, quinones, and dihydric phenols when present in quinone form also react with the sulfhydryl groups of urease. Heterocyclic sulfur compounds

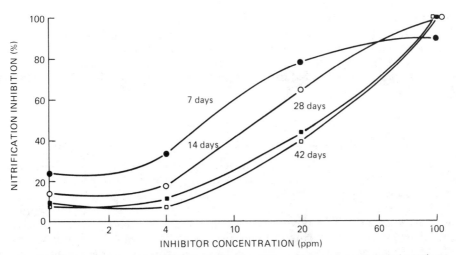

FIGURE 5-27.   Inhibition of nitrification in Rolling Hills soil by potassium ethyl xanthate. [Rowell and Akerboom, in *Effective Use of Nutrient Resources in Crop Production*, Proc. Alberta Soil Sci. Workshop, pp. 132–142. Edmonton, Alberta: Alberta Agriculture, (1979.)]

influence urease similarly by combining with sulfhydryl groups. The quinones, which are effective inhibitors of soil urease, are mild irritants to human beings. Offsetting the promising inhibitory action of the heterocylic sulfur compounds is their complete ineffectiveness under reducing conditions.

The second class of inhibitors are structural analogues of urea, such as thiourea, methylurea, and substituted ureas. They have structural similarities to urea and inhibit urease by competing for the same active site on the enzyme. These compounds are ineffective at high urea concentrations since they act competitively for the same sites on the enzyme.

Compounds that react with the nickel atom in the urease molecule comprise the third group of urease inhibitors. The hydroxamic acids are specific, noncompetitive inhibitors of urease and they are the most thoroughly studied of the known inhibitors. Caprylohydroxamic acid is the most potent member of this class of materials. It should be noted that not all compounds that react with nickel will inhibit urease.

Since 1980 there have been several reports that phenylphosphorodiamidate will effectively inhibit urease activity in flooded soils for about 4 days. It can be added to fluid fertilizers containing urea and can be cogranulated with urea.

Very few of the many substances capable of inhibiting urease meet the rather specific requirements of being (1) effective at low concentrations, (2) relatively nontoxic to higher forms of life, (3) inexpensive, (4) compatible with urea, and (5) movement in soil paralleling that of urea. Prospects for improving the effectiveness of urea through urease inhibition do not appear as promising as the more direct alternative of concentrating urea in the soil by using large particles or by some form of localized placement.

Moisture Supply and Response to Nitrogen

Crop response to fertilizer nitrogen is very dependent on moisture supplies during the growing season. In most cropping regions, especially in the low-rainfall areas of the western United States and Canada, the uncertain patterns of amount and distribution of precipitation make the use of a given amount of fertilizer nitrogen less profitable in some years than in others. The restricting effect of insufficient moisture on crop response to added nitrogen is readily apparent in Figures 2-7 and 2-8. Where adequate moisture conditions are assured by irrigation, there is little doubt about responses to nitrogen fertilization.

If growers can determine the probability of getting a profitable return from a given investment in fertilizer nitrogen, they are in a better position to plan their cropping programs for maximum net profit. It is also necessary for them to know the precipitation patterns before an expression can be developed to make such input–output calculations.

The Acidity and Basicity of Nitrogen Fertilizers

Some fertilizer materials produce an acid condition in the soil, others a basic reaction, and still others seemingly have no permanent influence on soil pH. Nitrogen carriers have a considerable effect on both the soil pH and on cation concentrations. Most of the popular nitrogen sources acidify soil and accelerate loss of cations by leaching. The development of acidity is illustrated by the nitrification equation given earlier in this chapter (see page 127). More information on the acidic and basic properties of commonly used fertilizers, particularly those containing nitrogen, is provided in Chapter 11.

Crop Responses to Various Sources of Fertilizer Nitrogen

Much research has been carried out to evaluate the effectiveness of various sources of nitrogen. It is not the purpose of this book to review individually the results of these various experiments. However, a summary of these findings, as well as of the principles involved in nitrogen fertilization, is given.

1. Nitrate is often the dominant form of nitrogen taken up by crops since it is generally present in higher concentrations in soil than is ammonium. In addition, its relative ease of movement through soil facilitates its absorption by plants.

2. Ammonium can be a significant source of nitrogen for plants, especially during the early growth stages of a number of important crop plants. Uptake of ammonium may produce several beneficial side effects, including stimulation of phosphorus uptake, acidification of high-pH soils near the root surface, and inhibition of certain root rots and other diseases.

3. Anhydrous ammonia, nonpressure UAN solutions, and urea have emerged as the principal nitrogen fertilizers in North America. Convenience of handling and application is a major advantage for solutions. The satisfactory properties of physically improved granular urea plus favorable economics of manufacturing, handling, storage, and transportation have contributed to its increased acceptance.

4. The nitrification pattern of both ammoniacal and natural organic materials provides little justification for the belief that these forms in warm, well-aerated, and moist soils release their nitrogen slowly, thus reducing excessive losses by leaching. In cold, fine-textured soils the water-insoluble forms may be expected to lose less of their nitrogen by leaching than the soluble forms because of reduced mineralization under cold conditions.

5. The principles of ion exchange generally influence the effectiveness of chemical sources of nitrogen. The ammoniacal form is retained briefly against leaching because of its adsorption by soil colloids. The nitrate form is not so retained. This difference will be greatest in fine-textured and least in coarse-textured soils. The downward movement is a factor to consider when top dressing with various sources of solid nitrogen fertilizers on soils of different texture. Differences in the leaching losses of these two forms are reduced when conditions favor rapid nitrification.

6. Localized placement of ammonium-containing or ammonium-forming fertilizers, such as is accomplished by injection of anhydrous ammonia and aqua ammonia, or in the case of dry materials, by banding or by applying very large particles, causes some drastic changes in soil conditions within or near the retention zone.

7. In or near the retention zone of ammoniacal fertilizers there is a temporary but dramatic rise in pH and greatly elevated concentrations of ammonia, ammonium, and nitrite. Osmotic suction in the zone is also increased. These conditions can greatly influence microbiological transformations of NH_4^+ and NO_2^-.

8. Nitrite as an intermediate in the microbial oxidation of ammonium to nitrate appears susceptible to nonenzymatic decomposition, resulting in emissions of oxides of nitrogen. There is new evidence that nitrite reactions with organic matter, causing the release of nitrogen and nitrous oxide gases, are more important than has been recognized previously. In addition, NO_2^- can be fixed by soil organic matter.

9. When conditions favor nitrification, the superiority of one form of nitrogen over the other may be related to the accompanying ion or to some element

contained as an impurity. This is illustrated by the use of a material such as ammonium sulfate. If applied on a sulfur-deficient soil, it would give an apparently better response than a non-sulfur-containing nitrogen carrier if sulfur were not included in some other component of the fertilizer. In such cases the way in which the limiting element should be supplied will be dictated by economic considerations.

10. Some nitrogen sources, such as anhydrous ammonia, nitrogen solutions, urea, and other ammoniacal materials, may lose ammonia by volatilization as a result of improper placement, surface application to alkaline soils, or, in the case of urea, surface application to soil or sod. In addition, if placed too close to seed or plants, injury may result from ammonia toxicity. These difficulties can most generally be corrected by proper placement and by adjusting the time of application.

11. Accumulations of mineral nitrogen can be lost in certain soils at the time of spring thaw in northern regions. Such losses are due mainly to denitrification. These losses from fall-applied ammoniacal fertilizers can be greatly reduced by some form of deep localized placement in the soil. Very large particles and clusters or nests of regular-sized fertilizer particles are techniques of localized placement which have successfully restricted denitrification. Inhibition of nitrification by compounds such as thiourea, carbon disulfide, and so on, has also decreased denitrification losses by lowering NO_3^- levels in the soil during the critical spring thaw period.

12. A number of chemical substances have been tested for control of nitrification and urease activity. Several substances are now commercially available for the inhibition of nitrification. These materials can be used in the management of fertilizer nitrogen where losses through leaching and biological and chemical denitrification are serious.

13. When differences in acid-forming properties, secondary or trace element content, and method and time of application and placement are recognized and handled accordingly, one source of fertilizer nitrogen is often as effective as any other in increasing crop yields. The determining factor in the selection of a source of nitrogen is then governed by reliability of supply, economics, and ease of application. Where factors such as supply, ease of application, and so on, are relatively unimportant, the nitrogen source selected should be the one that gives the maximum return on investment in fertilizer nitrogen.

14. The continued use of acid-forming fertilizer materials will lead to a decrease in pH with an accompanying decrease in crop yields unless lime sufficient to neutralize the acidity formed is applied to the soil.

Summary

1. Atmospheric nitrogen is fixed in soils by various free-living and symbiotic bacteria. The amounts fixed by these organisms are generally inadequate for the sustained high yields of crops in commercial farming.

2. The various forms of soil nitrogen and their turnover in soils were discussed. Important to the immobilization and release of nitrogen are such factors as the supply of carbon, phosphorus, and sulfur, soil aeration, and temperature.

3. Considerable attention was given to the reactions of inorganic nitrogen compounds in soils, especially those of ammonification and nitrification. The retention of the various forms of inorganic nitrogen in soils as well as gaseous losses of this element were covered.

4. The four general classes of nitrogen fertilizer were discussed: ammoniacal, nitrate, slowly available forms, and miscellaneous materials. Most of the ammoniacal forms are acid forming and their continued use

will lower soil pH values. The nitrate form is subject to loss by leaching. In coarse-textured soils, under high rainfall, such losses can be serious.

5. Crop responses to the various forms of nitrogen were discussed. It is generally concluded that when the nitrogen alone is considered, the results from one form are as good as another. However, method of application, accompanying elements in the carrier, and placement in the soil may cause differences in crop responses to the various carriers. The cost per unit of nitrogen applied to the land is an important item in determining the selection of the nitrogen fertilizer.

Questions

1. What are the ways, exclusive of synthetic nitrogen fixation, by which atmospheric nitrogen is made usable to higher plants?
2. What are the various microorganisms responsible for nitrogen fixation?
3. What soil property can exercise considerable influence on the survival and growth of rhizobia in soil? Describe at least two practical ways of improving the effectiveness of growth and performance of rhizobia.
4. Is it possible to distinguish between effective and noneffective nodules on the roots of legume plants? Describe the location and appearance of effective nodules.
5. Can nitrogen be fixed in association with crop plants such as corn, wheat, sorghum and rice, and so on? If so, how is it accomplished? What is the contribution of the crop plant, and does this contribution influence yields?
6. Define ammonification and nitrification. What are the factors affecting these reactions in soils?
7. Do crops utilize both NH_4^+ and NO_3^-? Which is the preferred form of nitrogen? Does stage of growth influence crop uptake of either NH_4^+ or NO_3^-?
8. Does uptake of either NH_4^+ or NO_3^- influence chemical composition of crops plants? Discuss the major differences in both organic and inorganic constituents.
9. Identify an important soil property that can be altered by uptake of NH_4^+ and NO_3^-. Describe at least two beneficial side effects resulting from NH_4^+ uptake.
10. Is energy consumed during the reduction of NO_3^- prior to protein formation in plants? Does NH_4^+ behave similarly?
11. Are high concentrations of NH_3 and NH_4^+ detrimental to crop growth? If so, briefly describe the harmful effects.
12. If leaching losses of nitrogen are to be minimized after the fall application of ammoniacal nitrogen, soil temperatures during these winter months should not rise above what point?
13. As a general rule, is the fall application of nitrate fertilizers a sound practice? Why?
14. Nitrification, specifically, is defined how? It is a two-step reaction. What are the two steps and what organisms are responsible for each?
15. Why is nitrification important? Would you consider this phenomenon a mixed blessing? Why? Be precise.
16. What is ammonia fixation? What are the soil conditions under which it occurs? Discuss the role that potassium and ammonia play, each

in the fixation or release of the other. How important do you consider this factor to be in the overall nitrogen fertilization picture?

17. Describe the environmental and soil conditions under which you would expect to get significantly lower leaching losses of NH_4^+ nitrogen as contrasted with NO_3^- nitrogen. Under what soil and environmental conditions would you expect not to get these differences?

18. You have disked in a large amount of barley straw just about a week before planting fall wheat. At planting time you applied fertilizer which supplied 20 lb of nitrogen, 20 lb of phosphorus, and 40 lb of potassium. The wheat germinates and shortly thereafter turns yellow. Tests show no NO_3^- nitrogen in the tissue. What is wrong with the wheat, and why? The farmer on whose field this is observed asks you, as County Farm Adviser, what to do. What is your answer?

19. In what forms may nitrogen as a gas be lost from soil? Discuss the conditions under which each form is lost and indicate the reactions that are thought to take place.

20. How would you prevent or minimize the various gaseous losses of nitrogen?

21. Describe a set of conditions that are now believed to result in serious denitrification losses in northern soils. Are there practical ways of largely eliminating such losses? If so, identify them.

22. Classify the various forms of nitrogen fertilizers.

23. What is the single most important *original* source of fertilizer nitrogen today?

24. What is the commercial nitrogen fertilizer with the highest percentage of nitrogen?

25. What solid nitrogenous material has the highest percentage of nitrogen?

26. What developments have resulted in the great increase in popularity of urea?

27. List the major changes in soil properties that occur in the injection zones of anhydrous and aqua ammonia. Also, identify the major changes in soil conditions within or near the retention zone of dry ammoniacal nitrogen sources.

28. What conditions favor NO_2^- accumulation? Describe harmful effects, if any, of NO_2^- on crops.

29. Can there be serious losses of nitrogen through NO_2^- as an intermediate substance? If so, list the pathways of nitrogen loss.

30. Identify at least three ways that NO_2^- will react with the organic matter fraction in soils.

31. Why is it sometimes unwise to apply urea to the surface of the soil?

32. What are the practical ways of improving the agronomic effectiveness of urea?

33. What is perhaps the most important factor governing the selection of the source of fertilizer nitrogen?

34. Is urea (46% nitrogen) at a price of $225.00 per ton more or less expensive per pound of nitrogen than anhydrous ammonia (82% nitrogen) at $325.00 per ton?

35. What precaution should be observed in applying ammoniacal nitrogen fertilizers to calcareous soils?

36. Intensive cultivation of land leads to a rapid decomposition of the organic matter and a more rapid rate of nitrification. Why?

37. What is the difference between nitrogen fixation and nitrification?

38. Why, specifically, have ammonia forms of nitrogen an acidifying effect on the soil?

39. In general, what types of nitrogen solutions can be dribbled onto the soil surface without losses of nitrogen to the atmosphere?
40. Describe the ideal source of fertilizer nitrogen. What are the major products and approaches used for controlling nitrogen fertilizer availability?
41. List some of the more important nitrification inhibitors. Do the xanthates have inhibitory properties? If so, what causes their inhibition of nitrification?
42. Describe the conditions where nitrification inhibitors have the greatest potential for increasing efficiency of nitrogen fertilizer management.
43. What is urease, and why is it important? List some of the chemicals that can control urease activity.
44. Can nitrogen fertilizers affect soil pH? What is the effect of the most widely used nitrogen products on soil reaction?

Selected References

Abruna, F., R. W. Pearson, and C. B. Elkins, "Quantitative evaluation of soil reaction and base status changes resulting from field application of residually acid forming nitrogen fertilizers." *SSSA Proc.*, **22:**539 (1958).

Adriano, D. C., P. F. Pratt, and S. E. Bishop, "Nitrate and salt in soils and ground waters from land disposal of dairy manure." *SSSA Proc.*, **35:**759 (1971).

Alexander, M., *Introduction to Soil Microbiology*, 2nd ed. New York: Wiley, 1977.

Allison, F. E., *Soil Organic Matter and Its Role in Crop Production* (Developments in Soil Science 3). Amsterdam: Elsevier Scientific, 1973.

Allison, F. E., J. N. Carter, and L. D. Sterling, "The effect of partial pressure of O_2 on denitrification in soil." *SSSA Proc.*, **24:**283 (1960).

Anderson, O. E., "The effect of low temperatures on nitrification of ammonia in Cecil sandy loam." *SSSA Proc.*, **24:**286 (1960).

Andrews, W. B., *The Response of Crops and Soils to Fertilizers and Manures*, 2nd ed. State College, Miss.: W. B. Andrews, 1954.

Ashworth, J., and G. A. Rodgers, "The compatability of the nitrification inhibitor dicyandiamide with injected anhydrous ammonia." *Can. J. Soil Sci.*, **61:**461 (1981).

Ashworth, J., G. G. Briggs, A. A. Evans, and J. Matala, "Inhibition of nitrification by nitrapyrin, carbon disulphide and trithiocarbonate." *J. Sci. Food Agr.*, **28:**673 (1977).

Baker, J. H., M. Peech, and R. B. Musgrave, "Determination of application losses of anhydrous ammonia." *Agron. J.*, **51:**361 (1959).

Bates, T. E., and S. L. Tisdale, "The movement of nitrate nitrogen through columns of coarse-textured soil materials." *SSSA Proc.*, **21:**525 (1957).

Bauder, J. W., and B. R. Montgomery, "Overwinter distribution and leaching of fall-applied nitrogen." *Soil Sci. Soc. Am. J.*, **43:**744 (1979).

Beaton, J. D., "Fertilizers and their use in the decade ahead." *Proc. 20th Annu. Meet. Agr. Res. Inst.* (October 1971).

Beaton, J. D., "Urea: its popularity grows as a dry source of nitrogen." *Crops Soils*, **30(6):**11 (1978).

Beaton, J. D., W. E. Janke, and S. S. Blair, "Fertilizer nitrogen." *Proc. Western Canada Nitrogen Symp.*, Proc. Alberta Soil Sci. Workshop. Edmonton, Alberta: Alberta Agriculture, 1976.

Bettany, J. R., J. W. B. Stewart, and E. H. Halstead, "Sulfur fractions and carbon, nitrogen, and sulfur relationships in grassland, forest, and associated transitional soils." *Soil Sci. Soc. Am. J.*, **37:**915 (1973).

Bixby, D. W., "Addition of sulfur in all types of fertilizer, including fluids." *Fert. Round Table*, Atlanta, Ga. (1978).

Blair, G. J., M. H. Miller, and W. A. Mitchell, "Nitrate and ammonium as sources of nitrogen for corn and their influence on the uptake of other ions." *Agron. J.*, **62:**530 (1970).

Blue, W. G., and C. F. Eno, "Distribution and retention of anhydrous ammonia in sandy soils." *SSSA Proc.*, **18:**420 (1954).

Breitenbeck, G. A., A. M. Blackmer, and J. M. Bremner, "Effects of different nitrogen fertilizers on emission of nitrous oxide from soil." *Geophys. Res. Lett.*, **7**(1):85 (1980).

Bremner, J. M., "Effects of soil processes on the atmospheric concentration of nitrous oxide," in D. R. Nielsen, Ed., *Nitrogen in the Environment*, Vol. 1: *Nitrogen Behavior in Field Soil*, pp. 477–491. New York: Academic Press, 1978.

Bremner, J. M., and L. A. Douglas, "Effects of some urease inhibitors on urea hydrolysis in soils." *SSSA Proc.*, **37:**225 (1973).

Bremner, J. M., and K. Shaw, "Denitrification in soil: II. Factors affecting denitrification." *J. Agr. Sci. (Camb.)*, **51:**40 (1958).

Bremner, J. M., G. A. Breitenbeck, and A. M. Blackmer, "Effect of anhydrous ammonia fertilization on emission of nitrous oxide from soils." *J. Environ. Qual.*, **10**(1):77 (1981).

Bundy, L. G., and J. M. Bremner, "Effects of substituted p-benzoquinones on urease activity in soils." *Soil Biol. Biochem.*, **5:**847 (1973).

Burford, J. R., and J. M. Bremner, "Relationships between the denitrification capacities of soils and total, water-soluble and readily decomposable soil organic matter." *Soil Biol. Biochem.*, **7:**389 (1975).

Campbell, C. A., E. A. Paul, and W. B. McGill, "Effect of cultivation and cropping on the amounts and forms of soil N." *Proc. Western Canada Nitrogen Symp.*, pp. 7–101. Edmonton, Alberta: Alberta Agriculture, 1976.

Cassman, K. G., and D. N. Munns, "Nitrogen mineralization as affected by soil moisture, temperature, and depth." *Soil Sci. Soc. Am. J.*, **44:**1223 (1980).

Chandra, P., "The effect of shifting temperatures on nitrification in a loam soil." *Can. J. Soil Sci.*, **42:**314 (1962).

Christianson, C. B., and C. M. Cho, "Denitrification under frozen conditions." *23rd Annu. Manitoba Soil Sci. Meet.*, pp. 105–109, Univ. of Manitoba (1979).

Christianson, C. B., R. A. Hedlin, and C. M. Cho, "Loss of nitrogen from soil during nitrification of urea." *Can. J. Soil Sci.*, **59:**147 (1979).

Cochran, V. L., L. F. Elliott, and R. I. Papendick, "Carbon and nitrogen movement from surface-applied wheat (*Triticum aestivum*) straw." *Soil Sci. Soc. Am. J.*, **44:**978 (1980).

Colliver, G. W., "Nitrification inhibitors can increase yields." *Farm Chem.*, **143**(4):114 (1980).

Cominco Ltd., *Chemicals and Fertilizers Product Data Manual*. Calgary, Alberta: Cominco Ltd., 1979.

Cooper, G. S., and R. L. Smith, "Sequence of products formed during denitrification in some diverse western soils." *Soil Sci. Soc. Am. J.*, **27:**659 (1963).

Cox, W. J., and H. M. Reisenauer, "Growth and ion uptake by wheat supplied nitrogen as nitrate, or ammonium, or both." *Plant Soil*, **38:**363 (1973).

Davis, C. H., "New developments in U.S. chemical fertilizer technology." *180th Natl Meet., Am. Chem. Soc.*, Las Vegas, Nev. (1980).

Diamond, R. B., and F. J. Myers, "Crop responses and related benefits from SCU." *Sulphur Inst. J.*, **8:**9 (1972).

Feagley, S. E., and L. R. Hossner, "Ammonia volatilization reaction mechanism between ammonium sulfate and carbonate systems." *Soil Sci. Soc. Am. J.*, **42:**364 (1978).

Fenn, L. B., "Ammonia volatilization from surface applications of ammonium compounds on calcareous soils: III. Effects of blending nonprecipitate forming with a precipitate forming ammonium compound." *Soil Sci. Am. J.*, **39:**366 (1975).

Fenn, L. B., and D. E. Kissel, "Ammonia volatilization from surface applications of ammonium compounds on calcareous soils: I. General theory." *Soil Sci. Soc. Am. J.*, **37:**855 (1973).

Fenn, L. B., and D. E. Kissel, "Ammonia volatilization from surface applications of ammonium compounds on calcareous soils: II. Effects of temperature and rate of NH_4^+-N application." *Soil Sci. Soc. Am. J.*, **38:**606 (1974).

Fenn, L. B., and D. E. Kissel, "Ammonia volatilization from surface applications of ammonium compounds on calcareous soils: IV. Effect of calcium carbonate content." *Soil Sci. Soc. Am. J.*, **39:**631 (1975).

The Fertilizer Institute, *Agricultural Anhydrous Ammonia Operator's Manual.* Washington, D.C.: The Fertilizer Institute, 1973.

Fisher, W. B., Jr., and W. L. Parks, "Influence of soil temperature on urea hydrolysis and subsequent nitrification." *SSSA Proc.*, **22:**247 (1958).

Freney, J. R., "Sulfur-containing organics," in A. D. McLaren and G. H. Peterson, Eds., *Soil Biochemistry*, pp. 229–259. New York: Marcel Dekker, 1967.

Giordano, P. M., and J. J. Mortvedt, "Release of nitrogen from sulfur-coated urea in flooded soil." *Agron. J.*, **62:**612 (1970).

Gould, W. D., "Inhibition of urease activity," in *Effective Use of Nutrient Resources in Crop Production*, Proc. Alberta Soil Sci. Workshop, pp. 143–158. Edmonton, Alberta: Alberta Agriculture, 1979.

Gould, W. D., F. D. Cook, and J. A. Bulat, "Inhibition of urease activity by heterocyclic sulfur compounds." *Soil Sci. Soc. Am. J.*, **42:**66 (1978).

Gould, W. D., F. D. Cook, and G. R. Webster, "Factors affecting urea hydrolysis in several Alberta soils." *Plant Soil*, **38:**393 (1973).

Hanawalt, R. B., "Environmental factors influencing the sorption of atmospheric ammonia by soils." *SSSA Proc.*, **33:**231 (1969).

Harapiak, J. T., and L. McCulley, "WCFL nitrogen studies on barley—1975." *Proc. 19th Annu. Manitoba Soil Science Meet.*, Winnipeg, Manitoba (1975).

Hauck, R. D., "Urea: soil chemistry and agronomic efficiency." *Proc., Fert. Ind. Round Table*, pp. 211–212, Washington, D.C. (November 4–6, 1975).

Hauck, R. D., and M. Koshino, "Slow-release and amended fertilizers," in R. A. Olson, T. J. Army, J. J. Hanway, and V. J. Kilmer, eds., *Fertilizer Technology and Use*, 2nd ed. Madison, Wis.: Soil Science Society of America, 1971.

Hays, J. T., "Methylolurea solutions—chemistry and fertilizer action." *180th Nat'l Meet., Am. Chem. Soc.*, Las Vegas, Nev. (1980).

Henry, J. L., T. J. Hogg, and E. A. Paul, "The effect of anhydrous ammonia on soil." *Proc. 1979 Soils Crops Workshop*, Publ. 403, pp. 34–57, Univ. of Saskatchewan, Saskatoon (1979).

Hinman, W. C., "Fixed ammonium in some Saskatchewan soils." *Can. J. Soil Sci.*, **44:**151 (1964).

Huber, D. M., "The use of fertilizers and organic amendments in the control of plant disease," in D. Pimentel, Ed., *Handbook Series in Agriculture: Pest Management.* pp. 357–394. Boca Raton, Fla.: CRC Press, 1980.

Huber, D. M., H. L. Warren, D. W. Nelson, and C. Y. Tsai, "Nitrification inhibitors—new tools for food production." *BioScience*, **27:**523 (1977).

Hutchinson, G. L., and A. R. Mosier, "Nitrous oxide emissions from an irrigated cornfield." *Science*, **205:**1125 (1979).

International Fertilizer Development Center and United Nations Industrial Development Organization, *Fertilizer Manual*, Reference Manual IFDC-R-1. Muscle Shoals, Ala.: IFDC, 1979.

Iritani, W. M., and C. Y. Arnold, "Nitrogen release of vegetable crop residues during incubation as related to their chemical composition." *Soil Sci.*, **89:**74 (1960).

Isfan, D., "Nitrogen rate–yield–precipitation relationships and N rate forecasting for corn crops." *Agron. J.*, **71:**1045 (1979).

Jansson, S. L., "Tracer studies on nitrogen transformations in soil with special attention to mineralization–immobilization relationships." *K. Lantbrukshoegsk, Ann.*, **24:**101 (1958).

Johnson, W. E., "Review of research results with anhydrous ammonia and aqua ammonia." *Proc. 1977 Soil Fert. Crops Workshop*, Publ. 328, pp. 100–109, Univ. of Saskatchewan, Saskatoon (1977).

Juma, N. G., and E. A. Paul, "Crop utilization and fate of fertilizer and residual nitrogen in soil," in *Effective Use of Nutrient Resources in Crop Production*, Proc. Alberta Soil Sci. Workshop, pp. 26–36. Edmonton, Alberta: Alberta Agriculture, 1979.

Kowalenko, C. G., and G. J. Ross, "Studies on the dynamics of 'recently' clay-fixed NH_4^+ using ^{15}N." *Can. J. Soil Sci.*, **60:**61 (1980).

Kresge, C. B., "Ammonia volatilization losses from nitrogen fertilizers when applied to soils." *Diss. Abstr.*, **20:**448 (1959).

Kresge, C. B., and D. P. Satchell, "Gaseous losses of ammonia from nitrogen fertilizers applied to soils." *Agron. J.*, **52:**104 (1960).

Leyshon, A. J., C. A. Campbell, and F. G. Warder, "Comparison of the effect of NO_3^- and NH_4-N on growth, yield, and yield components of Manitou spring wheat and Conquest barley." *Can. J. Plant Sci.*, **60:**1063 (1980).

Lindsay, W. L., M. Sadig, and L. K. Porter, "Thermodynamics of inorganic nitrogen transformations." *Soil Sci. Soc. Am. J.*, **45:**61 (1981).

Long, F. L., and G. M. Volk, "Availability of nitrogen from condensation products of urea and formaldehyde." *Agron. J.*, **55:**155 (1963).

McCants, C. B., E. O. Skogley, and W. G. Woltz, "Influence of certain soil fumigation treatments on the response of tobacco to ammonium and nitrate forms of nitrogen." *SSSA Proc.*, **23:**466 (1959).

Mahendrappa, M. K., "Ammonia volatilization from some forest floor materials following urea fertilization." *Can. J. For. Res.*, **5:**210 (1975).

Malhi, S. S., and M. Nyborg, "Effectiveness of several nitrification inhibitors and their effect on mineralization of soil nitrogen," in *Effective Use of Nutrient Resources in Crop Production*, Proc. Alberta Soil Sci. Workshop, pp. 113–131. Edmonton, Alberta: Alberta Agriculture, 1979.

Malhi, S. S., and M. Nyborg, "Rate of hydrolysis of urea as influenced by thiourea and pellet size." *Plant Soil*, **51:**177 (1979).

Malhi, S. S., and M. Nyborg, "An evaluation of carbon disulphide as a sulphur fertilizer and as a nitrification inhibitor." *Plant Soil*, **65:**203 (1982).

Malhi, S. S., F. D. Cook, and M. Nyborg, "Inhibition of nitrite formation by thiourea in pure cultures of *Nitrosomonas*," *Soil Biol. Biochem.*, **11:**431 (1979).

Mengel, K., and E. A. Kirkby, *Principles of Plant Nutrition*, 3rd ed. Bern: International Potash Institute, 1982.

Nielsen, R. F., and R. Carter, "Nitrogen deficiencies in fall fertilized winter wheat." *Proc. 31st Annu. Northwest Fert. Conf.*, pp. 125–127, Northwest Plant Food Association, Portland, Oreg. (1980).

Nommik, H., "Ammonium fixation and other reactions involving a nonenzymatic immobilization of mineral nitrogen in soil," in W. V. Bartholomew and F. E. Clark, Eds., *Soil Nitrogen*, pp. 198–258. Madison, Wis.: American Society of Agronomy, 1965.

Nommik, H., "The effect of pellet size on the ammonia loss from urea applied to forest soil." *Plant Soil*, **39**(2):309 (1973).

Nyborg, M., and R. H. Leitch, "Losses of soil and fertilizer nitrogen in northern Alberta," in *Effective Use of Nutrient Resources in Crop Production*, Proc. Alberta Soil Sci. Workshop, pp. 56–84. Edmonton, Alberta: Alberta Agriculture, 1979.

Nyborg, M., S. S. Malhi, and C. Monreal, "Placement of urea in big pellets or nests," in *Effective Use of Nutrient Resources in Crop Production*, Proc. Alberta Soil Sci. Workshop, pp. 99–112. Edmonton, Alberta: Alberta Agriculture, 1979.

Okon, Y., "Recent progress in research on biological nitrogen fixation with non-leguminous crops." *Phosphorus Agr.*, **82:**3 (1982).

Pang, P. C., R. A. Hedlin, and C. M. Cho, "Transformation and movement of band-applied urea, ammonium sulfate, and ammonium hydroxide during incubation in several Manitoba soils." *Can. J. Soil Sci.*, **53:**331 (1973).

Parish, D. H., L. L. Hammond, and E. T. Craswell, "Research on modified fertilizer materials for use in developing-country agriculture." *180th Nat'l Meet., Am. Chem. Soc.*, Las Vegas, Nev. (1980).

Parr, J. F., and J. C. Engibous, "Patterns of NH_3 distribution in the soil," in *Anhydrous Ammonia Agronomy Workshop*, pp. 4-1–4-8. Memphis, Tenn.: Agricultural Ammonia Institute, 1966.

Passioura, J. B., and R. Wetselaar, "Consequences of banding nitrogen fertilizers in soil: II. Effects on the growth of wheat roots." *Plant Soil*, **36:**461 (1972).

Patten, D. K., J. M. Bremner, and A. M. Blackmer, "Effects of drying and air-storage of soils on their capacity for denitrification of nitrate." *Soil Sci. Soc. Am. J.*, **44:**67 (1980).

Paulson, K. N., and L. T. Kurtz, "Locus of urease activity in soil." *Soil Sci. Soc. Am. J.*, **33:**897 (1969).

Pesek, J., G. Stanford, and N. L. Case, "Nitrogen production and use," in R. A. Olson, T. J. Army, J. J. Hanway, and V. J. Kilmer, Eds., *Fertilizer Technology and Use*, 2nd ed. Madison, Wis.: Soil Science Society of America, 1971.

Platou, J., "SCU—A progress report. *Sulphur Inst. J.*, **8:**9 (1972).

Porter, L. K., "Gaseous products produced by anaerobic reaction of sodium nitrite with oxime compounds and oximes synthesized from organic matter." *Soil Sci. Soc. Am. J.*, **33:**696 (1969).

Reisenauer, H. M., "Absorption and utilization of ammonium nitrogen by plants," in D. R. Nielsen, Ed., *Nitrogen in the Environment*, Vol. 2: *Soil-Plant-Nitrogen Relationships*, pp. 157–170. New York: Academic Press, 1978.

Rennie, D. A., J. D. Beaton, and R. A. Hedlin, "The role of fertilizer nutrients in western Canadian agricultural development." *Canada West Foundation 80-139*, Calgary, Alberta (1980).

Rennie, R. J., and R. I. Larson, "Dinitrogen fixation associated with disomic chromosome substitution lines of spring wheat in the phytotron and in the field," in P. B. Vose and A. P. Ruschel, Eds. *Associative Dinitrogen Fixation*, pp. 145–154. Boca Raton, Fla.: CRC Press, 1979.

Rennie, R. J., and R. I. Larson, "Dinitrogen fixation associated with disomic chromosome substitution lines of spring wheat." *Can. J. Bot.*, **57:**2771–2775 (1979).

Rice, W. A., "Inoculation, nodulation, and nitrogen fixation," in *Alfalfa Production in the Peace River Region*, Northern Res. Group Publ. 76-7, pp. D1–D12. Beaverlodge, Alberta: Agriculture Canada, 1976.

Rice, W. A., D. C. Penney, and M. Nyborg, "Effects of soil acidity on rhizobia numbers, nodulation, and nitrogen fixation by alfalfa and red clover." *Can. J. Soil Sci.*, **57:**197 (1977).

Rowell, M. J., and H. Akerboom, "Some new nitrification inhibitors and their potential use in agriculture and forestry," in *Effective Use of Nutrient Resources in Crop Production*, Proc. Alberta Soil Sci. Workshop, pp. 132–142. Edmonton, Alberta: Alberta Agriculture, 1979.

Russell, E. W., *Soil Conditions and Plant Growth*, 10th ed. London: Longmans, Green, 1973.

Ryden, J. C., "Denitrification loss from a grassland soil in the field receiving different rates of nitrogen as ammonium nitrate." *J. Soil Sci.*, **34:**355 (1983).

Savant, N. K., and S. K. DeDatta, "Movement and distribution of ammonium-N following deep placement of urea in a wetland rice soil." *Soil Sci. Soc. Am. J.*, **44:**559 (1980).

Scarsbrook, C. E., "Urea-formaldehyde fertilizer as a source of nitrogen for cotton and corn." *SSSA Proc.*, **22:**442 (1958).

Schrader, L. E., D. Domska, P. E. Jung, Jr., and L. A. Peterson, "Uptake and assimilation of ammonium-N and nitrate-N and their influence on the growth of corn (*Zea mays* L.)." *Agron. J.*, **64:**690 (1972).

Shiga, H., W. B. Ventura, and T. Yoshida, "Effect of deep placement of ball type fertilizer at different growth stages on yield and yield components of the rice plant in the Philippines." *Plant Soil*, **47:**351 (1977).

Skujins, J. J., "Enzymes in soil," in A. D. McLaren and G. H. Peterson, Eds., *Soil Biochemistry*, Vol. 1. New York: Marcel Dekker, 1967.

Smiley, R. W., "Rhizosphere pH as influenced by plants, soils, and nitrogen fertilizers." *Soil Sci. Soc. Am. J.*, **38:**795 (1974).

Smiley, R. W., "Forms of nitrogen and the pH in the root zone and their importance to root infections," in G. W. Bruehl, Ed., *Biology and Control of Soil-Borne Plant Pathogens*, Vol. 2. St. Paul, Minn.: American Phytopathology Society, 1974.

Smiley, R. W., "Colonization of wheat roots by *Gaeumannomyces graminis* inhibited by specific soils, microorganisms, and ammonium-nitrogen." *Soil Biol. Biochem.*, **10:**175 (1978).

Smith, C. J., and P. M. Chalk, "Gaseous nitrogen evolution during nitrification of ammonia fertilizer and nitrate transformations in soils." *Soil Sci. Soc. Am. J.*, **44**:277 (1980).

Smith, C. J., and P. M. Chalk, "Fixation and loss of nitrogen during transformations of nitrite in soils." *Soil Sci. Soc. Am. J.*, **44**:288 (1980).

Sowden, F. J., "Transformations of nitrogen added as ammonium and manure to soil with a high ammonium-fixing capacity under laboratory conditions." *Can. J. Soil Sci.*, **56**:319 (1976).

Spratt, E. D., and J. K. R. Gasser, "The effect of ammonium sulphate treated with a nitrification inhibitor, and calcium nitrate, on growth and N-uptake of spring wheat, ryegrass, and kale." *J. Agr. Sci. (Camb.)*, **74**:111 (1970).

Stanford, G., and E. Epstein, "Nitrogen mineralization–water relations in soils." *Soil Sci. Soc. Am. J.*, **38**:103 (1974).

Stanford. G., R. A. Vander Pol, and S. Dzienia, "Denitrification rates in relation to total and extractable soil carbon." *Soil Sci. Soc. Am. J.*, **39**:284 (1975).

Stanford, G., J. N. Carter, D. T. Westermann, and J. J. Meisinger, "Residual nitrate and mineralizable soil nitrogen in relation to nitrogen uptake by irrigated sugarbeets." *Agron. J.*, **69**:303 (1977).

Steen, W. C., and B. J. Stojanovic, "Nitric oxide volatilization from a calcareous soil and model aqueous solutions." *Soil Sci. Soc. Am. J.*, **35**:277 (1971).

Stevenson, F. J., "The nitrogen cycle in soils." Lecture notes, ICA Short Course, University of Illinois (1963).

Steyn, P. L., and C. C. Delwiche, "Nitrogen fixation by nonsymbiotic microorganisms in some California soils." *Environ. Sci. Technol.*, **4**(12):1122 (1970).

Tennessee Valley Authority, "Reduction of ammonia losses from urea applied to soil surfaces," in *12th Demonstration of New Developments in Fertilizer Technology*, pp. 39–41. Muscle Shoals, Ala.: TVA, 1978.

Tennessee Valley Authority, "Direct production of granular urea phosphate without removal of impurities," in *13th Demonstration of New Developments in Fertilizer Technology*, pp. 3–4. Muscle Shoals, Ala.: TVA, 1980.

Thomas, G. W., and D. E. Kissel, "Nitrate volatilization from soils." *SSSA Proc.*, **34**:828 (1970).

Tisdale, S. L., et al., "Sources of nitrogen in crop production." *North Carolina Agr. Exp. Sta. Tech. Bull. 96* (1952).

Tomlinson, T. E., "Urea—agronomic applications." *Proc. Fert. Soc. (London)*, 113 (1970).

Trappe, J. M., and W. B. Bollen, "Forest soil biology," in P. E. Heilman, H. W. Anderson, and D. M. Baumgartner, Eds., *Forest Soils of the Douglas-Fir Region*, pp. 145–151. Pullman: Washington State University, 1977.

Tuller, W. N., Ed., *The Sulphur Data Book: Freeport Sulphur Company*. New York: McGraw-Hill, 1954.

Volk, G. M., "Efficiency of fertilizer urea as affected by method of application, soil moisture, and lime." *Agron. J.*, **58**:249 (1966).

Volk, G. M., "Gaseous loss of ammonia from prilled urea applied to slash pine." *SSSA Proc.*, **34**:513 (1970).

Watkins, S. H., R. F. Strand, D. S. DeBell, and J. Esch, Jr., "Factors influencing ammonia losses from urea applied to northwestern forest soils." *Soil Sci. Soc. Am. J.*, **36**:354 (1972).

Welch, L. F., "Nitrogen use and behavior in crop production." *Illinois Agr. Exp. Sta. Bull. 761* (1979).

Welch, L. F., and A. D. Scott, "Nitrification of fixed ammonium in clay minerals as affected by added potassium." *Soil Sci.*, **90**:79 (1960).

Wetselaar, R., J. Passioura, and B. R. Singh, "Consequences of banding nitrogen fertilizers in soil: I. Effects on nitrification." *Plant Soil*, **36**:159 (1972).

Wullstein, L. H., "Soil nitrogen volatilization. A case for applied research." *Agr. Sci. Rev.*, **5**(2):8 (1967).

Wullstein, L. H., "Reduction of nitrite deficits by alkaline metal carbonates." *Soil Sci.*, **108**:222 (1969).

Wullstein, L. H., M. L. Bruening, and W. B. Bollen, "Nitrogen fixation associated with sand grain root sheaths (rhizosheaths) of certain xeric grasses." *Physiol. Plant.*, **46**:1 (1979).

Chapter 6

Soil and Fertilizer Phosphorus

THE importance of phosphorus in plant nutrition was discussed in Chapter 3. This element is present in plant tissues and in soils in smaller amounts than are nitrogen and potassium and in quantities about equal to that of sulfur. The generally small quantities of phosphorus in soils and its tendency to react with soil components to form relatively insoluble compounds, many of which have only limited availability to plants, make it a topic of major importance in soil fertility management.

Phosphorus Content of Soils

Phosphorus is taken up from soil by plants and unless the soil contains adequate phosphorus or it is supplied to soil from external sources, plant growth will be restricted. It does not occur as abundantly in soils as do other macronutrients, such as nitrogen or potassium. Total phosphorus concentration in most soils varies between about 90 and 2225 lb/A and averages about 890 lb/A in the surface 8 in. of soil. Young virgin soils in low-rainfall areas are usually high in total phosphorus. The average phosphorus content of virgin U.S. soils to a depth of 1 ft is shown in Figure 6-1. Low phosphorus levels in uncultivated soils of the humid Southeast are contrasted with the high level of native phosphorus in the soils of the prairie and western states.

The portion of soil phosphorus accessible for plant growth can be quite another story. Because little of this element is lost by leaching and because crop removals are generally small, phosphorus added as fertilizer can accumulate in the surface horizon of soils, particularly in those used for production of heavily fertilized crops such as potatoes, vegetables, and citrus. On the other hand, plant-accessible soil phosphorus levels can undergo a gradual decline where crop removal exceeds the amounts being returned to the soil in fertilizers, animal manures, and crop residues.

Although soils of the prairie and western states are often high in total phosphorus, many of them are characteristically low in the portion utilizable for crop growth. This condition is often aggravated by unfavorable environmental conditions such as low soil moisture, low soil temperatures early in the growing season, and high soil temperatures later when crops are heading and ripening.

Although the fraction of soil phosphorus utilizable for crop growth has been designated as "available phosphorus," this term is also used to refer to that portion of soil phosphorus extracted by various solvents. Extractants include water, dilute

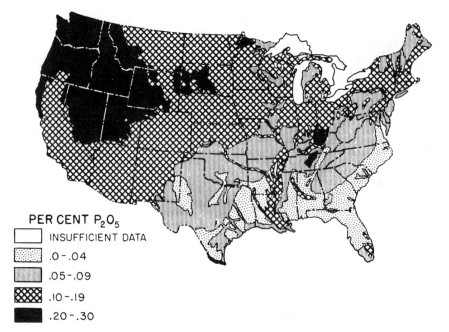

FIGURE 6-1. Phosphate content in the surface foot of soils in the United States. (Pierre and Norman, Eds., *Agronomy*, Vol. 4, p. 401, 1953. Reprinted with permission of Academic Press, Inc., New York.)

PER CENT P_2O_5

☐ INSUFFICIENT DATA

▒ .0 - .04

▥ .05 - .09

▩ .10 - .19

■ .20 - .30

acids or alkalies, and salt solutions. To be most effective in predicting the likelihood of a response to fertilizer phosphorus, the extractants should reflect contributions of the major factors affecting the supply of phosphorus to plants. Two of these factors are the amount of soil phosphorus (quantity) and the concentration of soil solution phosphorus (intensity). In practice, however, the commonly used extractants will usually characterize only one of these factors.

The various techniques of soil analysis are discussed in detail in Chapter 12, which deals with soil fertility evaluation. It is enough to mention at this point that the total quantities of soil phosphorus are much greater than those of the available phosphorus, but the latter is of greater importance to plant growth.

Forms of Soil Phosphorus

The various forms of phosphorus in soils and their dynamic interrelationships are summarized in Figure 6-2. Soil phosphorus can be classed generally as organic or inorganic, depending on the nature of the compounds in which it occurs. The organic fraction is found in humus and organic materials which may or may not be associated with humus. The levels of organic phosphorus in soils vary enormously, ranging from virtually zero to over 0.2%. Organic phosphorus content in the surface horizons of cultivated chernozemic soils in western Canada range from 25 to 55% of the total phosphorus content, with microbial phosphorus accounting for upward of 10% of the organic phosphorus content. Figure 6-3 is a simplified picture of the phosphorus cycle in soils which shows the interchange of organic and microbial phosphorus with solution and inorganic forms of phosphorus.

As shown in Figure 6-2, the inorganic fraction of soil phosphorus occurs in numerous combinations with iron, aluminum, calcium, flourine, and other elements. Solubility of these compounds in water varies from sparingly soluble to

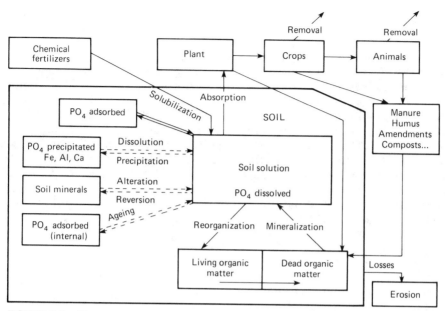

FIGURE 6-2. Phosphorus dynamics system in the soil. [Modified after Gachon, *Bull. Assoc. Fr. Etude Sol,* **4**:17 (1969).]

very insoluble. Phosphates also react with clays to form generally insoluble clay–phosphate complexes.

The inorganic phosphorus content of soils is frequently higher than that of organic phosphorus. An exception to this rule would, of course, be the phosphorus contained in predominantly organic soils. In addition, the organic phosphorus content of mineral soils is usually higher in the surface horizon than it is in the

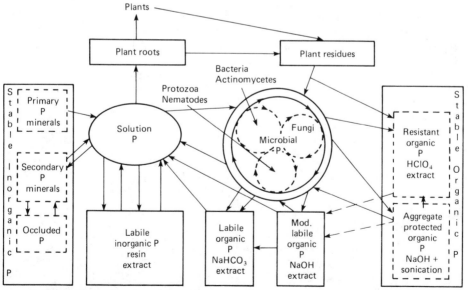

FIGURE 6-3. Schematic illustration of the measurable components of the phosphorus cycle in soils, showing the interchange of solution, organic, and microbial forms. [Chauhan et al., *Can. J. Soil Sci.,* **61**:373 (1981).]

subsoil because of the accumulation of organic matter in the upper part of the soil profile.

Soil Solution Phosphorus. It was mentioned in Chapter 3 that phosphorus is absorbed by plants largely as the primary and secondary orthophosphate ions ($H_2PO_4^-$ and HPO_4^{2-}), which are present in the soil solution. The amount of each form present depends largely on soil solution pH. At pH 7.22 there are approximately equal amounts of $H_2PO_4^-$ and HPO_4^{2-}. Below this pH, $H_2PO_4^-$ is the main form and it is predominant in many agricultural soils. The secondary orthophosphate ion (HPO_4^{2-}) becomes most important at pH values above 7.22. Plant uptake of HPO_4^{2-} is much slower than it is with $H_2PO_4^-$. Some soluble organic phosphate compounds may also be absorbed, but generally they are of minor importance under most soil conditions.

Concentration of the orthophosphate ions in soil solution and maintenance of suitable concentrations of them are of the greatest importance to plant growth. Absorption rates for $H_2PO_4^-$ by plant roots can be described by Michaelis–Menten kinetics. Figure 6-4 is an example of a relationship developed from Michaelis–Menten kinetics showing the dependence of phosphorus uptake by 18-day-old corn on phosphorus concentration in solution.

The actively absorbing surface of plant roots is the young tissue near the root tips. Relatively high concentrations of phosphorus accumulate in root tips followed by a zone of lesser accumulation where cells are elongating, then a second region of higher concentration where the root hairs are developed. Rapid replenishment of phosphorus in the soil solution is particularly important in the soil areas where roots are actively absorbing phosphorus.

The required concentration of phosphorus in the soil solution depends primarily on the crop species being grown and the level of production desired. Australian researchers indicate that concentrations of 0.2 to 0.3 ppm are adequate for a variety of crops. Fox of the University of Hawaii refers to the concentration of phosphorus in soil solution needed by crops as the external phosphorus requirement. Table 6-1 lists the external phosphorus requirement for several crops. Maximum corn grain yields may be obtained when solution concentrations are as low as 0.01 ppm if the yield potential is low, but high yield potential is associated with a level of about 0.025 ppm. The requirement for wheat is, on average, slightly greater than for corn. Sorghum is believed to have a requirement similar to that of corn. Soybean has a much higher requirement than corn.

Optimum solution concentrations of phosphorus are probably not constant for

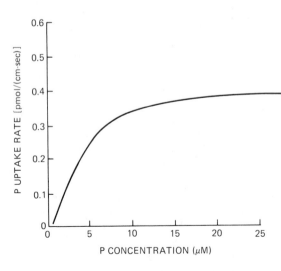

FIGURE 6-4. Relation between phosphorus concentration in solution and net phosphorus influx for 18-day-old corn plants. [Classen and Barber, *Plant Physiol.*, **42**:1407 (1974).]

a specific crop. Stage of growth and the occurrence of stress caused by disease and adverse climatic conditions are factors that are expected to substantially modify the desirable amounts of soil solution phosphorus.

Pathways responsible for movement of phosphorus to plant roots were discussed in Chapter 4. Mass or convective flow is considered generally to make only minor contributions to the phosphorus supply of plants. On the other hand, diffusion accounts for most of the phosphorus transport from soil to roots.

Mass Flow. The following example shows that mass flow of phosphorus to plants growing in low-phosphorus soils will normally provide only a small portion of the phosphorus requirements of crops. Water absorption by plants which results in convective or mass flow of phosphorus can be calculated by using the transpiration ratio or the weight of water transpired per unit weight of plant produced. Assuming a value of 400 for the transpiration ratio and a phosphorus concentration of 0.2% in the crop, an average phosphorus concentration of 5 ppm in the soil solution would be needed to provide all the phosphorus to the plant by mass flow.

Phosphorus concentration of the soil solution in fertile arable soils is about 0.3 to 3.0 ppm. In unfertilized soils it is around 0.05 ppm and is seldom higher than 0.3 ppm. If a value of 0.05 ppm is considered typical for soil solution phosphorus in unfertilized soils, mass flow would supply on the average only 1% of the phosphorus needed by plants. In fertilized soil with a soil solution concentration of 1 ppm of phosphorus, mass-flow contributions to the phosphorus nutrition of crops would approach 20% of the total requirement.

The very high phosphorus concentrations that exist temporarily in and near fertilizer bands are expected to encourage further phosphorus uptake by mass flow, as well as by diffusion. For example, a phosphorus concentration of 56 ppm has been measured near granules of monocalcium phosphate five weeks after application to a calcareous silty clay loam soil. Water-soluble phosphorus concentrations ranging between about 2 and 14 ppm have been found to occur in soil–fertilizer reaction zones.

Diffusion. The most important mechanism involved in the movement of phosphorus to absorbing roots is diffusion. With the exception of soils extremely high in phosphorus, it is responsible for the major part of the phosphorus reaching

TABLE 6-1. Estimated Concentration of Phosphorus in Soil Solution Associated with 75% and 95% of Maximum Yield of Selected Crops

Crop	Location(s) (Soil)	Approximate P in Soil Solution for Yield Indicated (ppm)	
		75% of Max.	95% of Max.
Cassava	Hawaii (Halii)	0.003	0.005
Peanuts	Hawaii (Halii)	0.003	0.01
Corn	Hawaii (Honokaa, Wahiawa), Nigeria (Ikenne), Southeastern United States (various)	0.008	0.025
Wheat *	India, Nebraska, Pakistan	0.009	0.028
Cabbage	Hawaii (Kula)	0.012	0.04
Potatoes	Bangladesh, Idaho, Hawaii (Kula), Ontario, Peru	0.02	0.18
Soybeans	Hawaii (Honokaa, Wahiawa)	0.025	0.20
Tomatoes	Hawaii (Kula)	0.05	0.20
Head lettuce	Hawaii (Kula, Wahiawa)	0.10	0.30

* Unpublished data of K. S. Memon, University of Hawaii.
Source: Fox, *Better Crops Plant Food,* **66:**24 (Winter 1981–1982).

roots. As noted in Chapter 4, a number of soil factors can profoundly influence phosphorus diffusion. The principal factors are (1) percentage by volume of the soil that is occupied by soil water, (2) tortuosity of the diffusion path, (3) the phosphate buffering capacity of the soil, and (4) temperature.

Maintenance of a suitable concentration of phosphorus in the soil solution (phosphorus intensity) depends on solid phase phosphorus going into solution to replace the amounts withdrawn by plant uptake. The quantity of solid phase phosphorus that acts as a reserve is frequently referred to as the capacity or quantity factor. The latter term is preferred since the present trend is to use capacity for describing gradients that relate quantity to intensity. Rate of organic matter formation and decomposition also influences phosphorus intensity or soil solution concentration. The equilibria that exist between the various components of the phosphorus quantity factor and phosphorus intensity are evident in Figure 6-2.

Description of soil phosphorus can be reduced to the following relationship, which simply states that

$$\text{soil solution} \rightleftharpoons \text{labile soil P} \rightleftharpoons \text{nonlabile P}$$

Labile soil phosphorus is the readily available portion of the quantity factor and it has a high dissociation rate, permitting rapid replenishment of solution phosphorus. Depletion of labile soil phosphorus usually causes nonlabile phosphorus to become labile again, but at a very slow rate. Thus it is apparent that the soil phosphorus quantity factor is comprised of both labile and nonlabile forms of phosphorus.

The equilibria shown in Figure 6-2 and in the relationship above can be temporarily disrupted by the addition of soluble phosphate fertilizers, by immobilization of soluble phosphorus by microorganisms, and by rapid mineralization of soil organic matter resulting from plowing and cultivation.

The native phosphorus in soils originated largely from the disintegration and decomposition of rocks containing the mineral apatite, $Ca_{10}(PO_4)_6(F, Cl, OH)_2$. Phosphorus is found in the soil as finely divided fluorapatite, hydroxyapatite, or chlorapatite, as iron or aluminum phosphates, as some of the compounds shown on page 199, or in combination with the clay fraction. It also occurs in combination with humus and other organic fractions.

The nature and reactions of phosphorus with the organic and inorganic soil components and the effect of added fertilizer phosphorus on the phosphate equilibria in soils are covered in the following sections.

Organic Soil Phosphorus.   The nature and reactions of organic soil phosphorus are not as well understood as are those of inorganic soil phosphorus. Nonetheless, some knowledge of this topic will give a better understanding of the overall role of phosphorus in soil fertility. It is in this light that the following brief summary is offered.

Forms of Organic Soil Phosphorus.   Most naturally occurring organic forms of phosphorus are esters of orthophosphoric acid and numerous mono- and diesters have been characterized. These organic phosphorus esters have been identified in five classes of compounds: inositol phosphates, phospholipids, nucleic acids, nucleotides, and sugar phosphates. The first three are the dominant groups. Black of Iowa State University, in summarizing what was known in 1968 of the chemical nature of organic phosphorus, concluded that about 2% of the total was present in nucleic acids, 1% in phospholipids, and 35% in inositol phosphates, leaving about 62% unaccounted for.

Inositol Phosphates.   Inositol is a homocyclic sugarlike compound, $C_6H_{12}O_6$, which can form a series of phosphate esters ranging from monophosphates up to a hexaphosphate. Phytic acid (myoinositol hexaphosphoric acid) is the most common ester of this group found in soils. The inositol phosphates are released from organic substances in soil at a much slower rate than many other esters, but they are quickly stabilized and can accumulate in some soils to the extent that they account for more than half of the organic phosphorus and about one-fourth of the total phosphorus. In a wide range of soils from the United States, Canada, and Australia, inositol hexaphosphates, or the penta- and hexaphosphates together, constituted up to nearly 20% of the soil organic phosphorus.

Inositol hexaphosphate forms a number of very insoluble salts through reactions with iron and aluminum under acid conditions and with calcium in alkaline solutions. It also forms strong complexes with proteins and with some metal ions. In these various precipitates and complexes, inositol hexaphosphate is more resistant to enzyme attack than are the more soluble ester salts.

Clay minerals such as montmorillonite and finely divided sesquioxides will strongly adsorb inositol hexaphosphate. Other myoinositol phosphates are sorbed, with the degree of sorption decreasing with declining numbers of phosphate groups. Sorption of inositol penta- and hexaphosphates can occur at the same active sites involved in sorption of inorganic orthophosphate ions.

Concentrations of inositol phosphates in soils vary widely. Values ranging from 2 to 54 ppm in tea soils to 100 to 400 ppm in Scottish agricultural soils have been recorded.

Nucleic Acids. Two distinct chemical forms of nucleic acids, ribonucleic acid and deoxyribonucleic acid, occur in all living things. Each consists essentially of a chain of sugar units, either ribose or deoxyribose, joined by phosphate ester bridges. A nitrogenous base derived from either purine or pyrimidine is attached to each sugar molecule. Nucleosides are units containing only one molecule of sugar linked to one molecule of the nitrogenous base. Phosphate derivatives of nucleosides are called nucleotides.

Nucleic acids are probably released into soil much more rapidly than the inositol phosphates and they are broken down more quickly. Because it has not been possible to isolate pure nucleic acids from soils, its measurement has usually been based on the amounts of nucleotides or purine and pyrimidine derivatives liberated by hydrolysis of soil organic-matter fractions. At one time it was believed that at least half of the organic phosphorus in soils was present in nucleic acids, but the introduction of specific methods of identification and measurement has resulted in much lower values.

Concentration of nucleic acid phosphorus in two Iowa soils was found to range between 1.2 and 6 ppm, which was equivalent to 0.2 and 1.8% of the total organic phosphorus. In a selection of Scottish soils the nucleic acid phosphorus concentration varied from 5 to 19 ppm and represented up to 2.4% of the soil organic phosphorus.

Phospholipids. Soil phospholipids are actual or potential esters of fatty acids containing phosphorus. Some of the most common phospholipids are derivatives of glycerol. Phosphatidylcholine (lecithin) and phosphatidylethanolamine are the predominant phospholipids in soils. The rate of release of phospholipids from organic sources in soils is rapid.

Phospholipid levels between 0.2 and 14 ppm were measured in Saskatchewan soils. These values constituted less than 5% of the organic phosphorus. Considerably more, as much as 14%, of the organic phosphorus in the B horizon of a Chernozem soil in Alberta occurred as phospholipids.

Other Esters. Much of the remainder of organic phosphorus in soils is believed to originate from microorganisms, especially from bacterial cell walls, which are known to contain a number of very stable esters.

All of the organic phosphorus in soils may not be intimately associated with the humus fraction of the organic matter. Workers in Australia, reviewing this complex topic of organic phosphorus, point out that, unlike carbon or nitrogen, all of the organic phosphorus can be fairly easily removed from soil by alkaline extracting reagents.

Organic Phosphorus Turnover in Soils. In soils in or near equilibrium with their environment the relationship between the amount of nitrogen and the amount of carbon present is fairly close. Workers in Australia have shown that a similar relationship exists between nitrogen and organic sulfur. This clear-cut relation, however, does not occur between nitrogen and organic phosphorus.

Decomposition of organic phosphorus substances is not too different from that of organic nitrogen compounds since there are active and inactive forms of both elements. The active substances are primarily the portions of the residues that have not yet been transformed into microbial substances. The inactive forms of phosphorus behave similarly to the resistant forms of nitrogen in humic acid. Resistance of organic phosphorus compounds to biological conversion is probably due to different reasons than those for nitrogen substances.

Studies on the mineralization of organic phosphorus in soils are not numerous because of the uncertainty of identifying its origin, that is, from organic or inorganic forms, since both exist simultaneously. Further complicating such investigations is the rapid reaction of released organic phosphorus with various soil constituents to form insoluble compounds and complexes. Many of these reactions were described earlier in the discussion on the forms of organic soil phosphorus.

Evidence for the mineralization of organic phosphate is of three kinds. The first is based on the lowering of the organic phosphorus level in soils as a result of long-continued cultivation. It has been observed that when virgin soils are brought under cultivation the content of organic matter decreases. With this decrease in organic matter there is an initial increase in the level of extractable inorganic phosphate in soils, but within a few years this also decreases.

The second type of evidence is based on the results of short laboratory experiments in which decreases in the organic phosphorus content of soils are related to increases in the dilute acid extractable inorganic phosphate.

A third type of evidence has been advanced, which is based on monitoring levels of soil organic phosphorus in the presence and absence of plants and by examining seasonal changes. An acceleration in mineralization of organic phosphorus during cropping was observed by researchers in Iowa and in Japan. Seasonal effects on accumulation of organic phosphorus have been found in New Zealand pastures and in Dark Brown Chernozems in southern Alberta. Examples of the seasonal patterns of organic phosphorus encountered in the southern Alberta soils are given in Table 6-2. Organic phosphorus content decreased with crop growth and development and increased again with the onset of dormancy.

Phosphatase enzymes play a major role in the mineralization of organic phosphates in soil. They are a broad group of enzymes that catalyze the hydrolysis of both esters and anhydrides of phosphoric acid. There exists in soil a wide range of microorganisms which through their phosphatase activities have the capability of dephosphorylating (mineralizing) all known organic phosphates of plant origin. Phosphatase activity of a soil is due to the combined functioning of the soil microflora and any free enzymes present. Tabatabai at Iowa State University has investigated the presence and activity of many of these phosphatases in soils.

Several studies show the behavior of organic phosphorus to be not entirely analogous to that of organic carbon and nitrogen. Experimental work carried out in Iowa indicated that mineralization of organic phosphorus increased with rises in soil pH but that mineralization of organic carbon and nitrogen did not. These

TABLE 6-2.   Seasonal Patterns of Total Organic Phosphorus in the Top Unfertilized 0- to 15-cm Layer of Two Irrigated Alfalfa Fields in Southern Alberta

		Organic P in Soil (ppm)						
	Year	*Apr.*	*May*	*June*	*July*	*Aug.*	*Sept.*	*Oct.*
Site 1	1	88	51	43	43	56	72	73
	2	242	153	146	134	116	122	138
	3	213						
Site 2	1	102	73	56	36	20	46	82
	2	226	158	130	110	100	120	146
	3	206						

Source: Dormaar, *Can. J. Soil Sci.,* **52:**107 (1972).

same pH effects on mineralization in the field are indicated by the fact that the ratios of total organic carbon and total nitrogen to total organic phosphorus increased with soil pH in the same samples.

The mineralization of organic phosphorus has been studied in relation to the C/N/P ratio in the soil. A C/N/P ratio of 100:10:1 for soil organic matter has been suggested, but values ranging from 229:10:0.39 to 71:10:3.05 have been found. It is obvious that no one set of figures will describe the ratio for all soils. It has been suggested that if the carbon/inorganic phosphorus ratio is 200:1 or less, mineralization of phosphorus will occur, and that if the ratio is 300:1, immobilization will occur. Some Australian workers believe the N/P ratio to be closely tied in with mineralization and immobilization of phosphorus and suggest that the decreased supply of one results in the increased mineralization of the other. Thus, if nitrogen were limiting, inorganic phosphate might accumulate in the soil and the formation of soil organic matter would be inhibited. The addition of fertilizer nitrogen under such conditions could result in the immobilization not only of some of the accumulated inorganic phosphorus but also of some of the added fertilizer nitrogen.

A concentration of about 0.2% phosphorus is critical in the mineralization of organic phosphorus substances. If the system contains less than this, net immobilization takes place, as both the plant and the native soil inorganic phosphates are utilized by microorganisms.

Much has yet to be learned about the immobilization and mineralization of phosphorus in soils and its relation to the supplies of carbon, nitrogen, and sulfur present. These reactions could have an important bearing on fertilizer practices, but the extent of this influence is not known. It is probably safe to assume the following:

1. If adequate amounts of nitrogen, phosphorus, and sulfur are added to soils to which crop residues are returned, some of the added elements may be immobilized in fairly stable organic combination with carbon compounds.
2. Continued cropping of soils without the addition of supplemental nitrogen, phosphorus, and sulfur will result in the mineralization of these elements and their subsequent depletion in such soils.
3. If nitrogen, phosphorus, or sulfur is present in insufficient amounts, the synthesis of soil organic matter may be curtailed. All of these reactions presuppose the presence of adequate carbon and conditions conducive to the synthesis and breakdown of soil organic matter.

In spite of the fact that in some mineral soils one-half to two-thirds of the total phosphorus is organic, this important fraction is usually ignored in measurements of available soil phosphorus. Most soil extractants measure a proportion of the labile inorganic phosphorus pool only. When organic phosphorus does appear in soil extracts, it is usually deliberately removed before carrying out the phosphorus analysis.

Inorganic Soil Phosphorus. Dissolved phosphorus from fertilizer materials, in wastewater, and from indigenous soil sources reacts with soil constituents to create less soluble forms. Phosphorus thus removed from the solution phase is said to be retained or fixed. An understanding of the changes that take place is important in the management of fertilizer phosphorus and for other purposes such as land treatment of liquid wastes.

A number of mechanisms have been proposed to explain phosphorus retention. These include precipitation–dissolution reactions, sorption–desorption reactions, and immobilization–mineralization reactions. The slow precipitation–dissolution reactions occurring in the soil affected by fertilizer phosphorus additions have been studied extensively. More recently, the research focus has switched to the relatively fast reactions by which phosphorus is either adsorbed by or desorbed from soil solids with charged surfaces. The later investigators tend to conclude that sorption–desorption reactions are of greater significance than precipitation–dissolution reactions.

Before initiating a discussion on the various phosphorus retention reactions, it is desirable to define several terms. The first is sorption, which means the removal of phosphorus from solution and its retention at soil surfaces. When phosphorus is held at the surface of a solid, it is said to be adsorbed. If the retained phosphorus penetrates more or less uniformly into the solid phase, it is considered to be absorbed or chemisorbed. The less specific overall term "sorption" is often preferred because of the difficulty in distinguishing between these two reactions. The reverse reaction, desorption, relates to the release of sorbed phosphorus into solution. "Fixation" is a term frequently used to collectively describe both sorption and precipitation reactions of phosphorus.

There is considerable evidence supporting a wide range of sorption and precipitation mechanisms as causes of phosphorus retention, with no explicit consensus as to the relative magnitudes of their contributions. Many researchers view phosphorus retention as a continuous sequence of precipitation, chemisorption, and adsorption. With low phosphorus solution concentrations, adsorption seems to be the dominant mechanism.

Under field conditions where phosphorus fertilizers are applied to soils as granules or highly concentrated droplets, the adjacent soil is contacted by soil solutions containing very high concentrations of phosphorus and accompanying cations. Phosphorus-retention reactions begin in this environment, which often initially favors in situ precipitation of phosphorus compounds.

Phosphorus forms difficulty soluble compounds with Fe^{3+} and Al^{3+} at low pH, more soluble compounds with Ca^{2+} and Mg^{2+} at pH values near neutrality, and difficultly soluble compounds with Ca^{2+} at higher pH values. There is a wide range in solubility of these various phosphate compounds and their availability to crops is usually greatest within the pH range of about 6 to 7 for most agricultural soils.

Soil retention or fixation of phosphorus is not usually considered to be a problem when applications of 40 to 100 lb of P_2O_5 per acre are sufficient to correct deficiencies. On the other hand, if rates of 600 to over 2000 lb of P_2O_5 per acre are required to produce a satisfactory level of available soil phosphorus, retention of added fertilizer phosphorus must receive due attention in soil fertility management.

Retention of phosphorus is most frequently a problem in acid soils high in finely divided sesquioxides. Olsen and Watanabe of the U.S. Department of Agriculture (USDA) reported that representative acid soils of the United States fixed 2.17 times more phosphorus per unit surface area of soil than did a similar group of neutral or calcareous soils. They also found that the phosphorus fixed was held with five times more bonding energy in acid soils than in calcareous soils.

Precipitates from Soil–Fertilizer Reactions. A summary is given in Table 6-3 of some of the initial reaction products that might precipitate in the soil zones affected by applications of common phosphatic fertilizers. This list is probably far from complete since only a limited number of soils and fertilizer salts were investigated. The chemical properties of fertilizer salts and their mixtures vary so widely that formation of a great variety of compounds in soil systems is to be expected.

These compounds probably do not occur under equilibrium conditions and most will gradually revert to more stable forms. However, some are relatively stable and will persist for considerable periods of time. Formation and stability of various soil–fertilizer reaction products are discussed more fully later in this chapter.

Many of these inorganic phosphates have been synthesized by scientists at the Tennessee Valley Authority (TVA) and other institutions and they have also been evaluated in terms of their availability to crop plants (Table 6-4). In the soil they are formed in part by the reaction of fertilizer salts with certain soil components,

TABLE 6-3.   Summary of Compounds Formed from the Reaction of Phosphate Fertilizers with Soils or Soil Constituents

Compound	Mineral Name	Compound	Mineral Name
$AlPO_4 \cdot 2H_2O$	Variscite	$FePO_4 \cdot 2H_2O$	Metastrengite
		$Fe_3(PO_4)_2 \cdot 8H_2O$	Vivianite
$AlPO_4 \cdot 2H_2O$	Metavariscite	$FeNH_4(HPO_4)_2$	—
$Al(NH_4)_2H(PO_4)_2 \cdot 4H_2O$	—	$Fe_3NH_4H_8(PO_4)_6 \cdot 6H_2O$	—
$Al_2(NH_4)_2H_4(PO_4)_4 \cdot H_2O$	—	$Fe_3KH_8(PO_4)_6 \cdot 6H_2O$	—
$Al_5(NH_4)_3H_6(PO_4)_8 \cdot 18H_2O$	NH_4-taranakite	$Fe_2K(PO_4)_2OH \cdot 2H_2O$	K-leucophosphite
$AlNH_4PO_4 \cdot 2H_2O$	—	$MgHPO_4 \cdot 3H_2O$	Newberryite
$AlNH_4PO_4OH \cdot 3H_2O$	—	$Mg_3(PO_4)_2 \cdot 4H_2O$	—
$Al_2NH_4(PO_4)_2OH \cdot 2H_2O$	—	$Mg_3(PO_4)_2 \cdot 22H_2O$	—
$Al_2NH_4(PO_4)_2OH \cdot 8H_2O$	—	$MgNH_4PO_4 \cdot 6H_2O$	Struvite
$AlKH_2(PO_4)_2 \cdot H_2O$	—	$Mg(NH_4)_2(HPO_4)_2 \cdot 4H_2O$	Schertelite
$Al_5K_3H_6(PO_4)_8 \cdot 18H_2O$	K-taranakite	$Mg_3(NH_4)_2(HPO_4)_4 \cdot 8H_2O$	Hannayite
$Al_2K(PO_4)_2OH \cdot 2H_2O$	Leucophosphite	$MgKPO_4 \cdot 6H_2O$	—
$AlKPO_4OH \cdot 0.5H_2O$	—	$Mg_2KH(PO_4)_2 \cdot 15H_2O$	—
$AlKPO_4OH \cdot 1.5H_2O$	—	$Al(NH_4)_2P_2O_7OH \cdot 2H_2O$	—
$Al_2K(PO_4)_2(F,OH) \cdot 3H_2O$	Minyulite	$Ca_2P_2O_7 \cdot 2H_2O$	—
$CaHPO_4$	Monetite	$Ca_2P_2O_7 \cdot 4H_2O$	—
$CaHPO_4 \cdot 2H_2O$	Brushite	$Ca_3H_2(P_2O_7)_2 \cdot 4H_2O$	—
$Ca_8H_2(PO_4)_6 \cdot 5H_2O$	Octocalcium phosphate	$Ca(NH_4)_2P_2O_7 \cdot H_2O$	—
$Ca_{10}(PO_4)_6(OH)_2$	Hydroxyapatite	$Ca_3(NH_4)_2(P_2O_7)_2 \cdot 6H_2O$	—
$Ca_{10}(PO_4)_6F_2$	Fluorapatite	$Ca_5(NH_4)_2(P_2O_7)_3 \cdot 6H_2O$	—
$CaAlH(PO_4)_2 \cdot 6H_2O$	—	$CaNH_4HP_2O_7$	—
$CaAl_6H_4(PO_4)_3 \cdot 20H_2O$	—	$Ca_2NH_4H_3(P_2O_7)_2 \cdot 3H_2O$	—
$CaNH_4PO_4 \cdot H_2O$	—	$CaK_2P_2O_7$	—
$Ca(NH_4)_2(HPO_4)_2 \cdot H_2O$	—	$Ca_3K_2(P_2O_7)_2 \cdot 2H_2O$	—
$Ca_2NH_4H_7(PO_4)_4 \cdot 2H_2O$	NH_4-Flatt's salt	$Ca_5K_2(P_2O_7)_3 \cdot 6H_2O$	—
$Ca_2(NH_4)_2(HPO_4)_3 \cdot 2H_2O$	—	$Ca_2KH_3(P_2O_7)_2 \cdot 3H_2O$	—
$CaKPO_4 \cdot H_2O$	—	$CaNa_2P_2O_7 \cdot 4H_2O$	—
$CaK_3H(PO_4)_2$	—	$Fe(NH_4)_2P_2O_7 \cdot 2H_2O$	—
$Ca_2KH_7(PO_4)_4 \cdot 2H_2O$	K-Flatt's salt	$Mg(NH_4)_2P_2O_7 \cdot 4H_2O$	—
$CaFe_2H_4(PO_4)_4 \cdot 5H_2O$	—	$Mg(NH_4)_6(P_2O_7)_2 \cdot 6H_2O$	—
$CaFe_2H_4(PO_4)_4 \cdot 8H_2O$	—	$Mg(NH_4)_2H_4(P_2O_7)_2 \cdot 2H_2O$	—
$Ca_3Mg_3(PO_4)_4$	—	$Ca(NH_4)_3P_3O_{10} \cdot 2H_2O$	—
$FePO_4 \cdot 2H_2O$	Strengite		

Source: Sample et al., in F. E. Khasawneh et al., Eds., *The Role of Phosphorus in Agriculture*, p. 284. Madison, Wis.: American Society of Agronomy, 1980.

TABLE 6-4.   Comparative Value of Various Phosphate Compounds As Sources of Phosphorus for Plants

Phosphate Compound	Test Conditions			Relative P Uptake from Neutral Soils (Mono-calcium Phosphate = 100)
	Particle Size Mesh (U.S. NBS)	Application Rate (mg P/3 kg soil)	Crop and Number of Harvests	
Products from acid water-soluble orthophosphate fertilizers				
$CaHPO_4 \cdot 2H_2O$	−100	60	Maize 3	123 *
dicalcium phosphate	−40 + 60	60	Maize 3	102 *
dihydrate	−20 + 40	60	Maize 3	66 *
$CaHPO_4$	−100	26	Ryegrass 1, sudangrass 1	110
anhydrous dicalcium	−60 + 100	60	Wheat 1, oats 1	65
phosphate	−10 + 14	60	Wheat 1, oats 1	27
$Ca_4H(PO_4)_3 \cdot 3H_2O$ octocalcium phosphate	−100	60	Ryegrass 1, sudangrass 1	77
$Ca_{10}(PO_4)_6(OH)_2$ hydroxyapatite	−100	60	Ryegrass 1, sudangrass 1	9
$Ca_{10}(PO_4)_6F_2$ fluorapatite	−325	200	Maize 3	2
$CaFe_2(HPO_4)_4 \cdot 5H_2O$ calcium ferric phosphate	−200	200	Maize 3	83
$K_3Al_5H_6(PO_4)_8 \cdot 18H_2O$ potassium taranakite	−200	200	Maize 3	79
	−200	150	Maize 1	40
$(NH_4)_3Al_5H_6(PO_4)_8 \cdot 18H_2O$ ammonium taranakite	−200	150	Maize 1	10
$KFe_3H_8(PO_4)_6 \cdot 6H_2O$ acid potassium ferric phosphate	−200	200	Maize 3	6
$(NH_4)Fe_3H_8(PO_4)_6 \cdot 6H_2O$ acid ammonium ferric phosphate	−325	200	Maize 3	38
Aluminum phosphate (colloidal)	−200	200	Maize 3	74
Ferric phosphate	−200	200	Maize 3	71
(colloidal)	−325	200	Maize 3	92
$FePO_4 \cdot 2H_2O$ strengite	−325	200	Maize 3	3
Products from alkaline water-soluble orthophosphate fertilizers				
$MgNH_4PO_4 \cdot 6H_2O$	−35 + 60	60	Oats 1	138 *
struvite	−20 + 35	60	Oats 1	124 *
$K_3CaH(PO_4)_2$	−35 + 60	60	Oats 1	123 *
	−20 + 35	60	Oats 1	136 *
$KAl_2(PO_4)_2 \cdot OH2H_2O$ basic potassium aluminum phosphate	−200	200	Maize 3	3
$KFe_2(PO_4)_2OH \cdot 2H_2O$ basic potassium ferric phosphate	−200	200	Maize 3	0
Products from soluble pyrophosphate fertilizers				
$Ca(NH_4)_2P_2O_7 \cdot H_2O$	−60 + 100	60	Wheat 1, oats 1	108
	−10 + 14	60	Wheat 1, oats 1	54
$Ca_3(NH_4)_2(P_2O_7)_2 \cdot 6H_2O$	−60 + 100	60	Wheat 1, oats 1	54
	−10 + 14	60	Wheat 1, oats 1	25
$Ca_2P_2O_7 \cdot 2H_2O$	−60 + 100	60	Wheat 1, oats 1	15
calcium pyrophosphate	−10 + 14	60	Wheat 1, oats 1	7

* Data relative to − 100-mesh anhydrous dicalcium phosphate = 100.
Source: Lindsay et al., *Trans. 7th Int. Congr. Soil Sci.,* **3:**580 (1960).

and many are termed *phosphate reaction products*. They are discussed more fully in a subsequent section of this chapter. The data showing the availability of these materials to plants were obtained by comparing the uptake of phosphorus by plants receiving this element as mono- or dicalcium phosphate to the uptake of phosphorus by plants receiving the same amount in the form of these various compounds. Phosphorus uptake by plants receiving the mono- or dicalcium phosphate is assigned a value of 100. Those materials giving values greater than 100 were more available to plants than mono- or dicalcium phosphate. Those with values less than 100 were not so available. Other phosphorus compounds have been identified in soils, but those listed here are among the most important.

Phosphorus is converted to much less soluble forms in acid soils by sorption reactions with iron and aluminum compounds and crystalline and X-ray amorphous colloids of low silica–sesquioxide ratios. When the solubility of various iron and aluminum phosphates is exceeded, precipitates of these compounds will also form. In basic or high-pH soils, soluble phosphorus is subject to sorption on calcite and to precipitation as stable calcium and/or magnesium phosphate minerals.

The relative importance of these various phosphorus retention mechanisms varies with soil properties. Important aspects of these mechanisms are reviewed in the next sections.

Sorption Reactions. The surfaces on which phosphate ions enter into sorption reactions are of two main types. The first group are surfaces of constant charge, such as the crystalline clay minerals, which interact with phosphorus principally through the cations held tightly to their platelike surfaces. The second category involves surfaces of variable charge including the Fe(III) and aluminum oxides and organic matter, for which H^+ and OH^- determine the surface charge, and calcite, for which Ca^{2+} and CO_3^{2-} are responsible for charge development.

In addition to these two distinctive groups there are some clay minerals such as kaolinite which have pH-dependent charges on their crystal edges. Also, amorphous clay minerals such as allophane are intimately associated with a hydroxyaluminum gel, which has a pH-dependent charge. Further, organic matter, with a pH-dependent charge, reacts with phosphorus through its cations held by coulombic forces.

Sorption on Surfaces of Constant Charge. The change in surface charge properties resulting from the occurrence of Ca^{2+}, Al^{3+}, or aluminum and iron polymers close to the surface of crystalline clay minerals will result in phosphorus adsorption at the clay surface.

Sorption on Surfaces of Variable Charge. Hydrated iron and aluminum oxides are the most important surfaces of variable charge in most soils excepting peats and highly calcareous soils. These oxides are characterized by a surface of negatively charged OH groups which take up and dissociate protons according to their acid strength and the ambient pH. Thus they are amphoteric, having either negative, zero, or positive charge, depending on the pH. The pH at which there are equal numbers of positive and negative charges on the surface defines the point of zero charge (PZC) of the oxide, which is approximately 8.5 and 9, respectively, for aluminum and iron oxides. The edge faces of kaolinite have a PZC of about 7.

At pH levels below the PZC, phosphorus and other anions, such as SO_4^{2-} and $H_3SiO_4^-$, are attracted to the positively charged oxide surface. The following two reactions are examples of how phosphorus ions can enter a metal oxide surface by exchanging with the metal ligands, OH or OH_2^+ (Figure 6-5). One fundamental principle that must be observed in the study of phosphorus adsorption mechanisms

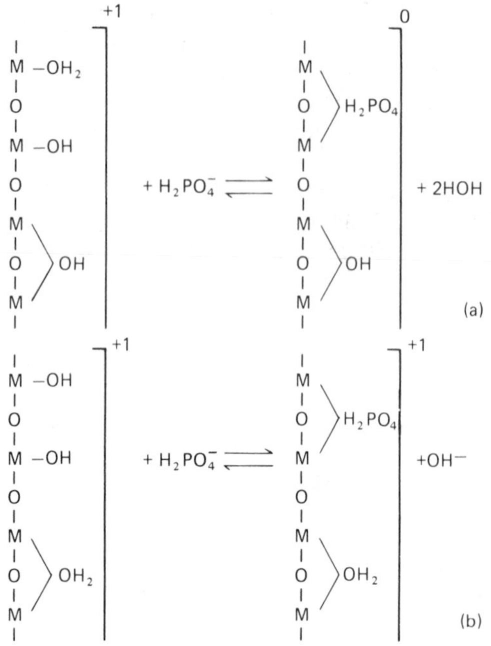

FIGURE 6-5. Displacement of (a) aquo and (b) hydroxy group from a metal oxide surface by phosphate. (Uehara and Gilman, *Chemistry, Physics and Mineralogy of Soil with Variable Charge Clays*, p. 44. Boulder, Colo.: Westview Press, 1981.)

is that charge balance must be maintained; that is,

$$eq/g \text{ P adsorbed} = eq/g \text{ OH}^- \text{ released (or g H}^+ \text{ adsorbed)}$$
$$+ \ eq/g \text{ positive charge lost}$$
$$+ \ eq/g \text{ negative charge gained by the surface}$$

Calcium Carbonate and Phosphorus Adsorption. Calcium carbonate occurring in soil primarily as the mineral calcite is a factor in phosphorus adsorption. This mineral develops a negative charge due to the greater tendency for Ca^{2+} ions than CO_3^{2-} ions to go into solution. Cations such as Mg^{2+} and the OH^- ion can be absorbed on calcite surfaces but more important, phosphorus ions become chemisorbed at the surface. This reaction of phosphorus with calcite surfaces apparently involves absorption of small amounts of phosphorus followed by precipitation of calcium phosphate at higher concentrations. Hydroxyapatite is formed in some adsorption reactions.

Adsorption of phosphorus by calcite can be described by the Langmuir isotherm, which will be discussed later. Only a small fraction of the calcite surface, in the order of 5%, is involved in reactions with phosphorus. As a consequence, calcite can continue to control solution pH while reacting with phosphorus.

The initial adsorption is thought to occur at certain sites where phosphate ions form surface clusters. These clusters of amorphous calcium phosphate act as nuclei for subsequent crystal growth. The kind of crystalline phosphate compounds that develop from these nuclei seems to depend on such factors as nature of the adsorbing calcite surface, solution conditions, and the time of reaction. Dicalcium phosphate, octacalcium phosphate, and hydroxyapatite are the principal crystalline phosphates that have been identified.

There are indications from research being reported by Freeman and Rowell of the University of Reading in England that there is a sequence in the adsorption and precipitation of phosphorus on to calcite. It appears that dicalcium phosphate first forms rapidly and then slowly changes to octacalcium phosphate. The octacalcium phosphate probably occurs as a coating on dicalcium phosphate.

Adsorption Relationships. Several adsorption equations have been developed to describe the removal of phosphorus from dilute solutions by soils or soil constitutents. The Freundlich, Langmuir, and Temkin equations, or modifications of these, have been used most frequently. These equations are helpful for understanding the processes involved and summarizing many results in a few numbers.

Adsorption isotherms are graphical representations of adsorption equations. A variety of isotherm shapes are possible, depending on the affinity of the soil or its components for phosphorus.

A brief description of the various equations follows. All of them are based on the fundamental equation

$$m = f(c) \tag{1}$$

where m is the quantity of phosphorus adsorbed at a phosphorus concentration equal to c.

The *Freundlich equation* is one of the first used in soil studies and it states that

$$m = AC^B \tag{2}$$

with m being the amount of phosphorus adsorbed per unit weight of soil, C denotes the phosphorus concentration in solution and A and B are constants that vary from soil to soil. According to the Freundlich equation, energy of phosphorus adsorption decreases as the amount of adsorption increases.

A second equation, the Langmuir equation, has been used to characterize phosphorus adsorption by soils and calcite. It and the Freundlich equation have been the most widely used for portraying phosphorus adsorption by soils. The most common form of the *Langmuir equation* is

$$m = \frac{ABc}{1 + Bc} \tag{3}$$

in which m and c are the same as in the Freundlich equation, A is the phosphorus adsorption maximum, and B is a constant related to bonding energy. This equation indicates that all increments of adsorbed phosphorus are held with the same bonding energy (constant B) and that there is a maximum adsorption capacity that will not be exceeded regardless of increasing concentrations. Neither of these conditions is usually met in soil–phosphorus systems.

The *Temkin equation* in the middle range of phosphorus adsorption is written as

$$\frac{m}{B} = \frac{RT}{E}\ln Fc \tag{4}$$

where E and F are constants and B, c, and m are the same as in the Langmuir equation. Similar to the Freundlich equation, this equation also implies that the energy of adsorption decreases with increasing amounts of adsorbed phosphorus.

In the application of these three equations it is assumed that equilibrium conditions exist, a state that seldom occurs in soil–phosphorus adsorption studies. It is also implied that some portion of the phosphorus adsorbed by soils is almost always irreversibly adsorbed. Despite these and other shortcomings, the three equations have been useful in characterizing the relationship between c and m over limited ranges in phosphorus concentration, usually below 100 ppm of phosphorus.

These adsorption equations, however, provide virtually no information about the mechanisms responsible for phosphorus adsorption. They are incapable of showing whether hydrous iron and aluminum oxides, silicate clays, calcite, or magnesite dominate the adsorption reactions. In addition, the equations do not indicate whether phosphorus adsorption involves replacement of either hydroxyl, silica, or bicarbonate.

Because of limited success with these equations for expressing phosphorus adsorption over a wide phosphorus concentration range, they have been modified to extend their scope. The "double" Langmuir equation,

$$m = \frac{A_1 B_1 c}{1 + B_1 c} + \frac{A_2 B_2 c}{1 + B_2 c} \tag{5}$$

was developed for use with soils that presumably contain two kinds of adsorption sites. The Freundlich equation was also extended by including a constant D to allow for phosphorus already present in the soil:

$$m = Ac^B - D \tag{6}$$

Recognizing that soils are complex with surfaces of mostly unknown heterogeneity and that it is unlikely for phosphorus adsorption to occur under well-defined conditions, Sibbesen of The Royal Veterinary and Agricultural University in Denmark proposed additional modifications to the Langmuir and Feundlich equations. In his proposed equations,

$$m = \frac{ABc^{-D}c}{1 + Bc^{-D}c} \tag{7}$$

$$m = Ac^{Bc^{-D}} \tag{8}$$

the "shape-governing" parameter, B, of the original equations is replaced by a shape-governing term

$$Bc^{-D}$$

the value of which decreases with increasing phosphorus concentration, c. D is a constant. Equation (8) was found to be the most suitable for describing phosphorus adsorption by soils. The earlier extension of the Freundlich equation [equation (6)] also agreed closely with experimental data.

Desorption Relationships. There is interest in knowing if phosphorus in firmly held or nonlabile forms can become available to plants. Desorption isotherms that characterize the release of adsorbed phosphorus have shown that the process is extremely slow. It is not usually completed in periods of only hours or days. The rate of desorption decreases with time according to the function $Kt^{0.3}$, where t is the period of desorption. Increasing the period of contact between soil and phosphorus reduces the value of K. In general, desorption appears to become very slow after about two days.

As might be expected, the extent of desorption is dependent on the nature of the adsorption complex at the surface of the hydrous iron and aluminum oxides. Formation of six-membered ring structures such as the one on the right in Figure 6-6 will prevent the desorption of phosphorus.

Solubility Diagrams. Many researchers have contended that the concentration of solution phosphorus in neutral and acid soils is controlled by the solubility equilibria of crystalline phosphorus compounds. Attempts to identify these discrete inorganic phosphates in soils by direct methods have been largely unsuccessful except in the case of soil–fertilizer reactions products or where soil components such as metal oxides and clay minerals were reacted with concentrated phosphorus solutions.

Solubility relationships is one of the indirect methods used to identify phosphate minerals in soils. Soluble phosphorus concentrations above a compound's isotherm

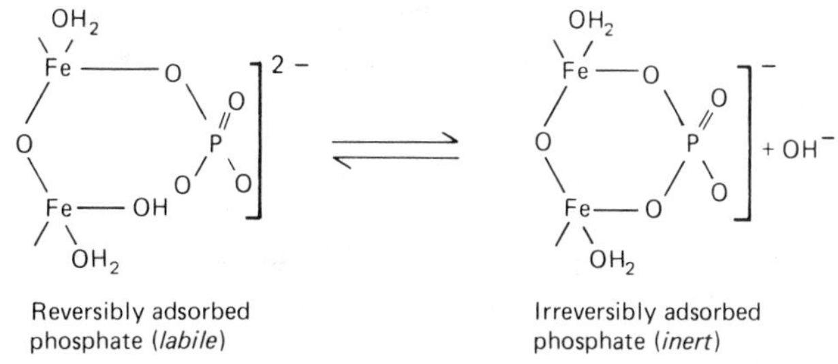

FIGURE 6-6. Reversible and irreversible adsorption of phosphorus on hydrous iron oxide. [Hingston et al., *J. Soil Sci.*, **25**:16 (1974).]

line signify supersaturation with respect to that compound, indicating that precipitation of the compound is possible. Levels of soil solution phosphorus below the compound's isotherm indicate undersaturation of the soil solution with respect to the compound, so that the compound, if present, would be expected to dissolve. Intersection of two isotherms indicates solutions containing both compounds.

The least soluble (or most stable) compounds at acid pH are variscite ($AlPO_4 \cdot 2H_2O$) and strengite ($FePO_4 \cdot 2H_2O$). They are expected to feature prominently in phosphorus retention in acid soils. Strengite is known to crystallize more rapidly than variscite and as a result phosphorus availability is expected to decline most rapidly when the iron phosphate is formed. The less crystalline aluminum phosphate has a greater surface area which is more favorable for release of phosphorus into the soil solution.

In soils containing large amounts of active magnesium, a number of insoluble magnesium phosphate compounds such as dimagnesium phosphate trihydrate, trimagnesium phosphate, and struvite may form. However, these magnesium phosphates are more soluble than dicalcium phosphate and octacalcium phosphate.

Factors Influencing Phosphorus Retention in Soils

Several of the factors influencing phosphorus retention in soils are apparent from the preceding discussion on the mechanisms involved in these reactions. Others are not. Because of the importance of retention and fixation in modifying the effectiveness of applied fertilizer phosphorus, these factors, and the extent to which they influence fixation, are briefly considered. The various factors fall into the following groupings: (1) nature and amount of soil components, (2) pH, (3) other ions, (4) kinetics, and (5) saturation of the sorption complexes.

Nature and Amount of Soil Components. Sorption–desorption reactions are affected by the type of surfaces contacted by phosphorus in the soil solution.

Hydrous Metal Oxides of Iron and Aluminum. These substances, particularly hydrous ferric oxide gel, have the capacity to sorb very large amounts of phosphorus. Although present in most all soils, they are most abundant in weathered soils.

Aluminum and iron oxides and their hydrous oxides can occur as discrete particles in soils or as coatings or films on other soil particles. They also exist as amorphous aluminum hydroxy compounds between the layers of expandable aluminum silicates.

It is generally accepted that in soils with significant contents of iron and aluminum oxide, the less crystalline the oxides are, the larger their phosphorus fixation capacity because of greater surface areas. However, crystalline hydrous metal oxides are usually capable of retaining more phosphorus than layer silicates.

Type of Clay. Phosphorus is retained to a greater extent by 1:1 than by 2:1 clays. Soils high in kaolinitic clays, such as those found in areas of high rainfall and high temperatures, will fix or retain larger quantities of added phosphorus than those containing the 2:1 type. The greater amount of phosphorus fixed by 1:1 clays is probably largely due to the higher amounts of hydrated oxides of iron and aluminum associated with kaolinitic clays. In addition, kaolinite develops pH-dependent charges on its edges which can enter into sorption reactions with phosphorus.

It is known that clays such as kaolinite with a low SiO_2/R_2O_3 ratio will fix larger quantities of phosphorus than will clays with a high ratio. Kaolinite, with a 1:1 silica/alumina lattice, has a larger number of exposed hydroxyl groups in the gibbsite layer which can be exchanged for phosphorus.

Figure 6-7 clearly shows the influence of clay mineralogy on phosphorus retention in three soils from Hawaii, all containing more than 70% clay. Very little phosphorus was retained by the Haplustoll soil, comprised mainly of montmorillonite and kaolinite, with only a minor component of iron and aluminum oxides.

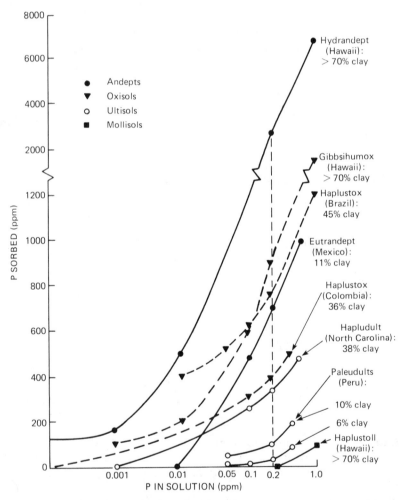

FIGURE 6-7. Examples of phosphorus sorption isotherms determined by the method of Fox and Kamprath. (Sanchez and Uehara, in F. E. Khasawneh, E. C. Sample, and E. J. Kamprath, Eds., *The Role of Phosphorus in Agriculture*, p. 480. Madison, Wis.: American Society of Agronomy, 1980.)

The Gibbsihumox soil, consisting primarily of well-crystallized iron and aluminum oxides, gibbsite, geothite, and hematite, had considerably more phosphorus retention capacity. Greatest phosphorus fixation occurred with the Hydrandept soil composed principally of X-ray amorphous colloids and finely divided gibbsite and geothite.

Clay Content. Soils containing large quantities of clay will fix more phosphorus than those comprised of small amounts of clay. In other words, the more surface area exposed with a given type of clay, the greater the tendency to retain phosphorus.

The effect of increasing clay content, in soils of similar clay mineralogy, on interaction with phosphorus is demonstrated in Figure 6-7. For example, compare the two Paleudult soils from Peru, one containing 10% clay and the other 6% clay, and the Hapludult soil from North Carolina with 38% clay. A similar relationship is evident in the following grouping of soils: Haplustox (Columbia) 36% clay, Haplustox (Brazil) 45% clay, and Gibbsihumox (Hawaii) >70% clay.

Amorphous Colloids. In young soils such as the Andepts, texture is often meaningless because of the presence of large quantities of X-ray amorphous colloids. Phosphorus retention in these soils is closely related to the content of X-ray amorphous colloids and with surface area. Volcanic ash is an important parent material of such soils and its weathering imparts a high phosphorus-sorption capacity to soils.

Amorphous aluminosilicate minerals like allophane have a large negative charge which is partly or entirely balanced by complex aluminum cations. Phosphorus becomes adsorbed by reacting with these aluminum cations.

Calcium Carbonate. A minor portion of the phosphorus sorption capacity of soils originates from calcium carbonate. Much of the adsorption attributed to it, however, may actually be due to hydrous ferric oxide impurities.

The amount and reactivity of calcium carbonate will influence phosphorus fixation. Impure calcites and those of high specific surface result in more adsorption of phosphorus and more rapid formation of calcium phosphate precipitates.

The activity of the phosphorus will be lower in those soils that have a high Ca^{2+} activity, a large amount of highly reactive calcium carbonate, and a large amount of calcium-saturated clay. Conversely, in order to maintain a given level of phosphate activity in the soil solution, it is necessary to add larger quantities of phosphate fertilizers to such soils. Workers at Colorado have studied phosphate equilibria in calcareous soils intensively. The results of one study, which illustrates the effect of texture and amount of added concentrated superphosphate (CSP) on phosphate activity, are shown in Figure 6-8. The phosphorus activity is expressed in terms of the mean activity of dicalcium phosphate in the soil solution. Two important points are illustrated in this figure. First, in any one soil the phosphate activity in the soil solution increased with additions of monocalcium phosphate. Second, larger additions of superphosphate were required to reach a given level of phosphate activity in fine-textured soils than were needed to reach the same activity in coarse-textured soils. Because a high level of phosphate activity is associated with a high level of phosphate availability to plants, the points illustrated in Figure 6-8 are of obvious practical concern.

Effect of pH. Soil pH has a profound influence on the amount and manner in which soluble phosphorus becomes fixed. Adsorption of phosphorus by iron and aluminum oxides declines with increasing pH. Gibbsite [γ-Al(OH)$_3$] adsorbs the greatest amount of phosphorus at between pH 4 and 5. Phosphorus adsorption by goethite (α-FeOOH) decreases steadily between pH 3 and 12.

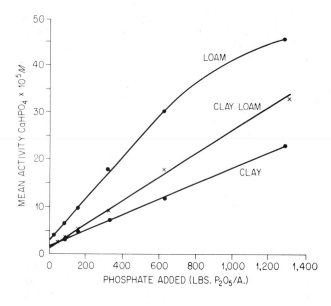

FIGURE 6-8. Phosphorus solubility (the mean activity of dicalcium phosphate in solution) as a function of the amounts of CSP added to three calcareous soils of different texture. [Cole et al., *SSSA Proc.*, **23**:119 (1959).]

Raising the pH of clays above 4 can increase phosphorus adsorption due to the increase in hydrolysis of Al^{3+}. Raising the pH may, however, reduce phosphorus adsorption where interlayer hydroxyaluminum polymers exist, as in many vermiculites.

Phosphorus availability in most soils is at a maximum in the pH range 6.0 to 6.5. At low pH values the retention results largely from the reaction with iron and aluminum and their hydrous oxides. As the pH increases, the activity of these reactants is decreased until, within the pH range just given, the activity of phosphorus is at a maximum. Above pH 7.0 the ions of calcium and magnesium, as well as the presence of the carbonates of these metals in the soil, cause precipitation of the added phosphorus, and its availability again decreases.

The pattern of phosphorus availability discussed is generally the one observed. If, however, a soil were basic because of the presence of cations such as sodium rather than calcium, a decrease in phosphorus availability would not necessarily be observed with a continuing increase in the soil pH. The presence of calcium or magnesium ions must accompany high pH values if there is to be a continued decrease in the solubility of soil phosphorus.

As will be discussed in Chapter 11, liming acid soils generally increases the solubility of phosphorus. Overliming can depress phosphorus solubility, however, due to the formation of more insoluble calcium phosphates similar to those occurring in basic soils naturally high in calcium.

Cation Effects. Adsorption of phosphorus by soils is influenced by the species and concentration of cations in the system. Divalent cations enhance phosphorus sorption relative to monovalent cations. For example, clays saturated with Ca^{2+} can retain greater amounts of phosphorus than those saturated with sodium or other monovalent ions. Current explanations for this effect of Ca^{2+} involve making positively charged edge sites of crystalline clay minerals more accessible to the phosphorus anions. This action of Ca^{2+} is possible at pH values slightly less than 6.5, but in soils more basic than this, dicalcium phosphate and other more basic calcium or magnesium phosphates would probably be directly precipitated from solution.

Concentration of exchangeable aluminum is also an important factor in phosphorus retention in soils since 1 meq of exchangeable aluminum per 100 g of soil when completely hydrolyzed may fix up to 102 ppm of phosphorus in solution. The following illustrates one of the possible ways that hydrolyzed aluminum can fix soluble phosphorus.

Cation Exchange:

$$\begin{bmatrix} \text{clay surface} \\ \end{bmatrix} \begin{matrix} Al^{3+} \\ \\ Al^{3+} \end{matrix} + 3Ca^{2+} \rightleftharpoons \begin{bmatrix} \text{clay surface} \\ \end{bmatrix} \begin{matrix} Ca^{2+} \\ Ca^{2+} \\ Ca^{2+} \end{matrix} + 2Al^{3+}$$

Hydrolysis:

$$Al^{3+} + 2H_2O \rightleftharpoons Al(OH)_2^{+} + 2H^{+}$$

Precipitation and/or Adsorption:

$$Al(OH)_2^{+} + H_2PO_4^{-} \rightleftharpoons \underline{Al(OH)_2H_2PO_4} \qquad (K_{sp} = 10^{-29})$$

Strong correlations between phosphorus adsorption and exchangeable aluminum have been reported when there is appreciable hydrolysis of aluminum. The extent to which phosphates may react with exchangeable aluminum is shown in Figure 6-9.

Anion Effects. Both inorganic and organic anions can compete in varying degrees with phosphorus for sorption sites, resulting in some cases in a decrease in the sorption of added phosphorus or a desorption of retained phosphorus. Weakly held inorganic anions such as nitrate and chloride are of little consequence, whereas specifically sorbed anions and acids such as hydroxyl, silicic acid, sulfate, and molybdate can be competitive. The strength of bonding of the anion with the sorption surface determines the competitive ability of that anion. For example, sulfate, even though it is considered to be specifically sorbed, is unable to desorb much phosphate. Apparently, phosphate is capable of forming a stronger bond at the surface than is sulfate.

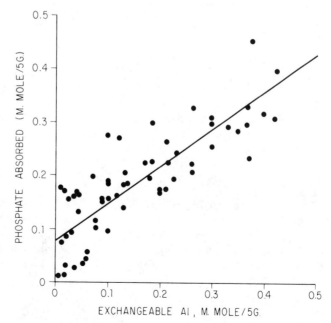

FIGURE 6-9. Effect of exchangeable aluminum on the amount of phosphorus adsorbed by suspended clay. [Coleman et al., *Soil Sci.*, **90**:1 (1960). Reprinted with permission of The Williams & Wilkins Company, Baltimore.]

Organic anions from various sources, such as organic waste materials and waste-water treatment, can affect the phosphorus sorption–desorption reactions in soils. The impact of organic anions on reduction of phosphorus sorption is related to their molecular structure and the pH of the system. It has been found that organic anions which form stable complexes with the iron and aluminum of soil components are particularly effective in reducing phosphorus sorption. Citrate has been found to form such complexes.

Anions of tricarboxylic acids are more effective in reducing phosphorus sorption than are those of dicarboxylic or monocarboxylic acids. Oxalate, citrate, and poly-galacturonate have been shown to be specifically sorbed at soil surfaces in a similar manner to phosphate.

Some of these effects of organic anions on phosphorus retention are probably partially responsible for the beneficial action of organic matter on phosphorus availability, which is discussed in a later section.

Saturation of the Sorption Complex. Amounts of phosphorus sorbed by soil are dependent on the saturation of the sorption complex or the number of sites available for reaction with the added phosphorus. Desorption of phosphorus is similarly strongly influenced by the extent of saturation of the sorption complex. Ease of desorption is greater at higher saturations because phosphorus is held less tightly with increasing surface coverage.

It has been found in North Carolina, for example, that there is a pronounced relationship between the amount of fixation of added fertilizer phosphorus and the R_2O_3/P_2O_5 ratio of the soil. The ratio R_2O_3/P_2O_5 is a measure of the amount of phosphorus present in relation to the iron and aluminum oxide content of the soil. A wide ratio indicates a small amount of phosphorus present or a low phosphorus saturation value. Under such conditions larger amounts of added phosphorus are fixed than when the ratio is narrow.

The practical implications of this relationship are quite important. On soils that have been heavily phosphated for several years it should be possible (1) to reduce the amount of phosphate currently applied in the fertilizer, (2) to utilize to a greater extent the phosphorus in the soil, or (3) to effect a combination of (1) and (2).

Organic Matter. It has been convincingly shown by California research-ers that organic phosphorus compounds can move in soils to a greater depth than can dissolved inorganic phosphorus. A range of organic phosphorus compounds moved four to six times deeper into a clay loam soil than did inorganic phosphorus.

Olsen* of the USDA at Colorado State University found appreciable downward movement of phosphorus following the field application of manure to a silty clay loam soil. By the second year after manuring there was phosphorus buildup in the 12- to 24-in. soil depth.

Campbell and Racz of the University of Manitoba observed greater extractable phosphorus levels in soil, to depths of 4 to 5 ft, from a feedlot site (with large accumulations of manure) than in soil from an adjacent field. These Manitoba researchers also found that organic and inorganic phosphorus from manure extract moved more rapidly in soil than equivalent amounts of phosphorus supplied as monopotassium phosphate.

The effect on phosphorus availability of other compounds arising from the de-composition of organic residues has received considerable attention. Numerous workers have reported that humus extracts from soils have increased the solubility of phosphorus. This has been variously described as resulting from (1) the for-mation of phosphohumic complexes which are more easily assimilated by plants, (2) anion replacement of the phosphate by the humate ion, and (3) the coating of

*S. R. Olsen, USDA, Fort Collins, Colorado; personal communication, 1981.

sesquioxide particles by humus to form a protective cover and thus reduce the phosphate fixing capacity of the soil.

Workers at the Massachusetts Experiment Station have suggested that certain organic anions arising from the decomposition of organic matter will form stable complexes with iron and aluminum, thus preventing their reaction with phosphorus. It was further stated that these complex ions release phosphorus previously fixed by iron and aluminum by the same mechanism. The anions that are most effective in replacing phosphates are citrate, oxalate, tartrate, malate, and malonate, some of which may also occur as degradation products during organic matter decay.

In some instances phosphorus adsorption has been found to correlate with the organic carbon content of soils. At first it was believed that this adsorption involved an exchange of phosphorus anions with OH groups in the organic matter. Other results indicate that it is the aluminum and to a lesser extent the iron adsorbed by the organic colloids which are active in phosphorus adsorption.

It is generally agreed that the turning under of stable or green manures results in better utilization of phosphorus by subsequent crops. Part of this favorable effect is likely related to the decomposition of organic residues which is accompanied by the evolution of carbon dioxide. This gas, when dissolved in water, forms carbonic acid, which is capable of decomposing certain primary soil minerals. It has been shown that in calcareous soils carbon dioxide production plays an important role in increasing phosphate availability. It has been shown in neutral soils, also, and evidence suggests the importance of carbon dioxide in increasing the availability of phosphorus in acid soils. The pH range over which carbonic acid may be important is rather wide but its greatest effect in dissolving soil phosphorus is probably under slightly acid to alkaline conditions.

On the basis of the available evidence, it is clear that the addition of organic materials to mineral soils may increase the availability of soil phosphorus.

Soils that consist largely of quartz sand and soils which are primarily muck and peat are subject to leaching losses of added phosphate fertilizer. This was demonstrated by laboratory work in North Carolina in which monocalcium phosphate was added to columns of two soils, one a muck and the other a soil containing 90% coarse sand and 10% organic matter. A portion of the results of this study is shown in Table 6-5.

It is apparent from these data that there was very little retention of added phosphate by either the surface or subsurface horizon of this soil. This is due to the absence of aluminum and iron compounds which are largely responsible for the retention of mineral phosphates in acid soils. In fact, the North Carolina workers found that by adding $AlCl_3$ to the two soils the leaching of added phosphorus in these studies was almost completely stopped.

TABLE 6-5.   Leaching of Fertilizer Phosphorus from an Acid Organic (Muck) Soil

P added (mg/column)	pH	Exchangeable Al (meq/100 g)	P Leached (mg/column)	Fertilizer P Absorbed (mg/column)
		Surface Soil		
0	4.6	0.02	0.48	—
2.5	4.6	0.02	2.95	0.03
10.0	4.6	0.02	8.67	1.81
		Subsurface Soil		
0	3.3	—	0.54	—
10.0	3.3	—	10.22	0.32

Source: Fox and Kamprath, SSSA Proc., **35**:154 (1971).

TABLE 6-6. Influence of Temperature and Source of Phosphorus on P Soluble in 1:15 Soil Water Extracts of Soil–Fertilizer Reaction Zones

Source of P (15 mg P/ 200 g soil)	Fertilizer P in 1:15 Soil Water Extracts (%) (Mean of Three Replicates Averaged over Soils and Soil–Fertilizer Reaction Periods)			Phosphorus Source Means *
	Temperature (°C)			
	5	20	35	
MKP	52.4	35.2	23.5	37.0
MAP	40.7	26.0	17.0	27.9 a
DAP	39.8	24.4	18.3	27.5 a
MCP	24.3	16.2	9.5	16.7
DCPD	3.3	4.4	3.4	3.7
Temperature means *	32.1	21.2	14.4	
	SEM of temperature means			= ± 0.42
	SEM of P source means			= ± 0.54
	SEM of temperature × P source means			= ± 0.93

* Any two means not having letters in common are significantly different at the 1% level by the Duncan test. Any two means with the same letter are not significantly different.
Source: Beaton et al., *Soil Sci. Soc. Am. J.*, **29:**194 (1965).

Temperature. Temperature affects most physical processes and the speed of chemical reactions generally increases with a rise in temperature. High temperatures are expected to slightly increase the molar solubility of compounds such as apatite, hydroxyapatite, octacalcium phosphate, variscite, and strengite. Mineralization of phosphorus from soil organic matter or crop residues is dependent on soil biological activity and increases in temperature are expected to stimulate biological activity up to the optima for the predominant biological systems.

The dissolution of granules of water-soluble phosphorus and resultant reactions with soil components to produce less soluble reaction products are hastened by higher temperatures. The data summarized in Table 6-6 show that there was a 33% reduction in concentration of water-soluble phosphorus for each 15°C increase in temperature. Most studies are in agreement that phosphorus retention increases at higher temperatures.

The soils in warm regions of the world are generally much greater fixers of phosphorus than the soils of more temperate regions. These warmer climates also give rise to soils with higher contents of the hydrous oxides of iron and aluminum.

At Purdue University, samples of Bedford silt loam were incubated for several months at −20.5 and 2.7°C. The soils were then leached with water at two temperature levels, 32 and 16°C. Even though the temperature of the leaching water affected the amount of phosphorus removed, the lower amount was removed in both cases from the soil incubated at the higher temperature. Similar findings were reported by workers from the Northern Plains Field Station at Mandan, North Dakota. Their studies showed that both the water- and $NaHCO_3$-soluble phosphate extracted from soils to which phosphate fertilizer had been added decreased when the soil incubation temperatures were above 59°F. However, on soils with no added phosphate the amounts of phosphorus extracted were not affected by soil incubation temperatures ranging from 45 to 80°F.

Time of Reaction. Phosphorus sorption by soils and many soil components follows two rather distinct patterns: an initial rapid reaction followed by a very much slower reaction. The adsorption reactions involving exchange of phosphorus for anions and the ligands of metal atoms on the surface of iron and aluminum oxides are extremely rapid. The very much slower continuing sorption

TABLE 6-7. Percentage of DCPD Hydrolyzed to OCP as a Function of Time and Temperature

Temperature (°C)	Percent OCP Present at:			
	1 Month	2 Months	4 Months	10 Months
10	< 5	20	20	70
20	< 5	40	75	100
30	< 5	30	80	100

Source: Sheppard and Racz, *Western Canada Phosphate Symp.*, p. 170 (1980).

reactions are believed to entail such changes as (1) diffusive penetration or chemisorption of surface-sorbed phosphorus into soil constituents (e.g., the incorporation of phosphorus into hydroxylaluminum or iron polymers and the occlusion of phosphorus in the surface of calcite), and (2) the precipitation of a phosphorus compound for which the solubility product has been exceeded. These slow reactions involve a shift in the form of phosphorus held at the surface from more loosely bound to more tightly bound types which are less accessible to plants.

The compounds precipitated during the reaction of phosphorus fertilizer salts in soils are metastable and will usually change with time into more stable and less soluble compounds. Table 6-7 shows the percentage of dicalcium phosphate dihydrate (DCPD) converted to octacalcium phosphate (OCP) as a function of time and temperature. An important practical consequence is the time after application during which the plant is best able to utilize the added fertilizer phosphorus. On some soils with a high fixing capacity this period may be short, whereas with other soils the period of utilization may last for months or even years. This time period will determine whether the fertilizer phosphorus should be applied at one time in the rotation or in smaller, more frequent applications. Also important is the placement of phosphorus in the soil. Band placement and broadcast applications of phosphates are discussed in a subsequent section of this chapter.

Placement of Added Fertilizer

The placement of fertilizer phosphorus in the soil is treated at greater length in a subsequent chapter, but because of the relationship of this topic to the subject matter covered here it will be considered briefly at this point. The finer the soil texture, the greater the retention of added fertilizer phosphorus. This would be predicted from a knowledge of the relation between the speed of a chemical reaction and the amount of the surface exposed by the reactants. If a finely divided soluble fertilizer phosphate is added to a soil by applying it broadcast and disking it in, the phosphate is exposed to a greater amount of surface; hence more fixation takes place than if the same amount of fertilizer had been applied in bands. Band placement reduces the surface of contact between the soil and fertilizer with a consequent reduction in the amount of fixation. Although this is not the only factor to consider in the placement of a phosphorus fertilizer, it is one of considerable importance when a crop is to be grown on a low phosphate soil with a high fixing capacity and when maximum return from the dollars spent on phosphorus is desired. Granulation of soluble phosphate fertilizers tends to have a similar effect, particularly in the case of larger granules such as those of 4 to 10 mesh.

Band placement generally increases the plant utilization of the water-soluble phosphates such as the superphosphates and ammonium phosphates. There are, however, other phosphatic fertilizers, which are classed as water insoluble, the plant utilization of which seems to be greatest when mixed with the soil rather than when applied in bands.

Phosphate Fertilizers

The development of the modern phosphate fertilizer industry began with the demonstration by Liebig in 1840 that the fertilizing value of bones could be increased by treatment with sulfuric acid. Shortly thereafter, in 1842, John B. Lawes patented a process by which phosphate rock was acidulated with this acid. He began the commercial production of this material in England in 1843. The first recorded sale of what is now commonly known as normal superphosphate manufactured in the United States did not take place until 1852.

From such beginnings has sprung the multimillion dollar fertilizer phosphate industry which uses essentially the same principle for the manufacture of soluble phosphates employed by John Lawes well over a hundred years ago. Ordinary superphosphate is a good fertilizer which is easily and inexpensively manufactured. However, new materials have been developed which, because of their higher content of phosphorus, their handling and storage properties, and the economics of manufacture, are replacing ordinary superphosphate. Much progress has been made during the last several years because of the cooperative efforts of scientists and engineers in both private and public agencies to improve the quality of fertilizers. The properties and behavior of some of these phosphates are discussed in the following sections of this chapter. The principles of the manufacture of phosphate fertilizers are considered in Chapter 10.

Calculation of Phosphorus Content of Fertilizers. Historically, the phosphorus content of fertilizers has been expressed in terms of its P_2O_5 equivalent. Attempts have been made to change this reporting procedure and to have the phosphorus content expressed in terms of its phosphorus equivalent, % P rather than % P_2O_5. There has also been interest in having potassium expressed as % K rather than as % K_2O, as has historically been the case. It must be admitted that percentage figures expressed on the elemental basis are easier to calculate—and to discuss—than on the oxide basis. As a matter of interest, nitrogen was formerly guaranteed as % NH_3 rather than as % N, as is now done.

The conversion of % P to % P_2O_5 and vice versa is simple; it is illustrated here with the salt, dicalcium phosphate ($CaHPO_4$). The percentage of phosphorus in this salt is

$31/136 \times 100 = 23\%$

To convert % P to % P_2O_5, and vice versa, the following expressions are used:

$\% P = \% P_2O_5 \times 0.43$

$\% P_2O_5 = \% P \times 2.29$

So the % P_2O_5 in the example of dicalcium phosphate is

$\% P_2O_5 = 23 \times 2.29 = 53$

Throughout the remainder of this book the phosphorus and potassium content of fertilizers will, wherever possible, be expressed in terms of % P or % K. However, because most fertilizer grades are listed, sold and discussed by the trade in terms of their P_2O_5 and K_2O content, this method will be used. The grade, or guarantee, often will be followed in parentheses by another figure stating it in terms of nitrogen, phosphorus, and/or potassium.

Phosphate Fertilizer Terminology. The solubility of the phosphorus in the different phosphate carriers is variable. The water solubility of fertilizer phosphates is not always the best criterion of the availability of this element to plants, though it is a good one. Perhaps the most accurate measure of the plant availability of any nutrient element is the extent to which it is absorbed by plants under conditions favorable to growth. Such determinations are not easily made when the availability of fertilizer elements has to be determined quickly on large numbers of samples, as in fertilizer control work. Chemical methods have been developed which permit a fairly rapid estimate of the water-soluble, available, and total phosphorus content of phosphate fertilizers. The terminology currently accepted in the United States to describe the availability of phosphate materials is frequently employed by persons dealing with fertilizers. A brief description of the analytical procedure as well as a definition of the terms follows.

Water-Soluble Phosphorus. The terms frequently encountered in describing the phosphate contained in fertilizers are *water-soluble, citrate-soluble, citrate-insoluble, available,* and *total phosphorus* (currently as P_2O_5). A small sample of the material to be analyzed is first extracted with water for a prescribed period of time. The slurry is then filtered, and the amount of phosphorus contained in the filtrate is determined. Expressed as a percentage by weight of the sample, it represents the fraction of the sample that is *water soluble*.

Citrate-Soluble Phosphorus. The residue from the leaching process is added to a solution of neutral $1 N$ ammonium citrate. It is extracted for a prescribed period of time by shaking, and the suspension is filtered. The phosphorus content of the filtrate is determined, and the amount present, expressed as a percentage of the total weight of the sample, is termed the *citrate-soluble* phosphorus.

Citrate-Insoluble Phosphorus. The residue remaining from the water and citrate extractions is analyzed. The amount of phosphorus found is termed *citrate insoluble*.

Available Phosphorus. The sum of the water-soluble and citrate-soluble phosphorus represents an estimate of the fraction available to plants and is termed *available* phosphorus.

Total Phosphorus. The sum of available and citrate-insoluble phosphorus represents the total amount present. The total phosphorus can, of course, be determined directly without resorting to the step-by-step process described.

Class of Phosphate Fertilizers

The original source of phosphorus in the early manufacture of phosphatic fertilizer was bones, but the supply was soon exhausted. Today, the only important source of fertilizer and industrial phosphorus is rock phosphate. Deposits of phosphate rock occur in several areas of the world. Their location and the extent of the deposits are discussed at greater length in Chapter 10. The phosphate compound in these deposits is apatite, which has the general formula $Ca_{10}(PO_4 \cdot CO_3)_6(F, Cl, OH)_2$. This mineral occurs in several forms, such as the carbonato-, fluoro-, chloro-, or hydroxyapatite. The rock is either heat-treated or acid-treated to break the apatite bond and to render the contained phosphate more soluble.

Acid-Treated Phosphates. Commercially important fertilizer phosphates are classed as either acid-treated or thermal-processed; the former constitute by far the more important group. Acid-treated phosphates are classified as: phosphoric acids; superphosphates; ammonium phosphates; ammonium polyphosphates; nitric phosphates; and miscellaneous and new phosphate materials.

Phosphoric Acid. Phosphoric acid (H_3PO_4) is manufactured by treating rock phosphate with sulfuric acid or by burning elemental phosphorus to phosphorus pentoxide and reacting it with water. That produced from sulfuric acid is known as *green* or *wet process acid* and is used largely in the fertilizer industry. Phosphoric acid made by burning is termed *white* or *furnace acid* and is used almost entirely by the nonfertilizer segment of the chemical industry. White acid has a much higher degree of purity than the green and is therefore more expensive. During periods of peak supply, however, some white acid may find its way into the fertilizer industry, for the manufacturers make it available at a lower price to eliminate expensive storage costs. Improved methods of producing wet acid, especially wet process superphosphoric acid, have led to the increased use of wet acid for processes formerly requiring white acid.

Agricultural grade phosphoric acid, which contain 24% phosphorus (55% P_2O_5), is used to acidulate phosphate rock to make triple superphosphate and is neutralized with ammonia in the manufacture of ammonium phosphates and liquid fertilizers. It can also be applied directly to the soil, particularly in alkaline and calcareous areas, by injection, but this method requires special equipment, so the acid is more frequently added to the irrigation water.

Phosphoric acid reacts with the soil components in essentially the same way as any other orthophosphate. In calcareous soils dicalcium phosphate, octacalcium phosphate, and the hydroxy- and carbonatoapatites will ultimately be formed. In acid soils complex phosphates of iron and aluminum will result. Phosphoric acid may, when applied in a band, bring temporarily into solution certain otherwise insoluble trace elements such as zinc, iron, and manganese. In addition, because it is in the liquid form, it may sometimes prove superior to solid sources of phosphorus when it is applied well after planting, a practice not generally recommended. This superiority was, in fact, demonstrated in Colorado. Band-placed phosphoric acid and triple superphosphate were applied to sugar beets as a late side dressing. The phosphoric acid was a better source of phosphorus than the triple superphosphate. In many other tests, when rates of applied phosphorus, method of placement, and time of application are comparable, phosphoric acid has been found to be equal to triple superphosphate as a source of phosphorus for most crops studied.

Superphosphoric Acid. Superphosphoric acid, another phosphoric acid, has been developed, largely through the efforts of the Tennessee Valley Authority. Originally, it was produced from white acid only, but it is now produced from the less expensive green acid, a development that can be very favorable to the fertilizer industry and the consumers of this acid.

Superphosphoric acid made from white acid contains 33 to 35% phosphorus (76 to 80% P_2O_5) in contrast to the 24% phosphorus for the unconcentrated acid. It is made essentially by dehydrating phosphoric acid, in addition to which the final product contains about 35 to 50% of such condensed phosphate radicals as tetra-, pyro-, and tripolyphosphates. Wet process super acid can be made by evaporating the water to concentrate orthophosphoric acid beyond the equivalent of 100% of H_3PO_4. Because of the impurities in wet process acid, however, the maximum concentration of P_2O_5 in the super acid so made is only about 72% (31% P). Superphosphoric acid is used for the manufacture of ammonium and calcium polyphosphates and liquid fertilizers, which are discussed in a subsequent section of this chapter.

Calcium Orthophosphates. The most important phosphate fertilizers, consisting of calcium orthophosphate, are ordinary superphosphate, 7 to 9.5% phosphorus (16 to 22% P_2O_5), triple or concentrated superphosphate, 19 to 23% phosphorus (44 to 52% P_2O_5), enriched superphosphates, 11 to 13% phosphorus (25 to 30% P_2O_5), and ordinary and triple superphosphates which have been reacted with different amounts of ammonia. The last group goes by the term *ammoniated superphosphates*.

The superphosphates are *neutral fertilizers* in that they have no appreciable effect on soil pH, as have phosphoric acid and the ammonium-containing fertilizers. The ammoniated superphosphates have a slightly acid reaction, depending, of course, on the extent to which they have been ammoniated.

Ordinary superphosphate (OSP) is manufactured by reacting sulfuric acid with rock phosphate. This product, which is essentially a mixture of monocalcium phosphate and gypsum, contains 7 to 9.5% phosphorus (16 to 22% P_2O_5), of which about 90% is water soluble and essentially all is classed as *available*. In addition, it contains about 8 to 10% sulfur as calcium sulfate. In areas in which the soils are deficient in that element the gypsum content has been an important contributor to satisfactory crop responses to this phosphate fertilizer. In fact, in numerous experiments conducted in the United States, in which various fertilizers were compared as sources of phosphorus for crops, the superiority of ordinary superphosphate to some of the other sources has frequently been associated with its sulfur content.

The phosphorus component of superphosphate reacts with the soil components, as does any water-soluble orthophosphate, in keeping with the reactions already discussed in this chapter. It is an excellent source of fertilizer phosphorus, but its low content of this element has resulted in its replacement in many areas by higher analysis materials. However, with the increasing incidence of sulfur deficiency in many soils throughout North America, there have been localized increases in the demand for this material.

Ordinary superphosphate is used largely in the production of mixed fertilizers in which other dry, powdered, or finely granular materials are blended to effect a product that contains nitrogen, phosphorus, and potassium. It is also used for direct application.

Triple or concentrated superphosphate (TSP or CSP) contains 19 to 23% phosphorus (44 to 52% P_2O_5), 95 to 98% of which is water soluble and nearly all of which is classed as available. It is essentially monocalcium phosphate and is manufactured by treating rock phosphate with *phosphoric* acid. CSP contains varying amounts of sulfur (usually less than 3%), depending on the manufacturing process. This low content of sulfur is insufficient to supply crop requirements on sulfur-deficient soils.

CSP is an excellent source of fertilizer phosphorus. Its high phosphorus content makes it particularly attractive when transportation, storage, and handling charges make up a large fraction of the total fertilizer cost. CSP is manufactured in pulverant and granular forms and is used in mixing and blending with other materials and in direct soil application.

Enriched superphosphates (ESP) are produced by treating rock phosphate with a mixture of phosphoric and sulfuric acids. Although any percentage of phosphorus between 8.5 and 19% can be obtained, the final product usually contains 11 to 13% (25 to 30% P_2O_5), of which 90 to 95% is water soluble. Nearly all of the contained phosphorus is available. The enriched superphosphates are mixtures of monocalcium phosphate and calcium sulfate. ESP obviously contains sulfur, the amount of which depends on the amount of sulfuric acid used in acidulating the rock.

ESP historically has never enjoyed the popularity in the United States that it has had in Europe. The phosphate component is as satisfactory a source of phosphorus as that in OSP and CSP. Further, the use of ESP is a simple way of upgrading the phosphorus component of the product and at the same time providing plant-nutrient sulfur on soils deficient in this element.

Ammoniated superphosphates are prepared by reacting anhydrous or aqua ammonia with ordinary or triple superphosphate. The total phosphorus content of the end product is decreased in proportion to the weight of the ammonia added. Ammoniation of superphosphates offers the advantage of inexpensive nitrogen but decreases the amount of water-soluble phosphorus in the product. The reduction in water-solubility is greater in OSP than in TSP.

The effect of excessive ammoniation of OSP on reducing crop yields has been studied in Mississippi (Wright et al., 1963). The results from this research were expressed in terms of superphosphate equivalents, which in effect is the extent, on a percentage basis, of the effectiveness of the variously ammoniated super-phosphate samples compared to the nonammoniated material. For crops responding to a high degree of water-soluble phosphorus a high degree of ammoniation of OSP will have a depressing effect on the plant availability of the contained phosphorus.

Ammonium Phosphates. Ammonium phosphates are produced by re-acting ammonia with phosphoric acid or a mixture of phosphoric and sulfuric acids. Some of the more widely used ammonium phosphate fertilizers are monoammon-ium phosphate (MAP), diammonium phosphate (DAP), and ammonium phosphate-sulfate (16-20-0) (8.6% P). Pure MAP contains 12% nitrogen and 61% P_2O_5 (26% P). The common MAP fertilizer grades vary from 11-48-0 (21% P) to 11-55-0 (24% P). Pure DAP has the grade 21-53-0 (23% P). Although small amounts of this relatively pure material are produced, other fertilizer grades of DAP are 16-48-0 (21% P) and 18-46-0 (20% P). Ammonium phosphate-sulfate is prepared by reacting ammonia with a mixture of phosphoric and sulfuric acids. The product contains 16% nitrogen and 20% P_2O_5 (8.6% P).

The ammonium phosphates are completely water soluble. Ultimately, they have an acid effect on soils because of the ammonia they contain, even though the initial reaction of DAP is alkaline, a point to be considered in more detail in a subsequent section of this chapter. The ammonium phosphates are usually offered in the granular form, though some crystalline materials are produced. They are used for formulating solid fertilizers, either by conventional mixing methods or in bulk blended goods (see Chapter 10) and (with the exception of 16-20-0) in manufac-turing suspension fertilizers. The ammonium phosphates are also used for direct application as starter fertilizers, monoammonium phosphate being particularly well suited for this purpose.

Care must be taken with row or seed placement of DAP since free NH_3 originating from this phosphorus source may cause seedling injury and inhibit root growth. This is especially true under basic soil conditions which favor the existence of free NH_3. Urea-ammonium phosphate (UAP) can also be toxic if it is used in large amounts close to the seed row. Adequate separation of seed from either DAP or UAP will usually be all that is required for satisfactory performance of these two phosphorus products.

Ammonium phosphates have the advantage of a high plant-food content which minimizes shipping, handling, and storage costs. As with the superphosphates, they have good handling properties, all of which combine to make them the fer-tilizer phosphates most rapidly increasing in popularity in the United States and abroad. Both mono- and diammonium phosphate, particularly the former, are now finding wide acceptance in making fluid-base suspensions.

Ammonium Polyphosphate. A relative newcomer to the field of am-monium phosphates is ammonium polyphosphate which was developed by TVA. Direct ammoniation of mixtures of electric furnace superphosphoric acid and up to 30% of the phosphorus as wet-process orthophosphoric acid resulted in a gran-ular product with a grade of 15-62-0 (27% phosphorus).

Ammonium polyphosphate is also produced by a two-stage ammoniation tech-nique employing merchant-grade (52 to 54% P_2O_5) wet-process acid. In this process, the heat of reaction is used to drive off free and combined water to form a reaction melt containing 75 to 80% of the phosphorus as polyphosphate. This melt can be dissolved to produce 11-34-0 or 11-37-0 (15 to 16% phosphorus) ammonium polyphosphate solutions with 70 to 80% of the P_2O_5 present in polyphosphate forms. Granulation of the reaction melt will result in a solid product analyzing 11-55-0 (24% phosphorus). Upon the addition of 99.5% urea solution, a granular urea-ammonium polyphosphate with a grade of 28-28-0 (12% phosphorus) can be made. An additional description of solid urea-ammonium polyphosphate fertilizers appears in Chapter 5.

Solid ammonium polyphosphates can be used for direct application or for bulk blending. Granular 15-62-0 can be shipped in bulk and then used to prepare polyphosphate liquid fertilizers. The 10-34-0 and 11-37-0 liquid materials can be applied directly or blended with other liquid fertilizer solutions.

Nitric Phosphates. Nitric phosphates are manufactured by reacting nitric acid with rock phosphate. One of the reaction products, calcium nitrate, is objectionable because of its hygroscopicity. When some sulfuric or phosphoric acid or a sulfate salt is added, most of the calcium nitrate is converted to calcium sulfate or phosphate. Another modification removes the excess calcium by introducing carbon dioxide to precipitate the calcium as calcium carbonate, and another employs refrigeration and centrifugation. The acidified slurry is ammoniated, so that the end products contain a complex assortment of salts such as ammonium phosphates, dicalcium phosphates, ammonium nitrate, calcium sulfate, and still others if potassium salts are added to make a complete fertilizer, as they frequently are.

The N/P_2O_5 ratio of materials produced by nitric phosphate methods ranges from 1:1 to 1:3. This is usually considered to be one of the principal objections to the production of nitric phosphates, for the possible number of grades that can be made is limited. In addition, most nitric phosphates generally have a lower degree of water-soluble phosphorus than have materials based on the neutralization of phosphoric acid with ammonia. Nitric phosphates are always produced in the granular form.

Nitric phosphates are not produced widely in the United States but are used extensively in certain European countries. They have been studied here in the United States, principally by the TVA, which in fact pioneered product development and agronomic testing of the materials in this country.

Results of numerous agronomic tests have shown that these materials are generally satisfactory sources of fertilizer phosphorus. When used on crops responding to water-soluble phosphorus, nitric phosphates may be inferior to those materials containing a high degree of water-soluble phosphorus. The water-soluble fraction of the total phosphorus in nitric phosphates will range from zero to perhaps 80%. A higher degree of water solubility may be obtained by replacing part of the nitric acid with phosphoric or sulfuric acids. A high degree of water solubility may also be obtained by adding nitric acid in amounts greater than that which is theoretically required to react completely with the phosphate rock. The excess acid is subsequently neutralized with ammonia.

Nitric phosphates, in general, will give the best results on acid soils and under crops with a relatively long growing season, such as turf and sod crops. It must be reemphasized, however, that when the degree of water-soluble phosphorus in these materials is kept high (60% or greater) they are usually just as effective as sources of phosphorus for most crops as the super- and ammonium phosphates.

Nitric phosphates are potentially important fertilizers in the United States because of the increasing supply of ammonia, hence nitric acid. The economics of producing these materials and the ability to increase their nitrogen content, in addition to the degree of water-soluble phosphorus, are the two factors that will probably determine the future of nitric phosphate fertilizers in this country.

Miscellaneous Phosphates. Several miscellaneous materials which have been used as phosphate fertilizers for several years and a few comparatively new compounds which show promise of becoming significant sources of fertilizer phosphorus include raw rock phosphate, potassium phosphate, dicalcium phosphate, ammonium phosphate-nitrate, magnesium ammonium phosphate, and ammonium polyphosphate.

Rock Phosphate. Untreated rock phosphate is stable and quite insoluble in water. Commercially important sources contain between 11.5% and 17.5% total phosphorus (27 to 41% P_2O_5). The citrate solubility varies from 5 to about 17% of the total phosphorus content and none of the contained phosphorus is water soluble.

Effectiveness of rock phosphate as a direct-application fertilizer is determined by its chemical reactivity, which in turn depends on the degree of carbonate substitution for phosphate in the apatite structure. Several laboratory solubility tests are suitable for estimating reactivity of phosphate rock. These include extraction with neutral ammonium citrate, 2% citric acid, 2% formic acid, and acid ammonium citrate, pH 3. Phosphate rocks with a high potential for direct application have more than 17% of their total phosphorus soluble in neutral ammonium citrate. Those with 12 to 17% citrate in soluble phosphorus are rated as having medium potential, while rocks with less than 12% citrate solubility are expected to have low potential. Depending on soil properties, crop type, and management, finely ground rock with high, medium, and low citrate solubilities will generally have effectiveness ranges of 80 to 100%, 50 to 80%, and 30 to 60%, respectively, compared to initial crop responses to triple superphosphate. It should be noted that the residual value of low-reactivity rocks can be considerable and as a result the initial differences between phosphate rock sources will often diminish with time.

Phosphate rock in its natural state is of only limited value to plants, unless extremely finely ground to at least 90% passing through 100-mesh screens (0.147 mm). It must also be thoroughly mixed into the soil and it is often necessary to apply quantities containing three to five times the phosphorus equivalent normally provided in conventional water-soluble fertilizers.

Finely ground apatitic phosphate rocks are effective only on acid soils, pH 6 or less. Calcined iron-aluminum phosphate ores with much higher citrate solubilities, 60 to 65% of total phosphorus, can be used successfully on neutral and calcareous soils. Calciphos prepared from C-zone Christmas Island iron-aluminum phosphate rock and Phosphal produced from the Thies deposit in Senegal are examples of two commercial iron-aluminum phosphate rock products that have been developed for direct application.

The effect of raw rock phosphate on crop yields has been studied in literally thousands of experiments conducted throughout the United States. It is not possible within the scope of this text to consider individually the data from these experiments, but a few summary statements concerning the effectiveness of raw rock as a source of fertilizer phosphorus, based on the results of these experiments, will be of interest. In these studies the rock was compared in effectiveness to that of OSP and CSP.

On acid soils, low in phosphorus, apatitic rock phosphate may be considered profitable, but the treated phosphates are generally more economical. In some tests rock phosphate produced greater yields of the test crop than did OSP, but *only* when the rock was supplied in quantities that furnished two to three times more phosphorus. Rock phosphate has been reported to give better residual effects than superphosphate, but whenever this was so it was found that the rates of applied rock phosphate were considerably in excess of those of superphosphate.

Rock phosphate should never be applied directly under any short-season row crop with the idea that phosphorus will be supplied. Its availability is low, and only when used in large amounts and under rotation, including red and sweet clover which are strong feeders on rock phosphate, should it be considered. Such a program was used in some states many years ago, notably in Illinois. If this system is initiated on a soil low in phosphorus, soluble phosphates, in addition to the rock phosphate, will be needed for the first few years to provide the plants with sufficient available phosphorus. When rock phosphate can be used, its choice should be based on the cost per unit of phosphorus compared with its cost per unit in the more available forms of fertilizer phosphate.

Workers in Australia have developed a granular material containing raw rock phosphate and finely ground elemental sulfur. The product, called Biosuper, is inoculated with the sulfur-oxidizing bacteria *Thiobacillus thiooxidans* to ensure the conversion of the sulfur to sulfuric acid. The acid in turn reacts with the phosphate rock, making the contained phosphorus more available to plants. Extensive field trials have shown that while the material is not as effective as superphosphate, it is a suitable product for use on pasture and range land. Its greatest use will probably be found in developing nations that do not have a sophisticated fertilizer industry.

Environmental conditions such as warm climates, moist soils, and long growing seasons will increase the effectiveness of rock phosphate. Ground phosphate rock is sometimes used for restoration of low-phosphorus soils on abandoned farms and on newly broken lands inherently low in available phosphorus. For these purposes a heavy initial application is recommended, such as 1 to 3 tons/ha, which may be repeated at 5- to 10- year intervals.

Factors limiting the use of rock phosphate include uncertain agronomic value; inconvenience of handling and applying the fine, dusty material; and relatively low P_2O_5 content compared with triple superphosphate or ammonium phosphate.

Partially Acidulated Rock Phosphate. In situations where the reactivity of available phosphate rock is inadequate for acceptable short-term yields and where the phosphorus-retention capacity of the soil quickly renders soluble phosphorus fertilizer unavailable to plants, partially acidulated rock phosphate can be a feasible alternative. The initial phosphorus availability of low reactivity phosphate rocks can be significantly improved by treatment with only 10 to 20% of the usual quantity of H_3PO_4 necessary for the manufacture of triple superphosphate. To produce a comparable material by acidulation with H_2SO_4, it takes 40 to 50% of the amount normally consumed in the manufacture of single superphosphate. The results to date indicate that fertilizer prepared with less than the conventional quantities of acid are effective sources of phosphorus.

Potassium Phosphate. Potassium phosphate is represented by two salts, KH_2PO_4 and K_2HPO_4, which have the grades 0-52-35 (22% P, 29% K) and 0-41-54 (18% P, 45% K), respectively. During the 1970s fertilizer-grade potassium phosphate with an analysis of 8-48-12 was manufactured on a small scale in California by Pennzoil. They are completely water soluble and find their greatest market in soluble fertilizers sold in small packets for home and garden use. Their high content of phosphorus and potassium makes them attractive possibilities for commercial application on a farm scale. Developments in the economics of producing these salts will determine whether they can be manufactured on a large scale for use as commercial fertilizers.

Other potassium phosphate fertilizers used on a limited basis include potassium polyphosphate solution with a grade of 0-26-27 and potassium metaphosphate containing about 60% P_2O_5 and 40% K_2O.

In addition to the high plant nutrient content of the potassium phosphates, they have other desirable characteristics. As they contain no chloride, potassium phosphates are ideally suited for "solanum"-type crops such as tobacco, potatoes, tomatoes, and many leafy vegetables which are sensitive to high levels of this element.

Their low salt index reduces the risk of injury to germinating seeds and to young seedlings when they are placed in or close to the seed row.

Dicalcium Phosphate. It is a component of many phosphate fertilizers that contain calcium. It arises from the ammoniation of solids or liquids in which there is monocalcium phosphate. The pure material has the formula $CaHPO_4$ and contains 23% phosphorus (53% P_2O_5). Although this material could be used as a fertilizer, it cannot compete with the less expensive phosphates presently available. Dicalcium phosphate is used as a mineral supplement for animals and pregnant women.

Dicalcium phosphate can be manufactured in several ways. One method used in Europe utilizes by-product hydrochloric acid. Phosphate rock is dissolved in hydrochloric acid and dicalcium phosphate is precipitated by stepwise addition of limestone and slaked lime. Another development involves acidulation with nitric acid, ammoniating, and carbonating the slurry with carbon dioxide. Because of the increasing availability of lower-cost nitric acid, this process holds promise for future commercial exploitation.

The Tennessee Valley Authority has reported successful production of dicalcium phosphate by reaction of phosphate rock with SO_2, H_2O, and acetone.

Ammonium Phosphate Nitrate. This fertilizer is essentially a homogeneous mixture of ammonium phosphate and ammonium nitrate. It is completely water soluble and there are several processes that can be used to manufacture products containing different amounts of nitrogen and phosphorus as well as potassium. Sample grades are (on an N-P_2O_5-K_2O basis) 23-23-0, 24-24-0, 25-25-0, 30-10-0, 8-16-32, and 15-15-15. TVA operated a demonstration-scale plant for several years producing 25-25-0 and 30-10-0 grades.

Ammonium Phosphate-Urea, Ammonium Polyphosphate-Urea, and Magnesium Ammonium Phosphate. These three materials were covered in Chapter 5.

Heat-Treated Phosphates. Heat-treated or thermal phosphate fertilizers are manufactured by heating phosphate rock to varying temperatures with or without additives, such as silica. In the United States, thermal-process phosphates do not account for a large segment of the total fertilizer phosphates manufactured. Some of them are locally important in various parts of the world. Their chief drawbacks are the following:

1. They are generally more expensive to produce than acid-derived phosphates.
2. They contain no water-soluble phosphorus and the available phosphorus is frequently considerably less than 100% of the total present.
3. They do not lend themselves to ammoniation, hence are of no value in the manufacture of N-P-K fertilizers.

Heat-treated rock phosphate is frequently referred to as defluorinated phosphate. Defluorination may be effected by calcination or fusion. By calcination is meant the heating of phosphate rock, usually in the presence of steam and silica, to temperatures *below* the melting point of the mixture. Fusion is the term applied when similar mixtures are heated *above* the melting point so that the charge fuses or runs together, thus forming a glassy product. The calcined material is sintered and porous in appearance.

Several thermal-process phosphates are of sufficient importance to warrant some discussion. These materials, with their total and citrate-soluble contents, are defluorinated phosphate rock, 9% total phosphorus (21% P_2O_5), 8% citrate-soluble phosphorus (18% P_2O_5); phosphate rock-magnesium silicate glass, 10% total (22.5%

P_2O_5), 8% citrate-soluble (19% P_2O_5); Rhenania phosphate, 12% total, 11.8% citrate soluble (28% and 27.5% P_2O_5, respectively); calcium metaphosphate, 27.5% total, 27% citrate soluble (64% and 63% P_2O_5, respectively); and basic slag, 1.0 to 7.8% total (2.3 to 18% P_2O_5), of which 60 to 80% is citrate soluble.

Defluorinated phosphate rock, commonly referred to as Coronet phosphate, is made by combining finely ground phosphate rock, high silica tailings from the phosphate rock mining process, and enough water to form a slurry. This slurry is passed into an oil-fired rotary kiln in which the temperatures in the burning zone are between 1480 and 1590°C. Heating at these temperatures for about 30 minutes produces a porous mass that is quenched in water. The product is ground so that 60 per cent will pass through a 200-mesh screen. Most of this material goes into animal feeds. However, it is also a satisfactory source of fertilizer phosphorus except on alkaline or calcareous soils.

Phosphate rock-magnesium silicate glass is formed by fusing rock phosphate and either olivine or serpentine in a furnace at 1550°C. It finds its greatest use in areas in which the supplies of sulfuric or other acids are inadequate. When finely ground, this material is a satisfactory source of plant-nutrient phosphorus on acid soils. It is not suitable for use on alkaline or calcareous soils.

Rhenania phosphate was first developed in Germany in 1917. It is prepared by calcining a mixture of soda ash, silica, and phosphate rock at 1100 to 1200°C. It is quenched with water and ground to pass a 180-mesh screen. Rhenania phosphate is not used to any extent in this country, though it is popular in some areas of Europe. It is a very satisfactory source of phosphorus when used on acid soils, but is generally unsatisfactory when applied to soils that are alkaline or calcareous.

Calcium metaphosphate was developed in Germany in 1929. Work on this compound in the United States has been done almost entirely by TVA. Its successful manufacture requires a relatively cheap source of electric power, which is perhaps the chief factor limiting its commercial production in this country.

Calcium metaphosphate is prepared by burning elemental phosphorus in the presence of finely divided phosphate rock. The melt is not quenched but poured onto a water cooled drum flaker and then ground to the desired degree of fineness.

Subsequent studies by TVA have shown that calcium metaphosphate can be partly hydrolyzed by treatment with sulfuric acid. It can then be ammoniated and used as an orthophosphate in the manufacture of complete N-P-K fertilizers.

Calcium metaphosphate is a suitable source of phosphorus for most crops grown on acid soils but variable results have been obtained on alkaline or calcareous soils.

Basic slag, or Thomas slag, is a by-product of the basic open-hearth method of making steel from pig iron. In the United States it is produced only in the area around Birmingham, Alabama, and its use is confined to a radius of several hundred miles around that city. However, it is a very popular phosphate fertilizer in Europe. Both in the United States and on the European continent its production is dependent entirely on the production of steel by the open-hearth process and supplies vary accordingly.

Basic slag in the United States runs around 3% P_2O_5, whereas that in Europe is usually much higher, in a range of 14 to 18%, because of the higher phosphate content of its iron ore. In addition to its phosphate content, basic slag has a neutralizing value of 60 to 80%, which makes it particularly valuable on acid soils, for it will act like lime to neutralize soil acidity.

Basic slag is a good source of phosphorus on acid soils. It has not been used in the United States on alkaline soils, for, like other thermal phosphates, its application to such soils could not be recommended. In Europe, however, where the soil phosphate levels are frequently quite high because of centuries of intensive fertilization, basic slag is said by some to be a satisfactory source of phosphorus on neutral to slightly alkaline as well as on acid soils.

All of the thermal-process phosphates must be finely ground to be effective sources of fertilizer phosphate. Although the fluoroapatite structure is broken by the heat treatment, the phosphate remains in a completely water-insoluble form. Therefore fine grinding is essential to ensure that a large total surface is exposed for contact by the soil and root systems. The availability of these materials to plants is related directly to their specific surface, which is, of course, inversely related to the particle size of the material. These fertilizers should not be banded but applied broadcast and disked in for the same reason that fine grinding is necessary.

Bacterial Phosphate Fertilization

In the USSR and several eastern European countries soils are inoculated with bacteria which apparently increase the plant availability of native and applied soil phosphorus. Several species of microorganisms are effective in this respect, but the one principally employed is *Bacillus megatherium* var. *phosphaticum*. Cultures of these organisms, which the Soviets term *phosphobacterins*, are prepared commercially and distributed to the farmers. In 1958 it was estimated that about 10 million hectares (about 25 million acres) were treated with phosphobacterin. Azotobacterin, discussed in Chapter 5, is also popular and is used in Soviet agriculture.

The increase in available soil phosphorus results primarily from the decomposition of organic phosphorus compounds. The treatment is most effective on neutral to somewhat alkaline soils and on those high in organic matter. Some organisms, which produce considerable acidity as a result of their metabolic activity, will also dissolve some of the plant-unavailable soil mineral phosphates, such as hydroxyapatite. Some work has been reported in which the availability of rock phosphate and other water-insoluble phosphates inoculated with phosphobacterin was increased.

Yield increases from the use of phosphobacterin are reported to range from 0 to 70%. The average, however, is about 10%. Crops responding to this treatment are sugar beets, potatoes, cereals, vegetables, some grasses, and legumes.

The effectiveness of phosphobacterin has been investigated by workers at the U.S. Department of Agriculture, who showed that the culture readily decomposes glycerophosphates. It was also found that the yield of tomatoes grown in greenhouse tests was increased 7.5%. There was no increase in the yield of wheat. However, neither phosphorus concentration nor total phosphate uptake was favorably influenced by phosphobacterin.

The use of phosphobacterin is undoubtedly of value in increasing crop yields in the USSR, where use of chemical fertilizers is still rather limited. How popular it will remain when the supply of mineral fertilizers increases is a matter of conjecture. It is fairly certain that the inoculation of soil or fertilizers with phosphobacterin is not likely to be practiced on a large scale in the United States in the foreseeable future.

A summary of the properties and plant-nutrient contents of the most important phosphate fertilizers is given in Table 6-8.

Behavior of Phosphate Fertilizers in Soils

Effectiveness of phosphorus fertilizers is determined by the properties of both the phosphorus salt and the soil being fertilized and by the reactions which occur between the phosphorus fertilizer and various soil constituents. Dissolution of granules of water-soluble phosphorus fertilizer is fairly rapid, even under conditions of low soil moisture. Water sufficient to initiate dissolution moves to the granule by either capillarity or vapor transport. A nearly saturated solution of the

TABLE 6-8. Composition of Phosphatic Fertilizer Materials

Material	Total Nitrogen (%)	Total Potassium (%)	Total Sulfur (%)	Total Calcium (%)	Total Magnesium (%)	Phosphorus Total (%)	Phosphorus Available * (% of Total)
Ammonium phosphates †							
21-53-0	21	—	—	—	—	23	100
21-61-0	21	—	—	—	—	27	100
11-48-0	11	—	0–2	—	—	21	100
16-48-0	16	—	0–2	—	—	21	100
18-46-0	18	—	0–2	—	—	20	100
16-20-0	16	—	14	—	—	8.7	100
Ammoniated OSP	2–5	—	10–72	17–21	—	6.1–8.7	96–98
Ammoniated CSP	4–6	—	0–1	12–14	—	19–21	96–99
Ordinary super-phosphate	—	—	11–12	18–21	—	7–9.5	97–100
Conc. (triple) super-phosphate	—	—	0–1	12–14	—	19–23	96–99
Enriched super-phosphate	—	—	7–9	16–18	—	11–13	96–99
Dicalcium phosphate	—	—	—	29	—	23	98
Superphosphoric acid	—	—	—	—	—	34	100
Phosphoric acid	—	—	0–2	—	—	23	100
Potassium phosphate	—	29–45	—	—	—	18–22	100
Ammonium phosphate nitrate	30	—	—	—	—	4	100
Ammonium poly-phosphate	15	—	—	—	—	25	—
Magnesium ammonium phosphate	8	—	—	—	14	17	—
Raw rock phosphate	—	—	—	33–36	—	11–17	14–65
Basic slag	—	—	0.2	32	3	3.5–8	62–94
Defluorinated phosphate rock	—	—	—	20	—	9	85
Phosphate rock–magnesium silicate glass	—	—	—	20	8.4	10	85
Rhenania phosphate	—	—	—	30	0.3	12	97
Calcium metaphosphate	—	—	—	19	—	27	99
Potassium meta-phosphate	—	29–32	—	—	—	24–25	—

* By neutral 1.0 N ammonium citrate procedure.
† Ammonium phosphate grades expressed as % N, % P_2O_5, % K_2O.

phosphorus fertilizer material forms in and around fertilizer granules, droplets, and bands.

Formation of these nearly saturated solutions of phosphorus fertilizer salts causes an osmotic potential gradient to be established between the concentrated fertilizer solution and the soil water. While water is drawn into this fertilizer zone by vapor transport, the fertilizer solution moves into the surrounding soil. This movement of water inward and fertilizer solution outward continues to maintain a nearly saturated solution as long as any of the original salt remains. Even after the original addition of fertilizer phosphorus has disappeared, such osmotic gradients will persist until dilution or reactions between fertilizer nutrients and soil constituents restore the soil solution to its original composition.

As the virtually saturated solutions of phosphorus salts move into the first increments of soil, the chemical environment is dominated by the solution properties rather than by the soil properties. The composition of saturated solutions of the major phosphorus fertilizer materials used in North America is summarized

TABLE 6-9. **Phosphate Compounds Commonly Found
in Fertilizers and Compositions of Their Saturated Solutions**

Compound	Formula	Solution Symbol	pH	P (mol/liter)	Accompanying Cation (mol/liter)	
					Composition of Saturated Solution	
Highly water-soluble						
Monocalcium phosphate	$Ca(H_2PO_4)_2 \cdot H_2O$	TPS	1.0	4.5	Ca	1.3
		MTPS	1.5	4.0	Ca	1.4
Monoammonium phosphate	$NH_4H_2PO_4$	MAP	3.5	2.9	NH_4	2.9
Monopotassium phosphate	KH_2PO_4	MKP	4.0	1.7	K	1.7
Triammonium pyrophosphate	$(NH_4)_3HP_2O_7 \cdot H_2O$	TPP	6.0	6.8 $(3.4\ P_2O_7)$	NH_4	10.2
Diammonium phosphate	$(NH_4)_2HPO_4$	DAP	8.0	3.8	NH_4	7.6
Dipotassium phosphate	K_2HPO_4	DKP	10.1	6.1	K	12.2
Sparingly soluble						
Dicalcium phosphate	$CaHPO_4$ $CaHPO_4 \cdot 2H_2O$	DCP	6.5	~0.002	Ca	0.001
Hydroxyapatite	$Ca_{10}(PO_4)_6(OH)_2$	HAP	6.5	$\sim 10^{-5}$	Ca	0.001

Source: Sample et al., in F. E. Khasawneh et al., Eds., *The Role of Phosphorus in Agriculture,*
p. 275. Madison, Wis.: American Society of Agronomy, 1980.

in Table 6-9. It can be seen that solutions formed in the group of water-soluble
phosphorus fertilizers have pH values between 1.0 and 10.1 and contain from 1.7
to 6.1 mol/liter of phosphorus. The concentration of accompanying cations ranges
from 1.3 to 12.2 mol/liter. Because of the wide variation in chemical properties of
these phosphorus sources alone and in mixtures with other fertilizer materials,
there are marked differences in their reactions with soil and its components.

When the concentrated phosphorus solution leaves the granule, droplet, or band
site and moves into the surrounding soil, the soil components are altered by the
solution and at the same time the solution composition is changed by its contact
with soil. Some soil minerals may actually be dissolved by the concentrated phosphorus solution resulting in the release of large quantities of reactive cations such
as Fe^{3+}, Al^{3+}, Mn^{2+}, Ca^{2+}, and Mg^{2+}. Cations from exchange sites may also be
displaced by these concentrated solutions. Phosphorus in the concentrated solutions reacts with these cations to form specific and often identifiable compounds,
many of which were listed previously in Table 6-3. These compounds are referred
to as soil–fertilizer reaction products. In addition, phosphorus can be adsorbed on
soil constituents which were relatively unaltered during their exposure to the
concentrated phosphorus solutions.

Precipitation reactions are favored by the very high phosphorus concentrations
existing in close proximity to granules, droplets, and bands of fertilizer phosphorus.
Adsorption reactions are expected to be most important at the periphery of the
soil–fertilizer reaction zone, where phosphorus concentrations are much lower.
Although both precipitation and adsorption occur at the application site, precipitation reactions usually account for most of the phosphorus being retained in that
vicinity. The precipitation of dicalcium phosphate at the application site of monocalcium phosphate monohydrate is readily apparent in Figure 6-10. From 20 to
34% of the applied phosphorus will remain as this reaction product at the granule
site.

Although the initial reaction products are metastable and are usually trans-

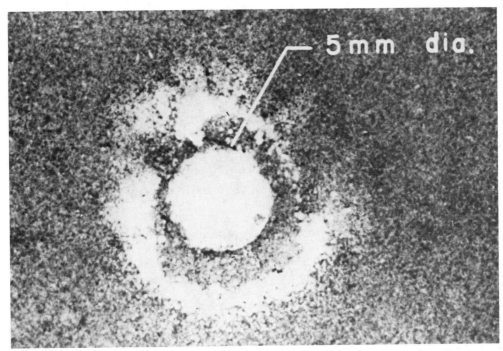

FIGURE 6-10. Distribution of monocalcium phosphate reaction products after 14 days' reaction at 5°C in the Bradwell very fine sandy loam. [Hinman et al., *Can. J. Soil Sci.*, **42**:229 (1962).]

formed with time into more stable but less-water-soluble compounds, they will have a favorable influence on the phosphorus nutrition of crops. Some of the initial reaction products will provide phosphorus concentrations in solution 1000 times those in untreated soil. An example of the alteration of dicalcium phosphate to much-less-soluble octacalcium phosphate is given in Table 6-7. The rate of this change of the initial reaction products is influenced by soil properties and environmental factors. Residual value of fertilizer phosphorus is dependent on the nature and reactivity of long-term reaction products.

Monocalcium phosphate monohydrate, the major phosphorus component of superphosphates, monoammonium phosphate, and diammonium phosphate are the most widely used orthophosphate fertilizers. Mono- and dipotassium phosphate are potentially useful orthophosphates which up until now have not been produced on a large scale. Brief descriptions of the reactions of these phosphorus carriers in soils follow.

Monocalcium Phosphate. Reactions of fertilizers containing this form of phosphorus in acid soils will produce a variety of substances, including colloidal (Fe, Al, X)$PO_4 \cdot nH_2O$, dicalcium phosphate (both the dihydrate and anhydrate), $CaFe_2(HPO_4)_4 \cdot 8H_2O$, $CaAlH(PO_4)_2 \cdot 6H_2O$, and $CaAl_6H_4(PO_4)_8 \cdot 20H_2O$. Under less acidic conditions, dicalcium phosphate dihydrate, $K(AlFe)_3H_8(PO_4)_6 \cdot 6H_2O$, and $K_3Al_5H_6(PO_4)_8 \cdot 18H_2O$ will form.

In calcareous soils, dicalcium phosphate is the dominant initial reaction product and in the presence of extremely large amounts of calcium carbonate, octacalcium phosphate may also form. When there are substantial quantities of $MgCO_3$ present in basic soils, $MgHPO_4 \cdot 3H_2O$ (newberryite) will form as well as dicalcium phosphate (dihydrate and anhydrate).

Inclusion of other fertilizer salts such as $(NH_4)_2SO_4$, NH_4NO_3, NH_4Cl, KNO_3,

K_2SO_4, and KCl with monocalcium phosphate will significantly reduce the amount of phosphorus precipitated in the vicinity of the application site.

Contrasted with monocalcium phosphate monohydrate, ammonium ortho-, and polyphosphate fertilizers containing little or no calcium or micronutrient cations ordinarily leave little or no residue at the original site. Instead, the resultant reaction products tend to be distributed throughout the soil some distance away.

Monoammonium Phosphate. Dicalcium phosphate dihydrate and ammonium taranakite are the major products formed when this phosphorus salt reacts in soils. Formation of the former reaction product is favored by increasing amounts of calcium. Colloidal $(Fe, Al, X)PO_4 \cdot nH_2O$ may also be produced in acidic soils.

When soil Mg levels are high, $Mg_3(NH_4)_2(HPO_4)_2 \cdot 8H_2O$ (hannayite) and $MgNH_4PO_4 \cdot 6H_2O$ (struvite) are formed.

Monopotassium Phosphate. Potassium taranakite is a major compound identified following the reaction of monopotassium phosphate in soils. Racz and Soper at the University of Manitoba in Canada observed that dicalcium phosphate dihydrate and dimagnesium phosphate trihydrate were initial reaction products when this compound was added to neutral and calcareous soils. In soils with a water-soluble Ca/Mg ratio of approximately 1.5 or greater, dicalcium phosphate dihydrate formed. Dicalcium phosphate dihydrate and/or dimagnesium phosphate trihydrate formed in soils having a water-soluble Ca/Mg ratio of less than 1.5.

Diammonium Phosphate. The following compounds have been identified following the reaction of this basic phosphorus salt with soil and its constituents: $Ca_4H(PO_4)_3 \cdot 3H_2O$, $Ca_2(NH_4)_2(HPO_4)_3 \cdot 2H_2O$, $Ca(NH_4)_2(HPO_4)_2 \cdot H_2O$, $CaNH_4PO_4 \cdot H_2O$, $CaHPO_4 \cdot 2H_2O$, $MgNH_4PO_4 \cdot 6H_2O$, $NH_4Al_2(PO_4)_2OH \cdot 8H_2O$, $Ca_8H_2(PO_4)_6 \cdot 5H_2O$, colloidal apatite, and $Ca_{10}(PO_4)_6(OH)_2$. Dicalcium phosphate dihydrate is one of the most abundant of these reaction products in high-calcium soils and octacalcium phosphate $[Ca_8H_2(PO_4)_6 \cdot 5H_2O]$ may also dominate in such soils.

Dipotassium Phosphate. The principal substance resulting from reactions of this phosphorus source was observed to be $CaK_2H(PO_4)_2$.

Reactions of Polyphosphates. Commercial polyphosphate fertilizers contain both orthophosphate and condensed forms of phosphorus. The various phosphorus species present in a typical granular ammonium polyphosphate with an analysis of 15% nitrogen and 62% P_2O_5 are 41% ortho-, 54% pyro-, 4% tripoly-, and 1% tetrapoly- and more highly condensed phosphates. Ammonium polyphosphate liquid fertilizers may contain, as indicated in Table 6-10, an even higher proportion of tripoly- and more highly condensed phosphates. Triammonium pyrophosphate, $(NH_4)_3HP_2O_7 \cdot H_2O$, is the most abundant nonorthophosphate species in ammonium polyphosphate fertilizers.

Less is known about the behavior of polyphosphate fertilizers in soil than the orthophosphates. When polyphosphate fertilizers are applied to soils there are opportunities for both precipitation and adsorption of the polyphosphate species. In addition, the orthophosphate present initially plus that formed by hydrolysis of the polyphosphate component is expected to undergo reactions with soil and its constituents similar to those described earlier for orthophosphate compounds.

Hydrolysis or the combination of water with condensed phosphates results in a stepwise breakdown, producing orthophosphates and various shortened polyphosphate fragments. The shortened polyphosphate fragments then undergo further hydrolysis. Reactions of polyphosphates in soil and the nature of the substances formed

TABLE 6-10. Typical Phosphate Species Distribution in Liquid Ammonium Phosphate Fertilizers Made from Wet-Process Superphosphoric Acid by the Pipe-Reactor Process

P_2O_5 Species	Formula	Percent of Total P_2O_5
Ortho	H_3PO_4	21
Pyro	$H_4P_2O_7$	31
Tri	$H_5P_3O_{10}$	23
Tetra	$H_6P_4O_{13}$	12
Penta	$H_7P_5O_{16}$	7
Higher	$H_{n+2}P_nO_{3n+1}$	6

Source: NFSA Liquid Fertilizer Manual.

are dependent on the rate of their reversion back to orthophosphates. Slow hydrolysis rates would permit condensed phosphates to sequester or form soluble complexes with soil cations and thus avoid or reduce phosphorus retention reactions.

Transformation of polyphosphates back to orthophosphates occurs by two principal pathways, either chemically or biologically. Chemical hydrolysis of condensed phosphates proceeds very slowly in sterile, neutral solutions at room temperature. Clay minerals and hydrous oxides, particularly iron oxide, are reported to make minor contributions to the chemical hydrolysis of polyphosphates. In soils, where both mechanisms can function, hydrolysis is usually rapid.

Several factors control hydrolysis rates in soils, with enzymatic activity provided by plant roots and microorganisms probably being the most important. Phosphatases associated with plant roots and rhizosphere organisms are now believed to be the group of enzymes responsible for the biological hydrolysis of pyro- and polyphosphates.

The various conditions, such as temperature, moisture, available sources of carbon, pH, and nutritional levels, which encourage microbial and root development will also favor phosphatase activity and the resultant hydrolysis of polyphosphates. It should be noted that this hydrolysis can be inhibited by the presence of orthophosphate which is known to repress the synthesis of certain phosphatases.

Temperature is probably the most important environmental factor influencing rate of hydrolysis of polyphosphate in soil. As seen in Figure 6-11, the extent of hydrolysis of pyro- and tripolyphosphate was increased substantially by elevating the temperature from 5°C to 35°C.

Retention of pyrophosphate by soil constituents will substantially lower hydrolysis rates by decreasing its accessibility for conversion by phosphatase. There are indications that pyrophosphates are adsorbed more strongly on clays and soils than are orthophosphates.

Khasawneh and his co-workers at the Tennessee Valley Authority have conducted studies on the reactions of phosphorus from triammonium pyrophosphate, ammonium polyphosphate, and diammonium phosphate in soil. They found that the diffusive movement of phosphorus from the application site was accompanied by two major reactions: (1) hydrolysis of pyro- and polyphosphates to the orthophosphate form, and (2) precipitation reaction of all phosphate anions with soil. Hydrolysis of the water-soluble fraction of the condensed forms of phosphorus was rapid, even faster than diffusion. As a result, movement of the pyro- and polyphosphates occurred largely in the orthophosphate form. The precipitation reactions of pyro- and polyphosphate were virtually complete in the first week, were irreversible, and localized within well-defined zones, and accounted for nearly one-fourth of the added phosphorus.

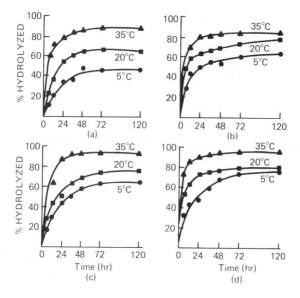

FIGURE 6-11. Effects of temperature on hydrolysis of water-soluble sodium pyrophosphate and sodium tripolyphosphate. (a) Lakeland soil with 200 ppm phosphorus pyrophosphate. (b) Newdale soil with 200 ppm phosphorus pyrophosphate. (c) Lakeland soil with 200 ppm phosphorus tripolyphosphate. (d) Newdale soil with 200 ppm phosphorus tripolyphosphate. [Chang and Racz, *Can. J. Soil Sci.*, **57**:271 (1977).]

Sufficient work has been done to establish that polyphosphates are as effective as orthophosphates as sources of phosphorus for crops. Plants can absorb and utilize the polyphosphates [which are primarily pyrophosphates with the formula $(NH_4)_3HP_2O_7 \cdot H_2O$] directly.

Because polyphosphates have the ability to form complex ions with some metals, it has been suggested that they may be effective in mobilizing zinc in soils in which deficiencies of this element have been induced by high pH or high phosphate levels. Giordano and his research colleagues at the Tennessee Valley Authority found only slight influence of pyrophosphates on zinc concentrations in the soil solution. For one day after the addition of a high rate (2000 ppm of phosphorus) of triammonium pyrophosphate, there was an increase in initial concentration of zinc in the soil solution. This short-lived effect probably resulted from either sequestering of zinc by the pyrophosphate or solubilization of soil organic matter. Any complexing of zinc by this condensed phosphate source could only be transitory because hydrolysis of pyrophosphate was very rapid in their studies.

Movement and Redistribution of Fertilizer Phosphorus. Initial movement of phosphorus away from fertilizer application sites seldom exceeds 3 to 5 cm. The extent of phosphorus movement of diammonium orthophosphate (DAP), triammonium pyrophosphate monohydrate (TPP), and ammonium polyphosphate (APP) in a fine sandy loam soil (pH 6.0) is portrayed in Figure 6-12.

Movement of phosphorus fertilizer salts is dependent on initial moisture content of the soil. Salt diffusion processes are favored by higher moisture content. At the Tennessee Valley Authority, it was shown in studies conducted under the leadership of Khasawneh that triammonium pyrophosphate moved 18, 25, and 34 mm when the moisture content in the treated soil was 6.7, 9.6, and 19%, respectively.

There are differences in the distribution of phosphorus in the reaction zones of phosphorus fertilizers. This aspect was referred to briefly in the discussion on behavior of phosphate fertilizers in soils. Visible differences can be observed in Figure 6-10 and there is more evidence in Figure 6-12.

Extensive reaction zones combined with thorough distribution and a fine state of subdivision of the reaction products are all factors that should enhance absorption of phosphorus by plant roots encountering reaction zones. These conditions may, in fact, offset the low water solubility of many of the phosphorus compounds which are precipitated in reaction zones.

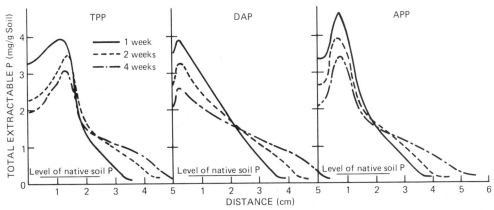

FIGURE 6-12. Phosphorus distribution profiles in columns treated with TPP, DAP, or APP. [Khasawneh et al., *Soil Sci. Soc. Am. J.*, **38**:446 (1974).]

Considering the root zone as a whole, the initial distribution of phosphorus fertilizers is heterogeneous or nonuniform. In time, phosphorus tends to become more evenly dispersed. This is due partly to slow diffusion of phosphorus away from its original site, but it can also be due to destruction of a fertilizer band by subsequent cultivation.

Biocycling or redistribution of phosphorus by the action of plant roots is another pathway that is often overlooked. Read at the Swift Current, Saskatchewan, Research Station has observed a gradual downward movement of phosphorus in soils receiving large batch applications of between 205 to 820 lb of P_2O_5 per acre. He found a significant increase in the amount of $NaHCO_3$- extractable phosphorus at all sampling depths from 15 to 120 cm, where the heaviest initial rate was used. The two lower rates of 205 and 410 lb of P_2O_5 per acre also resulted in some increase in extractable phosphorus at all depths. This transport from surface to lower soil depths was explained by cycling of phosphorus taken up by crops back into the entire rooting media through decomposition of roots. Researchers in Montana have found similar increases in available soil phosphorus to depths of 12 in. after the addition of heavy initial applications of phosphorus fertilizer.

Dissolution of Sparingly Soluble Compounds. The principles of dissolution of water-soluble phosphorus fertilizer salts also apply to phosphorus reaction products of low water solubility, such as variscite, strengite, dicalcium phosphate, dimagnesium phosphate, octacalcium phosphate, hydroxyapatite, and struvite. Soil moisture moves to particles of these compounds and saturated solutions form in and around them. The principal difference between the two groups of phosphorus containing substances is that the saturated solutions associated with sparingly soluble compounds are very dilute and as a consequence the osmotic gradient between the saturated phosphorus solution and the soil solution is low. Water movement into the zone of higher phosphorus concentration is very slow and the concentrated solution influences only small volumes of soil surrounding the dissolving particles.

Interaction of Nitrogen with Phosphorus. The addition of nitrogen fertilizers in combination with phosphorus enhances plant uptake of phosphorus. Since nitrogen is the principal constituent of plants, accounting for at least one-half of the total number of ions absorbed, it is reasonable that phosphorus uptake is influenced by the presence of fertilizer nitrogen. There is evidence that nitrogen promotes phosphorus uptake by plants by (1) increasing top and root growth, (2) altering plant metabolism, and (3) increasing solubility and availability of phosphorus.

A number of researchers have reported a synergistic effect of supplying nitrogen and phosphorus together on root growth. The greater root mass is believed to be responsible for increased crop uptake of phosphorus. Ammoniacal fertilizers have a greater stimulating effect on phosphorus absorption than nitrate.

Increased absorption of phosphate through physiological changes in plants receiving NH_4^+-N has been attributed to (1) stimulated uptake of $H_2PO_4^-$ or HPO_4^{2-} to balance a greater cation uptake, (2) to enlarged sinks in the higher protein of NH_4^+ nourished plants, and (3) to effects on the phosphorus-carrier complex. There is a further suggestion that a generally higher level of nitrogen metabolites produced by previous exposure to either nitrate or ammonium will stimulate subsequent phosphorus uptake and translocation. The pH is also reduced.

Greater effectiveness of fertilizer phosphorus has been reported from many areas of the United States and Canada when modern fertilizer application systems place phosphorus in close association with ammoniacal nitrogen sources. For example, it was demonstrated by Murphy and others in Kansas that there are some definite agronomic advantages, often resulting in 5 to 6 bu/A yield increases of winter wheat, to be gained by using the technique of simultaneously knifing anhydrous ammonia and ammonium polyphosphate solution into the soil. This and other application methods are discussed at greater length in Chapter 13.

Other Factors Affecting Phosphate Fertilizers

A number of other fertilizer properties and soil conditions that affect the efficacy of phosphate fertilizers should be mentioned. In the following brief discussions of several of these factors, frequent reference will be made to granules or particles of dry phosphorus sources. It should be noted that in many instances fine sprays and large droplets of fluid phosphorus fertilizers are analogous to small and regular-sized granular solid phosphorus sources.

Effect of Size of Granule. The size and type of fertilizer granule, and the manner in which it is placed in the soil, have a marked influence on the relative effectiveness of phosphatic fertilizers. Small granules that possess a large surface area per unit of mass and permit intimate contact with the soil and better distribution through the soil are normally assumed to be more effective than large particles. Particle surface area is of great importance.

As a general rule, best results are achieved with water-insoluble or slightly soluble phosphates on both acid and calcareous soils when they are applied in powdered form or in very fine granules (less than 35 mesh) and mixed thoroughly with the soil of the root zone. This is more important, however, on calcareous than on acid soils. Decreasing granule size increased plant response to fertilizers containing dicalcium phosphate, a sparingly soluble phosphorus compound.

Water-soluble phosphates give best results on acid soils when they are granulated and applied in a band. Granule size is generally satisfactory in the range of 12 to 50 mesh, but the present trend is toward 8 to 12 mesh. Granular forms of water-soluble materials give good results on calcareous soils, although experimental evidence indicates that the best results (as in water-insoluble materials) will probably be obtained from pulverized materials thoroughly mixed with the soil.

Granular materials are preferred to the powdered or pulverized forms because of their greater ease of handling and spreading. Nitric phosphates are generally manufactured in a granular form and are coated and shipped in special moistureproof bags because of their hygroscopicity. As a result, granular nitric phosphates with their low water solubility (less than 50%) are not suitable for use on calcareous soils, whereas granulated materials of high water solubility, such as monoammonium phosphate and triple superphosphate, are quite satisfactory.

Soil Moisture. Moisture content of the soil has a decided effect on the effectiveness and rate of availability of applied phosphorus in various forms. Experimental work has shown that when the soil water content is at field capacity, 50 to 80% of the water-soluble phosphorus can be expected to move out of the fertilizer granule within a 24-hour period. Even in soils with only 2 to 4% moisture, 20 to 50% of the water-soluble phosphorus will move out of the granule within the same time.

Under wet conditions response to granular phosphates of high water solubility is superior to that from powdered materials. Under dry conditions, however, powdered materials are likely to give better results.

Granule Distribution. Distribution of fertilizer granules in the soil, granule diameter, and phosphorus content all affect crop response to applied phosphates. Poor distribution may be caused by inadequate mixing with the soil or by relatively light applications, so that the few granules applied are widely spaced in the soil. Poor distribution, particularly unfavorable in materials of low water solubility, is explained simply by the decreased probability of a plant root reaching a fertilizer granule.

The effects of poor distribution may be partly overcome by the use of highly water-soluble granular materials because of their ability to disperse into soil zones surrounding the granule and to react quickly with the soil constituents.

Rate of Application. At low rates of applied phosphorus water solubility may be much more important than at high rates. This is due to the same factors that influence crop response under conditions of poor distribution and suggests that when optimum application rates cannot be used for some time it is important, to get full benefit from the limited amounts of fertilizer applied, that materials of high water solubility be used. The effect of the degree of water-soluble phosphorus on corn yields in Iowa is shown in Figure 6-13.

Residual Phosphorus. When the amount of phosphorus added to soil as fertilizers exceeds removal by cropping, the phosphorus residues gradually increase with a corresponding rise in phosphorus concentration in the soil solution. Because there is generally a rapid decline in effectiveness of soluble phosphorus fertilizers over the first few months, or years after application, the potential contributions of residual phosphorus to succeeding crops are frequently considered unimportant. The fact that rate of reversion of plant available phosphorus to less soluble and ultimately unavailable forms tends to decline with time is often overlooked. There can be long lasting benefits from residual phosphorus, the duration and magnitude of which depend largely on the rate of initial applications, crop removal, and the buffering capacity of soil for phosphorus.

The performance of initial applications of phosphorus cannot be fully assessed from one or two consecutive crops. On both acidic and basic soils, substantial benefits from residual phosphorus can persist for as long as 5 to 10 years or more. The duration of response will, of course, be influenced by the amount of residual phosphorus.

Initial applications of at least 350 to 360 lb of P_2O_5 per acre are usually required on acid soils with high phosphorus fixation capacity. In some soils such as Oxisols in Hawaii, with extremely high capacity for phosphorus retention, amounts of between 1000 to 2000 lb of P_2O_5 per acre are needed to establish an effective reserve of residual phosphorus.

Field experiments conducted on calcium-dominated soils in Idaho, Montana,

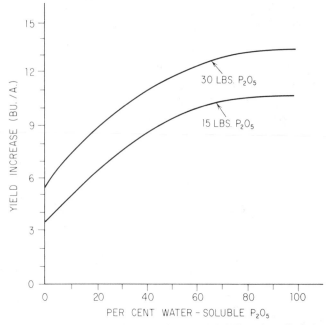

FIGURE 6-13. Effect of rate and water solubility of applied phosphate fertilizer on the yield increase of corn. [Webb et al., *SSSA Proc.*, **22**:533 (1958).]

North and South Dakota, Oregon, New Mexico, Alberta, Saskatchewan, and Manitoba have clearly shown that large batch applications of from about 175 to 250 lb of P_2O_5 per acre can remain effective for periods of 8 to 10 years or longer. The long-term benefits of an initial application of 184 lb of P_2O_5 per acre for seven consecutive winter wheat crops in Montana can be seen in Figure 6-14. Researchers

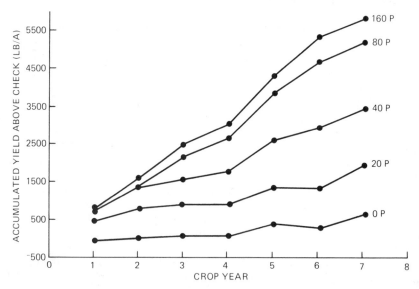

FIGURE 6-14. Average accumulated yield increase above that of the check (no nitrogen or phosphorus) treatment for two study sites for each crop year with 40 lb/Acre of nitrogen. [Halvorson and Black, *Better Crops Plant Food*, **64**:33 (Winter 1981–1982).]

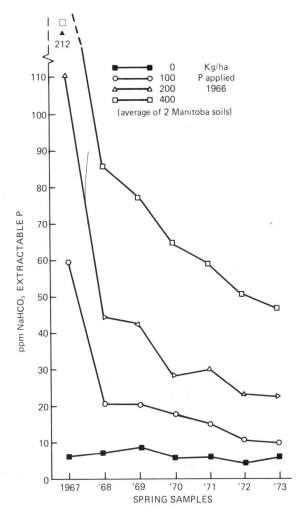

FIGURE 6-15. Effect of single applications of phosphorus in 1966 on the NaHCO₃ extractable phosphorus levels in the soils while being cropped alternately with wheat and flax from 1967 to 1973. [Spratt, *Better Crops Plant Food*, **62**:24 (1978).]

in Montana contend that high levels of residual phosphorus are desirable because they assure that soil phosphorus levels are high enough to meet crop requirements over a wide range of climatic conditions and yield levels.

The effect of large initial additions of phosphorus on available soil phosphorus is demonstrated in Figure 6-15. At all three rates there was a substantial improvement in the level of available phosphorus in these deficient soils. Following a rapid decline in available phosphorus in the first year there was a gradual and steady drop of between 5 to 8 ppm per year until reaching the 10- to 15-ppm range. Phosphate fertilizer is usually recommended when soil test levels are below 15 ppm and large economical yield increases are expected from applied phosphorus when soil tests are below 10 ppm.

Olsen and his associates in Colorado have observed that residual phosphorus from fertilizer or manure occurred mainly as octacalcium phosphate in a group of 23 alkaline and calcareous soils from the eastern part of that state. They found that this compound has a very high availability coefficient for plants. Therefore, accumulation of fertilizer phosphorus in the form of octacalcium phosphate seems to be beneficial since this phosphorus is potentially all available to crops.

There is some question about the need for additions of fresh phosphorus even when residual phosphorus levels are high. Low rates of phosphorus in starter fertilizers placed in or near the seed row are potentially beneficial on high-phosphorus soils when the crop is stressed by cold, wet conditions and by diseases such as root rots. For example, Alessi and Power found that although residual phosphorus contributed significantly to yields of dryland spring wheat at Mandan, North Dakota, additional banding of phosphorus was required to maximize grain production. Annual banding of 30 lb of P_2O_5 per acre with the seed raised yields about 10% over and above the increases resulting from residual phosphorus. This favorable effect of seed placed phosphorus has been attributed to the stimulatory effect it had on wheat growth in the cool springs of the northern Great Plains. Similar observations have been made for other cropping situations such as corn production in Iowa and Wisconsin.

Comparative Fertilizer Value of Various Phosphate Materials

The suitability of various materials as sources of fertilizer phosphorus has been evaluated in thousands of greenhouse and field tests throughout North America and Europe. It is not the purpose of this book to include a discussion of the results of even a fraction of these experiments, for references which themselves include extensive bibliographies of the literature on this subject are cited at the end of this chapter.

Since OSP was almost the sole source of fertilizer phosphorus for many decades, it was only natural that it serve as a yardstick for the evaluation of more recently developed materials. OSP, however, contains sulfur, while most of the newer phosphate materials contain only minor amounts of this nutrient. Under conditions of sulfur deficiency OSP will give superior results, but this is not a valid test of the materials as carriers of *phosphorus*. For that reason CSP has become the preferred standard. If sulfur deficiency is suspected in the test area, it can be uniformly corrected by the addition of a suitable sulfur carrier.

On the basis of published experimental data, the following general conclusions may be drawn concerning the effectiveness of various phosphates under different circumstances:

1. Both water-soluble and citrate-soluble phosphates are *available* to plants. However, there may be a considerable difference in crop response under different circumstances.
2. For maximum yields short-season fast-growing crops and those with restricted root systems generally require a fertilizer containing a high proportion of water-soluble phosphorus. Such crops frequently give only limited response to applications of citrate-soluble materials.
3. A high degree of water solubility (greater than 60%) is less important on long-season crops and perennials with extensive root systems such as permanent pastures and meadows.
4. A high degree of water solubility may be desirable for early growth and stand establishment in crops such as small grains and corn (maize).
5. When the amount of phosphate to be applied is limited, the greatest crop response will almost always be obtained when a large proportion of the fertilizer phosphorus is in water-soluble form and when the fertilizer is applied in a band with or near seed or in a band near transplants. This is particularly true on soils that are depleted or naturally low in phosphorus.
6. On acid to neutral soils granular fertilizers with a high degree of water solubility are more effective than powdered fertilizers containing the same proportion of water-soluble phosphorus when the fertilizer is to be mixed with the soil. Within limits, the larger the fertilizer granule, the greater its effectiveness under these conditions.
7. On acid to neutral soils band application of powdered fertilizers with a high

degree of water solubility will give better results than mixing the fertilizer with the soil.

8. On calcareous soils granular forms of highly water-soluble phosphates will generally give good results, although experimental evidence indicates that better results are sometimes obtained with pulverized materials thoroughly mixed with the soil. Granular nitric phosphates of low water solubility (less than 50%) are not suitable for use on calcareous soils.

9. Best results may be achieved with materials of low water solubility when they are applied in powdered form or in very small granules and mixed thoroughly with the soil. This is somewhat more important, however, on calcareous than on acid soils.

10. Monoammonium phosphate will generally give better results than diammonium phosphate on calcareous soils, although both are water soluble.

11. With phosphates of low water solubility, effectiveness decreases with an increase in particle size.

12. Effectiveness of rock phosphate as a direct-application fertilizer depends on its chemical reactivity which can be estimated by solubility in neutral ammonium citrate. Phosphate rocks with a high potential for direct application have more than 17% of their total phosphorus soluble in neutral ammonium citrate.

13. Partially acidulated phosphate rock may be useful in situations where the reactivity of phosphate rock is inadequate for acceptable short-term yields and where the phosphorus-retention capacity of soil quickly renders soluble phosphorus fertilizers unavailable to crops.

14. The thermal-process phosphates, when finely ground, are satisfactory sources of phosphorus for most crops on acid soils, but they have failed generally to give favorable results on neutral and alkaline soils.

15. Maximum response will not be obtained from an applied phosphatic fertilizer, whether water soluble or water insoluble, unless adequate quantities of the other plant nutrients, including the secondary and micronutrient elements, are present. Experimental evidence indicates that phosphorus utilization by plants may be improved by the presence of nitrogen, especially ammonium ions.

16. Fertilizer application systems which simultaneously place water-soluble phosphorus and ammoniacal sources of nitrogen in the same band are usually more effective than those where nitrogen and phosphorus are applied separately.

In general, it may be concluded that when the granule size and the degree of water solubility of the phosphorus in fertilizers are comparable there is likely to be little if any difference in the effectiveness of materials based on sulfuric, phosphoric, or nitric acid acidulation. However, highly water-soluble materials, such as ammonium phosphate and the superphosphates, will give satisfactory results under all circumstances in which crop response to an applied phosphate is possible. Although water-insoluble phosphates and those of low water solubility may give equally good results under some conditions, they are not as suitable as the water-soluble materials.

Summary

1. Phosphorus occurs in soils in both inorganic and organic forms. The concentration of the inorganic forms ($H_2PO_4^-$, HPO_4^{2-}) in the soil solution is the most important single factor governing the availability of this element to plants. Uptake of $H_2PO_4^-$ is more rapid than HPO_4^{2-}, with the former being most abundant at pH values below 7.2.

2. The concentration of phosphate ion in the soil solution is influenced by the rate and extent to which this element is immobilized by biological factors and by reaction with the mineral fraction of soils. Soils high in soluble iron and aluminum react with ortho- and polyphosphates to form a variety of insoluble compounds, including variscite and strengite, which are largely unavailable to plants. Soluble phosphates also undergo reactions in soils high in clays (especially those of the 1:1 type and accompanying hydrated oxides of iron and aluminum) which convert them to forms of limited availability to plants.

3. Availability of soluble phosphates is reduced by the high calcium activity prevailing in most basic soils. In such soils phosphate is precipitated as relatively insoluble dicalcium phosphate, and other more basic calcium phosphates such as octacalcium phosphate and hydroxyapatite. Insoluble magnesium phosphate compounds including dimagnesium phosphate trihydrate, trimagnesium phosphates, and struvite may form in basic soils containing large amounts of active magnesium.

4. The availability of added water-soluble fertilizer phosphates can be considerably extended by placing them in a band in the soil. Similar results can be obtained by granulating the phosphate materials.

5. The terminology peculiar to phosphate fertilizers was discussed and the terms *water-soluble*, *citrate-soluble*, *available*, and *total phosphorus* were defined. The adequacy of the current method for assaying the availability of phosphate fertilizers was covered.

6. Phosphate fertilizers are classed generally on the basis of their manufacture as heat- or acid-treated phosphates. Heat-treated phosphates are either calcined or fused. Calcining is heat treatment below the melting point of the furnace charge, and fusing results from heating the charge above its melting point. The contained phosphorus is water insoluble. Acid-treated phosphates are those in which the phosphate rock is treated with a strong acid such as sulfuric, phosphoric, or nitric. The phosphorus in unammoniated acid-treated phosphates is largely water soluble.

7. The phosphobacterins, so widely publicized by Soviet scientists, were discussed. Although these bacterial cultures may be required to increase native soil phosphates under conditions of Soviet agriculture, they do not appear to be helpful in areas in which large amounts of inorganic phosphate fertilizers are used.

8. The accumulation and benefits of residual phosphorus were reviewed. Formation of octacalcium phosphate is an important factor in the value of residual phosphorus in basic soils.

9. The behavior and properties of the various phosphate fertilizers in the soil were discussed. The reaction of soluble phosphate fertilizers with various soil components gives rise to what is termed *fertilizer–soil reaction products* and it is the solubility of these compounds that largely governs plant availability of added phosphate fertilizers.

10. The suitability of the various materials as sources of fertilizer phosphorus was covered. As a general rule, those materials that have a high percentage of the contained phosphate in the water-soluble form are more generally acceptable than are those with none or only a small amount. There are, however, certain crops and certain soil conditions with and on which the less water-soluble forms perform as well as those that are water soluble. As with nitrogen, the cost per unit of contained phosphorus should loom large as one of the determining factors in the selection of a phosphate fertilizer, but this decision should be tempered by the response of the crop to water-soluble forms.

Questions

1. What is the original source of soil phosphorus?
2. Is the soil phosphorus in organic combinations available to plants?
3. What are the factors affecting the retention of phosphorus in soils?
4. How is phosphorus availability influenced by soil pH?
5. What are soil phosphate reaction products?
6. What are probably the two most important factors that influence the uptake of phosphorus by plants?
7. What is phosphate retention or fixation? Why is it important agriculturally? Is fixed phosphorus totally lost to plants?
8. What are the various mechanisms of phosphate retention in acid mineral soils?
9. What soil properties influence the retention or fixation of added fertilizer phosphorus?
10. What can be done to reduce the amount of fixation of fertilizer phosphorus?
11. What is the original source of most fertilizer phosphorus?
12. A fertilizer contains 46% phosphorus pentoxide. To what percent of phosphorus does this correspond?
13. Derive the conversion factor:

$$\%P = \frac{\%P_2O_5}{2.29}$$

14. What is meant by ammoniating superphosphates? What is the effect on the water- and citrate-soluble contents of ammoniating OSP and CSP? Why the difference?
15. What acids are commonly used to acidify phosphate rock? Why, specifically, does acid treatment of phosphate rock render the phosphorus more plant available?
16. Describe the soil conditions under which you might expect an appreciable downward movement of phosphorus through the soil profile.
17. What is the significance of a wide R_2O_3/P_2O_5 ratio? A narrow ratio? Is this important to the grower? Why?
18. Under what soil conditions would the band placement of phosphorus result in its greatest utilization by the plant? If there were no such thing as phosphorus fixation, what method of fertilizer placement would probably result in the greatest utilization of this element by plants? Why?
19. Phosphates held in organic combination are generally considered to be of little value to plants during cold weather. Why?
20. A soil was reported to contain 20 ppm of available phosphorus. To how many pounds per acre of ordinary superphosphate (8.5% P) does this correspond?
21. Chemically speaking, on what is the stability of rock phosphate based?
22. Under what types of soil and cropping conditions might the use of rock phosphate give satisfactory results? Explain.
23. With what type of crop would the use of a high-water-soluble phosphate be particularly recommended?
24. What types of phosphate fertilizer are not recommended for use on alkaline and calcareous soils?
25. On what basis should phosphate fertilizers be purchased—total or available phosphorus?

26. What advantages are offered by the high-analysis phosphates such as DAP, MAP, and CSP? What disadvantages?
27. What is residual phosphorus? Why is it important agriculturally?
28. Why are dicalcium phosphate and octacalcium phosphate important reaction products?
29. What is an adsorption isotherm? Write the equations for three of the most widely used adsorption isotherms.
30. What information is not provided by adsorption isotherms? What factor is mainly responsible for the shape of isotherms?
31. What is known about desorption rates?
32. Give brief descriptions of the main pathways of transporting soil phosphorus to plant roots. Can phosphate fertilization alter the importance of any of these pathways?
33. Compare typical soil solution concentrations of phosphorus in unfertilized soil with the required soil solution levels for top yields of corn, wheat, and head lettuce.
34. Define phosphorus intensity and quantity factors. What is labile soil phosphorus?
35. What are polyphosphates? Are they stable in biologically active soils?
36. Briefly describe the sequence of events that takes place during the dissolution of water-soluble phosphate fertilizers.
37. What is meant by biocycling of phosphorus?
38. Describe how the presence of nitrogen improves plant utilization of phosphate fertilizers. Which of the two forms, NH_4^+ or NO_3^-, is most beneficial?
39. What occurs during the hydrolysis of polyphosphates, and what agents are responsible for this reaction?
40. What are typical distances for the initial movement of phosphorus from fertilizer application sites? Will phosphorus in the reaction zones eventually become more uniformly distributed in the soil?

Selected References

Adriano, D. C., G. M. Paulsen, and L. S. Murphy, "Phosphorus–iron and phosphorus–zinc relationships in corn (*Zea mays* L.) seedlings as affected by mineral nutrition." Agron. J. ,**63**:36 (1971).

Alessi, J., and J. F. Power, "Effects of banded and residual fertilizer phosphorus on dryland spring wheat yield in the Northern Plains." *Soil Sci. Soc. Am. J.*, **44**:792 (1980).

Allison, F. E., *Soil Organic Matter and Its Role in Crop Production*. Amsterdam: Elsevier Scientific, 1973.

Allred, S. E., and A. J. Ohlrogge, "Principles of nutrient uptake from fertilizer bands: VI. Germination and emergence of corn as affected by ammonia and ammonium phosphate." *Agron. J.*, **56**: 309 (1964).

Anderson, G., "Assessing organic phosphorus in soils," in F. E. Khasawneh, E. C. Sample, and E. J. Kamprath, Eds., *The Role of Phosphorus in Agriculture*, pp. 411–413. Madison, Wis.: American Society of Agronomy, 1980.

Armiger, W. H., and M. Fried, "The plant availability of various sources of phosphate rock." *SSSA Proc.*, **21**:183 (1957).

Barber, S. A., "Soil–plant interactions in the phosphorus nutrition of plants," in F. E. Khasawneh, E. C. Sample, and E. J. Kamprath, Eds., *The Role of Phosphorus in Agriculture*, pp. 591–615. Madison, Wis.: American Society of Agronomy, 1980.

Barrow, N. J., "Phosphorus in soil organic matter." *Soils Fert.*, **24**:169 (1961).

Barrow, N. J., "Evaluation and utilization of residual phosphorus in soils," in F. E. Khasawneh, E. C. Sample, and E. J. Kamprath, Eds., *The Role of Phos-*

phorus in Agriculture, pp. 333–359. Madison, Wis.: American Society of Agronomy, 1980.

Beaton, J. D., "Fertilizers and their use in the decade ahead." *Proc. 20th Annu. Meet., Agr. Res. Inst.*, St. Louis, Mo. (1971).

Beaton, J. D., and N. A. Gough, "The influence of soil moisture regime and phosphorus source on the response of alfalfa to phosphorus." *SSSA Proc.*, **26:**265 (1962).

Beaton, J. D., and D. W. L. Read, "Phosphorus uptake from a calcareous Saskatchewan soil treated with mono-ammonium phosphate and its reaction products," *Soil Sci.*, **94:**404 (1962).

Beaton, J. D., and D. W. L. Read, "Effect of temperature and moisture on phosphate uptake from a calcareous Saskatchewan soil treated with several pelleted sources of phosphorus." *SSSA Proc.*, **27:**61 (1963).

Beaton, J. D., R. C. Speer, and G. Brown, "Effect of soil temperature and length of reaction period on water solubility of phosphorus in soil fertilizer reaction zones." *Soil Sci. Soc. Am. J.*, **29:**194 (1965).

Beaton, J. D., R. C. Speer, D. W. L. Read, and A. Dueck, "Distribution of monocalcium and monoammonium phosphate reaction products in a calcareous Saskatchewan soil." *Nature (London)*, **198:**813 (1963).

Bell, L. C., and C. A. Black, "Crystalline phosphates produced by interaction of orthophosphate fertilizers with slightly acid and alkaline soils." *Soil Sci. Soc. Am. J.*, **34:**735 (1970).

Bixby, D. W., D. L. Rucker, and S. L. Tisdale, "Phosphatic fertilizers, properties and processes," *Tech. Bull. 8* (rev.). Washington, D.C.: The Sulphur Institute, 1968.

Black, C. A., *Soil–Plant Relationships*, 2nd ed. New York: Wiley, 1968.

Blair, G. J., M. H. Miller, and W. A. Mitchell, "Nitrate and ammonium as sources of nitrogen for corn and their influence on the uptake of other ions." *Agron. J.*, **62:**530 (1970).

Bohn, H. L., B. L. McNeal, and G. A. O'Connor, *Soil Chemistry*. New York: Wiley, 1979.

Bouldin, D. R., and E. C. Sample, "Laboratory and greenhouse studies with monocalcium, monoammonium, and diammonium phosphates." *SSSA Proc.*, **23:**338 (1959).

Bouldin, D. R., J. D. DeMent, and E. C. Sample, "Interaction between dicalcium and monoammonium phosphates granulated together." *J. Agr. Food Chem.*, **8:**470 (1960).

Bouldin, D. R., J. R. Lehr, and E. C. Sample, "The effect of associated salts on transformations of monocalcium phosphate monohydrate at the site of application." *SSSA Proc.*, **24:**464 (1960).

Bouma, D., "The effect of ammonium sulfate usage on the availability of soil phosphorus to citrus." *Australian J. Agr. Res.*, **11:**292 (1960).

Bromfield, S. M., "Some factors affecting the solubility of phosphates during the microbial decomposition of plant material." *Australian J. Agr. Res.*, **11:**304 (1960).

Brown, A. L., and B. A. Krantz, "Source and placement of zinc and phosphorus for corn (*Zea mays* L.)." *SSSA Proc.*, **3:**086 (1966).

Burns, G. R., D. R. Bouldin, C. A. Black, and W. L. Hill, "Estimation of particle size effects of water-soluble phosphate fertilizer in various soils." *SSSA Proc.*, **27:**556 (1963).

Caldwell, A. G., and C. A. Black, "Inositol hexaphosphate: II. Synthesis by soil microorganisms." *SSSA Proc.*, **22:**293 (1958).

Caldwell, A. G., and C. A. Black, "Inositol hexaphosphate: III. Content in soils." *SSSA Proc.*, **22:**296 (1958).

Calvert, D. V., H. F. Massey, and W. A. Seay, "The effect of exchangeable calcium on the retention of phosphorus by clay fractions of soils of the Memphis catena." *SSSA Proc.*, **24:**333 (1960).

Chang, C., and G. J. Racz, "Effects of temperature and phosphate concentration on rate of sodium pyrophosphate and sodium tripolyphosphate hydrolysis in soil." *Can. J. Soil Sci.*, **57:**27 (1977).

Chauhan, B. S., J. W. B. Stewart, and E. A. Paul, "Effect of labile inorganic phosphate status and organic carbon additions on the microbial uptake of phosphorus in soils." *Can. J. Soil Sci.*, **61**:373 (1981).

Classen, N., and S. A. Barber, "A method for evaluating the influence of concentration on the uptake rate of nutrients." *Plant Physiol.*, **54**:564 (1974).

Cole, C. V., and S. R. Olsen, "Phosphorus solubility in calcareous soils: II. Effects of exchangeable phosphorus and soil texture on phosphorus solubility." *SSSA Proc.*, **23**:119 (1959).

Cole, C. V., G. S. Innis, and J. W. B. Stewart, "Simulation of phosphorus cycling in semiarid grasslands." *Ecology*, **58**:1 (1977).

Coleman, N. T., J. T. Thorup, and W. A. Jackson, "Phosphate-sorption reactions that involve exchangeable Al." *Soil Sci.*, **90**:1 (1960).

Cook, R. L., K. Lawton, L. S. Robertson, et al., "Phosphorus solubility, particle size and placement as related to the uptake of fertilizer phosphorus and crop yields." *Commer. Fert.*, **94**:41 (1957).

Cooper, R., "Bacterial fertilizers in the Soviet Union." *Soils Fert.*, **22**:327 (1959).

Cosgrove, D. J., "Metabolism of organic phosphates in soil," in A. D. McLaren and G. H. Peterson, Eds., *Soil Biochemistry*, Vol. 1, pp. 216–228. New York: Marcel Dekker, 1967.

De Datta, S. K., R. L. Fox, and G. D. Sherman, "Availability of fertilizer phosphorus in Latosols of Hawaii." *Agron. J.*, **55**:311 (1963).

Dick, W. A., and M. A. Tabatabai, "Inorganic pyro-phosphatase activity of soils." *Soil Biol. Biochem.*, **10**:59 (1978).

Dormaar, J. F., "Distribution of inositol phosphates in some chernozemic soils of southern Alberta." *Soil Sci.*, **104**:17 (1967).

Dormaar, J. F., "Phospholipids in chernozemic soils of southern Alberta." *Soil Sci.*, **110**:136 (1970).

Dormaar, J. F., "Seasonal pattern of soil organic phosphorus." *Can. J. Soil Sci.*, **52**:107 (1972).

Duncan, W. G., and A. J. Ohlrogge, "Principles of nutrient uptake from fertilizer bands: II. Root development in the band." *Agron. J.*, **50**:605 (1958).

Duncan, W. G., and A. J. Ohlrogge, "Principles of nutrient uptake from fertilizer bands: III. Band volume, concentration, and nutrient composition." *Agron. J.*, **51**:103 (1959).

Eivazi, F., and M. A. Tabatabai, "Phosphatases in soils." *Soil Biol. Biochem.*, **9**:167 (1977).

Englestad, O. P., and S. E. Allen, "Ammonium pyrophosphate and ammonium orthophosphate as phosphorus sources: effects of soil temperature, placement, and incubation." *SSSA Proc.*, **35**:1002 (1971).

Ensminger, L. E., "Response of crops to various phosphate fertilizers." *Alabama Agr. Exp. Sta. Bull. 270* (1950).

Follett, R. H., L. S. Murphy, and R. L. Donahue, *Fertilizers and Soil Amendments*, Englewood Cliffs, N.J.: Prentice-Hall, 1981.

Fox, R. L., "Using phosphate sorption curves to determine P requirements." *Better Crops Plant Food*, **66**:24 (Winter 1981–1982).

Fox, R. L., and E. J. Kamprath, "Adsorption and leaching of P in acid organic soils and high organic matter sand." *SSSA Proc.*, **35**:154 (1971).

Franklin, W. T., and H. M. Reisenauer, "Chemical characteristics of soils as related to phosphorus fixation and availability." *Soil Sci.*, **90**:192 (1960).

Freeman, J. S., and D. L. Rowell, "The adsorption and precipitation of phosphate onto calcite." *J. Soil Sci.*, **32**:75 (1981).

Gachon, L., "Les méthodes d'appréciation de la fertilité phosphorique des sols." *Bull. Assoc. Fr. Etude Sol*, **4**:17 (1969).

Ganiron, R. B., D. C. Adriano, G. M. Paulsen, and L. S. Murphy, "Effect of phosphorus carriers and zinc sources on phosphorus–zinc interaction in corn." *SSSA Proc.*, **33**:306 (1969).

Gerretsen, F. C., "The influence of microorganisms on the phosphate intake by the plant." *Plant Soil*, **1**:51 (1949).

Gilliam, J. W., "Hydrolysis and uptake of pyrophosphate by plant roots." *SSSA Proc.*, **34**:83 (1970).

Gilliam, J. W., and E. C. Sample, "Hydrolysis of polyphosphate in soils: pH and biological effects." *Soil Sci.*, **106:**352 (1968).

Giordano, P. M., E. C. Sample, and J. J. Mortvedt, "Effect of ammonium ortho- and pyrophosphate on Zn and P in soil solution." *Soil Sci.*, **111:**101 (1971).

Gough, N. A., and J. D. Beaton, "Influence of phosphorus source and soil moisture on the solubility of phosphorus." *J. Sci. Food Agr.*, **14:**224 (1963).

Grunes, D. L., "Effect of nitrogen on the availability of soil and fertilizer phosphorus to plants." *Adv. Agron.*, **11:**369 (1959).

Hagin, J., and J. Berkovits, "Efficiency of phosphatic fertilizers of varying water solubility." *Can. J. Soil Sci.*, **41:**68 (1961).

Halvorson, A. D., and A. L. Black, "Long-term benefits of a single broadcast application of phosphorus." *Better Crops Plant Food*, **64:**33 (Winter 1981–1982).

Hashimoto, I., and J. R. Lehr, "Mobility of polyphosphates in soil." *SSSA Proc.*, **37:**36 (1973).

Hashimoto, I., and Z. T. Wakefield, "Hydrolysis of pyrophosphate in soils: response to temperature and effect on heavy-metal uptake by plants." *Soil Sci.*, **118:**90 (1974).

Hemwall, J. B., "The fixation of phosphorus in soils." *Adv. Agron.*, **9:**95 (1957).

Hignett, P. T., and J. A. Brabson, "Phosphate solubility, evaluation of water insoluble phosphorus in fertilizers by extraction with alkaline ammonium citrate solutions." *J. Agr. Food Chem.*, **9:**272 (1961).

Hingston, F. J., A. M. Posner, and J. P. Quirk, "Anion adsorption by goethite and gibbsite: II. Desorption of anions from hydrous oxide surfaces." *J. Soil Sci.*, **25:**16 (1974).

Hinman, W. C., J. D. Beaton, and D. W. L. Read, "Some effects of moisture and temperature on transformation of monocalcium phosphate in soil." *Can. J. Soil Sci.1*, **42:**229 (1962).

Huffman, E. O., "The reactions of fertilizer phosphate with soils." *Outlook Agr.*, **5:**202 (1968).

Huffman, E. O., "Behaviour of fertilizer phosphates." *Trans. 9th Int. Congr. Soil Sci.*, **2:**745 (1968).

Humphreys, F. R., and W. L. Pritchett, "Phosphorus adsorption and movement in some sandy forest soils." *SSSA Proc.*, **35:**495 (1971).

International Fertilizer Development Center, *Fertilizer Manual*. Muscle Shoals, Ala.: International Fertilizer Development Center, 1979.

Jacob, K. D., Ed., *Fertilizer Technology and Resources in the United States*. Vol. 3 of *Agronomy*. New York: Academic Press, 1953.

Kamprath, E. J., "Residual effect of large applications of phosphorus on high phosphorus fixing soils." *Agron. J.*, **59:**25 (1967).

Khasawneh, F. E., I. Hashimoto, and E. C. Sample, "Reactions of ammonium ortho- and polyphosphate fertilizers in soil: II. Hydrolysis and reactions with soil." *Soil Sci. Soc. Am. J.*, **43:**52 (1979).

Khasawneh, F. E., E. C. Sample, and I. Hashimoto, "Reactions of ammonium ortho- and polyphosphate fertilizers in soil: I. Mobility of phosphorus." *Soil Sci. Soc. Am. J.*, **38:**446 (1974).

Khasawneh, F. E., E. C. Sample, and E. J. Kamprath, Eds., *The Role of Phosphorus in Agriculture*. Madison, Wis.: American Society of Agronomy, 1980.

Larsen, J. E., R. Langston, and G. F. Warren, "Studies on the leaching of applied labeled phosphorus in organic soils." *SSSA Proc.*, **22:**558 (1958).

Larsen, J. E., G. F. Warren, and R. Langston, "Effect of iron, aluminum, and humic acid on phosphorus fixation by organic soils." *SSSA Proc.*, **23:**438 (1959).

Lathwell, D. J., J. T. Cope, Jr., and J. R. Webb, "Liquid fertilizers as sources of phosphorus for field crops." *Agron. J.*, **52:**251 (1960).

Lawton, K., and J. A. Vomocil, "The dissolution and migration of P from granular superphosphate in some Michigan soils." *SSSA Proc.*, **18:**26 (1954).

Lawton, K., C. Apostolakis, and R. L. Cook, "Influence of particle size, water solubility and placement of fertilizers on the nutrient value of phosphorus in mixed fertilizers." *Soil Sci.*, **82:**465 (1956).

Lewis, E. T., and G. J. Racz, "Phosphorus movement in some calcareous and noncalcareous Manitoba soils." *Can. J. Soil Sci.*, **49:**305 (1969).

Lindsay, W. L., *Chemical Equilibria in Soils*. New York: Wiley, 1979.

Lindsay, W. L., and E. C. Moreno, "Phosphate phase equilibria in soils." *SSSA Proc.*, **24:**177 (1960).

Lindsay, W. L., and H. F. Stephenson, "Nature of the reactions of monocalcium phosphate monohydrate in soils: I. The solution that reacts with the soil." *SSSA Proc.*, **23:**12 (1959).

Lindsay, W. L., and H. F. Stephenson, "Nature of the reactions of monocalcium phosphate monohydrate in soils: II. Dissolution and precipitation reactions involving iron, aluminum, manganese, and calcium." *SSSA Proc.*, **23:**18 (1959).

Lindsay, W. L., and H. F. Stephenson, "Nature of the reactions of monocalcium phosphate monohydrate in soils: IV. Repeated reactions with metastable triple-point solution." *SSSA Proc.*, **23:**440 (1959).

Lindsay, W. L., and A. W. Taylor, "Phosphate reaction products in soil and their availability to plants." *Trans. 7th Int. Congr. Soil Sci.*, **3:**580 (1960).

Lindsay, W. L., and P. L. G. Vlek, "Phosphate minerals," in J. B. Dixon and S. B. Weed, Eds., *Minerals in Soil Environments*, pp. 639–672. Madison, Wis.: Soil Science Society of America, 1977.

Lindsay, W. L., A. W. Frazier, and H. F. Stephenson, "Identification of reaction products from phosphate fertilizers in soils." *SSSA Proc.*, **26:**446 (1962).

McLean, E. O., and T. J. Logan, "Sources of phosphorus for plants grown in soils with differing phosphorus fixation tendencies." *SSSA Proc.*, **34:**907 (1970).

Martin, W. E., J. Vlamis, and J. Quick, "Effect of ammoniation on availability of phosphorus in superphosphates as indicated by plant response." *Soil Sci.*, **75:**41 (1959).

Mattingley, G. E. G., "The agricultural value of some water and citrate soluble fertilizers: an account of recent work at Rothamsted and elsewhere." *Proc. Fert. Soc. (London)*, **75:**57 (1963).

Miller, M. H., and A. J. Ohlrogge, "Principles of nutrient uptake from fertilizer bands: I. Effect of placement of nitrogen fertilizer on the uptake of band-placed phosphorus at different soil phosphorus levels." *Agron. J.*, **50:**95 (1958).

Miller, M. H., and V. N. Vij, "Some chemical and morphological effects of ammonium sulfate in a fertilizer phosphorus band for sugar beets." *Can. J. Soil Sci.*, **42:**87 (1962).

Miner, G., and E. Kamprath, "Reactions and availability of banded polyphosphate in field studies." *SSSA Proc.*, **35:**927 (1971).

Mishustin, E. N., and A. N. Naumova, "Bacterial fertilizers, their effectiveness and mechanism of action." *Mikrobiologiya*, **31:**543 (1962) [*Soils Fert.*, **25:**382 (1962)].

National Fertilizer Solutions Association, *Liquid Fertilizer Manual*. Peoria, Ill.: National Fertilizer Solutions Association, 1980.

Neller, J. R., "Effect on plant growth of particle size and degree of solubility of phosphorus labeled in 12-12-12 fertilizer." *Soil Sci.*, **94:**413 (1962).

Norland, M. A., R. W. Starostka, and W. L. Hill, "Crop response to phosphorus fertilizers as influenced by level of phosphorus solubility and by time of placement prior to planting." *SSSA Proc.*, **22:**529 (1958).

Nye, P. H., and W. N. M. Foster, "A study of the mechanism of soil phosphate uptake in relation to plant species." *Plant Soil*, **9:**338 (1958).

Olsen, S. R., and A. D. Flowerday, "Fertilizer phosphorus interactions in alkaline soils," in R. A. Olson, T. J. Army, J. J. Hanway, and V. J. Kilmer, Eds., *Fertilizer Technology and Use*, 2nd ed. Madison, Wis.: Soil Science Society of America, 1971.

Olsen, S. R., and F. E. Khasawneh, "Use and limitations of physical–chemical criteria for assessing the status of phosphorus in soils," in F. E. Khasawneh, E. C. Sample, and E. J. Kamprath, Eds., *The Role of Phosphorus in Agriculture*, pp. 361–410. Madison, Wis.: American Society of Agronomy, 1980.

Olsen, S. R., R. A. Bowman, and F. S. Watanabe, "Behavior of phosphorus in the soil and interactions with other nutrients." *Phosphorus Agr.* **70:**31 (1977).

Olsen, S. R., F. S. Watanabe, and R. A. Bowman, "Evaluation of soil phosphate residues by plant uptake and extractable phosphorus." *Proc. 1978 Soils Crops*

Workshop, Publ. 390, pp. 48–75, Univ. of Saskatchewan, Saskatoon, (1978).

Olsen, S. R., F. S. Watanabe, and R. E. Danielson, "Phosphorus absorption by corn roots as affected by moisture and phosphorus concentration." *SSSA Proc.*, **25:**289 (1961).

Olson, R. A., A. C. Drier, and G. W. Lowrey, et al., "Availability of phosphoric acid to small grains and subsequent clover in relation to: I. Nature of soil and method of placement." *Agron. J.*, **48:**106 (1956).

Olson, R. A., T. J. Army, J. J. Hanway, and V. J. Kilmer, Eds., *Fertilizer Technology and Use*, 2nd ed. Madison, Wis.: Soil Science Society of America, 1971.

Ozanne, P. G., "Phosphate nutrition of plants—a general treatise," in F. E. Khasawneh, E. C. Sample, and E. J. Kamprath, Eds., *The Role of Phosphorus in Agriculture*, pp. 559–589. Madison, Wis.: American Society of Agronomy, 1980.

Ozanne, P. G., D. J. Kirton, and T. C. Shaw, "The loss of phosphorus from sandy soils." *Australian J. Agr. Res.*, **12:**409 (1961).

Parish, D. H., L. L. Hammond, and E. T. Craswell, "Research on modified fertilizer materials for use in developing-country agriculture." *180th Nat'l Meet., Am. Chem. Soc.*, Las Vegas, Nev. (1980).

Philen, O. D., Jr., and J. R. Lehr, "Reactions of ammonium polyphosphates with soil minerals." *SSSA Proc.*, **31:**196 (1967).

Phillips, A. B., and J. R. Webb, "Production, marketing, and use of phosphorus fertilizers," in R. A. Olson, T. J. Army, J. J. Hanway, and V. J. Kilmer, Eds., *Fertilizer Technology and Use*, 2nd ed. Madison, Wis.: Soil Science Society of America, 1971.

Phillips, A. B., R. D. Young, F. G. Heil, and M. M. Norton, "Fertilizer technology, high analysis superphosphate by the reaction of phosphate rock with superphosphoric acid." *J. Agr. Food Chem.*, **8:**310 (1960).

Pierre, W. H., and A. G. Norman, Eds., *Soil and Fertilizer Phosphorus in Crop Nutrition*. Vol. 4 of *Agronomy*. New York: Academic Press, 1953.

Potash and Phosphate Institute, *Phosphorus for Agriculture: A Situation Analysis*. Atlanta, Ga.: Potash and Phosphate Institute, 1978.

Power, J. F., D. L. Grunes, W. O. Willis, and G. A. Reichman, "Soil temperature and phosphorus effects upon barley growth." *Agron. J.*, **55:**389 (1963).

Power, J. F., D. L. Grunes, G. A. Reichman, and W. O. Willis, "Soil temperature effects on phosphorus availability." *Agron. J.*, **56:**545 (1964).

Power, J. F., W. O. Willis, D. L. Grunes, and G. A. Reichman, "Effect of soil temperature, phosphorus, and plant age on growth analysis of barley." *Agron. J.*, **59:**231 (1967).

Racz, G. J., and R. J. Soper, "Reaction products of orthophosphates in soils containing varying amounts of calcium and magnesium." *Can. J. Soil Sci.*, **47:**223 (1967).

Ragland, J. L., and W. A. Seay, "The effects of exchangeable calcium on the retention and fixation of phosphorus by clay fractions of soil." *SSSA Proc.*, **21:**261 (1957).

Read, D. W. L., "Bio-cycling of phosphorus in soil." *Better Crops Plant Food*, **66:**24 (1982).

Read, D. W. L., and R. Ashford, "Effect of varying levels of soil and fertilizer phosphorus and soil temperature on the growth and nutrient content of bromegrass and Reed canarygrass." *Agron. J.*, **60:**680 (1968).

Reisenauer, H. M., "Absorption and utilization of ammonium nitrogen by plants," in D. R. Nielsen, Ed., *Nitrogen in the Environment*, Vol. 2: *Soil–Plant–Nitrogen Relationships*, pp. 157–170. New York: Academic Press, 1978.

Riley, D., and S. A. Barber, "Effect of ammonium and nitrate fertilization on phosphorus uptake as related to root-induced pH changes at the root–soil interface." *Soil Sci. Soc. Am. J.*, **35:**301 (1971).

Robinson, R. R., V. G. Sprague, and C. F. Gross, "The relation of temperature and phosphate placement to growth of clover." *SSSA Proc.*, **23:**225 (1959).

Rogers, H. T., "Crop response to nitrophosphate fertilizers." *Agron. J.*, **43:**468 (1951).

246 SOIL FERTILITY AND FERTILIZERS

Russell, E. W., *Soil Conditions and Plant Growth*, 10th ed. London: Longmans, Green, 1973.

Sample, E. C., F. E. Khasawneh, and I. Hashimoto, "Reactions of ammonium ortho- and polyphosphate fertilizers in soil: III. Effects of associated cations." *Soil Sci. Soc. Am. J.*, **43**:58 (1979).

Sample, E. C., R. J. Soper, and G. J. Racz, "Reactions of phosphate fertilizers in soils," in F. E. Khasawneh, E. C. Sample, and E. J. Kamprath, Eds., *The Role of Phosphorus in Agriculture*, Chapter 11. Madison, Wis.: American Society of Agronomy, 1980.

Sanchez, P. A., and G. Uehara, "Management considerations for acid soils with high phosphorus fixation capacity," in F. E. Khasawneh, E. C. Sample, and E. J. Kamprath, Eds., *The Role of Phosphorus in Agriculture*, pp. 471–514. Madison, Wis.: American Society of Agronomy, 1980.

Sauchelli, V., Ed., *Chemistry and Technology of Fertilizers*. Am. Chem. Soc. Monogr. 148. New York: Reinhold, 1960.

Sauchelli, V., *Manual on Fertilizer Manufacture*, 3rd ed. Caldwell, N.J.: Industry Publications, 1963.

Savant, N. K., and G. J. Racz, "Effect of added orthophosphate on pyrophosphate hydrolysis in soil." *Plant Soil*, **36**:719 (1972).

Savant, N. K., and G. J. Racz, "Hydrolysis of sodium pyrophosphate and tripolyphosphate by plant roots." *Soil Sci.*, **113**:18 (1972).

Schenk, M. K., and S. A. Barber, "Root characteristics of corn genotypes as related to P uptake." *Agron. J.*, **71**:921 (1979).

Shapiro, R. E., W. H. Armiger, and M. Fried, "The effect of soil water movement vs. phosphate diffusion on growth and phosphorus content of corn and soybeans." *SSSA Proc.*, **24**:61 (1960).

Sheard, R. W., G. J. Bradshaw, and D. L. Massey, "Phosphorus placement for the establishment of alfalfa and bromegrass." *Agron. J.*, **63**:922 (1971).

Sheppard, S. C., and G. J. Racz, "Phosphorus nutrition of crops as affected by temperature and water supply." *Proc. Western Canada Phosphate Symp.*, pp. 159–199. Edmonton, Alberta: Alberta Soil and Feed Testing Laboratory, 1980.

Sibbesen, E., "Some new equations to describe phosphate sorption by soils." *J. Soil Sci.*, **32**:67 (1981).

Simpson, K., "Factors influencing uptake of phosphorus by crops in southeast Scotland." *Soil Sci.*, **92**:1 (1961).

Skujins, J. J., "Enzymes in soil," in A. D. McLaren and G. H. Peterson, Eds., *Soil Biochemistry*, pp. 371–414. New York: Marcel Dekker, 1967.

Smith, J. H., and F. E. Allison, "Phosphobacterin as a soil inoculant. Laboratory, greenhouse, and field evaluation." *USDA Tech. Bull. 1263* (1962).

Smith, J. H., F. E. Allison, and D. A. Soulides, "Evaluation of phosphobacterin as a soil inoculant." *SSSA Proc.*, **25**:109 (1961).

Soon, Y. K., and M. H. Miller, "Changes in the rhizosphere due to NH_4^+ and NO_3^- fertilization and phosphorus uptake by corn seedlings (*Zea mays* L.)." *Soil Sci. Soc. Am. J.*, **41**:77 (1977).

Soper, R. J., and G. J. Racz, "Reactions and behavior of phosphorus fertilizer in soil." *Proc. Western Canada Phosphate Symp.*, pp. 65–91. Edmonton, Alberta: Alberta Soil and Feed Testing Laboratory, 1980.

Spiers, G. A., and W. B. McGill, "Effects of phosphorus addition and energy supply on acid phosphatase production and activity in soils." *Soil Biol. Biochem.*, **11**:3 (1979).

Spratt, E. D., "Residual fertilizer P benefits wheat and oilseed production." *Better Crops Plant Food*, **62**:24 (1978).

Stanberry, C. O., W. H. Fuller, and N. R. Crawford, "Comparison of phosphate sources for alfalfa on a calcareous soil." *SSSA Proc.*, **24**:304 (1960).

Stanford, G., and D. R. Bouldin, "Biological and chemical availability of phosphate–soil reaction products." *Trans. 7th Int. Congr. Soil Sci.*, **2**:388 (1960).

Stewart, J. W. B., "The importance of P cycling and organic P in soils." *Better Crops Plant Food*, **64**:16 (1980–1981).

Striplin, M. M., Jr., D. McKnight, and G. H. Megar, "Fertilizer materials, phosphoric acid of high concentration." *J. Agr. Food Chem.*, **6**:298 (1958).

Swaby, R. J., and J. Sherber, "Phosphate dissolving microorganisms in the rhizosphere of legumes." *Proc. Univ. Nottingham 5th Easter School Agr. Sci.* 289 (1958).

Syers, J. K., and I. K. Iskander, "Soil–phosphorus chemistry," in I. K. Iskander, Ed., *Modeling Wastewater Renovation Land Treatment*, pp. 571–599. New York: Wiley, 1981.

Tabatabai, M. A., and W. A. Dick, "Distribution and stability of pyrophosphates in soils." *Soil Biol. Biochem.*, **11**:655 (1979).

Taylor, A. W., W. L. Lindsay, and E. O. Huffman, "Potassium and ammonium taranakites, amorphous aluminum phosphate, and variscite as sources of phosphate for plants." *SSSA Proc.*, **27**:148 (1963).

Terman, G. L. "Agronomic results with polyphosphate fertilizers." *Phosphorus Agr.* **65**:21 (1975).

Terman, G. L., D. R. Bouldin, and J. R. Lehr, "Calcium phosphate fertilizers: I. Availability to plants and solubility in soils varying in pH. *SSSA Proc.*, **22**:25 (1958).

Terman, G. L., D. R. Bouldin, and J. R. Webb, "Phosphorus availability, crop response to phosphorus in water-soluble phosphates varying in citrate solubility and granule size." *J. Agr. Food Chem.*, **9**:166 (1961).

Terman, G. L., J. D. DeMent, and O. P. Engelstad, "Crop response to fertilizers varying in solubility of the phosphorus as affected by rate, placement and seasonal environment." *Agron. J.*, **53**:221 (1961).

Terman, G. L., J. D. DeMent, L. G. Clements, and J. A. Lutz, Jr., "Crop response to ammoniated superphosphates and DCP as affected by granule size, water solubility, and time of reaction with the soil." *J. Agr. Food Chem.*, **8**:13 (1960).

Thien, S. J., and W. W. McFee, "Influence of nitrogen on phosphorus absorption and translocation in *Zea mays*." *SSSA Proc.*, **34**:87 (1970).

Tisdale, S. L., and D. L. Rucker, "Crop response to various phosphates." *Tech. Bull. 9*. Washington, D.C.: The Sulphur Institute, 1964.

Tisdale, S. L., and E. Winters, "Crop response to calcium metaphosphate on alkaline soils." *Agron. J.*, **45**:228 (1953).

Van Burg, P. F. J., "The agricultural evaluation of nitrophosphates with particular reference to direct and cumulative phosphate effects and to interaction between water solubility and granule size." *Proc. Fert. Soc. (London)*, **75**:7 (1963).

Van Wazer, J. R., Ed., *Phosphorus and Its Compounds*, Vol. II. New York: Interscience, 1961.

Waggaman, W. A., *Phosphoric Acid, Phosphates, and Phosphatic Fertilizers*, 2nd ed. New York: Reinhold, 1952.

Walsh, L. M., and J. D. Beaton, Eds., *Soil Testing and Plant Analysis*, rev. ed. Madison, Wis.: Soil Science Society of America, 1973.

Watanabe, F. S., S. R. Olsen, and R. E. Danielson, "Phosphorus availability as related to soil moisture." *Trans. 7th Int. Congr. Soil Sci.*, **3**:450 (1960).

Webb, J. R., and J. T. Pesek, "An evaluation of phosphorus fertilizers varying in water solubility: I. Hill applications for corn." *SSSA Proc.*, **22**:533 (1958).

Webb, J. R., and J. T. Pesek, "An evaluation of phosphorus fertilizers varying in water solubility: II. Broadcast applications for corn." *SSSA Proc.*, **23**:381 (1959).

Webb, J. R., K. Eik, and J. T. Pesek, "An evaluation of phosphate fertilizers applied broadcast on calcareous soils for corn." *SSSA Proc.*, **25**:232 (1961).

Webb, J. R., J. T. Pesek, and K. Eik, "An evaluation of phosphorus fertilizers varying in water solubility: III. Oat fertilization.' *SSSA Proc.*, **25**:22 (1961).

Welch, L. F., D. L. Mulvaney, L. V. Boone, G. E. McKibben, and J. W. Pendleton, "Relative efficiency of broadcast versus banded phosphorus for corn." *Agron. J.*, **58**:283 (1966).

White, R. E., "Retention and release of phosphate by soil and soil constituents," in P. B. Tinker, Ed., *Soils and Agriculture*, pp. 71–114. Oxford: Blackwell Scientific, 1980.

Williams, C. H., and A. Steinbergs, "Sulphur and phosphorus in some eastern Australian soils." *Australian J. Agr. Res.*, **9**:483 (1958).

Williams, C. H., E. G. Williams, and N. M. Scott, "Carbon, nitrogen, sulphur, and phosphorus in some Scottish soils." *J. Soil Sci.*, **11:**334 (1960).

Wright, B., J. D. Lancaster, and J. L. Anthony, "Availability of phosphorus in ammoniated ordinary superphosphate." *Mississippi Agr. Exp. Sta. Tech. Bull.* *52* (1963).

Chapter 7

Soil and Fertilizer Potassium

POTASSIUM is absorbed by plants in larger amounts than any other mineral element except nitrogen. Although the total potassium content of soil is usually many times greater than the amount taken up by a crop during a growing season, in most cases only a small fraction of it is available to plants. Potassium–soil mineral relationships are consequently of major significance and they will be considered in this chapter.

Potassium Content of Soils

Unlike phosphorus, potassium is present in relatively large quantities in most soils. Phosphorus concentration of the earth's crust is only 0.11%, whereas that of potassium averages about 1.9%. Concentration of potassium in soil normally varies between about 0.5 to 2.5% and typically is about 1.2%.

The potassium content of soils is variable and may range from only a few hundred pounds per acre furrow slice in coarse-textured soils formed from sandstone or quartzite to 50,000 lb or more in fine-textured soils formed from rocks high in the potassium-bearing minerals.

Soils of the southeastern and southern coastal plain areas of the United States are formed from marine sediments which have been highly leached and are generally low in their content of plant nutrients. Soils of the Midsouth are formed from igneous, sedimentary, and metamorphic rocks. Because of their age and the climate under which they were formed, these soils are low in potassium even though the parent rocks are frequently high in potassium-bearing minerals. The soils of the middle and far western states are formed from geologically young parent materials and under conditions of lower rainfall. The low content of potassium in the coastal soils of the Pacific Northwest is accounted for by high rainfall in that area.

In tropical soils the total content of potassium may be quite low because of their origin, high rainfall, and continued high temperatures. The last two factors have hastened the release and leaching of soil potassium over the years.

Origin of Soil Potassium

Exclusive of that added in fertilizers, the potassium contained in soils originates from the disintegration and decomposition of rocks containing potassium-bearing minerals. The minerals that are generally considered to be original sources of potassium are the potassium feldspars orthoclase and microcline ($KAlSi_3O_8$), muscovite [$KAl_3Si_3O_{10}(OH)_2$], biotite [$K(Mg,Fe)_3AlSi_3O_{10}(OH)_2$], and phlogopite

$[KMg_2Al_2Si_3O_{10}(OH)_2]$. The nature and mode of weathering of these potassium-bearing minerals depend to a large extent on their properties and the environment. As far as plant response is concerned, the availability of potassium in these minerals, although slight, is of the order biotite > muscovite > potassium feldspars.

Potassium is also found in the soil in the form of secondary or clay minerals: (1) illites or hydrous micas, (2) vermiculites, (3) chlorites, and (4) interstratified minerals in which two or more of the preceding types occur in more or less random arrangement in the same particle.

Forms of Soil Potassium

Soil potassium is generally believed to exist in four categories based on their availability to crops. These groups in increasing order of availability along with estimates of the approximate amounts in each are: mineral (structural), 5000 to 25,000 ppm; nonexchangeable (fixed or difficultly available), 50 to 750 ppm; exchangeable, 40 to 600 ppm; and solution, 1 to 10 ppm.

The principal components of each of these categories and the transfer of potassium between them are depicted in Figure 7-1. The relative importance of the four groupings depends on the mineralogical composition of the soil. Distinctions between the four forms are more arbitrary than implied in Figure 7-1 since there is a progression of forms of potassium in soils.

Movement of potassium in dynamic soil systems produces equilibria between the various potassium fractions. It is signified in Figure 7-1 that exchangeable and solution potassium equilibrate rapidly, whereas difficultly available or fixed potassium equilibrates very slowly with the exchangeable and solution forms. Transfer of potassium from the mineral or structural fraction to any of the other three forms is extremely slow in most soils and as a result it is considered essentially unavailable to crops during a single growing season.

Potassium is held tightly in feldspars and micas which are very resistant to weathering. Fixed or difficultly available or nonexchangeable potassium is present mainly within clay minerals such as illite, vermiculite, and chlorite. The small particle size of clays facilitates potassium release. Exchangeable potassium, held by electrostatic forces on clay and organic matter in the solid phase of soil, is easily transferred to the soil solution by exchange with other cations.

Because of the continuous removal of potassium by crop uptake and leaching, a static equilibrium probably never occurs. There is a continuous but slow transfer of potassium in the primary minerals to the exchangeable and slowly available forms. Under some soil conditions, including applications of large amounts of fertilizer potassium, some reversion to the slowly available form will occur. The unavailable form accounts for 90 to 98% of the total soil potassium, the slowly available form, 1 to 10%, and the readily available form, 0.1 to 2%.

Soil Solution Potassium. Plants take up potassium as the K^+ ion mainly from, or via, the soil solution. The concentration of potassium needed in the soil solution will vary considerably depending on type of crop and the amount of growth wanted. German workers report that optimum potassium level in the soil solution appears to be 20 to 60 ppm, depending on the nature of crop, soil structure, general fertility level, and moisture supply. Singh and Jones at the University of Idaho using a sorption-isotherm technique found that 14.5 ppm potassium in the equilibrium solution was sufficient for optimum yields of celery and potato crops having a high potassium requirement. For crops such as tomatoes and beans, which have intermediate potassium needs, they observed that 8.7 ppm potassium in sorption equilibrium solution would maximize growth.

Levels of water-soluble potassium in soils of humid regions commonly range

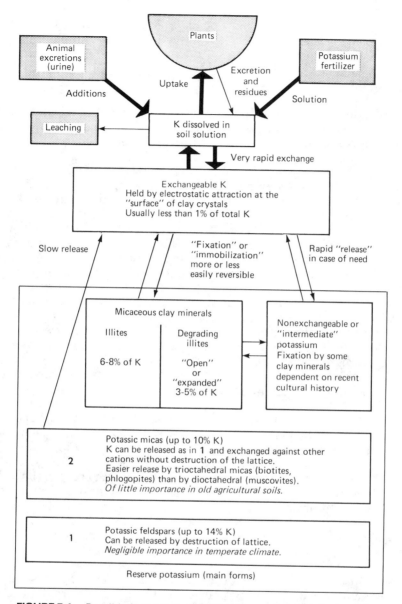

FIGURE 7-1. Possible forms and equilibria of potassium in soils. (Modified from Quémener, *IPI Res. Topics No. 4*. Bern: International Potash Institute, 1979.)

between 1 and 80 ppm, with a value of 4 ppm being representative. Potassium concentration in soil saturation extracts usually varies from 3 to 156 ppm and the higher figures are found in arid or saline soils. Concentration of water-soluble potassium may be as low as 8 ppm in deficient soils. Under field conditions the potassium concentration of the soil solution varies considerably due to concentration and dilution processes brought about by evaporation and precipitation, respectively.

Effectiveness of soil solution K^+ for crop nourishment is influenced by the presence of other cations, particularly calcium and magnesium. It may also be desirable to consider aluminum ions in very acidic soils and sodium ions in salt-affected soils. The activity ratio (AR_e^K),

$$\frac{\text{activity of potassium}}{\sqrt{\text{activity of calcium and magnesium}}} \quad \text{or} \quad \frac{(a_K)}{\sqrt{a_{Ca\,+\,Mg}}}$$

in a solution in equilibrium with a soil, has provided a satisfactory estimate of the availability or the potential of potassium. This ratio is a measure of the "intensity" of labile potassium in the soil. It indicates the potassium that is immediately available to crop roots.

A single AR_e^K measurement will only characterize momentary availability. Soils with similar AR_e^K values may have quite differing capacities for maintaining AR_e^K while K^+ is being depleted by plant uptake or by leaching. Thus to describe fully the potassium status of soils it is necessary to specify not only the current potential of potassium in the labile pool but also the way in which the intensity depends on the quantity of labile potassium present. More discussion of quantity–intensity relationships (Q/I) appears in the following section on exchangeable potassium since this fraction has the principal role in replenishing solution potassium.

Electroultrafiltration (EUF), a procedure involving both electrodialysis and ultrafiltration, is proposed by West German researchers to satisfactorily characterize the availability of soil solution potassium to crops. The principle of the EUF method consists of utilizing the acceleration imposed upon ions by an electric field for the separation of ions from soil colloids. It measures intensity of soil solution potassium and, as might be expected, it is closely correlated with the activity ratio AR_e^K. There are indications that EUF is better suited than AR_e^K for distinguishing between soils of different potassium availabilities.

Absorption by Plants. Only a small proportion (6 to 10%) of the total potassium required by plants is obtained by direct contact between roots and soil particles. Transport of potassium from various soil regions toward plant roots is therefore important in the potassium nourishment of plants. This potassium transport is achieved mainly by mass flow and diffusion which occur in the soil solution.

Intensity of potassium is of great significance in plant nutrition because of the effect it has on diffusion and mass flow. The total amount of potassium that can diffuse through the soil solution is directly related to the intensity of potassium in the soil solution at any given time. Transfer of potassium in the flow of water to roots is also controlled by potassium intensity.

Mass Flow. The amount of potassium conveyed by mass flow depends on the amount of water used by plants and the potassium content of the water flowing through soil to the root. It is normally considered to be a less important mechanism than diffusion for supplying potassium to root surfaces.

An estimate of the relative contribution of mass flow transport of potassium to the total needs of a plant can be made by assuming that the normal potassium concentration in tissue of the crop is 2.5% and that each unit weight of the crop results in transpiration of 400 times its weight of water. On this basis water moving to the root would need to contain in excess of 60 ppm potassium. Since most soils, particularly those in humid regions, have only about one-tenth this concentration, mass flow is considered to be of only minor importance in satisfying the potassium requirements of crops. It is apparent, however, that mass flow could be a major factor in the potassium nourishment of crops grown in soils naturally high in water-soluble potassium or where fertilizer potassium has elevated potassium levels in the soil solution. Involvement of mass flow in transport of potassium from around fertilizer sources is identified in Table 7-1.

TABLE 7-1.   Mechanisms and Speed of Potassium Transport in Soils

Situation	Mechanism	Speed (cm/day)
In profile	Mainly mass flow	Up to 10
Around fertilizer source	Mass flow and diffusion	~0.1
Around root	Mainly diffusion	0.01–0.1
Out of clay interlayers	Diffusion	10^{-7}

Source: Tinker, in G. S. Sekhon, Ed., *Potassium in Soils and Crops.* New Delhi: Potash Research Institute of India, 1978.

Diffusion. Diffusion is the movement of potassium in response to a concentration gradient resulting in potassium transport from a zone of high concentration to one of lower concentration. It is a slow process in comparison to mass flow, as is evident from the values in Table 7-1. This pathway of potassium transport takes place in the moisture films surrounding soil particles and it is favored by conditions that permit ready passage of the migrating K^+ ions.

Potassium diffusion to roots is limited to very short distances in soil, usually only 1 to 4 mm from the root surface during a growing season. Some understanding of the general nature of the diffusion of potassium to corn roots can be gained from examination of the autoradiographs appearing in Figure 7-2. These autoradiographs were made using ^{86}Rb, an element very similar to potassium, but having a more suitable radioactive isotope. The behavior of Rb in soil closely resembles that of potassium. Figure 7-2, in effect, shows that potassium within only a few millimeters of the root can be absorbed. Potassium farther away may be in available form, but it is not available positionally.

According to Barber and his colleagues at Purdue University, diffusion in many soils will furnish from 88 to 96% of the potassium presented to roots. German investigators showed that 93% of the 496 lb of potassium per acre taken up by a sugar beet crop was accounted for by diffusion. Because diffusion is prominent in providing potassium to crops, factors that affect the rate of supply by diffusion will also influence the availability of potassium to crops. Amounts of potassium diffusing to root systems are dependent on the potassium concentration gradient, the rate of diffusion, and the surface area of roots.

Potassium concentration gradients depend on soil potassium levels and the lowering of potassium concentration in the rhizosphere by active potassium uptake by roots. The latter varies with plant species and is influenced by root metabolism and rate of potassium absorption by the root.

The major soil and environmental conditions, including moisture content, tortuosity of the diffusion path, temperature, and so on, which influence diffusion rates of ions such as K^+, were reviewed briefly in Chapters 4 and 6.

The amounts of potassium present in soil solution are insufficient to meet crop requirements throughout a growing season. Therefore, for satisfactory potassium nutrition of plants soil solution potassium must be continuously replenished from the other fractions identified in Figure 7-1.

Exchangeable Potassium. Like other exchangeable cations, the K^+ ion is held around negatively charged soil colloids by electrostatic attraction. Cations held in this manner are easily displaced or exchanged when the soil is brought into contact with neutral salt solutions. The amount of potassium exchanged varies with the cation used in the measurement. In many laboratories neutral normal ammonium acetate solution is the standard solution used for determining the amount of exchangeable potassium in soils. Usually, less than 1% of the total potassium in soils occurs in this form.

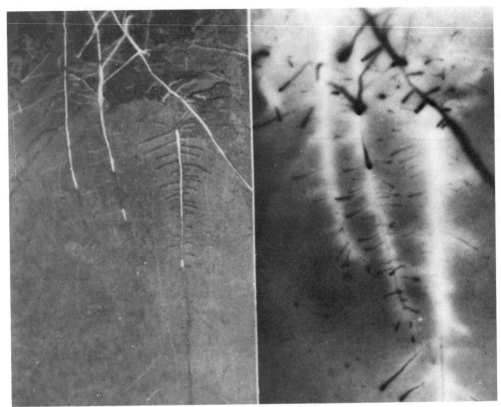

FIGURE 7-2. Autoradiograph showing the depletion of ^{86}Rb about a corn root due to diffusion of Rb to the root. The dark areas are where ^{86}Rb is present, the light area, are areas of depletion. (Barber et al., in "Production, marketing, and use of potassium fertilizers," in R. A. Olson, T. J. Army, J. J. Hanaway, and V. J. Kilmer, Eds., *Fertilizer Technology and Use*, 2nd ed. Madison, Wis.: Soil Science Society of America, 1971.)

The distribution of potassium between negatively charged sites on soil colloids and the soil solution is a function of the kinds and amounts of complementary cations, the anion concentration, and the properties of the soil cation exchange materials. Calcium is commonly the major cation in the soil solution and on the exchange complex. Consequently, calcium–potassium relationships have been the object of considerable study.

Some of the principles of K^+ exchange from soil colloids are summarized by the two equations that follow. Consider first the reaction

$$\text{clay} \begin{bmatrix} \text{Al} \\ \text{K} \\ \text{K} \\ \text{K} \\ \text{K} \end{bmatrix} + \text{CaSO}_4 \rightarrow \begin{bmatrix} \text{Al} \\ \text{Ca} \\ \text{K} \\ \text{K} \end{bmatrix} \text{clay} + \text{K}_2\text{SO}_4$$

If a soil colloid is saturated with potassium and a neutral salt such as calcium sulfate is added, there will be a replacement of some of the adsorbed potassium ions by the calcium ions. The amount of replacement taking place will depend on the nature and amount of the added salt as well as that of the ionic population adsorbed on the clays. On some soils used for production of perennial crops, calcium sulfate is applied to encourage potassium movement into the subsoil, where it becomes available to roots deeper in the profile.

Suppose that a soil condition is represented by the following equation:

$$\text{clay}\begin{array}{c}\text{Ca}\\\text{Al}\\\text{Al}\end{array}\Big] + \text{KCl} \rightarrow \begin{array}{c}{}^{1\!/_2}\text{Ca}\\\text{K}\\\text{Al}\\\text{Al}\end{array}\Big[\ \text{clay} + {}^{1\!/_2}\text{CaCl}_2$$

This soil clay contains adsorbed calcium and aluminum ions to which potassium chloride has been added. Because the calcium is more easily replaced than the aluminum, the added potassium will replace some of the calcium and will itself be adsorbed onto the surface of the clay. This reaction illustrates an important point—the greater the degree of calcium saturation, the greater the adsorption by clay of potassium from the soil solution. This is not inconsistent with the point previously illustrated in which the calcium from calcium sulfate replaced potassium from the colloid. Calcium, when added as a neutral salt, replaces aluminum only with great difficulty, and, if a soil clay contains, in the adsorbed form, ions such as K^+, Na^+, and NH_4^+ in addition to Al^{3+}, these ions, rather than the aluminum, will be replaced. In such cases there will be a net transfer of potassium to the soil solution.

Sandy soils with a high degree of base saturation lose less of their exchangeable potassium by leaching than soils with a low degree of base saturation. Liming is the common means by which the base saturation of soils is increased, and it must follow that liming decreases the loss of exchangeable potassium.

Exchange Sites. Exchangeable potassium on soil colloids is not homogeneous. It is held at three types of exchange sites or binding positions as shown in Figure 7-3. The planar position (p position) on the outside surfaces of some phyllosilicates such as mica is rather unspecific for potassium. On the other hand, the edge position (e) and the inner position (i) particularly have a rather high specificity for potassium.

The K^+ ions bound by a p position are in equilibrium with fairly high potassium concentration of the soil solution, whereas concentrations of soil solution potassium in equilibrium with potassium held on i and e sites is rather low. Under field conditions, soil solution potassium concentrations are likely the net result of the three possible equilibria. A high proportion of potassium adsorption by clay minerals tends to saturate the specific binding sites resulting in higher potassium concentrations in the equilibrated soil solution.

Because of the major role of exchangeable potassium in replenishing soil solution potassium removed by cropping or lost by leaching, there is much interest in defining the relationship between exchangeable potassium (Q for quantity) and

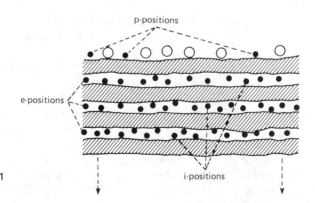

FIGURE 7-3. Binding sites for potassium on 2:1 clay minerals such as illite, vermiculite, and chlorite. [Mengel and Haeder, *Potash Rev.*, **11**:1 (1973).]

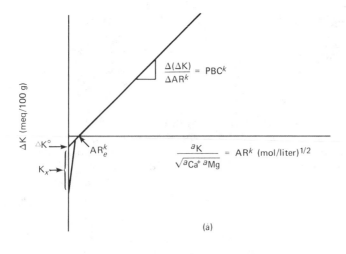

$$\frac{\Delta(\Delta K)}{\Delta AR^k} = PBC^k$$

$$\frac{{}^{a}K}{\sqrt{{}^{a}Ca^{+}\,{}^{a}Mg}} = AR^k \ (\text{mol/liter})^{1/2}$$

(a)

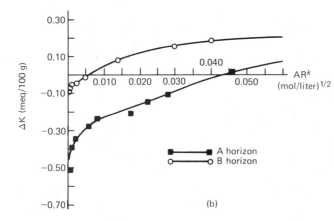

(b)

FIGURE 7-4. (a)Typical quantity/intensity (Q/I) plot and (b) the Q/I relationship in Kalmia soil with 2.24 metric tons/ha of lime and O kg/ha of potassium. [Sparks and Liebhardt, *Soil Sci. Soc. Am. J.*, **45**:786 (1981).]

the activity of soil solution potassium (I for intensity). The Q/I concept developed by Beckett in England is used by many as an aid in predicting the potassium status of soils. A number of researchers use the Q/I relationship to describe subsequent (Q) and immediate (I) availability of potassium to plants.

A typical Q/I curve along with one for a specific Delaware soil is represented in Figure 7-4. Identification and interpretation of the parameters involved in a plot of this type follow.

ΔK = quantity by which the soil gains or loses potassium in reaching equilibrium or the quantity (Q) factor

AR^K = activity ratio for potassium or the intensity (I) factor

ΔK^0 = labile or exchangeable potassium

AR^K_e = equilibrium activity ratio for potassium

K_x = specific potassium sites

PBC^K = potential buffering capacity

The potassium intensity factor (AR^K) is computed from the measured concentrations of calcium, magnesium, potassium, and sodium corrected to the appropriate activities by application of extended Debye–Hückel theory.

The AR_e^K value is a measure of availability or intensity of labile potassium in soil and it can be increased by potassium fertilization. There are suggestions that liming can either increase or decrease this value.

Labile soil potassium (held in p positions) may be more reliably estimated by ΔK^o than by the usual measurement of exchangeable potassium. Higher values of labile potassium ($-\Delta K^o$) have indicated a greater potassium release into soil solution, resulting in a larger pool of labile potassium. Fertilizer potassium will increase the labile potassium pool and liming of cropped soils has also increased it.

The PBC^K value is a measure of the ability of the soil to maintain the intensity of potassium in the soil solution and is proportional to the cation exchange capacity. A high PBC^K signifies good potassium supplying power while a low figure suggests a need for frequent potassium fertilization. Liming has increased this value, presumably as a result of the increase in pH-dependent cation exchange capacity.

The linear portion of Q/I curves such as is illustrated in Figure 7-4(a) is attributed to nonspecific sites for potassium, whereas the curved part is ascribed to specific sites with a high potassium affinity. As pointed out earlier, the nonspecific sites are associated with planar surfaces while the specific sites are believed to occur on edges of clay crystals and in interlayer or wedge zones of weathered micas.

If PBC^K is low, small changes in exchangeable potassium produce large differences in soil solution potassium. Potential buffering capacity is extremely small in sandy soils in which the cation exchange capacity is due mainly to organic matter. In such soils, intense leaching or rapid plant growth can seriously deplete available potassium in just a few days.

In general, the relation between exchangeable potassium and soil solution potassium is a good measure of the availability of the more labile potassium in soils to plants. The ability of a soil to maintain the activity ratio against depletion by plant roots and leaching is governed partly by the character of the labile potassium pool and also by the rate of release of fixed potassium, and by the diffusion and transport of potassium ions in the soil solution.

Nonexchangeable and Mineral Potassium. Soil solution and exchangeable potassium, the two forms described so far, are regarded as readily available but in most soils they comprise only a very small proportion of the total potassium, usually less than one percent. The remainder is generally referred to as (1) nonexchangeable and (2) mineral or reserve potassium. Because mineral or reserve potassium is largely nonexchangeable, it will be discussed in this section along with more active nonexchangeable potassium occurring in illitic clay and other 2:1 types of intergrade minerals. Potassium associated with these clay minerals is generally considered to be slowly available.

Although nonexchangeable potassium reserves are not always immediately available, they can contribute significantly to maintenance of the labile potassium pool in soil. In some soils, nonexchangeable potassium becomes available as the exchangeable and water-soluble potassium is removed by cropping or lost by leaching. In others, release from nonexchangeable reserves is too slow to meet crop requirements.

This form of potassium is a constituent of various minerals, and as indicated in Figure 7-1, two main groups are involved:

1. Primary minerals, feldspars, and micas (phlogopites, muscovites, biotites), originating from the parent rock.
2. Secondary minerals (clays of the illite family) formed by alteration of micas.

Nonexchangeable potassium in clays of the illite family, especially in vermiculite and 2:1 intergrade clay minerals, is often measured by extraction with a strong acid such as 1 N boiling nitric acid. With certain soils, nonexchangeable potassium release helps to explain crop responses or absence of responses to fertilizer potassium applied at various levels of exchangeable potassium.

The potassium release characteristics of nonexchangeable potassium fractions depend on the type of the potassium-containing minerals as well as their amount and particle size distribution. Past cropping history and the nature and amount of potassium fertilizer residues also influence the liberation of potassium from soil and its components.

Rate of potassium replenishment of the labile pool in soils is largely governed by the nature and weathering processes of micas and feldspars in which soil potassium principally exists. Feldspars have a three-dimensional crystal structure with potassium located at the interstices throughout the mineral lattice. Release from such a three-dimensional tektosilicate is more difficult than from the interlamellar surface of micas. Potassium can only be liberated from feldspars by destruction of the mineral. In micas, interlayer potassium can be released by exchange with other cations without fundamental alteration of the mineral.

Feldspars. The potassium feldspars are the largest natural reserve of potassium in many soils. In moderately weathered soils, there are usually considerable quantities of potassium feldspars. They often occur in much smaller amounts or may even be absent in strongly weathered soils such as those in humid tropical areas. Potassium feldspars exist in the light mineral fraction of soils. These potassium-bearing minerals may be present in the clay fraction as well as in the silt and sand fractions.

Figure 7-5 shows the percent distribution of feldspar and mica potassium in clay, silt, and sand fractions of two Saskatchewan soils. Most of the potassium feldspar was found in the silt and sand fractions. The observed partition of both potassium-containing minerals is due to the effects of soil-forming processes and topography.

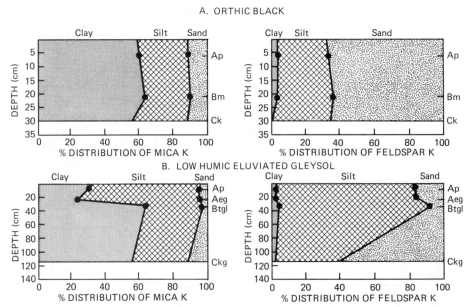

FIGURE 7-5. Distribution of mica potassium and feldspar potassium among clay, silt, and sand fractions of the Orthic Black (A) and Low Humic Eluviated Gleysol (B) profiles in the Oxbow catena. [Somasiri et al., *Soil Sci. Soc. Am. J.*, **35**:500 (1971).]

TABLE 7-2. Apparent Rate Constant and Arrhenius Heat of Activation for the Release of Lattice Potassium from Potassium Minerals

Mineral	Rate Constant (h^{-1}) at:		Arrhenius Heat of Activation (kcal/mol)
	28°C	38°C	
Biotite	1.46×10^{-2}	3.09×10^{-2}	14.00
Phlogopite	9.01×10^{-4}	2.44×10^{-3}	18.57
Muscovite	1.39×10^{-4}	4.15×10^{-4}	20.41
Microcline	7.67×10^{-5}	2.63×10^{-4}	22.97

Source: Huang et al., *Trans. 9th Int. Congr. Soil Sci.*, **2:**705 (1968).

Micas. The micas are 2:1 layer-structured silicates composed of a sheet of alumina octahedra between two sheets of silica tetrahedra. In muscovite (dioctahedral) only two out of the three octahedral positions are occupied by trivalent ions (Al^{3+}). Biotite and phlogopite, the other major types of mica in soils, are classified as trioctahedral and have all three octahedral positions filled by cations such as Mg^{2+} and Fe^{2+}. Potassium ions reside mainly in between the silicate layers.

Bonding of potassium is stronger in dioctahedral than in trioctahedral structure. As a consequence, members of the mica group of potassium-bearing minerals have varying abilities to supply potassium (Table 7-2). The rate of potassium liberation from biotite is 13 to 16, 75 to 105, and 118 to 190 times faster than from phlogopite, muscovite, and microcline feldspar, respectively. Rate of potassium release from phlogopite is approximately six to seven times greater than from microcline. Muscovite is capable of liberating its potassium at about twice the rate possible from microcline.

The readiness with which potassium is released from soil minerals has been estimated by calculating heat of activation values from the Arrhenius equation. It is apparent in Table 7-2 that minerals with the highest heats of activation release less potassium per unit time.

The gradual release of potassium from positions in the mica lattice results in the formation of illite (hydrous mica) and eventually vermiculite with accompanying gain of water or OH_3^+ and swelling of the lattice (Figure 7-6). There is also an increase in specific surface charge and cation exchange capacity of the clay minerals formed during the weathering and transformation of mica.

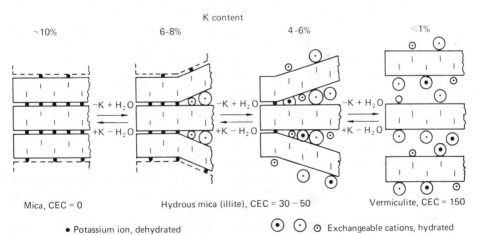

FIGURE 7-6. Schematic weathering of micas and their transformation into clay minerals: a matter of potassium release and fixation. (McLean, E. O. in G. S. Sekhon, Ed., *Potassium in Soil and Crops*, pp. 1–13. New Delhi: Potash Research Institute of India, 1979.)

Potassium release from mica (and the reverse reaction, fixation) is both a cation exchange and a diffusion process requiring time for the exchanging cation to reach the site and for the exchanged ion (K^+) to diffuse from it. Low potassium concentration or activity in the soil solution favors the liberation of interlayer potassium. Thus depletion of potassium by the plant or leaching may induce release of potassium from nonexchangeable interlayer positions.

It is possible for K^+ to be progressively released and diffused from all interlayer locations, or it may come only from alternate interlayers leading to formation of interstratified mica-vermiculite.

Conversion of freshly applied potassium and/or soil solution potassium to fixed or nonexchangeable forms can also occur. The various factors affecting potassium fixation will be summarized later in this chapter.

Potassium Fixation

It was observed in the early 1920s, at the Rothamsted Research Station in England, that potassium could be fixed and released by soils. Since this early finding these reactions of potassium in soils have become widely recognized.

Potassium fixation does not occur to the same extent in all soils or under all conditions. It reaches its maximum, however, in soils high in 2:1 clays and with large amounts of illite. Both fixation and release of potassium are shown in Figure 7-6. Fixation of potassium is the result of reentrapment of K^+ ions between the layers of the 2:1 minerals, especially those such as illite.

Other cations, excepting NH_4^+, are apparently too large for ready entry into openings in the oxygen network of the silica sheets. Potassium ions are sufficiently small to gain entrance into the silica sheets where they become held very firmly by electrostatic forces. The NH_4^+ ion has nearly the same ionic radius as the K^+ ion and is subject to similar fixation by 2:1 clays. Cations such as Ca^{2+} and Na^+ have different ionic radii and are not susceptible to entrapment by expanding-type clays.

The interrelationship between the fixation of K^+ and NH_4^+ was touched on briefly in the section on ammonium fixation in Chapter 5. Because NH_4^+ can be fixed by clays in a manner similar to that of K^+, its presence will alter both the fixation of added potassium and the release of fixed potassium. Even as the presence of K^+ can block the release of fixed NH_4^+, as pointed out in Chapter 5, so can the existence of NH_4^+ block the release of fixed potassium. This is illustrated in Figure 7-7, which shows a reduction in the release of nonexchangeable potassium with increasing amounts of added NH_4^+. The NH_4^+ ions evidently are held in openings in the oxygen network of the silica sheets, thus closing the adjacent sheets and further trapping the K^+ ions already present. This phenomenon may be of some importance in fine-textured agricultural soils which have a high fixation capacity for both K^+ and NH_4^+. However, it is not generally considered to be a serious factor in limiting crop response to either applied NH_4^+ or K^+.

Clay Minerals. As indicated previously, potassium fixation is controlled mainly by clay minerals, with both their quantity and nature being important. Illite, weathered mica, vermiculite, smectite, and interstratified minerals take part in fixation reactions. The 1:1-type minerals such as kaolinite do not fix potassium.

Soil Reaction. Potassium fixation capacity can be reduced by the presence of Al^{3+} and aluminum-hydroxide cations and their polymers which form under acid conditions. These Al^{3+} cations will occupy the potassium selective binding sites.

The occurrence of hydroxyl aluminum-iron interlayer groups under acidic con-

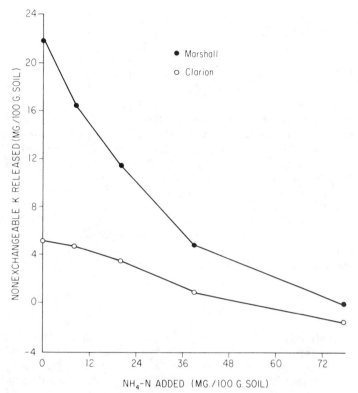

FIGURE 7-7. Nonexchangeable potassium released by Marshall and Clarion surface soils during a 10-day cropping period as influenced by the amount of added NH_4^+. [Welch and Scott, *SSSA Proc.*, **25**:102 (1961).]

ditions will prevent the collapse of silica layers in highly expanded clays. This separation prevents potassium ions from being trapped and subsequently fixed by collapsing silica layers.

Concentration of Added Potassium. Increasing the concentration of K^+ in soils with a high fixation capacity will obviously encourage greater fixation.

Wetting and Drying. Air drying of some soils high in exchangeable potassium will result in fixation and a decline in exchangeable potassium. On the other hand, drying of field-moist soils with low to medium levels of potassium, particularly subsoils, will frequently increase exchangeable potassium. In some cases the exchangeable potassium in subsoils will be increased several-fold by drying. Scott and his co-workers at Iowa State University have explained the release of potassium upon drying as being caused by cracking of edge-weathered micas and exposing interlayer potassium, which can then be released to exchange sites.

Investigations conducted with standard minerals at Ohio State University by Richards and McLean indicated that most of the potassium fixation under moist conditions was associated with illite or minerals containing mica as an impurity. Vermiculite or other minerals containing expanded mica (beidellite, illite) were mainly responsible for potassium fixed during drying. Soil clay fractions of mixed mineral compositions fixed more potassium, especially in a dry–rewet sequence, than did similar mixtures prepared from standard clay minerals.

The effects of wetting and drying on the availability of potassium under field conditions are not known. They are important, however, in soil testing. Frequently, soil test procedures call for the air drying of samples before analysis. It is apparent from the foregoing discussion that this drying treatment can substantially modify soil test potassium values—and subsequent recommendations for potassium fertilization.

Freezing and Thawing. The freezing and thawing of moist soils may also be important in release of fixed potassium and in fixation of the exchangeable form. Controlled laboratory studies have shown that, with alternate freezing and thawing, certain soils will release to the exchangeable form a fraction of the fixed potassium. In other soils, particularly those high in exchangeable potassium, no such release was observed, and in fact, some of the exchangeable potassium became fixed in less readily available form. This phenomenon was observed in soils that contained appreciable quantities of illite. It seems probable that freezing and thawing may play a significant role in the potassium supply of certain soils depending on their clay mineralogy and degree of weathering.

Retention of potassium in less available or fixed forms is of considerable significance to the practical aspects of farming. As with phosphorus, the conversion of potassium to slowly available or fixed forms reduces its immediate value as a plant nutrient. However, it must not be assumed that potassium fixation is completely unfavorable.

Potassium fixation, in the first place, results in a conservation of this element which otherwise might be lost by leaching on sandy soils. In the second place, fixed potassium tends to become available over a long period of time and is thus not entirely lost to plants, although crop plants do vary in their ability to utilize slowly available potassium.

Soil Factors Affecting Potassium Availability to Plants

Kinds of Clay Minerals. The greater the proportion of clay minerals high in potassium, the greater the potential potassium availability in a soil. For example, soils containing vermiculite or montmorillonite will have more potassium than soils containing predominately kaolinitic clays, which are more highly weathered and very low in potassium. However, human-made differences appear when soils with vermiculite and montmorillonite are intensively cropped over a period of time without supplemental potassium. They may also be low in potassium.

Occasionally, sandy soils low in potassium, such as in Delaware, may show little yield response from applied potassium. This lack of response is attributed to release of potassium from fixed and mineral forms rapidly enough to maintain adequate exchangeable and solution potassium. Similar behavior has been observed in soils in other areas, including Nebraska and Kansas.

Cation Exchange Capacity (CEC). This is related to amount and type of clay and the amount of organic matter. Finer-textured soils usually have a higher CEC and can hold a greater amount of exchangeable potassium. However, a higher level of exchangeable potassium does not always mean that a higher level of potassium will be maintained in the soil solution. In fact, soil solution potassium in the finer-textured soils (loams and silt loams) may be considerably lower than that in a coarse-textured soil (sandy) at any given level of exchangeable potassium.

In Ohio, sufficiency levels of potassium are considered to increase with soil CEC according to this formula:

sufficiency exchangeable potassium (pp2m or lb/A) for corn = $220 + (5 \times \text{CEC})$

Based on this formula a soil with a CEC of 15 would have a sufficiency level of 295 lb exchangeable potassium. There are indications that the optimum level for soybeans is higher than for corn, 280 rather than 220 in the formula above.

The Amount of Exchangeable Potassium. Determination of exchangeable potassium is the universal measure for predicting potassium availability and potassium needs. It is a helpful tool and many studies show the relationship between soil test potassium and response to applied potassium.

It is not to be implied that additions of some nutrient should be stopped because of its already high soil level. Maximum economic returns from a farming enterprise are almost always obtained only when high soil fertility is maintained, not only for potassium but for the other nutrients as well. What is implied, however, is that fertilizer applications of potassium can be adjusted downward with increasing levels of available soil potassium. If the soil level of an element begins to decrease, fertilizer applications can again be increased. Soil testing will provide the farm operator with a measure of the fertility level of his soil and fertilizer applications of the various plant nutrients can be adjusted to maintain or increase crop production.

In the preceding section, mention was made of the lack of response on soils testing low in potassium. On the other hand, studies in North Dakota, Montana, and the prairie provinces indicate a small but nevertheless profitable response of small grain and other crops to applied potassium on soils with test values of over 800 lb/A. In Illinois, profitable responses by corn to potassium on soils considered to be high in potassium have been obtained.

Capacity to Fix Potassium. This has been mentioned earlier. Investigations in Illinois indicate that on the average 4 lb/A of K_2O will increase the exchangeable potassium 1 lb. However, the amount of K_2O needed may vary from 2 to 50 lb/A or more depending on the soil. Fortunately, some of the potassium that is fixed may be subsequently released to crops. However, the release may be too slow for high levels of crop production.

Subsoil Potassium and Rooting Depth. The exchangeable levels of potassium in subsoils vary with soil type. However, little progress has been made in relating subsoil potassium to crop need for and responses to potassium.

In Montana, it has been shown that subsoil physical factors affect potassium availability. Average annual temperature of the subsoil at 20 in., subsoil consistence or force required to break peds, and amounts of clay in the zone of free calcium carbonates in the soil were related to potassium availability and crop response to applied potassium. These three variables accounted for 80% of the variability in yield response to applied potassium.

Soil Moisture. There was some discussion of the involvement of soil moisture in potassium diffusion and uptake by plants in the section on diffusion. It is important to remember that with low soil moisture, water films around soil particles are thinner and discontinuous, resulting in a more tortuous path for K^+ movement. This slows down movement of potassium to roots by diffusion. With increased potassium levels or higher moisture contents in the soil, potassium diffusion is accelerated.

Evidence of the substantial effects that soil moisture and temperature have on potassium transport in soil has been obtained by Skogley and co-workers at Montana State University (Figure 7-8). In this Montana study, hydrogen-saturated cation exchange resin sinks were placed in the center of soil packed into cylinders. Diffusion of potassium from soil with varying moisture and temperature conditions

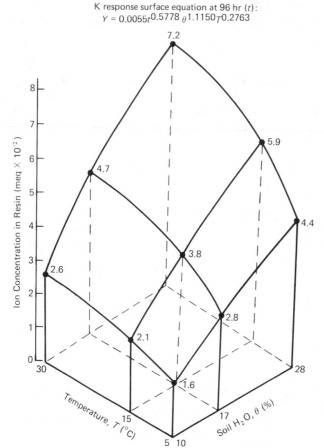

K response surface equation at 96 hr (t):
$$Y = 0.0055t^{0.5778}\theta^{1.1150}T^{0.2763}$$

FIGURE 7-8. Diffusion of potassium to an ion exchange resin sink in Bozeman silt loam during 96 hours as influenced by temperature and soil moisture. [Skogley, *Proc. 32nd Annu. Northwest Fert. Conf.*, Billings, Mont. (1981).]

into the resin sinks was measured over a 96-hour period. Increasing soil moisture from 10 to 28% increased total potassium transport by up to 175%.

In Figure 7-9 impeded potassium diffusion and the resultant reduction in growth of corn caused by low soil moisture content can be recognized. A split-root technique was used in this study which by means of a nutrient solution provided a part of the root with a full supply of water and all mineral nutrients except potassium. Potassium was supplied by growing the other part of the root in soil with different contents of exchangeable potassium maintained at several moisture levels. Corn yields declined as the moisture content of the soil dropped from 38% to 22%, but these yield depressions were largely offset by high levels of exchangeable potassium. It is concluded that the yield reduction was due to insufficient potassium diffusion from soil to the corn roots because an increase in exchangeable potassium increased the yield.

Aeration. Respiration and the normal functioning of roots are very dependent on an adequate oxygen supply. Under high levels of moisture or in compact soils, oxygen supply is lowered and absorption of potassium and other mineral elements is slowed. It can be seen in Table 7-3 that the inhibitory action of poor aeration on nutrient uptake is most pronounced with potassium.

Cropping systems that are detrimental to soil tilth and which cause reduced soil porosity and increased compaction have been found to impair potassium uptake. It was observed in an earlier study in Montana that marked decreases in potassium

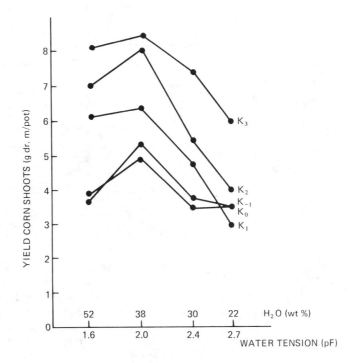

FIGURE 7-9. Effect of the soil moisture and the potassium status of the soil on growth of corn. [Mengel and Haeder, *Potash Rev.*, **11**:1 (1973).]

levels in sugar beet petioles were associated with high soil moisture and consequent reduced soil air space.

Applications of potassium fertilizer to increase the concentration of potassium in soil have helped overcome the depressive influence of poor aeration on potassium absorption by plants.

Soil Temperature. Temperature as an environmental factor strongly affects the uptake of potassium by plants. There is uncertainty about just how much the effect of temperature on potassium uptake is due to changes in availability of soil potassium or to modifications in root activity and rate of plant physiological processes. Another possibility is alteration of the structure of water and aqueous solutions. It is generally agreed that reduced temperature slows down plant processes, plant growth and rate of potassium uptake.

Research conducted at Purdue University has shown that potassium influx into corn roots at 15°C (59°F) was only about one-half of that at 29°C (84°F). This

TABLE 7-3. Uptake of Nutrients by Corn Grown in Nonaerated and Aerated Cultures of a Silt Loam Soil Containing 50% Water

Component Measured	Relative Uptake: Nonaerated/Aerated
Potassium	0.3
Nitrogen	0.7
Magnesium	0.8
Calcium	0.9
Phosphorus	1.3
Dry matter	0.6

Source: Lawton, *Soil Sci. Soc. Am. J.*, **10**:263 (1946).

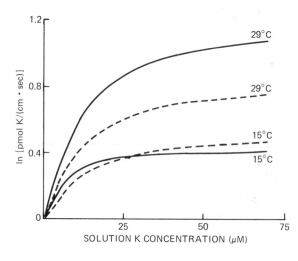

FIGURE 7-10. Rate of potassium influx into young corn roots is increased by higher temperature and potassium concentration in solution. Solid lines are for 11-day-old and dashed lines for 16-day-old corn roots. [Ching and Barber, *Agron. J.*, **71**:1040 (1979).]

difference is shown graphically in Figure 7-10. In the same study, root length increase over a 6-day period was eight times greater at 29°C than at 15°C. Potassium concentration in the shoot was 8.1% at 29°C and 3.7% at 15°C.

Additionally, the Purdue University researchers investigated the interaction of the rate of applied potassium and root temperature on total potassium uptake by corn plants. Some of this work is summarized in Figure 7-11. From the sloping part of each curve it can be seen that potassium uptake per unit of potassium applied was 2.6 times higher at 29°C than at 15°C. The soil ceased to limit uptake at approximately the same potassium level at both temperatures. In spite of the restricted potassium influx into roots at 15°C, shown previously in Figure 7-10, a low level of potassium in the soil was insufficient to maximize total potassium uptake at 15°C. The supplemental potassium needed to increase potassium uptake at this low temperature is thought to overcome the adverse effect that low temperature has on rate of diffusion.

As indicated above, capacity of the soil to supply potassium to roots is reduced by the effect that low temperature has on diffusion. The potassium diffusion study represented in Figure 7-8 is further evidence of the large effect that temperature has on this phenomenon. Increasing temperature from 5°C to 30°C increased potassium accumulation in the resin sink by about 65% at each level of soil moisture.

Also influencing the capability of soil to satisfy plant requirements for potassium is the concentration of potassium in soil solution. The values in Table 7-4 show less potassium in soil solution when soils were equilibrated at 15°C than at 29°C. This effect of temperature was greatest at low potassium levels. Also of special

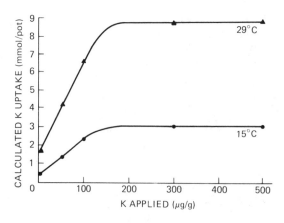

FIGURE 7-11. Total potassium uptake by corn plants growing at 15 and 29°C root temperatures as a function of potassium level in Raub soil. Predicted potassium uptake was highly correlated with observed potassium uptake ($r^2 = .98$). [Ching and Barber, *Agron. J.*, **71**:1040 (1979).]

TABLE 7-4. Effect of the Soil Potassium Level and Temperature on the Initial Concentration of Potassium in Soil Solution (C_{li}), Buffer Power of Potassium on the Solid Phase for Potassium in Solution (*b*), and Effective Diffusion Coefficient ($\bar{D}e$) in the Raub Silt Loam

K Added (μg/g)	C_{li}(μmol/ml) at:		b at:		$\bar{D}e \times 10^7$ (cm²/sec) at:	
	15°C	29°C	15°C	29°C	15°C	29°C
0	0.046	0.089	32	23	0.15	0.39
50	0.174	0.256	12	9.5	0.50	0.94
100	0.355	0.516	8.7	3.3	0.69	2.7
300	1.97	2.66	2.2	1.8	2.7	5.0
500	6.26	8.10	1.2	1.2	5.0	7.5
700	11.90	13.90	1.2	1.2	5.0	7.5

Source: Ching and Barber, *Agron. J.,* **71**:1040 (1979).

interest is the fact that in the situation where no potassium was added, the effective diffusion coefficient at 15°C was only 0.4 of what it was at 29°C. The divergences in effective diffusion coefficients at the two temperatures tended to diminish at the higher soil potassium levels.

Providing high levels of potassium is seen as a practical way of overcoming some of the problems of low temperature. Transport of potassium from fertilizer sources was identified in Table 7-1 as being appreciably faster than movement around roots or out of clay interlayers. Temperature effects are probably a major reason crop responses are seen to row-applied fertilizer for early-planted crops such as corn. The beneficial effect of potassium fertilization for early seedings of barley grown on Montana soils high in available potassium is mainly attributed to improving potassium supply under cool soil conditions (Table 7-5).

Soil pH. In very acid soils toxic amounts of exchangeable aluminum and manganese create an unfavorable root environment for the uptake of potassium or any other element. Plant uptake of magnesium and other basic cations is usually restricted in such soils. Impairment of magnesium absorption will in itself reduce plant growth, as well as potassium uptake and utilization.

Liming very acid soils to nearly neutral pH levels has had an inconsistent effect on potassium concentration in the soil solution, with increases, decreases, and no effects having been observed. Some of these differences may be related to the considerable effect that liming has on the forms and reactivity of soil aluminum.

TABLE 7-5. Effect of Potassium Fertilization on Early Seedings of Barley Grown on Montana Soils High in Available Potassium, 1968–1973

Seeding Date	K_2O* (lb/A)	Yield (bu/A)
April 6	0	48
	20	55
May 6	0	36
	20	42
June 3	0	30
	20	33

* N at 60 lb and P_2O_5 at 25 lb/A per year.
Source: Dubbs, *Better Crops Plant Food,* **65**:27 (1981).

When acid soils are limed, exchangeable Al^{3+} and hydroxyaluminum cations such as $Al(OH)_2^+$ are converted to insoluble $Al(OH)_3$. This change removes the Al^{3+} from cation exchange competition with K^+ and it frees blocked binding sites so that K^+ can compete with Ca^{2+} for them. As a consequence, much greater amounts of K^+ can be held by clay colloids and removed from the soil solution. Leaching losses of potassium will also likely be reduced.

Raising soil pH from 5.5 to 7.0 will favor the collapse of silicate layers of expanded clays and trap K^+ already present in the interlayers. Hydroxyaluminum cations keep the clay layers wedged apart but lose this ability when they are changed to $Al(OH)_3$. Potassium trapped in this manner is unaccessible to plants.

It is generally agreed that the use of calcitic liming materials on acid soils low in exchangeable potassium can induce a potassium deficiency through ion competition effects. Liming soils already at pH 6.0 to 7.5 will usually decrease exchangeable and water-soluble potassium levels in the soil and decrease potassium uptake by plants. Removal by liming of the restrictive effect of aluminum on root growth and vigor is expected to more than offset the competitive effect of the added calcium on potassium uptake by plants.

When acid soils are limed there is usually a substantial increase in pH-dependent cation exchange capacity. Raising soil pH from 5 to 6 will increase the effective cation exchange capacity by as much as 50%. Potential effects of this pH-related change on the ability of soils to retain potassium were referred to in the section on Q/I relationships. Greater retention of exchangeable potassium should reduce leaching losses and generally improve the availability of potassium to crops. The competitive effects of calcium and magnesium will increase, however, if potassium levels are not increased accordingly.

Applications of high rates of KCl to acid soils can result in large increases in the concentration of potentially toxic elements such as aluminum and manganese in the soil solution. The toxicity of aluminum and manganese can be intensified and the benefits of increasing potassium supply nullified. The extremely high rates of KCl used to induce these changes will, under field conditions, most likely occur in a fertilizer band or in the soil immediately adjacent to a granule or droplet of fertilizer potassium.

Calcium and Magnesium. Both Ca^{2+} and Mg^{2+} compete with potassium for entry into plants. Thus it is expected that soils high in one or both of these basic cations will require high levels of potassium for satisfactory nutrition of crops.

According to the activity ratio defined earlier in this chapter, potassium uptake would be reduced as calcium and magnesium are increased and conversely uptake of these two cations would be reduced as the available supply of potassium is increased. This concept also indicates that the availability of potassium is more dependent on its concentration relative to that of calcium and magnesium than on the total quantity of potassium present. Because the activity ratio does not always agree with potassium, calcium, and magnesium plant uptake measurements, it appears that due consideration must also be given to the absolute amount of potassium at hand.

Although the depressive action of calcium in calcareous soils on potassium uptake such as that shown in Table 7-6 has been recognized widely, many soil testing laboratories fail to make appropriate adjustments in potassium fertilizer recommendations. Consistent economic responses in yield of wheat and barley have been reported in Manitoba from the addition of potassium to calcareous soils which by most standards would be rated as potassium sufficient.

An indication of the interference of high magnesium on potassium uptake in corn is apparent in Figure 7-12. Such effects of excessive amounts of magnesium on potassium availability should receive attention in modern crop production.

TABLE 7-6. **Effect of Potassium Uptake and Growth on Young Corn Plants in a High-Lime Iowa Soil**

Soil Condition	Uptake (meq/100 g of Dry Matter)			Weight of Corn Plants (g)
	K	Ca	Mg	
High Lime	23	55	101	1.2
Normal	107	32	39	12.0

Source: Pierre and Bower, *Soil Sci.*, **55**:23 (1943).

Relative Amounts of Other Nutrients—Native or Applied. If nutrients other than potassium are limiting, the soil potassium is not put under stress and its status will not be fully evaluated. An example is given in Figure 7-13 where, with low phosphorus, added nitrogen had little effect on potassium uptake by wheat even though 289 lb of K_2O per acre had been applied. Contrasted with this is a dramatic increase in potassium uptake when nitrogen was used in combination with 131 lb of P_2O_5 per acre.

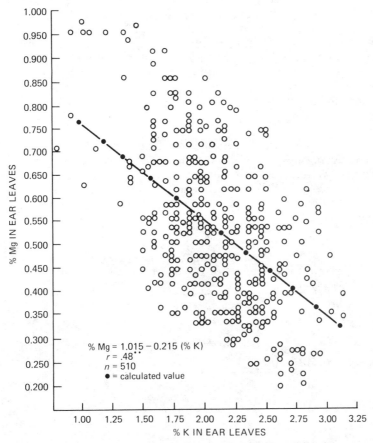

% Mg = 1.015 − 0.215 (% K)
r = .48**
n = 510
● = calculated value

FIGURE 7-12. High levels of magnesium depress potassium concentration in ear leaves of corn. [Spratt, in *Tech. Sci. Papers, Manitoba Agron. Ann. Conf.*, pp. 40–56, Winnipeg, Manitoba (December 12–13, 1979).]

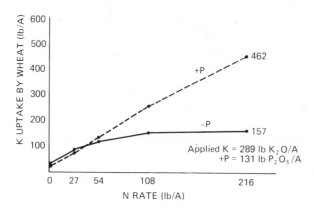

FIGURE 7-13. Potassium uptake by wheat increases with nitrogen rates and potassium. [Schwartz and Kafkafi, *Agron. J.*, **70**:227 (1978).]

In situations where other plant nutrients are low, applications of potassium may reduce yields. This is illustrated by the example in Table 7-7 of corn production in Iowa. With adequate nitrogen, however, supplemental potassium increased corn yields.

Yields of soybeans in Brazil have been raised from 39.5 bu/A to 53.3 bu/A by increasing the rates of phosphorus. Concentration of potassium in the harvested beans was increased from 2.1% to 2.3% by phosphorus fertilization. Removal of potassium in the beans was extended from 51 lb to 77 lb of K_2O per acre. It is obvious that maintenance rates of nutrients, potassium in this case, will often need to be boosted when limitations imposed by other nutrients are removed.

Plants receiving high NH_4^+ and inadequate potassium may develop toxicity symptoms remedied by potassium applications. Potassium concentration may be as high in the NH_4^+-nourished plants as it is in those supplied with NO_3^-. Apparently, additional potassium is needed for proper utilization of high levels of NH_4^+.

Tillage. Most fertilizer recommendations originally were based on a plow furrow $6\frac{2}{3}$ in. deep. Plowing to 10 to 12 in. and chiseling to 12 to 15 in. is becoming more extensive. When tillage is increased to 10 in., 50% more soil is involved. Where the practice of soil buildup is used, potassium requirement will thus be increased about 50%, depending on the potassium status of soil deeper in the tilled layer.

TABLE 7-7. Effect of Potassium Fertilization on Yields When the Nitrogen Supply for Corn Is Low

N (lb/A)	Yield of Corn (bu/A)		Yield Change (bu/A)
	K_2O (lb/A)		
	0	96	
0	78	71	− 7
40	99	94	− 5
80	127	130	+ 3
160	139	143	+ 4
320	144	157	+ 13

*Source: Better Crops Plant Food, **63**:19 (Spring 1979).*

TABLE 7-8.   Effect of Added Potassium on the Uptake of
Potassium by Corn and the Yield Loss from No-Till on a Medium
Potassium Soil in Wisconsin (1976 Results)

K_2O Applied Annually, 1972–1976 (lb/A)	K in Leaves at Silking (%)		Yield Loss from Not Plowing (bu/A)
	Plowed	No-Till	
0	0.73	0.59	51
80	1.40	1.04	29
160	1.71	1.42	18

Source: Schulte, Better Crops Plant Food, **63**:25 (Fall 1979).

Tillage practices influence potassium availability by modifying other factors, such as oxygen or aeration, temperature, soil moisture, and positional availability of applied potassium. Residues left on the soil surface may lower soil temperature, reduce evaporation, and enhance moisture movement into soil, thus changing nutrient–moisture relations.

Minimum tillage and no-till are becoming more common and there is evidence that potassium availability under these systems is reduced because of increased compaction, less aeration, lower temperature, and positional availability of potassium applied on the soil surface. In a number of studies comparing the effects of plowed and unplowed till-planted operations on corn yields and potassium concentration in leaves, there was little doubt that omission of plowing decreased yields and potassium availability to the crop. An example of reductions in yield and potassium uptake in no-till corn grown in Wisconsin is given in Table 7-8. Higher corn yields are usually associated with leaf concentrations above 2.0%. Fertilizer potassium had a major role in counteracting these undesirable effects of no-till, and this aspect of crop management should receive adequate attention, since in some instances low soil fertility has been blamed for lower yields under no-till.

Plant Factors Affecting Potassium Availability

It is widely recognized that crops differ in their capacity to take up potassium from a soil with a given level of exchangeable potassium. Also, various crops have different internal or critical levels of potassium that are necessary to meet the metabolic and osmotic needs of the crop.

Cation Exchange Capacity of Roots.   One popular explanation for varying abilities of crops to utilize soil potassium is related to cation exchange properties of plant roots. It is suggested that those plants with roots of relatively low cation exchange values are best able to extract soil potassium. This low cation exchange capacity seems to be associated with a low calcium requirement. Grasses and cereals with roots of low exchange capacities respond least to applications of fertilizer potassium on many soils, in contrast to plants such as clovers, which have higher root cation exchange capacities. Additional information on this subject was covered in Chapter 4.

Although the cation exchange capacity of roots may be important in determining the ability of plants to absorb the more slowly available forms of soil potassium, it is only one of the factors involved. Some of the other aspects of plant growth that influence potassium uptake are reviewed briefly in the following section.

TABLE 7-9. Potassium Absorption Properties for Corn, Onion, and Wheat

Parameter	Maize (16 days)	Onion (20 days)	Wheat (15 days)
Maximum influx, I_{max} [pmol/(cm · sec)]	4.07	3.97	2.93
Average influx, I_n [pmol/(cm · sec)]	1.42	0.53	0.33
Absorption cylinder (mm^2)	2.00	0.18	1.50
Depletion zone (mm)	11.00	7.00	7.00

Source: Baligar and Barber, *11th Congr. Int. Soc. Soil Sci. Abstr.*, **1:**309 (1978).

Root System and Crop. Researchers at Purdue University have identified the following five plant factors as having considerable influence on potassium uptake.

1. Ion flux (movement) and concentration.
2. Root radius.
3. Rate of water uptake.
4. Root length.
5. Rate of root growth.

Root type and density are two of the major characteristics affecting potassium availability to crops. The ability of crops to take up water will also influence their uptake of potassium. Potassium absorption properties of three crops are compared in Table 7-9. Not shown in this table is root density, which was very much lower for onions than for corn. Because of differences in root morphology and potassium influx characteristics, corn removed 47% of the exchangeable and solution potassium, whereas onions removed only 6% of these forms of potassium from the same soil volume.

Grasses tend to have a more fibrous root system with many lateral branches, whereas alfalfa, especially under low potassium levels, tends to be tap rooted with

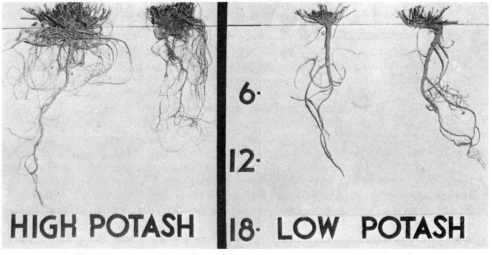

Figure 7-14. High available potassium encourages root development and density of taprooted legumes such as alfalfa. (Munson, in *Potassium for Agriculture—A Situation Analysis.* Atlanta, Ga.: Potash & Phosphate Institute, 1980.)

FIGURE 7-15. Adequate K on corn increases root branching and density, size of stalks, and size of ear. (Courtesy of Potash & Phosphate Institute, Atlanta, Ga.)

less root renewal. High availability of potassium actually enhances root development, producing more branching and lateral roots in both alfalfa (Figure 7-14) and grasses such as corn (Figure 7-15).

Differences in rate of potassium uptake between crop species can result in competition for potassium. A stable productive association between white clover and ryegrass can be difficult to maintain because of the ability of the latter species to absorb potassium at rates two to five times faster than white clover. Figure 7-16 shows that over this particular range of potassium supply, the specific rate of potassium absorption for ryegrass is twice the value for red clover. This is in addition to the advantage provided by the more extensive root system of ryegrass.

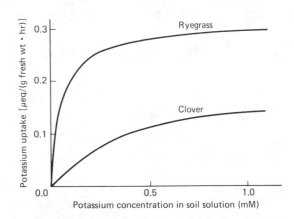

FIGURE 7-16. Potassium influx isotherms for ryegrass and white clover plants growing under a potassium stress. [Dunlop et al., *New Phytol.,* **83**:365 (1979).]

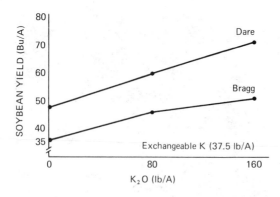

FIGURE 7-17. Soybean varieties have different capacities to respond to soil and fertilizer potassium. [Terman, *Agron. J.*, **69**:234 (1977).]

It thus does seem that variations in root architecture are not solely responsible for the competitive advantage that ryegrass has over white clover under potassium deficient conditions.

Variety or Hybrid. There are clear indications that ion absorption by plants is under genetic control and that considerable differences exist both between and within genera. Corn hybrids have different capacities to take up potassium. Florida workers found that under low fertility, hybrids had a range of potassium uptake from only 23 to 41 lb of potassium per acre. Under high fertility, however, the range increased to between 164 to 279 lb of potassium per acre.

Soybean varieties have varying capacities to absorb and utilize potassium. An indication of differences between the varieties Dare and Bragg is given in Figure 7-17. Dare was the more responsive to additions of fertilizer potassium.

Growth of five barley cultivars grown in competition for five different levels of available potassium supply is summarized in Table 7-10. The data reveal a consistent pattern of superior performance by Herta and Fergus and generally poor performance by Conquest.

Following the measurement of fairly apparent differences in rates of potassium uptake among selected barley varieties, Glass and co-workers at the University of British Columbia recorded a strong correlation between potassium uptake and hydrogen ion efflux from the roots of barley varieties. A summary of this relationship for 24 barley varieties is illustrated in Figure 7-18.

Plant Population and Spacing. There is a trend with several of the principal crops for growers to use higher plant populations per acre (corn) or closer rows (soybeans). Narrower row spacing of 3 to 4 in. has been shown in England, Indiana, and Wisconsin to be advantageous for small grains such as wheat and

TABLE 7-10.   Capacities of Barley Cultivars to Respond to Potassium

	Fresh Weight per Gram of Plant with a Nutrient Solution of Potassium Concentration (μM):				
Cultivar	0.1	1.0	10	100	500
Herta	0.256	0.294	0.287	0.474	1.121
Fergus	0.227	0.278	0.256	0.494	1.116
Carlsberg	0.236	0.209	0.227	0.379	0.987
Olli	0.130	0.189	0.208	0.547	0.946
Conquest	0.143	0.139	0.180	0.408	0.820

Source: Glass and Perley, *Plant Physiol.*, **65**:160 (1980).

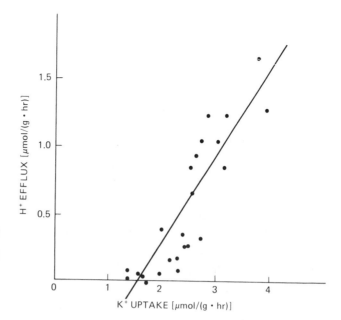

FIGURE 7-18. Plot of net H$^+$ efflux against K$^+$ uptake for 24 barley varieties. Plants, previously grown in a balanced inorganic medium containing 60 μM K$^+$ were exposed to 1 mM K$_2$SO$_4$ plus 0.5 mM CaSO$_4$ for 24 hours. Correlation coefficient = 0.88 (significant at $P < 0.001$). [Glass et al., *Plant Physiol.*, **68**:457 (1981).]

barley. Higher plant populations and closer row spacings affect nutrient needs, including potassium. There is considerable evidence that potassium soil fertility needs are being affected. In an experiment in Kentucky, the response of corn to potassium fertilization was 22 and 39 bu/A at low and high plant populations, respectively. As the plant population of two corn hybrids in Florida was raised from 19,000 plants to 39,000 plants per acre, the potassium uptake by one hybrid increased from 150 lb to 228 lb of potassium per acre and from 128 lb to 170 lb of potassium per acre for the other.

Where plant populations are increased without appropriate increases in potassium fertilization or availability, there can be greater potassium uptake but yield may actually be reduced. For example, increasing corn population from 12,146 to 36,438 plants per acre on an irrigated, low-potassium soil in Florida decreased the yield from 110 bu/A to 56 bu/A but increased the potassium uptake from 117 lb/A to 160 lb/A. However, inclusion of 240 lb of K$_2$O per acre in combination with these same plant populations boosted yield from 110 bu/A to 231 bu/A and potassium uptake from 117 lb/A to 259 lb/A.

Because of these relationships potassium fertilizer recommendations are often raised for higher stand and yield levels. Potassium rates 10 to 20% heavier than usual may be advised for soybeans in 7-in. or solid seedings.

Yield Level. As yield level increases, the amount of potassium available usually must be augmented to compensate for the greater potassium uptake. For example, in Minnesota, a 255-bu/A yield of corn removed a total of 304 lb of potassium per acre. Comparing these values with yield and potassium uptake by a 145-bu/A crop, the yield was 1.76 greater while potassium removal was 2.2 times more.

The yield goal or level of production being sought is not considered by all soil test laboratories in making recommendations for fertilizer potassium. At the Missouri Soil Testing Laboratory yield goal is taken into account when advising on rates of potassium. The basic equation given below is used for determination of potassium buildup and maintenance requirements when the potassium soil test level is less than the desired or optimum soil test plus five times CEC.

$$X_k = 75.5 \frac{(X_d + 5CEC)^{1/2} - X_o^{1/2}}{4} + (Y_g)(K_r)$$

where: X_k = annual potassium fertilizer rate, lb of K_2O per acre

X_d = desired or optimum potassium soil test level, lb of potassium per acre

CEC = cation exchange capacity, meq per 100 g

X_o = observed soil test level, lb of potassium per acre

Y_g = yield goal

K_r = potassium removal, lb of K_2O per yield unit

If the measured soil test potassium exceeds the optimum plus five times CEC, the following equation is used instead of the previous one:

$$X_k = 75.5[(X_d + 5CEC)^{1/2} - X_o^{1/2}] + (Y_g)(K_r)$$

Values of X_k less than 20 but greater than 0 are are rounded to 20. If X_k is less than 0, it is recorded as 0.

As is the case in the Missouri approach, it is important with higher yields to adjust maintenance rates of potassium in accordance with optimum soil potassium levels and the yield attained.

Time Factor. One aspect of the time factor involves numbers of crops grown and intensity of cropping practices such as multiple cropping. Initially, some soils are well supplied with available potassium and they may not show much response to applied potassium. However, on soils that do not have a large quantity of potassium that can be released at a fairly constant rate (capacity), potassium levels may be drawn down in a short time and large crop responses to applied potassium will be obtained after several cropping seasons.

Figure 7-19 illustrates long-term development of a potassium requirement for rice grown on a soil classified as unresponsive to potassium. Exchangeable soil potassium was 356 lb/A initially. By the end of the third year of cropping, addition of potassium was beneficial and the response continued to increase with time. At the end of 8 years, soil test potassium on the NP only plots had declined to 102

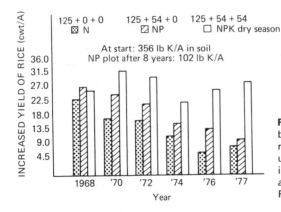

FIGURE 7-19. Potassium fertilization becomes beneficial after several years of rice production on a soil initially unresponsive to potassium. (von Uexkull, in G. S. Sekhon, Ed., *Potassium in Soils and Crops*, pp. 241–259. New Delhi: Potash Research Institute of India, 1978.)

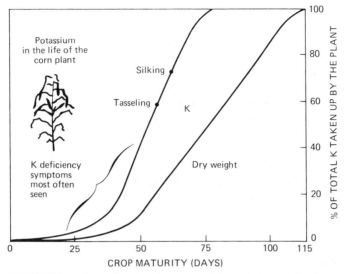

FIGURE 7-20. Potassium is taken up rapidly early in the life of the corn plant. (Aldrich et al., *Modern Corn Production*, 2nd ed. A. & L. Publications, Champaign, Ill.: 1975.)

lb of potassium per acre, a decrease of 254 lb of potassium per acre or 32 lb of potassium per acre yearly.

Another consideration within the framework of time is the maximum demand for available potassium as plant growth progresses. Crop requirements for potassium obviously vary at different growth stages, as exhibited by corn in Figure 7-20. The young corn seedling does not need much potassium, but the rate of uptake rises rapidly to a peak in the 3 weeks prior to tasseling. Thus, for satisfactory growth, rate of supply of potassium from soil must at least match crop needs during critical periods of maximum uptake.

The actual amounts of potassium taken during the growing season by corn yielding 187 bu/A are compiled in Table 7-11. These values clearly indicate the changing demands for potassium supplies.

Where multiple cropping is practiced, both total yield and potassium crop removal per unit of time are increased. This intensifies the demand for soil potassium, accelerating the rate of potassium drawdown and expanding the amounts of potassium needed for maintenance programs.

TABLE 7-11. Plant Nutrient Requirements of Corn Yielding 187 bu/A Change During the Growing Season

Plant Age (days)	Nutrients Absorbed (lb/A/day)		
	N	P	K
20–30	1.5	0.15	1.3
30–40 (knee-high)	6.0	0.60	7.4
40–50	7.4	0.90	8.6
50–60 (tasseling)	4.7	0.80	3.3

Source: Spies, Purdue Univ. (1973).

Luxury Consumption. The term *luxury consumption* has been grossly misused. It means that plants will continue to absorb an element in amounts in excess of that required for optimum growth. It results in an accumulation of the element in the plant without a corresponding increase in growth and suggests inefficient and uneconomical use of that particular element. However, with higher crop yields, a much greater concentration of potassium, and other nutrients as well, is required. As an example, 1.0 to 1.2% potassium in alfalfa was formerly thought to be adequate. Now a potassium level of 2 to 3% is considered necessary to maintain consistently high yields and good stands and to enable alfalfa plants to survive stress periods with a minimum decrease in growth.

Placement of Fertilizer Potassium

The factors determining the most effective placement of fertilizer potassium are those, generally, that influence the potassium equilibria in the soils covered in this chapter. If a soil is high in potassium-fixing clay minerals and limited quantities of potassium fertilizers are to be applied, best results will be obtained if the material is applied in bands rather than broadcast and mixed with the soil. This results from the reduced area of contact between the fertilizer potassium and the soil. In fact, banding of potassium fertilizers on high potassium-fixing river clay soils was 3.65 times more effective than broadcast application.

On soils low in potassium, row or band placement of potassium is often more effective than broadcast dressings. However, on soils high in potassium and at higher levels of applied potassium, one method is as satisfactory as the other. With high-yield crops and on soils requiring considerable potassium, broadcasting is the most widely used method of application.

Small grains frequently benefit from seed placed potassium, particularly in low-potassium soils and when low rates of fertilizer potassium are being used (Table 7-12). Potential germination and seedling damage limit the amount of potassium that can be applied in this manner. Therefore, broadcasting sufficient potassium for at least 2 or 3 years of production may be a practical alternative. Limited root systems, shorter growing seasons, and cooler temperatures in some years result in greater response from potassium placed with seed than from that which is broadcast and incorporated.

If high rates of potassium are applied during planting operations, the soluble potassium fertilizer salts should be placed at a safe distance from the seed row, usually at least 1 to 2 in. away. Sidebanding about 2 in. to the side and 1 in. below the seed or transplant crown has been widely used for such placement of potentially

TABLE 7-12.   Comparison of Potassium Placement Methods for Spring Barley in Central Alberta

	Placement Method	Yield Increase (bu/A) in:	
		1974 (6 tests)	1976 (13 tests)
15 lb K$_2$O/A	Broadcast	8.6	—
	Banded	12.8	6.2
	With seed	18.8	10.7
30 lb K$_2$O/A	Broadcast	17.0	—
	Banded	18.8	8.0
	With seed	21.0	12.2

Source: Walker, *Better Crops Plant Food,* **62**:13 (Summer 1978).

injurious rates of potassium. Failure to provide adequate separation will frequently result in salt injury to the seedlings, reduced and uneven stands, and lowered yields. This point is treated at greater length in the chapter dealing with fertilizer application.

Loss of Potassium by Leaching

It is often thought that considerable potassium can be lost by leaching. On the contrary, in most soils, excepting those that are quite sandy or subject to flooding, losses by this mechanism are small. Work carried out in Illinois indicates that losses from the silt loam soil studied were equivalent to only about 1.4 lb of potassium per acre per year. Other studies in Florida, however, showed that large amounts of potassium would be lost from coarse sandy soils with low cation exchange capacities. Passage of 16 in. of water through such soils under experimental conditions resulted in removal from the soil of 126 lb of potassium, or over 93% of the total supply of this nutrient.

The wide differences illustrated by these two examples can be explained on the basis of the total cation exchange capacities of these soils and their relative degrees of potassium saturation. The silt loam soil had a much higher exchange capacity than the sandy soil. It was therefore able to hold a much larger amount of potassium. Additions of fertilizer potassium would result in adsorption of a much higher proportion of the added potassium than it would in the case of the sandy soil. The unadsorbed part, of course, remains in the soil solution and is removed in leaching waters, which explains why such a large quantity was lost from the sandy soil with the low CEC.

While organic soils, such a mucks, have high exchange capacities, the bonding strength for cations such as potassium is not great, and the exchangeable potassium level tends to vary somewhat with the intensity of rainfall. Thus emphasis should be placed on annual applications rather than on buildup of soil potassium. Since crops grown on organic soils characteristically have a high potassium need, it is important to monitor the fertility level with soil tests.

Losses of potassium from surface soils can be reduced by liming the soil to maintain it at a favorable pH level. In addition, potassium in percolating waters is usually readsorbed by clays in the subsoil, so that a buildup of this element in the lower profile is frequently observed. In the final analysis leaching losses of potassium are of consequence only in coarse-textured or organic soils in areas of high rainfall.

Fertilizers Containing Potassium

Extensive deposits of soluble potassium salts are found in many areas of the world, most of them well beneath the surface of the earth but some in the brines of dying lakes and seas. Many of these deposits and brines are of a high degree of purity and therefore lend themselves to mining operations for the production of agricultural and industrial potassium salts, usually termed potash by the fertilizer trade.

Location of the major potash deposits in North America is shown in Figure 7-21. In the United States there are production locations in New Mexico, California, and Utah. Sustained production of potash began in the 1960s in Saskatchewan, Canada, and two new mines are scheduled to come on stream in New Brunswick in 1983–1984. Commercial production is being considered for the late 1980s at one other location in New Brunswick and at a site in southwestern Manitoba near the Saskatchewan border.

The Saskatchewan deposit is the world's largest known high-grade potash deposit. It extends in a broad belt 450 miles long and up to 150 miles wide into Manitoba on the east and southward into North Dakota and Montana. Depth of

FIGURE 7-21. Major potash production areas in North America.

this deposit increases from 3000 ft at the northern edge to 7000 ft at the International Boundary.

As with nitrogen and phosphorus, there have been impressive increases in potassium consumption in recent years. As more knowledge of crop production is acquired, made urgent by the rise in world population and the need for higher and more profitable yields, usage not only of potassium but of all fertilizers will continue to increase and at a significant rate.

Like phosphorus, the potassium content of fertilizers is presently guaranteed in terms of its potassium oxide equivalent. This is determined analytically by measuring the amount of potassium salt that is soluble in an aqueous solution of ammonium oxalate. As mentioned in Chapter 6, there is some interest in expressing the plant nutrient content of fertilizers in terms of the element instead of the oxide. Converting % K to % K_2O and the reverse can be accomplished by the two following expressions.

$$\% \text{ K} = \% \text{ K}_2\text{O} \times 0.83$$

$$\% \text{ K}_2\text{O} = \% \text{ K} \times 1.2$$

Practically all of the potassium fertilizers are water soluble. They consist essentially of potassium in combination with chloride, sulfate, nitrate, or polyphosphate. Some double salts exist, such as potassium-magnesium sulfate. The properties of the individual potassium carriers are covered in the following sections and their mining and manufacturing are described in Chapter 10.

Potassium Chloride (KCl). This salt is sold under the commercial term *muriate of potash.* The term *muriate* is derived from muriatic acid, a common name for hydrochloric acid. Fertilizer-grade muriate contains 50 to 52% potassium (60 to 63% K_2O) and varies in color from pink or red to white, depending on the mining and recovery process used. There is no agronomic difference among the products.

Muriate of potash is generally marketed in five particle sizes: special standard, white soluble, standard, coarse, and granular. Typical particle size distributions for these size grades in sieve mesh openings are $-28 + 200$ for special standard and white soluble, and $-14 + 65$, $-8 + 28$, and $-6 + 20$, respectively for standard, coarse, and granular. A steadily increasing demand for the coarse and granular types has been noted because of their suitability for blending with other products as well as for direct application. The white soluble grade is popular in the fluid fertilizer market.

Muriate of potash is by far the most widely used potassium fertilizer. It is employed for direct application to the soil and for the manufacture of N-P-K fertilizers. When added to the soil, it dissolves in the soil moisture. The resulting K^+ and Cl^- ions behave in keeping with the pattern already discussed in this chapter.

Potassium Sulfate (K_2SO_4). *Sulfate of potash* is the term usually applied to this salt by the fertilizer trade. It is a white material containing 42 to 44% potassium (50 to 53% K_2O) and 17% sulfur. It is produced by different processes some of which involve reactions of other salts with potassium chloride and some of which involve the reaction with sulfur or sulfuric acid.

Potassium sulfate finds its greatest use on potatoes and tobacco, which are sensitive to large applications of chlorides. Its behavior in the soil is essentially the same as that of muriate, but it has the advantage of supplying plant-nutrient sulfur, which is deficient in many soils.

Potassium Magnesium Sulfate (K_2SO_4, $MgSO_4$). This is a double salt of magnesium sulfate and potassium sulfate with a small amount of sodium chloride, which is largely removed in processing. The material contains 18% potassium (22% K_2O), 11% magnesium, and 22% sulfur. It has the advantage of supplying both magnesium and sulfur and is frequently included in mixed fertilizers for that purpose on soils deficient in these two elements. It reacts as would any other neutral salt when applied to the soil.

Potassium Nitrate (KNO_3). This fertilizer material is also known as saltpeter or niter. It contains 13% nitrogen and 37% potassium (44% K_2O). Agronomically, it is an excellent source of fertilizer nitrogen and potassium. Its once high production cost restricted its use to crops of high acre value, but in 1963 commercial production was started in the United States. Before that time most of the fertilizer grade was imported, some of it from Chile, where it was manufactured in conjunction with the production of nitrate of soda.

Potassium nitrate is being marketed largely for use on fruit trees and on crops such as tobacco, cotton, and vegetables. If production costs can be lowered, it might compete with other sources of nitrogen and potassium for use on crops of a lower value per acre.

Potassium Phosphates (KPO_3, $K_4P_2O_7$, KH_2PO_4, K_2HPO_4). The Tennessee Valley Authority, Scottish Agricultural Industries, and others have produced potassium metaphosphate (KPO_3) in pilot plants. This water-insoluble material contains 33% potassium (40% K_2O) and 27% phosphorus (60% P_2O_5). It hydrolyzes in water and in the soil to the orthophosphate form. Although an excellent source of both phosphorus and potassium and possessing the unique advantage of having no salt effect, potassium metaphosphate has not been available commercially because of high cost.

A potassium polyphosphate solution with a grade of 0-26-26 was produced for a short time in the United States. It was made by reacting superphosphoric acid and potassium hydroxide and contained a mixture of ortho-, pyro-, and higher polyphosphates. Solid condensed potassium polyphosphates with analyses ranging from 0-42-42 to 0-47-47 have been made in the pilot plant of one United States firm.

Some high-analysis speciality fertilizers contain the ortho potassium phosphates (KH_2PO_4 or K_2HPO_4) produced from potassium hydroxide or potassium carbonate and phosphoric acid. High-purity electric furnace phosphoric acid is normally used for the manufacture of mono- and dipotassium phosphates used to formulate liquids for foliar application.

Experimental quantities of a mixture of mono- and dipotassium phosphate were made in the late 1950s in western Canada and successfully test-marketed in California. This product, analyzing 0-36-36, was made by neutralizing wet-process phosphoric acid with potassium hydroxide.

Development activities in Israel demonstrated the feasibility of making monopotassium phosphate and its ammoniated double salt, monopotassium-monoammonium phosphate. Monopotassium phosphate contains 52% P_2O_5 and 32% K_2O and the ammoniated double salt has a grade of 6-56-18.

Novel technology was developed simultaneously in Ireland and in the United States for the production of fertilizer-grade monopotassium phosphate (KH_2PO_4) by the reaction of potassium bisulfate and sulfuric acid on phosphate rock. Approximate grades of solid products were 9-48-16, 5-46-30, and 0-47-31. Liquid fertilizers resulting from this process assayed 0-30-11, 0-20-20 (clear), and 0-25-20 (clear).

Further processing of the monopotassium phosphate yielded low-molecular-weight, highly water soluble polymers of the $KO(KPO_3)_nK$ type. Typical analyses of the solid products ranged from 0-60-30 to 0-50-40.

Advantages of the various potassium phosphates are:

1. High analysis.
2. Low salt index.
3. Adapted to preparation of clear liquid fertilizers high in K_2O.
4. Polyphosphates can be formulated with controlled solubility.
5. Fluorine and chlorine free, making them well suited for use on tobacco, potatoes, and other crops sensitive to excessive amounts of chlorine.

Potassium Carbonate (K_2CO_3), Potassium Bicarbonate ($KHCO_3$), and Potassium Hydroxide (KOH). These salts are used on a limited scale, primarily for the production of high-purity fertilizers for foliar application or other speciality uses. Up to the present time their high cost of manufacture has precluded widespread use as commercial fertilizers. Studies on the agronomic suitability of potassium bicarbonate suggests that its use on acid soils will reduce loss of cations by leaching. There were also indications that this material will increase the effectiveness of phosphate fertilizers. The composition of important potassium fertilizers and other potassium salts is shown in Table 7-13.

Agronomic Value of Various Potassium-Containing Fertilizers

Potassium fertilizers have been compared in numerous field and greenhouse trials and the following points summarize the results:

1. In general, if the material is being used for its potassium content alone, one material is as good as any other. The selection should be based on the cost per pound of potassium applied to the soil.
2. Sources such as potassium nitrate and potassium polyphosphate, which contain

TABLE 7-13.   Plant Nutrient Content of Common Potassium Fertilizers and Other Potassium Salts

Material	N (%)	P_2O_5 (%)	K_2O (%)	S (%)	Mg (%)
Potassium chloride	—	—	60–62	—	—
Potassium sulfate	—	—	50–52	17	—
Potassium magnesium sulfate	—	—	22	22	11
Potassium nitrate	13	—	44	—	—
Potassium and sodium nitrate	15	—	14	—	—
Manure salts	—	—	22–27	—	—
Potassium hydroxide	—	—	83	—	—
Potassium carbonate	—	—	< 68	—	—
Potassium orthophosphates	—	30–60	30–50	—	—
Potassium polyphosphates	—	40–60	22–48	—	—
Potassium metaphosphate	—	55–57	38	—	—
Potassium calcium pyrophosphate	—	39–54	25–26	—	—

the other major elements, are fully as effective as sources of potassium as potassium chloride and must be evaluated on the basis of economy of supplying potassium as well as nitrogen and phosphorus. Materials such as these may be completely absorbed by plants, and no anions such as chloride or sulfate will remain. Use of similar forms in combination with other fertilizer materials in greenhouse culture permits the maintenance of adequate nitrogen, phosphorus, and potassium without danger of accumulation of excess salts. The same principle would apply at least in part under some field conditions in which heavy fertilization is required.

3. Accompanying elements, such as sulfur, magnesium, chlorine, and sodium, are agronomically important on many soils. The value of the accompanying element must be considered in choosing among the various sources. In addition, the economics of applying the elements in question in one material or in separate materials must be kept in mind.

4. Tobacco is a crop that is extremely sensitive to excessive amounts of chloride. Although rates of up to 20 lb/A are beneficial, quantities in excess of 30 to 40 lb will impair burning quality. In some potato, sweet potato, and citrus areas high quantities of chloride are avoided. Instead, sources such as potassium sulfate or potassium nitrate may furnish the major portion of the potassium.

Summary

1. The potassium content of soils is variable, but many soils contain large amounts of this element. The total amount present, however, is no criterion of the amount available to plants.

2. The availability of potassium to plants is governed by the equilibria in the soil system among four forms of potassium which vary in their availability. These fractions, together with estimates of the approximate amounts in each, are mineral or structural 5000 to 25,000 ppm; nonexchangeable (fixed or difficultly available) 50 to 750 ppm; exchangeable 40 to 600 ppm; and solution 1 to 10 ppm.

3. Fixed or difficultly available potassium is nonexchangeable and it occurs mainly within the lattice of clay minerals such as illites or hydrous mica, vermiculite, chlorite, and interstratified minerals, in which two or more of the preceding types occur in more or less random arrangement in the same particle.

4. Potassium present in the soil solution and held in an exchangeable state by soil colloids is classified as readily available. About 0.1 to 2%

of the total potassium in soils will normally be accounted for by solution and exchangeable forms.

5. The activity ratio (AR_o^K) is useful for estimating the effectiveness of soil solution potassium in the presence of competing ions such as Ca^{2+} and Mg^{2+}. It may also be desirable to take into account the influence of Al^{3+} in very acid soils and of Na^+ in salt-affected soils. This ratio is a measure of the intensity of labile potassium in the soil or the amount of potassium that is immediately available to plant roots.

6. Transport of potassium from soil to absorbing plant roots is achieved mainly by mass flow and diffusion, which both occur from, or via, the soil solution. The former is the less important mechanism for satisfying potassium requirements of crops except in soils naturally high in solution potassium or where fertilizer potassium has elevated potassium levels in the soil solution.

7. A number of soil and environmental conditions will influence rates of potassium diffusion. Some of these are moisture content, temperature, clay content, salt content, and potassium concentration.

8. Exchangeable potassium is held around negatively charged soil colloids by electrostatic attraction. Potassium held in this manner is easily displaced or exchanged when the soil is placed in contact with neutral salt solutions.

9. Exchangeable potassium is bound at three types of exchange sites: planar, edge, and inner positions. The planar position on outside surfaces of clay minerals is rather unspecific for potassium, while the edge and inner binding positions have high specificity for potassium.

10. Additions of fertilizer potassium, as well as potassium in the soil solution, can be fixed in some soils, especially those high in 2:1 clays and illite. Potassium fixation capacities are affected by a number of factors, including type and amount of clay minerals, soil reaction, potassium concentration, wetting and drying, and freezing and thawing.

11. Potassium fixation is not completely unfavorable since it helps to conserve the element, which might otherwise be lost from soils where leaching is a problem. Also, fixed potassium tends to become available over a long period of time. In most soils potassium fixation is not a serious crop production problem.

12. Plant uptake of available soil potassium can be impaired significantly by various soil and environmental conditions such as low moisture, low temperature, restricted aeration, compaction and unsatisfactory tilth, low pH, and high concentrations of basic cations such as Ca^{2+} and Mg^{2+}, and other interfering cations such as Al^{3+} in very acidic soils.

13. Liming acidic soils will improve potassium availability by increasing effective cation exchange capacity, thus increasing the ability of soils to retain exchangeable potassium. The competitive effect of Al^{3+} on plant uptake of K^+ is reduced when insoluble aluminum hydroxide is formed following the addition of lime.

14. Crops vary in their capability of utilizing soil potassium. They may also have different internal requirements to satisfy metabolic and osmotic needs of the crop.

15. Root type and density are the two major factors influencing availability among crops. Grasses tend to have a more fibrous root system and legumes tend to be tap rooted. High availability of potassium enhances root development.

16. When rates of such nutrients as nitrogen and phosphorus are more adequate and as yields increase, the demand for available potassium is greatly increased.

17. With today's intensive agriculture, which demands the production of high-yielding crops, considerable potassium is required to fulfill the needs of these crops. Under such conditions, the potassium released from slowly available forms in the soil will be insufficient in many soils.

18. The principal fertilizer compounds containing potassium are potassium chloride, potassium sulfate, potassium magnesium sulfate, and potassium nitrate. When soils are low in available potassium or have the capacity to fix large amounts of potassium, band placement of these fertilizer salts is often more effective than broadcast dressings. However, on soils high in potassium and at higher levels of applied potassium, one method is as satisfactory as the other.

19. Agronomic effectiveness of the principal potassium fertilizers is about equal in cropping situations where only supplemental potassium is needed. Obviously, there will be differences among sources if there is also a need for other plant nutrients such as chlorine, sulfur, or magnesium. As with nitrogen and phosphorus, an important factor governing selection of a potassium fertilizer source where only potassium is needed is the cost per unit of potassium applied to the land.

Questions

1. Why are soils with much clay generally more fertile than sandy soils? Are they *always* more productive? Why?

2. Under what soil conditions is there most likely to be reversion of available or added potassium to less available forms?

3. A soil was found to contain 1.5 meq per 100 g of exchangeable potassium. To how many pounds of muriate of potash per acre does this correspond?

4. What effect will the liming of an acid soil have on the retention of potassium?

5. Why does not the addition of gypsum to an acid soil result in an increased conservation of potassium?

6. What is the original source of soil potassium? Is finely ground granite dust a suitable source of potassium? Why?

7. Is potassium released more readily from feldspar than from the potassium-bearing micas?

8. Describe the general nature of the micas.

9. Do members of the mica group have similar abilities to supply potassium?

10. In which soil particle-size fractions are feldspar and micas usually found?

11. Describe the changes and end results that occur when mica minerals weather in soils.

12. Is soil solution potassium important in the nourishment of crops? Summarize its role.

13. What is the activity ratio, and what does it measure?

14. Is exchangeable potassium available to crops? Are some fractions of soil potassium nonexchangeable?

15. Which forms of potassium are not readily available to plants?

16. How does cation exchange capacity affect the amount of potassium in solution at a given level of exchangeable potassium?

17. Are there different binding sites for exchangeable potassium? Name the mechanism responsible for potassium retention by soil colloids.

18. What mechanisms are thought to account for potassium fixation in soil?

19. Does fixed potassium tend to become available to plants?
20. From your knowledge of cation exchange in soils, would you predict that the addition of sodium would tend to deplete or conserve the supply of soil potassium? Why?
21. By what processes is potassium transported to the plant root surface? What factors govern this movement?
22. What factors control the amount of potassium present in the soil solution?
23. Under what conditions (soil and environmental) would you expect to obtain the least and greatest crop response to surface-applied top dressings of a soluble potassium fertilizer?
24. Why does applied potassium increase the rate of diffusion to the plant root?
25. Low moisture, low temperature, or more clay in a soil decrease the rate of diffusion of potassium to the root. Why does applied potassium increase the rate of diffusion?
26. Under what conditions is band-placed potassium likely to be superior to broadcast applications?
27. Can plant uptake of available soil potassium be impaired by soil and environmental factors? If so, list the principal factors.
28. Do crops vary in their ability to use soil potassium? If there are differences, explain why they occur.
29. Are there differences in the capability of crop varieties and hybrids to absorb potassium? Summarize any examples.
30. What are the two major root characteristics that cause differences among crops in uptake of potassium?
31. List important plant factors that can influence the availability of and need for potassium.
32. Why might continuous cropping at high yield levels deplete available potassium over time and increase the probability of a response to potassium?
33. What is *luxury consumption* of potassium? Is it a serious problem under most soil and cropping conditions? How can it be minimized?
34. Name the common sources of fertilizer potassium.
35. A fertilizer is guaranteed to contain 30% potassium oxide. To what percentage of potassium does this correspond?
36. Under what soil conditions might you prefer to use potassium magnesium sulfate rather than muriate and dolomite or muriate alone?
37. Are there situations where potassium chloride will be more effective than potassium sulfate or potassium nitrate?
38. Are there any incentives for undertaking commercial production of potassium phosphate fertilizers? If so, list them.
39. How many pounds of potassium sulfate would be required to supply the same amount of potassium in 350 lb of 60% muriate of potash?
40. In what form is potassium absorbed by plants?

Selected References

Adams, F., and J. B. Henderson, "Magnesium availability as affected by deficient and adequate levels of potassium and lime." *SSSA Proc.*, **23**:65 (1962).

Agarwal, R. R., "Potassium fixation in soils." *Soils Fert.*, **23**:375 (1960).

Aldrich, S. R., W. O. Scott, and E. R. Leng, *Modern Corn Production*, 2nd ed. Champaign, Ill.: A & L Publications, 1975.

Attoe, O. J., "Potassium fixation and release in soils occurring under moist and drying conditions." *SSSA Proc.*, **11**:145 (1946).

Baligar, V. C., and S. A. Barber, "Mechanism of K supply to plant roots characterized by K/Rb ratio." *11th Congr. Int. Soc. Soil Sci.*, **1**:309 (1978).

Baligar, V. C., and S. A. Barber, "Genotypic difference of corn for ion uptake." *Agron. J.*, **71**:870 (1979).

Barber, S. A., "Relation of fertilizer placement to nutrient uptake and crop yield: II. Effects of row potassium, potassium soil level, and precipitation." *Agron. J.*, **51**:97 (1959).

Barber, S. A., "Properties of the plant root that influence fertilizer practice." *Proc. 7th Int. Coll. Plant Anal. Fertilizer Prob.* (J. Wehermann, Ed.), German Society of Plant Nutrition (1974).

Barber, T. E., and B. C. Matthews, "Release of non-exchangeable soil potassium by resin-equilibration and its significance for crop growth." *Can. J. Soil Sci.*, **42**:266 (1962).

Barber, S. A., R. D. Munson, and W. B. Dancy, "Production, marketing, and use of potassium fertilizers," in R. A. Olson, T. J. Army, J. J. Hanway, and V. J. Kilmer, Eds., *Fertilizer Technology and Use*, 2nd ed., Chapter 10. Madison, Wis.: Soil Science Society of America, 1971.

Barshad, I., "Cation exchange in micaceous minerals: I. Replacement of interlayer cations of vermiculite with ammonium and potassium ions." *Soil Sci.*, **77**:463 (1954).

Barshad, I., "Cation exchange in micaceous minerals: II. Replaceability of ammonium and potassium from vermiculite, biotite, and montmorillonite." *Soil Sci.*, **78**:57 (1954).

Bartlett, R. J., and T. J. Simpson, "Interaction of ammonium and potassium in a potassium-fixing soil." *SSSA Proc.*, **31**:219 (1967).

Bates, T. E., and A. D. Scott, "Control of potassium release and reversion associated with changes in soil moisture." *SSSA Proc.*, **33**:566 (1969).

Beckett, P. H. T., "Studies on soil potassium: I. Confirmation of the ratio law: measurement of potassium potential." *J. Soil Sci.*, **15**:1 (1964).

Beckett, P. H. T., "Studies of soil potassium: II. The 'immediate' Q/I relations of labile potassium in the soil." *J. Soil Sci.*, **15**:9 (1964).

Black, C. A., *Soil–Plant Relationships*, 2nd ed. New York: Wiley, 1968.

Bohn, H., B. McNeal, and G. O'Connor, *Soil Chemistry*. New York: Wiley, 1979.

Bolt, G. H., M. E. Summer, and A. Kamphorst, "A study of the equilibria between three categories of potassium in an illitic soil." *SSSA Proc.*, **27**:294 (1963).

Bolton, E. F., V. A. Dirks, and W. I. Findlay, "Effect of lime on corn yield, soil tilth, and leaf nutrient content for five cropping systems on Brookston clay soil." *Can. J. Soil Sci.*, **59**:225 (1979).

Boswell, F. C., and O. E. Anderson, "Potassium movement in fallowed soils." *Agron. J.*, **60**:688 (1968).

Bray, R., "A nutrient mobility concept of soil–plant relationships." *Soil Sci.*, **78**:9 (1954).

Bray, R. H., and E. E. DeTurk, "The release of potassium from non-replaceable forms in Illinois soils." *SSSA Proc.*, **3**:101 (1938).

Buckman, H. O., and N. C. Brady, *The Nature and Properties of Soils*, 7th ed. New York: Macmillan, 1969.

Burns, A. F., and S. A. Barber, "The effect of temperature and moisture on exchangeable potassium." *SSSA Proc.*, **25**:349 (1961).

Ching, P. C., and S. A. Barber, "Evaluation of temperature effects on K uptake by corn." *Agron. J.*, **71**:1040 (1979).

Claassen, M. E., and G. E. Wilcox, "Comparative reduction of calcium and magnesium composition of corn tissue by NH_4N and K fertilization." *Agron. J.*, **66**:521 (1974).

Cruse, R. M., D. R. Linden, J. K. Radke, W. E. Larson, and K. Larnitz, "A model to predict tillage effects on soil temperature." *Soil Sci. Soc. Am. J.*, **44**:378 (1980).

Dennis, E. J., and R. Ellis, Jr., "Potassium in fixation equilibria and lattice changes in vermiculite." *SSSA Proc.*, **26**:230 (1962).

Dowdy, R. H., and T. B. Hutcheson, Jr., "Effect of exchangeable potassium level and drying on release and fixation of potassium by soils as related to clay mineralogy." *SSSA Proc.*, **27**:31 (1963).

Dowdy, R. H., and T. B. Hutcheson, Jr., "Effect of exchangeable potassium level and drying upon availability of potassium to plants." *SSSA Proc.*, **27:**521 (1963).

Drake, M., J. Vengris, and W. G. Colby, "Cation exchange capacity of plant roots." *Soil Sci.*, **72:**139 (1951).

Drechsel, E. K., "Potassium phosphates: the new generation of super phosphates." Paper presented to Am. Chem. Soc., Div. Fert. Soil Chem. (August 28, 1973).

Dubbs, A. L., "Nitrogen's effect on seeding date of barley and spring wheat." *Better Crops Plant Food*, **65:**27 (1981).

Dunlop, J., A. D. M. Glass, and B. D. Tomkins, "The regulation of K^+ uptake by ryegrass and white clover roots in relation to their competition for potassium." *New Phytol.*, **83:**365 (1979).

Farina, M. P. W., M. E. Sumner, C. O. Plank, and W. S. Letzsch, "Exchangeable aluminum and pH as indicators of lime requirement for corn." *Soil Sci. Soc. Am. J.*, **44:**1036 (1980).

Fried, M., and H. Broeshart, *The Soil–Plant System.* New York: Academic Press, 1967.

Glass, A. D. M., and J. E. Perley, "Varietal differences in potassium uptake by barley." *Plant Physiol.*, **65:**160 (1980).

Glass, A. D. M., M. Y. Siddiqi, and K. I. Giles, "Correlations between potassium uptake and hydrogen efflux in barley varieties." *Plant Physiol.*, **68:**457 (1981).

Goulding, K. W. T., "Potassium retention and release in Rothamsted and Saxmundham soils." *J. Sci. Food. Agr.*, **32:**667 (1981).

Goulding, K. W. T., and O. Talibudeen, "Potassium reserves in a sandy clay loam soil from the Saxmundham experiment: kinetics and equilibrium thermodynamics." *J. Soil Sci.*, **30:**291 (1979).

Graham, E. R., and D. H. Kampbell, "Soil potassium availability and reserve as related to the isotopic pool and calcium exchange equilibria." *Soil Sci.*, **106:**101 (1968).

Grava, J., G. E. Spalding, and A. C. Caldwell, "Effect of drying upon the amounts of easily extractable potassium and phosphorus in Nicollet clay loam." *Agron. J.*, **53:**219 (1961).

Gray, B., M. Drake, and W. G. Colby, "Potassium competition in grass–legume associations as a function of root cation exchange capacity." *SSSA Proc.*, **17:**235 (1953).

Grimme, H., K. Nemeth, and L. C. von Braunschweig, "Some factors controlling potassium availability in soils." *Proc. Int. Symp. Soil Fert. Eval. New Delhi*, **1:**5 (1971).

Hamm, J. W., and J. L. Henry, "A review of current potassium research in Saskatchewan." *Proc. 1973 Saskatchewan Soil Fert. Workshop*, pp. 168–179, Saskatoon (February 8–9, 1973).

Hammond, L. C., W. H. Allaway, and W. E. Loomis, "Effects of oxygen and carbon dioxide levels upon absorption of K in plants." *Plant Physiol.*, **30:**155 (1955).

Hanson, R. G., "Soybeans need P for adequate nodulation." *Better Crops Plant Food*, **62:**26 (1979).

Hanway, J. J., and A. D. Scott, "Soil potassium–moisture relations: II. Profile distribution of exchangeable K in Iowa soils as influenced by drying and rewetting." *SSSA Proc.*, **21:**501 (1957).

Hanway, J. J., and A. D. Scott, "Soil potassium–moisture relations: III. Determining the increase in exchangeable soil potassium on drying soils." *SSSA Proc.*, **23:**22 (1959).

Henderson, R., "The application of potassic fertilizers to pasture and the incidence of hypomagnesemia." *Potash Ltd. Tech. Ser.*, **1:**23 (1960).

Hood, J. T., N. C. Brady, and D. J. Lathwell, "The relationship of water-soluble and exchangeable potassium to yield and potassium uptake by Ladino clover." *SSSA Proc.*, **20:**228 (1956).

Huang, P. M., "Feldspars, olivenes, pyroxenes, and amphiboles," in J. B. Dixon and S. B. Weed, Eds., *Minerals in Soil Environments*, Chapter 15. Madison, Wis.: Soil Science Society of America, 1977.

Huang, P. M., "Soil potassium in relation to crop response to potash." *Proc. Workshop, Potash Phosphate Inst. Canada*, pp. 41–54 (1979).

Huang, P. M., L. S. Crosson, and D. A. Rennie, "Chemical dynamics of potassium release from potassium minerals common in soils." *Trans. 9th Int. Congr. Soil Sci.*, **2:**705 (1968).

Jackson, J. E., and G. W. Burton, "An evaluation of granite meal as a source of potassium for coastal Bermuda grass." *Agron. J.*, **50:**307 (1958).

Jacob, K. D., Ed., *Fertilizer Technology and Resources in the United States*. New York: Academic Press, 1953.

James, D. W., "Relationships of K in arid western soils." *Better Crops Plant Food*, **54:**(1)22 (1970).

Kafkafi, U., B. Bar-Yosef, and A. Hadas, "Fertilizer decision model—a synthesis of soil and plant parameters in a computerized programme," in *Improved Use of Plant Nutrients*, FAO Soils Bull. 37, pp. 11–31 (1978).

Kilmer, V. J., S. E. Younts, and N. C. Brady, Eds., *The Role of Potassium in Agriculture*. Madison, Wis.: American Society of Agronomy, 1968.

Larson, W. E., "Response of sugar beets to potassium fertilization in relation to soil physical and moisture conditions." *Soil Sci. Soc. Am. J.*, **18:**313 (1954).

Lawton, K., "The influence of soil aeration on the growth and absorption of nutrients by corn plants." *Soil Sci. Soc. Am. J.*, **10:**263 (1945).

Liebhardt, W. C., N. Spoljaric, W. W. Hsu, and L. Cotnoir, "Potassium release characteristics and mineralogical characteristics in some Delaware soils." *Commun. Soil Sci. Plant Anal.*, **10**(1):1025 (1979).

McLean, E. O., "Exchangeable K levels for maximum crop yield on soils of different exchange capacities." *Commun. Soil Sci. Plant Anal.*, **7:**823 (1976).

McLean, E. O., "Influence of clay content and clay composition on potassium availability," in G. S. Sekhon, Ed., *Potassium in Soils and Crops*, Chapter 1. New Delhi: Potash Research Institute of India, 1978.

Matthews, B. C., and C. G. Sherrell, "Effect of drying on exchangeable potassium of Ontario soils and the relation of exchangeable K to crop yield." *Can. J. Soil Sci.*, **40:**35 (1960).

Mengel, K., "Potassium availability and its effect on crop production." *Proc. Jap. Potash Symp.*, Tokyo, Japan (1971).

Mengel, K., and H. E. Haeder, "Potassium availability and its effect on crop production." *Potash Rev.*, **11:**1 (1973).

Mortland, M. M., K. Lawton, and G. Uehara, "Fixation and release of potassium by some clay minerals." *SSSA Proc.*, **21:**381 (1957).

Mortland, M. M., K. Lawton, and G. Uehara, "Alteration of biotite to vermiculite by plant growth." *Soil Sci.*, **82:**477 (1958).

Moss, P., "Some aspects of the cation status of soil moisture: I. The ratio law and soil moisture content." *Plant Soil*, **18:**99 (1963).

Munson, R. D., "Potassium availability and uptake," in *Potassium for Agriculture—A Situation Analysis*. Atlanta, Ga.: Potash and Phosphate Institute, 1980.

Nair, P. K. R., and H. Grimme, "Q/I relations and electroultrafiltration of soils as measures of potassium availability to plants." *Z. Pflanzenernaehr. Bodenkd.*, **142:**87 (1979).

Nelson, W. L., "Plant factors affecting potassium availability and uptake," in V. J. Kilmer, S. E. Younts, and N. C. Brady, Eds., *The Role of Potassium in Agriculture*, Chapter 17. Madison, Wis.: American Society of Agronomy, 1968.

Nemeth, K., and H. Grimme, "Effect of soil pH on the relationship between K concentration in the saturation extract and K saturation of soils." *Soil Sci.*, **114:**349 (1972).

Nemeth, K., K. Mengel, and H. Grimme, "The concentration of K, Ca, and Mg in the saturation extract in relation to exchangeable K, Ca, and Mg." *Soil Sci.*, **109:**179 (1970).

Nuttall, W. F., B. P. Warkentin, and A. L. Carter, " 'A' values of potassium related to other indexes of soil potassium availability." *SSSA Proc.*, **31**:344 (1967).

Olson, R. A., et al., Eds., *Fertilizer Technology and Use*, 2nd ed. Madison, Wis.: Soil Science Society of America, 1971.

Page, A. L., F. T. Bingham, T. J. Ganje, and M. J. Garber, "Availability and fixation of added potassium in two California soils when cropped to cotton." *SSSA Proc.*, **27**:323 (1963).

Page, A. L., W. D. Burge, T. J. Ganje, and M. J. Garber, "Potassium and ammonium fixation by vermiculitic soils." *SSSA Proc.*, **31**:337 (1967).

Parks, W. L., and W. M. Walker, "Effect of soil K, K fertilizer and method of fertilizer placement upon corn yields." *Soil Sci. Soc. Am. J.*, **33**:427 (1969).

Pope, A., and H. B. Cheney, "The K-supplying power of several western Oregon soils." *SSSA Proc.*, **21**:75 (1957).

Pratt, P. F., "Potassium removal from Iowa soils by greenhouse and laboratory procedures." *Soil Sci.*, **72**:107 (1951).

Pretty, K. M., "Factors influencing potassium requirements and response to potassium fertilization, North American experience," in G. S. Sekhon, Ed., *Potassium in Soils and Crops*, pp. 147–163. New Delhi: Potash Research Institute of India, 1978.

Quémener, J., "The measurement of soil potassium." *IPI Res. Topics 4*. Bern: International Potash Institute, 1979.

Renger, M., and O. Strebel, "Water and nutrient transport to plant roots as a function of depth and time under field conditions," in *Soils in Mediterranean Type Climates and Their Yield Potential*, pp. 65–77, 14th Colloq. IPI. Bern: International Potash Institute, 1979.

Rich, C. I., "Mineralogy of soil potassium," in V. J. Kilmer, S. E. Younts, and N. C. Brady, Eds., *The Role of Potassium in Agriculture*, Chapter 5. Madison, Wis.: American Society of Agronomy, 1968.

Richards, G. E., and E. O. McLean, "Release of fixed potassium from soils by plant uptake and chemical extraction techniques." *SSSA Proc.*, **25**:98 (1961).

Robertson, W. K., L. C. Hammond, and L. G. Thompson, Jr., "Yield and nutrient uptake by corn for silage on two soil types as influenced by fertilizer, plant population and hybrids." *Soil Sci. Soc. Am. J.*, **29**:551 (1965).

Roy, R. N., S. Sectharaman, and R. N. Singh, "Soil and fertilizer K in crop nutrition." *Fert. News*, **23**(6):3 (1978).

Sauchelli, V., Ed., *The Chemistry and Technology of Fertilizers*. ACS Monogr. 148. New York: Reinhold, 1960.

Sauchelli, V., *Manual on Fertilizer Manufacture*, 3rd ed. Caldwell, N.J.: Industry Publications, 1963.

Schulte, E. E., "Build up soil K levels before shifting to minimum tillage." *Better Crops Plant Food*, **63**(3):25 (1979).

Schwartz, S., and U. Kafkafi, "Mg, Ca and K status of corn silage and wheat at periodic stages of growth in the field." *Agron. J.*, **70**:227 (1978).

Scott, A. D., and T. E. Bates, "Effect of organic additions on the changes in exchangeable potassium observed in drying soils." *SSSA Proc.*, **26**:209 (1962).

Scott, T. W., and F. W. Smith, "Effect of drying upon availability of potassium in Parsons silt loam surface soil and subsoil." *Agron. J.*, **49**:377 (1957).

Singh, B. B., and J. P. Jones, "Use of sorption-isotherms for evaluating potassium requirements of plants." *Soil Sci. Soc. Am. J.*, **39**:881 (1975).

Singh, B. B., and J. P. Jones, "Sorption isotherms for predicting potassium requirements of some Iowa soils." *Commun. Soil Sci. Plant Anal.*, **7**:197 (1976).

Smith, D., "Levels and sources of K for alfalfa as influenced by temperature." *Agron. J.*, **63**:497–500 (1971).

Somasiri, S., and P. M. Huang, "The nature of K-feldspars of selected soils in the Canadian prairies." *Soil Sci. Soc. Am. J.*, **37**:461 (1973).

Somasiri, S., S. Y. Lee, and P. M. Huang, "Influence of certain pedogenic factors on potassium reserves of selected Canadian prairie soils." *Soil Sci. Soc. Am. J.*, **35**:500 (1971).

Sparks, D. L., "Chemistry of soil potassium in Atlantic coastal plain soils: a review." *Commun. Soil Sci. Plant Anal.*, **11**:435 (1980).

Sparks, D. L., and W. C. Liebhardt, "Effect of long-term lime and potassium applications on quantity-intensity (Q/I) relationships in sandy soil." *Soil Sci. Soc. Am. J.*, **45**:786 (1981).

Sparks, D. L., D. C. Martens, and L. W. Zelazny, "Plant uptake and leaching of applied and indigenous potassium in Dothan soils." *Agron. J.*, **72**:551 (1980).

Sparks, D. L., L. W. Zelazny, and D. C. Martens, "Kinetics of potassium exchange in a Paleudult from the coastal plain of Virginia." *Soil Sci. Soc. Am. J.*, **44**:37 (1980).

Spencer, K., "Potassium fertilizers, soil interactions, plant uptake, losses and efficient usage," in *Fertilizers and the Environment*. Sydney: Australian Institute of Agricultural Science, 1974.

Spies, C., "Crops have big appetites." *Purdue Agron. Crops Soils Notes 218* (July 1973).

Spratt, E. D., "Nutritional status of Manitoba corn, 1973–77." *Tech. Sci. Papers, Manitoba Agron. Annu. Conf.*, pp. 40–56, Winnipeg, Manitoba (December 12–13, 1979).

Stevenson, F. J., and A. P. S. Dhariwal, "Distribution of fixed ammonium in soils." *SSSA Proc.*, **23**:121 (1959).

Terman, G. L., "Yields and nutrient accumulation by determinant soybeans as affected by applied nutrients." *Agron. J.*, **69**:234 (1977).

Thomas, G. W., and B. W. Hipp, "Soil factors affecting potassium availability," in V. J. Kilmer, S. E. Younts, and N. C. Brady, Eds., *The Role of Potassium in Agriculture*, Chapter 13. Madison, Wis.: American Society of Agronomy, 1968.

Thorne, D. W., and H. B. Peterson, *Irrigated Soils*. New York: Blakiston, 1954.

Thorup, R. M., and A. Mehlich, "Retention of potassium meta- and ortho-phosphates by soils and minerals." *Soil Sci.*, **91**:38 (1961).

Tinker, P. B., "Transport processes in the soil as factors in K availability," in G. S. Sekhon, Ed., *Potassium in Soils and Crops*, pp. 21–33. New Delhi: Potash Research Institute of India, 1978.

Von Uexkull, H. R., "Agronomic and economic evaluation of crop responses to K fertilizer—rice: the Far Eastern and S.E. Asian experience," in G. S. Sekhon, Ed., *Potassium in Soils and Crops*, pp. 241–259. New Delhi: Potash Research Institute of India, 1978.

Walker, D. W., "Potassium fertilization in central Alberta," *Better Crops Plant Food*, **62**:13 (Summer 1978).

Walsh, L. M., and J. D. Beaton, Eds., *Soil Testing and Plant Analysis*, rev. ed. Madison, Wis.: Soil Science Society of America, 1973.

Walsh, L. M., and J. T. Murdock, "Native fixed ammonium and fixation of applied ammonium in several Wisconsin soils." *Soil Sci.*, **89**:183 (1960).

Welch, L. F., "Availability of non-exchangeable potassium and ammonium to plants and micro-organisms." *Diss. Abstr.*, **19**:1885 (1958).

Welch, L. F., and A. D. Scott, "Nitrification of fixed ammonium in clay minerals as affected by added potassium." *Soil Sci.*, **90**:79 (1960).

Welch, L. F., and A. D. Scott, "Availability of non-exchangeable soil potassium to plants as affected by added potassium and ammonium." *SSSA Proc.*, **25**:102 (1961).

Westermann, D. T., T. L. Jackson, and D. P. Moore, "Effect of potassium salts on extractable soil manganese." *Soil Sci. Soc. Am. J.*, **35**:43 (1971).

Williams, D. E., "The absorption of potassium as influenced by its concentration in the nutrient medium." *Plant Soil*, **15**:387 (1961).

Wood, L. K., and E. E. DeTurk, "The absorption of potassium in soils in non-replaceable forms." *SSSA Proc.*, **5**:152 (1941).

Woodruff, C. M., "The energies of replacement of calcium by potassium in soils." *SSSA Proc.*, **19**:167 (1955).

York, E. T., Jr., R. Bradfield, and M. Peech, "Calcium–potassium interactions in soils and plants: I. Lime-induced potassium fixation in Mardin silt loam." *Soil Sci.*, **76**:379 (1953).

Chapter 8

Soil and Fertilizer Sulfur, Calcium, and Magnesium

SULFUR, calcium, and magnesium, often referred to as secondary elements in the past, are macronutrients that are required in relatively large amounts for good crop growth. Sulfur and magnesium are needed by plants in about the same quantities as phosphorus, an element traditionally classed as a major plant nutrient. For many plant species, the calcium requirement is greater than that for phosphorus.

The occurrence and reactions of sulfur in soil are different from those of calcium and magnesium. The SO_4^{2-} ion is relatively mobile in the soil solution and sulfur, like nitrogen, is subject to biological and chemical oxidation–reduction reactions.

Calcium and magnesium are much more stable than sulfur. They exist as the cations Ca^{2+} and Mg^{2+} and are associated with the soil colloidal fraction in much the same way as potassium.

Reactions of sulfur, calcium, and magnesium in soils and the factors influencing their availability to plants are considered in this chapter. Ways of applying these elements in fertilizers are also covered.

Sulfur

Sources in Soils. The average sulfur content of the earth's crust is estimated to be between 0.06 and 0.10%. It is usually ranked as the thirteenth most abundant element. Sulfur occurs in the elemental form, as well as sulfides, sulfates, and in organic combinations with carbon and nitrogen.

The main sulfur-bearing minerals in rocks and soils are gypsum ($CaSO_4 \cdot 2H_2O$), anhydrite ($CaSO_4$), epsomite ($MgSO_4 \cdot 7H_2O$), mirabilite ($Na_2SO_4 \cdot 10H_2O$), pyrite and marcasite (FeS_2), sphalerite (ZnS), chalcopyrite ($CuFeS_2$), and cobaltite ($CoAsS$). Other important sulfides, including pyrrhotite ($Fe_{11}S_{12}$), galena (PbS), arsenopyrite ($FeS_2 \cdot FeAs_2$), and pentlandite ($Fe,Ni)_9S_8$, are found throughout the world.

Elemental sulfur occurs in deposits over salt domes, in volcanic deposits, and in deposits associated with calcite, gypsum, and anhydrite. Minor amounts of sulfur occur as gaseous oxides released during volcanic activity and as hydrogen sulfide of volcanic, hydrothermal, and biological origin. Hydrogen sulfide, an important commercial source of sulfur, is also a contaminant in many natural gas fields. Organic compounds containing sulfur occur in crude oil, coal, and tar sands.

Silicate minerals generally contain less than 0.01% sulfur; however, it may be much more abundant in biotites, chlorites, and layer-type clay minerals. The sulfur content of igneous rocks usually ranges from 0.02% in acidic and plutonic types to 0.07% in ultrabasic and volcanic types. The concentration of sulfur in sedimentary rocks varies from 0.02 to 0.22%, and thus they are an important source of this element in many soils.

The original source of soil sulfur was doubtless the sulfides of metals in plutonic rocks. As these rocks were exposed to weathering, the minerals decomposed and sulfide was oxidized to sulfate. These sulfates were then precipitated as soluble and insoluble sulfate salts in arid or semiarid climates, absorbed by living organisms, or reduced by other organisms to sulfides or elemental sulfur under anaerobic conditions. Some of the released sulfates, of course, found their way to the sea in drainage waters.

Sulfate is not only prevalent in minerals in rocks and soils. Ocean waters contain approximately 2700 ppm sulfate. Its normal range in other natural waters ranges from 0.5 to 50 ppm but may reach 60,000 ppm (6%) in highly saline lakes and sediments. A number of solid mineral sulfates were listed earlier.

The sulfur in most arable land today is in the form of organic matter, soluble sulfates in the soil solution, or adsorbed on the soil complex. It will be recalled that sulfur is a component of proteins and that when these materials are returned to the soil and converted to humus a large fraction of the sulfur remains in organic combination. Much of the total sulfur found in the surface of humid-region soils is in the organic form. In arid soils, of course, the sulfates of calcium, magnesium, sodium, and potassium are frequently precipitated in large quantities in the soil profile. Appreciable amounts of exchangeable SO_4^{2-} may be present in subsoils that contain 1:1 clays and hydrous oxides of iron and aluminum.

Another source of soil sulfur is the atmosphere. Around centers of industrial activity, in which coal and other sulfur-containing products are burned, sulfur dioxide is released into the air, and much of this gas is later brought back to earth by the rain. Plants may also absorb sulfur dioxide by gaseous diffusion into the leaves, and this sulfur is utilized by the plant in its normal metabolic processes. If the concentration in the air is too great, however, there may be injury to plants. Exposure to as little as 0.5 ppm of SO_2 for 3 hours can cause visible injury to the foliage of sensitive vegetation.

In the United States more than 26 million tons of SO_2 (13 million tons of sulfur) were estimated to have been emitted into the atmosphere in 1980. This is a noteworthy decline from the approximately 32 million tons ejected in 1972. Emissions of SO_2 in Canada in 1980 were about 6.5 million metric tons, with some 8 to 9% of this national figure occurring in Alberta. Liberation of SO_2 to the atmosphere by European nations plus American and Canadian contributions reached almost 100 million tons in 1980. Most of this resulted from the combustion of fossil fuels, but industrial processes such as ore smelting, petroleum refining operations, and others contributed about 20% of the total emitted. Much of this SO_2 is returned to the soil in rainfall and within a relatively small radius around the point of emission. Available data indicate that the amounts of sulfur brought down in rainfall in the United States range from about 1 lb/A per year in rural areas to 100 lb/A per year near areas of industrial activity.

Such sulfur emissions are partly responsible for the acid rain and snow which falls in northern Europe, northeastern United States and in the provinces of Ontario and Quebec in eastern Canada. Strong acids including sulfuric acid have lowered the pH of precipitation in much of these areas to between 4 and 5. Precipitation in nonindustrial areas is characterized by pH values in the range 5.7 to 7.0, whereas rain in highly populated industrial regions always has a pH lower than 5.7.

In addition to the deposition of SO_2 in rain and snow on soil, this pollutant can reach soils by several other mechanisms: rain interception by trees, dry particulates, and direct adsorption. The proportion of SO_2 deposited in each way varies from region to region. Direct adsorption may be the most important way that SO_2 enters soils.

Because of the growing concern over air pollution, legislation may ultimately be enacted to require the cleaning and scrubbing of all waste gases. Many industrial companies are presently required to clean effluent gases, and it is likely that this practice will spread and, in turn, will reduce the amounts of sulfur brought down by the rain and absorbed from the atmosphere directly by plants and soil. While the emission of sulfur compounds into the atmosphere by industrial activity has received a lot of unfavorable publicity, the fact remains that 70% of the total content of sulfur compounds in the atmosphere is of nonmanmade origin. Volatile sulfur compounds are released in large quantities from volcanic activity, from tidal marshes, from decaying organic matter, and from other sources.

Forms and Behavior of Sulfur in Soils. Sulfur is present in the soil in many different forms, both organic and inorganic. It exists in multiple oxidation states ranging from $+6$ in H_2SO_4 and its derivatives (hexavalent or oxidized sulfur) to -2 in H_2S and its derivatives (divalent or reduced sulfur). Also, sulfur occurs in solid, liquid, or gaseous phases.

The inorganic forms are readily-soluble sulfate, adsorbed sulfate, insoluble sulfate coprecipitated with calcium carbonate, and reduced inorganic sulfur compounds. Since plants obtain sulfur primarily from soil as dissolved sulfate, easily soluble sulfate plus adsorbed sulfate represent the readily available fraction of soil sulfur which is utilized by plants.

Many surface soils contain most of their sulfur in organic combinations. Researchers agree that in excess of 90% of the total sulfur in most noncalcareous surface soils exists in organic forms.

The dynamic nature of sulfur in soils and the relationships between the variety of forms are shown schematically in Figure 8-1. There are similarities between

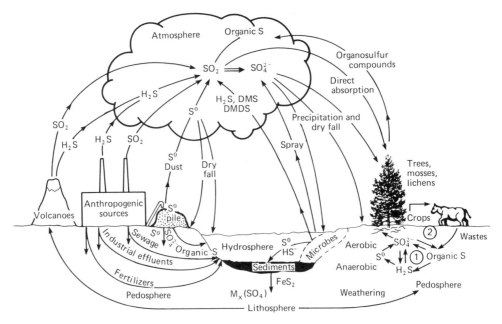

FIGURE 8-1. Simplified version of the overall sulfur cycle in nature. The encircled 1 refers to the microbiological cycle and the encircled 2 depicts a food cycle with higher animals. (Krouse and McCready, in P. A. Trudinger and D. J. Swaine, Eds., *Studies in Environmental Science*, Vol. 3: *Biogeochemical Cycling of Mineral-Forming Elements*. Amsterdam: Elsevier Scientific, 1979.)

the nitrogen and sulfur cycles in that both have gaseous components and their occurrence in soils is associated with organic matter.

Easily Soluble Sulfate. Sulfur is normally taken up by plants as the SO_4^{2-} ion. Concentrations of 3 to 5 ppm of SO_4^{2-} in the soil or solution cultures have proved adequate for the growth of many plant species. Rape and alfalfa appear to require higher concentrations.

Sulfate ions reach roots by diffusion and mass flow. In soils containing 5 ppm or more of soluble SO_4^{2-}, virtually all of the sulfur requirement of most crops can be supplied by mass flow.

Except for soils in dry areas which may have accumulations of sulfate salts, most soils are generally considered to contain less than 25% of their total sulfur as inorganic SO_4^{2-}. In surface soils the figure is more likely to be of the order of 10% or less. Readily soluble SO_4^{2-} in many North American soils represents between 1 to 10% of the total sulfur content. The easily soluble SO_4^{2-} fraction of most well-drained surface soils of Australia is usually less than 5% of the total sulfur.

Concentrations of less than 20 ppm of readily soluble SO_4^{2-} are common in North American soils. In sulfur-deficient soils or soils of low sulfur supplying capacity, the amount of readily soluble SO_4^{2-} is frequently between 5 and 10 ppm. Sandy sulfur-deficient soils often contain less than 5 ppm.

Levels of readily soluble SO_4^{2-} vary greatly with depth of soil and fluctuate considerably within the profile from location to location. It may sometimes reach a maximum in subsurface horizons but it can also be very low in sandy subsoils. Accumulations of easily soluble SO_4^{2-} occur most commonly in calcareous or gypsiferous horizons and in impervious horizons where moisture movement and leaching are restricted.

Adsorbed Sulfate. Adsorbed sulfate is an important fraction of some soils, particularly those containing large amounts of the hydrous oxides of aluminum and iron. Many Ultisol (Red-Yellow Podzol) and Oxisol (Latosol) soils contain appreciable amounts of adsorbed sulfate or have the ability to absorb this form of sulfur. These soils are highly weathered and occur in regions of high rainfall in the southeastern United States, Pacific Northwest, Central and South America, Asia, and Africa. After cropping for a number of years the organic-matter content of such soils approaches a stable equilibrium level and the amounts of organic sulfur released will be quite small.

This fraction is believed to contribute significantly to the sulfur needs of plants growing in highly weathered soils because it is usually readily available. Also, release of organic sulfur is likely to be low in these soils. In certain soils, adsorbed SO_4^{2-} may not be as rapidly available as soluble SO_4^{2-} and it may be released over longer periods of time.

Although crops can utilize adsorbed SO_4^{2-} in subsoils, they might experience sulfur deficiency in the early growth stages until root development is sufficient to reach the retention zones. Deep-rooted crops such as alfalfa and sericea lespedeza are unlikely to have such temporary shortages of available sulfur.

The analyses reported in Table 8-1 are indicative of the amounts of adsorbed SO_4^{2-} in various southeastern U.S. soils. It should be noted that these values are only an approximate measure of this fraction because of limitations in the extraction method and since no corrections were made for easily soluble SO_4^{2-}. The occurrence of adsorbed SO_4^{2-} generally follows a characteristic pattern with surface soils containing very little while there is a definite accumulation at deeper depths. These zones of high concentration of adsorbed SO_4^{2-} are found at depths ranging from 6 to 30 in. below the surface.

TABLE 8-1. Concentration of Adsorbed Sulfate Sulfur in Selected Soils of the Southeastern United States

Soil Depth (in.)	Concentration of Sulfate Sulfur Extracted with Sodium Acetate * (ppm)				
	Norfolk Sandy Loam A.	Norfolk Sandy Loam B.	Chewacla Silt Loam	Fannin Clay Loam	Magnolia Fine Sandy Loam
0–6	0	2	1	3	1
6–12	1	7	1	83	54
12–18	63	114	1	100	95
18–24	93	131	6	58	106
24–30	97	135	24	44	146
30–36	110	127	47	28	87

* pH 4.8.

Source: Ensminger, *Alabama Agr. Exp. Sta. Bull. 312* (1958); and Jordan, *U.S. Dept. Agr. Tech. Bull. 1297* (1964).

Reserves of adsorbed SO_4^{2-} in subsoils are believed to be the result of eluviation or leaching of sulfates from the upper part of soil profiles followed by their retention at lower depths.

Adsorbed SO_4^{2-} can account for up to one-third of the total sulfur in subsoils. In surface soils it usually represents less than 10% of the total sulfur present.

The mechanisms causing SO_4^{2-} retention and the various factors influencing its retention will be reviewed briefly in a later section of this chapter. Sulfate adsorption by soils is considered to be a beneficial phenomenon since it will help to conserve sulfur fertilizers and native supplies of soil sulfur by reducing leaching losses in regions of high rainfall.

Sulfate Coprecipitated with Natural Calcium Carbonates. Sulfate associated with calcium carbonate is often an important fraction of the total sulfur in calcareous soils. It most likely occurs as a coprecipitated or co-crystallized impurity with calcium carbonate. This form of sulfate is thought to be relatively unavailable to plants, particularly when the calcium carbonate is present as coarse particles. As much as 93% of the total sulfur in an Australian soil and 75% in a Scottish soil occurred as SO_4^{2-} coprecipitated with calcium carbonate. If this fraction contributes significantly to the sulfur needs of crops growing in low sulfur surface soils, some degree of deficiency will likely occur until their roots are deep enough to extract SO_4^{2-}.

Solubility and availability of SO_4^{2-} coprecipitated with calcium carbonate are believed to be influenced by several factors, including particle size of the calcium carbonate, soil moisture content, common ion effects, and ionic strength. The common ion effect probably also plays an important role in formation of this fraction.

The grinding that is normally done in soil sample preparation will render sulfate in this fraction accessible to chemical extraction. Consequently, more sulfur will likely be classified as being available by a particular soil-test procedure than is the actual case under field conditions.

Other Precipitated Forms of Sulfate. Selenite, a crystalline form of gypsum, has been found in poorly drained subsoils. Extremely insoluble barium and strontium sulfates may occasionally account for substantial amounts of SO_4^{2-} in some soils. Jarosite $[KFe_3(OH)_6(SO_4)_2]$ and coquimbite $[Fe_2(SO_4)_3 \cdot 9H_2O]$ minerals were isolated from efflorescent coatings on clods of drained and cultivated tidal marsh soils.

Reduced Inorganic Sulfur. Although SO_4^{2-} is the stable form of sulfur in well-drained upland soils, minor amounts of sulfide forms are known to occur. Under waterlogged conditions, sulfide is the principal stable form of sulfur. Some elemental sulfur and organic compounds containing reduced sulfur may be found in certain natural anaerobic environments.

Sulfides. Under anaerobic conditions in waterlogged soils, there may be accumulations of H_2S formed by the decay of organic matter. Also, SO_4^{2-} present in the soil serves as an electron acceptor for sulfate-reducing bacteria and it is usually reduced to H_2S. This reduction of SO_4^{2-} is both redox potential- and pH-dependent. Little or no S^{2-} accumulates at redox potentials above -150 mV, or with a pH outside the range 6.5 to 8.5.

Sulfide accumulation is limited primarily to coastal regions influenced by seawater. In normal submerged soils well supplied with Fe, the H_2S liberated from organic matter and from SO_4^{2-} is almost completely removed from solution by reaction with Fe^{2+}. The first iron sulfide precipitate is usually amorphous FeS, which undergoes conversion to either pyrite or marcasite, both with the formula FeS_2. Mackinawite (FeS) and greigite (Fe_3S_4) may be intermediates in this conversion.

The principal insoluble sulfides formed in flooded rice soils are believed to be FeS_2 and Fe_2S_3. Occurrence of FeS in such soils is probably limited and does not become appreciable unless redox potentials are -400 mV or less. In some rice paddy soils, high in organic matter and low in active metallic elements such as iron, free H_2S may prove harmful to the rice roots.

Reactivity of sulfides in soils varies greatly according to the manner in which they are formed. Presumably, it depends chiefly on particle size and degree of crystallinity of the precipitates. Oxidation of FeS precipitates may be complete after only a few hours exposure to the atmosphere, and will probably in most cases be terminated within a few days. Amorphous and crystalline forms of marcasite are considerably more reactive than pyrite and oxidize at a much greater rate.

Pyrite will persist in soils with appreciable amounts of it still present in previously waterlogged land many years after drainage and cultivation. Resistance of pyrite to oxidation has been ascribed to inhibitory effects of phosphorus.

Occurrence of the minerals pyrite and marcasite is widespread in coal and allied materials. They may be finely dispersed and invisible to the naked eye or in a large-size range of single crystals and crystal aggregates. When pyritic spoil is exposed to the atmosphere, sulfuric acid is formed resulting in acidification of mine waters and mine spoil areas. The acidic level in some waters may make them unsatisfactory for use as a water supply or for recreational purposes. The acidic soil environment can create problems in revegetation and use of mine spoil areas.

Sulfide and sulfite are oxidized rapidly by strictly chemical means. The initial stage in the oxidation of pyrite, with the formation of elemental sulfur as an intermediate, may be largely chemical. Higher temperature, oxygen tension, and specific surface all increase the rate of oxidation of pyrite. The presence of Fe^{3+}, and possibly also Fe^{2+}, accelerates the oxidation of pyrite. A slightly acid reaction and adequate moisture favor the oxidation process. Perfect soil drainage is required before oxidation will be complete.

Elemental Sulfur. It is not a direct product of sulfate reduction under reducing conditions in soil but is instead an intermediate formed during the oxidation of sulfides by largely chemical processes. Accumulation of elemental sulfur may occur, however, in soils of delta regions where the complete oxidation of reduced forms of sulfur is interrupted by periodic flooding.

Organic Forms. Most of the sulfur in surface horizons of well-drained agricultural soils of the humid, semiarid, temperate, and subtropical regions is present in organic forms. It is generally agreed that sulfur associated with these organic compounds accounts for well over 90% of the total sulfur in most noncalcareous surface soils.

The proportion of total sulfur existing in organic forms varies considerably according to soil type and depth in the soil profile. It is usually somewhat lower in subsoils than in surface horizons.

As described in Chapter 5, there is a close relationship between organic carbon, total nitrogen, and total sulfur in soils. In addition to the examples of C/N/S ratios referred to in Chapter 5, it is of interest to note that several Scottish soils had ratios of total C/N/nonsulfate sulfur of 113:10:1.3 for calcareous soils and 147:10:1.4 for humid noncalcareous ones. The N/S ratio in these studies was less variable than for C/S. In Iowa soils studied by Tabatabai and Bremner, the association of carbon, nitrogen, and sulfur occurred approximately in the ratio of 110:10:1.5, respectively.

Several Australian workers have suggested that the C/N/S ratios are often remarkably similar for different groups of soils. Stewart and Bettany at the University of Saskatchewan thoroughly reviewed this subject and found marked differences in the mean C/N/S ratios, among and within types of world soils. These differences have been ascribed to variations in parent material and other soil-forming factors, such as climate, vegetation, leaching intensity, and drainage.

A close association frequently exists between the nitrogen and sulfur constituents of soil organic matter. Total nitrogen and organic sulfur are often more closely correlated than organic carbon and organic sulfur. The N/S ratio in most soils falls within the narrow range 6 to 8:1.

The nature and properties of the organic sulfur fraction in soils are important since they govern the release of plant-available sulfur as illustrated in Figure 8-2. While much of the organic sulfur in soils remains uncharacterized, three broad groups of sulfur compounds are recognized. These are HI-reducible sulfur, carbon-

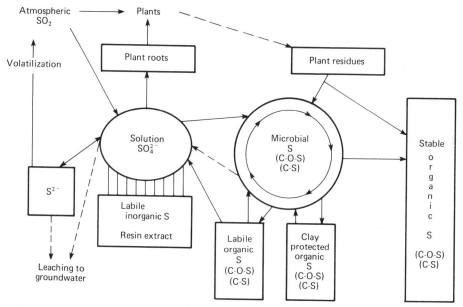

FIGURE 8-2. Involvement of organic sulfur fractions in release of plant available sulfur. [Stewart and Bettany, *Proc. Alberta Soil Sci. Workshop*, Edmonton, Alberta, p. 184 (February 23–24, 1982).]

TABLE 8-2.   Fractionation of Organic Sulfur in Surface Soils

Location *	HI-Reducible S as Percent of Total		C-Bonded S † as Percent of Total		Residual, Inert S as Percent of Total	
	Range	Mean	Range	Mean	Range	Mean
Quebec, Canada (3)	44–78	65	12–32	24	0–44	11
Alberta, Canada (15)	25–71	49	12–32	21	7–45	30
Saskatchewan, Canada (54)	28–59	45	n.d. ‡		n.d.	
Australia (21)	—	52	n.d.		n.d.	
Australia (15)	32–63	47	22–54	30	3–31	23
Iowa, U.S. (24)	36–66	52	5–20	11	21–53	37
Brazil (6)	36–70	51	5–12	7	24–59	42

* Figures in parentheses refer to number of samples.
† Determined by reduction with Raney nickel.
‡ n.d., not determined.
Source: Biederbeck, in M. Schnitzer and S. U. Khan, Eds., *Soil Organic Matter,* Chapter 6. New York: Elsevier, 1978.

bonded sulfur, and residual or inert sulfur. The relative importance of these three categories is shown in Table 8-2.

HI-Reducible. This fraction is comprised of organic sulfur that is reduced to H_2S by hydriodic acid. Its sulfur is not bonded directly to carbon and is believed to be largely in the form of sulfate esters and ethers with C—O—S linkages. Examples of substances in this grouping include arylsulfates, alkylsulfates, phenolic sulfates, sulfated polysaccharides, choline sulfate, sulfated lipids, and so on. About 50% of the organic sulfur occurs in this fraction, but it can range from about 27 to 59%. Values as high as 94.5% have been reported for Iowa subsoils.

Carbon-Bonded. Sulfur directly bonded to carbon is a major component of this grouping. It is determined by reduction to sulfide with Raney nickel. This analytical procedure has been criticized because it fails to reduce all carbon-bonded sulfur compounds and it can produce artifacts. Nevertheless, it is still a useful method for characterizing a significant portion of the sulfur bonded in this manner.
     The sulfur-containing amino acids, cystine and methionine, are principal components of this fraction, which accounts for about 10 to 20% of the total organic sulfur. More oxidized sulfur forms, including sulfoxide, sulfones, sulfenic, sulfinic, and sulfonic acids and sulfur-containing heterocyclic compounds, are also constituents of this fraction.

Inert or Residual. Organic sulfur that is not reduced by either hydriodic acid or Raney nickel is considered to be inert or residual. This unidentified fraction generally represents approximately 30 to 40% of the total organic sulfur. It is exceptionally stable since it resists degradation by drastic chemical treatments. Consequently, it is probably of little importance as a potential source of sulfur for plants.

Behavior of Sulfur in Soils.
     Inorganic Sulfate. Almost all of the inorganic sulfur in well-drained arable soils exists as the sulfate ion in combination with cations such as Ca^{2+}, Mg^{2+}, K^+, Na^+, or NH_4^+. The amounts of sulfate typically occurring in its three main forms, water-soluble salts, adsorbed by soil colloids, and as insoluble compounds, were discussed earlier in this chapter.

TABLE 8-3. Sulfate Levels in Two Soil Depths at Locations in Southern Alberta

Site and Year	Previous Crop *	Precipitation (mm)	SO_4-S (kg/ha)	
			0–15 (cm)	0–30
Barons				
1976	Smf	204	6	9
1977	St	116	3	7
1978	St	160	6	11
Milk River				
1977	St	81	2	39
1978	Smf	292	4	11
Jefferson				
1977	Smf	110	5	24
Welling				
1976	St	132	14	23

* Smf, summer fallow; St, stubble (recrop).
Source: Bole and Pittman, in *Effective Use of Nutrient Resources in Crop Production,* Proc. Alberta Soil Sci. Workshop, p. 335 (1979).

Soluble Sulfate. The quantities of soluble sulfate in soils vary widely both between soil types and within soil profiles. Large seasonal and year-to-year fluctuations in levels of soluble sulfate in the surface soil and at greater depths can occur, as is evident in Table 8-3. This variability is caused by the interaction of environmental and seasonal conditions on the mineralization of organic sulfur, either downward or upward movement of sulfate salts in soil moisture, and sulfate uptake by plants.

Sulfate content of soils is also affected by the application of sulfur-containing fertilizers, and the sulfate present in precipitation and irrigation waters. In localized areas near centers of industrial activity, the sulfate content of soils can be increased by direct absorption of SO_2 and the fallout of dry particulates.

One of the reasons that soil sampling to 24 in. is recommended in dry regions of western Canada is to better evaluate the sulfur-supplying power of soil. Table 8-3 shows there can be appreciable amounts of sulfate below the top 6 in. of soil.

Because of its anionic nature and the solubility of most of its common salts, sulfate like nitrate can, under conditions of large amounts of percolating water, be readily leached from surface soil. However, its tendency to disappear from soils varies widely. As an example, workers at the University of Georgia showed that cotton, grown for five years on two texturally different soils, responded differently to applications of sulfur at 0, 4, 8, 16, and 32 lb/A per year. On the silt loam soil no responses to added sulfur were observed at the end of the 5-year experiment. On the sandy loam soil, however, a sulfur deficiency developed during the fourth cropping year at the zero level of added sulfur.

Another piece of work conducted in California illustrates the rapidity with which sulfur can be lost from soils under conditions of heavy rainfall. Sulfur as gypsum was applied to a stand of clover growing on a field plot of sandy loam soil. Rates of applied gypsum were 100, 200, and 300 lb/A and rainfall contributed about 21 lb/A. Measurements showed that in the growing season following fertilization 77% of the sulfur applied at the 100-lb rate of gypsum and 78% of that applied at the 300-lb rate were accounted for in the percolating water collected from these plots. There was a significant yield response to the added sulfur, and about 31 and 57%, respectively, of the tissue sulfur in these plants was accounted for by the 100- and 300-lb applications of gypsum.

The relation between the amount of percolating water and the downward movement of sulfate was determined with radioactive ^{35}S by workers in Oregon. A given

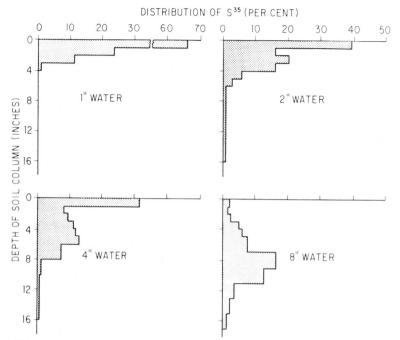

FIGURE 8-3. Distribution of sulfur throughout columns of a Willamette soil as a function of the amount of added water. [Chao et al., *SSSA Proc.*, **26**:27 (1962).]

amount of sulfate sulfur was added to columns of soil. Different amounts of water were then added to these columns and the distribution pattern of the sulfur was measured. The results are shown in Figure 8-3. The greater the amount of added water, the greater the net downward movement of the sulfate. The soil used in this study contained 29% clay and is classed as a silt or silt loam.

Another factor influencing the loss of sulfates is the nature of the cation population of the soil solution. This is discussed more fully in the following section which deals with sulfate retention, but because the relationship between sulfate retention and sulfate leaching is generally an inverse one it is mentioned here briefly. Leaching losses of sulfate are greatest when monovalent ions such as potassium and sodium predominate; next in order are the divalent ions such as calcium and magnesium; and leaching losses are least when soils are acid and appreciable amounts of exchangeable aluminum and iron are present.

Adsorbed Sulfate. Soils vary widely in their capacity to adsorb sulfate. Many possess little or no sulfate adsorption capacity while in others this form of sulfur plays an important part in contributing to the sulfur nourishment of crops, conserving sulfur from excessive leaching, and in determining sulfur distribution in soil profiles.

Sulfate adsorption is readily reversible and is influenced by a number of soil properties, the most important of which are identified below.

1. *Clay content and type of clay mineral.* Adsorption of sulfate usually increases with the clay content of soils. Kaolin minerals retain more sulfate than the montmorillonite group of clays. Capacities of hydrogen saturated clays for sulfate adsorption are kaolinite > illite > bentonite. When saturated with aluminium, adsorption is about the same for kaolinite and illite but much lower for bentonite.

2. *Hydrous oxides.* Hydrous oxides of aluminum and to a lesser extent iron show marked tendencies to retain sulfates, especially the former in certain soils. These compounds are probably responsible for most of the sulfate adsorption in many soils.
3. *Soil horizon or depth.* Most soils have some capacity to adsorb sulfate. The amounts of sulfate in surface horizons may be low but are often greater in lower soil horizons. Capacity for sulfate adsorption is often greater in subsoils, due to the presence of more clay and iron and aluminum oxides.
4. *Effect of pH.* Adsorption of SO_4^{2-} in soil systems is favored by strongly acid conditions. It becomes almost negligible at pH values above 6.5.
5. *Sulfate concentration.* The amount of sulfate adsorbed is concentration dependent. Adsorbed SO_4^{2-} is in kinetic equilibrium with SO_4^{2-} in solution. Adsorption maxima have not been reached in many laboratory studies, particularly under acidic conditions.
6. *Effect of time.* Sulfate retention increases with the length of time it is in contact with the adsorbing substances.
7. *Presence of other anions.* Sulfate is generally considered to be weakly held with the strength of retention decreasing in the order hydroxyl > phosphate > sulfate = acetate > nitrate = chloride. Phosphate will displace or reduce the adsorption of sulfate but sulfate has little effect on phosphate. There is little, if any, effect of chloride on sulfate retention. Molybdates will depress sulfate adsorption.
8. *Effect of cations.* The amount of sulfate retained is affected by the associated cation of the salt or by the exchangeable cation. This effect follows the lyotropic series: that is, $H^+ > Sr^+ > Ba^{2+} > Ca^{2+} > Mg^{2+} > Rb^+ > K^+ > NH_4^+ > Na^+ > Li^+$. Both the cation and the sulfate from a salt may be retained but tenacity of adsorption of anion and cation will likely differ.
9. *Organic matter.* In some soils organic matter may contribute to retention of sulfate.

Of all these factors, amount and type of soil colloids, pH, concentration of sulfate, and presence of other ions in the equilibrium solution appear to influence sulfate adsorption the most.

Harward of Oregon State University and Reisenauer of the University of California have described and summarized several possible mechanisms of sulfate retention. These are:

1. Anion exchange caused by positive charges developed on hydrous iron and aluminum oxides or on the crystal edges of clays, especially kaolinite, at low pH values.
2. Retention of sulfate ions by hydroxyaluminum complexes by coordination.
3. Salt adsorption resulting from attraction between the surface of soil colloids and the salt.
4. Amphoteric properties of soil organic matter which develop positive charges under certain conditions.

Workers in Virginia have shown that chlorides as well as sulfates were adsorbed by a Cecil soil and a kaolinite with which they worked. They have suggested a mechanism which they claim accounts for the adsorption of sulfate. In explaining their postulate, they assume a homoionic aluminum-saturated clay with coatings of the hydrated oxides R (iron and aluminum). Then

$$yK + Al_x[clay] + yH_2O \rightarrow \quad Al_x(OH)_y{}^{Ky}[clay] + yH$$

$$SO_4^{2-} + R_x(OH)_y[clay] \rightarrow R_x[(OH)_{y-z}(SO_4)_z]clay + zOH^-$$

It is assumed that the K^+ adsorption sites develop from the exchange and/or hydrolysis of aluminum on the clay surface. As a result of this hydrolysis, some

H^+ ions go into solution. At the same time SO_4^{2-} replaces OH^- ions from $R(OH)$ coatings on clay and substitutes for them. The replaced OH^- ions in turn react with the H^+ ions. Whether the pH of the system increases or decreases depends on the relative rates of the two reactions: hydrolysis and OH^- exchange.

The mechanism proposed explains the several observed phenomena as follows: sulfate adsorption is increased as the pH is lowered because the replaced OH^- ions are more effectively neutralized. Increased cation affinity causes the replacement of more aluminum and results in still further hydrolysis. It has been shown that sulfate adsorption is a function of time as well as of the other factors discussed. In explaining the time effect, it is suggested that the continuation of the aluminum hydrolysis, which produces H^+ ions, neutralizes OH^- ions and thus carries the reaction to completion.

Elemental Sulfur and Sulfides. Elemental sulfur is not found in well-drained upland soils. Under waterlogged conditions, in which bacterial reduction is taking place, sulfides are formed and in some instances elemental sulfur is deposited.

In its pure form elemental sulfur is a yellow, inert, water-insoluble crystalline solid. Commercially, it is stored in the open, where it remains unaltered by moisture and temperature. When sulfur is finely ground and mixed with soil, however, it is oxidized to sulfate by soil microorganisms. Because of this property, sulfur has been used for many years in the reclamation of alkali soils (which also contain free calcium carbonate).

The oxidation of sulfur in the soil, plus the fact that it is in the elemental form (meaning, of course, that it provides the highest amount of plant nutrient sulfur for the least bulk), has suggested its use as a source of this element in fertilizers. As mentioned in Chapters 5, 6, and 7, many of the higher analysis fertilizers currently being manufactured contain no sulfur. When used for several years on soils low in sulfur, crop deficiencies of this element have appeared. In some areas the deficiency is severe. Work is being carried out to develop fertilizers containing sulfur in the elemental form, and the agronomic suitability of these products is being tested. Sulfur has been successfully introduced into such materials as urea, anhydrous ammonia, CSP, ammonium phosphate, and solid and fluid N-P-K materials. Its usefulness as a plant nutrient depends, of course, on the rate at which it is oxidized to sulfate.

Factors Affecting Sulfur Oxidation in Soils. Elemental sulfur, sulfides, and several other inorganic sulfur compounds can be oxidized in the soil by purely chemical means but these are usually much slower, and therefore of less importance than microbial oxidation. The rate of biological oxidation of sulfur depends on the interaction of three groups of factors. These factors are (1) the microfloral population in soil, (2) characteristics of the sulfur source, and (3) environmental conditions in the soil.

Soil Microflora. Two classes of bacteria are specially adapted for sulfur oxidation. These are chemolithotrophic sulfur bacteria which utilize energy released from the oxidation of the inorganic sulfur for the fixation of CO_2 into organic matter. Their activity is described in the following general equation:

$$CO_2 + S + \frac{1}{2}O_2 + 2H_2O \rightarrow [CH_2O] + SO_4^{2-} + 2H^+$$

The *Thiobacilli* are typical chemolithotrophic bacteria. Many of them are strict autotrophic aerobes but some are facultative autotrophs and mixotrophic.

The second class of sulfur oxidizers is photolithotrophic sulfur bacteria, which carry out photosynthetic carbon fixation using sulfide and other sulfur compounds as "oxidant sinks." The following equation summarizes their behavior:

$$CO_2 + 2H_2S \xrightarrow{\text{light}} [CH_2O] + H_2O + S^o$$

In this class of microorganisms, there are two principal groups, the *Chlorobium*, or green bacteria, and *Chromatium*, or purple bacteria. They are obligate anaerobes and their habitat is usually H_2S-containing muds and stagnant waters exposed to light.

The best known, and usually considered to be the most important, group of sulfur-oxidizing organisms are the autotrophic bacteria belonging to the genus *Thiobacillus*. This group has received much more attention than the heterotrophic fungi, actinomycetes, and bacteria whose role in the conversion of elemental sulfur in the soil has been largely ignored. The most common form is *T. thiooxidans*, but others are *T. thioparus*, *T. copraliticus*, and *T. ferrooxidans*. A new acidophilic species, *T. kabobis*, was isolated by E. J. Laishley and his co-workers at the University of Calgary. Swaby and Vitolins of CSIRO in Australia found that in 228 soils, *T. intermedius* was more numerous than all the other species of *Thiobacilli* together.

There is considerable variability in sulfur oxidation rates among soils (Table 8-4). These variations are usually attributed to differences in numbers of *Thiobacillus*. In the Australian soils referred to in this table, *T. thiooxidans* was not detected in one-third of the soils examined and was seldom abundant in the remaining two-thirds. The most numerous sulfur oxidizers were heterotrophic bacteria, followed by facultative, then obligate autotrophic *Thiobacilli*, and finally green and purple bacteria.

The initial rate of sulfur oxidation in pot and incubation studies can be greatly increased by inoculation, usually with *Thiobacilli*, but also with heterotrophs. Figure 8-4 demonstrates the stimulatory effect of inoculation with two species of *Thiobacilli* on oxidation rate of elemental sulfur in a soil lacking these organisms. A considerable difference occurred in the activity of these two species, with *T. thioparus* outperforming *T. thiooxidans*.

These favorable effects of inoculation are frequently short-lived under most conditions, and less benefit is usually obtained in field trials. Addition of sulfur and some of its compounds to soil will encourage greater populations of sulfur-oxidizing microorganisms and the improved oxidizing power of the soil. This enrichment of sulfur-oxidizing capacity and the overriding influence of environmental factors could be the reason for limited success from field-scale inoculations. An example is given in Table 8-5 of how the population of *Thiobacilli* increased following the burial side by side of two elemental sulfur cylinders (each 6 in. in diameter by 12 in. in length) in a field soil that was initially devoid of these organisms.

Where inoculation is being attempted, addition of the inoculum to the soil will

TABLE 8-4. Sulfur Oxidation Rates in Soils

Location and Number of Soils	Percent S Oxidation in Various Incubation Periods
Australia	
51	0– 0.4
64	0.5– 4.0
58	4.1–12.9
100	13.0–61.6
Nebraska (1)	14.0–56.0
Oregon (1)	2.0–50.0
Saskatchewan (4)	20.0–44.0
Wisconsin (54)	3.0–73.0

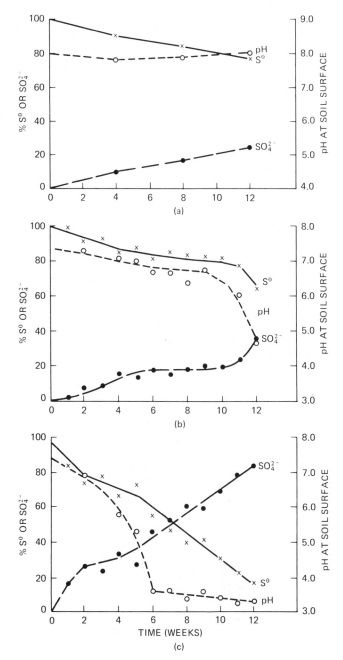

FIGURE 8-4. Changes in sulfur and sulfate content, and pH during the oxidation of sulfur in Solonetzic soils: (a) uninoculated (b) inoculated with *Thiobacillus thiooxidans*; (c) inoculated with *T. thioparus*. [McCready and Krouse, *Can. J. Soil Sci.*, **62**:105 (1982).]

probably be more successful than direct inoculation of sulfur fertilizers. Poor microbial survival has generally been experienced with the latter method.

The possible involvement of heterotrophic organisms in conversion of elemental sulfur should not be overlooked. For example, *Fusarium solani* is capable of oxidizing about 2% of the available sulfur per week. Soil bacteria such as the *Arthrobacter* spp. have oxidized elemental sulfur to SO_4^{2-} in pure culture studies. Many fungi and some of the common soil actinomycetes, *Alternaria tenuis, Aureobasidium pullulans, Epicoccum nigrum, Penicillium* spp., and *Streptomyces* spp.,

TABLE 8-5. Changes in pH and *Thiobacillus* Population in Field Soil Surrounding Large Cylinders of Elemental Sulfur

Sampling Time	pH of Soil	Number of Bacteria per Gram Dry Weight of Soil	Number of Thiobacillus per Gram Dry Weight of Soil	
			Less Acidic	Acidic
1974				
Oct.	7.7	$86 \times 10^5 (\pm 1.9)$ *	Nil	Nil
1975				
May	7.0	$14 \times 10^5 (\pm 6.6)$	38	Nil
July	7.5	$19 \times 10^6 (\pm 3.8)$	96	Nil
Sept.	7.6	$66 \times 10^5 (\pm 3.0)$	2.8×10^2	Nil
1976				
May	7.4	$19 \times 10^7 (\pm 7.0)$	3.3×10^5	Nil
July	8.0	$29 \times 10^6 (\pm 5.0)$	1.3×10^4	296
Sept.	8.1	$29 \times 10^6 (\pm 7.0)$	1.6×10^4	5

* Standard error of the mean (SEM).
Source: Laishley et al., in W. E. Krumbein, Ed., *Environmental Biogeochemistry and Geomicrobiology,* Vol. 2: *The Terrestrial Environment,* Chapter 56. Ann Arbor, Mich.: Ann Arbor Science Publishers, 1978.

have the ability to oxidize elemental sulfur. The Australian researchers Vitolins and Swaby found that heterotrophic yeasts and several genera of heterotrophic and facultative autotrophic bacteria were far more numerous than the strict autotrophs, and could play an important role in sulfur oxidation in soils where conditions are unfavorable for the autotrophs.

Temperature. As in most biological reactions, an increase in temperature increases the rate at which sulfur is oxidized in the soil. This is illustrated in Figure 8-5. The data indicate an increasing rate of oxidation up to 30°C. In other studies a marked increase in activity was observed from 23 to 30°C, with further increases to 40°C. At temperatures above 55 to 60°C the sulfur-oxidizing organisms

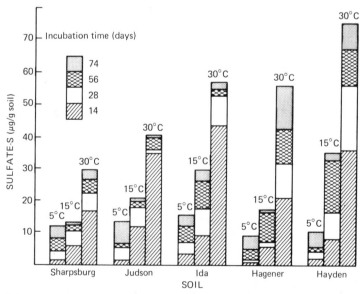

FIGURE 8-5. Effects of temperature and time of incubation on oxidation of elemental sulfur (100μg of sulfur per gram of soil) in soils. [Nor and Tabatabai, *Soil Sci. Soc. Am. J.,* **41**:739 (1979).]

are killed. Microbial oxidation will occur at soil temperatures as low as 4°C, but the process is slow below 10°C.

Optimum temperatures will naturally vary for the different sulfur-oxidizing organisms. Temperatures between about 25 and 40°C will be close to ideal for most.

Soil Moisture and Aeration. Sulfur-oxidizing bacteria are mostly aerobic and their activity will decline if oxygen is lacking due to waterlogging. It is apparent in Figure 8-6 that sulfur oxidation is favored by soil moisture levels near to field moisture capacity (FMC). Also evident is the impedance in oxidizing activity when the soils were either excessively wet or dry. Thus the moisture conditions most suitable for plant growth will also encourage a high rate of sulfur oxidation.

Dry soils retain their ability to oxidize sulfur, but there can be a lag period following rewetting before they regain full capacity for this process.

Soil pH. *Thiobacillus thiooxidans* is capable of surviving at extremely low pH values and grows best in the pH range 1.0 to 4.0. As the initial soil pH of 8.4 dropped during the incubation study referred to in Figure 8-4, oxidation of sulfur was enhanced. This effect was most pronounced with *Thiobacillus thioporus*, which has an optimum pH range of 7.2 to 4.5. Other sulfur-oxidizing organisms appear to have different pH requirements, but in general oxidation of added sulfur proceeds most rapidly under acidic soil conditions.

Soil Type and Properties. The large differences observed in the initial rate of sulfur oxidation by soils have been found to be unrelated to soil type or to various soil properties, including texture, organic-matter content, field moisture

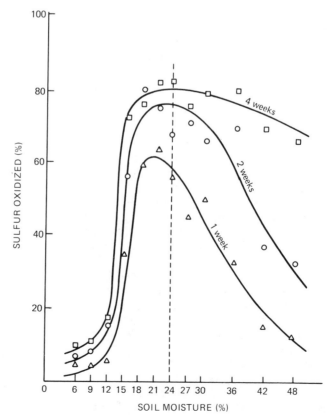

FIGURE 8-6. Percentage of added sulfur oxidized in a Miami silt loam incubated at various moisture levels after 1-, 2-, and 4- week periods. Dashed line, field moisture capacity. [Kittams and Attoe, *Agron. J.*, **57**:331 (1965).]

capacity, pH, initial sulfate sulfur, or degree of sulfur deficiency of the soil. Liming effects on this process are variable.

Microorganisms responsible for sulfur oxidation are believed to require most of the same nutrients needed by plants plus perhaps a few others. They can compete with plants for nutrient supply and temporary nitrogen deficiencies have been reported in plants following stimulation of sulfur oxidizing activity by the addition of sulfur. More rapid sulfur oxidation has been observed in fertilized soil than in one low in both phosphorus and potassium.

Thiobacilli require the NH_4^+ form of nitrogen, and NO_3^- can be injurious. These organisms can withstand high concentrations of ammonia in injection zones of anhydrous ammonia. High Cl^- concentrations will cause some depression of sulfur oxidation, but the effect is less than for most other biological transformations in the soil. Some strains of sulfur-oxidizing organisms have remarkable tolerance to heavy metals encountered in the oxidation of sulfur compounds in metal ores.

Organic matter is not essential for activity of autotrophic sulfur bacteria, but heterotrophs require organic carbon as a source of energy. Results reported from research on the influence of organic matter additions on sulfur oxidation rates have been inconsistent.

Fertilizer Interactions. Sulfur as an integral part of fertilizer granules may be oxidized more rapidly than when it is added separately. Bloomfield at the Rothamsted Experimental Station found that sulfur in granulated triple super-phosphate and diammonium phosphate fertilizers oxidized faster than sulfur alone, in both acid and calcareous soils. Researchers at the University of Minnesota obtained a more rapid oxidation of sulfur combined with N-P fertilizers than when it was supplied with triple superphosphate. There are several possible reasons for this enhancement, including the effects of the nutrients, nitrogen, phosphorus, or calcium and the existence of more favorable moisture conditions around the fertilizer granule. The temporary low pH resulting from dissolution of certain fertilizer materials such as triple superphosphate might also improve growth of acidophilic sulfur oxidizers such as *T. thiooxidans*.

Characteristics of the Elemental Sulfur Source. Properties of elemental sulfur and its rate and method of placement can influence sulfur oxidation rates. A brief discussion of these points follows.

Form and Particle Size.

Solid sulfur exists in three distinct crystalline forms: orthorhombic, monoclinic, and cubic. It also occurs as an elastomer—an amorphous black, rubberlike solid structurally identical to liquid sulfur.

A mixture of the orthorhombic, monoclinic, and elastomeric forms may be produced during the cooling of liquid sulfur. Although rhombic sulfur is the stable form at room temperatures and pressure, others, such as the elastomer, probably coexist for indefinite periods. Oxidation rates of ground and sublimed sulfur have been compared and the results revealed that surface area or particle size was more important than surface shape.

Many researchers have observed that initial sulfur oxidation rates increase as particle size is reduced. The tremendous impact of particle-size distribution of applied sulfur on its rate of transformation to SO_4^{2-} is illustrated in Table 8-6. Oxidation is slow in the coarse sulfur fraction, 30 to 50 mesh and coarser. These results and others are the basis of a general rule of thumb for satisfactory use of elemental sulfur as a source of plant-nutrient sulfur. According to this rule, 100% of the sulfur material must pass a 16-mesh screen, and 50% of that should in turn pass through a 100-mesh screen.

The finer the particle size of a given mass of sulfur, the greater the specific surface and the faster is sulfate formation. Because of the inverse relationship

TABLE 8-6.   **Effect of Granule Size on the Rate of Oxidation of Elemental Sulfur Added to a Dorset Sandy Loam and Incubated at 30°C for Increasing Periods of Time**

Particle Size of Sulfur (meshes/in.)	Incubation Time (days)									
	10		30		60		90		180	
	Application Rate (ppm)									
	10	50	10	50	10	50	10	50	10	50
	Percent Sulfur Oxidized									
10–20	1.0	0.4	1.0	0.6	0	0.8	—	—	—	—
30–50	2.0	1.2	2.0	2.8	5.0	3.2	—	—	—	—
60–80	3.0	2.2	5.0	4.2	10.0	12.6	9.0	11.4	20.0	25.6
< 100	15.0	13.4	29.0	45.4	58.0	56.2	43.0	—	53.0	—

Source: Li and Caldwell, *Soil Sci. Soc. Am. J.*, **30**:370 (1966).

between surface area and particle diameter, oxidation rate increases exponentially with decreasing particle diameter. This relationship was recognized by Fox and his co-workers in the mid-1960s. Figure 8-7 is a recent example of the relationship between surface area and the availability of elemental sulfur for canola in pot experiments.

Rate and Placement.
Application of heavier rates of sulfur will increase the amount of surface area exposed to sulfur oxidizing organisms, which should result in a linear increase in the release of available sulfur. This relationship is evident in Table 8-6. In poorly buffered soils the reverse can occur, with the percentage of sulfur oxidized declining with higher rates of application.

Placement of sulfur can often affect its oxidation rate, and broadcasting, followed by incorporation, is generally thought to be superior to banding. Uniform distribution of sulfur particles throughout the soil will (1) provide greater exposure of sulfur particles to oxidizing microorganisms, (2) minimize any potential problems caused by excessive acidity, and (3) provide more favorable moisture relationships.

Sulfides and Polysulfides.   Sulfides and polysulfides do not exist in well-drained upland soils. Sulfides of heavy metals and other ions are found in soils

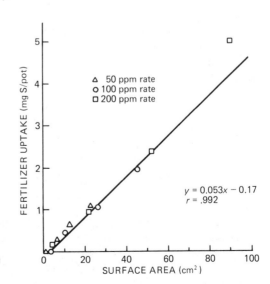

FIGURE 8-7.   Influence of surface area of applied elemental sulfur on the uptake of sulfur by canola. [Janzen et al., *Proc. Alberta Soil Sci. Workshop*, p. 229, Edmonton, Alberta (February 23–24, 1982).]

$y = 0.053x - 0.17$
$r = .992$

TABLE 8-7.   Effect of Time and Added Organic Matter on the Production of Hydrogen Sulfide in a Waterlogged Soil

Time after Submergence (days)	Concentration of Hydrogen Sulfide in Soil for Three Treatments		
	Control (ppm)	Green Manure (ppm)	Straw (ppm)
0	Nil	Nil	Nil
7	Nil	12.5	10.5
14	3.5	23.8	18.5
21	5.0	32.9	28.8
28	6.8	52.8	40.5
38	8.2	57.6	54.8
51	10.5	60.3	62.5
63	13.0	62.5	65.6
78	15.0	65.0	66.0

Source: Mandal, *Soil Sci.,* **91:**121 (1961). Reprinted with the permission of The Williams & Wilkins Co., Baltimore.

under waterlogged conditions. In fact, the deep color of the shore of the Black Sea is caused by the accumulation of iron sulfide.

When sulfates are added to waterlogged soils from which oxygen is excluded, the sulfate is reduced to hydrogen sulfide. If hydrogen sulfide is not precipitated by iron and other similar metals, it escapes to the air aboveground. The effect of waterlogging on the production of hydrogen sulfide in a paddy soil was studied in the laboratory. Its production as affected by time of submergence and added organic matter is shown in Table 8-7. It is obvious that it increased with both time and added organic matter.

In some paddy soils sulfide production is believed to cause Akiochi. The soils on which this disease of rice is observed are old and degraded and low in iron content. On soils containing adequate iron, it does not occur.

In some tidal marsh lands large quantities of reduced sulfur compounds accumulate and the soil pH is raised. When the areas are drained, the sulfur compounds are oxidized to sulfates with a considerable lowering of soil pH. A classic example is found in Kerala State on the southwestern coast of India. During the monsoon season these soils are under water and the pH is about 7.0. When the monsoon season passes, the soil is no longer inundated. The sulfur compounds are oxidized and the pH drops to around 3.5. This cycle is repeated annually.

Polysulfides are used as sulfur fertilizers, soil conditioners, and for treatment of irrigation waters to improve water penetration into soil. Ammonium polysulfide is used for all three purposes. Calcium polysulfide is marketed in limited quantities in the southwestern United States, particularly in California, as a soil conditioner and for treatment of irrigation waters. In each case the polysulfide is changed to colloidal sulfur and a sulfide when it is added to the soil or irrigation water. The reaction of calcium polysulfide is shown in the following equation:

$$CaS_5 + H_2O + CO_2 \rightarrow H_2S + CaCO_3 + 4S$$

Because of the colloidal nature of sulfur deposited from polysulfides, its conversion to sulfates takes place quite rapidly (Figure 8-8).

Organic Sulfur Transformations. The nature of organic sulfur forms in soils was described earlier in this chapter, but little was said about the changes and reactions of this major source of plant-available sulfur. It was shown previously in Figures 8-1 and 8-2 that the transformations of sulfur in soil are many and

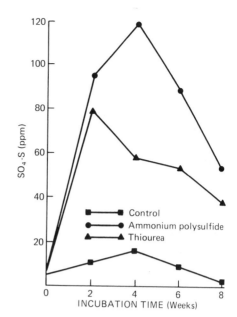

FIGURE 8-8. Rate of oxidation of ammonium polysulfide and thiourea added to the Cooking Lake sandy loam and incubated at 20°C. [Harron and Malhi, *Can. J. Sci.*, **58**:109 (1978).]

varied with changes often being cyclic. Also, the element converts back and forth from inorganic to organic forms due to the presence of living organisms.

Biological cycling among the major sulfur pools is shown in Figure 8-9. The principal classes of organisms responsible for these reactions are: dissimilatory reducers (*Desulfovibrio, Desulfotomaculum*)—reaction 2; *Desulfuromonas*—reaction 5; assimilatory reducers (bacteria, fungi, algae, plants)—reactions 1 and 3; chemolithotrophs (*Thiobacillus, Beggiatoa*)—reactions 4, 6, and 8; photolithotrophs (*Chlorobium* and *Chromatium*)—reactions 4, 6, and 8; and heterotrophic microorganisms—reactions 7 and 9 with some involved in 4, 6, and 8.

When plant and animal residues are returned to the soil they are attacked by microorganisms, thus releasing some of the sulfur as sulfate. Most of the sulfur, however, remains in organic form and eventually becomes part of the soil humus. In contrast to the relatively rapid decomposition of fresh organic residues in soil, degradation and release of sulfur from the large humus fraction is limited and slow.

Although some of the sulfur requirement of plants can be derived from other sources, it is largely dependent on the sulfate released from the organic soil fraction and from plant and animal residues. Approximately 4 to 13 lb of sulfur per acre as sulfate is mineralized each year from the organic fraction, which in most surface

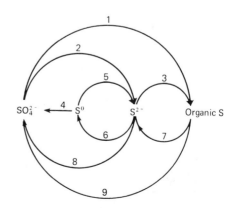

FIGURE 8-9. Biological cycling of the major sulfur pools. (Trudinger, "The biological sulfur cycle." in P. A. Trudinger and D. J. Swaine, Eds., *Studies in Environmental Science*, Vol. 3: *Biogeochemical Cycling of Mineral-Forming Elements*, Chapter 6.1. Amsterdam: Elsevier Scientific, 1979.)

soils may typically contain several hundred pounds of sulfur per acre. Some simple organic sulfur compounds, such as the sulfur-containing amino acids which can be assimilated by plant roots, may also be released. They are, however, of minor significance because of their instability in soil.

Mineralization. The mechanisms of mineralization of sulfur (defined as for nitrogen in Chapter 5) from the decomposition of a wide range of organic materials are not completely understood. Many organisms are involved in the oxidation of organic sulfur compounds to SO_4^{2-} and the exact pathways are difficult to trace (Figure 8-9).

It is believed that conversion of organic sulfur to inorganic SO_4^{2-} is mainly the result of microbial activity. Any factor that affects the growth of microorganisms is expected to alter mineralization of sulfur. Mineral content of organic matter, temperature, moisture, pH, presence or absence of plants, time and cultivation, and availability of food supply are all known to affect the mineralization of sulfur.

Mineral Content of Organic Matter. Mineralization of sulfur depends on the sulfur content of the decomposing material in much the same way that the mineralization of nitrogen depends on the nitrogen content. This is illustrated by the data in Figure 8-10. It is apparent that smaller amounts of sulfate were liberated from the materials containing the smaller percentages of this element and that an analogous situation exists for the mineralization of nitrogen. In the samples containing less than about 0.15% sulfur, there was, in fact, a reduction in the level of soil sulfate at the end of the incubation period, which may suggest immobilization of sulfur.

Sulfur may be immobilized in soils in which the ratio of either C or N to S is too wide. At or below a C/S weight ratio of approximately 200:1, only mineralization of sulfur occurs. Above this ratio immobilization or tie-up of SO_4^{2-} in various organic forms is favored, particularly if the ratio is greater than 400:1. The immobilized sulfur is bound in soil humus, in microbial cells, and in by-products of microbial synthesis. Immobilization occurs during the mineralization of organic fractions with wide C/S ratios because of conversion of a larger portion of carbon into microbial biomass with a resultant higher need for sulfur than when the C/S ratio is low. Fresh organic residues commonly have C/S ratios of about 50:1.

The practical implication of maintaining an appropriate balance in the soil between nitrogen and sulfur is well illustrated by some work done in Colorado. Wheat straw with a low sulfur content was incorporated into soils fertilized with N-P-K

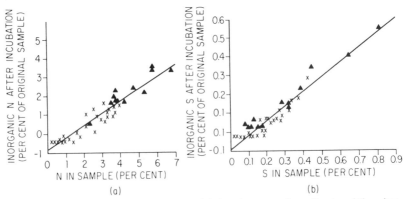

FIGURE 8-10. Relationships between: (a) the nitrogen mineralized and the nitrogen content of the original material; (b) the inorganic sulfur present after incubation and the sulfur content of the original material. [Barrow, *Australian J. Agr. Res.,* **11**:960 (1960).]

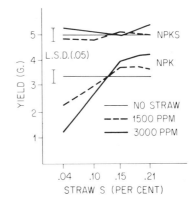

FIGURE 8-11. Growth of winter wheat as affected by incorporating different straws of varying sulfur contents into N-P-K fertilized soil, with and without fertilizer sulfur. [Stewart et al., *SSSA Proc.*, **30**:355 (1966).]

or N-P-K-S. Winter wheat was then planted on these soils. A portion of the results of this study are shown in Figure 8-11. Two points are clear from these data. The addition of wheat straw in increasing amounts to soil to which no fertilizer sulfur was added progressively decreased the growth of wheat plants. The addition of sulfur in the fertilizer overcame the limiting effect of the straw, and the growth of wheat was the same on the treatments receiving straw as it was on those receiving no straw.

Apparently the addition of straw with a low sulfur content to the soil used in this study tied up the available soil sulfur because of the immobilization by soil microorganisms during decomposition of the straw. This situation was aggravated by the addition of fertilizer nitrogen which further widened the N/S ratio of the soil, resulting in the immobilization of any available SO_4. As a result the wheat plants had insufficient sulfur available to them to permit proper growth. The addition of fertilizer sulfur to this soil, however, overcame this unfavorable situation.

In practical farming operations where large amounts of straw, stover, or other organic materials are to be returned to the soil, the grower should take steps to ensure that adequate nitrogen and sulfur are available to promote rapid decomposition of the added straw. Otherwise, a temporary nitrogen or sulfur deficiency may be induced in the following crop.

Temperature. Williams in Australia found in incubation studies of moist soil for 64 days that mineralization of sulfur was either negligible or severely impeded at 10°C. It increased with increasing temperatures from 20°C to 40°C but was less at 50°C than at 40°C. Other investigators have reported no effect of increasing soil temperatures between 10 and 20°C on sulfur mineralization. Tabatabai and Al-Khafaji at Iowa State University showed that in samples representing 12 major soil series in Iowa much more sulfur was released during incubation at 35°C than at 20°C (Figure 8-12). They reported an average Q_{10} of sulfur mineralization of 1.9 for these soils.

A restricting influence of low temperatures on mineralization is anticipated because of greatly reduced intra- and extracellular enzymatic activity. This effect of low temperature is consistent with the relatively greater sulfur content of soils formed under northern climates.

Moisture. Soil moisture is known to affect the activity of sulfatases, the rate of sulfur mineralization, the form in which sulfur is released from organic matter, and the movement of sulfates in soil. Mineralization of sulfur in Australian soils incubated at low (< 15%) and high (> 40%) moisture levels was impaired, whereas it was enhanced at 60% of field moisture-holding capacity.

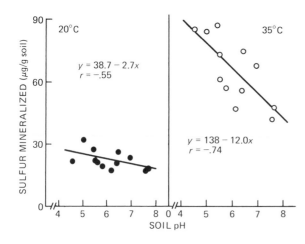

FIGURE 8-12. Relationship between total sulfur mineralized and pH of soils incubated at 20 or 35°C for 26 weeks. [Tabatabai and Al-Khafaji, *Soil Sci. Soc. Am. J.*, **44**:1000 (1980).]

Drying. Part of the soil organic sulfur appears to be very unstable and is readily converted to SO_4^{2-} by physical treatments such as heating, air drying, or grinding. Soils prepared for laboratory and greenhouse studies usually receive some or all of these treatments.

Air drying alone causes considerable SO_4^{2-} formation, perhaps resulting from the splitting of organic sulfates. Some simple organic sulfur compounds may also be released during drying. Increases of 20 to 80% in extractable SO_4^{2-} have been found upon air-drying soils at room temperature.

The SO_4^{2-} liberated by air drying is readily available to plants and it can be just as effective as fertilizer sulfur. It is not known if the air drying effect is significant under field conditions, but it is conceivable that it may contribute to the seasonal and year-to-year fluctuations in soil SO_4^{2-} levels.

Wetting and Drying. Gradual moisture changes in the range between field capacity and wilting point have little influence on sulfur mineralization. However, drastic differences in soil moisture conditions can produce a flush of sulfur mineralization in some soils. The source of this sulfur is thought to be primarily from ester sulfates in the soil organic matter.

Increased availability of sulfur due to soil wetting and drying is believed to be important under field conditions. Barrow in Australia considers it to be a major factor contributing to the peak period of growth common on sulfur-deficient soils after dry periods. There is a possibility in some regions that SO_4^{2-} release associated with periodic drastic soil moisture changes may be more significant in supplying sulfur to crops than the slow gradual release by microbial decomposition.

Soil pH. The effect of pH on mineralization of sulfur is not clear. Rates of sulfur mineralization in 12 Iowa soils were found to be negatively correlated with soil pH (Figure 8-12). In soils from other regions the amount of sulfur released is directly proportional to pH up to a value of 7.5. At pH values above 7.5 in these soils, mineralization increased more rapidly suggesting that an additional factor such as chemical hydrolysis might be involved. Formation of SO_4^{2-} in some soils has been found to be proportional to the amount of calcium carbonate applied and not to the resulting pH. Soil pH in the nearly neutral range is normally expected to encourage microbial activity, including sulfur mineralization.

Presence or Absence of Plants. Soils generally mineralize more sulfur in the presence of growing plants than in their absence. In some cases more than twice as much sulfur is mineralized when soils are cropped than when uncropped.

This increase in SO_4^{2-} liberation has been explained as being due to the "rhizosphere effect" brought about by the excretion by plant roots of amino acids and sugars and the resultant stimulation of microbial activity.

Appreciable immobilization of added sulfate has been observed in uncropped soils. This suggests that applications of sulfate fertilizer to fallowed soils could be largely tied up in the organic form and not immediately available to the following crop.

Time and Cultivation. As is the case with nitrogen, when land is first cultivated the sulfur content of the soil tends to decline rapidly. With time an equilibrium level is reached which is characteristic of climate, cultural practices, and soil type. At the equilibrium level soil humus virtually ceases to further liberate SO_4^{2-}. Before reaching this point, the rate of mineralization of sulfur will gradually diminish and become inadequate to meet plant needs.

In western Canada, the C/N/S ratios of virgin soils are wider than the corresponding cultivated surface soils. Narrowing of this ratio upon cultivation suggests that sulfur is relatively more resistant to mineralization than carbon and nitrogen, or the losses of organic carbon and nitrogen are proportionately greater than sulfur.

Mineralization Patterns. In the many investigations of sulfur mineralization four rather definite patterns have emerged. These are (1) immobilization of sulfur during the initial stages of incubation, followed by mineralization of sulfur in the later stages; (2) a steady, linear, release with time over the whole period of incubation; (3) a rapid release of SO_4^{2-} during the first few days followed by a slower linear release; and (4) a rate of release that decreased with time. They appear to be unrelated to any particular soil property, but are probably due to the chemical nature of the decomposing fraction of the soil organic matter.

Sulfatase Activity. As much as 50% of the total sulfur in surface soils may be present as organic sulfate esters. Sulfatase enzymes that hydrolyze these esters and release inorganic sulfate may play an important role in the mineralization process. The general action of sulfatases is:

$$R \cdot O \cdot SO_3^- + H_2O \overset{\text{sulfatase}}{\rightleftharpoons} R \cdot OH + HSO_4^-$$

The HI-reducible ester sulfates are considered to be the natural substrates for sulfatase enzymes in soil.

Tabatabai and Bremner at Iowa State University were the first to detect sulfatase enzymes in soils. They found arylsulfatases, which are enzymes capable of hydrolyzing arylsulfates. Arylsulfatase activity has been observed in a variety of U.S., African, and Canadian soils. Most of this activity was due to extracellular, particle-adsorbed arylsulfatases. Levels of arylsulfatase activity vary according to soil type, depth, and climate. The activity of this enzyme decreases markedly with soil depth. This decrease is associated with lower contents of organic carbon. Further research in Iowa revealed that activity of this enzyme was significantly correlated with organic carbon content but not with pH, sulfur content, or texture of the soils.

In some regions there are wide seasonal variations in arylsulfatase activity. Seasonal weather conditions may influence activity of this enzyme more than the soil properties with which it has been correlated.

Source of Mineralizable Sulfur. Because of the vital role that sulfur mineralization has in providing plant-available SO_4^{2-}, there is interest in characterizing the labile sulfur reserve in soil organic matter. Freney in Australia demonstrated that over a 9-month plant growth period most of the available sulfur removed by plants came from the ester sulfate fraction, although there were changes in all sulfur fractions. Changes in the carbon-bonded sulfur fraction were due mainly to decomposition of the sulfur-containing amino acids.

Studies of the soil organic sulfur fraction require suitable solvents for extracting high proportions of organic materials without significantly altering their composition. An improved technique, employing an alkali-pyrophosphate extraction followed by peptization and ultrasonic dispersion, has been developed at the University of Saskatchewan. It extracts up to 25% more organic matter than the traditional 0.1 M NaOH procedure. Bettany and co-workers at the University of Saskatchewan using this more effective method reported that the contributions of six soil organic-matter fractions to sulfur mineralization during 65 years of cultivation were, in decreasing order, clay-associated humic acid (36%), conventional humic acid (26%), humin <2 μm (18%), conventional fulvic acid (14%), clay-associated fulvic acid (4%), and humin >2 μm (3%).

The distribution of HI-reducible sulfur among the various soil organic matter fractions was determined in the Saskatchewan research (Table 8-8). This ester sulfate fraction was much lower in the cultivated soil than in the pasture soil. During the 65 years of cultivation some fractions lost a greater percentage of this form of sulfur than others. Major differences ranging from 57 to 33% were found in the HA-A, humin <2 μm, HA-B, and humin >2 μm fractions.

Sulfur Volatilization. Volatile sulfur compounds are produced through microbial transformations of soil sulfur compounds under both aerobic and waterlogged conditions. From a collection of surface samples of 25 Iowa soils treated with soluble sulfate, volatilization of sulfur was observed in 14 soils when incubated under waterlogged conditions. Four of these soils, however, also released volatile sulfur compounds when incubated aerobically. Where volatilization occurred, the volatile sulfur detected was dimethyl sulfide (CH_3SCH_3) alone or dimethyl sulfide accompanied by smaller quantities of carbonyl sulfide (COS), carbon disulfide (CS_2), methyl mercaptan (CH_3SH), and/or dimethyl disulfide (CH_3SSCH_3). Dimethyl sulfide accounted for 55 to 100% of all sulfur volatilized.

No release of volatile sulfur was detected from this group of Iowa soils when there was less than 2.0% organic matter. However, volatilization occurred in five out of the six soils containing more than 5 to 7% organic matter. The actual amounts of sulfur volatilized were very small and did not represent more than 0.05% of the total sulfur present in soil. Such losses are probably insignificant under field conditions.

TABLE 8-8.   Distribution of HI-Reducible Sulfur Among Soil Organic-Matter Fractions of a Well-Drained Black Chernozemic Soil (Udic Haploboroll)

| | HI-Reducible S * (μg/g soil) in: | | |
	Pasture	Cultivated Soil	Percent Difference †
HA-A	28 (1.3)†	12 (7.3)	57
HA-B	71 (2.1)	42 (3.6)	41
FA-A	85 (1.6)	76 (1.1)	11
FA-B	11 (1.3)	10 (1.2)	9
Humin > 2 μm	12 (4.4)	8 (14.1)	33
Humin < 2 μm	37 (1.6)	18 (3.8)	51
Total fractions	244	166	
Percent recovery	98.1	96.9	

* Figures in parentheses represent the coefficients of variations (%) in the replicates.
† The concentration of HI-S of the cultivated soil was subtracted from that of the pasture soil and expressed as a percentage of the pasture HI-S soil fractions.
Source: Bettany et al., *Soil Sci. Soc. Am. J.,* **44:**70 (1980).

The same five volatile sulfur compounds distinguished in the Iowa study discussed above also have been identified as products of microbial decomposition of sulfur-containing amino acids in soils under aerobic and waterlogged conditions and as products of anaerobic decomposition of animal manures. Methyl mercaptan, dimethyl sulfide, and dimethyl disulfide are formed during decomposition of methionine, and carbon disulfide is released in the breakdown of cystine or cysteine.

Volatilization of sulfur as methyl mercaptan, dimethyl sulfide, and dimethyl disulfide has also occurred from soils treated with residues of a variety of cruciferous crops. Such compounds may represent 30% of the total organic sulfur in these crops but not over 5% in alfalfa, clover, and beets.

Grundon in Australia, in an extensive study involving 25 plant species, found that volatile sulfur compounds were evolved from intact plants and from oven drying of top and root samples. Losses in the drying treatment ranged in magnitude from about 0.3 to 6.0% of the total sulfur content of tops or roots. This work indicated that the presence of volatile sulfur compounds is much more widespread than was believed previously. It occurred in a large number of forage species, including pasture grasses and legumes, and field crops such as wheat, sorghum, cotton, rape, and sunflower.

Even though potential losses of soil and plant sulfur by volatilization are probably only of minor consequence, the various volatile compounds may have important side effects. Those substances released from soil during the decomposition of cruciferous crop residues have been reported to control root rot in peas, beans, and sesame. Carbon disulfide is a potent inhibitor of nitrification and methyl mercaptan, dimethyl sulfide, and dimethyl disulfide are also capable of retarding it. The offensive and sometimes toxic odors from decomposing animal manures are caused by volatile sulfur compounds such as H_2S, CH_3SH, CH_3SCH_3, CH_3SSCH_3, COS, and CS_2.

Volatile sulfur compounds released by intact plants may affect the palatability and acceptability of forage plants to grazing animals. Sulfur losses from forage species when they are dried in haymaking or pelleting might also influence quality and palatability.

Acidulating the Soil. It is occasionally necessary to increase soil acidity. Acidification may be needed when land is inherently high in carbonates, as in the arid western regions of the United States and Canada. Farmers in humid regions may overlime, or dust from limestone-graveled roads may blow onto field borders, causing a localized and excessively high pH. In other areas, moderately acid soils may need further acidification for the growth of such plants as potatoes, blueberries, cranberries, azaleas, rhododendrons, camellias, or conifer seedlings.

The fundamental soil chemistry of the acidification of soils is the same as that of liming soils. The pH is decreased, however, and different materials are employed. The agents used to reduce soil pH are elemental sulfur, sulfur dioxide, sulfuric acid, aluminum sulfate, iron sulfate, and ammonium polysulfide. Ammonium sulfate, ammonium phosphate, and similar compounds, though primarily used as fertilizers, are also quite effective in decreasing the soil pH, as pointed out in Chapter 11.

Elemental Sulfur. Pound for pound, elemental sulfur is the most effective of the soil acidulents. Its conversion to sulfuric acid in moist, warm, well-aerated soils by the autotrophic and heterotrophic organisms was discussed at considerable length earlier in the chapter. In calculating the amount to apply to the soil, reference must be made to the buffer curve of that soil. If it is assumed that all of the sulfur is to be converted to sulfuric acid, the calculation of the amount needed is a simple matter, for 1 meq of sulfuric acid, 0.049 g, will be formed by the oxidation of 1 meq of sulfur, 0.016 g. Theoretically, then, 1000 lb/A of

TABLE 8-9.   Effect of Soil Acidification on the Yield of Irrigated Sorghum Grown on a Calcareous Cut Soil in Rush County, Kansas

Year	Sulfur Source	Application Rate (lb S/A)	Yield Without S (bu/A)	Yield With S (bu/A)	Yield Increase Due to S (bu/A)	Percent Increase
1966	Elemental S	1000	15	53	38	253.3
	Elemental S	4000	15	78	63	420.0
1967	Elemental S	1000	36	51	15	41.6
	Elemental S	4000	36	56	20	55.5

Source: J. D. Beaton and R. L. Fox, "Production, marketing, and use of sulfur products," Chapter 11. Reprinted from *Fertilizer Technology and Use*, 2nd ed., published and copyrighted by Soil Science Society of America, 1971, by permission of copyright owners.

limestone could be neutralized by 320 lb of elemental sulfur if it were completely transformed to sulfuric acid by sulfur-oxidizing organisms.

Ordinary ground sulfur can be applied broadcast and disked in several weeks before planting the crop, for the initial velocity of the reaction, particularly in cold alkaline soils, may be somewhat slow. Under some conditions it may be advisable to acidulate a zone near the plant roots to increase water penetration or increase phosphorus and micronutrient availability. Both of these conditions frequently need to be corrected on extensive areas of dryland and irrigated saline-alkali soils of the United States in the southwestern, intermountain, and western states and in western Canada. Elemental sulfur can be applied in bands either as dry ground sulfur or granular sulfur bentonite. Sulfur suspensions can also be employed. When elemental sulfur is applied in a band, much smaller amounts are required than are needed when it is applied broadcast.

Land leveling to facilitate irrigation and for other purposes often exposes calcareous and high-pH subsoils that are unfavorable for optimum plant growth. The beneficial effect of elemental sulfur on production of irrigated grain sorghum on a cut soil in Kansas is illustrated in Table 8-9. Yield was increased from 42% to 420% by the sulfur treatment on these high-pH calcareous soils.

Problems of high soil pH are not confined to arid and semiarid areas. Applications of relatively high rates of granular sulfur and elemental sulfur have brought about striking improvements in rice yields in Arkansas, Louisiana, and Texas in the United States and in Western Australia. In several instances the yield increases were related to increased availability of micronutrients.

One major effect of the acidity produced by additions of elemental sulfur and other sulfur-containing compounds to high pH and calcareous soils is increased availability of soil nutrients. The influence of soil-applied SO_2 on uptake by sorghum of four essential nutrients is shown in Figure 8-13. Only 25 to 50% neutralization of the basicity was necessary to produce substantial improvement in uptake of these nutrients. There was no significant increase in yield between 25 and 75% neutralization.

Sulfuric Acid. Researchers in a number of states, including Arizona, California, Montana, and Texas, have clearly demonstrated that sulfuric acid is beneficial in arid-land agriculture. Benefits reported from its use by Stroehlein and others at the University of Arizona are: reclaiming sodium- or boron-affected soils, increasing availability of phosphorus and micronutrients, reducing ammonia volatilization, increasing water penetration, controlling certain weeds and soil-borne pathogens, and enhancing the establishment of range grasses.

An example is given in Table 8-10 of the favorable influence of sulfuric acid and other soil ameliorants on yield of rice in Western Australia. Increased availability of nitrogen, phosphorus, potassium, iron, manganese, and zinc appeared to con-

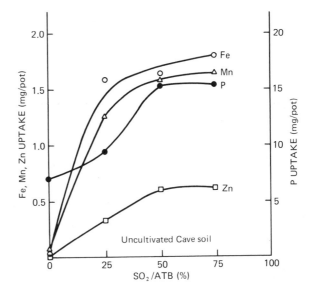

FIGURE 8-13. Effect of SO_2 treatment as percentage of acid-titratable basicity (ATB) on uptake by sorghum of phosphorus, iron, manganese, and zinc. Growth period of approximately $2^1/_2$ months on an uncultivated calcareous Cave soil. [Miyamoto et al., *Sulphur Inst. J.,* **10**(2):14 (1974).]

tribute to the higher yields obtained from additions of either sulfuric acid or sulfur.

Sulfuric acid can be added directly to the soil, but it is unpleasant to work with and requires the use of special acid-resistant equipment. It can be dribbled on the surface or applied with a knifeblade applicator, similar to the way in which anhydrous ammonia is treated. It can also be applied in ditch irrigation water. Sulfuric acid has the advantage of reacting instantaneously with the soil. In some areas it can be applied by custom suppliers who have the equipment necessary for handling this acid.

Aluminum Sulfate. Aluminum sulfate is a popular material among floriculturists for acidulating the soil in which azaleas, camellias, and similar acid-tolerant ornamentals are grown. When this material is added to water, it hydrolyzes as follows:

$$Al_2(SO_4)_3 + 6H_2O \rightarrow 2Al(OH)_3 + 3H_2SO_4$$

TABLE 8-10.   Effect of Sulfur, Sulfuric Acid, and Gypsum Soil Ameliorants on the Mean Grain Yield (14% Moisture) of Bluebonnet 50 and IR661-1-170-1-3 Rice on Cununurra Clay (Averaged over Levels and Methods of Application), Kimberley Research Station, 1975–1976 Wet Season

	Mean Yield (tons/ha) of:	
Soil Ameliorant	*Bluebonnet 50*	*IR661-1-170-1-3*
Control	2.69	5.85
Gypsum	2.70	6.00
Sulfur	3.24	6.72
Conc. H_2SO_4	3.69	6.96
LSD		
$P = 0.05$	0.31	0.86
$P = 0.01$	0.48	n.s.

Source: Chapman, *Australian J. Exp. Agr. Anim. Husb.,* **20:**724 (1980).

This solution is quite acid. When the salt is added to the soil, in addition to hydrolysis in the soil solution, the aluminum replaces any exchangeable hydrogen on the soil colloid and drives the pH even lower:

$$Al_2(SO_4)_3 + [clay]\frac{4H}{Ca} \rightarrow \frac{Al}{Al}[clay] + CaSO_4 + 2H_2SO_4$$

Aluminum sulfate, which is largely a specialty material, is not widely used in general agriculture.

Iron Sulfate. Iron sulfate $(FeSO_4)$ is also applied to soils for acidification. Its behavior is similar to that of aluminum.

Ammonium Polysulfide. Liquid ammonium polysulfide, mentioned previously in this chapter, is also used to lower soil pH and to increase water penetration in irrigated saline-alkali soils. It can be applied in a band 3 or 4 in. to the side of the seed or it can be metered into the ditch irrigation systems. Band application would be more effective in correcting micronutrient deficiencies than application through irrigation water. The polysulfide decomposes into ammonium sulfide and colloidal sulfur when applied. The sulfur and sulfide are subsequently oxidized to sulfuric acid.

Acidification in Fertilizer Bands. Because of the high buffering capacity of many calcareous and high-pH soils, it is usually too expensive to use enough acidifying material for complete neutralization of soil alkalinity. It is known from Arizona work that it is unnecessary to neutralize the alkalinity of the entire soil mass because soil zones more favorable for root growth and nutrient uptake can be created by confining the acid-forming materials to bands, furrows, auger holes, and so on.

Banding of acid-forming fertilizer materials has improved the growth of lettuce and other vegetable crops in Arizona, lettuce and potatoes in the San Luis Valley of Colorado, and corn and winter wheat in Kansas. The acidifying effect of a banded mixture of ammonium thiosulfate and ammonium polyphosphate fertilizer solution is shown in Figure 8-14. Accompanying this soil acidification was an increase in soil levels of iron and manganese. Increased plant uptake of nutrients such as phosphorus and manganese due to the acidifying effect of sulfur-containing fertilizers like ammonium thiosulfate and ammonium sulfate has been measured with corn in Kansas.

Some Practical Aspects. Plant roots absorb sulfur almost entirely as the sulfate ion. The concentration of this ion in the soil is important to growing

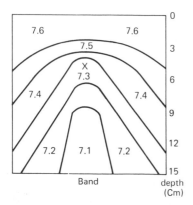

FIGURE 8-14. Application of a phosphorus–sulfur fertilizer solution lowers soil pH in the vicinity of fertilizer band (× denotes point of application). [Leiker, M.Sc. thesis, Kansas State Univ. (1970).]

plants. It will be governed largely by the factors affecting its retention in and removal from soils, items which have been discussed in the preceding sections.

Leaching losses of sulfur are greatest on coarse-textured soils under high rainfall. Under such conditions sulfur-containing fertilizers may have to be applied more frequently than on fine-textured soils and under lighter rainfall. In some areas a fertilizer containing both sulfate sulfur and elemental sulfur may be required. This practice is followed in some of the more humid regions of Australia and New Zealand. Elemental sulfur is mixed with ordinary superphosphate and applied to the land. Plant response to sulfur is thus extended over a longer period of time.

Sulfates are retained by certain clays and the hydrous oxides of iron and aluminum usually found in the B or C horizons of some humid region soils. Deep-rooted crops, such as alfalfa, are able to utilize this adsorbed sulfur. Even in these soils, however, the surface horizons may be deficient in sulfur, and a supply of this element should be applied to the young seedling until its roots have penetrated the sulfur-rich zones.

Large seasonal and year-to-year fluctuations in the levels of soluble sulfate often occur in surface soils and also at greater depths. This variability is caused by the interaction of environmental and seasonal conditions on the mineralization of organic sulfur, movement of sulfate salts in soil moisture, and sulfate uptake by plants. The quantities of sulfur in precipitation and in the atmosphere that reach soils will also vary from season to season and from year to year.

There are substantial differences in the sulfur fertilizer requirement of crops. Spencer in Australia uses three categories, high, moderate, and low, to describe the fertilizer sulfur needs of a variety of crops. His classification and the recommended rates of sulfur fertilization are given in Table 8-11. The actual amount of sulfur needed will depend on the balance between all soil additions of this nutrient via precipitation, air, irrigation water, crop residues, fertilizers and other agricultural chemicals, and all soil losses through crop removal, leaching, and erosion.

In addition to the observed differences in sulfur needs among crop species, there are indications of varietal or cultivar differences as well. Jones and his co-workers at the University of California's Hopland Station found that differences in response of annual clovers to sulfur were as great among varieties of a given species as

TABLE 8-11.   Tentative Classification of Crops According to Their Sulfur Fertilizer Requirement

Crop	Fertilizer Required in Deficient Areas * (kg S/ha)	Crop	Fertilizer Required in Deficient Areas * (kg S/ha)
Group I (high)		Group III (low)	
Cruciferous forages	40–80	Sugar beet	15–25
Lucerne	30–70	Cereal forages	10–20
Rapeseed	20–60	Cereal grains	5–20
Group II (moderate)		Peanuts	5–10
Coconuts	–50		
Sugarcane	20–40		
Clovers and grasses	10–40		
Coffee	20–40		
Cotton	10–30		

* Figures cited for the high end of the range apply where the potential yield is high, accessions in rainfall are low, the soil is low in available sulfur and there is considerable loss in effectiveness of applied sulfur. Figures cited for the low end refer to the opposite situation.

For perennials, a further consideration is whether the requirement refers to a corrective fertilizer dressing or to an annual maintenance dressing. The former is typically about four times the latter in amount.

Source: Spencer, in K. D. McLachlan, Ed., *Sulphur in Australasian Agriculture.* Sydney: Sydney University Press, 1975.

between species. Researchers in British Columbia observed with five alfalfa cultivars that 'duPuit' grew the poorest without supplemental sulfur and responded the most to sulfur additions. There are unpublished reports from Minnesota of differences in sulfur response among three corn hybrids. Two hybrids did not benefit from sulfur fertilization while production by the third one increased up to 24 bu/A.

Grasses are better able to utilize sulfate than the legumes. In grass-legume meadows the grasses can absorb the available sulfate at a faster rate than the legumes. Unless an adequate soil level of this element is maintained, the legumes will be forced out of the mixture, for sulfur is required for nitrogen fixation by the rhizobia. This condition has been demonstrated in New Zealand by Walker and his co-workers.

There may be immobilization of added sulfur in some soils because of its conversion to organic forms. This will be particularly true on soils that have large amounts of carbon and nitrogen but only limited quantities of sulfur. On other soils containing larger amounts of sulfur in relation to carbon and nitrogen, there will be the release of sulfate sulfur.

Elemental sulfur and polysulfides must be converted to the sulfate form before they can be absorbed by plants. The rate of conversion of elemental sulfur will be influenced by numerous and previously mentioned factors. Elemental sulfur-sulfate combinations will likely be preferable to solely elemental sulfur on soils with low sulfur-oxidizing capacity and where environmental conditions are unfavorable for sulfate formation.

With the exception of ammonium, aluminum, and iron sulfates, sulfate salts have no effect on soil pH. These three salts, elemental sulfur, and polysulfides are acid forming and their continued use will lower the soil pH. Availability of phosphorus and micronutrients, including boron, copper, iron, manganese, and zinc, is often improved in the acidified soil zone, even though the effect may only be temporary.

Where there is an undesirable reduction in pH, it can be corrected by an adequate liming program. One pound of elemental sulfur will produce an amount of acid that will require 3 lb of calcium carbonate for neutralization.

At the low yield levels of the past, some of the sulfur needs were often satisfied by incidental additions of this nutrient via precipitation and air. The importance of these two sources depends on the composition of fuels, distance from emitting sources, pollution control measures, and various climatic factors.

Withdrawal of sulfur from soils is accelerated and increased by numerous management factors, including high-yielding varieties, multiple cropping, irrigation, and heavier rates of other plant nutrients. Prolonged use of sulfur-free fertilizers will eventually induce a sulfur deficiency of crops. This deficiency will appear sooner on coarse-textured, humid-region soils and in areas in which the amounts of sulfur in the rainfall and atmosphere are low. In addition, the production of high-yielding crops, most of which are removed from the soil, will further aggravate a deficiency of sulfur.

Need for sulfur fertilizers is closely related to the amounts of nitrogen being applied to crops. Because both nitrogen and sulfur are involved in protein synthesis, the full benefit from the addition of one is dependent on an ample supply of the other. The strong association between nitrogen and sulfur is readily apparent in Figure 8-15. Here the magnitude of response to sulfur increases with the rate of nitrogen added—averaging 3 bu/A for 24 to 48 lb of nitrogen per acre, 6 bu/A for nitrogen rates in the range 72 to 96 lb/A and 10 bu/A for 144 to 168 lb of nitrogen per acre.

Sulfur-Containing Fertilizers. There are numerous sulfur-containing fertilizer materials, some of which have already been mentioned in Chapters 5, 6,

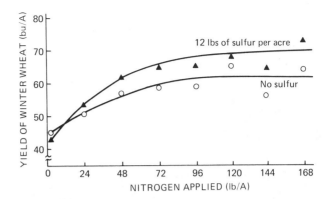

FIGURE 8-15. Sulfur increases the response of winter wheat to nitrogen on a low-sulfur Oregon soil 0–24 in. = 2 ppm, 24–48 in. = 1.3 ppm, and 48–72 in. = 1.4 ppm). [Rasmussen, *Proc. Pendleton–Walla Walla Fert. Conf.*, Walla Walla, Wash. (January 6, 1976).]

and 7, dealing with nitrogen, phosphorus, and potassium. Table 8-12 lists the most common sulfur-containing fertilizer materials and gives their typical content of sulfur and other plant nutrients.

Most of these materials are old and established sources of fertilizer sulfur. Their behavior in the soil is determined by the nature of the sulfur, and the reactions they undergo are described in the preceding section. Studies comparing the effectiveness of sources of sulfur are by no means as numerous as are those that compare the agronomic effectiveness of nitrogen, phosphorus, and potassium fertilizers. The available data suggest that when the sulfate itself is considered, one source of sulfate sulfur is generally equal to any other (provided the accompanying cation is not zinc, copper, or manganese, which salts must be applied sparingly) and that the factor determining the selection should be the cost per unit of sulfur applied to the land.

Elemental Sulfur. When elemental sulfur is compared with sulfate sulfur, the results will depend on several factors related to the former source. These are particle size, rate, method, and time of application; sulfur-oxidizing characteristics of the soil; and environmental conditions. When the sulfur is finely ground and mixed with soil possessing a high oxidizing capacity, it is usually just as effective as sulfate sources. Time of application is especially important with elemental sulfur products and they should be worked into the soil as far ahead of planting as possible.

If elemental sulfur is placed on the surface of the soil and compared with a soluble sulfate similarly placed, the sulfate may give initially better responses. Because of its solubility it can move into the root zone with percolating waters. The elemental sulfur must first be oxidized to sulfate, and this is not a rapid process when it is surface applied. For this reason, topdresssing elemental sulfur is not recommended. High rates of elemental sulfur may be required in situations where conversion to sulfate is slow. These high rates may need to be repeated only once every 3 or 4 years.

Sulfur-Bentonite. A product consisting of 90% elemental sulfur and 10% bentonite has been manufactured in the United States and Canada since 1970. Particles of this sulfur source are sized for blending with solid nitrogen, phosphorus, and potassium materials. Dust problems have occurred with it due to breakdown of its irregular, nonspherical particles, particularly when handled excessively with mechanical augers. This fertilizer is made by adding bentonite to molten sulfur; the molten mass cools and solidifies, and is then crushed and screened. When the product is applied to soil, the bentonite component imbibes moisture, causing granules to disintegrate, thereby forming more finely divided sulfur which is more rapidly converted to sulfate. This material has gained wide acceptance as

TABLE 8-12. Sulfur-Containing Fertilizer Materials

Material	Formula	Nitrogen	P_2O_5	K_2O	Sulfur	Other	Sulfur Content (lb/ton)
Aluminum sulfate	$Al_2SO_4 \cdot 18H_2O$	0	0	0	14.4	11.4 (Al)	288
Alunite	$K_2Al_6(OH)_{12}(SO_4)_4$	0	0	10.5	14.1	17.9 (Al)	282
Ammonia-sulfur solution	$NH_3 + S$	74	0	0	10		200
		70.5	0	0			280
Ammonium bisulfite	NH_4HSO_3	14.1	0	0	32.3		646
Ammonium bisulfite solution	$NH_4HSO_3 + H_2O$	8.5	0	0	17		340
Ammonium nitrate-sulfate	[a]	30	0	0	5		100
Ammonium phosphate	MAP (crude)	11	48	0	2.2		44
Ammonium phosphate-sulfate	MAP, DAP + $(NH_4)_2SO_4$	16.5	20.5	0	15.5		310
		13	39	0	7		140
Ammonium polysulphide	NH_4S_x	20.5	0	0	45		900
Ammonium polysulphide solution	NH_4S_x	20	0	0	40		800
Ammonium sulfate	$(NH_4)_2SO_4$	21	0	0	24.2		484
Ammonium sulfate-nitrate	$(NH_4)_2SO_4 \cdot NH_4NO_3$	26	0	0	12.1		242
Ammonium thiosulfate	$(NH_4)_2S_2O_3$	18.9	0	0	43.3		866
Ammonium thiosulfate solution	$(NH_4)_2S_2O_3 + H_2O$	12	0	0	26		520
Aphthitalite	$(K,Na)_3(NaSO_4)_2$	0	0	42.5	15.1		302
Aqua-sulfur solution	$NH_3 + NH_3S_x + H_2O$	20	0	0	5		100
Basic slag (Thomas)		0	15.6	0	3		60
Cement flue dust		0	0.6	6	3	3 (Mg)	60
Cobalt sulfate	$CoSO_4 \cdot 7H_2O$	0	0	0	11.4	21 (Co)	228
Copper sulfate	$CuSO_4 \cdot 5H_2O$	0	0	0	12.8	25.5 (Cu)	256
Ferrous ammonium sulfate	$Fe(NH_4)_2(SO_4)_2$	6	0	0	16	16 (Fe)	320
Ferrous sulfate	$FeSO_4 \cdot H_2O$	0	0	0	18.8	32.8 (Fe)	376
Ferrous sulfate (copperas)	$FeSO_4 \cdot 7H_2O$	0	0	0	11.5	20 (Fe)	230
Glaubers salt	$Na_2SO_4 \cdot 10H_2O$	0	0	0	10		200
Guano (Peruvian)		0	0	11	1.1		22
Gypsum (anhydrite)	$CaSO_4$	0	0	0	23.5	41.1 (CaO)	470
Gypsum (hydrated)	$CaSO_4 \cdot 2H_2O$	0	0	0	18.6	32.6 (CaO)	372
Gypsum (by-product)[b]		0	2.5	0	17.2	21.6 (CaO)	344
Gypsum (impure)[c]		0	0	0	13.6	23.8 (CaO)	272
Kainit[d]	$MgSO_4 \cdot KCl \cdot 3H_2O$	0	0	19	12.9	9.7 (Mg)	258
Kalinite	$K_2SO_4 \cdot Al_2(SO_4)_3 \cdot 24H_2O$	0	0	9.9	13.5	5.7 (Al)	270
Kieserite	$MgSO_4 \cdot H_2O$	0	0	0	23	17.5 (Mg)	460
Krugite	$K_2SO_4 \cdot MgSO_4 \cdot 4CaSO_4 \cdot 2H_2O$	0	0	10.7	21.9	2.8 (Mg)	438
Leonite	$K_2SO_4 \cdot MgSO_4 \cdot H_2O$	0	0	25.5	20.5	7.8 (Mg)	410
Lime sulfur (dry)	CaS_x	0	0	0	57	43 (Ca)	1140
Lime sulfur (solution)	$CaS_5 + Ca_2SO_3 \cdot 5H_2O +$ $CaS_4 + CaSO_3 \cdot 2H_2O$	0	0	0	23-24	9 (Ca)	480

Material	Formula						
Magnesium sulfate	$MgSO_4$	0	0	0	30	20 (Mg)	600
Magnesium sulfate (Epsom salt)	$MgSO_4 \cdot 7H_2O$	0	0	0	13	9.8 (Mg)	260
Manganese sulfate	$MnSO_4 \cdot 4H_2O$	0	0	0	14.5	25 (Mn)	290
Polyhalite	$2CaSO_4 \cdot MgSO_4 \cdot K_2SO_4 \cdot 2H_2O$	0	0	15.6	21.2	4.0 (Mg)	424
Potassium sulfate	K_2SO_4	0	0	50	17.6		352
Pyrites	FeS_2	0	0	0	53.5[e]	46.5 (Fe)	1070
Schoenite (picromerite)	$K_2SO_4 \cdot MgSO_4 \cdot 6H_2O$	0	0	23.3	15.9	6.0 (Mg)	318
Sodium bisulfate (nitre cake)	$NaHSO_4$	0	0	0	26.5		530
Sodium sulfate (salt cake)	Na_2SO_4	0	0	0	22.6		452
Sulfate of potash-magnesia (Langbeinite)	$K_2SO_4 \cdot 2MgSO_4$	0	0	22	22	11 (Mg)	440
Sulfuric acid (100%)	H_2SO_4	0	0	0	32.7		654
Sulfuric acid (66° Bé = 93%)	H_2SO_4	0	0	0	30.4		608
Sulfuric acid (60° Bé = 77.7%)	H_2SO_4	0	0	0	25.4		508
Sulfuric acid (56° Bé = 71.17%)	H_2SO_4	0	0	0	23.2		465
Sulfur	S	0	0	0	100		2000
Sulfur dioxide	SO_2	0	0	0	50		1000
Superphosphate, concentrated	[f]	0	54	0	1.5		30
Superphosphate, normal	[g]	0	20	0	13.9		278
Superphosphate, 20% normal, ammoniated	[h]	4.6	19	0	12.0		240
Superphosphate, triple	[i]	0	46	0	1.5		30
Superphosphate, triple, ammoniated	[j]	6.9	42	0	1.4		28
Syngenite	$K_2SO_4 \cdot CaSO_4 \cdot H_2O$	0	0	28.8	19.5	12.2 (Ca)	390
Urea-gypsum	$CaSO_4 \cdot 4CO(NH_2)_2$	17.3	0	0	14.8		296
Urea-sulfur	$CO(NH_2)_2 + S$	40	0	0	10		200
Urea-sulfuric acid	$CO(NH_2)_2 \cdot H_2SO_4$	10–28	0	0	9–18		
Zinc sulfate	$ZnSO_4 \cdot H_2O$	0	0	0	17.8	36.4 (Zn)	356

[a] NH_4NO_3 (39%), $3NH_4NO_3 \cdot (NH_4)_2SO_4$ (49%), using an ammoniator-granulator.

[b] Average content of byproduct from wet process phosphoric acid plant.

[c] Average purity of agricultural gypsum (73% $CaSO_4 \cdot 2H_2O$).

[d] Often has considerable NaCl content, and may have as low as 12% K_2O.

[e] Commercial pyrites average 48–50 % sulfur.

[f] Heat treated triple superphosphate (i.e., not made with superphosphoric acid). Analysis will be that of regular triple superphosphate as affected by the amount of water removed.

[g] $Ca(H_2PO_4)_2 \cdot 2H_2O$, 3%; $Ca(H_2PO_4)_2$ anhydrous, 17%; $CaSO_4 \cdot 2H_2O$, 9%; $CaSO_4$ anhydrous, 41%; H_3PO_4, 8%; other 22%.

[h] $NH_4H_2PO_4$, 5.5%; $(NH_4)_2SO_4$, 28%; $Ca_3(PO_4)_2$, 17.5%; $CaCO_3$, 37%; other 12% (ammoniated at 6 lb of NH_3 per unit of P_2O_5).

[i] $Ca(H_2PO_4)_2 \cdot H_2O$, 63–73%; $CaHPO_4$ and other phosphates, 17–29%; $CaSO_4 \cdot 2H_2O$, 3–6%, other 6–12%.

[j] $CaHPO_4$, 35%; $NH_4H_2PO_4$, 30%; $Ca_3(PO_4)_2$, 10%; $(NH_4)_2HPO_4$, 8%; $(NH_4)_2SO_4$, 2%, other 15% (ammoniated at 4 lb of NH_3 per unit of P_2O_5).

Source: Bixby and Beaton, "Sulphur containing fertilizers: properties and applications." *Tech. Bull. 17.* Washington, D.C.: The Sulphur Institute, 1970.

a source of plant nutrient sulfur for high-analysis bulk-blend formulations. Alternative methods of making a less dusty, uniform, spherical product are being investigated.

Because of the uncertainty of adequate availability of sulfur-bentonite during the first growing season after application, it should be incorporated into soil at least 4 or 5 months prior to planting. When applied just before seeding and on severely sulfur-deficient soils, some sulfate should also be provided. In areas where conversion of elemental sulfur is slow, it may take several repeated applications of sulfur-bentonite before the rate of sulfate formation becomes sufficient to satisfy crop requirements.

Ammonia-Sulfur and Sulfur Dioxide. There has been some interest in ammonia-sulfur and liquid SO_2 as fer. .izer materials. Although both have been shown to be satisfactory sources of plant available sulfur, they are not widely used because of either equipment and application difficulties or incompatibility with existing application systems. However, SO_2 produced from elemental sulfur in specially designed field burners is added to irrigation waters in some areas of Arizona and California. The SO_2 increases infiltration rates of water in some soils, especially those irrigation waters of low electrolyte content. This material and other acidifying sulfur-containing substances, including elemental sulfur, sulfuric acid, ammonium polysulfide, ammonium thiosulfate, ammonium bisulfite, ammonium sulfate, and so on, will also improve the availability of plant nutrients in high pH and calcareous soils.

Ammonium Thiosulfate. Ammonium thiosulfate is a clear liquid material with no appreciable vapor pressure. It contains 12% nitrogen and 26% sulfur and is the most popular sulfur-containing product used in the fluid fertilizer industry. It is essentially noncorrosive and may be stored in mild steel or aluminum containers. This sulfur source is compatible with nitrogen solutions and complete (N-P-K) liquid mixes, which are neutral to slightly acid in reaction. It cannot be used with acidic (pH 5.8 or less) materials or mixes. In addition to its wide adaptability for clear liquid mixtures, it is commonly used in suspensions.

Ammonium thiosulfate can be applied to the soil directly or in mixtures or it can be added to irrigation water, both sprinkler and open-ditch systems. In some areas it is used both as a fertilizer and for water treatment.

When applied to the soil it decomposes to form colloidal elemental sulfur and ammonium sulfate. The sulfate is immediately available, whereas the elemental sulfur is converted to sulfate by bacterial oxidation, thus extending the availability of this element to the growing crop. The acidifying effect of this material in fertilizer bands and the resultant increase in availability of phosphorus and micronutrients in high-pH and calcareous soils were described in an earlier part of the chapter. calcium saturations of 20% or less and aluminum saturations of 68% or more.

Ammonium Polysulfide. This substance is a red to brown to black solution having a hydrogen sulfide odor. It contains approximately 20% nitrogen and 45% sulfur. Its three main uses are for supplying plant-available nitrogen and sulfur, for reclaiming high-pH soils, and for treatment of irrigation water. It finds its greatest use in irrigated arid-land agriculture.

Ammonium polysulfide is recommended for mixing with anhydrous ammonia and it is also compatible with aqua ammonia and urea-ammonium nitrate solutions. For stability with the latter solutions, not less than 10% by volume of ammonium polysulfide should be used. It normally is considered incompatible with phosphate-containing liquids. This material has a low vapor pressure and it should be stored at a pressure of 0.5 psig to prevent loss of ammonia and subsequent precipitation of sulfur.

It may be applied in open-ditch irrigation systems with or without previous dilution. Application through sprinkler systems is not usually recommended. It is applied to soil by injection with anhydrous ammonia or aqua ammonia. By diluting ammonium polysulfide with water to lower the nitrogen concentration to less than 5%, it can be applied directly to the soil surface. The simultaneous application of ammonium polysulfide and anhydrous ammonia is a common way of providing both nitrogen and sulfur in the small grain-producing areas of eastern Oregon, eastern Washington, and northern Idaho.

Ammonium polysulfide is a good source of plant nutrient sulfur but is not as convenient or pleasant to handle as is ammonium thiosulfate. It produces acidification effects similar to those already noted in discussions of other sulfur sources and in the acidification section of this chapter.

Urea-Sulfuric Acid. A simple, rapid, and economic batch process has been developed in California for mixing urea and sulfuric acid to produce this concentrated liquid nitrogen sulfur fertilizer. Measures must be taken to dissipate the heat evolved in the process. Two typical grades used as acidifying amendments as well as sources of nitrogen contain 10% nitrogen and 18% sulfur and 28% nitrogen and 9% sulfur, respectively. They can be applied directly to the soil or added for control of water quality in drip and overhead sprinkler systems.

Because these urea-sulfuric acid formulations have pH values between 0.5 and 1.0, it is necessary to construct all equipment items from stainless steel and other noncorrosive materials. Workers must be extremely safety conscious and wear protective clothing. Careful monitoring of the pH of irrigation water applied through aluminum equipment is necessary to avoid acidification below pH 5.0 with resultant deterioration of aluminum components.

Ammonium Bisulfite. This low analysis (8.5% nitrogen and 17% sulfur) clear liquid product with a strong odor of sulfur dioxide has been marketed to a limited extent for many years in the Pacific Northwest. However, its use has now declined to virtually nil in this region. It is well suited to mixing with aqueous ammonia and nitrogen solutions and like ammonium polysulfide it should not be used in an acidic medium. In addition to its use in liquid fertilizer blends, it can be applied in irrigation water similar to and for the same purposes as ammonium thiosulfate.

Urea-Sulfur. A description of this sulfur source was given in Chapter 5. The agronomic effectiveness of sulfur in the most recently developed version of this fertilizer is uncertain. Sulfur responses from 36-0-0-20(S) in the prairie provinces have been variable, perhaps due to differences in population and activity of oxidizing organisms in soils at the experimental sites. There are indications from several more southerly locations in the United States that elemental sulfur in this product becomes available sufficiently fast to meet crop needs. This product's excellent physical properties coupled with satisfactory performance in most evaluations suggest that it has a promising future.

Elemental Sulfur Suspensions. The addition of finely ground elemental sulfur to water containing 2 to 3% attapulgite clay results in a suspension containing 40 to 60% sulfur. These suspensions can be applied directly to the soil or they can be combined with suspension fertilizers to supply plant-nutrient sulfur. They are easy to handle and have the added advantage of being non-dusty, a great drawback to the handling of run-of-pile agricultural grade sulfur.

Phosphate-Elemental Sulfur Materials. Elemental sulfur can be readily incorporated into mono- and diammonium phosphates using several different methods. Any desired amount of sulfur may be added, but the usual concentration in pilot-plant and laboratory products has been 5 to 20% sulfur. The limited amount of work done with such fertilizers indicates that they are suitable sources of both phosphorus and sulfur.

A granular product based on concentrated superphosphate with an approximate analyses of 0-40-0-20(S) has been prepared by the TVA. Elemental sulfur and concentrated superphosphate are granulated using steam and water in a granulating drum. Commercial production of a similar material analyzing 0-35-0-28(S) was undertaken during the late 1960s by one manufacturer in the western United States. In spite of excellent results and farmer acceptance, production ceased after just a few years because of manufacturing and labor difficulties.

Sulfur-fortified normal superphosphates are popular in some countries such as Australia and New Zealand. Ordinary superphosphate is enriched with elemental sulfur to make mixtures containing from 18 to 35% sulfur. The added elemental sulfur is superior in its residual effect to the calcium sulfate already in the ordinary superphosphate.

There are small amounts of sulfur, normally between about 1 and 3%, in the ammonium phosphates and concentrated superphosphates which probably make minor contributions to the sulfur nutrition of fertilized crops.

Fertilizer Use Guidelines. For purposes of convenience, recommendations for the use and proper application of common sulfur-containing fertilizers are summarized in Table 8-13.

Calcium

Form Utilized by Plants. Calcium is absorbed by plants as the ion, Ca^{2+}. This takes place from the soil solution and possibly by root interception or contact exchange. The quantities of calcium required by plants can be readily transported to root surfaces by mass flow in most soils, excepting highly leached and unlimed acid soils. In soils abundant in this nutrient, excesses of it probably accumulate in the vicinity of roots.

Representative levels of calcium in the soil solution of temperate region soils are 30 to 300 ppm. In soils of higher-rainfall areas, soil solution calcium concentration will range from 8 to 45 ppm, and usually averages about 33 ppm. A level of 15 ppm Ca^{2+} in the soil solution was found by Barber at Purdue University to be adequate for high corn yields. Levels of from 100 to 300 ppm are commonly used in solution cultures. Loneragan and Snowball of the University of Western Australia demonstrated that Ca^{2+} concentrations of between 100 to 400 ppm in flowing culture solutions were sufficient for high yields of 30 grasses, cereals, legumes, and herbs.

The presence of concentrations of calcium in the root medium higher than are necessary for proper plant growth will normally have little effect on its uptake. This is because the calcium content of plants is largely genetically controlled. Although the Ca^{2+} concentration of the soil solution is frequently about 10 times greater than that of K^+, its uptake is usually lower than that of K^+. Capacity of plants for uptake of Ca^{2+} is limited because it can be absorbed only by young root tips in which the cell walls of the endodermis are still unsuberized.

Source of Soil Calcium. The calcium concentration of the earth's crust is about 3.64% and it is more common than most other plant nutrients. Soils vary widely in calcium content with extremely sandy soils of humid regions containing very low amounts of this nutrient. Its concentration in calcium carbonate free soils of humid temperate regions of the world normally ranges from approximately 0.7

to 1.5%. Highly weathered soils of the humid tropics may contain as little as 0.1 to 0.3% calcium. Calcium levels in calcareous soils vary from less than 1 to more than 25%. Values of more than approximately 3% indicate the presence of calcium carbonate.

The calcium present in soils, exclusive of that added as lime or in fertilizer materials, has its origin in the rocks and minerals from which the soil was formed. The plagioclase mineral, anorthite ($CaAl_2Si_2O_8$), is the most important primary source of calcium. Other minerals in this group, including impure albite, are of less significance. Pyroxenes (augite) and amphiboles (hornblende) are fairly common in soils and also contain calcium. Small amounts of calcium may also originate from biotite, epidote minerals, apatite, and certain borosilicates.

Calcite ($CaCO_3$) is often the dominant source of calcium in soils of semiarid and arid regions. Dolomite [$Ca,Mg(CO_3)_2$] may also be present in association with calcite. In some arid-region soils, calcium sulfate or gypsum ($CaSO_4 \cdot 2H_2O$) may be present. Calcium is liberated when these various minerals disintegrate and decompose.

The fate of released calcium is less complex than that of potassium. Calcium ions set free in solution may (1) be lost in drainage waters, (2) absorbed by organisms, (3) adsorbed onto surrounding clay particles, or (4) reprecipitated as a secondary calcium compound, particularly in arid climates. As far as is known, there is no conversion in the soil of calcium to a form comparable to fixed or slowly available potassium.

As a general rule, coarse-textured, humid-region soils formed from rocks low in calcium-containing minerals are low in their content of this element. Soils that are fine-textured and formed from rocks high in the calcium-containing minerals are much higher in their content both of exchangeable and total calcium. However, in the humid region even soils formed from limestones are frequently acid in the surface layers because of the removal of calcium and other basic cations by excessive leaching. As water containing dissolved carbon dioxide percolates through the soil, the carbonic acid so formed displaces calcium (and other basic cations) in the exchange complex. If considerable percolation of such water through the soil profile takes place, soils gradually become acid.

The calcium content of soils of the arid regions is generally high, regardless of texture, as a result of low rainfall and little leaching. Many of the soils of the arid regions actually have within their profiles secondary deposits of calcium carbonate or calcium sulfate.

Deficiencies of calcium leading to yield reductions are rather rare in agricultural soils, since most acid soils usually contain sufficient amounts of this nutrient for plant growth. More common are indirect calcium deficiencies caused by its shortage in fruit and storage organs which grow rapidly but with restricted internal supplies of calcium.

Behavior of Calcium in Soil. The calcium in acid, humid-region soils occurs largely in the exchangeable form and as undecomposed primary minerals. In most of these soils calcium, aluminum and hydrogen ions are present in the greatest quantity on the exchange complex. Like any other cation, the exchangeable and solution forms are in dynamic equilibrium. If the activity of calcium in the solution phase is decreased, as it might be by leaching or plant removal, there tends to be replacement from the adsorbed phase. Conversely, if the activity of calcium in the soil solution is suddenly increased, there tends to be a shift of equilibrium in the opposite direction, with subsequent adsorption of some of the calcium by the exchange complex.

In soils not containing calcite, dolomite, or gypsum, the amount of calcium in the soil solution is dependent on the amount of exchangeable calcium present. Soil

TABLE 8-13. Recommendations for Use of Fertilizers Containing Sulfur

Solid Materials	Recommended Use	Remarks
Ammonium phosphate–sulfur; ammonium polyphosphate–sulfur; ammonium phosphate–urea phosphate–sulfur; concentrated superphosphate–sulfur, urea–sulfur	For direct application and bulk blends apply materials several months before beginning of growing season	If used in starter fertilizer or shortly before beginning of growing season, some readily available sulfate should be included (15 to 20% of the total sulfur applied)
Water-degradable elemental sulfur; flake sulfur and porous granular sulfur	For direct application and bulk blends apply materials several months before beginning of the growing season; fall applications should be encouraged	Where feasible incorporate into soil 4 or 5 months prior to planting; when applied just in advance of planting or on severely sulfur deficient soils, some readily available sulfate should be included—see above; heavy initial rate may be required
Ammonium sulfate	For direct application and to some extent for bulk blending; should be effective at almost any time	Tends to segregate in bulk blends unless physical properties are improved by granulation; where significant leaching losses are expected, apply shortly before planting or the beginning of the growing season
Ammonium nitrate–sulfate; ammonium phosphate-sulfate; ammonium phosphate–sulfate–gypsum; ordinary superphosphate; potassium sulfate; potassium magnesium–sulfate	For direct application and bulk blends; should be effective at almost any time	Where significant leaching losses of sulfate are expected, apply shortly before planting or the beginning of the growing season
Calcium sulfate (gypsum)	For direct application; should be effective at almost any time	Difficulties may be encountered in application (dustiness, clogging)
Fluid Materials	**Recommended Use**	**Remarks**
Ammonium thiosulfate	For direct application and blending with most fluid fertilizer products; can be broadcast prior to planting or applied in starter fertilizers; can be topdressed on certain growing crops; can also be added through open-ditch and sprinkler irrigation systems; should be effective almost any time	Can be blended with all neutral fluid phosphate products now available, all nitrogen solutions (except anhydrous ammonia) and most micronutrient solutions

Ammonium bisulfite	For direct application and blending with most other nitrogen and nitrogen-phosphate fluids; can be sprayed or dribbled on prior to planting; can also be added through open-ditch and sprinkler irrigation systems; should be effective at almost any time	Can be blended with all neutral fluid phosphate products now available and all nitrogen solutions except anhydrous ammonia; in many instances it is applied simultaneously with anhydrous ammonia, the two products from different tanks are taken to the ground through separate metering systems and injected into soil on common applicator knives
Ammonium polysulfide	For direct application and blending with other nitrogen solutions; frequently injected into soil; broadcast spray applications are possible following dilution with water; single preplant applications are effective; repeated applications at low rates are often made to growing crops through open-ditch irrigation systems; should be effective at almost any time	Ammonium polysulfite is generally not considered suitable for mixing with phosphate-containing fluids
Sulfuric acid	For mixing with wet process ammonium polyphosphate and anhydrous ammonia in the preparation of clear liquid blends; should be effective at almost any time	Sulfuric acid has been applied directly to crops such as onions and garlics for weed control purposes
Suspensions containing elemental sulfur	For direct application and for simultaneous application with other fertilizers, the suspensions should be applied several months before beginning of the growing season	If used in starter fertilizer or shortly before beginning of growing season, readily available sulfate should be included (15 to 20% of the total sulfur applied)
Suspensions containing one or more sulfate salts such as ammonium sulfate, potassium sulfate, potassium magnesium–sulfate, and ammonium phosphate–sulfate	Should be effective at almost any time	Where significant leaching losses are expected, apply shortly before planting or the beginning of the growing season

Source: Bixby and Beaton, "Sulphur containing fertilizers: properties and applications." *Tech. Bull. 17.* Washington, D.C.: The Sulphur Institute. 1970.

TABLE 8-14. **Effect of Calcium Concentration and pH in Subsurface Nutrient Solution on Soybean Taproot Elongation in the Nutrient Solution**

		Experiment 2					Experiment 3	
pH	Ca Concentration Added (ppm)	Taproot Elongation Rate * (mm/hr)	Taproot Harvest Length † (mm)	Oven Dry Wt/mm (mg)		pH	Ca Concentration Added (ppm)	Taproot Elongation Rate (mm/hr)
5.6	0.05	2.66	461	0.20		4.75	0.05	0.11
	0.50	2.87	453	0.23			0.50	0.91
	2.50	2.70	455	0.32				
4.5	0.05	0.04	24	0.54		4.0	2.50	0.44
	0.50	1.36	270	0.26			5.00	1.26
	2.50	2.38	422	0.31				
LSD	5%	0.35	62					0.75
	1%	0.54	97					1.13

* Elongation rate during first 4 hr in solution.
† Harvested 7$\frac{1}{2}$ days after entering the solution.
Source: Lund, *Soil Sci. Soc. Am. J.,* **34**:457 (1970).

factors of the greatest importance in determining the availability of calcium to plants are the following:

1. Total calcium supply.
2. Soil pH.
3. Cation exchange capacity.
4. Percent calcium saturation of soil colloids.
5. Type of soil colloid.
6. Ratio of calcium to other cations in solution.

Total supply of calcium in very sandy acid soils with low cation exchange capacities can be too low to provide sufficient available calcium to crops. On such soils supplemental calcium will be needed to supply calcium, as well as to correct the acidity.

High H^+ activity occurring at low soil pH will impede calcium uptake. Lund of the USDA at Auburn University found in short-term split-root experiments that much higher calcium concentrations were required for soybean root growth as the pH was lowered from 5.6 to 4.0 (Table 8-14).

In acid mineral soils calcium is not readily available to plants at low base saturation. For example, a soil having only 2000 lb of exchangeable calcium per acre but representing a high degree of saturation of a low cation exchange capacity might well supply plants with more of this element than one containing 8000 or 9000 lb of exchangeable calcium per acre but saturating a lower percentage of a higher cation exchange capacity. The degree of calcium saturation is of considerable importance in this respect, for as the amount of this element held in the exchangeable form by soil colloids decreases in proportion to the total exchange capacity of the colloidal complex, the amount of calcium absorbed by plants decreases.

An exchange complex dominated by calcium is usually associated with higher crop yields. Many crops will respond to calcium applications when the degree of calcium saturation of the exchange capacity falls below 25%. High calcium saturation is indicative of a favorable pH for growth of most plants and for microbial activity. Also, a prominence of calcium will usually mean low concentrations of undesirable exchangeable cations such as Al^{3+} in acidic soils and Na^+ in sodic soils.

Calcium saturations of below 40 to 60% and aluminum saturations of 40 to 60% have lowered cotton yields. Soybeans are reported to suffer calcium deficiency at calcium saturations of 20% or less and aluminum saturations of 68% or more. Near normal growth of sugarcane in Hawaii is possible with 12% calcium saturation of exchange complexes in volcanic soils.

The type of clay influences the degree of calcium availability; 2:1 clays require a much higher degree of saturation for a given level of plant utilization than 1:1 clays. Montmorillonitic clays require a calcium saturation of 70% or more before this element is released sufficiently rapidly to growing plants. Kaolinitic clays, on the other hand, are able to satisfy the Ca^{2+} requirements of most plants at saturation values of only 40 to 50%.

Increasing aluminum concentration in the soil solution has been found to reduce calcium uptake by various plants, including corn, cotton, soybeans, and wheat. The depressing action of aluminum on rate of calcium uptake by wheat is evident in Figure 8-16. Evans and Kamprath at North Carolina State University showed that the concentration of aluminum in the soil solution of mineral soils is determined by the percent saturation of the effective cation exchange capacity with exchangeable aluminum. With organic soils, they observed that the amount of exchangeable aluminum rather than the percent saturation was more important in determining the amount of aluminum in the soil solution.

Calcium availability and its uptake by plants are also influenced by the ratios between calcium and other cations in the soil solution. A calcium/total cation ratio of 0.10 to 0.15 is desirable for development of cotton roots. Elongation of soybean taproots proceeded most satisfactorily when the $Ca/(Ca + Mg + K)$ ratios in nutrient cultures were between 0.10 and 0.20. Blossom-end rot, a calcium-deficiency disorder in tomatoes, can be prevented by maintaining calcium/soil solution salt ratios in the range 0.16 to 0.20.

Form of Inorganic Nitrogen. While Ca^{2+} uptake is depressed by NH_4^+, K^+, Mg^{2+}, Mn^{2+}, and Al^{3+}, its absorption is increased when plants are supplied

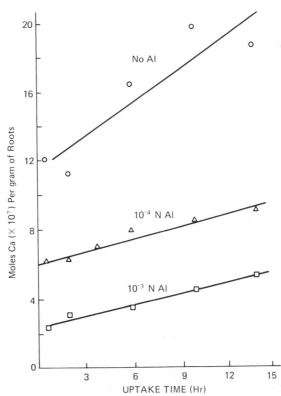

FIGURE 8-16. Influence of $AlCl_3$ on the rate of calcium uptake from $10^{-3}N$ $CaCl_2$ by excised wheat roots as determined by Ca^{45} measurements. [Johnson and Jackson, *Soil Sci. Soc. Am. J.*, **28**:381 (1964).]

with NO_3^--N. A high level of NO_3^- nutrition stimulates organic anion synthesis and the resultant accumulation of cations, particularly Ca^{2+}.

Leaching Losses of Calcium. Where leaching occurs, sodium is lost most readily from soil but the quantities of calcium lost are much greater. Calcium is often the dominant cation in drainage waters, springs, streams, and lakes. Leaching losses of calcium usually vary from 75 to 200 lb/A per year.

Plant Factors. Mention was made earlier of the restricted zone of calcium uptake at root tips. Conditions impairing the growth of new roots will reduce access of plant roots to calcium and thus induce deficiency. Also, problems related to inadequate calcium uptake are more likely to occur with plants that naturally have poor or small root systems than with those possessing more highly developed rooting systems.

Special attention must be given to the calcium requirements of certain crops, including peanuts, tomatoes, and celery, which are often unable to obtain sufficient calcium from soils supplying adequate calcium for most other crops. Proper calcium nourishment is also important for crops such as alfalfa, cabbage, potatoes, and sugar beets, which are known to have high requirements for this nutrient.

Calcium Fertilizers. Calcium is not normally formulated as such into mixed fertilizers but rather is present as a component of the materials supplying other nutrients, particularly phosphorus. Normal superphosphate and triple superphosphate contain 18 to 21 and 12 to 14% calcium, respectively. Calcium concentration in calcium nitrate is about 19%. Synthetic chelates such as CaEDTA contain approximately 3 to 5% calcium, while some of the natural complexing substances used as micronutrient carriers have varying calcium levels usually ranging from 4 to 12%.

The primary sources of calcium are the liming materials such as calcite, dolomite, hydrated lime, precipitated lime, and blast furnace slag which are applied to neutralize soil acidity. In situations where calcium is required without the need for correcting soil acidity, gypsum is a favorite source of calcium. Chelated calcium can also be foliarly applied to crops lacking this element but not suffering from soil acidity.

Gypsum ($CaSO_4 \cdot 2H_2O$) has been used as a fertilizer for a long time. It was applied in the early Greek and Roman times and was also used extensively in Europe in the eighteenth century. Deposits are found at several locations in the United States and Canada and in many other areas of the world. Large amounts of by-product gypsum are on hand as a result of the manufacture of phosphoric acid.

Gypsum is a source of calcium for peanuts in the United States and is applied directly to the plant in early bloom. In several African countries, for example, Nigeria, Senegal, and Upper Volta, the gypsum contained in the superphosphate applied to this crop is as valuable for its sulfur content as it is for calcium. Large acreages of soil in these countries are severely sulfur deficient. Gypsum has little effect on soil reaction, hence may have some value on crops that demand an acid soil, yet need considerable calcium. It is widely used on the alkali soils of the West. The calcium replaces sodium on the exchange complex, and the sodium sulfate is carried out in the drainage water. This replacement serves to flocculate the soil and make it more permeable to water.

Magnesium

Form Utilized by Plants. Magnesium is absorbed by plants from the soil solution as the ion Mg^{2+}. Like calcium, plant magnesium needs of plants in most soils can be satisfied by the amounts transported by mass flow. Lesser amounts of magnesium may also reach plant roots by interception. The quantities of magnesium taken up by plants are usually less than calcium or potassium.

There are wide ranges in the Mg^{2+} concentration of soil solutions. Although levels of 5 to 50 ppm are considered typical in temperate region soils, values of from 120 to 2400 ppm have been recorded. Its concentration in nutrient solutions usually varies between about 30 to 100 ppm, with about 24 ppm being the approximate level needed for proper nutrition of most plants.

Source of Soil Magnesium. Magnesium constitutes 1.93% of the earth's crust. As with similar data for calcium and potassium, this figure represents the average of a wide range of values. The total magnesium content of soils is variable, ranging from only 0.1% in coarse, sandy soils in humid regions to perhaps 4% in fine-textured, arid, or semiarid soils formed from high magnesium parent materials.

Magnesium in the soil originates from the decomposition of rocks containing primary minerals such as biotite, dolomite, hornblende, olivene, and serpentine. It is also found in the secondary clay minerals chlorite, illite, montmorillonite, and vermiculite. Substantial amounts of epsomite ($MgSO_4 \cdot 7H_2O$) and perhaps also hexahydrite ($MgSO_4 \cdot 6H_2O$) and bloedite [$Na_2Mg(SO_4)_3 \cdot 4H_2O$] may occur in arid or semiarid soils. On decomposition of the primary minerals, the magnesium is released into surrounding waters. It may then be (1) lost in these percolating waters, (2) absorbed by living organisms, (3) adsorbed by surrounding clay particles, or (4) reprecipitated as a secondary mineral. This last phenomenon would be expected to take place most readily in an arid climate.

Magnesium in the secondary clay minerals is slowly available and can be released by leaching and by exhaustive cropping. Vermiculite has an abnormally high affinity for magnesium and it can be a significant source of this nutrient.

Magnesium deficiencies are not common. Conditions where soil supplies of it are likely to be insufficient for satisfactory crop growth include acid, sandy, highly leached soils with low cation exchange capacities; calcareous soils with inherently low magnesium levels; acid soils receiving high rates of liming materials low in magnesium; high rates of NH_4 or potassium fertilization; and cropping with plants having high magnesium requirements.

Excesses of magnesium can occur in certain situations where soils are formed on serpentine bedrock or are influenced by groundwaters high in this element. Normal calcium nutrition can be disrupted when exchangeable magnesium exceeds calcium on an equivalent weight basis.

Behavior of Magnesium in the Soil. The soil magnesium available to plants is in the exchangeable and/or water-soluble forms. The same general principles apply to its behavior as apply to calcium and potassium. The absorption of magnesium by plants depends on the amount present, soil pH, the degree of magnesium saturation, the nature of the other exchangeable ions, and the type of clay.

Like potassium, but to a lesser extent perhaps, magnesium may occur in soils in a somewhat slowly available form, in which it is in equilibrium with exchangeable magnesium. The formation of these relatively unavailable forms in acid soils would be favored by the presence of large quantities of soluble magnesium compounds and a 2:1 clay. Presumably there could be an entrapment of magnesium ions between the expanding and contracting sheets of the mineral. Such a reaction has been suggested, but the extent to which it takes place is not actually known.

Sumner and his co-workers at the University of Georgia have observed radical differences in exchangeable Mg^{2+} following the addition of a magnesium liming material. At first, Mg^{2+} levels increased but as the pH approached neutrality they decreased. This reduction was attributed to magnesium fixation through reactions with soluble silica or aluminum chlorite and to coprecipitation with $Al(OH)_3$.

The coarse-textured soils of the humid regions are those in which a deficiency of magnesium is generally manifested. These soils normally contain only small amounts of both total and exchangeable Mg^{2+}. Soils are probably deficient when they contain less than 25 to 50 ppm of exchangeable Mg^{2+}.

Exchangeable magnesium normally accounts for 4 to 20% of the cation exchange capacities of soils, but in soils derived from serpentine rock exchangeable magnesium ions can exceed those of calcium. The critical magnesium saturation of soils found necessary for optimum plant growth coincides closely with this range. Actual percent saturation values are dependent on the nature of the soil, the crop, and other factors. In most instances, magnesium saturation should not be less than 10%.

The impaired plant uptake of magnesium in many strongly acid soils is caused by high levels of exchangeable aluminum. Aluminum saturation percentages of 65 to 70% are often associated with magnesium nutritional problems. High aluminum in plants is thought by some to be a factor responsible for grass tetany, a nutritional disorder in cattle caused by lack of magnesium. The availability of soil magnesium can also be adversely affected by high H^+ activity in acid soils where exchangeable Al^{3+} is not a major cause of the acidity.

Plant deficiencies of magnesium can occur in soils with wide ratios of exchangeable Ca/Mg. This ratio should ideally not be greater than 7:1. On many humid-region, coarse-textured soils the continued use of high calcic liming materials may result in an unfavorable calcium:magnesium balance and the consequent development of magnesium deficiency symptoms on certain crops.

High levels of exchangeable potassium can interfere with magnesium uptake by crops. This K/Mg antagonism is of major concern in low-magnesium soils. Guidelines developed in the United Kingdom for satisfactory ratios of exchangeable potassium to magnesium are of interest. On a weight basis, the recommended K/Mg ratios are <5:1 for field crops, 3:1 for vegetables and sugar beets, and 2:1 for fruit and greenhouse crops. On an equivalent weight basis (K = 39.1 and Mg = 12.16) these ratios become about 1.5:1, 1.0:1.0, and 0.6:1, respectively.

There is also competition between NH_4^+ and Mg^{2+} which can lower the availability of the latter nutrient to crops. Its effects are expected to be greatest when high rates of ammoniacal fertilizers are applied to soils having low levels of exchangeable magnesium. This interaction is sometimes thought to contribute to grass tetany problems. The mechanism of this interaction is not fully understood. It probably involves both the H^+ released when NH_4^+ is absorbed by roots, as well as the direct effect of NH_4^+ itself.

Leaching Losses of Magnesium. Magnesium, like calcium, can be easily leached from soils. These losses are on the order of 5 to 60 lb of magnesium per acre per year. The amounts lost depend on the interaction of several factors, including the magnesium content of soil, rate of weathering, intensity of leaching, and uptake by plants. Leaching of magnesium is often a severe problem in sandy soils, particularly following the addition of fertilizer salts such as potassium chloride and to a lesser extent potassium sulfate (Table 8-15). Very little displacement of magnesium occurred when equivalent amounts of potassium were applied as either carbonate, bicarbonate, or phosphate. Apparently, magnesium in such coarse-textured soils is released by ion exchange and its removal in percolating moisture is enhanced by the presence of anions such as chlorides and sulfates.

Plant Factors. Plant species and varieties differ in their magnesium requirement. Pastures, corn, potatoes, oil palm, cotton, citrus, tobacco, and sugar beets are among the crops that are highly responsive to magnesium. Deficiencies of magnesium often become apparent on Newtown, McIntosh, and Spartan apple varieties in heavy crop years.

TABLE 8-15.   Percentage of Exchangeable Magnesium Displaced by Various Potassium Salts in the First Week of Leaching

	Soil Type and Initial Exchangeable Mg (meq %)		
	Taupo Sandy Silt (0.56 meq %)	Te Kopuru Sand (0.59 meq %)	Patea Sand (0.22 meq %)
	% displaced		
KCl (0.06 g)	12.1	6.8	31.4
K₂SO₄	4.3	6.1	30.8
KHCO₃	< 0.1	< 0.1	1.6
K₂CO₃	< 0.1	< 0.1	1.6
KH₂PO₄	< 0.1	< 0.1	2.6
KCl + super (0.14 g)	15.8	16.0	58.3
KHCO₃ + super	5.7	2.7	29.2
KHCO₃ + double super	< 0.1	—	2.1

Source: Hogg., *New Zealand J. Sci.*, **5**:64 (1962).

Corn inbreds and hybrids differ genetically in their uptake of and response to magnesium. Figure 8-17 illustrates the difference in magnesium efficiency of two corn inbreds, B57 and Oh40B. At magnesium concentrations below 1.2 mM (29 ppm) in the nutrient solution, inbred B57 grew markedly better and produced higher amounts of leaf and root material than did Oh40B. However, at higher levels of magnesium nutrition, the reverse occurred, with Oh40B outperforming B57.

Seasonal and environmental conditions will interact with plant varieties to produce magnesium deficiency. An example of the occurrence of magnesium deficiency in certain corn varieties in a year characterized by a cool, cloudy, and wet growing season is given in Table 8-16. Severe magnesium deficiencies were evident in all varieties except Pioneer 3965. Magnesium deficiency symptoms and reduced growth

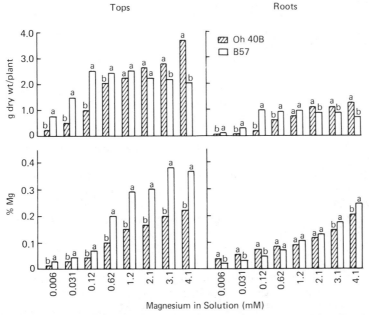

FIGURE 8-17.   Dry matter yields and magnesium concentrations of Oh40B and B57 tops and roots grown at 8 levels of magnesium. Standard errors of the mean were determined for values making up each bar. The values of bars with different letters differ significantly at the 0.05 level. [Clark, Soil Sci. Soc. Am. J., **38**:488 (1975).]

TABLE 8-16. Magnesium Deficiency Rating, Total Plant Yield, and Selected Nutrient Composition for Selected Corn Varieties

	Symptom Rating in 1976 *	Dry Matter Yield † (tons/ha) **			Nutrient Concentration (% in Dry Matter)					
					Magnesium		Nitrogen		Potassium	
		1975	1976†	1977	1976	1977	1976	1977	1976	1977
Idahybrid BX110	3	17.8	7.4	18.3	0.32	0.56	3.27	2.94	2.50	1.77
Pride R097	3	—	8.1	18.8	0.22	0.52	2.41	2.37	2.10	1.90
Pioneer 3965	6	19.7	18.6	17.5	0.32	0.37	3.13	2.66	2.77	1.90
Pioneer 3960	2	13.9	3.6	—	0.18	—	2.40	—	2.30	—
Buckerfields BX4331	3	16.4	8.5	—	0.18	—	2.69	—	2.93	—

* Ranking: 1, very severe deficiency; 6, no deficiency.
** Tons per hectare × 0.446 = tons per acre.
† May to August 1976—cool, wet, and cloudy.
Source: Broersma and van Ryswyk, in *Research 1977 Roundup.* Kamloops, B.C.: Range Research Station, Agriculture Canada, 1977.

in certain corn varieties have occurred in previous years at this location when growing conditions were similar to those which prevailed in 1976.

Magnesium Sources. In contrast to calcium, few of the carriers of primary nutrients contain large amounts of magnesium. Potassium magnesium sulfate, which was described in Chapter 7 as a potassium source, is a very important exception. Dolomite is commonly applied to acid soils low in magnesium and more will be said about it in Chapter 11 on liming.

When magnesium deficiencies exist in soils having a satisfactory pH, sources such as potassium magnesium sulfate, magnesium sulfate (epsom salt), and kieserite are used (Table 8-12). Other materials containing significant amounts of magnesium are magnesia (MgO, 55% Mg), magnesium nitrate [$Mg(NO_3)_2$, 16% Mg], magnesium silicate (basic slag, 3 to 4% Mg; serpentine, 26% Mg), magnesium chloride solution ($MgCl_2 \cdot 10H_2O$, 8 to 9% Mg), synthetic chelates (2 to 4% Mg), and natural organic complexing substances (4 to 9% Mg).

Potassium magnesium sulfate, magnesium sulfate (epsom salts and kieserite), and magnesia are the most widely used materials in dry fertilizer formulations. Magnesium silicate as basic slag or in ground up serpentine are other dry sources which are sometimes used for direct application.

Magnesium sulfate, magnesium chloride, magnesium nitrate, and synthetic and natural magnesium chelates are well suited for application in clear liquids and foliar sprays. Magnesium deficiency of citrus trees in California is frequently corrected by foliar applications of magnesium nitrate. In some tree-fruit growing areas, sprays of epsom salts are applied annually to maintain magnesium levels, and in seriously deficient orchards several annual applications are necessary.

Magnesium silicate, potassium magnesium sulfate, magnesium chelates, magnesium nitrate, magnesium sulfate, magnesium chloride, and magnesium oxide provide up to 3% MgO in most suspension formulas. In some instances greater amounts of these sources can be used to raise the MgO content to as high as 4 or 5%. Potassium magnesium sulfate is probably the most widely used magnesium additive in suspensions. A special suspension grade, 100% passing through a 20-mesh screen, of this material is available commercially.

Magnesium is a required component of fertilizers for certain crops grown where soil conditions are conducive to magnesium deficiencies. The fertilizer for flue-cured tobacco produced in some areas is a case in point and must contain a guaranteed percentage of magnesium expressed as MgO. Tobacco fertilizers in the province of Ontario must contain water-soluble magnesium percentages equal to 40% of the total nitrogen percentage. Significant amounts of magnesium are recommended in fertilizers formulated for potatoes in the state of Maine and in the adjoining provinces of New Brunswick and Prince Edward Island.

Grass Tetany. A problem of considerable importance in some areas is that of a low magnesium content of forage crops, particularly grass forages. Cattle consuming such forages may suffer from hypomagnesemia, more commonly known as grass tetany, which is an abnormally low level of blood magnesium. High rates of applied nitrogen fertilizers, especially ammoniacal forms, or high levels of potassium may depress the magnesium level in the plant tissue. It has been shown, for example, that the magnesium level of young corn plants is markedly reduced when NH_4^+ rather than NO_3^- is the source of applied nitrogen (Figure 12-13). As grass tetany often occurs in the spring, the nitrogen may still be in the NH_4^+ form, particularly if cool weather has prevailed.

Another factor of importance is that high protein content of ingested forages (and other feeds, too) will depress the absorption of magnesium by the animal. Aluminum is also thought to be involved.

Levels of soil magnesium may be increased through the use of dolomitic limestone, if liming is advisable, or through the use of magnesium-containing fertilizers, such as potassium-magnesium sulfate. Also, the inclusion of legumes in the forage program is advisable as these plants have a higher content of magnesium than do grasses. Cattle can also be fed a magnesium salt to help prevent grass tetany.

Although some workers maintain that hypomagnesemia is the result of excessive nitrogen or potassium fertilization, it seems more reasonable to class it as a magnesium deficiency and to treat it accordingly. An extensive review of the causes and management practices to help prevent and correct grass tetany is the subject of a Special Publication of the American Society of Agronomy.

Summary—Sulfur

1. In most humid region soils a large proportion of the sulfur occurs in organic combination. With soils containing appreciable amounts of hydrous oxides, some adsorbed sulfate sulfur will be found. In arid regions, of course, soluble sulfate salts accumulate in the soil profile.

2. Plants require concentrations of about 3 to 5 ppm of sulfate in the soil solution. Sulfate, which is the form absorbed by plants, reaches plant roots mainly by mass flow, especially when solution concentrations are 5 ppm or greater.

3. In sulfur-deficient soils the level of soluble sulfate is usually less than 5 to 10 ppm.

4. There are large seasonal and year-to-year fluctuations in sulfate concentrations in surface soils. This variability is due to the interaction of environmental and seasonal conditions on release of sulfate from organic forms, movement of sulfate in soil moisture, and sulfate uptake by plants.

5. Sulfate is subject to leaching but appreciable quantities can be retained by adsorption in soils high in iron and aluminum oxides and in 1:1 clays.

6. The mobilization and immobilization of soil sulfur depends on the supply of organic carbon, nitrogen, and phosphorus as well as the activity of soil microorganisms.

7. Factors affecting the growth of microorganisms will alter mineralization of sulfur. Some of the most important of these are the mineral content of organic matter, temperature, pH, presence or absence of plants, time and cultivation, and availability of food supply.

8. Although not all of the organic sulfur in soils has been characterized, three broad groups of sulfur compounds are recognized. These are (a) HI-reducible or sulfate esters and ethers, (b) carbon-bonded sulfur compounds such as cystine and methionine, and (c) residual or inert sulfur, which is very stable since it resists degradation by drastic chemical treatment.

9. Sulfides, polysulfides, and elemental sulfur are converted to sulfates by soil microorganisms. The speed with which this conversion takes place depends on the temperature, moisture, organisms present, and the soil pH. Of great importance, particularly with elemental sulfur, are the fineness of the material, temperature, and the population and activity of sulfur-oxidizing organisms.

10. The best known group of sulfur-oxidizing organisms are the autotrophic bacteria belonging to the genus *Thiobacillus.*

11. Additions of elemental sulfur and its compounds to soil will encourage greater populations of sulfur-oxidizing microorganisms and improved oxidizing power of the soil. Inoculation of soil will improve sulfur oxidation rates under controlled conditions in the laboratory or greenhouse.

12. Sulfatase enzymes have been detected in soil. These enzymes hydrolyze organic sulfate esters releasing inorganic sulfate. Activity of the sulfatases is dependent on the organic carbon content of soils and it decreases with soil depth.

13. The acidity produced by elemental sulfur and some of its compounds is beneficial in the treatment of calcareous and high-pH soils, particularly under irrigated conditions. Some of the benefits reported are reclaiming sodium- or boron-affected soils, increasing availability of phosphorus and micronutrients, reducing ammonia volatilization from irrigation water, increasing water penetration, controlling weeds and soilborne pathogens, and enhancing the establishment of range grasses.

14. Localized zones of temporary soil acidification such as in fertilizer bands can improve the availability of phosphorus and micronutrients.

15. There are substantial differences in the sulfur requirements of crops. The actual amount of sulfur fertilizer needed depends on the balance between all soil additions of this nutrient from precipitation, air, irrigation water, crop residues, fertilizers and other agricultural chemicals, and all soil losses through crop removal, leaching, and erosion.

16. In addition to differences in sulfur needs among crop species, there are varietal or cultivar differences as well.

17. The need for sulfur fertilizers on low-sulfur soils is closely related to the amounts of nitrogen being applied to crops.

18. The effectiveness of elemental sulfur fertilizers depends on several factors. These are particle size; rate, method, and time of application; sulfur oxidizing capacity of the soil; and environmental conditions.

19. When granular elemental sulfur fertilizers are applied shortly before a crop is to be grown on severely sulfur-deficient soils, some readily available sulfate should also be provided. The same guideline for sulfate supplementation should be followed when the sulfur-oxidizing ability of soil is low or uncertain.

Summary—Calcium and Magnesium

1. Calcium is absorbed by plants as the ion Ca^{2+} and magnesium is taken up as the ion, Mg^{2+}. Both are readily transported to root surfaces by the mechanism of mass flow. Root interception can also provide these two nutrients in some soils.

2. In spite of ample supplies of calcium in the soil solution, plant uptake of this nutrient is limited because it can be absorbed only by young root tips where the cell walls of the endodermis are unsuberized.

3. Calcium deficiencies causing reductions in crop growth are rare in agricultural soils. More common are disorders related to calcium shortage in fruit and storage organs caused by restrictions in its movement within plant structures.

4. Calcium and magnesium, which are somewhat similar in their behavior in soils, are held as exchangeable ions by electrostatic attraction around negatively charged soil colloids. Magnesium may under some soil conditions become fixed in the lattice structure of certain clay minerals and by other soil constituents. Calcium is not subject to fixation reactions.

5. A high degree of calcium saturation of soil exchange complexes is desirable because of the resultant favorable pH for the growth of most plants and for microbial activity. Also, when calcium is prevalent there will usually be low concentrations of such undesirable exchangeable cations as Al^{3+} in acidic soils and Na^+ in sodic soils.

6. Magnesium deficiencies are not widespread and the conditions where they are most likely to occur include acid, sandy, highly leached soils with low cation exchange capacities; calcareous soils with inherently low magnesium levels; acid soils receiving high rates of ammoniacal and potassium fertilizers; and cropping with plants of high-magnesium requirements.

7. There are important genetic differences in the magnesium requirement and efficiency of plants.

8. The occurrence of magnesium deficiency can be brought on by climatic factors (e.g., cool, wet, and cloudy weather). Lack of calcium can result from conditions such as low temperatures, drought, poor aeration, and so on, which impair growth of new roots and reduce the access of plants to calcium.

9. Major sources of magnesium are potassium magnesium sulfate, magnesium sulfate, magnesium oxide, magnesium nitrate, and magnesium chloride. Calcium is supplied by both dolomitic and calcitic limestone as well as gypsum. Some calcium is supplied incidentally in calcium nitrate and in the calcium phosphate fertilizers.

Questions—Sulfur

1. What are the forms of sulfur found in soils?

2. Which of the sulfur forms is of special importance in plant nutrition?

3. How does the plant available form of sulfur reach plant roots? What is the approximate concentration of this form needed in the soil solution?

4. What factors influence the release of inorganic sulfate from organic matter?

5. What effect has drying the soil on the availability of organic sulfur?

6. What is the importance of the C/N/P/S ratio to the availability of soil sulfur?
7. What are the soil conditions under which losses of sulfur by leaching would be expected?
8. What are the soil conditions under which little loss of sulfur by leaching would be expected?
9. Describe the soil and climatic conditions under which sulfur deficiencies in the field are most likely to occur.
10. Discuss sulfate adsorption by soils with emphasis on the factors affecting this phenomenon.
11. What are the three broad groups of organic sulfur compounds?
12. Which of the above groups seems to be an active reserve of mineralizable sulfur?
13. What are the factors affecting the oxidation of elemental sulfur in soils?
14. What soil microorganisms are responsible for oxidation of elemental sulfur? Which group is the best known and most thoroughly investigated?
15. Are there active populations of sulfur-oxidizing organisms in all soils?
16. Will the addition of elemental sulfur to soil influence the population of sulfur-oxidizing microorganisms?
17. Does inoculation with sulfur-oxidizing organisms improve sulfur-oxidation rates? What is the preferable method of inoculation?
18. What is sulfatase? What is its importance?
19. Is sulfur volatilized from soil, crop residues, or animal manures?
20. Can sulfur volatilize from intact plants and during oven drying of top and root samples?
21. What side effects in soil and changes in plants are possible because of sulfur volatilization?
22. Does acidity produced by the oxidation of elemental sulfur and its compounds have beneficial effects? What are they? Can there be detrimental effects produced by this acidity?
23. Will acidification in fertilizer bands improve nutrient availability? If so, under what soil conditions?
24. What factors influence the amount of sulfur fertilizer needed?
25. Are there differences in sulfur needs among crop species and varieties?
26. When using granular elemental sulfur fertilizer, are there conditions making it desirable to apply some fertilizer sulfate also? What are the conditions?

Questions—Calcium and Magnesium

1. In what forms are calcium and magnesium absorbed by plants? What is the primary transport mechanism of these forms to the root surface?
2. Are deficiencies of calcium and magnesium common? What conditions are conducive to shortages of these two nutrients in soils?
3. Is magnesium fixed in soils?
4. Why is it desirable to have a high degree of calcium saturation of the cation exchange capacity of soils?
5. Why is soil acidity usually associated with impaired uptake of calcium and magnesium?

6. Does calcium influence magnesium uptake by plants? Does the reverse take place?
7. What other fertilizer nutrients influence magnesium availability?
8. By what common agricultural practice is magnesium most easily added to soils?
9. Under what soil conditions might you prefer to use potassium magnesium sulfate rather than muriate and dolomite or muriate alone?
10. What are some incidental sources of plant-nutrient calcium?
11. How is a low-calcium soil condition usually corrected?
12. Are there genetic differences in the requirement of plants for magnesium and in their efficiency of using it?
13. Which crops have a high calcium requirement?
14. Can climate influence availability and uptake of calcium and magnesium?
15. What are some common sources of fertilizer magnesium?

Selected References

Adams, F., and J. B. Henderson, "Magnesium availability as affected by deficient and adequate levels of potassium and lime." *SSSA Proc.*, **26:**65 (1962).

Adams, F., and R. W. Pearson, "Crop response to lime in the southeastern United States and Puerto Rico," in R. W. Pearson and F. Adams, Eds., *Soil Acidity and Liming*, Chapter 4. Agronomy Series 12. Madison, Wis.: American Society of Agronomy, 1967.

Allen, B. L., "Mineralogy and soil taxonomy," in J. B. Dixon, S. B. Weed, J. A. Kittrick, M. H. Milford, and J. L. White, Eds., *Minerals in Soil Environments*, Chapter 22. Madison, Wis.: Soil Science Society of America, 1977.

Allen, V. G., D. L. Robinson, and F. G. Hembry, "Aluminum is linked to grass tetany." *Better Crops Plant Food*, **65:**14 (Spring 1981).

Anderson, A. J., "The significance of sulphur deficiency in Australian soils." *J. Australian Inst. Agr. Sci.*, **18:**135 (1952).

Anderson, D. W., E. A. Paul, and R. J. St. Arnaud, "Extraction and characterization of humus with reference to clay associated humus." *Can. J. Soil Sci.*, **54:**317 (1974).

Anderson, D. W., S. Saggar, J. R. Bettany, and J. W. B. Stewart, "Particle size fractions and their use in studies of soil organic matter: I. The nature and distribution of forms of carbon, nitrogen, and sulfur." *Soil Sci. Soc. Am. J.*, **45:**767 (1981).

Atkinson, W. T., M. H. Walker, and R. G. Weir, "The phosphorus and sulphur needs of pastures in New South Wales." *9th Int. Grassland Conf.*, Brazil (January 1965).

Banwart, W. L., and J. M. Bremner, "Formation of volatile sulfur compounds by microbial decomposition of sulfur-containing amino acids in soils." *Soil Biol. Biochem.*, **7:**395 (1975).

Banwart, W. L., and J. M. Bremner, "Volatilization of sulfur from unamended and sulfate-treated soils." *Soil Biol. Biochem.*, **8:**19 (1976).

Banwart, W. L., and J. M. Bremner, "Evolution of volatile sulfur compounds from soils treated with sulfur-containing organic materials." *Soil Biol. Biochem.*, **8:**439 (1976).

Bardsley, C. E., Jr., and H. V. Jordan, "Sulfur availability in seven southeastern soils as measured by growth and composition of white clover." *Agron. J.*, **49:**310 (1957).

Barrow, N. J., "A comparison of the mineralization of nitrogen and of sulfur from decomposing organic materials." *Australian J. Agr. Res.*, **11:**960 (1960).

Barrow, N. J., "Studies on the mineralization of sulfur from soil organic matter." *Australian J. Agr. Res.*, **12:**306 (1961).

Barrow, N. J., "Studies on the adsorption of sulfate by soils." *Soil Sci.*, **104:**342 (1967).

Beaton, J. D., "Fertilizers and their use in the decade ahead." *Proc. 20th Annu. Meet., Agr. Res. Inst.* (October 13, 1971).

Beaton, J. D., "Sulphur: one key to high yields in small and coarse grains." *Solutions*, **24**(6):16 (November–December 1980).

Beaton, J. D., and R. L. Fox, "Production, marketing, and use of sulfur products," in R. A. Olson et al., Eds., *Fertilizer Technology and Use*, 2nd ed., Chapter 11. Madison, Wis.: Soil Science Society of America, 1971.

Beaton, J. D., S. L. Tisdale, and J. Platou, "Crop responses to sulphur in North America." *Tech. Bull. 18.* Washington, D.C.: The Sulphur Institute, December 1971.

Bettany, J. R., S. Saggar, and J. W. B. Stewart, "Comparison of the amounts and forms of sulfur in organic matter fractions after 65 years of cultivation." *Soil Sci. Soc. Am. J.*, **44**:70 (1980).

Biederbeck, V. O., "Soil organic sulfur and fertility," in M. Schnitzer and S. U. Khan, Eds., *Soil Organic Matter*, Chapter 6. New York: Elsevier Scientific, 1978.

Bixby, D. W., and J. D. Beaton, "Sulfur-containing fertilizers, properties and applications." *Tech. Bull. 17.* Washington, D.C.: The Sulphur Institute, December 1970.

Blair, G. J., "The sulphur cycle." *J. Australian Inst. Agr. Sci.*, **37**:113 (1971).

Bledsoe, R. W., and R. E. Blaser, "The influence of sulfur on the yield and composition of clovers fertilized with different sources of phosphorus." *J. Am. Soc. Agron.*, **39**:146 (1947).

Bloomfield, C., "Effect of some phosphate fertilizers on the oxidation of elemental sulfur in soil." *Soil Sci.*, **103**:219 (1967).

Bohn, H., B. McNeal, and G. O'Conner, *Soil Chemistry*. New York: Wiley, 1979.

Bole, J. B., and U. J. Pittman, "Crop response to applied sulphur in southern Alberta and the uptake of subsoil sulphates." *Proc. Alberta Soil Sci. Workshop*, Lethbridge, Alberta (February 26–27, 1979).

Bornemisza, E., and R. Llanos, "Sulfate movement, adsorption, and desorption in three Costa Rican soils." *SSSA Proc.*, **31**:356 (1967).

Broersma, K., and A. L. van Ryswyk, "Magnesium deficiencies observed in corn at the Kamloops Research Station 1976 Variety Trial," in *Research 1977 Roundup*, pp. 55–57. Kamloops, B.C.: Agriculture Canada, 1977.

Burns, G. R., "Oxidation of sulphur in soils." *Tech. Bull. 13.* Washington, D.C.: The Sulphur Institute, June 1968.

Chang, M. L., and G. W. Thomas, "A suggested mechanism for sulfate adsorption by soils." *SSSA Proc.*, **27**:281 (1963).

Chao, T. T., M. E. Harward, and S. C. Fang, "Movement of S^{35} tagged sulfate through soil columns." *SSSA Proc.*, **26**:27 (1962).

Chao, T. T., M. E. Harward, and S. C. Fang, "Adsorption and desorption phenomena of sulfate ions in soils." *SSSA Proc.*, **26**:234 (1962).

Chao, T. T., M. E. Harward, and S. C. Fang, "Soil constituents and properties in the adsorption of sulfate ions." *Soil Sci.*, **94**:276 (1962).

Chao, T. T., M. E. Harward, and S. C. Fang, "Cationic effects on sulfate adsorption by soils." *SSSA Proc.*, **27**:35 (1963).

Chapman, A. L., "Effect of sulphur, sulphuric acid, and gypsum on the yield of rice on the Cununurra soils of the Ord Irrigation Area, Western Australia." *Aust. J. Exp. Agr. Anim. Husb.*, **20**:724 (1980).

Chapman, F. M., J. L. Mason, and J. E. Miltimore, "Response of alfalfa cultivars to sulfur." *Can. J. Plant Sci.*, **52**:493 (1972).

Chapman, H. D., "Calcium," in H. D. Chapman, Ed., *Diagnostic Criteria for Plants and Soils*, Chapter 6. Univ. of California, Division of Agricultural Sciences, 1966.

Clark, R. B., "Differential magnesium efficiency in corn inbreds: I. Dry-matter yields and mineral element composition." *Soil Sci. Soc. Am. J.*, **39**:488 (1975).

Clement, L., "Sulphur increases availability of phosphorus in calcareous soils." *Sulphur Agr.*, **2**:9 (1978).

"Control techniques for sulfur oxide air pollutants." *Natl. Air Pollut. Control Admin. Publ. AP-52* (January 1969).

Corey, R. B., E. E. Schulte, and R. A. Swanson, "A comparison of sulfur carriers." *Proc. 1974 Fert. Aglime Conf.*, Vol. 13, p. 42, Univ. of Wisconsin, Madison (1974).

Couto, W., D. J. Lathwell, and D. R. Bouldin, "Sulfate sorption by two Oxisols and an Alfisol of the tropics." *Soil Sci.*, **127**:108 (1979).

Doll, E. C., and R. E. Lucas, "Testing soils for potassium, calcium, and magnesium," in L. M. Walsh and J. D. Beaton, Eds., *Soil Testing and Plant Analysis*, rev. ed., Chapter 10. Madison, Wis.: Soil Science Society of America, 1973.

Doner, H. E., and W. C. Lynn, "Carbonate, halide, sulfate, and sulfide minerals," in J. B. Dixon, S. B. Weed, J. A. Kittrick, M. H. Mlford, and J. C. White, Eds., *Minerals in Soil Environments*, Chapter 3. Madison, Wis.: Soil Science Society of America, 1977.

During, C., "Recent research work: sulphur." *New Zealand J. Agr.*, **93**:549 (1956).

Embleton, T. W., "Magnesium," in H. D. Chapman, Ed., *Diagnostic Criteria for Plants and Soils*, Chapter 18. Univ. of California, Division of Agricultural Sciences, 1966.

Ensminger, L. E., "Sulfur in relation to soil fertility." *Alabama Agr. Exp. Sta. Bull. 312* (1958).

Ensminger, L. E., "Some factors affecting the adsorption of sulfate by Alabama soils." *SSSA Proc.*, **18**:259 (1964).

Evans, C. E., and E. J. Kamprath, "Lime response as related to percent Al saturation, solution Al, and organic matter content." *Soil Sci. Soc. Am. J.*, **34**:893 (1970).

Fox, R. L., A. D. Flowerday, F. W. Hosterman, H. F. Rhoades, and R. A. Olson, "Sulfur fertilizers for alfalfa production in Nebraska." *Nebraska Res. Bull. 214* (1964).

Freney, J. R., "The oxidation of cysteine to sulfate in soil." *Australian J. Biol. Sci.*, **13**:387 (1960).

Freney, J. R., "Some observations on the nature of organic sulfur compounds in soils." *Australian J. Agr. Res.*, **12**:424 (1961).

Freney, J. R., and K. Spencer, "Soil sulfate changes in the presence and absence of growing plants." *Australian J. Agr. Res.*, **11**:339 (1960).

Fried, M., "The absorption of sulfur dioxide by plants as shown by the use of radioactive sulfur." *SSSA Proc.*, **13**:135 (1949).

Greenwood, M., "Sulphur deficiency in groundnuts in northern Nigeria." *Trans. 5th Int. Congr. Soil Sci.*, **3**:245 (1955).

Grundon, N. J., "Release of volatile sulphur compounds by plants: development of technique for studying release by intact plants and oven-drying plant material." Ph.D. Thesis, Univ. of Queensland (1975).

Harron, W. R., and S. S. Malhi, "Release of sulphate from the oxidation of thiourea and ammonium polysulphide." *Can. J. Soil Sci.*, **58**:109 (1978).

Hassan, N., and R. A. Olson, "Influence of applied sulfur on availability of soil nutrients for corn (*Zea mays* L.) nutrition." *Soil Sci. Soc. Am. J.*, **30**:284 (1966).

Hatch, L. F., "What makes sulfur unique?" *Hydrocarbon Process.*, **51**:75 (1972).

Hilder, E. J., "Some aspects of sulfur as a nutrient for pastures in New England soils." *Australian J. Agr. Res.*, **5**:39 (1954).

Hogg, D. E., "Studies on soil magnesium: I. A laboratory investigation into the displacement of magnesium in soils." *New Zealand J. Sci.*, **5**:64 (1962).

Janzen, H. H., J. R. Bettany, and J. W. B. Stewart, "Sulphur oxidation and fertilizer sources." *Proc. Alberta Soil Sci. Workshop*, p. 229, Edmonton, Alberta (February 23–24 1982).

Jensen, J., "Some investigations of plant uptake of sulfur." *Soil Sci.*, **95**:63 (1963).

Jones, M. B., and J. E. Ruckman, "Effect of particle size on long-term availability of sulfur on annual-type grasslands." *Agron. J.*, **61**:936 (1969).

Jones, M. B., P. W. Lawler, and J. E. Ruckman, "Differences in annual clover responses to phosphorus and sulfur." *Agron. J.*, **62**:439 (1970).

Jones, M. B., W. E. Martin, and W. A. Williams, "Behavior of sulfate sulfur and elemental sulfur in three California soils in lysimeters." *SSSA Proc.*, **32**:535 (1968).

Jordan, H. V., "Sulfur as a plant nutrient in the southern United States." *USDA Tech. Bull. 1297* (1964).

Jordan, H. V., and C. E. Bardsley, "Response of crops to sulfur in southeastern soils." *SSSA Proc.*, **22:**254 (1958).

Jordan, H. V., and L. E. Ensminger, "The role of sulfur in soil fertility." *Adv. Agron.*, **10:**408 (1958).

Kamprath, E. J., "Possible benefits from sulfur in the atmosphere." *Combustion*, **44:**16 (October 1972).

Kamprath, E. J., and C. D. Foy, "Lime–fertilizer–plant interactions in acid soils," in R. A. Olson, T. J. Army, J. J. Hanway, and V. J. Kilmer, Eds., *Fertilizer Technology and Use*, 2nd ed., Chapter 5. Madison, Wis.: Soil Science Society of America, 1971.

Kamprath, E. J., W. L. Nelson, and J. W. Fitts, "The effect of pH, sulfate, and phosphate concentrations on the adsorption of sulfate by soils." *SSSA Proc.*, **20:**463 (1956).

Kamprath, E. J., W. L. Nelson, and J. W. Fitts, "Sulfur removed from soils by field crops." *Agron. J.*, **49:**289 (1957).

Kittams, H. A., and O. J. Attoe, "Availability of phosphorus in rock phosphate–sulfur fusions." *Agron. J.*, **57:**331 (1965).

Kowalenko, C. G., and L. E. Lowe, "Effects of added nitrogen on the net mineralization of soil sulphur from two soils during incubation." *Can. J. Soil Sci.*, **58:**99 (1978).

Krouse, H. R., and R. G. L. McCready, "Biogeochemical cycling of sulfur," in P. A. Trudinger and D. J. Swaine, Eds., *Studies in Environmental Science*, Vol. 3: *Biogeochemical Cycling of Mineral-Forming Elements*, Chapter 6.4. Amsterdam: Elsevier Scientific, 1979.

Laishley, E. J., "Implications of microbial attack on sulphur bonded construction materials." *Alberta Sulphur Research Ltd. Q. Bull.*, **15:**26 (1978).

Laishley, E. J., M. G. Tyler, and R. G. McCready, "Environmental assessment of soils in contact with sulfur-based construction material," in W. E. Krumbein, Ed., *Environmental Biogeochemistry and Geomicrobiology*, Vol. 2: *The Terrestrial Environment*, Chapter 56. Ann Arbor, Mich.: Ann Arbor Science Publishers, 1978.

Lawton, K., "Chemical composition of soils," in F. E. Bear, Ed., *Chemistry of the Soil*, Chapter 2. ACS Monogr. 126. New York: Reinhold, 1955.

Li, P., and A. C. Caldwell, "The oxidation of elemental sulfur in soil." *Soil Sci. Soc. Am. J.*, **30:**370 (1966).

Lindell, D. L., and R. C. Sornsen, "Sulfur and manganese release from soils treated with prilled sulfur and sulfur-bentonite." *Soil Sci. Soc. Am. J.*, **38:**368 (1974).

Liu, M., and G. W. Thomas, "Nature of sulfate retention by acid soils." *Nature (London)*, **192:**384 (1961).

Lobb, W. R., "Sulphur investigations in North Otago." *New Zealand J. Agr.*, **89:**434 (1954).

Loneragan, J. F., and K. Snowball, "Calcium requirements of plants." *Aust. J. Agr. Res.*, **20:**465 (1969).

Loneragan, J. F., I. C. Rowland, A. D. Robson, and K. Snowball, "The calcium nutrition of plants." *Proc. 11th Int. Grasslands Congr.*, Surfers Paradise, Australia, pp. 358–367 (1970).

Lowe, L. E., and W. A. DeLong, "Aspects of the sulphur status of three Quebec soils." *Can. J. Soil. Sci.*, **41:**141(1961).

McClung, A. C., L. M. DeFreitas, and W. L. Lott, "Analyses of several Brazilian soils in relation to plant responses to sulphur." *SSSA Proc.*, **23:**221 (1959).

McCready, R. G. L., and H. R. Krouse, "The potential use of stable isotopes in evaluating the effect of oxidized sulfur compounds on soil microorganisms." *Proc. Alberta Soil Sci. Workshop*, Lethbridge, Alberta (February 26–27, 1979).

McCready, R. G. L., and H. R. Krouse, "Sulfur isotope fractionation during the oxidation of elemental sulfur by *Thiobacilli* in a solonetzic soil." *Can. J. Soil Sci.*, **62:**105 (1982).

McGill, W. B., and C. V. Cole, "Comparative aspects of cycling of organic C, N, S, and P through soil organic matter." *Geoderma*, **26:**267 (1981).

McKell, C. M., and W. A. Williams, "A lysimeter study of sulfur fertilization of an annual-range soil." *J. Range Manag.*, **13:**113 (1960).

McLachlan, K. D., "The occurrence of sulfur deficiency on a soil of adequate phosphorus status." *Australian J. Agr. Res.*, **3:**125 (1952).

McLachlan, K. D., "Phosphorus, sulfur, and molybdenum deficiencies in soils from eastern Australia in relation to the nutrient supply and some characteristics of soil and climate." *Australian J. Agr. Res.*, **6:**673 (1955).

Martin, W. E., "Sulfur deficiency widespread." *Calif. Agr.*, **11:**10 (1958).

Mathers, A. C., "Effect of ferrous sulfate and sulfuric acid on grain sorghum yields." *Agron. J.*, **62:**555 (1970).

Mengel, K., and E. A. Kirkby, *Principles of Plant Nutrition*, 3rd ed. Bern: International Potash Institute, 1982.

Metson, A. J., "Sulphur in forage crops." *Tech. Bull. 20.* Washington, D.C.: The Sulphur Institute, January 1973.

Miyamoto, S., J. Ryan, and J. L. Stroehlein, "Effects of SO_2 on calcareous soils." *Sulphur Inst. J.*, **10**(2):14 (1974).

Mohammed, E. T. Y., J. Letey, and R. Branson, "Sulphur compounds in water treatment—effect on infiltration rate." *Sulphur Agr.*, **3:**7 (1979).

Mortvedt, J. J., and H. G. Cunningham, "Production, marketing and use of secondary and micronutrient fertilizers," in R. A. Olson, T. J. Army, J. J. Hanway, and V. J. Kilmer, Eds., *Fertilizer Technology and Use,* 2nd ed. Madison, Wis.: Soil Science Society of America, 1971.

Muth, O. H., and J. E. Oldfield, Eds., *Symposium: Sulfur in Nutrition.* Westport, Conn.: AVI, 1970.

Nearpass, D. C., M. Fried, and V. J. Kilmer, "Greenhouse measurement of available sulfur using radioactive sulfur." *SSSA Proc.*, **25:**287 (1961).

Nemeth, K., K. Mengel, and H. Grimme, "The concentration of K, Ca, and Mg in the saturation extract in relation to exchangeable K, Ca, and Mg." *Soil Sci.*, **109:**179 (1970).

Newton, J. D., C. F. Bentley, J. A. Toogood, and J. A. Robertson, "Grey wooded soils and their management." *Univ. Alberta Bull. 21*, 5th ed. rev. (March 1959).

Nicolson, A. J., "Soil sulfur balance studies in the presence and absence of growing plants." *Soil Sci.*, **109:**345 (1970).

Nor, Y. M., and M. A. Tabatabai, "Oxidation of elemental sulfur in soils." *Soil Sci. Soc. Am. J.*, **41:**736 (1977).

Parr, J. F., and P. M. Giordano, "Agronomic effectiveness of anhydrous ammonia–sulfur solutions: 2." *Soil Sci.*, **106:**448 (1968).

Pepper, I. L., and R. H. Miller, "Comparison of the oxidation of thiosulfate and elemental sulfur by two heterotrophic bacteria and *Thiobacillus thiooxidans.*" *Soil Sci.*, **126:**9 (1978).

Ralph, B. J., "Oxidative reactions in the sulfur cycle," in P. A. Trudinger and D. J. Swaine, Eds., *Studies in Environmental Science*, Vol. 3: *Biogeochemical Cycling of Mineral-Forming Elements,* Chapter 6.3. Amsterdam: Elsevier Scientific, 1979.

Rasmussen, P. E., "Update-nitrogen and sulfur research on wheat." *Proc. Pendleton–Walla Walla Fert. Dealers Conf.*, Walla Walla, Wash. (January 6, 1976).

Reynolds, D. M., E. J. Laishley, and J. W. Costerton, "Physiological and ultrastructural characterization of a new acidophillic *Thiobacillus* species (*T. kabobis*)." *Can. J. Microbial.*, **27:**151 (1981).

Roberts, S., and F. E. Koehler, "Extractable and plant-available sulfur in representative soils of Washington." *Soil Sci.*, **106:**53 (1968).

Scholfield, P. E., P. E. H. Gregg, and J. K. Syers, " 'Biosuper' as a phosphate fertiliser: a glasshouse evaluation." *New Zealand J. Exp. Agr.*, **9:**63 (1981).

Schulte, E. E., K. A. Kelling, and C. R. Simson, "Too much magnesium in soil?" *Solutions,* **24**(6):106 (1980).

Shear, C. B., Ed., "International symposium on calcium nutrition of economic crops." *Commun. Soil Sci. Plant Anal.*, **10:**1 (1979).

Spencer, K., "Sulphur requirements of plants," in K. E. McLachlan, Ed., *Sulphur in Australasian Agriculture.* Sydney, Australia: Sydney University Press, 1975.

Stewart, B. A., L. K. Porter, and F. G. Viets, Jr., "Effect of sulfur content of straws on rates of decomposition and plant growth." *SSSA Proc.*, **30:**355 (1966).

Stewart, J. W. B., and J. R. Bettany, "Dynamics of soil organic phosphorus and sulfur." *Publ. R290*, Saskatchewan Institute of Pedology, Saskatoon (1982).

Stewart, J. W. B., and J. R. Bettany, "Dynamics of organic sulphur." *Proc. Alberta Soil Sci. Workshop*, p. 184, Edmonton, Alberta (February 23–24, 1982).

Stroehlein, J. L., S. Miyamoto, and J. Ryan, "Sulfuric acid for improving irrigation waters and reclaiming sodic soils." *Agr. Eng. Soil Sci. Bull. 78-5,* Univ. of Arizona, Tucson (1978).

Stromberg, L. K., and S. L. Tisdale, "Treating irrigated arid-land soils with acid-forming sulphur compounds." *Tech. Bull. 24.* Washington, D.C.: The Sulphur Institute, March 1979.

Sumner, M. E., and F. C. Boswell, "Alleviating nutrient stress," in G. F. Arkin and H. M. Taylor, Eds., *Modifying the Root Environment to Reduce Crop Stress,* Chapter 4. ASAE Monogr. 4. St. Joseph, Mo.: American Society of Agricultural Engineers, 1981.

Sumner, M. E., P. M. W. Farina, and V. J. Hurst, "Magnesium fixation—a possible cause of negative yield responses to lime applications." *Commun. Soil Sci. Plant Anal.*, **9:**995 (1978).

Tabatabai, M. A., and J. M. Bremner, "Factors affecting soil arylsulfatase activity." *Soil Sci. Soc. Am. J.*, **34:**427 (1970).

Tabatabai, M. A., and J. M. Bremner, "Distribution of total and available sulfur in selected soils and soil profiles." *Agron. J.*, **64:**40 (1972).

Tabatabai, M. A., and A. A. Al-Khafaji, "Comparison of nitrogen and sulfur mineralization in soils." *Soil Sci. Soc. Am. J.*, **44:**1000 (1980).

Tisdale, S. L., "The use of sulphur compounds in irrigated, aridland agriculture." *Sulphur Inst. J.*, **6:**2 (1970).

Tisdale, S. L., and B. R. Bertramson, "Elemental sulfur and its relationship to manganese availability." *SSSA Proc.*, **14:**11 (1949).

Trudinger, P. A., "The biological sulfur cycle," in P. A. Trudinger and D. J. Swaine, Eds., *Studies in Environmental Science*, Vol. 3: *Biogeochemical Cycling of Mineral-Forming Elements*, Chapter 6.1. Amsterdam: Elsevier Scientific, 1979.

Venema, K. C. W., "Some notes regarding the function of the sulfate anion in the metabolism of oil producing plants, especially oil palms. Part I." *Potash Trop. Agr.*, **5** (July 1962).

Wainwright, M., "Microbial sulphur oxidation in soil." *Sci. Prog.*, **65:**459 (1978).

Wainwright, M., and K. Killham, "Sulphur oxidation by *Fusarium solani.*" *Soil Biol. Biochem.*, **12:**555 (1980).

Walker, T. W., "Sulfur responses on pastures in Australia and New Zealand." *Soils Fert.*, **18:**185 (1955).

Walker, T. W., "The use of sulphur as a fertilizer." *Int. Conf. Sulfur Agr.*, Palermo (1964).

Walker, T. W., and A. F. R. Adams, "Competition for sulfur in a grass-clover association." *Plant Soil,* **9:**353 (1958).

Walsh, L. M., and J. D. Beaton, *Soil Testing and Plant Analysis*, rev. ed. Madison, Wis.: Soil Science Society of America, 1973.

Weir, R. G., "The oxidation of elemental sulphur and sulphides in soils," in K. E. McLachlan, Ed., *Sulphur in Australasian Agriculture.* Sydney: Sydney University Press, 1975.

White, J. G., "Mineralization of nitrogen and sulfur in sulfur-deficient soils." *New Zealand J. Agr. Res.*, **2:**225 (1959).

Wieringa, K. T., "Solid media with elemental sulphur for detection of sulphur-oxidizing microbes." *Antonie van Leeuwenhoek J. Microbiol. Serol.*, **32:**183 (1966).

Williams, C. H., "The chemical nature of sulphur compounds in soil," in K. D. McLachlan, Ed., *Sulphur in Australasian Agriculture.* Sydney: Sydney University Press, 1975.

Williams, C. H., and A. Steinbergs, "Sulphur and phosphorus in some eastern Australian soils." *Australian J. Agr. Res.*, **9:**483 (1958).

Williams, C. H., E. G. Williams, and N. M. Scott, "Carbon, nitrogen, sulfur, and phosphorus in some Scottish soils." *J. Soil Sci.*, **11:**334 (1960).

Woodruff, C. M., "The energies of replacement of calcium by potassium in soils." *SSSA Proc.*, **19:**167 (1955).

York, E. T., Jr., R. Bradfield, and M. Peech, "Calcium–potassium interactions in soils and plants: I. Lime-induced potassium fixation in Mardin silt loam." *Soil Sci.*, **76:**379 (1953).

Chapter 9

Micronutrients and Other Beneficial Elements in Soils and Fertilizers

$\mathbf{T}$HE nature and behavior in soil of the plant-nutrient elements boron, cobalt, copper, iron, manganese, molybdenum, and zinc, traditionally classed as micronutrients, will be discussed in this chapter. Similar information will be reviewed for sodium, chlorine, and silicon which are less well known but nevertheless are beneficial for growth of some plants. Information on selenium is also provided since grazing animals can obtain much of their requirements for this element through the soil–plant system.

Micronutrients were first recognized as a limiting factor in crop production in the United States in Florida during the 1920s. Since then deficiencies of one or more micronutrients have been reported in most of the states in the United States and in nearly all of the Canadian provinces.

Materials used to supply micronutrients and the other beneficial elements, as well as the customary methods of applying them, will be considered in the following discussions.

General Relationships of Micronutrient Cations

The various steps and processes involved between the weathering of soil minerals or decomposition of organic residues and plant uptake of micronutrient cations are represented in Figure 9-1. This diagram implies that soluble complexing and chelating agents, such as those known to be secreted from roots, have a prominent role in increasing the transport of heavy metal micronutrients to roots. Knowledge is limited about the stage in the uptake process where breakdown of these complexes results in release of the protected micronutrient. This figure also indicates that mass flow and diffusion are the main processes responsible for movement of micronutrient cations from the soil solution to plant roots.

Boron

Boron is the only nonmetal among the micronutrient elements. It has a constant valence of $+3$ and has a very small ionic radius. Boron occurs in low concentrations in the earth's crust and in most igneous rocks and soils. Its concentration in the

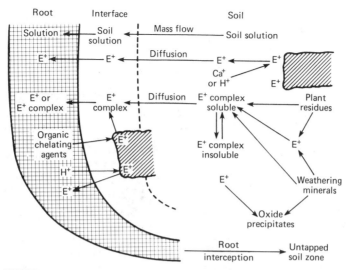

FIGURE 9-1. Principal processes operative in the transfer of trace elements from the soil to the plant root. The general symbol E^+ is used to represent any trace-element cation that can participate in the particular process illustrated. A complementary diagram could be produced to illustrate anionic processes. [Mitchell, *Geol. Soc. Am. Bull.*, **83**:1069 (1972). Reprinted with permission of the Macaulay Institute for Soil Research, Aberdeen, Scotland.]

earth's crust is 10 ppm and it ranges from 5 to 15 ppm in igneous rocks. Among the usual sedimentary rocks, shales have the highest boron concentrations of up to 100 ppm, with boron present mainly in the clay minerals.

The total concentration of boron in most soils varies between 2 and 200 ppm and it more frequently ranges from 7 to 80 ppm. Less than 5% of the total soil boron is generally available to plants. Extraction with hot water is the most reliable evaluation of soil boron availability.

Tourmaline, a complex borosilicate, is the main boron-containing mineral found in most soils. It is quite insoluble and resistant to weathering. Consequently, release of boron from this mineral is quite slow and the increasing frequency of boron deficiencies suggests that it is incapable of supplying plant requirements under prolonged heavy cropping. Boron reserves are different in the soils of arid climates, where borates of alkali and alkaline earth elements predominate.

There are four regions in the United States in which the soils seem to be particularly low in boron: the Atlantic coastal plain, the Pacific coastal area, the Pacific Northwest, and northern Michigan, Wisconsin, and Minnesota. Areas of boron deficiency in the United States are shown in Figure 9-2. These deficient soil areas extend northward across the International Boundary into Canada.

Large amounts of boron can be growth retarding and toxic. Fortunately, boron does not normally occur in toxic quantities in most arable soils unless it has been added in excessive amounts in commercial fertilizers. In arid regions, however, toxic concentrations of boron may occur naturally or they may develop because of irrigation with waters containing high amounts of boron.

Forms of Soil Boron. Boron exists in four major forms in soil: in rocks and minerals, adsorbed on surfaces of clays and hydrous iron and aluminim oxides, combined with organic matter, and as free nonionized boric acid (H_3BO_3) and $B(OH)_4{}^-$ in the soil solution. The relationships among these forms are demon-

FIGURE 9-2. Areas of boron deficiency in the United States. (Turner, *Boron in Agriculture*. New York: U.S. Borax, 1976.)

strated in Figure 9-3. Distribution of boron between the solid and solution phases is very important because of the narrow range in solution concentration, resulting in deficiency and toxicity.

Soil Solution Boron. Undissociated boric acid (H_3BO_3) is the predominant species expected in soil solution at pH values ranging from less than 5 to 9. The other soluble boron species $H_2BO_3^-$ is of little importance at pH values below 8, and only at a pH near 9.0 does it equal boric acid.

Although the process of boron uptake by plants is not fully comprehended, it appears that undissociated boric acid is the most effective form. Boron can be transported from the soil solution to absorbing plant roots by both mass flow and diffusion. The former mechanism is apparently the most important factor in the uptake of boron by crops. Passage of boron into the plant is probably the result of both passive and active processes.

Adsorbed Boron. Adsorbed boron constitutes a labile pool or reserve that maintains boron concentration in the soil solution, and it helps to conserve soil supplies of boron by reducing leaching losses. It is a major form of boron in alkaline, high-boron soils.

Adsorbed boron is basically constituted of complexes of either molecular boric acid or the hydrated borate ion $B(OH)_4^-$ with various minerals. Four main adsorption sites for boron have been identified: (1) broken Si—O and Al—O bonds at the edges of aluminosilicate minerals, (2) amorphous hydroxide structures, (3) magnesium hydroxide clusters, and (4) iron and aluminum oxy and hydroxy compounds.

Adsorption of the $B(OH)_4^-$ anion on the edges of broken bonds is facilitated by breakage and fracturing of the edges of clay minerals. Boron adsorption by allophane in weathered soils is an example of adsorption on an amorphous mineral. Retention of boron by magnesium hydroxide is associated with arid-zone soils

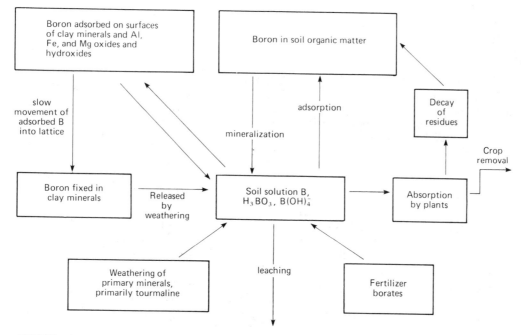

FIGURE 9-3. Boron cycle. [Adapted from Berger by Mengel, *U.S. Borax Plant Food Borate Meet.*, Lafayette, Ind. (1980).]

where coatings of this compound form on ferromagnesian minerals and on micaeous layer silicates.

Increasing pH and the presence of aluminum compounds rather than analogous iron substances both favor hydroxylborate reactions. Other factors enhancing boron adsorption are fineness of soil texture and organic-matter content.

Adsorption of borate by soils may be described within certain concentration limits by adsorption equations. For example, Okazaki and Chao found that when Hawaiian soils were equilibrated with solutions containing from 0 to 50 ppm boron, adsorption of boron was characterized by the Langmuir equation.

Organically Complexed Boron. A significant quantity of plant available boron in soils is held in the organic-matter fraction. Soils high in organic matter are also frequently high in boron. The exact form of the boron–organic matter combinations is not known, but diol-type compounds are believed to form as a consequence of boron reacting with α-hydroxyaliphatic acids and ortho-dihydroxy derivatives of aromatic compounds.

$$=\!C\!-\!O \diagdown \atop =\!C\!-\!O \diagup B\!-\!OH$$

This ring compound can be transformed into a boron-containing organic acid such as the following:

$$H^+ \left| \begin{array}{c} =\!C\!-\!O \diagdown \quad O\!-\!C\!= \\ =\!C\!-\!O \diagup B \diagdown O\!-\!C\!= \end{array} \right|^-$$

Diol-type structures generated during the microbial breakdown of soil polysaccharides can readily react with boron to form similar organic complexes.

Boron in Minerals. Boron can substitute for Al^{3+} and/or Si^{4+} ions in silicate minerals. Following its adsorption on clay surfaces, boron will slowly diffuse into interlayer positions.

Factors Affecting Availability and Movement of Boron. There are several factors that influence the availability and movement of soil boron. These are soil texture, amount and type of clay, pH and liming, organic matter, interrelationships with other elements, and environmental conditions.

Soil Texture. It has generally been found that coarse-textured, well-drained, sandy soils are low in boron and crops with a high requirement such as alfalfa, respond to applications of 3 to 5 lb/A of boron. Sandy soils with fine-textured subsoils generally do not respond as much to the addition of boron as do those with coarse-textured subsoils. Leaching studies have shown that boron added to soils remains soluble and may move out of the upper horizons. The removal of added boron depends on the quantity of water added and the texture of the soil. The finer-textured soils retain the added boron for longer periods than the coarse-textured soils.

In low organic matter, sandy soils up to 85% of the boron added as fertilizer can be leached by only 5 in. of water. Undissociated boric acid and borate anions in the soil solution are relatively free to move in soil water and can be quickly leached from soil. Movement is less rapid in heavy-textured soils and in ones higher in organic matter because of increased boron fixation in the clay and organic matter fractions.

The fact that clays retain boron more effectively than sands does not necessarily imply that plants will absorb this element from clays in greater quantities than from sands when equal concentrations of water-soluble boron are present. In fact, plants will take up much larger quantities of boron from sandy soils than they will from fine-textured soils at equal concentrations of water-soluble soil boron.

Amount and Type of Clay. Available boron concentrations are usually greater in heavy soils than in coarse-textured soils. On a weight basis, micaceous-type clays such as illite have the greatest boron-adsorption capacities. Montmorillonite follows illite in ability to retain boron, while kaolinite has the lowest boron-adsorption capacity.

Soil pH and Liming. Soil pH and liming strongly influence the availability of boron to plants. This element generally becomes less available to plants with increasing soil pH. There is often a dramatic drop in boron availability and plant uptake at pH levels greater than 6.3 to 6.5. This is illustrated by the results in Table 9-1 for five cuttings of tall fescue.

Liming strongly acid soils frequently induces at least a temporary boron deficiency in susceptible plants. The severity of the deficiency depends on several factors, including the moisture status of the soil, the nature of the crop, and the period of time elapsed following liming.

Freshly precipitated $Al(OH)_3$ adsorbs large quantities of boron, much more than soil, even though aging of $Al(OH)_3$ greatly reduces its boron retention capacity. The marked reduction in boron availability following liming is believed to be caused mainly by boron adsorption on $Al(OH)_3$, which forms from exchangeable Al^{3+} and hydroxyl-aluminum cations. The surface area of $Al(OH)_3$ available for adsorption is an important consideration. Adsorption of boron by iron and aluminum hydroxides is very pH dependent, with maximum adsorption at pH 7 for $Al(OH)_3$ and pH 8 to 9 for $Fe(OH)_3$.

Moderate liming can be used as a corrective treatment on soils containing excesses of boron by depressing availability and plant uptake of this nutrient. It

**TABLE 9-1.   Boron Uptake and Percentage Recovery of Added
Boron by Five Harvests of Tall Fescue at Five Soil pH Levels**

| | Amount of Added B (mg/pot) | | | | | | |
| | *0* | *4.5* | *8.9* | *17.8* | *4.5* | *8.9* | *17.8* |
Soil pH	B Uptake (mg/pot)				% Recovery		
4.7	0.47	1.85	4.15	9.40	30.7	41.3	50.2
5.3	0.45	1.92	4.45	9.51	32.7	44.9	50.9
5.8	0.44	1.98	4.14	9.10	34.2	41.6	48.7
6.3	0.45	1.98	4.03	9.37	34.0	40.2	50.1
7.4	0.22	0.80	1.40	3.76	12.9	13.3	19.9
LSD 0.05	0.030	0.38	0.25	1.63			
LSD 0.01	0.043	0.55	0.36	2.37			

Source: Peterson and Newman, *Soil Sci. Soc. Am. J.,* **40:**280 (1976).

should be noted that heavy liming does not always lead to greater boron adsorption and reduced plant uptake. Higher pH resulting from liming of soils high in organic matter may encourage organic matter decomposition and release of boron.

Organic Matter. In addition to the effect that organic matter can have on complexing boron, it is one of the main sources of boron in acid soils. The greater availability of boron in surface soils compared with subsurface soils is undoubtedly related to the greater quantities of organic matter present in the former. Applications of organic materials to soils can raise substantially the concentration of boron in plants and even cause phytotoxicity.

Interrelationships with Other Elements. Uptake of boron is dependent on the concentration of other ions in the growth medium.

Calcium. There appears to be a functional relationship between calcium and boron in plants. Low tolerance to boron occurs when plants have a low calcium supply. When calcium nutrition is high, there is a greater requirement for boron. The occurrence of free calcium ions in alkaline and recently overlimed soils will restrict boron availability. Free calcium is reported to protect crops from boron excesses.

The Ca/B ratio in leaf tissues has been used to assess the boron status of crops. To avoid misleading results, it is essential that neither of these nutrients is deficient or toxic. Boron deficiency is indicated by ratios greater than 1370:1 in barley sampled at the boot stage; 3300:1 in rutabaga leaf tissue; and 1200 to 1500:1 in tobacco.

Potassium. A balance between potassium and boron seems to exist in plants. At low levels of boron nutrition, increased rates of applied potassium may accentuate boron-deficiency symptoms. Boron deficiency in alfalfa has been aggravated by potassium applications. The effect of potassium may be indirect because of its influence on calcium absorption.

Increased potassium rates have been reported to accentuate boron toxicity at high levels of boron supply. Boron uptake by tomato and corn was enhanced by high potassium in combination with high boron.

Nitrogen. Liberal nitrogen applications have been useful in controlling excess boron in citrus and other crops.

Soil Moisture. Boron deficiency is often associated with dry weather and low-soil-moisture conditions. It is generally accepted that boron availability declines under dry soil conditions. This behavior seems to be related to restricted release of boron from organic complexes and to impaired ability of plants to extract boron from soil due to lack of moisture in the root zone. Although boron levels in soil may be high, low soil moisture impairs transport of boron to absorbing root surfaces.

Plant Factors. Plants differ in their boron requirements. The crops most sensitive to lack of boron include sugar beet, mangolds, and celery. Others with a high boron requirement include apple, asparagus, broccoli, brussels sprouts, cabbage, cauliflower, clovers, kale, alfalfa, radish, rapeseed, red beet, spinach, sunflower, and turnip. It is apparent that members of the Leguminosae and Brassicaceae are prominent in this list of crops with a high sensitivity to insufficient boron. By way of contrast, the boron needs of the Graminiae (cereals, grasses) are low.

Genetic variability contributes to differences in boron uptake and leaf concentration. Investigations with tomatoes revealed that susceptibility to boron deficiency is controlled by a single recessive gene. Tomato variety T3238 is boron-inefficient, while the variety Rutgers is boron-efficient. Corn hybrids have similarly exhibited genetic variability related to boron uptake and leaf concentration.

Because of the narrow range between sufficient and toxic levels of available soil boron, the sensitivity of crops to excess boron is of interest. Some of the most sensitive crops to boron toxicity are peach, grapes, kidney beans, soybeans, and figs. Semitolerant plants include barley, peas, corn, potato, alfalfa, tobacco, and tomato. The most tolerant crops are turnips, sugar beet, and cotton.

Boron Fertilizers. Boron is one of the most widely applied micronutrients. Sodium tetraborate is the most commonly used boron source. Various levels of hydration of the tetraborate salts result in boron concentrations ranging from 11 to 20%. Table 9-2 lists the principal boron fertilizers with their chemical formulas and percentages of boron.

Borates. Fertilizer borates, applied either directly to the soil, to the plants, or blended with commercial fertilizers, are economical for correction of boron deficiencies. Fertilizer Borate-48 Fine Mesh is used mainly in mixed fertilizers and suspensions, Fertilizer Borate-68 Coarse is well suited for blending with mixed fertilizers, and Fertilizer Borate-68 Granular is designed for direct application and addition to blended and granular mixed fertilizers.

Solubor is a highly concentrated, completely soluble source of boron which can be applied as a spray or dust directly to foliage of fruit trees, vegetables, and other crops. It is also used in liquid and suspension fertilizer formulations. Solubor is preferred to borax because it dissolves more readily and causes minimum changes in crystallization temperatures.

Borosilicate Glasses. Several years ago it was found that the salts of boron, and other trace elements as well, could be fused with glass, shattered and applied to soils, into which the salt is slowly released as the glass dissolves. These materials are referred to as *frits* and they have been successfully employed to extend the plant availability of highly soluble and reactive microelement salts. Borosilicate glass is one example.

The boron content of these frits varies, but is generally on the order of 2 to 11%. As would be anticipated in a low water-soluble material of this type, its availability is influenced by the particle-size distribution of the finished product. The more coarsely divided materials will be less effective per unit mass of material (or per

TABLE 9-2. Principal Boron Fertilizers and Their Formulas and Boron Percentages

Source	Formula	Percent B (approx.)
Borax	$Na_2B_4O_7 \cdot 10H_2O$	11
Boric acid	H_3BO_3	17
Colemanite (Portabor)	$Ca_2B_6O_{11} \cdot 5H_2O$	10–16
Sodium pentaborate	$Na_2B_{10}O_{16} \cdot 10H_2O$	18
Sodium tetraborate		
Fertilizer Borate-48 (Agribor, Tronabor)	$Na_2B_4O_7 \cdot 5H_2O$	14–15
Fertilizer Borate-68	$Na_2B_4O_7$	21
Solubor	$Na_2B_4O_7 \cdot 5H_2O +$ $Na_2B_{10}O_{16} \cdot 10H_2O$	20–21
Ulexite	$NaCaB_5O_9 \cdot 8H_2O$	9–10
Boron Frits	Complex borosilicates	
FTE 11		11
FTE 115		11
FTE 171		2
FTE 181		2
176 E		2
176 F		6
501		6

Sources: Fleming, in B. E. Davies, Ed., *Applied Soil Trace Elements,* p. 171. New York: Wiley, 1980; Gupta, *Adv. Agron.,* **31**:273 (1979); and Turner, *Boron in Agriculture.* New York: U.S. Borax, 1970.

pound of contained boron, assuming the same percentage of boron in coarse and fine materials) than the more finely divided materials. This, together with their availability compared with that of borax, is shown in Figure 9-4. The lower boron content of the alfalfa grown on soils treated with frits is apparent from these curves. The impact of particle size is also obvious. The use of borosilicate glasses offers certain advantages in terms of extended availability not offered by the more readily soluble borax. These advantages are most obvious on sandy soils and under conditions of high rainfall. It is likely that under such soil conditions the use of borosilicate glasses will increase.

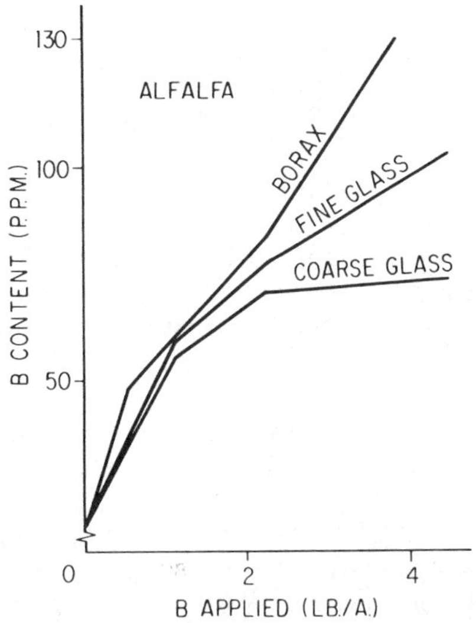

FIGURE 9-4. Effect of borax and two particle sizes of a boron glass on the uptake of boron by alfalfa. [Holden et al., *J. Agr. Food Chem.*, **10**:188 (1962). Copyright 1962 by the American Chemical Society and reproduced by permission of the copyright owner.]

Colemanite. This naturally occurring calcium borate is often used on sandy soils because it is less soluble and less subject to leaching than the sodium borates.

Ulexite. Commercial supplies of this boron source became available a few years ago. Release of boron is slower from it than from the more soluble sodium borates.

Application Methods and Rates of Boron Fertilization. The most common methods of boron application are broadcast, banded, or applied as a foliar spray or dust. In the first two methods the boron fertilizer source is usually mixed with N-P-K-S products and applied to soil. Finely divided boron salts can also be coated on dry fertilizer materials.

Boron fertilizers should be applied uniformly to soil because of the narrow range between deficiency and toxicity. Segregation of granular boron sources in dry fertilizer blends must be avoided. Application of boron with fluid fertilizers eliminates the segregation problem.

Foliar application of boron is practiced for perennial tree fruit crops, often in combination with pesticides other than those formulated in oils and emulsions. Boron may also be included in summer sprays of zinc chelate, magnesium, manganese, and urea. Foliar applications of boron with insecticides are also used in cotton.

Rates of boron fertilization depend on plant species, soil cultural practices, rainfall, liming, and soil organic matter as well as other factors. Application rates of 0.5 to 3 lb of boron per acre are generally recommended.

Method of application has an important bearing on the amounts of boron recommended. Mortvedt of the Tennessee Valley Authority pointed out, for example, that the boron rate for vegetable crops is 0.4 to 2.7 lb/A if broadcast, 0.4 to 0.9 lb/A if banded, and 0.09 to 0.4 lb/A if foliar applied. Rutabaga requires more boron than other vegetables to control deficiency and the usual rate of 3.6 lb/A broadcast can be reduced by half when boron is banded or applied to foliage.

Cobalt

It is well established that cobalt is essential in symbiotic N_2 fixation, but there is still some question as to whether or not it is required for higher plants. The importance of cobalt is most evident when sickness develops in ruminant animals receiving inadequate amounts of dietary cobalt. Rumen microorganisms require it for the synthesis of vitamin B_{12}. Cobalt levels in soil are of significance because soil is the source of plant cobalt for animals.

The average total cobalt concentration in the earth's crust is 40 ppm. Acidic rocks, including granites, containing large amounts of iron-rich ferromagnesian minerals are low in cobalt, with levels ranging from 1 to 10 ppm. Much higher levels from 100 to 300 ppm of cobalt may be present in magnesium-rich ferromagnesian minerals. Sandstones and shales are normally low in cobalt, with concentrations frequently below 5 ppm. Other sedimentary rocks, such as shale, may contain considerably more cobalt, usually from 20 to 40 ppm.

Content in Soils. Total cobalt content of soils typically ranges from 1 to 70 ppm and averages about 8 ppm. Cobalt deficiencies in ruminants are often associated with forages produced on soils containing less than 5 ppm of total cobalt.

Small quantities of cobalt minerals often occur in combination with nickel minerals. In Australian soils most of the total cobalt is associated with the manganese minerals birnessite and lithiophorite. There also seems to be a relationship between the occurrence of cobalt in soil and the presence of magnesium minerals in parent rock.

Deficient Areas. A geographical pattern of low and adequate cobalt areas has been identified in the United States. Extremely low cobalt soil areas are found in the lower Atlantic coastal plain from North Carolina to Florida. Soils in this region are primarily sandy groundwater podzols or Humaquods containing 1 ppm or less of total cobalt. Soils in glaciated regions of the Northeast and the Great Lakes states as far west as Wisconsin are generally low in total cobalt. The low-cobalt soils in New Hampshire and southern Maine have formed on granitic glacial drift.

In addition to low-cobalt soils in the United States, deficiencies of this element occur in other countries, including Australia, Canada, Ireland, Kenya, New Zealand, Norway, and Scotland. The low-cobalt soils identified in Canada are located mainly in the eastern provinces of Ontario, Quebec, and Nova Scotia. Spodosols formed in coarse-textured deposits and organic (Histosol) soils are implicated with cobalt deficiency in Quebec.

Other kinds of soils in which cobalt deficiency most commonly occurs are (1) acidic, highly leached, sandy soils with low total cobalt; (2) some highly calcareous; and (3) some peaty soils.

Behavior and Availability of Cobalt in Soil. Cobalt is retained in soil mainly in specifically adsorbed exchangeable forms or as clay–organic matter complexes similar to the example in Figure 9-5. It is very strongly held in these forms and, like Cu^{2+}, its concentration in the soil solution is very low. German workers indicate that soil concentrations of available cobalt extracted with 0.1 N HCl should not be below 0.2 to 0.3 ppm.

Cobalt's behavior in soil resembles that of the other heavy metals and like iron, manganese, and zinc, it has a strong tendency to form chelates. It is capable of interfering in the uptake and pattern of action of other heavy metals. Excess cobalt nutrition produces growth symptoms similar to iron and manganese deficiencies.

Among the several factors that influence cobalt availability is the presence of crystalline manganese oxide minerals. These minerals have a high adsorption capacity for heavy metals, especially for cobalt. They are capable of retaining almost all of the cobalt in soil, leading to deficiencies in crops and to fixation of soil applied cobalt fertilizer. An average of 79% of the total cobalt in certain Australian soils is associated with manganese oxide minerals. Cobalt appears to replace manganese in the surface layers of these minerals.

Availability of cobalt to plants is decreased by soil treatments that allow conversion of soluble Mn^{2+} ions to insoluble oxides. Conversely, it is increased by those which transform manganese from oxides to mobile Mn^{2+} ions. Thus cobalt availability is favored by increasing acidity and waterlogging or reducing conditions. Liming and drainage are practices that reduce cobalt availability.

Polish investigators determined that the nature of the clay had a pronounced influence on the adsorption of cobalt from solutions. The order of adsorption was muscovite > hematite > bentonite = kaolin. Workers at Cornell University obtained similar results which showed that the expanding-lattice clays have a greater capacity for the adsorption of cobalt than has the nonexpanding kaolinite.

Cobalt Fertilization. Cobalt deficiency of ruminants can be corrected by administration of the element in any one of the following ways: by adding it to feed, salt licks, or drinking water; by drenching; and by use of cobalt bullets. Instead of providing cobalt directly to animals, it can be supplied indirectly by fertilizing forage crops with small amounts of cobalt compounds to raise plant levels of this element. For this purpose, rates of between about 1.5 And 3 oz/A of cobalt as the sulfate salt are used.

There are soils in Australia that are too low in available cobalt for satisfactory nodulation and nitrogen fixation by subterranean clover and alfalfa. Applications of only 0.5 to 2 oz/A of cobalt, as cobalt sulfate, can correct this condition. Cobaltized superphosphate, which contains small amounts of $CoSO_4$, is a fertilizer source of cobalt used in Australia and New Zealand. Sprays of $CoSO_4$ equivalent to slightly less than 4 oz/A of cobalt have effectively increased the cobalt concentration in subterranean clover. Seed treatment with very small quantities of $CoSO_4$ has been attempted on a limited scale.

Copper

The essentiality of copper for the growth of higher plants was first recognized in 1931. Its concentration in the earth's crust typically averages about 55 to 70 ppm. Copper levels in igneous rocks vary from 10 to 100 ppm, while they range between 4 and 45 ppm in sedimentary rocks. The copper concentration in soils of the United States ranges from nearly 1 to over 40 ppm and averages about 9 ppm. Total copper may fall to 1 or 2 ppm in deficient soils.

Sulfides are the predominant minerals of copper in the earth's crust, with strong covalent bonds formed between reduced copper (Cu^+) and sulfide (S^{2-}) anions. Chalcopyrite ($CuFeS_2$) is the most widely occurring copper mineral. Chalcocite (Cu_2S) and bornite ($CuFeS_4$) are other important copper-containing sulfide minerals. Relatively soluble secondary minerals of Cu^{2+}, including oxides, carbonates, silicates, sulfates, and chlorides, are also formed, but most of them do not persist under strong leaching conditions.

Copper deficiency is often the first nutritional disease to appear in plants grown on newly reclaimed acid Histosols and this condition is often referred to as "reclamation disease." Precautions should be taken to avoid copper deficiency whenever new areas of organic soils are brought into production. The occurrence of copper deficiency in many areas of Histosols is complicated by the presence of underlying deposits of marl, phosphatic limestone, or other calcareous materials that adversely affect the availability of copper.

Copper deficiencies in the United States are less common than deficiencies of other micronutrients. The geographical pattern of deficiency tends to be quite localized and is most often associated with plants grown on peats and mucks (Histosols). The incidence of copper deficiencies is highest in Florida, Wisconsin, Michigan, and New York, where high-value speciality crops are intensively grown on peats and mucks. This nutritional problem occurs generally in the Florida citrus area predominantly in organic and very sandy soils. Reports of copper deficiency have also come from the Pacific Coast states.

Vegetable crops such as onions and carrots have exhibited copper deficiency when grown on Histosols in the province of Manitoba. Small grains have frequently shown copper deficiency on peat soils in Alberta and Manitoba and in several eastern Canadian provinces. There are also instances of inadequate copper in well-drained sandy and calcareous sandy soils in Manitoba and in sandy calcareous soils in northeastern and northwestern Saskatchewan.

Use of copper-containing fungicides in the past has probably prevented or corrected copper deficiencies. Excessive use of such fungicides has created copper toxicity problems in some instances. Copper toxicity has also been encountered in soils affected by mine wastes.

Forms of Soil Copper. In addition to the copper in the lattice structure of primary and secondary minerals already mentioned, it exists in many other forms:

1. In the soil solution—ionic and complexed.
2. On normal cation exchange sites of clays and organic matter—held electrostatically in response to coulombic forces.
3. Occluded and coprecipitated in soil oxide material.
4. On specific adsorption sites—when held in this form copper cannot be removed by the reagents normally used for replacing exchangeable ions.
5. In biological residues and living organisms.

Most of the copper in soils is very insoluble and can only be extracted by strong chemical treatments which dissolve various mineral structures or solubilize organic matter. There is, however, a significant "pool" of diffusible copper, probably mainly in the form of organic complexes, which is in equilibrium with the very low level of copper in soil solution.

Soil Solution Copper. The concentration of copper in soil solutions is usually very low, in the range 10^{-8} to $10^{-6} M$ (0.6 to 63 ppb). At pH values below 6.9, divalent Cu^{2+} is the dominant species. Above pH 6.9, $Cu(OH)_2^0$ is the principal solution species and $CuOH^+$ assumes some importance near pH 7. Hydrolysis reactions of copper ions are shown in the following equations:

$$Cu^{2+} + H_2O \rightleftharpoons CuOH^+ + H^+$$

$$CuOH^+ + H_2O \rightleftharpoons Cu(OH)_2^0 + H^+$$

The complexes $CuSO_4^0$ and $CuCO_3^0$ are also important forms of copper. Solubility of Cu^{2+} is very pH dependent and it increases 100-fold for each unit decrease in pH.

Because of the immobility of copper in soil, it is supplied to plant roots mainly by root interception. Oliver and Barber at Purdue University determined that root interception represented on average 87% of the copper absorbed by soybeans from soil in pots. Other researchers have concluded that movement of copper into plants is largely dependent on the exploitation of soil by roots.

Copper concentration in soil solutions is much too low to be controlled by the solubility of copper minerals occurring in soil. Rather, it is governed both by adsorption reactions of copper on the surfaces of iron, aluminum, and manganese oxides and by its binding with organic matter.

McLaren and Crawford at the University of Nottingham in England postulated the following relationship between plant available forms and available reserves of copper:

$$\text{exchangeable and soluble Cu} \rightleftharpoons \text{specifically adsorbed Cu} \rightleftharpoons \text{organically bound Cu}$$

Occluded and Coprecipitated Copper. A significant fraction of soil copper is nondiffusible and is occluded or buried in various mineral structures, such as layer silicate soil clays and iron and manganese oxides. Copper is capable of isomorphous substitution in octahedral positions of crystalline silicate clays. It is also probably present as an impurity within carbonate minerals, especially magnesium and iron carbonate. Copper may also be trapped within oxide structures since it readily coprecipitates in aluminum and iron hydroxides.

Adsorbed Forms of Copper. The Cu^{2+} ion is specifically or chemically adsorbed by layer silicate clays, organic matter, and oxides of either iron, aluminum, or manganese. Minor amounts of these oxides as impurities may be responsible for much of the adsorption attributed to clay minerals. With the exception of Pb^{2+}, Cu^{2+} is the most strongly adsorbed of all the divalent transition and heavy metals on iron and aluminum oxides and oxyhydroxides.

Clay Minerals and Oxides. The mechanism of adsorption by oxides, unlike the loose electrostatic attraction occurring with negatively charged clay particles, appears to involve formation of Cu-O-Al or Cu-O-Fe surface bonds. This chemisorption process is thus likely controlled by the quantity of surface hydroxyl groups. Manganese oxides also specifically adsorb Cu^{2+}, and synthetic preparations of them have a stronger affinity for Cu^{2+} than that possessed by iron or aluminum oxides.

Adsorption of copper by aluminum, iron, and manganese oxides increases as a function of pH. As the pH is raised, hydrolysis of ionic species such as $Cu(H_2O)_6^{2+}$ on the surfaces of clay minerals is promoted, resulting in concomitant release in protons and a reduction in exchangeability of the bound Cu^{2+}.

Copper specifically adsorbed by clay minerals and by iron and aluminum oxides is exchangeable with other cations, but it is probably nonexchangeable when bonded to manganese oxides and organic matter.

Organic Matter. In many soils one-fifth to one-half of the copper occurs in organically bound forms. Copper is more strongly bound to organic matter than is any other micronutrient. Small organic acid molecules solubilize Cu^{2+} by chelation and complexation and most of the soluble copper in surface soils is organically complexed. As a result of this organic binding, there is more dissolved copper in the soil solution than normally occurs in the absence of organics. They are believed to aid in copper uptake by plants under conditions of low availability, (e.g., high pH).

The Cu^{2+} ion is directly bonded to two or more organic functional groups, chiefly carboxyl, carbonyl, or phenol. Trace levels of Cu^{2+} in peats may also be strongly bound in porphyrin complexes. Very little is known about the effect of these various reactions on the availability of copper to plants but it is known that both soluble and insoluble complexes are formed.

Humic and fulvic acids contain multiple binding sites for copper, with COO^- carboxyl groups playing a prominent role. In most mineral soils, organic matter is intimately associated with clay, probably as a clay–metal–organic complex. Figure 9-5 shows how the interaction of organic matter with clay still provides an organic surface for complexing copper. Stevenson at the University of Illinois has suggested that at soil organic-matter levels up to about 8%, both organic and mineral surfaces are involved in copper adsorption, while at higher concentrations of organic matter, binding of copper takes place mostly on organic surfaces. For soils having similar clay and organic-matter contents, the contribution of organic matter to the complexing of copper will be highest when the predominant clay mineral is kaolinite and lowest when montmorillonite is the principal clay mineral.

Factors Affecting Availability and Movement of Copper. The availability and movement of soil copper are affected by a number of soil properties, including texture, pH, cation exchange capacity, contents of organic matter, and hydrous oxides.

Texture. The supply of copper to the soil solution is usually lower in excessively leached podzolic sands and calcareous sands than in other soil types.

Soil pH. The mobility of copper in soil solutions often decreases with increasing pH and its supply to plants is correspondingly reduced because of greatly diminished solubility and increased sorption onto mineral colloid surfaces. The general effect of increasing adsorption with increasing pH is believed to be due to (1) an increased generation of pH-dependent sites on the colloids, (2) reduced competition with hydrogen ions, and (3) a change in the hydrolysis state of copper in solution.

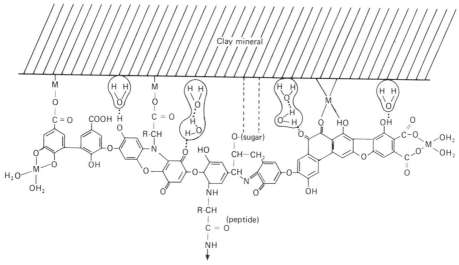

FIGURE 9-5. Schematic diagram of clay–organic matter–metal (M) complex. (Stevenson and Fitch, in J. F. Loneragan, Ed., *Copper in Soils and Plants*, p. 70. New York: Academic Press, 1981.)

As soil pH changes, the quantity and nature of the organic binding agents present in the soil solution will likely be affected. The degree of complexing of copper might also be altered.

Interrelationships with Other Elements. There are numerous interactions involving copper and the supply of plant nutrients. Applications of nitrogen fertilizer can aggravate copper deficiencies and increasing use of N-P-K fertilizers has been implicated in greater incidence of copper deficiencies. The occurrence of copper deficiencies following the use of acid-forming nitrogen fertilizers may in part be due to increased aluminum levels in the soil solution. Furthermore, increased growth resulting from the application of nitrogen or other fertilizer nutrients may be proportionally greater than copper uptake, which in effect dilutes copper concentration in plants. Increasing the nitrogen supply to crops can adversely affect mobility of copper in plants since it is known that a high nitrogen level in plants impedes translocation of copper from older leaves to growing points, where it is most needed.

High concentrations of zinc and aluminum in soil solution depress absorption of copper by plant roots and may intensify copper deficiency. Additions of zinc fertilizers to low copper soils have been reported to induce copper deficiencies in cereal crops in Western Australia. Both phosphorus and iron have been found to restrict copper absorption by plants.

Incorporation of Crop Residues. Ploughing unharvested *Brassica* root crops into soil has been observed by scientists in England to aggravate copper deficiency in the following crop. It is possible that the large amounts of sulfur released during decomposition of these crops and others such as rapeseed/canola may immobilize copper in the soil solution.

Severe copper deficiency has appeared in crops planted in haystack residues in Western Australia. Explanations for this reduction in effectiveness of fertilizer copper include (1) chemical reactions of copper with organic compounds and other substances originating from decomposing straw, (2) competition for available copper by stimulated microbial populations, and (3) inhibition of root development and ability to absorb copper.

Plant Factors. Some plant species exhibit copper deficiency more frequently than others. Crops that are highly responsive to copper include wheat, rice, alfalfa, carrots, lettuce, spinach, table beets, sudangrass, citrus, and onions. On the other hand, crops with the greatest apparent tolerance to low copper are beans, peas, potatoes, asparagus, rye, pasture grasses, *Lotus* spp., soybeans, lupine, rape, and pines.

There are striking genotypic differences in the copper nutrition of plants. Some of the possible mechanisms resulting in expression of genotypic differences are: (1) differences in the rate of copper absorption by roots, (2) better exploration of soil through greater root length per plant or per unit area, (3) better contact with soil through longer root hairs, (4) modification of availability of copper in soil adjacent to roots by root exudation, (5) acidification or change in redox potential, (6) more efficient transport of copper from roots to shoots, and/or (7) lower tissue requirement for copper.

Among small-grain species, rye has exceptional tolerance to low levels of soil copper supply. It will be healthy where wheat fails completely without the application of copper. Rye can extract up to twice as much copper as wheat under the same conditions. This genetic advantage over wheat is inherited by the wheat–rye hybrid, triticale. The usual order of sensitivity of the small grains to copper deficiency in the field is wheat > barley > oats > rye.

Varietal differences in tolerance to low copper supply are important and sometimes they can be as large as those among crop species. An Australian researcher, Nambiar, has found substantial difference in the sensitivity of wheat varieties to copper deficiency. Studies conducted in the 1980s in Manitoba clearly showed that there is wide variation in the ability of barley varieties to endure low copper nutrition. Other investigators have observed large differences in the copper utilization efficiency of two varieties of oats.

Copper Toxicity. Like most of the microelements, copper in large amounts is toxic to plants. Toxicity symptoms commonly include reduced shoot vigor, poorly developed and discolored root systems, and leaf chlorosis. The chlorotic condition in shoots superficially resembles an iron deficiency, but the mechanism responsible for this symptom is unclear. Toxicities are uncommon, occurring in limited areas of high natural availability of copper; after additions to soil of materials containing considerable concentrations of copper, such as sewage sludge, municipal composts, pig and poultry manures, and mine wastes; and from repeated use of copper-containing pesticides such as Bordeau mixture, $CuSO_4$ alone, and copper oxychloride.

Copper Fertilizers. As shown in Table 9-3, a wide range of compounds can be used to supply copper in solid, suspension and liquid macronutrient fertilizers. The usual copper source is $CuSO_4 \cdot 5H_2O$, although CuO, mixtures of $CuSO_4$ and $Cu(OH)_2$, and copper chelates are also used. Copper sulfate salt ($CuSO_4 \cdot 5H_2O$) contains 25.5% sulfur. This salt is quite soluble in water and is compatible with most fertilizer materials.

Copper ammonium phosphate can also be used either for soil application or as a foliar spray. It is only slightly soluble in water but can be suspended in water and sprayed on the plants. It contains 30% copper and, like the other metal ammonium phosphates (recall the discussion of magnesium ammonium phosphate in Chapter 5), is a slowly available material.

Copper salts are also contained in frits and in this form are suitable for soil application. Another relatively recent development is that of the chelates of copper and other micronutrients. The nature and development of these compounds are covered in detail in a subsequent section of this chapter. It is sufficient at the moment to indicate that chelates are metal–organic complexes which, though soluble themselves, do not ionize to any degree. They retain the copper (and similar

TABLE 9-3. Copper Compounds Used as Fertilizers

Source	Formula	Percent Cu	H_2O Solubility
Cu metal	Cu	100	Insoluble
Cuprite	Cu_2O	89	Insoluble
Tenorite	CuO	75	Insoluble
Covellite	CuS	66	Insoluble
Chalcocite	Cu_2S	80	Insoluble
Chalcopyrite	$CuFeS_2$	35	Insoluble
Malachite	$CuCO_3 \cdot Cu(OH)_2$	57	Insoluble
Azurite	$2CuCO_3 \cdot Cu(OH)_2$	55	Insoluble
Chalcanthite	$CuSO_4 \cdot 5H_2O$	25	Soluble
Copper sulfate monohydrate	$CuSO_4 \cdot H_2O$	35	Soluble
Basic copper sulfates	$CuSO_4 \cdot 3Cu(OH)_2$ (general formula)	13–53	Insoluble
Copper nitrate	$Cu(NO_3)_2 \cdot 3H_2O$		Soluble
Copper acetate	$Cu(C_2H_3O_2)_2 \cdot H_2O$	32	Slightly
Copper oxalate	$CuC_2O_4 \cdot 5H_2O$	40	Insoluble
Copper oxychloride	$CuCl_2 \cdot 3CuO \cdot 4H_2O$	52	Insoluble
Copper ammonium phosphate	$Cu(NH_4)PO_4 \cdot H_2O$	32	Insoluble
Copper chelates	Na_2Cu EDTA	13	Soluble
Copper chelates	NaCu HEDTA	9	Soluble
Copper polyflavanoids	—	5–7	Soluble
Copper–sulfur frits	—	0.5–20	Varies
Copper–glass fusions	—	Varies	Varies

Source: Gilkes, in J. F. Loneragan et al., Eds., *Copper in Soils and Plants,* p. 98. New York: Academic Press, 1981.

metals) in a soluble form, permitting their absorption by plants, yet preventing their conversion to insoluble forms in the soil.

Copper can be applied as organic compounds in the form of CuEDTA, copper ligninsulfonates, and copper polyflavonoids. These compounds can be applied either to the soil at rates of about 1 to 5 lb/A or as a foliar spray at rates considerably less than this.

Soil and foliar applications are both effective, but soil applications are more common, with rates of from 0.6 to about 20 lb/A of copper needed to correct deficiencies. Accessibility of soil applied copper to plant roots is important and utilization is enhanced by reducing particle size and increasing the number of granules or droplets per unit soil volume. Effectiveness is also increased by thoroughly mixing copper fertilizers into the root zone or by banding them in the seed row. When band-applied, copper rates may have to be reduced to prevent possible root injury. Additions of copper can be ineffective when root activity is restricted by excessively wet or dry soil, root pathogens, toxicities, and deficiencies of other nutrients.

Application of copper in foliar sprays is confined mainly to emergency treatment of deficiencies identified after planting. In some areas, however, copper is included in regular foliar spraying programs.

Copper dusts or solutions can be used for seed treatment, but this method of providing copper tends to be less effective than soil or foliar applications. Separation in the case of dusts and sensitivity of germinating seeds to copper at rates high enough to eliminate deficiency both limit the effectiveness of seed treatment. Similarly to foliar application, this method of treatment must be repeated for several years in succession.

Soil applied copper will generally have long-lasting residual effects. Beneficial effects from 1.3 to 2.7 lb/A of copper have persisted undiminished for up to 35 years under Western Australian conditions. It has also been reported in Australia that residual effects of copper treatments were still reflected in the copper status of pasture and sheep 13 and 17 years, respectively, after fertilization. Australian research has shown in several instances that the effectiveness of copper 2 to 12 years after application may be two to three times better than current applications banded in the seed row.

Iron

Iron comprises about 5% of the earth's crust and is the fourth most abundant element in the lithosphere. Common minerals of iron are olivene [$(Mg, Fe)_2SiO_4$], pyrite (FeS), siderite ($FeCO_3$), hematite (Fe_2O_3), goethite ($FeOOH$), magnetite (Fe_3O_4), and limonite [$FeO(OH)\cdot nH_2O + Fe_2O_3\cdot nH_2O$].

Since iron may be either concentrated or depleted during soil development, its normal concentration in soil varies widely from 0.7 to 55%. The average iron level in soils is estimated at 3.8%. Most of this soil iron is found in primary minerals, clays, oxides, and hydroxides. Hematite and goethite are the most common iron oxides in soils.

Deficient Soil Areas. Deficiency of iron has been reported in one or more crops in at least 25 states of the United States with most of them located in the western half of the nation. Matocha and Cunningham of Texas A & M University indicate that more than 12 million acres in 22 western states are deficient in available iron and that approximately 300,000 acres of grain sorghum are affected in South Texas alone. Iron deficiency of soybeans is common on high-pH soils in the upper Midwest.

Iron deficiencies have been confirmed in western Canada, principally in ornamental shrubs and fruit trees. These deficiencies are associated with basic soils which are sandy or organic. Although reports of iron-deficiency problems in field crops are rare, there are a few isolated instances where growth of crops such as barley was improved by supplemental iron.

Iron deficiencies have also been reported in Hawaii, having been induced by high levels of manganese on certain manganiferous soils in that state.

Soil Solution Iron. Iron is taken up by plant roots as Fe^{2+} and it can be transported to root surfaces as iron chelates. The low solubility of compounds containing Fe^{3+} severely limits availability and uptake of Fe^{3+}. Diffusion and mass flow are believed to be the two mechanisms responsible for the transfer of iron from soil to root surfaces. Investigations at Purdue University showed that diffusion was the main way in which iron was transported to soybean roots.

The solubility of iron in soils is principally controlled by ferric oxides. Hydrolysis, pH, chelation, and redox are important modifying factors. Inorganic Fe(III) in the soil solution may hydrolyze to form $Fe_2(OH)_2^{4+}$, Fe^{3+}, $FeOH^{2+}$, $Fe(OH)_2^{+}$, $Fe(OH)_3^{0}$, and $Fe(OH)_4^{-}$. The first four species are prevalent under acidic conditions, while the latter two are dominant above pH 7. Plant absorption of any one of these ions will cause the others to dissociate, thereby restoring equilibrium relationships among all species.

The solubility of iron in the soil solution is very dependent upon pH as is readily apparent in Figure 9-6. For each unit increase in pH there are 1000-fold and 100-fold decreases, respectively, in the solubility of Fe^{3+} and Fe^{2+}. Only at pH 3 does the total soluble iron concentration become sufficient to supply iron adequately to roots by mass flow. This figure also shows that at normal soil pH values the amount of available iron falls far short of plant requirements, even when allowances are made for its transfer to roots by other processes, such as diffusion and root interception.

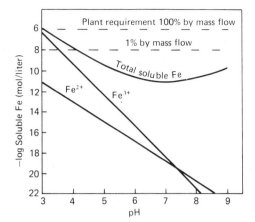

FIGURE 9-6. Solubility of iron in relation to pH and concentration of iron in solution necessary for mass flow to be important in the movement of iron to plants. (Lindsay, in E. W. Carson, Ed., *The Plant Root and Its Environment*, Chapter 17. Charlottesville, Va.: University Press of Virginia, 1974.)

Soluble organic complexes, many of which have chelating properties, react with iron in solution. These natural iron chelates maintain much higher iron concentrations in soil solutions than those normally occurring in solutions in equilibrium with only inorganic iron compounds. Fulvic acid and low-molecular-weight humic acids are capable of complexing and transporting metallic ions such as iron. The influence of synthetic chelates on increasing iron concentration in soil solution and resultant plant uptake of iron is shown in Figure 9-7. Diffusion of iron to the absorbing sorghum roots was encouraged by the higher concentration of mobile iron.

These organic substances capable of complexing and/or chelating iron originate as root exudates, from organic matter, as metabolic products from microbial activity, or as iron chelate fertilizers. Matocha and his colleagues at Texas A & M University found that substantial quantities of plant-complexed iron were cycled to the soil through sorghum stubble residue and remained available for the succeeding crop. They found that iron chelated by sorghum plant metabolites increased iron uptake 100% on a moderately deficient soil and in excess of 200% on severely deficient soil.

Oxidation–reduction reactions, normally the result of changes in O_2 partial pressure, exert considerable influence on the amount of soluble iron in the soil solution. The insoluble Fe^{3+} form predominates in well-drained soils, while levels of soluble Fe^{2+} increase significantly when soils become oxygen deficient due to excess water. Lindsay at Colorado State University, using the pe + pH parameter to describe redox relationships, showed that lowering redox increases Fe^{2+} solubility 10-fold for each unit decrease in pe + pH. Amounts of Fe^{2+} increase rapidly at redox

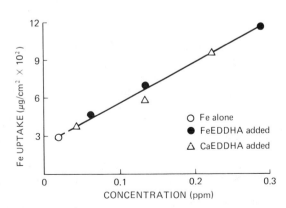

FIGURE 9-7. Uptake of iron at the end of 18 hours by sorghum roots as a function of the concentration of iron in the soil solution and chelate treatment. [O'Connor et al., *Soil Sci. Soc. Am. J.*, **35**:407 (1971).]

potentials below 200 mV. There is little effect of redox on total soluble Fe until the pe + pH declines to less than 12. In soils of near neutral reaction $FeOH^+$ is the major Fe^{2+} species and it is slightly more abundant than $Fe(OH)_2^+$, the principal Fe^{3+} species. Below pH 6.75, Fe^{2+} is the dominant ionic form of ferrous iron.

Factors Affecting Availability and Movement of Iron. Low total iron content of soil is seldom the basis for plant deficiencies. In most cases the problem is one of soil and environmental conditions impairing the availability of soil iron. Jones of the University of Georgia has brought attention to at least 16 different factors contributing to or causing iron deficiency in plants. Brief discussions of those factors appearing to have considerable influence on availability and movement of iron follow.

Ion Imbalance. The effect of imbalance of ions such as copper, iron, and manganese was covered in the section dealing with copper in soils. The same situation exists for iron. Iron deficiencies observed on many Florida soils probably result from an accumulation of copper in these soils after long years of application in sprays and fertilizers. Pineapples in Hawaii have exhibited iron chlorosis when grown on soils high in manganese, and other plants growing on soils developed from serpentine have exhibited iron deficiency because of excess nickel. Work carried out by scientists of the U.S. Department of Agriculture has shown that iron deficiencies on soybeans grown on two soils occurred because of a low ratio in the plants of Fe/(Cu + Mn).

Effects of pH, Bicarbonate, and Carbonates. Iron deficiency is most often encountered on high-pH and calcareous soils in arid regions, but it may also occur on acid soils that are extremely low in total iron. Solubility of iron in the soil solution reaches a minimum between pH 7.4 and 8.5. It is within this pH range in soils that iron deficiencies usually occur. The significance of high pH on iron availability is obvious when it is realized that Fe^{3+} and Fe^{2+} activities in solution decrease 1000-fold and 100-fold, respectively, for each unit increase in pH.

Irrigation waters and soils high in the bicarbonate ion (HCO_3^-) may aggravate iron deficiencies. The undesirable effects of HCO_3^- may simply be due to the high pH levels accompanying its presence.

The pH of most soils containing calcium carbonate falls in the range 7.3 to 8.5, which coincides with greatest incidence of iron deficiency and lowest solubility of soil iron. Similar conditions of pH and iron availability result from the presence of mixtures of calcite and dolomite (calcium carbonate–magnesium carbonate) in soils. Bicarbonate ion can be formed in calcareous soils by the following reaction of carbon dioxide and water on calcite:

$$CaCO_3 + CO_2 + H_2O \rightarrow Ca^{2+} + 2HCO_3^-$$

Although the presence of lime alone does not necessarily induce iron deficiency, this substance in combination with certain environmental conditions appears to be responsible for the problem in some plant types. Several of these soil and environmental conditions are discussed in the next section.

Excessive Water and Poor Aeration. The reaction shown in the preceding section is promoted by the accumulation of CO_2 such as will occur when soils are excessively wet and poorly drained. Consequently, any compact heavy-textured soil that is calcareous and basic in reaction is potentially iron deficient. Iron chlorosis is often associated with cool, rainy weather when soil moisture is high and soil aeration poor.

The effect of lack of O_2 in poorly drained soils on root development and absorption of H_2O and nutrients should not be ignored as a factor contributing to iron deficiency. It is known that lime-induced chlorosis will often disappear when problem soils are allowed to dry out.

Flooding and submergence of soils in which HCO_3^- formation is of no concern can improve iron availability by increasing concentrations of the active Fe^{2+} ionic form. There can actually be a buildup of toxic concentrations of Fe^{2+} in the soil solution of latosols.

It will be recalled that in the discussion on sulfur in submerged soil it was stated that the hydrogen sulfide released from the reduction of sulfur compounds was precipitated as iron sulfide. If the soil were low in iron, the hydrogen sulfide was not precipitated and was thought to cause Akiochi. This disease of rice occurs on soils that have been in paddy rice for years. The data shown in Table 9-4 suggest what happens to the iron. After years of reduction it is converted to the ferrous form, in which state it is probably slowly leached from the soil.

Organic Matter. Additions of organic matter to well-drained soils have produced varying effects on iron availability. Some researchers have found that manure corrected iron deficiency. Organic materials such as manure may supply chelating agents that aid in maintaining the solubility of micronutrients. Ligands with iron chelating properties have been extracted from poultry manure.

Improved soil structure resulting from applications of organic manures should increase iron availability because of better soil aeration. However, this ameliorating effect may be negated by higher levels of CO_2 and associated HCO_3^- produced as a result of greater microbial activity.

The presence of organic matter can have a profound influence on iron solubility in waterlogged soils, as the data in Table 9-4 show. Iron reduction was greatly

TABLE 9-4. Effect of Time and Added Organic Matter on the Reduction of Ferric Iron in a Submerged Soil

Time of Submergence (days)	Fe^{2+} in Solution (ppm)	Exchangeable Fe^{2+}
	Control	
0	Nil	Nil
7	Nil	Nil
14	Nil	Trace
21	Nil	10
28	Nil	20
38	Nil	31
51	Trace	40
63	2.0	46
78	4.0	50
	Straw	
0	Nil	Nil
7	Nil	26.0
14	30.0	108.0
21	90.4	162.0
28	132.0	200.0
38	192.0	446.0
51	184.0	450.0
63	128.0	482.0
78	104.0	523.0

Source: Mandal, *Soil Sci.*, **91**:121 (1961). Reprinted with the permission of The Williams & Wilkins Co., Baltimore.

accelerated by the addition of organic matter, and the longer the period of submergence, the greater the amount of soluble plus exchangeable iron.

Interrelationships with Other Nutrients. In addition to iron deficiencies evoked by excesses of essential nutrients such as cobalt, copper, manganese, and zinc, it has been observed that too much phosphorus or molybdenum will encourage iron deficiency. At least two iron–phosphorus relationships are known in plants. First, high phosphorus levels usually aggravate iron deficiency through some kind of inactivation reaction. The proportion of shoots to roots is increased by high phosphorus levels, especially when the iron status is low. The second relationship indicates that plants can be more tolerant of low iron when phosphorus is also low.

Nitrogen nutrition can also influence the occurrence of iron chlorosis. Plants receiving NO_3^- are more likely to develop this condition than those nourished with NH_4^+. Nitrate uptake leads to an alkalization effect in the root zone and in the plant, a condition that can markedly lower iron solubility and availability. Iron solubility and availability are favored by the acidity that develops when NH_4^+ is utilized by plants.

Deficiencies of potassium or zinc can disrupt movement of iron within plants. Lack of either of these two nutrients causes iron to accumulate in the stem nodes of corn.

Plant Factors. Plant genotypes differ in their ability to take up iron from a growth medium. Differential responses to iron deficiency occur in a wide variety of plants, including corn, peanuts, sorghum, soybeans, and tomatoes. Genetic control has been used widely in citrus and grape production where scions with desirable fruit quality are grafted on iron-efficient rootstocks.

The ability of plants to absorb and translocate iron appears to be a genetically controlled adaptive process that responds to iron deficiency or stress. Roots of iron-efficient plants alter their environment to improve the availability and uptake of iron. Some of the biochemical reactions and changes enabling iron-efficient plants to tolerate and adapt to iron stress are:

1. Excretion of H^+ ions from roots.
2. Excretion of various reducing compounds from roots.
3. Rate of reduction (Fe^{3+} to Fe^{2+}) increases at the root.
4. Organic acids, particularly citrate, increase in the root sap.
5. Adequate transport of iron from roots to tops.
6. Less accumulation of phosphorus in roots and shoots, even in the presence of relatively high phosphorus in the growth medium.

Table 9-5 rates a number of plants according to their sensitivity or tolerance to low levels of available iron. Some crops appear in more than one category because of variations in soil, growing conditions, and differential response of varieties of a given crop. Iron-efficient varieties should be selected for conditions where iron deficencies are likely to occur.

Iron Fertilizers. Iron chlorosis is one of the most difficult micronutrient deficiencies to correct in the field. Table 9-6 lists the iron-containing materials that are commonly used to treat iron deficiencies. In general, soil applications of ionizable ferrous salts, such as ferrous sulfate, have not been satisfactory because of their rather rapid oxidation to much less soluble ferric iron.

Correction of iron deficiencies is done mainly with foliar sprays. One dressing of a 2–3% ferrous sulfate solution at a rate of 15 to 30 gal/A is usually sufficient to alleviate mild chlorosis. However, several sprays 7 to 14 days apart may be

TABLE 9-5. **Sensitivity of Crops to Low Levels of Available Iron in Soil** *

Sensitive	Moderately Tolerant	Tolerant
Berries	Alfalfa	Alfalfa
Citrus	Barley	Barley
Field beans	Corn	Corn
Flax	Cotton	Cotton
Forage sorghum	Field beans	Flax
Fruit trees	Field peas	Grasses
Grain sorghum	Flax	Millet
Grapes	Forage legumes	Oats
Mint	Fruit trees	Potatoes
Ornamentals	Grain sorghum	Rice
Peanuts	Grasses	Soybeans
Soybeans	Oats	Sugar beets
Sundangrass	Orchard grass	Vegetables
Vegetables	Ornamentals	Wheat
Walnuts	Rice	
	Soybeans	
	Vegetables	
	Wheat	

* Some crops are listed under two or three categories because of variations in soil, growing conditions, and differential response of varieties of a given crop.

Source: Mortvedt, *Farm Chem.*, **143**:42 (1980).

needed to remedy more severe iron deficiencies. Inclusion of several drops per gallon of a mild detergent to serve as a wetting agent is normally necessary.

Injections of iron salts directly into trunks and limbs of fruit-tree species such as pears and plums have been very effective in controlling iron chlorosis. Treatments in California orchards typically consist of pressure injection at 200 psi of between 1 to 2 pints to 1 to 2 quarts per tree of 1–2% solutions of ferrous sulfate.

TABLE 9-6. **Some Sources of Fertilizer Iron**

Source	Formula	Percent Fe (approx.)
Ferrous sulfate	$FeSO_4 \cdot 7H_2O$	19
Ferric sulfate	$Fe_2(SO_4)_3 \cdot 4H_2O$	23
Ferrous oxide	FeO	77
Ferric oxide	Fe_2O_3	69
Ferrous ammonium phosphate	$Fe(NH_4)PO_4 \cdot H_2O$	29
Ferrous ammonium sulfate	$(NH_4)_2SO_4 \cdot FeSO_4 \cdot 6H_2O$	14
Iron frits	Varies	Varies
Iron ammonium polyphosphate	$Fe(NH_4)HP_2O_7$	22
Iron-sul	Mixture $FeO(OH)$, $KFe_3(OH)_6(SO_4)_2$, FeS_2, and $CuFeS_2$	20
Iron chelates	NaFeEDTA	5–14
	NaFeHEDTA	5–9
	NaFeEDDHA	6
	NaFeDTPA	10
Iron polyflavonoids	—	9–10
Iron ligninsulfonates	—	5–8
Iron methoxyphenylpropane	FeMPP	5

Source: Mortvedt et al., Eds., *Micronutrients in Agriculture*, p. 357. Madison, Wis.: Soil Science Society of America, 1972.

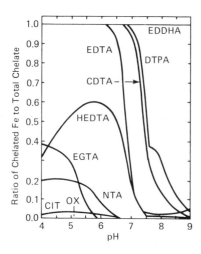

FIGURE 9-8. Iron chelate stabilities in soil solutions equilibrated with 10^{-4} concentrations of H^+, Ca^{2+}, Mg^{2+}, Al^{3+}, and Fe^{3+}. (Norvell, in J. J. Mortvedt, P. M. Giordano, and W. L. Lindsay, Eds., *Micronutrients in Agriculture,* p. 126. Madison, Wis.: Soil Science Society of America, 1972.)

With the exception of ferrous sulfate, perhaps the most widely used iron sources are the synthetic chelates and natural organic complexes identified in Table 9-6. These materials are water soluble and can be applied to the soil or foliage. Iron in chelated forms and in organic complexes is protected from the usual soil reactions, which result in formation of insoluble compounds such as iron hydroxide, iron phosphate, and iron carbonate. Spraying 0.10 to 0.15 lb/A of iron in chelates in a 15-in. band over the soybean row is effective about 80% of the time.

The stability of iron chelates in soils is of interest because of the need to supply iron to plants under widely different pH and other soil conditions. Stabilities of a number of iron chelates over a pH range of 4 to 9 are compared in Figure 9-8. It is apparent that HEDTA is effective in acid soils, whereas EDTA is useful only in slightly acid soils. Above pH 7.0, the chelates become increasingly effective in the order CDTA, DTPA, and EDDHA. The latter is unique since its stability remains constant in the pH range 4 to 9.

Local acidification of small portions of the root zone is an effective method for correcting iron deficiencies in calcareous and high-pH soils. Several sulfur products, such as elemental sulfur, ammonium thiosulfate, sulfuric acid, ammonium bisulfite, sulfur dioxide, and ammonium polysulfide, will lower soil pH and simultaneously act as reducing agents to convert ferric iron to the more readily available ferrous form. It is apparent in Figure 8-13 that SO_2 injected into a calcareous Arizona soil markedly increased the uptake of iron by sorghum.

Complexing with polyphosphate fertilizers also increases the plant availability of fertilizer iron. The TVA workers, Mortvedt and Giordano, reported this effect with both sulfate and chelate sources of iron, but iron EDDHA was more effective than ferrous or ferric sulfate at the same iron rates.

Applications of gypsum may improve iron availability by reducing the interference of molybdenum on plant uptake of iron. This effect may account for some of the benefits of gypsum additions on basic soils.

Manganese

Manganese concentration in the earth's crust averages 1000 ppm. It is an element that is widely distributed and at least traces of it are found in most rocks, particularly in ferromagnesian materials. Manganese, when released through weathering of primary rocks, will combine with O_2, CO_3^{2-}, and SiO_2 to form a number of secondary minerals, including pyrolusite (MnO_2), hausmannite (Mn_3O_4), manganite ($MnOOH$), rhodochrosite ($MnCO_3$), and rhodonite ($MnSiO_3$). The oxides, pyrolusite and manganite, are the most abundant.

Amounts of total manganese in soils generally range between 20 and 3000 ppm and average about 600 ppm. The usual forms of manganese in soils are various oxides and hydroxides. They occur as coatings on soil particles, deposited in cracks and veins, and mixed with iron oxides and other soil constituents in nodules. Individual crystallites are small and have large surface areas.

Deficient Soil Areas. It has been estimated that 13 million acres in 30 states may be low in manganese. Reported areas of manganese-deficient soils in the United States are more widespread in humid regions of the East, especially around the Great Lakes, than in the arid calcareous soils of the West. Most neutral or basic soils are potentially manganese deficient, but naturally wet fields that have been drained and put into crop production seem to be even more prone to this problem. These drained soils are characteristically dark colored and high in organic matter, and usually they are derived from calcareous parent materials. Manganese-deficient Oregon soils matching the above description contain from 20 to 30% organic matter.

Other soil conditions identified with manganese deficiency are:

1. Thin, peaty soils overlying calcareous subsoils.
2. Alluvial silt and clay soils and marsh soils derived from calcareous materials.
3. Poorly drained calcareous soils high in organic matter.
4. Calcareous black sands and reclaimed acid heath soils.
5. Calcareous soils freshly broken up from old grassland.
6. Old black garden soils where manure and lime have been applied regularly for many years.
7. Very sandy acid mineral soils that are low in native manganese content and where the limited quantities of available manganese may have been leached from the root zone.

Manganese deficiency has been identified in Canada's prairie provinces on neutral or alkaline soils of abundant organic matter and on sandy and peat soils underlain with calcareous subsoils. It also occurs in British Columbia in a few orchards situated on alluvial fans. Low manganese is the most common micronutrient deficiency in both soybeans and small grains in Ontario.

Forms of Soil Manganese. Manganese in soil is usually thought to exist as exchangeable Mn^{2+}, as water-soluble Mn^{2+}, as both water-soluble and insoluble organically bound manganese, as easily reducible manganese, and as various manganese oxides (Figure 9-9). These various forms are in a state of equilibrium with one another and they differ in their degree of availability to plants. It is generally agreed that for satisfactory manganese nutrition of crops, levels of water-soluble, exchangeable, and easily reducible manganese should be of the order 2 to 3 ppm, 0.2 to 5 ppm, and 25 to 65 ppm, respectively.

Two major processes are operative in this cycle. The first is oxidation–reduction, while the second is production and decomposition of natural chelating agents that can complex manganese in both soluble and insoluble forms. The continuous flux of organic matter and decomposition of plant residues is thought to contribute significantly to the solubilization of inert forms of manganese and to the maintenance of water-soluble manganese.

Factors influencing the solubility of soil manganese include pH, redox, and complexation. Soil moisture, aeration, and microbial activity influence redox, while complexation is affected by organic matter and microbial activity.

Soil Solution Manganese. The principal ion species in solution is Mn^{2+} and its concentration decreases 100-fold for each unit increase in pH. Other species of only minor importance are $MnSO_4(aq)$, $MnHCO_3^+$, and $MnOH^+$.

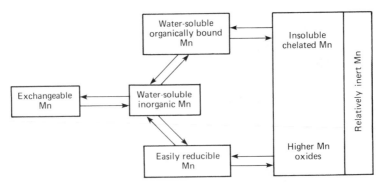

FIGURE 9-9. Manganese cycle. [Ghanem et al., *Plant Soil*, **34**:653 (1971).]

Dissolved Mn^{2+} in the soil solution is of direct importance to plants since it is usually assumed that they obtain all of their requirement in this form. Concentrations of Mn^{2+} measured in the soil solution of acid and neutral soils have varied from less than 0.01 ppm to almost 13 ppm, with levels commonly in the range 0.01 to 1 ppm. Organic-matter reactions with manganese were important in these soils since between 84 and 99% of the Mn^{2+} was complexed.

There may be large seasonal changes in Mn^{2+} levels in the soil solution. Fluctuations of at least 100-fold were found in a mixed hardwood forest soil from the state of New York. Similar effects have resulted from prolonged storage of soil samples.

Studies at Purdue University suggest that diffusion is the most important mechanism for transport of Mn^{2+} from soil to the surface of soybean roots. This process is enhanced by the formation of organic manganese complexes which greatly increase both Mn^{2+} in solution and in turn the strength of the requisite concentration gradient.

Manganese in the soil solution is greatly increased under acid, low-redox conditions. Transformation of Mn^{4+} to Mn^{2+} takes place when the redox potential is in the range $+200$ to $+400$ mV.

In extremely acid soils, Mn^{2+} solubility can be sufficiently great to cause toxicity problems in sensitive plant species. Because of the mobility of Mn^{2+}, it can leach from soils, particularly from acid podzols. The frequent occurrence of manganese deficiency in poorly drained mineral and organic soils is often attributed to low manganese levels resulting from intense leaching of soluble Mn^{2+}.

Factors Affecting Availability and Movement of Manganese. Many soil, seasonal, and management factors influence manganese availability and movement. The most important of these factors are reviewed briefly in this section.

Imbalance of Heavy Metal Ions. High levels of copper, iron, or zinc in the growth medium will impede manganese uptake by plants. Similar effects of ion imbalances on availability of cobalt, copper, and iron have already been described.

Effect of pH and Carbonates. Consistent with the known sensitivity of Mn^{2+} concentrations to pH, management practices that change soil pH will also greatly influence Mn^{2+} availability and uptake. Cheng and Quellette in Quebec showed that liming very acid soils resulted in a threefold decrease in the amount of exchangeable Mn^{2+} in the soil and a twofold decrease in the manganese content of the plant. Reductions in exchangeable Mn^{2+} in a North Carolina soil following liming are apparent in the data in Table 9-7.

TABLE 9-7.   Effect of Added Peat on the Acid-Extractable Manganese Content of a Limed and Unlimed Norfolk Sandy Loam After 3 Weeks' Incubation

	Acid-Extracted Mn (ppm)	
Treatment	No Lime	Limed
54 ppm. Mn	48.0	64.8
54 ppm. Mn + 2% peat	50.2	50.6
54 ppm. Mn + 4% peat	50.0	39.0
54 ppm. Mn + 10% peat	53.9	32.4

Source: Sanchez et al., *SSSA Proc.,* **23**:302 (1959).

On the other hand, low manganese availability in high pH and calcareous soils and in overlimed, poorly buffered, coarse-textured soils can be largely overcome by acidification through the use of acid-forming nitrogen or sulfur materials.

High pH also favors the formation of less available organic complexes of manganese. Activity of the soil microorganisms which oxidize soluble manganese to unavailable forms reaches a maximum near pH 7.

Excessive Water and Poor Aeration. Soil submergence and waterlogging lower redox potential and increase the amount of soluble Mn^{2+} in soils. This is especially true in acid soils high in active manganese. As soluble Mn^{2+} concentration increases under submerged conditions, there is a corresponding decline in O_2 levels.

Manganese availability can be increased by poor aeration in compact soils and by local accumulations of CO_2 within root masses and other soil microsites. The resulting low-redox conditions will render manganese more available without appreciably affecting redox potential or pH of the bulk soil.

Organic Matter. Availability of Mn^{2+} can be affected in several ways by the presence of organic materials. The low availability of manganese in basic soils high in organic matter is attributed to the formation of unavailable chelated Mn^{2+} compounds. It may also be held in unavailable organic complexes in peats or muck soils.

Additions of certain natural organic materials such as peat moss, compost, and wheat and clover straw have increased the manganese in water-soluble plus exchangeable, exchangeable, and easily reducible fractions. The stimulatory effect of peat moss additions on exchangeable Mn^{2+} in an unlimed North Carolina soil is illustrated by the data in Table 9-7.

Interrelationships with Other Nutrients. Nitrogen sources can influence manganese availability in several ways. Addition of physiologically acidic nitrogen materials such as NH_4Cl, $(NH_4)_2SO_4$, NH_4NO_3, $NH_4H_2PO_4$, and $CO(NH_2)_2$ to the growth medium will enhance plant uptake of manganese.

Neutral chloride-containing salts including KCl, NaCl, and $CaCl_2$ have increased the manganese concentration of plants and levels of available manganese in acidic soils. The relative order of the salt effect on increasing extractable manganese was $KBr > KCl > KNO_3 > K_2SO_4$. Jackson and his coworkers at Oregon State University observed that KCl increased the manganese concentration in bush beans and sweet corn more than did K_2SO_4 or K_2CO_3. The stimulatory effect of KCl on manganese uptake was so strong that it produced toxicity symptoms in trifoliate bean leaves.

Phosphate fertilization has been reported to both enhance and decrease manganese availability. The action of phosphorus sources probably depends on the nature of the soil, fertilizer properties, and plant characteristics. Several investigators have observed an inverse relationship between phosphorus and manganese in plant tissue.

Seasonal and Climatic Effects. Very pronounced seasonal variations in manganese availability have been observed. It is believed that wet winter weather favors the presence of Mn^{2+}, whereas warm, dry summer conditions encourage the formation of less available oxidized forms of manganese.

Several workers have indicated that dry weather either induced manganese deficiency or aggravated this nutritional problem, particularly in fruit trees. Wet weather is one of the conditions usually associated with a high incidence of gray speck of oats, a manganese deficiency disorder, in western Canada. Increasing soil temperature during the growing season has been found to improve manganese uptake in some cropping situations, presumably because of greater plant growth and root activity.

Soil Microorganisms. Deficiencies of manganese caused by soil organisms oxidizing Mn^{2+} to Mn^{4+} have been reported. A number of bacterial and fungal genera are recognized as being capable of performing this transformation. It is unclear, however, to what extent the effect is indirect. For example, following a microbially produced rise in pH, there could be precipitation of the oxide through standard chemical pathways.

Plant Factors. Several plant species, including soybeans, have been reported to show differences in sensitivity to manganese deficiency. Among the cereal grains, oats are especially sensitive, although some cultivars are less sensitive than others. European barley varieties are sensitive to manganese deficiency, while those of wheat are somewhat more tolerant. A study conducted by Nyborg in the province of Alberta revealed that barley was less susceptible to manganese deficiency than was wheat; there were differences in sensitivity of wheat and oats cultivars to manganese stress; and relative susceptibility to manganese deficiency in the crops tested was: Olli and Parkland barley < Thatcher wheat < Saunders wheat < Glen oats < Exeter oats < Victory and Abegweit oats.

These differences in response of manganese-efficient and manganese-inefficient plants are believed to be due to internal factors rather than to the effects of the plants on the growth medium. Reductive capacity at the root may be the factor restricting uptake and translocation of manganese. There may also be significant differences in the amounts and properties of root exudates generated by plants. These substances can materially influence supplies of available Mn^{2+}. Further, dissimilarity between manganese-efficient and manganese-inefficient oat plants has been attributed to calcium substituting for manganese at nonspecific sites in manganese-efficient plants, thus freeing adequate manganese for essential reactions. It is possible that some or all of the plant characteristics possessed by iron-efficient plants may similarly influence manganese uptake in plants tolerant of manganese stress.

Major crops are grouped in Table 9-8 according to their sensitivity to low manganese supply. Although it is difficult to generalize, small grains, soybeans, some vegetables, and tree fruits are usually considered to be most affected by inadequate manganese nutrition. Some agriculturists have expressed the view that lack of manganese is the most common micronutrient problem in soybean production.

Manganese Fertilizers and Correction of Deficiencies. Manganese sulfate is widely used for correction of manganese deficiency. It may be applied to

TABLE 9-8. Sensitivity of Crops to Low Levels of Available Manganese in Soil *

Sensitive	Moderately Tolerant	Tolerant
Alfalfa	Barley	Barley
Citrus	Corn	Corn
Fruit trees	Cotton	Cotton
Oats	Field beans	Field beans
Onions	Fruit trees	Fruit trees
Potatoes	Oats	Rice
Soybeans	Potatoes	Rye
Sugar beets	Rice	Soybeans
Wheat	Rye	Vegetables
	Soybeans	Wheat
	Vegetables	
	Wheat	

* Some crops are listed under two or three categories because of variation in soil, growing conditions, and differential response of varieties of a given crop.
Source: Mortvedt, *Farm Chem.,* **143**:42 (1980).

soil or directly to the crop as a foliar spray. Other manganese fertilizers and their manganese concentrations are noted in Table 9-9. In addition to the simple inorganic compounds, fertilizer manganese is commercially available in chelated, organically complexed, and fritted forms. The natural organic complexes and chelate sources of manganese are best suited for spray applications.

Manganese oxide is only slightly water soluble but it is usually a satisfactory source of manganese. Particle size of this material is important and it must be finely ground to be effective. The manganese frits, $MnCO_3$, and MnO_2 are less popular manganese carriers.

Rates of manganese application range from 1 to 25 lb/A. The quantities at the top end of this range are recommended for broadcast application, while those at the lower end are typically used in foliar sprays. Band treatments are usually only about one-half of the broadcast rates. Applications at the higher rates may be required on organic soils. Band application of manganese sources in combination with N-P-K fertilizers is commonly practiced and it seems to be a superior way of providing manganese.

TABLE 9-9. Sources of Manganese Used for Fertilizer

Source	Formula	Percent Mn (approx.)
Manganese sulfate	$MnSO_4 \cdot 4H_2O$	26–28
Manganous oxide	MnO	41–68
Manganese carbonate	$MnCO_3$	31
Manganese chloride	$MnCl_2$	17
Manganese oxide	MnO_2	63
Manganese frits	Fritted glass	10–35
Natural organic complexes	—	5–9
Synthetic chelates	MnEDTA	5–12
Manganese methoxyphenylpropane	MnMPP	10–12

Sources: Mortvedt et al., Eds., *Micronutrients in Agriculture,* p. 363. Madison, Wis.: Soil Science Society of America, 1972; and Mortvedt, *Farm Chem.,* **143**:42 (1980).

When manganese products are applied alone without the benefit of other fertilizer materials, banding is more effective than broadcasting. Oxidation to less available forms of manganese is apparently delayed in bands.

Broadcast application of manganese chelates and natural organic complexes is not normally advised because soil calcium or iron can replace manganese in these fertilizers and the freed manganese is usually converted to unavailable forms. Meanwhile, the more available complexed calcium or iron probably accentuates the manganese-deficiency condition.

Either manganese sulfate or one of the various natural organic complexes or synthetic chelates is used successfully in foliar application. One-fifth to one-half pound of actual manganese in 20 gal of water per acre spray applied until the crop foliage is wet to dripping will correct most deficiencies. Where deficiencies are pronounced, multiple foliar applications may be necessary.

Lime or high-pH-induced manganese deficiencies can be rectified by acidification resulting from the use of elemental sulfur and its compounds. The same principles apply as were discussed in the foregoing section on iron.

Molybdenum

Molybdenum, which occurs in the earth's crust and in soils in extremely small quantities, is usually found in concentrations of less than 1 ppm in plants. The average concentration of this element in the lithosphere is about 2 ppm and in soils it typically ranges from 0.2 to 5 ppm and averages approximately 2 ppm. Its median concentration in U.S. soils is estimated to be about 1.2 to 1.3 ppm and may vary from 0.1 to 40 ppm.

The main forms of molybdenum in soil include: in nonexchangeable positions in the crystal lattice of primary and secondary minerals, as an exchangeable anion, bound to iron and aluminium oxides, as water-soluble molybdenum in the soil solution, and as organically bound molybdenum.

Deficient Soil Areas.
Deficiencies of molybdenum in the United States occur largely on the acid sandy soils of the Atlantic and Gulf coasts, although responses to this element have also been reported in California and the Pacific Northwest, Nebraska, and the states bordering the Great Lakes. Soils of eastern Canada are acidic, coarse textured, and highly leached in character. In keeping with these properties they are molybdenum deficient. Large soil areas in New Zealand and Australia are also deficient in molybdenum.

This nutrient may also be lacking in highly podzolized soils low in total molybdenum, acid soils high in hydrous oxides of iron and aluminum or those derived from calcareous parent materials, and certain neutral and calcareous soils.

Soil Solution Molybdenum.
The major solution species of molybdenum in decreasing order of importance are: $MoO_4^{2-} > HMoO_4^- > H_2MoO_4^-$. Above pH 4.2, the first-named species is dominant. Concentration of the first two species increases dramatically with increasing soil pH.

Plants absorb molybdenum as MoO_4^{2-}. The extremely low concentrations of molybdenum in soil solution, reported to be of the order of 2 to 8 ppb in saturation extracts, is reflected in the customary low molybdenum content of plant material. At concentrations above 4 ppb in the soil solution, molybdenum is transported to plant roots by mass flow. Diffusion becomes the dominant means for molybdenum transfer to plant roots when the concentration is less than 4 ppb.

Factors Affecting Availability and Movement of Molybdenum.
Two of the most important factors affecting the availability of molybdenum are soil pH and the amount of aluminum and iron oxides in soil. The effects of these and

several other soil properties, including levels of sulfate and phosphate on molybdenum availability, are considered next.

Soil pH and Liming. Plant availability of molybdenum, unlike that of other micronutrients, increases with decreasing soil acidity. Figure 9-10 demonstrates the pH dependence of molybdenum activity in four Colorado soils. This relationship can be written according to the following equation, which shows that there is a 10-fold increase in MoO_4^{2-} activity per unit increase in soil pH.

$$\text{soil} + MoO_4^{2-} \rightleftharpoons \text{soil} - MoO_4^- + OH^-$$

An even greater sensitivity of molybdate activity to soil reaction occurred in one of the soils. Below pH 7, soil 2 exhibited a 100-fold increase in the molybdenum solubility per unit increase in pH. Changes of such large magnitude are expected when the mineral wulfenite ($PbMoO_4$) is present in soil.

Liming to correct soil acidity will thus improve the molybdenum nutrition of plants. In most instances this practice is enough to correct or prevent molybdenum deficiency. Numerous studies have established that molybdenum uptake by plants is increased by the application of alkaline materials such as limestone and basic slag, but is decreased by physiologically acid fertilizers such as ammonium sulfate and nitrate.

Reaction with Iron and Aluminum. Molybdenum is adsorbed strongly by iron and aluminum oxides. A portion of the adsorbed molybdenum becomes unavailable to the plant, while the remainder is in equilibrium with the soil solution molybdenum. As plant roots remove molybdenum from solution, more molybdenum is desorbed into the solution by simple mass action. Because of adsorption reactions, soils that are high in iron, especially noncrystalline iron on clay surfaces, tend to be low in available molybdenum.

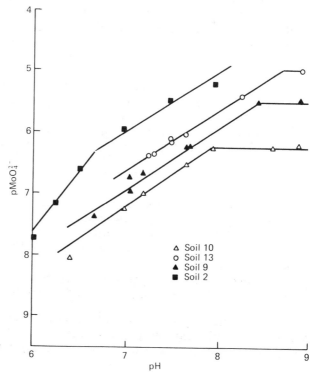

FIGURE 9-10. pH dependence of molybdenum solubility in four Colorado soils. [Vlek and Lindsay, *Soil Sci. Soc. Am. J.*, **41**:42 (1977).]

TABLE 9-10. Effect of Sulfur and Molybdenum on the Yield and Sulfur and Molybdenum Concentration of Brussels Sprouts *

		Aboveground Tissue	
Treatment	Yield (g/pot)	Mo (ppm)	S (%)
Sulfur †			
No S added	12.6 a	5.09 c	0.25 w
50 ppm S	13.8 b	0.88 d	0.60 x
100 ppm S	13.8 b	0.50 e	0.70 y
Molybdenum			
No Mo added	12.7 a	0.08 f	0.53 xz
Seed treated with Mo	13.6 b	0.16 g	0.49 z
2.5 ppm Mo	13.9 b	6.23 h	0.51 z

* Values followed by a common letter are not significantly different at $P = 0.01$.

† Sulfur treatments did not alter soil pH or the exchangeable molybdenum content.

Source: Gupta, *Sulphur Inst. J.,* **5**(1):4 (1969).

Interrelationships with Other Nutrients. Phosphorus enhances the absorption and translocation of molybdenum by plants. Some researchers attribute this beneficial effect to the release of adsorbed MoO_4^{2-}, thus making it more available to plants. An alternative explanation involves plant biochemical processes and stimulation of molybdenum release from root cells into the translocation system.

In contrast to phosphorus, high levels of SO_4^{2-} in the rooting media depress molybdenum uptake by plants. Sulfate applied as the salt of ammonium, calcium, or potassium has a similar influence on plant absorption of molybdenum. The restricting effect of sulfur application on molybdenum content of Brussels sprouts is evident in Table 9-10. On soils with borderline molybdenum deficiencies, the application of heavy rates of sulfate-containing fertilizers may induce a molybdenum deficiency in plants. In such situations the inclusion of molybdenum in the fertilizer at locally recommended rates may be advisable.

Both copper and manganese have been observed to act antagonistically on molybdenum uptake, and if present in excess they may induce molybdenum deficiency. Magnesium has been reported to have an opposite effect and will encourage molybdenum absorption by plants.

The nitrate form of nitrogen apparently encourages molybdenum uptake by plants, while ammoniacal sources act oppositely. This beneficial effect of nitrate nutrition is perhaps related to the release of OH^- ions and an accompanying increase in solubility of soil molybdenum.

Environmental Effects. Molybdenum deficiency has been found to be more severe under dry soil conditions in which the soil is unable to supply sufficient molybdenum for the growing plant. Transport of molybdenum from soil to root surfaces by either mass flow or diffusion would probably be impaired by low soil moisture content.

A favorable influence of high soil temperature on molybdenum solubility has been reported with a very large increase occurring between 26 and 65°C.

Plant Factors. There are varietal differences in the susceptibility of crops to molybdenum deficiency. Molybdenum-efficient and molybdenum-inefficient varieties of alfalfa, cauliflower, corn, and kale have been identified. The differential susceptibility of cauliflower varieties to molybdenum deficiency is seemingly unrelated to requirement for the nutrient, but rather in their ability to extract soil molybdenum.

Crops that are very sensitive to insufficient molybdenum are the legumes, crucifers (broccoli, Brussels sprouts, cauliflower, rapeseed, etc.), and citrus. Other crops that are also sensitive to a low molybdenum supply are beet, cotton, lettuce, spinach, sweet corn, sweet potatoes, and tomatoes. Subterranean clover is one of the most sensitive legumes to a lack of molybdenum. Small grains and some row crops tend to be more tolerant to low levels of available molybdenum in soil.

Molybdenum Toxicities. Excessive amounts of molybdenum are toxic, especially to grazing cattle or sheep. There are areas of molybdenum toxicity in the United States, principally in western states such as California, Oregon, Washington, Idaho, Nevada, Colorado, and Montana. Molybdenum toxicity is also recognized in Florida. In Canada, molybdenum toxicity has been reported in the western provinces of Manitoba, Saskatchewan, and British Columbia. Molybdenum toxicity also exists in certain areas of Australia. Soils in the problem areas are locally quite high in their content of this element.

Molybdenosis, as this disease of cattle is called, is actually caused by an imbalance of molybdenum and copper in the diet of the ruminant. This is customary when the molybdenum content of the forage is greater than 5 ppm. Molybdenum toxicity causes stunted growth and bone deformation in the animal. It can be corrected by oral feeding of copper, injections of copper glycinate suspensions, or the application of copper sulfate to the soil. Other practices used to decrease molybdenum toxicity are application of sulfur or manganese and improving soil drainage.

The soil conditions on which high-molybdenum forage may be found are wet soils mostly neutral to alkaline in reaction, often with either a thick A_1 horizon or an A_1 horizon capped by a thin surface layer of peat or muck. Pockets of peats may also be present in these problem areas. Molybdenum toxicity on the broad deltas of central California and broad basinlike lowlands in the Klamath area in Oregon and the Florida Everglades is generally associated with Histosols.

Molybdenum Fertilizers and Correction of Deficiencies. Sources of molybdenum used for fertilizer are listed in Table 9-11. Rates of molybdenum application are very low, only 0.5 to 5 oz/A. Molybdenum may be applied to soil, sprayed on foliage, or put on seed prior to planting. The optimum molybdenum rate depends on the application method with lower rates used in the latter two methods.

Seed treatment, involving soaking seed in a solution of sodium molybdate before seeding, is widely used because of the low application rates needed. A slurry or dust treatment of the seed is also effective. Another related method of incorporating MoO_3 in a seed pellet has been successful for pasture legumes.

To obtain satisfactory distribution of the small quantities of molybdenum applied to soil, molybdenum sources are sometimes combined with N-P-K fertilizers. In Australia, MoO_3 has been incorporated into rock phosphate pellets.

Foliar spray applications with ammonium or sodium molybdate also are effective in correcting molybdenum deficiencies. A comparison of the effectiveness of seed treatment versus foliar sprays is shown in Table 9-12.

It has been demonstrated in Australia, New Zealand, southwestern Oregon, and elsewhere that the application of molybdenum to clovers will in some cases produce yield increases equivalent to those obtained from the addition of several tons of limestone. In inaccessible areas, particularly with rough terrain, transportation and application of large tonnages are difficult and expensive. As a consequence, under these conditions molybdenum fertilization is often preferred to liming.

TABLE 9-11. Sources of Molybdenum Used for Fertilizer

Sources	Formula	Percent Mo (approx.)
Ammonium molybdate	$(NH_4)_6Mo_7O_{24} \cdot 2H_2O$	54
Sodium molybdate	$Na_2MoO_4 \cdot 2H_2O$	39
Molybdenum trioxide	MoO_3	66
Molybdenum frits	Fritted glass	1–30

Source: Mortvedt, *Farm Chem.,* **143**:42 (1980).

Zinc

The zinc content of the lithosphere has been estimated to be about 80 ppm. Its total content in soils ranges from 10 to 300 ppm and averages approximately 50 ppm. Total zinc concentration in soil is no more a criterion of its availability to plants than is the case for other plant nutrients.

The igneous rocks basalt and granite contain on the average 100 and 40 ppm of zinc, respectively. Among sedimentary rocks, shale contains more zinc (95 ppm) than either limestone (20 ppm) or sandstone (16 ppm). Sphalerite (ZnS), smithsonite ($ZnCO_3$), and hemimorphite [$Zn_4(OH)_2Si_2O_7 \cdot H_2O$] are common zinc-containing minerals. Zinc has a strong tendency to combine with sulfides and it occurs most frequently in the lithosphere as sphalerite.

Deficient Soil Areas. Zinc deficiencies are widespread in the United States. As a result of extensive research conducted since the 1960s on detection and correction of these deficiencies, most states now recognize the need for supplemental zinc for one or more crops. Large acreages of zinc-deficient soils have been identified in at least eight states.

Occurrences of zinc deficiency are also known in Canada, western Europe, Great Britain, Israel, New Zealand, Australia, Central and South Africa, and Brazil. This nutritional problem in Canada has long been delineated in the provinces of Ontario and British Columbia and has now been confirmed in isolated cases in the three Prairie Provinces.

Soil conditions most often associated with zinc deficiencies are: acid sandy soils low in total zinc; neutral or basic soils, especially calcareous soils; soils with a

TABLE 9-12. Effect of Seed and Foliar Application of Molybdenum on the Yield of Canning Peas

Treatment *	Yield (kg/ha)	
	Vines	Peas
Check	1070	1520
Seed application		
Slurry	2150	2770
Dust	1500	1950
Foliar spray	2370	2500
LSD 0.05	427	410

* Molybdenum was applied as $Na_2MoO_4 \cdot 2H_2O$ at a rate of 56 g/ha.

Source: Hagstrom and Berger, *Soil Sci.,* **100**:52 (1963). Reprinted with the permission of The Williams & Wilkins Co., Baltimore.

high content of fine clay and silt; soils high in available phosphorus; some organic soils; and subsoils exposed by land-leveling operations or by wind and water erosion.

Forms of Soil Zinc. The forms of zinc in soils considered to influence in varying degree its supply to plants are: water-soluble Zn^{2+}; exchangeable Zn^{2+}; adsorbed Zn^{2+} on surfaces of clay, organic matter, carbonates, and oxide minerals; organically complexed Zn^{2+}; and Zn^{2+} substituted for Mg^{2+} in the crystal lattices of clay minerals. Identification of these various phases has been difficult because of the small amounts of zinc involved. Also, the specific minerals controlling the solubility of Zn^{2+} in soils are not known.

Most of the zinc in many soils is located in ferromagnesian minerals such as augite, hornblende, and biotite. The occurrence of Zn^{2+} in these minerals is due to its isomorphous replacement of some of the original Fe^{2+} and Mg^{2+}.

Soil Solution Zinc. Reported amounts of Zn^{2+} in the soil solution are very low. Concentrations of about 75 ppb were found in New York soils, while in a large group of Colorado soils the Zn^{2+} levels were all less than 2 ppb. On average 60% of the Zn^{2+} in solution in these soils was complexed, presumably by organic matter.

The oxidation state of zinc in soils is exclusively Zn^{2+}. Several zinc hydrolysis species exist in solution with Zn^{2+} predominating at soil reactions below pH 7.7. Above this pH, $ZnOH^+$ becomes the most abundant species until it is supplanted by $Zn(OH)_2^0$ at pH 9.1.

Complexes with nitrate, chloride, and phosphate can form, but with the exception of $ZnHPO_4^0$ their contributions to total zinc in solution are considered insignificant. In neutral and calcareous soils it is possible, depending on pH and phosphorus activity, for $ZnHPO_4^0$ to equal $ZnOH^+$ concentration.

Zinc solubility is highly pH dependent and decreases 100-fold for each unit increase in pH. This important theoretical relationship is represented by the following equation developed by Lindsay and Norvell at Colorado State University:

$$\text{soil} - Zn + 2H^+ \rightleftharpoons Zn^{2+} \quad \log K^0 = 5.8$$

which can be expressed

$$\log Zn^{2+} = 5.8 - 2pH$$

Thirty-fold reductions of zinc concentration in soil solutions for every unit of pH increase in the pH range 5 to 7 have been observed in studies at Cornell University.

Plants take up zinc as the Zn^{2+} ion, and mobility of this ion in soil has an important bearing on proper nutrition of plants. Diffusion is believed to be the dominant mechanism for transporting Zn^{2+} to plant roots. Figure 9-11 demonstrates how complexing agents or acids from root exudates or from decomposing organic residues facilitate the diffusion of Zn^{2+} to a simulated root.

Factors Affecting Availability and Movement of Zinc. The plant availability of Zn^{2+} is conditioned by a number of soil and environmental factors: pH, adsorption on surfaces of clay, organic matter, carbonates, and oxide minerals; complexation by organic matter; interactions with other nutrients; and climatic conditions.

Soil pH. The availability of soil zinc to plants decreases with increased soil pH, as would be expected by reason of the sensitivity of Zn^{2+} solubility to pH. Most pH-induced zinc deficiencies occur within the range 6.0 to 8.0 and calcareous soils are particularly prone to this nutritional problem. It should be noted that not all basic soils are zinc deficient because of mechanisms such as chelation of Zn^{2+}

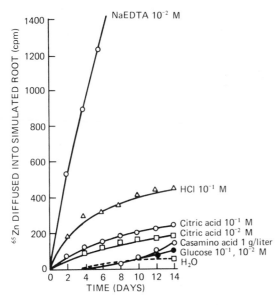

FIGURE 9-11. Effect of various complexing agents and acids on the accumulative diffusion of zinc into a simulated root over a 14-day period. [Elgawhary et al., *Soil Sci. Soc. Am. J.*, **34**:211 (1970).]

by naturally occurring organic substances which may compensate for the low solubility of Zn^{2+} at high pH.

At high pH, zinc forms insoluble compounds such as $Zn(OH)_2$ and $ZnCO_3$ which can reduce the available zinc in soils to lower levels. Formation of these substances alone is not responsible for zinc deficiencies in plants. The drastic decline in Zn^{2+} solubility with increasing pH is the overruling factor.

Liming acid soils, especially ones low in zinc, will reduce uptake of Zn^{2+}. This depressive action is usually attributed to the effect that increasing pH has on lowering Zn^{2+} solubility. It is possible, however, that some Zn^{2+} could be adsorbed on the surface of freshly added particles of liming agents such as $CaCO_3$.

It is generally accepted that adsorption of Zn^{2+} by soil constituents, including clay minerals and various oxide minerals of aluminum, iron, and magnesium, is highly pH dependent with the amounts fixed increasing markedly with increasing pH. The strength of Zn^{2+} adsorption by hydrated Al_2O_3 and Fe_2O_3 was reported by researchers at the University of Manitoba to increase with increasing pH.

Adsorption of Zn^{2+} by organic matter is also influenced by pH. For example, the amounts of Zn^{2+} complexed by humic acids increases with rising pH. Also with increasing pH, the stabilities of zinc–organic complexes increase up to a point, after which the complexes break up and hydroxides form.

Adsorption by Oxide Minerals. Clay minerals, sesquioxides, carbonates, and soil organic matter are all recognized as being able to adsorb Zn^{2+}.

The exact mechanisms of Zn^{2+} adsorption on oxide surfaces are not clear but several have been proposed. Two Australian researchers, Quirk and Posner, viewed this process as a bridging ligand between two neutral sites:

In their scheme, an olation bridge and ring structure is postulated. Such adsorption is considered a growth or an extension of the surface resulting in specific or irreversible retention of Zn^{2+}. In addition to the bridging between two neutral sites shown here, it has been suggested by others that Zn^{2+} could be adsorbed to two positive sites or to a positive and neutral site.

A second mechanism functioning at lower pH and resulting in nonspecific adsorption of Zn^{2+} has been advanced by investigators at the University of Manitoba. They proposed the following adsorption reaction:

$$
\begin{array}{c}
\text{OH}_2 \quad \text{ZnCl}^+ \\[2pt]
\diagup \\
\text{Fe or Al} \qquad + \quad \text{or} \qquad \qquad \rightleftharpoons \text{Fe or Al} \!-\! \begin{array}{c} \text{H} \\ \text{O} \\ \text{OH}_2 \end{array} \!-\! \text{ZnCl} + \text{H}^+ \\[2pt]
\diagdown \\
\text{OH}_2 \quad \text{Zn}^{2+} + \text{Cl}^-
\end{array}
$$

In this process only one H^+ is released for each Zn^{2+} and the retained Zn^{2+}, being less firmly held, can be replaced by other cations, such as Ca^{2+}, Mg^{2+}, and Ba^{2+}.

Adsorption by Clay Minerals. Fixation of Zn^{2+} by clay minerals such as bentonite, illite, and kaolinite has been reported by several investigators. The presence of iron and aluminum oxides influenced this adsorption. Removal of iron oxides increased the capacity of clays to adsorb Zn^{2+} more frequently than it decreased their retentiveness. Zinc adsorption capacity is usually directly related to the cation exchange capacity of clays and hydrous oxides.

Zinc reversibly bound by silicate clay minerals is an exchangeable cation and should be of value to plants. On the other hand, it appears that the effectiveness of soil and fertilizer zinc can be impaired to some extent when Zn^{2+} is specifically adsorbed by clays.

Adsorption by Carbonate Minerals. Zinc adsorption by carbonates or precipitation of $Zn(OH)_2$ or $ZnCO_3$ is believed to be partly responsible for the unavailability of Zn^{2+} in calcareous soils. Detailed analysis of the Zn^{2+} adsorption curves presented in Figure 9-12 revealed that $CaCO_3$ content was the principal factor contributing to the zinc adsorption maximum. Although $ZnCO_3$ was precipitated at higher Zn^{2+} concentrations, zinc availability to plants will probably not be seriously impaired because the solubility of this compound is too high for it to persist in soils.

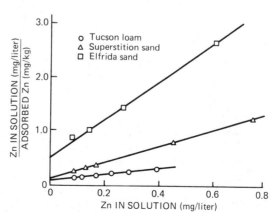

FIGURE 9-12. Adsorption of low concentrations of zinc by calcareous Arizona soils. [Udo et al., *Soil Sci. Soc. Am. J.*, **34**:405 (1970).]

Zinc is most strongly adsorbed by magnesite ($MgCO_3$), to an intermediate degree by dolomite [$CaMg(CO_3)_2$], and least of all by calcite ($CaCO_3$). In magnesite and dolomite it appears that zinc is actually adsorbed into the crystal surfaces at sites in the lattice normally occupied by magnesium atoms.

Complexation by Soil Organic Matter. The high capacity of organic soils to fix zinc and other micronutrients has brought attention to the potential role of the organic-matter fraction in soils on zinc complexation. It is well known that zinc forms stable complexes with soil organic-matter components. The humic and fulvic acid fractions are prominent in zinc adsorption.

Three classes of reactions of organic matter with micronutrients such as zinc have been distinguished:

1. Immobilization by high-molecular-weight organic substances such as lignin.
2. Solubilization and mobilization by short-chain organic acids and bases.
3. Complexation by initially soluble organic substances which then form insoluble salts.

Thus the action of organic matter on zinc can be expected to be variable depending on the characteristics and amounts of the organic materials involved. When reactions 1 and/or 3 prevail, availability of soil zinc will be adversely affected such as likely occurs in zinc deficient peats and humic gley soils. On the other hand, formation of soluble chelated compounds of zinc will enhance availability by shielding the retained zinc from fixation reactions.

Substances present in or derived from freshly applied organic supplements have the capacity to chelate Zn^{2+}. However, the increased solubilization and extractability of zinc is not always reflected in enhanced zinc uptake by plants.

Interaction with Other Nutrients. Other metal cations, including Cu^{2+}, Fe^{2+}, and Mn^{2+}, inhibit plant uptake of Zn^{2+}, possibly because of competition for the same carrier site. The antagonistic effect of several cations, especially Cu^{2+} and Fe^{2+}, on zinc uptake by rice is clearly demonstrated in Figure 9-13.

Phosphorus. There are widespread reports that high phosphorus availability in soil induces zinc deficiency. This interaction occurs mainly when plants are grown without sufficient zinc. Many of the results have been confusing because the accentuated symptoms of zinc deficiency were not always accompanied by reductions in zinc concentration in plant tops.

Loneragan and his colleagues in Australia and Jackson and Christensen at Oregon State University appear to have solved this intriguing problem. It seems that when plants are deficient in zinc their ability to regulate phosphorus accumulation is either lost or severely impaired. As a consequence, phosphorus is absorbed by roots and transported to plant tops in such excess that it becomes toxic and produces symptoms resembling zinc deficiency, whereas zinc concentrations in plant tops remain relatively unaffected. These workers also established that increasing phosphorus supply to zinc-deficient plants further enhances accumulation and toxicity of phosphorus in old leaves.

Serious concerns have been expressed about inducing zinc deficiencies, particularly in sensitive crops such as corn, flax, and beans, as a result of systematically building up available phosphorus levels in soil. This new evidence on phosphorus toxicity being associated with marginal zinc nutrition reveals that the problem can be corrected or avoided by ensuring that zinc supplies are also adequate. Usually, modest applications of zinc are sufficient.

The influence of high levels of available phosphorus in soil on mycorrhizal uptake of zinc might also be a factor in faulty zinc nutrition of crops. Mycorrhizae are known to increase the uptake of zinc and copper by many plants. Research con-

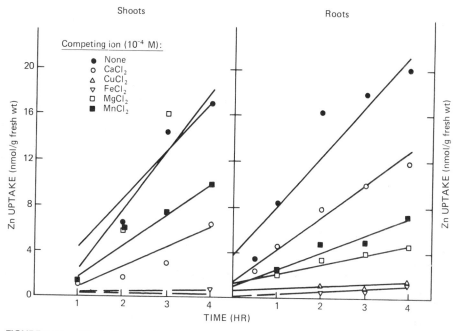

FIGURE 9-13. Effect of competing ions on uptake of zinc in shoots and roots of rice seedlings immersed in 5×10^{-8} M $Zn^{65}Cl_2$. [Giordano et al., *Plant Soil*, **41**:637 (1974).]

ducted by Lambert and others at the Pennsylvania State University showed that phosphorus fertilization suppressed mycorrhizal uptake of zinc and copper in corn and soybeans. Two corn lines were affected slightly differently by this phosphorus × mycorrhizal interaction.

The popularly held belief that P-Zn reactions in soil, such as the formation of insoluble $Zn_3(PO_4)_2 \cdot 4H_2O$, are responsible for phosphorus-induced zinc deficiency should be largely discounted. Solubility of this compound is sufficiently high that it will readily provide zinc to plants.

Sulfate. The highly mobile $ZnSO_4^0$ complex is an important species in soils and contributes significantly to total zinc in solution. Solubility and mobility of Zn^{2+} in soils are believed to be increased by the presence of SO_4^{2-} and subsequent formation of this complex.

Olsen and Watanabe of the USDA showed that a Fe × Mo × S interaction existed. In this three-way interaction, molybdenum decreased iron uptake in the absence of gypsum; however, SO_4^{2-} supplied as gypsum decreased molybdenum uptake, which, in turn, increased concentrations of zinc as well as iron and manganese in sorghum grown on six basic Colorado soils. The slight lowering of pH that occurred when gypsum was used may also have contributed to higher availability of the three metal cations. Additional work at this USDA laboratory demonstrated that gypsum added to a slightly acid soil increased the zinc and iron concentrations in leaves of sorghum plants. It has also been reported that gypsum applications markedly decreased the pH of flooded, strongly basic rice soils in the Punjab, India, and significantly increased the yield and zinc uptake of rice.

Nitrogen. Liberal applications of nitrogen fertilizer can stimulate plant growth and increase zinc requirements beyond the available supply. The amount and properties of the nitrogen source and its placement in relation to the zinc fertilizer has a notable effect on zinc availability. Nitrogen fertilizers that are acid forming will increase the uptake of both native and supplemental zinc. On the other hand, products with a neutral-to-basic effect are known to depress zinc uptake.

Flooding. At one time zinc deficiency in flooded rice soils was believed to be confined to those with high pH or those containing $CaCO_3$. However, this problem is now also known to occur in acid soils. When soils are submerged the concentration of most nutrient elements in the soil increases, but this is not true for zinc. In acid soils the decrease in zinc levels may be attributed to the increase in pH following reduction. The lowering of pH that occurs when calcareous soils are submerged is normally expected to increase zinc solubility. Observed reductions in zinc concentration could be the result of formation of a less soluble mineral such as franklinite ($ZnFe_2O_4$) or sphalerite (ZnS). Kittrick at Washington State University has demonstrated that if zinc levels in soils are controlled by the latter mineral, the likelihood of zinc deficiency is greater the higher the soil pH and the poorer the aeration. He suggests that some of the deleterious effects of H_2S on plants under anaerobic conditions may actually be due to zinc deficiency.

Climatic conditions. Zinc deficiencies are generally most pronounced during cool, wet spring seasons and often disappear with the coming of warmer weather. Apparently, climatic conditions that prevail during early spring contribute to zinc deficiency in a complex manner, possibly involving poor light as well as cool temperature and excessive moisture.

Increases in soil temperature have been shown to increase the availability of zinc to crops. The growth and zinc uptake responses to zinc fertilizer have been shown to be greater at low temperatures than at high temperatures, suggesting that native soil zinc becomes more available in warmer soil and more nearly satisfies crop requirements. Relative responses to fertilizer zinc have also tended to be greater in the presence of supplemental iron.

Movement of Zinc in Soil. It is obvious from the foregoing that zinc is relatively immobile in most soils. Its movement has been studied by workers in California. Zinc sulfate and zinc oxide were applied to the surface of soil placed in leaching columns. Water was passed through these columns and the distribution of zinc throughout the length of the columns was determined. The results are shown in Figure 9-14. It is apparent that most of the zinc was retained in the surface inch of soil and that only little if any downward movement occurred. These researchers indicated that even though the zinc was retained in a water-insoluble form much of it was removable with a dithizone extractant. Zinc so removed has been well correlated with plant uptake of this element, at least by workers in California.

Plant Factors. Species and varieties of plants differ characteristically in their susceptibility to zinc deficiency. Sensitivity of various crops to low levels of soil zinc is rated in Table 9-13. Corn and beans are very susceptible to low zinc. Fruit trees in general, and citrus and peach in particular, are also sensitive.

Corn cultivars and inbred lines differ in their ability to take up and use zinc. A comparison of the performance of inbreds A635 and H84 under varied zinc nutrition showed the latter to be more zinc-efficient. It produced more total dry matter and more dry matter per unit zinc, developed fewer zinc-deficiency symptoms, had more zinc in tops compared to roots, and had fewer imbalances of other mineral elements when grown at low levels of zinc. Differential responses between these two inbreds for zinc appeared to be caused by differences in translocation, requirements, and utilization of zinc, and imbalanced accumulations of mineral elements that interact with zinc. In addition to these characteristics of the zinc-efficient

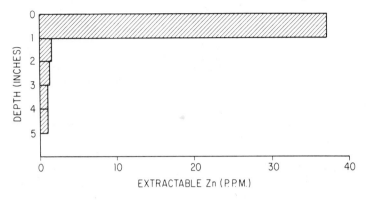

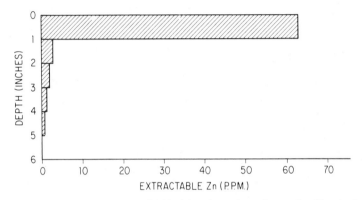

FIGURE 9-14. Movement of added zinc through a column of soil leached with demineralized water. [Brown et al., *SSSA Proc.*, **26**:167 (1963).]

TABLE 9-13. Sensitivity of Crops to Low Levels of Available Zinc in Soil

Very Sensitive	Mildly Sensitive	Insensitive
Beans, lima and pea	Alfalfa	Asparagus
Castor beans	Barley	Carrots
Citrus	Clovers	Forage grasses
Corn	Cotton	Mustard and other crucifers
Flax	Potatoes	Oats
Fruit trees (deciduous)	Sorghum	Peas
Grapes	Sugar beets	Peppermint
Hops	Tomatoes	Rye
Onions	Wheat	Safflower
Pecans		
Pine		
Rice		
Soybeans		
Sudangrass		

Sources: Adapted from *Zinc in Crop Nutrition.* New York: International Lead Zinc Research Organization, Inc., and Zinc Institute, Inc., 1974; and Mortvedt, *Farm Chem.*, **143**(11):56 (1980).

corn inbred, other investigators have reported substantial differences among corn plant roots to exploit the soil for zinc.

Striking differences in the susceptibility of navy bean varieties to zinc deficiency were reported some years ago. The Saginaw variety is both more zinc efficient and more tolerant to excess zinc than is Sanilac. Genotypic differences in sensitivity to zinc deficiency or responsiveness to zinc fertilization are known to occur in several other crops, including oats, potatoes, and soybeans.

Zinc Fertilizers and Correction of Deficiencies. A number of zinc-containing sources can be used for correction of deficiencies (Table 9-14). Zinc sulfate, containing about 35% zinc, continues to be a popular fertilizer material. In addition to zinc sulfate and the other inorganic salts, zinc can also be supplied in frits, synthetic chelates, and natural organic complexes.

Although $Zn(OH)_2$ minerals are not specified in Table 9-14, they as well as ZnO and $ZnCO_3$ make good fertilizers in soils because they dissolve sufficiently to maintain adequate Zn^{2+} levels for plants. Zinc phosphate, $Zn_3(PO_4)_2$, is less soluble than the oxides, hydroxides, or carbonates, but in soils it can be expected to supply available zinc to plants over extended periods of time. Zinc-iron-ammonium sulfate made by ammoniating spent sulfuric acid containing iron and zinc impurities is also commercially available.

Recommended amounts of zinc depend on crop, zinc source, method of application, and severity of zinc deficiency. Rates usually range from 3 to 20 lb/A when inorganic zinc salts are used and from 0.5 to 2.0 lb/A when the zinc source is either a chelate or an organic complex. For most field and vegetable crops, 10 lb/A is recommended in clay and loam soils, and 3 to 5 lb/A in sandy soils. In most cropping situations, applications of 10 lb/A of zinc can be effective for 3 to 5 years.

The methods of providing zinc fertilizers for control or prevention of deficiencies range from soil and foliar applications to tree injections. Because the mobility of zinc can be very limited in soils, soil applications of zinc should be worked thoroughly and deeply into the soil.

In the case of long-term perennials such as hops, grapes, and tree fruits, preplant soil applications of zinc are effective. The rates of zinc for establishing hops and grapes is 20 lb/A with a higher amount of 100 lb/A of zinc for tree fruits. Soil applications are only of limited value after these crops have been established.

Because of the immobility of zinc in soil, especially those that are basic or calcareous, spatial distribution of zinc fertilizer granules or droplets plays an

TABLE 9-14.   Some Sources of Fertilizer Zinc

Source	Formula	Percent Zn (approx.)
Zinc sulfate monohydrate	$ZnSO_4 \cdot H_2O$	35
Zinc sulfate heptahydrate	$ZnSO_4 \cdot 7H_2O$	23
Basic zinc sulfate	$ZnSO_4 \cdot 4Zn(OH)_2$	55
Zinc oxide	ZnO	78
Zinc carbonate	$ZnCO_3$	52
Zinc sulfide	ZnS	67
Zn frits	(silicates)	Varies
Zinc phosphate	$Zn_3(PO_4)_2$	51
Zinc chelates	$Na_2ZnEDTA$	14
	NaZnNTA	13
	NaZnHEDTA	9
Zinc polyflavonoid	—	10
Zinc ligninsulfonate	—	5

Source: Mortvedt et al., Eds., *Micronutrients in Agriculture*, p. 371. Madison, Wis.: Soil Science Society of America, 1972.

important role in supplying sufficient zinc to plant roots. Properties that increase the number of fertilizer particles or droplets applied per unit area of soil at a given application rate are low zinc concentration, low particle or droplet density, small particle or droplet size, high solubility, and special properties such as acidulation and chelation.

Foliar applications are effective and they are used primarily for tree crops. Sprays containing up to 50 lb of zinc sulfate in 100 gal of water are usually applied at the rate of 10 to 15 lb/A of zinc to dormant orchards. Spray treatments of up to 10 lb of $ZnSO_4$ per 100 gal of water are sometimes applied directly to the foliage of growing crops, but they can be harmful. Damage to foliage can be prevented by adding half as much hydrated lime or soda ash to the solution or using less soluble materials like zinc oxide or carbonate.

Other plant methods include seed coatings, root dips, and tree injections. The former treatment may not supply enough zinc for small seeded crops, but dipping potato seed pieces in a 2% ZnO suspension is said to be satisfactory.

Chelates and natural organic complexes are particularly suitable for foliar sprays when quick recovery of young seedlings is needed. These zinc sources can be used in high-analysis liquid fertilizers, where their high solubility and compatability with other components are great advantages. Chelates such as ZnEDTA are mobile and can be surface applied under irrigated cropping conditions.

The efficiency of banded zinc applications can often be improved by the presence of nitrogen fertilizers, particularly those that acidify the soil (Figure 9-15). Availability of indigenous soil zinc was also increased by treatment with the acid-forming nitrogen fertilizers ammonium sulfate and ammonium nitrate. Enhancement of zinc availability through acidification by elemental sulfur and other sulfur compounds was mentioned in Chapter 8.

Sodium

Although sodium constitutes an appreciable fraction of the earth's crust (2.8%), soils contain lower amounts, ranging from 0.1 to 1%. Its estimated average concentration in soils is 0.63%. This lower sodium amount indicates a weathering

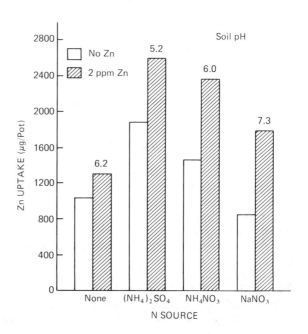

FIGURE 9-15. Plant uptake of zinc in cropping sequence of two crops of sorghum followed by four clippings of Ladino clover and final crop of sorghum as affected by application of $ZnSO_4$, several nitrogen sources, and final pH in noncalcareous fine sandy loam soil with initial pH of 7.2. [Viets et al., *Soil Sci. Soc. Am. J.*, **21**:197 (1957).]

away of sodium from sodium-containing substances in soil. Sodium is normally found in very small amounts in soils of humid regions, whereas it can be a major component of soils in arid and semiarid regions. Occasionally, sodium will be present in measurable quantities in soils of humid areas where sodium nitrate fertilizer has been used for many years.

Forms in Soil. Three forms of sodium are found in soils: fixed in insoluble silicates, exchangeable in the structures of other minerals, and soluble in the soil solution. In the majority of soils, most of the sodium is present in silicates. Sodium in highly-leached soils occurs in high-albite plagioclases and in small amounts in perthite, micas, pyroxenes, and amphiboles, which exist mainly in the fine sand and silt fractions. In arid and semiarid soils, sodium exists in silicates, as well as in quantity as $NaCl$, Na_2SO_4, and sometimes as Na_2CO_3 and other soluble salts.

The amounts of soluble and exchangeable Na^+ vary greatly according to soil type. In most productive humid-region soils the relative importance of exchangeable Na^+ in relation to other major cations is $Ca^{2+} > Mg^{2+} > K^+ = Na^+$.

Exchangeable Na^+ is present in two forms in soils. The first is held loosely on clay sheets and the second is held tightly on specific sites, probably on the broken edges of clay sheets. This more strongly bound sodium is probably the major form of exchangeable Na^+ in humid soils.

The loosely held Na^+ is utilized by annual crops. Sugar beets respond to sodium fertilization when readily exchangeable Na^+ in soil is less than 0.05 meq per 100 g.

In arid regions and if soils are irrigated with sodic waters, exchangeable Na^+ levels generally exceed those of K^+. Furthermore, sodium salts accumulating in poorly drained soils of the arid and semiarid regions will be major contributors to soil salinity.

Effect of Sodium on Soil Properties. The dispersing action of Na^{2+} on clay and organic matter with resultant breakdown of soil aggregates and impairment of permeability to air and water is well known. High sodium-affected soils become almost impervious to water and air because of the loss of large pores. Root penetration is impeded, clods are hard, and preparation of a good seedbed becomes difficult. Surface crusting will result in poor germination and uneven stands. For these reasons the presence of large quantities of sodium in fine-textured soils is undesirable. There is concern over the effects of sodium when it occurs in excess of 10 to 20% of the cation exchange capacity.

The detrimental effect of high levels of exchangeable Na^+ is conditioned by soil texture and type of soil clays. With fine-textured soils containing a large proportion of swelling clays only 10% exchangeable Na^+ can be tolerated, whereas in sandy soils the upper limit is 30%. Soils with these excessive amounts of exchangeable Na^+ are termed alkali soils and they present quite a problem in certain western and midwestern states and in large areas in the Prairie Provinces of Canada.

Growth of most crop plants is severely reduced on alkali soils (Table 9-15). Although high concentrations of sodium are toxic to some plants and the associated high pH can create deficiencies of the micronutrient cations, the main impediment to growth is the loss of soil permeability due to the detrimental effect of sodium on soil structure. Tree fruits and berries, however, are sensitive to sodium and will often develop toxicity symptoms before being harmed by unsatisfactory physical properties.

Reclamation of sodic soils involves replacement of the unwanted exchangeable Na^+ with Ca^{2+}, followed by removal in leaching water of the displaced Na^+ as sodium sulfate. Calcium-containing substances such as gypsum are used to supply Ca^{2+} when soils are low in this element. On soils containing calcium carbonate,

TABLE 9-15. Typical Reductions in Crop Yields at Various Exchangeable Sodium Percentages (ESP)

Type of Soil	ESP (%)	Average Decrease in Crop Yield (%)
Slightly sodic	7–15	20–40
Moderately sodic	15–20	40–60
Very sodic	20–30	60–80
Extremely sodic	> 30	> 80

Source: Velasco, *Sulphur Agri.*, **5**:2 (1981).

any of the acid-forming sulfur compounds can be used to solubilize this natural source of calcium.

Effect on Plant Growth. On the positive side, sodium is beneficial for the growth of some plants, as explained in Chapter 3. Its effect on plants ranges from that of an essential element for halophytic species to that of being relatively toxic to other plants, particularly stone fruits. As mentioned in Chapter 3, sodium is, in fact, now recognized as being essential for those plants having C-4 dicarboxylic photosynthetic pathways.

The beneficial effects of sodium on plant growth tend to be greatest when potassium nutrition is poorest. There is evidence that Na^+ can partially substitute for K^+, as also mentioned in Chapter 3. The extent of this substitution is closely related to the characteristic ability of plants to take up Na^+. Crops are listed in Table 9-16 according to their uptake potential for sodium. Growth of those crops with high and medium ratings will be favorably influenced by Na^+. Crops in the high category will be recognized as the so-called sodium-loving plants.

Sodium is absorbed by plants as the Na^+ ion, largely from the soil solution. Concentrations of between 0.5 and 5 ppm Na^+ in soil solution are representative of temperate-region soils.

High and harmful sodium levels in plants are often accompanied by low potassium concentrations. Potassium concentrations above 3.0 to 3.5% generally tend to keep sodium levels below 3%. Sodium concentrations as little as 1 to 2% in leaves of sensitive crops will, of course, lower yields. Calcium levels in plants containing excessive sodium may also be low to the point of being deficient.

Sodium Fertilizers. Responses to sodium have been observed in crops with a high uptake potential (Table 9-16) or a definite sodium requirement. To

TABLE 9-16. Sodium Uptake Potential of Various Crops

High	Medium	Low	Very Low
Fodder beet	Cabbage	Barley	Buckwheat
Sugar beet	Coconut	Flax	Maize
Mangold	Cotton	Millet	Rye
Spinach	Lupins	Rape	Soya
Swiss chard	Oats	Wheat	Swede
Table beet	Potato		
	Rubber		
	Turnips		

Source: Marschner, *Proc. 8th Colloq. Int. Potash Inst.*, pp. 50–63 (1971); cited by Mengel and Kirkby, *Principles of Plant Nutrition*, 3rd ed. Bern: International Potash Institute, 1982.

achieve maximum yield of these crops their sodium demands must be satisfied. The sodium demand of such crops appears to be independent of, and perhaps even greater than, their potassium demand.

In some parts of Europe, fertilization of grass forage crops with sodium is considered desirable. Sodium concentrations of between 1 and 2% in pasture grasses will improve palatability and will provide part of the animals requirements for this mineral.

The important sodium-containing fertilizers identified by Finck of the University of Kiel, West Germany, are:

Potassium fertilizers with various NaCl contents.

Sodium nitrate or Chile saltpeter (about 25% Na).

Rhenania phosphate (about 12% Na).

Multiple-nutrient fertilizers with sodium (e.g., pasture fertilizers with 15% nitrogen, etc., and 3% sodium).

Sodium chloride (NaCl) or common salt that contains approximately 40% sodium is in practice used only as a constituent of potassium fertilizer salts.

Chlorine

On the basis of an estimated average concentration of merely 0.05% in the earth's crust, chlorine is considered to be a minor constituent of the lithosphere. For example, igneous and metamorphic rocks contain 0.05% chlorine, while its concentration in limestones is considerably less at 0.02%. Chlorine is generally absent in shales and only trace quantities occur in sedimentary sandstones.

The majority of Cl^- in soils is believed to originate from salts trapped in soil parent material, from marine aerosols, and from volcanic emissions. Nearly all of the soil Cl^- has been in the oceans at least once, being returned to the land surface either by uplift and subsequent leaching of marine sediments, or by oceanic salt spray carried in rain or snow. Most of the soil Cl^- commonly occurs as soluble salts such as NaCl, $CaCl_2$, and $MgCl_2$. Chlorine is sometimes the principal anion in extracts of saline soils. The quantity of Cl^- in soil solutions may range from 0.5 ppm or less to over 6000 ppm.

External Sources of Chlorine.
Extremely variable amounts of chlorine reach soils from the atmosphere by fertilization, by crop protection programs, and in irrigation waters. Annual depositions in precipitation of 12 to 35 lb/A of chlorine are customary and in coastal areas these depositions may increase to over 100 lb/A. In addition to this source of chlorine, undetermined amounts are absorbed directly from the atmosphere, probably mainly as gaseous HCl released during volcanic eruptions. It is believed that most of the chlorine in seawater got there by washout of this HCl from volcanic activity.

Levels of 2 ppm of Cl^- or more are typical in the precipitation near seacoasts. The actual quantities depend on the amount of sea spray, which in turn is related to such factors as the temperature; the foam formation on tops of waves; the strength and frequency of winds sweeping inland from the sea; the topography of the coastal region; and the amount, frequency, and intensity of precipitation. Salty droplets or dry salt dust may be whirled to very great heights by strong air currents and carried for a long time over great distances.

Concentration of Cl^- in precipitation drops off rapidly in the interior direction, with areas about 500 miles inland averaging about 0.2 ppm. In the Midwest and Great Plains regions of the United States, the values are very uniform over large areas and range from 0.1 to 0.2 ppm.

Behavior in Soil. The Cl^- anion is bound very slightly by most soils in the mildly acid to neutral pH range and becomes negligible at pH > 7. Appreciable amounts can be adsorbed, however, with increasing acidity, particularly by kaolinitic soils, which have significant pH-dependent positive charge.

Nonspecifically adsorbed anions such as NO_3^-, Cl^-, and SO_4^{2-} are readily exchangeable because they are held in solution some distance away from the mineral surface. The divalent SO_4^{2-} ion is adsorbed more strongly and completely than the monovalent Cl^- and NO_3^- ions. Chloride is often used as an indicator of NO_3^- mobility because the former is not subject to biological and chemical transformations, and in most other aspects these ions behave similarly.

Because of the great mobility of Cl^- in all but extremely acid soils, it can be rapidly cycled through soil systems. The rate and extent of migration and accumulation of Cl^- in soils is very dependent on water circulation. It will accumulate where the internal drainage of soils has been or is restricted. Chloride in groundwaters within a few meters of the soil surface can be moved by capillarity into the root zone and deposited at or near the soil surface.

Problems of excess Cl^- occur in some irrigated areas. They are usually the result of interactions of two or more of the following factors:

1. Significant amounts of Cl^- in the irrigation water.
2. Failure to apply sufficient water to adequately leach out Cl^- accumulations in the root zone.
3. Unsatisfactory soil physical properties and drainage conditions for proper leaching.
4. High water table and capillary movement of chloride into the root zone.

Environmental damage in localized areas from high concentrations of Cl^- has resulted from sources related to human activities such as road deicing, water softening, salt-water spills associated with the extraction of oil and natural gas deposits, and disposal of feedlot wastes and various industrial brines.

The principal effect of too much Cl^- is to increase the osmotic pressure of soil water and thereby lower the availability of water to plants. In addition, some woody plants, including most fruit trees, berry and vine crops, and ornamental shrubs, are specifically sensitive to Cl^- ions and develop leaf-burn symptoms when Cl^- concentration reaches about 0.50% on a dry matter basis. Leaves of tobacco and tomatoes thicken and begin to roll.

Although most of the attention by far has been directed to plant damage caused by excesses of Cl^-, the possibility of deficiencies of this nutrient should not be ignored. Water-soluble chlorine levels of less than 2 ppm in soil are considered low.

Nutrient Interactions. Chloride uptake and accumulation in plants is depressed by high concentrations of NO_3^- and SO_4^{2-}. A mutual antagonism between Cl^- and NO_3^- has been observed for a number of plants, including potatoes, *Nitella*, beans, tomatoes, buckwheat, sugar beets, and perennial ryegrass. Lowering of NO_3^- uptake by increasing levels of Cl^- has also been observed in barley, corn, and wheat. This strong negative interaction between Cl^- and NO_3^- has been attributed to competition for carrier sites at root surfaces.

Chloride is beneficial for a number of plants especially the "salt"- or "Cl^- loving" ones, such as beet, spinach, cabbage, and celery. This was discussed in Chapter 3. Growth responses of coconut to dressings of KCl have been found to be closely related to the Cl^- content of leaves and negatively correlated with K^+ levels in the crop. Such findings help to explain the often observed positive effect of NaCl or even seawater on coconuts.

Another example of the favorable effect of increasing Cl^- concentration in crop

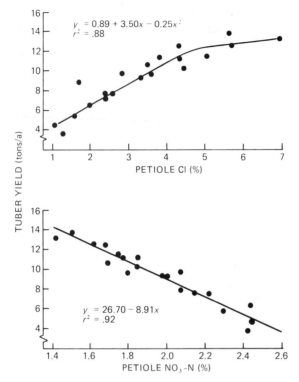

$y = 0.89 + 3.50x - 0.25x^2$
$r^2 = .88$

$y = 26.70 - 8.91x$
$r^2 = .92$

TUBER YIELD (tons/a)

PETIOLE Cl (%)

PETIOLE NO_3-N (%)

FIGURE 9-16. Relationship between yield of potatoes in central Oregon and Cl^- and NO_3^- concentrations in petiole samples taken on August 5, 1980. [Jackson et al., unpublished data, Oregon State Univ. (1981).]

plants and yield of a particular crop appears in Figure 9-16. Here potato yields increased as the Cl^- levels in petioles increased from 1.1% to 6.9%. Also shown is that NO_3^- decreased as Cl^- increased. Although not indicated here, uptake of K^+ increased also. In some situations decreased NO_3^- with added Cl^- is thought to be desirable.

Although these beneficial effects of Cl^- on plant growth are not fully understood, there are strong indications from research at Oregon State University led by Christensen and Jackson that improved plant water relationships and inhibition of plant diseases are two of the factors. These factors and their interactions were reviewed in Chapter 3.

Fertilizer Sources and Fertilization. When additional Cl^- is desirable, it can be supplied by

Ammonium chloride	66% Cl
Calcium chloride	65% Cl
Magnesium chloride	74% Cl
Potassium chloride	47% Cl
Sodium chloride	60% Cl

Rates of chloride will vary depending on a number of conditions, including crop, method of application, and purpose of addition (i.e., for correction of nutrient deficiency, for disease suppression, or for improved plant water status). Where take-all root rot of winter wheat is suspected in western Oregon, banding 35 to 40 lb/A of Cl^- with or near the seed at planting is recommended. Broadcast dressings of between 75 to 125 lb/A of Cl^- in February or March have effectively reduced crop stress from take-all and they have also reduced the severity of attack by leaf and head diseases (i.e., stripe rust and septoria).

Silicon

This element is the second most abundant element in the earth's crust. Its average concentration in the lithosphere is 27.6% and in soils it normally ranges between 23 and 35% and typically averages 32% by weight. Silicon has a central role in rock weathering and soil development. It is a major soil component lost during weathering and the conversions of silicon to secondary minerals are important aspects of soil development.

Sandy soils can contain as much as 40% silicon contrasted with as little as 9% silicon in highly weathered tropical soils. The laterite deposits in the tropics, consisting largely of hydrous aluminum and ferric oxides, are left after silicon is removed during intense weathering of soils.

Major sources of silicon include primary silicate minerals, secondary alumino-silicates, and several forms of silica (SiO_2). Silica occurs as six distinct minerals: quartz, tridymite, cristobalite, coesite, stishovite, and opal. The first three of these minerals plus opal comprise the major weight and volume percentage of most soils. Quartz is overwhelmingly the most common mineral species in most soils where it often makes up 90 to 95% of all sand and silt fractions.

Deficient Soil Areas.

Low-silicon soils are present in intensively weathered high-rainfall areas of Hawaii. Certain peat and sandy soils in south Florida are lacking sufficient silicon for top yields of sugarcane. Silicon inadequacies have also been reported in Australia, Ceylon, India, Japan, Mauritius, Puerto Rico, Saipan, and South Africa.

Properties of the silicon-deficient Hawaiian soils include low total silicon and high total aluminum contents, low base saturation, and a correspondingly low pH. Further, they all have extremely high phosphorus-fixing capacity. Some of the soils are bauxitic in nature, almost to the point of being potential sources of industrial aluminum. Supplies of plant available Fe^{2+} and Mn^{2+} may also be high in these soils.

Soil Solution Silicon.

In the usual pH range of soils, $H_4SiO_4^0$ or $[Si(OH)_4]$ is the principal silicate species in solution. At pH values above 8.5, ionic species such as H_3SiO_4 contribute significantly to total silicon in solution. At high concentrations of about 28 ppm of silicon in solution, the monomer $H_4SiO_4^0$ polymerizes to form precipitates of amorphous silica. Conditions favoring this reaction are limited, and it probably takes place in basic soil solutions, in interstitial solutions between expanded 2:1 lattice silicates, and perhaps on the surface of actively weathering minerals.

The solubility of silicon in water is unaffected by pH in the range 2 to 9. However, this independence does not apply in soils since the Australian workers Jones and Handreck reported that silicon concentration decreased from 33 to 11 ppm as pH increased from 5.4 to 7.2. Other scientists have demonstrated that silicon concentration increases on either side of a minimum lying between pH 8 and 9.

There is substantial evidence that concentration of $H_4SiO_4^0$ or monosilic acid in soil solutions is largely controlled by a pH-dependent adsorption reaction. Although silicon is adsorbed on the surfaces of various kinds of inorganic substances in soils, iron and aluminum oxides are believed to be mainly responsible for this reaction. The adsorptive capacity of aluminum oxides declines markedly with greater degree of crystallinity. The mechanism for this reversible retention of $H_4SiO_4^0$ is uncertain.

Silicon concentrations of less than 0.9 to 2 ppm in soil solution were shown by Fox and his colleagues at the University of Hawaii to be insufficient for proper nutrition of sugarcane. Results from a sandy soil in south Florida support the reliability of these Hawaiian values for predicting the need for supplemental sil-

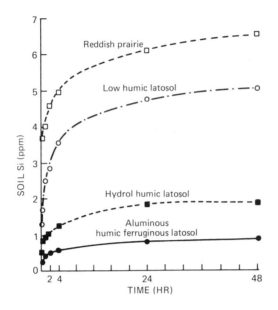

FIGURE 9-17. Concentration of water soluble silica in four Hawaii soils. The Reddish Prairie and Hydrol Humic Latosol are developed from ash with low and high rainfall, respectively. The Low Humic Latosol and Humic Ferruginous Latosol are from basalt, with low and high rainfall, respectively. [Fox et al., *Soil Sci. Soc. Am. J.*, **31**:775 (1967).]

icon. By way of comparison, levels of from 3 to 37 ppm silicon in solution have been reported for a wide range of normal soils. Figure 9-17 illustrates the large differences in soluble silicon observed in several Hawaiian soils.

Factors Affecting Availability and Movement of Silicon. Among the soil and management conditions influencing plant uptake of silicon are content of iron and aluminum oxides, liming, flooding, and nutrient supply.

Increasing amounts of iron and aluminum oxides will lower both supply of silicon in solution and plant uptake. Liming has frequently decreased silicon uptake by various plants, including barley, oats, red clover, rice, ryegrass, sorghum, and sugarcane. Conversely, acidification has increased silicon uptake by oats.

Higher water content of soil encourages silicon uptake, particularly by crops such as rice. Concentration of silicon in the soil solution is increased with time of submergence, in one case increasing from 11 to 19 ppm in 50 days. The reasons for this increase are not fully understood because the higher pH resulting from the flooding of acidic rice soils is expected to decrease silicon concentration in the soil solution.

Heavy applications of nitrogen are known to make rice plants more susceptible to fungal attack because of decreases in silicon concentration in the straw. To correct or avoid this problem, recommendations are made in Japan and elsewhere to add silicon-bearing materials when high rates of nitrogen fertilizer are used.

Plant Factors. Plants take up different amounts of silicon according to their species. It is generally accepted that Gramineae contain 10 to 20 times the silicon concentration normally found in legumes and other dicotyledons. Lowland or paddy rice commonly contains between 4.6 and 7.0% silicon in the straw.

Fertilizer Sources and Fertilization . The major materials that have successfully increased silicon supplies to crops are:

Calcium silicate slag	
[mostly $CaAl_2Si_2O_8$ with some $CaSiO_3$	
(by-product of electric furnaces)]	18 to 21% Si
Calcium silicate ($CaSiO_3$)	31% Si
Sodium metasilicate ($NaSiO_3$)	23% Si

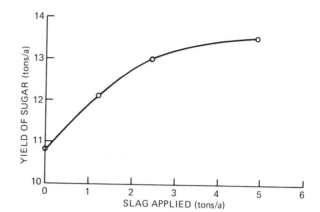

FIGURE 9-18. Effect of electric furnace slag on yield of sugar from sugarcane grown on an aluminous humic ferruginous latosol in Hawaii. Mean results for plant and ratoon crops combined. [Ayres, *Soil Sci.*, **101**:216 (1966).]

Minimum rates of 5000 lb/A of calcium silicate are applied to sugarcane and in some areas greater quantities of 8000 lb/A or more may be beneficial. Broadcasting followed by mixing into the soil before planting is an acceptable method of application. Annual calcium silicate applications of between 500 and 1000 lb/A applied in the cane row have also improved yields of sugarcane. The beneficial effect of increasing rates of electric furnace slag on sugar yields in Hawaii is evident in Figure 9-18. Additions of lime that increase calcium levels and decrease soil acidity do not produce similar dramatic improvements in growth of sugarcane.

Selenium

This element is apparently not needed by plants, but it must be present in fodder and crop plants since it is essential for animals. An understanding of the forms and behavior of selenium in soil is important because of its need in livestock diets and the potential hazards from excessive dietary intake. Selenium ranks about 70th in terrestrial abundance and occurs in minute amounts in nearly all materials of the earth's crust. The amount of this element in crustal rocks averages only 0.09 ppm. It tends to concentrate in sedimentary rocks, particularly those containing organic matter. Even when selenium is consolidated in lithospheric materials, it rarely exceeds 100 to 500 ppm.

Selenium exhibits some of the same chemical and physical behavior as sulfur. Unlike sulfur it is seldom found in the native state. Instead, it is a major constituent of 39 mineral species and a minor component of 37 others, chiefly sulfides. Selenium is most common in chalcopyrite, bornite, and pyrite minerals. It has five important oxidation states -2, 0, $+2$, $+4$, and $+6$.

Seleniferous Soils. The total selenium concentration in most soils lies between 0.1 and 2 ppm and averages about 0.3 ppm. Seleniferous soils developed from Cretaceous shale tend to have much higher concentrations, ranging from 2 to 10 ppm. There are extensive areas of high-selenium soils in South Dakota, Wyoming, Montana, North Dakota, Nebraska, Kansas, Colorado, Utah, Arizona, and New Mexico which produce vegetation toxic to livestock. Large areas of similar seleniferous soils also occur in the provinces of Alberta, Saskatchewan, and Manitoba.

Selenium-toxic soils are also encountered in Colombia, Ireland, South Africa, the state of Queensland in Australia, the United Kingdom, the USSR, Mexico, and Israel. The parent materials of these soils are sedimentary deposits with shales predominating.

Toxic seleniferous soils are usually basic in reaction and contain free calcium carbonate. They occur in regions of low rainfall, usually less than 20 in. of total annual precipitation. Another important characteristic of these soils is the presence of water-soluble selenium.

Selenium-rich acid soils occur in the Hawaiian Islands and in Puerto Rico. They do not support toxic seleniferous vegetation. Selenium in these highly ferruginous soils is relatively unavailable to plants.

Low Selenium Soils. Several areas in the United States have been identified where the crop plants contain low concentrations of selenium. The main areas are in the Pacific Northwest, the southern Atlantic seaboard, and the Great Lakes region, extending south and east to include large parts of Wisconsin, Illinois, Indiana, Michigan, Ohio, West Virginia, Maryland, Pennsylvania, and the New England states. Smaller areas also occur in Arizona and New Mexico.

Many of the low-selenium soils in the United States are acid, mostly Spodosols and Inceptisols and contiguous areas of Alfisols. Insufficient plant uptake of selenium is usually caused by one or both of the following soil factors: low total selenium in the soil parent material or low availability of selenium in acid and poorly drained soils.

Low selenium soils occur in Canada in the interior of British Columbia, west central Alberta, northern Ontario, the Eastern Townships and Lower St. Lawrence regions of Quebec, northern New Brunswick, and elsewhere in the Maritime Provinces. Selenium deficiency of livestock is also known to be a problem in certain areas of New Zealand, Australia, Scotland, Finland, Sweden, Austria, Germany, France, western USSR, Turkey, and Greece.

Soil Selenium. The forms of selenium generally considered to be present in soil are (1) selenides (pyritic), (2) elemental selenium, (3) selenites, (4) selenates, and (5) organic selenium compounds. These forms and their relationships in soils and weathering sediments are shown in Figure 9-19. According to this schematic summary, the chemical forms of selenium in soils and sediments are closely related to redox potential, pH, and solubility.

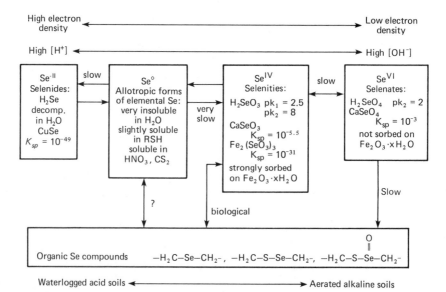

FIGURE 9-19.   Generalized chemistry of selenium in soils and weathering sediments. [Allaway, in D. E. Hemphill, Ed., *Trace Substances in Environmental Health—II*, Univ. of Missouri (1968).]

Selenides (Se^{2-}). Selenides are largely insoluble and they are associated with sulfides in soils of semiarid regions where weathering is limited. They contribute little to selenium uptake because of their insolubility. Some selenium may become available to plants, however, as a result of weathering and slow hydrolysis.

Elemental Selenium (Se^0). Elemental selenium is present in small amounts in some soils. It may be either an intermediate in the oxidation of the element to a soluble form or a transitory constituent of neutral and acid soils during the reduction of selenites under acid conditions. Significant amounts of elemental selenium may be oxidized to selenites and selenates by microorganisms in neutral and basic soils. There are indications that some oxidation of selenium may proceed by nonbiological pathways.

Selenites (SeO_3^{2-}). A large fraction of the selenium in acid soils may occur as stable complexes of selenites with hydrous iron oxides. A team of USDA researchers at Ithaca, New York, showed that the thermodynamically stable selenium compound in acid-to-neutral soils is a ferric selenite–ferric hydroxide adsorption complex. The low solubility of iron–selenite complexes is apparently responsible for the nontoxic levels of selenium in plants growing on acid ferruginous soils having very high total selenium contents.

Plants will absorb selenite but generally to a lesser extent than selenate. However, both forms are taken up equally by some *Astragalus* species which accumulate selenium.

Selenates (SeO_4^{2-}). Selenates are frequently associated with sulfates in arid-region soils. This form of selenium is stable in a basic, oxidizing environment such as that found in many well-aerated, semiarid seleniferous soils. Other forms of selenium will be oxidized to selenates under these conditions. Only limited quantities of selenate occur in acid and neutral soils, except possibly for a few weeks following the addition of selenites or selenates to the soil.

They are highly soluble and readily available to plants, thus being largely responsible for toxic accumulations in plants grown on high-pH soils. Most of the water-soluble selenium in soils probably occurs as selenates.

Organic Forms of Selenium. Very little is currently known about the nature of organic forms of selenium in soils. It can be an important fraction since up to 40% of the total selenium in some soils is present in humus.

Soluble organic selenium compounds are liberated through the decay of seleniferous plants. Such substances derived from accumulator or indicator plants are readily taken up by other plants. The soluble selenium of plant origin is quite stable in semiarid areas, and much of it remains available in soil. Organic forms of selenium are believed to be more soluble under basic than acidic soil conditions, which would enhance their availability to plants in basic soils of semiarid regions.

Factors Affecting Movement and Availability of Selenium. Plant availability of selenium is influenced by a number of conditions, including pH and liming, interrelationships with other elements, and climatic conditions. Selenium uptake by plants has generally been reported to be greater at high pH values than at acid reactions. Solubility of selenium in the soil solution is lowest at slightly acid-to-neutral pH, while it increased under both more acidic and basic pH values. High soil pH will facilitate the oxidation of selenites to the more readily available selenates.

Increased yields in response to nitrogen fertilization may lower selenium concentrations in crops through dilution of limited supplies of soil selenium. Higher yields resulting from sulfur fertilization will have this same effect on selenium levels in crops, which is sometimes mistakenly thought to be the result of strictly ion antagonism. However, it should be noted that SO_4^{2-} and SeO_4^{2-} ions can act antagonistically toward one another. In some areas of the western United States and Canada where sulfur fertilization is practiced, there has been much concern about increased incidence and severity of selenium deficiencies in cattle due to the restrictive action of sulfur on selenium uptake by crops.

Phosphatic fertilizers have had variable effects on selenium uptake by plants, depending on the form of selenium present and on the chemical composition of the phosphorus source. Straight phosphate salts may increase selenium availability, while those products containing other substances such as gypsum may be inconsistent in their action.

A greater frequency of livestock nutritional disorders caused by low selenium has been observed in Sweden after cold and rainy summers than after hot dry ones. High summer temperatures are amenable to increased selenium concentration in feedstuffs.

Plant Factors. Plants differ in their uptake of available selenium from species to species, with stage of growth, and season. Certain species of *Astragalus* absorb many times as much selenium as do other plants growing in the same soil. These *Astragalus* plants, often referred to as selenium-accumulators, utilize selenium in an amino acid peculiar to the species. Other seleniferous plants include species of *Machaeranthera*, *Haplopappus*, and *Stanleya*.

Plants such as the cruciferae (cabbage, mustard, etc.) and onions, which require large amounts of sulfur, absorb intermediate amounts of selenium. Grasses and grain crops absorb low-to-moderate amounts of selenium.

Selenium Fertilization. Although selenium-deficiency disorders such as "muscular dystrophy" or "white muscle disease" in cattle and sheep can be corrected or prevented by approved therapeutic measures, there is interest in fertilizing with selenium to produce forages of adequate selenium status for grazing animals rather than to satisfy any particular plant requirements. Selenium fertilization is acceptable if proper precautions are taken:

1. At no stage should herbage become toxic to grazing animals; topdressing of growing plants must be avoided.
2. Undesirably high levels of selenium in edible animal tissue should not result.
3. Protection against selenium deficiency should be provided for at least one grazing season following application during the dormant season.

Fertilization with selenites is preferred because they are slower acting and thus less likely to produce excessive levels of selenium in plants than the rapidly available selenates. The latter are effective for obtaining fast initial increases in selenium concentration, but they are usually more expensive and perhaps more difficult to procure. Soil additions of elemental selenium are not always effective in raising selenium levels in plants.

The New Zealand researcher Grant demonstrated that the addition of sodium selenite at low rates up to 1 oz of actual selenium per acre was a satisfactory means of increasing the selenium status of pasture forages on normal soils or, at worst, ones only marginally deficient in selenium. Allaway and his team of USDA investigators showed that a solution of this same selenium source injected to a depth of 10 cm in a low-selenium soil in Oregon elevated selenium uptake by alfalfa sufficiently to prevent white muscle disease in lambs for up to 2 years postapplication. The rate of application in this Oregon study was 1 ppm of selenium in the surface soil.

Foliar application of sodium selenite at rates of up to 15 g/ha of selenium in 359 liters/ha of water was shown by Cary of the USDA to be an efficient way to increase selenium in field corn. Recovery of the applied selenium varied from about 7.5% for treatment when the fourth to sixth leaves were fully expanded to 18% later when the crop was in the early tassel stage of growth.

Selenium is present in phosphate rocks and in superphosphate produced from them. Superphosphate containing 20 ppm or more selenium may provide sufficient selenium to the plants in selenium-deficient areas to protect livestock from selenium-deficiency disorders.

Chelates

In the preceding discussion of the microelements, the terms *chelates* or *chelating agents* were frequently employed. Very simply, metal chelates are defined as cyclic structures of a metal atom and an organic component in which the two components are held together with varying degrees of strength, varying from a rather loose bonding force to the strong metal-organic bond typical of metal porphyrins.

Metal chelates are soluble in water. Those commonly used for agriculture dissociate to only a very slight degree. Hence, although they decrease the *activity* of metallic ions in aqueous solution, the *solubility* of these metals is greatly increased in combination with the chelating agent. The term *sequestering agent* is sometimes used in lieu of chelating agent.

Numerous substances have the ability to chelate or sequester metallic ions. Several compounds are commercially important in agriculture, four of which are ethylenediaminetetraacetic acid (EDTA), diethylenetriaminepentaacetic acid (DTPA), cyclohexanediaminetetraacetic acid (CDTA), and ethylenediaminedi(o-hydroxyphenylacetic acid) (EDDHA).

The metallic ions commonly sequestered commercially are iron, copper, zinc, and manganese. Just how these chelated metals are absorbed and utilized by the plant is not known. It was once thought that the intact molecule was absorbed by the root hairs, for some of the early work showed a 1:1 ratio of metal:chelate in the root cells. Later work indicated that the metal and the chelate were not absorbed uniformly and that at low pH values appreciable amounts of the chelates remained in the solution external to the root even though the metal itself was absorbed. It must be admitted that at this writing the mechanism of metal:chelate absorption and utilization by plants is not completely understood.

The chelates without the metals iron, copper, manganese, and zinc exist as acids or sodium salts. When applied to soils, they have the ability to sequester the heavy metals from their insoluble forms in the soil. In fact, it has been possible to correct iron deficiencies simply by applying sodium-chelate directly to the soil around the root zone.

These chelated compounds are quite effective, for only small amounts are required to correct deficiency of the various metals. Chelates can be applied either to the soil or to plant leaves in the form of an aqueous spray.

Summary

1. The solubility and availability of the micronutrient cations are affected by the presence of complexing and chelating agencies as well as the oxidation–reduction (redox) potential in the soil.

2. The availability and movement of boron in the soil are influenced by a number of factors, the most important of which are climatic conditions, soil texture, pH, and liming. Boron reaches plant roots primarily by mass flow and to a lesser extent by diffusion. Boron deficiencies can be overcome through the application of a number of boron-containing fertilizers.

3. Cobalt is needed to improve legume nodulation and for the synthesis of vitamin B_{12} in ruminant animals. Soil areas low in cobalt have been identified in the United States, Canada, and a number of other countries.

4. Copper deficiencies are most frequently found in organic soils that have just been brought into production and with high-value specialty crops which are grown intensively in peat and muck soils. It is strongly bound to soil organic matter, and its solubility in the soil solution decreases 100-fold for each unit increase in soil pH.

5. In soil solutions of pH values below 6.9, the divalent ion Cu^{2+} is the dominant form of this element. Above this pH value, $Cu(OH)_2^0$ is the major species. In addition to soil texture, pH, and organic matter the plant availability of copper is influenced by its interaction with other micronutrients.

6. Copper deficiencies can be corrected by the application of a wide variety of commercially available fertilizer materials applied at rates ranging from 0.6 to 20 lb/A of copper.

7. Plant-nutrient deficiencies of iron have been reported in 25 states in the United States and on a limited scale in western Canada. The solubility and availability of iron in soils are controlled by pH, chelation, redox potential of the soil, and the presence of carbonates and bicarbonates.

8. Iron is taken up by plants principally as the ion Fe^{2+} and is transported to the roots as iron chelates. Because of low solubility, only small amounts of Fe^{3+} are available to plant roots.

9. Plant genotypes differ widely in their ability to absorb and utilize iron from the growth medium. Under field conditions, iron chlorosis is one of the most difficult of the micronutrient element deficiencies to correct. Its correction is accomplished primarily with foliar sprays and to a lesser extent by band application of some acidifying agent in the root zone.

10. Crop deficiencies of manganese have been reported in a number of states in the United States and in the three prairie provinces in Canada.

11. The solubility of soil manganese is influenced by the soil pH, redox potential, and the formation of ion complexes with organic matter. Mn^{2+} is the principal ion species in the soil solution and diffusion is the principal mechanism for its movement to plant root surfaces. The formation of manganese organic complexes enhances its movement to root surfaces. Like iron, plant susceptibility to manganese deficiency varies greatly with its genetic makeup.

12. Manganese deficiencies can be corrected by foliar sprays of simple inorganic salts or chelated manganese compounds. Soil applications of manganese compounds as well as lowering of the soil pH are effective in correcting manganese deficiencies.

13. Unlike most of the micronutrients, deficiencies of molybdenum increase with an increase in soil pH.

14. Plants absorb molybdenum from the soil as the ion MoO_4^{2-}, and its movement to plant roots is largely through mass flow. At low concentrations, however, diffusion becomes the main transfer mechanism.

15. Deficiencies of molybdenum can be prevented or overcome through seed treatment prior to planting, applied directly to the soil in the fertilizer, or sprayed on the foliage. Rates of application are extremely low, 0.5 to 5.0 oz/A of molybdenum.

16. Zinc deficiencies have been reported in most of the states in the

United States and in at least five Canadian provinces. Deficiencies of this element are found most frequently in sandy soils low in total zinc and neutral or basic soils, especially those that are calcareous, and exposure to subsoils resulting from land-leveling and wind and water erosion.

17. Zinc availability is highly pH dependent, decreasing 100-fold for each unit increase in pH. The Zn^{2+} ion is the form absorbed by plants, and diffusion is the main process of transporting this ion to plant root surfaces. The formation of zinc–organic complexes facilitates this diffusion.

18. Plant-nutrient zinc can be supplied from several fertilizer sources directly to the soil. Tree injections and foliar applications of this element have also been successfully utilized. The effectiveness of banded applications of zinc in the soil have been increased by the presence of acid-forming nitrogen fertilizers and certain acid-forming sulfur compounds.

19. Sodium is an element that is beneficial to the growth of some plants and it is essential for certain halophytic species. It is detrimental, however, to the growth of many other plants. The benefits of sodium generally tend to be greatest when potassium nutrition is poorest.

20. There is evidence that sodium can partially substitute for potassium in plant nutrition, which in turn is closely related to the plants' ability to absorb sodium. Harmful levels of sodium in plants are usually accompanied by low concentrations of potassium and calcium.

21. Chlorine is a minor constituent of the lithosphere, and the variable amounts of this element in soils are thought to come from marine aerosols, volcanic emissions, fertilization, crop protection programs, and in irrigation waters. The mobility of the chlorine ion is great in most soils, and the strength with which it is held is very weak, its absorption decreasing with increases in soil pH.

22. While chlorine has been found to be beneficial for many plants, large amounts of this ion increase the osmotic pressure of the soil solution and thereby lower the availability of water to plants.

23. Some plants, notably rice and sugarcane, have responded to applications of soluble silicate fertilizers. The major sources of silicon in soils are primary silicon minerals, secondary aluminosilicates, and several forms of silica. Soils low in silicon are usually those found in the tropics occurring under highly weathered conditions. Plant intake of silicon is influenced by soil content of iron and aluminum, liming, flooding, and nutrient supply.

24. Selenium is not an essential plant nutrient, but it is essential for animals. A deficiency of it results in white muscle disease or muscular dystrophy.

25. There are large areas of soils in the United States and Canada that are low in selenium, and there are also significant soil areas in each where the selenium level is high and plants absorb amounts sufficient to be toxic to livestock.

26. Selenium deficiencies in animals can be overcome through selenium fertilization of the forages ingested by these animals. Several precautions must be strictly observed when fertilizing forages with this element to ensure that toxicities to the animals do not result.

Questions

1. What plant nutrient elements are classified as micronutrients?
2. Name other nutrients discussed in this chapter that may also be beneficial for some plants. Are they micronutrients?
3. What micronutrient elements exist in soil solution as cations? Which are present in anionic form?
4. Give the principal form(s) in which each element discussed in this chapter occurs in the soil solution.
5. Report the primary mechanism(s) responsible for movement from the soil to root surfaces of each nutrient reviewed in this chapter.
6. Indicate how the solubility of each nutrient is influenced by pH.
7. The theoretical solubility of the heavy metal micronutrients in water seems in most cases to be too low for satisfactory crop nutrition. In spite of this, soil solution concentrations are often much higher and plant nutrition better than expected. Discuss the mechanism(s) responsible for higher solubility and improved plant uptake.
8. Availabilities of the plant nutrient elements discussed in this chapter are altered by liming. What ones are decreased by liming? What ones are increased by liming?
9. What elements in the micronutrient grouping react in various ways with carbonate minerals? Is their availability altered by these reactions?
10. Which of the various micronutrients and other beneficial elements react with oxides of iron and aluminum? Is their availability affected by these reactions?
11. The solubility and availability of certain of the micronutrients and other beneficial elements are influenced by flooding and submergence. Which elements are affected by these conditions? What happens to their solubility and availability?
12. Describe at least three major roles that organic-matter additions can have on micronutrient supplies to plants.
13. Can microbial activity influence solubility and availability of micronutrients? Explain the reasons for your answer.
14. Why are nutrient interrelationships important in plant uptake of micronutrients and the other beneficial elements? Give examples of strong interactions (a) among heavy metal cations, (b) between nitrogen and at least five other elements, (c) between phosphorus and at least five other elements, (d) between potassium and at least two other elements, (e) between SO_4^{2-}-containing substances and at least two other elements, and (f) between molybdenum and one other micronutrient.
15. What is meant by a mutually antagonistic interaction? Give an example of such an interaction.
16. Manganese availability can be greatly increased by the addition of which neutral fertilizer salts?
17. Describe briefly the nature of the latest findings related to the interaction of phosphorus and zinc.
18. Is the sensitivity of plants to deficiencies of micronutrients and the other beneficial elements related in any way to plant genetic differences? Explain your answer.
19. What is meant by iron-efficient and iron-inefficient plants? Give examples of plants in these two categories.
20. Outline the biochemical reactions and changes associated with iron-efficient plants. Do these plant characteristics have any bearing on the efficiency of plant uptake of other heavy metal micronutrients?

21. Availability and plant absorption of micronutrients and other beneficial elements can be influenced by climatic and environmental conditions. Describe the effect of climatic factors on the occurrence of boron, manganese, molybdenum, and zinc deficiencies.
22. Why is copper deficiency often referred to as "reclamation disease"?
23. The deficiency of which micronutrient is generally considered one of the most difficult to correct in the field?
24. What nutrient deficiencies are best controlled by foliar treatments of micronutrient fertilizers?
25. Correction of what nutrient deficiencies is best accomplished by soil applications of micronutrient fertilizers?
26. Acidification of high-pH and calcareous soils in localized zones such as fertilizer bands can be helpful in the treatment of what micronutrient deficiencies?
27. In the treatment of certain micronutrient deficiencies, it is necessary to thoroughly incorporate the fertilizer additives deeply into the soil. This is true for what micronutrients? Why?
28. Name the important external sources of Cl^- reaching soils. Why are some contributions typically much higher near seacoasts?
29. How is the availability of copper affected by the level of soil organic matter and soil pH?
30. How is a cobalt deficiency in animals commonly remedied?
31. What are some common copper fertilizers?
32. What are the reasons that have been suggested to explain iron chlorosis of plants?
33. What are the ways commonly employed to overcome iron chlorosis?
34. In what forms is manganese believed to exist in soils?
35. Under what soil conditions has a manganese toxicity been observed?
36. What are some common manganese fertilizer materials?
37. In what way is the behavior of molybdenum in soils different from the behavior of the other microelements?
38. What soil factors influence the availability of zinc in soils?
39. In what ways is selenium important in soils?
40. What are chelates? Which of the microelements are frequently applied as chelates?
41. What are frits? How is their availability influenced by their particle size?
42. What soil form of selenium is readily available to plants?
43. What soil conditions favor the existence of the soil form of selenium readily available to plants?
44. Why doesn't vegetation grown on seleniferous soils in Hawaii and Puerto Rico contain toxic levels of selenium?
45. Are there practical ways of increasing the selenium concentration of plants? List them.

Selected References

Adams, F., and J. I. Wear, "Manganese toxicity and soil acidity in relation to crinkle leaf of cotton." *SSSA Proc.*, **21**:305 (1957).

Allaway, W.H., "Control of the environmental levels of selenium," in D. E. Hemphill, Ed., *Trace Substances in Environmental Health, II.* Columbia: Univ. of Missouri, 1968.

Amer, F., A. I. Rezk, and H. M. Khalid, "Fertilizer zinc efficiency in flooded calcareous soil." *Soil Sci. Soc. Am. J.*, **44**:1025 (1980).

Anderson, A. J., "Molybdenum deficiency on a South Australian ironstone soil." *J. Australian Inst. Agr. Sci.*, **8**:73 (1942).

Anderson, A. J., "The significance of sulphur deficiency in Australian soils." *J. Australian Inst. Agr. Sci.*, **18**:135 (1952).

Anderson, A. J., "Molybdenum as a fertilizer." *Adv. Agron.*, **8**:164 (1956).

Aubert, H., and M. Pinta, *Developments in Soil Science*, Vol. 7: *Trace Elements in Soils*. Amsterdam: Elsevier Scientific, 1977.

Ayres, A. S., "Calcium silicate slag as a growth stimulant for sugarcane on low-silicon soils." *Soil Sci.*, **101**:216 (1966).

Banerjee, D. K., R. H. Bray, and S. W. Melsted, "Some aspects of the chemistry of cobalt in soils." *Soil Sci.*, **75**:421 (1953).

Barrie, L. A., and A. Sirois, "An analysis and assessment of precipitation chemistry measurements made by CANSAP (The Canadian Network for Sampling Precipitation): 1977–1980." *Rep. AQRB-82-003-T*. Downsview, Ontario: Environment Canada, Atmospheric Environment Service, 1982.

Beaton, J. D., "Fertilizers and their use in the decade ahead." *Proc. 20th Annu. Meet., Agr. Res. Inst.* (October 13, 1971).

Biggar, J. W., and M. Fireman, "Boron adsorption and release by soils." *SSSA Proc.*, **24**:115 (1960).

Boawn, L. C., "Comparison of zinc sulfate and zinc EDTA as zinc fertilizer sources." *Soil Sci. Soc. Am. J.*, **37**:111 (1973).

Boawn, L. C., F. G. Viets, Jr., and C. L. Crawford, "Effects of nitrogen carrier, nitrogen rate, zinc rate, and soil pH on zinc uptake by sorghum, potatoes, and sugar beets." *Soil Sci.*, **90**:329 (1960).

Boswell, F. C., K. Ohki, M. B. Parker, L. M. Shuman, and D. O. Wilson, "Methods and rates of applied manganese for soybeans." *Agron. J.*, **73**:909 (1981).

Bowman, R. A., and S. R. Olsen, "Effect of calcium sulfate on iron and zinc uptake in sorghum." *Agron. J.*, **74**:923 (1982).

Bradford, G. R., "Boron," in H. D. Chapman, Ed., *Diagnostic Criteria for Plants and Soils*, Chapter 4. Univ. of California, Division of Agricultural Sciences, 1966.

Brown, J. C., and W. D. Bell, "Iron uptake dependent upon genotype of corn." *Soil Sci. Soc. Am. J.*, **33**:99 (1969).

Brown, J. C., and W. E. Jones, "Differential response of oats to manganese stress." *Agron. J.*, **66**:624 (1974).

Brown, J. C., R. S. Holmes, and L. O. Tiffin, "Hypotheses concerning iron chlorosis," *SSSA Proc.*, **23**:231 (1959).

Brown, J. C., J. E. Ambler, R. L. Chaney, and C. D. Foy, "Differential responses of plant genotypes to micronutrients," in J. J. Mortvedt, P. M. Giordano, and W. L. Lindsay, Eds., *Micronutrients in Agriculture*, Chapter 16. Madison, Wis.: Soil Science Society of America, 1972.

Brownell, P. E., and C. J. Crossland, "The requirement for sodium as a micronutrient by species having the C_4 dicarboxylic photosynthetic pathway." *Plant Physiol.*, **49**:794 (1972).

Broyer, T. C., C. M. Johnson, and R. P. Huston, "Selenium and nutrition of *Astragalus*: I. Effects of selenite or selenate supply on growth and selenium content." *Plant Soil*, **36**:635 (1972).

Broyer, T. C., C. M. Johnson, and R. P. Huston, "Selenium and nutrition of *Astragalus*: II. Ionic sorption interactions among selenium, phosphate, and the macro- and micronutrient cations." *Plant Soil*, **36**:651 (1972).

Cary, E. E., and W. H. Allaway, "Selenium content of field crops grown on selenite-treated soils." *Agron. J.*, **65**:922 (1973).

Cary, E. E., and M. Rutzke, "Foliar application of selenium to field corn." *Agron. J.*, **73**:1083 (1981).

Chapman, H. D., "Zinc," in H. D. Chapman, Ed., *Diagnostic Criteria for Plants and Soils*, Chapter 33. Univ. of California, Division of Agricultural Sciences, 1966.

Cheng, B. T., "Dynamics of soil manganese." *Agrochimica*, **17**:84 (1972).

Christensen, N. W., and T. L. Jackson, "Potential for phosphorus toxicity in zinc-stressed corn and potato." *Soil Sci. Soc. Am. J.*, **45**:904 (1981).

Christensen, N. W., T. L. Jackson, and R. L. Powelson, "Suppression of take-all root rot and stripe rust diseases of wheat with chloride fertilizers." *Proc. 9th*

Int. Plant Nutr. Colloq., Vol. 1, pp. 111–116, Warwick University, Warwick, England (Commonwealth Agricultural Bureaux) (August 22–27, 1982).

Christensen, N. W., R. G. Taylor, T. L. Jackson, and B. L. Mitchell, "Chloride effects on water potentials and yield of winter wheat infected with take-all root rot." *Agron. J.*, **73:**1053 (1981).

Clark, R. B., "Differential response of maize inbreds to Zn." *Agron. J.*, **70:**1057 (1978).

Davies, E. B., "Factors affecting molybdenum deficiency in soils." *Soil Sci.*, **81:**209 (1956).

Donald, C. M., and K. Spencer, "The control of molybdenum deficiency in subterranean clover by presoaking the seed in sodium molybdate solution." *Australian J. Agr. Res.*, **2:**295 (1951).

Edwards, I. K., Y. P. Kalra, and F. G. Redford, "Chloride determination and levels in the soil–plant environment." *Environ. Pollut. Ser. B*, **2:**109 (1981).

Esty, J. C., A. B. Onken, L. R. Hossner, and R. Matheson, "Iron use efficiency in grain sorghum hybrids and parental lines." *Agron. J.*, **72:**589 (1980).

Fleming, G. A., "Essential micronutrients: I. Boron and molybdenum," in B. E. Davies, Ed., *Applied Soil Trace Elements*, Chapter 5. New York: Wiley, 1980.

Fleming, G. A., "Essential micronutrients: II. Iodine and selenium," in B. E. Davies, Ed., *Applied Soil Trace Elements*, Chapter 6. New York: Wiley, 1980.

Fox, R. L., J. A. Silva, O. R. Younge, D. L. Plucknett, and G. D. Sherman, "Soil and plant silicon and silicate response by sugarcane." *Soil Sci. Soc. Am. J.*, **31:**775 (1967).

Gartrell, J. W., "Distribution and correction of copper deficiency in crops and soils," in J. F. Loneragan, A. D. Robson, and R. D. Graham, Eds., *Copper in Soils and Plants*, Chapter 14. New York: Academic Press, 1981.

Garvin, J. P., V. A. Haby, and P. O. Kresge, "Effect of fertilizer N, P, K, Cl, and S on yield, protein percentage, nutrient content and root rot on barley." *Proc. 32nd Annu. Northwest Fert. Conf.*, pp. 87–96, Billings, Mont. (July 14–16, 1981).

Gascho, G. J., "Response of sugarcane to calcium silicate slag. I. Mechanisms of response in Florida." *Soil Crop Sci. Soc. Florida Proc.*, **37:**55 (1977).

Gilkes, R. J., "Behaviour of Cu additives-fertilizers," in J. F. Loneragan, A. D. Robson, and R. D. Graham, Eds., *Copper in Soils and Plants*, Chapter 5. New York: Academic Press. 1981.

Giordano, P. M., and J. J. Mortvedt, "Agronomic effectiveness of micronutrients in macronutrient fertilizers," in J. J. Mortvedt, P. M. Giordano, and W. L. Lindsay, Eds., *Micronutrients in Agriculture*, Chapter 20. Madison, Wis.: Soil Science Society of America, 1972.

Giordano, P. M., J. C. Noggle, and J. J. Mortvedt, "Zinc uptake by rice as affected by metabolic inhibitors and competing cations." *Plant Soil*, **41:**637 (1974).

Giordano, P. M., E. C. Sample, and J. J. Mortvedt, "Effect of ammonium ortho- and pyrophosphate on Zn and P in soil solution." *Soil Sci.*, **111:**101 (1971).

Gissel-Nielsen, G., "Selenium content of some fertilizers and their influence on uptake of selenium in plants." *J. Agr. Food Chem.*, **19:**564 (1971).

Gissel-Nielsen, G., "Influence of pH and texture of the soil on plant uptake of added selenium." *J. Agr. Food Chem.*, **19:**1165 (1971).

Gupta, U. C., "S × Mo interaction in plant nutrition." *Sulphur Inst. J.*, **5(1):**4 (1969).

Hader, R. J., M. E. Harward, D. D. Mason, and D. P. Moore, "An investigation of some of the relationships between Cu, Fe, and Mo in the growth and nutrition of lettuce: I. Experimental design and statistical methods for characterizing the response surface." *SSSA Proc.*, **21:**59 (1957).

Hiatt, A. J., and J. L. Ragland, "Manganese toxicity of burley tobacco." *Agron. J.*, **55:**47 (1963).

Hodgson, J. F., K. G. Tiller, and M. Fellows, "Effect of iron removal on cobalt sorption by clays." *Soil Sci.*, **108:**391 (1969).

Holden, E. R., and A. J. Engel, "Boron supplements. Response of alfalfa to applications of a soluble borate and a slightly soluble borosilicate glass." J. Agr. Food Chem., **5:**275 (1957).

Jackson, T. L., R. L. Powelson, and N. W. Christensen, "Combating take-all root rot of winter wheat in western Oregon." *Oregon State Univ. Ext. Serv. FS250* (September 1981).

Jackson, T. L., D. T. Westermann, and D. P. Moore, "The effect of chloride and lime on the manganese uptake by bush beans and sweet corn." *Soil Sci. Soc. Am. J.*, **30:**70 (1966).

Jackson, T. L., M. J. Johnson, S. James, and D. Sullivan, "A new view of potassium chloride fertilization of potatoes." *Better Crops Plant Food*, **66:**6 (Fall 1982).

James, D. W., W. H. Weaver, and R. L. Reeder, "Chloride uptake by potatoes and the effects of potassium chloride, nitrogen and phosphorus fertilization." *Soil Sci.*, **109:**48 (1970).

Johnson, C. M., "Molybdenum," in H. D. Chapman, Ed., *Diagnostic Criteria for Plants and Soils*, Chapter 20. Univ. of California, Division of Agricultural Sciences, 1966.

Johnson, C. M., "Selenium in soils and plants: contrasts in conditions providing safe but adequate amounts of selenium in the food chain," in D. J. D. Nicholas and A. R. Egan, Eds., *Trace Elements in Soil–Plant–Animal Systems*, pp. 165–180. New York: Academic Press, 1975.

Johnson, J. W., "Boron supply and availability in soils." *U.S. Borax Plant Food Borate Meet.*, Lafayette, Ind. (1980).

Jones, L. H. P., and K. A. Hendreck, "Studies of silica in the oat plant: III. Uptake of silica from soils by the plant." *Plant Soil*, **23:**79 (1965).

Khalid, R. A., J. A. Silva, and R. L. Fox, "Residual effects of calcium silicate in tropical soils. I. Fate of applied silicon during five years cropping." *Soil Sci. Soc. Am. J.*, **42:**89 (1978).

Kilmer, V. J., "Minerals and agriculture," in P. A. Trudinger and D. J. Swaine, Eds., *Studies in Environmental Science*, Vol. 3: *Biogeochemical Cycling of Mineral-Forming Elements*, Chapter 9. Amsterdam: Elsevier Scientific, 1979.

Kubota, J., "Regional distribution of trace element problems in North America," in B. E. Davies, Ed., *Applied Soil Trace Elements*, Chapter 12. New York: Wiley, 1980.

Lakin, H. W., "Selenium accumulation in soils and its absorption by plants and animals." *Geol. Soc. Am. Bull.*, **83:**181 (1972).

Leeper, G. W., "Forms and reactions of manganese in soil." *Soil Sci.*, **63:**79 (1947).

Lindsay, W. L., "Role of chelation in micronutrient availability," in E. W. Carson, Ed., *The Plant Root and Its Environment*, Chapter 17. Charlottesville, Va.: University Press of Virginia, 1974.

Loganathan, P., R. G. Burau, and D. W. Fuerstenau, "Influence of pH on the sorption of Co^{2+}, Zn^{2+}, and Ca^{2+} by a hydrous manganese oxide." *Soil Sci. Soc. Am. J.*, **41:**57 (1977).

Löhnis, Marie P., "Manganese toxicity in field and market garden crops." *Plant Soil*, **3:**193 (1951).

Loneragan, J. F., "The availability and adsorption of trace elements in soil–plant systems and their relation to movement and concentration of trace elements in plants," in D. J. D. Nicholas and A. R. Egan, Eds., *Trace Elements in Soil–Plant–Animal Systems*, pp. 109–134. New York: Academic Press, 1975.

Ludwick, A. E., "Manganese availability in manganese-sulfur granules." M.S. thesis, Univ. of Wisconsin (1964).

Lunt, O. R., "Sodium," in H. D. Chapman, Ed., *Diagnostic Criteria for Plants and Soils*, Chapter 27. Univ. of California, Division of Agricultural Sciences, 1966.

McLachlan, K. D., "Phosphorus, sulfur, and molybdenum deficiencies in soils from eastern Australia in relation to the nutrient supply and some characteristics of soil and climate." *Australian J. Agr. Res.*, **6:**673 (1955).

McLaren, R. G., and D. V. Crawford, "Studies on soil copper: II. The specific adsorption of copper by soils." *J. Soil Sci.*, **24:**444 (1973).

Magat, S. S., V. L. Cadigal, and J. A. Habana, "Yield improvement of coconut in elevated inland area of Davao (Philippines) by KCl fertilization." *Oléagineux*, **30**(10):413 (1975).

Mandal, L. N., "Transformations of iron and manganese in water-logged rice soils." *Soil Sci.*, **91**:121 (1961).

Matocha, J. M., and D. Pennington, "Effects of complexed iron materials on iron chlorosis of grain sorghum grown on calcareous soils." *Agron. Abstr. 1978 Annu. Meet.*, Chicago (December 3–8, 1978).

Mengel, D. B., "Boron in soils and plant nutrition: A summary." *U.S. Borax Plant Food Borate Meet.*, Lafayette, Ind. (1980).

Mengel, K., and E. A. Kirkby, *Principles of Plant Nutrition*, 3rd ed. Bern: International Potash Institute, 1982.

Mikkelson, D. S., and D. M. Brandon, "Zinc deficiency in California rice." *Calif. Agr.*, **29**(9):8 (1975).

Moore, D. P., M. E. Harward, D. D. Mason, R. J. Hader, W. L. Lott, and W. A. Jackson, "An investigation of some of the relationships between copper, iron and molybdenum in the growth and nutrition of lettuce: II. Response surfaces of growth and accumulations of Cu and Fe." *SSSA Proc.*, **21**:65 (1957).

Mortvedt, J. J., and F. R. Cox, "Production, marketing, and use of calcium, magnesium and micronutrient fertilizers," in O. P. Engelstad, F. C. Boswell, T. C. Tucker, and L. F. Welch, Eds., *Fertilizer Technology and Use*, 3rd ed., Chapter 13. Madison, Wis.: Soil Science Society of America, In press.

Mortvedt, J. J., P. M. Giordano, and W. L. Lindsay, Eds., *Micronutrients in Agriculture*. Madison, Wis.: Soil Science Society of America, 1972.

Murphy, L. S., and L. M. Walsh, "Correction of micronutrient deficiencies with fertilizers," in J. J. Mortvedt, P. M. Giordano, and W. L. Lindsay, Eds., *Micronutrients in Agriculture*, Chapter 15. Madison, Wis.: Soil Science Society of America, 1972.

Nelson, W. L., "Boron needs as related to maximum yields." *Better Crops Plant Food*, **64**:3 (Winter 1980–1981).

Nyborg, M., "Sensitivity to manganese deficiency of different cultivars of wheat, oats and barley." *Can. J. Plant Sci.*, **50**:198 (1970).

Okhi, K., D. O. Wilson, and O. E. Anderson, "Manganese deficiency and toxicity sensitivities of soybean cultivars." *Agron. J.*, **72**:713 (1980).

Olsen, S. R., and F. S. Watanabe, "Interaction of added gypsum in alkaline soils with uptake of iron, molybdenum, manganese, and zinc by sorghum." *Soil Sci. Soc. Am. J.*, **43**:125 (1979).

Ouellette, G. J., "Effects of lime, nitrogen, and phosphorus on the response of ladino clover to molybdenum." *Can. J. Soil Sci.*, **43**:117 (1963).

Ozanne, P. G., "Chlorine deficiency in soils." *Nature (London)*. **182**:1172 (1958).

Parker, M. B., and H. B. Harris, "Soybean response to molybdenum and lime and the relationship between yield and chemical composition." *Agron. J.*, **54**:480 (1962).

Peterson, L. A., and R. C. Newman, "Influence of soil pH on the availability of added boron." *Soil Sci. Soc. Am. J.*, **40**:280 (1976).

Prather, R. J., "Sulfuric acid as an amendment for reclaiming soils high in boron." *Soil Sci. Soc. Am. J.*, **41**:1098 (1977).

Quirk, J. P., and A. M. Posner, "Trace element adsorption by soil minerals," in D. J. D. Nicholas and A. R. Egan, Eds., *Trace Elements in Soil–Plant–Animal Systems*, pp. 95–107. New York: Academic Press, 1975.

Racz, G. J., and P. W. Haluschak, "Effects of phosphorus concentration on Cu, Zn, Fe, and Mn utilization by wheat." *Can. J. Soil Sci.*, **54**:357 (1974).

Rana, S. K., and G. J. Ouellette, "Cobalt status in Quebec soils." *Can. J. Soil Sci.*, **47**:83 (1967).

Randall, G. W., E. E. Schulte, and R. B. Corey, "Effect of soil and foliar-applied manganese on the micronutrient content and yield of soybeans." *Agron. J.*, **67**:502 (1975).

Reid, J. M., and G. J. Racz, "Effect of Cu, Mn, and fertilization on yield and chemical composition of wheat and barley." *Proc. 24th Annu. Manitoba Soil Sci. Meet.* (December 3–4, 1980).

Reil, W. O., J. A. Beutel, C. L. Hemstreet, and W. S. Seyman, "Trunk injection corrects iron and zinc deficiency in pear trees." *Calif. Agr.*, **32**(10):22 (1978).

Reisenauer, H. M., "Relative efficiency of seed- and soil-applied molybdenum fertilizer." *Agron J.*, **55**:459 (1963).

Reisenauer, H. M., A. A. Tabikh, and P. R. Stout, "Molybdenum reactions with soils and the hydrous oxides of iron, aluminum, and titanium." *Soil Sci. Soc. Am. J.*, **26**:23 (1962).

Reisenauer, H. M., L. M. Walsh, and R. G. Hoeft, "Testing soils for sulphur, boron, molybdenum, and chlorine," in L. M. Walsh and J. D. Beaton, Eds., *Soil Testing and Plant Analysis*, rev. ed., Chapter 12. Madison, Wis.: Soil Science Society of America, 1973.

Robson, A. D., and D. J. Reuter, "Diagnosis of copper deficiency and toxicity," in J. F. Loneragan, A. D. Robson, and R. D. Graham, Eds., *Copper in Soils and Plants*, Chapter 13. New York: Academic Press, 1981.

Safaya, N. M., and A. P. Gupta, "Differential susceptibility of corn cultivars to zinc deficiency." *Agron. J.*, **71**:132 (1979).

Sanchez, C., and E. J. Kamprath, "The effect of liming and organic matter content on the availability of native and applied manganese." *SSSA Proc.*, **23**:302 (1959).

Sims, J. L., J. E. Leggett, and U. R. Pal, "Molybdenum and sulfur interaction effects on growth, yield, and selected chemical constituents of burley tobacco." *Agron. J.*, **71**:75 (1979).

Singh, M., and N. Singh, "The effect of forms of selenium on the accumulation of selenium, sulfur, and forms of nitrogen and phosphorus in forage cowpea." *Soil Sci.*, **127**:264 (1979).

Stephens, C. G., and C. M. Donald, "Australian soils and their responses to fertilizers." *Adv. Agron.*, **10**:168 (1958).

Sumner, M. E., and P. M. W. Farina, "Phosphorus interactions with other nutrients and lime in field cropping systems," in Keith Syers, Ed., *Phosphorus in Agricultural Systems*. New York: Elsevier Scientific, 1984.

Sumner, M. E., H. R. Boerma, and R. Isaac, "Differential genotypic sensitivity of soybeans to P–Zn–Cu imbalances." *Proc. 9th Int. Plant Nutr. Colloq.*, pp. 652–657, Warwick Univ., Warwick, England, (Commonwealth Agricultural Bureaux) (1982).

Thomas, J. D., and A. C. Mathers, "Manure and iron effects on sorghum growth on iron-deficient soil." *Agron. J.*, **71**:793 (1979).

Turner, J. R., *Boron in Agriculture*. New York: U.S. Borax, 1976.

Turner, J., "Boron for crop production." *Fert. Prog.*, **11**(3):24 (1980).

Udo, E. J., H. L. Bohn, and T. C. Tucker, "Zinc adsorption by calcareous soils." *Soil Sci. Soc. Am. J.*, **34**:405 (1970).

van der Elst, F. H., and R. Tetley, "Selenium studies on peat: I. Selenium uptake of pasture after incorporation of sodium selenite with peat soil." *New Zealand J. Agr. Res.*, **13**:945 (1970).

Velasco, I., "Improving the sodic soils of Spain." *Sulphur Agr.*, **5**:2 (1981).

von Uexkull, H. R., "Response of coconuts to (potassium) chloride in the Philippines." *Oléagineux*, **27**(1):13 (1972).

Walker, T. W., A. F. R. Adams, and H. D. Orchiston. "The effects and interactions of molybdenum, lime, and phosphate treatments on the yield and composition of white clover grown on acid, molybdenum responsive soils." *Plant Soil*, **6**:20 (1955).

Wallace, A., and R. T. Mueller, "Complete neutralization of a portion of calcareous soil as a means of preventing iron chlorosis." *Agron. J.*, **70**:888 (1978).

Wallace, A., E. M. Romney, and G. V. Alexander, "Lime-induced chlorosis caused by excess irrigation water." *Commun. Soil Sci. Plant Anal.* **7**(1):47 (1976).

Wallihan, E. F., "Iron deficiency in California crops." *Calif. Agr.*, **30**(3):4 (1976).

Walsh, L. M., and J. D. Beaton, Eds., *Soil Testing and Plant Analysis*, rev. ed. Madison, Wis.: Soil Science Society of America, 1973.

Wear, J. I., and R. M. Patterson, "Effect of soil pH and texture on the availability of water-soluble boron in the soil." *SSSA Proc.*, **26**:344 (1962).

Welch, R. M., M. J. Webb, and J. F. Loneragan, "Zinc in membrane function and its role in phosphorus toxicity." *Proc. 9th Int. Plant Nutr. Colloq.*, Vol. 1,

p. 710, Warwick University, Warwick, England (Commonwealth Agricultural Bureaux) (August 22–27, 1982).

Westermann, D. T., and C. W. Robbins, "Effect of SO_4-S fertilization on Se concentration of alfalfa (*Medicago sativa* L.)." *Agron. J.*, **66:**207 (1974).

Westermann, D. T., T. L. Jackson, and D. P. Moore, "Effect of potassium salts on extractable soil manganese." *Soil Sci. Soc. Am. J.*, **35:**43 (1971).

Westfall, D. G., W. B. Anderson, and R. J. Hodges, "Iron and zinc response of chlorotic rice grown on calcareous soils." *Agron. J.*, **63:**702 (1971).

Williams, C., and I. Thornton, "The effect of soil additives on the uptake of molybdenum and selenium from soils from different environments." *Plant Soil*, **36:**395 (1972).

Wilson, C. M., R. L. Lovvorn, and W. W. Woodhouse, "Movement and accumulation of water soluble boron within the soil profile." *Agron. J.*, **43:**363 (1951).

Yoshikawa, F. T., W. O. Reil, and L. K. Stromberg, "Trunk injection corrects iron deficiency in plum trees." *Calif. Agr.*, **36**(2–3):13 (1982).

Zinc in Crop Nutrition. New York: International Lead Zinc Research Organization, Inc., and Zinc Institute, Inc., 1974.

Chapter 10

Fertilizer Manufacture

THE properties and use of fertilizer materials containing nitrogen, phosphorus, and potassium were discussed in Chapters 5, 6, and 7. It is not the purpose of this book to cover in detail the manufacturing processes of these materials, for this is a topic more properly covered in courses on fertilizer technology. Some knowledge, however, of the basic reactions in fertilizer production is of value in better understanding the essential roles assumed by these materials in the overall picture of soil fertility and its place in the agribusiness enterprise.

In the following sections are discussed the reactions and processes encountered in the manufacture of nitrogen, phosphorus, and potassium materials. The production of only a few compounds is necessary for the synthesis of the many discussed in Chapters 5, 6, and 7. The information contained in the present chapter therefore deals with the reactions involved in the manufacture of these fundamental materials.

Processes for combining the basic fertilizer materials into solid and fluid multinutrient fertilizer products are also described. Many aspects of blending solid and fluid fertilizers are covered in these discussions. Fertilizer–pesticide mixtures are reviewed and information on fertilizer control and regulation is included to indicate the extent of consumer protection.

Nitrogen Fertilizer Materials

Basic Synthetic Processes. The fixation of atmospheric nitrogen is required for the production of all synthetic nitrogen fertilizer materials. The supply of this element is for all intents and purposes inexhaustible. It constitutes 80% of the atmosphere.

Until the end of the nineteenth century this limitless store of nitrogen was virtually unavailable to man in a combined form. Some, of course, was fixed by symbiotic and nonsymbiotic bacteria and by discharge of electricity in the atmosphere, but large-scale commercial fixation has been undertaken only in recent years.

There are three basic methods for the fixation of elemental nitrogen: (1) the synthetic ammonia process, in which nitrogen and hydrogen form ammonia gas; (2) the cyanamide process, in which nitrogen is reacted with calcium carbide; and (3) direct oxidation of nitrogen.

Synthetic Ammonia Production. All processes used are essentially modifications of the original Haber–Bosch process, which was developed in Germany in 1910. Ammonia synthesis is based on the reaction of nitrogen and hy-

TABLE 10-1.   Major Feedstocks Used for Ammonia Production in the World in 1971 and 1975 and Forecasts for Future Years

Feedstock	Percent of Ammonia Production					
	1971	1975	1980	1985	1990	2000
Natural gas	60	62	71.5	71.0	69.5	68.0
Naphtha	20	19	15.0	13.0	8.5	6.5
Fuel oil	4.5	5	7.0	8.5	10.0	12.0
Coal	9	9	5.5	6.5	7.5	10.5
Total	93.5	95.0	99.0	99.0	95.5	97.0

Source: International Fertilizer Development Center and United Nations Industrial Development Organization, *Fertilizer Manual*, Reference Manual IFDC-R-1. Muscle Shoals, Ala.: IFDC, 1979.

drogen in the presence of a catalyst, which is currently comprised of magnetite (Fe_3O_4) with additions of potassium, alumina, and calcium, at temperatures up to 1200°C. The pressure required varies from 200 to 1000 atm, depending on the modification employed. The reaction is expressed by the following equation:

$$3H_2 + N_2 \longrightarrow 2NH_3$$

Conversion of the two reactants to ammonia usually range between 17 and 28%.

Hydrogen for this reaction may be obtained from several sources. The principal feedstock sources of hydrogen are listed in Table 10-1. Other hydrogen-containing feedstocks are mentioned later in the discussion following steam reforming. Also, in locations where low-cost hydroelectric power is available, hydrogen can be provided by the electrolysis of water. As a matter of interest, for 40 years (until 1971) electrolytic hydrogen was used for ammonia production at Trail, British Columbia, Canada. All of the nitrogen needed for the ammonia synthesis reaction comes from air.

The chemical composition of natural gas, which currently accounts for over 70% of the world's ammonia production, is summarized in Table 10-2.

Methane and the other common feedstocks must be decomposed to release hydrogen for subsequent combination with nitrogen. Steam-reforming processes are

TABLE 10-2.   Typical Analysis of Natural Gas Feedstock for Ammonia Production

Constituent	Percent by Volume	
	At Wellhead	As Delivered by Pipeline
CH_4	75.9	93.3
N_2	—	2.0
Argon	—	0.4
CO_2	7.3	0.01
H_2S	8.9	< 5 ppm
Hydrocarbons		
C_2H_6	3.3	3.3
C_3H_8	1.2	0.9
C_4H_{10}	0.8	0.2
C_5H_{12}	0.5	0.01
$C_6H_{14}+$	2.3	—

Source: International Fertilizer Development Center and United Nations Industrial Development Organization, *Fertilizer Manual*, Reference Manual IFDC-R-1. Muscle Shoals, Ala.: IFDC, 1979.

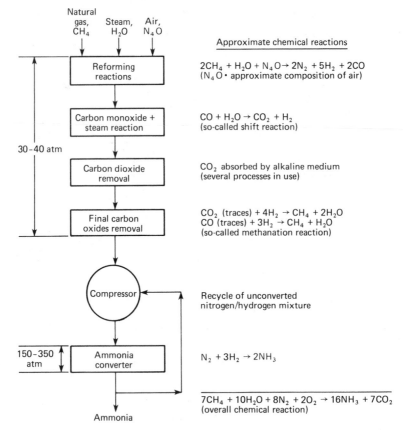

Natural gas, CH_4 Steam, H_2O Air, N_4O

Approximate chemical reactions

Reforming reactions

$2CH_4 + H_2O + N_4O \rightarrow 2N_2 + 5H_2 + 2CO$
($N_4O \cdot$ approximate composition of air)

Carbon monoxide + steam reaction

$CO + H_2O \rightarrow CO_2 + H_2$
(so-called shift reaction)

30–40 atm

Carbon dioxide removal

CO_2 absorbed by alkaline medium
(several processes in use)

Final carbon oxides removal

CO_2 (traces) $+ 4H_2 \rightarrow CH_4 + 2H_2O$
CO (traces) $+ 3H_2 \rightarrow CH_4 + H_2O$
(so-called methanation reaction)

Compressor

Recycle of unconverted nitrogen/hydrogen mixture

150–350 atm

Ammonia converter

$N_2 + 3H_2 \rightarrow 2NH_3$

$7CH_4 + 10H_2O + 8N_2 + 2O_2 \rightarrow 16NH_3 + 7CO_2$
(overall chemical reaction)

Ammonia

FIGURE 10-1. Block flowchart of ammonia synthesis by steam reforming of natural gas. (International Fertilizer Development Center and United Nations Industrial Development Organization, *Fertilizer Manual*, Reference Manual IFDC-R-1. Muscle Shoals, Ala.: IFDC, 1979.)

employed for this purpose in over 80% of the world's ammonia production. Figure 10-1 shows the principal chemical reactions and operating pressures for a typical steam-reforming process. There are three distinct steps in the manufacture of ammonia: synthesis gas production, purification, and ammonia synthesis.

Other hydrogen feedstocks that can be treated by steam-reforming processes include liquefied petroleum gas or LPG, which is mainly butane and propane; liquefied natural gas (LNG); naphtha; refinery gases; coke-oven gas; and heavy oil. Other suitable feedstocks not currently in commercial use include methanol, ethanol, and methane derived from organic wastes or from biomass specifically grown to produce cellulosic materials.

The various steps of ammonia production are carried out at temperatures varying from 0 to 1200°C. Fuel is thus needed for heating the reactants. Although measures are usually taken to recover as much of the heat as possible, a substantial portion is lost. Some form of energy is also utilized for driving machinery.

Prices of natural gas greatly affect the cost of manufacturing ammonia. Natural gas feedstock costs now represent about 75 to 80% of the variable or cash costs of producing ammonia. At one time natural gas costs accounted for much less than 50% of the total ammonia manufacturing costs. In 1981, U.S. ammonia producers paid an average of $2.33 per 1000 ft³ of gas, a dramatic rise from $0.28 in 1970.

The effect of natural gas costs on the total manufacturing cost of ammonia is shown in the following table.

Cost of Natural Gas ($/1000 ft³)	Percent of Total Ammonia Manufacturing Cost Attributable to Natural Gas Cost
0.30	27.5
0.60	42.3
0.90	53.0
1.80	68.6
2.70	77.2
3.00	79.0

It is noteworthy that ammonia manufacturing consumes only about 3% of the world production of natural gas. Usage of oil products and coal for this purpose represents approximately 0.5% of global production.

The Cynamide Process. The cyanamide process was developed in Germany in 1898 by Frank and Caro. Essentially, it requires the reaction at 1100°C of highly purified nitrogen gas with calcium carbide to form calcium cyanamide:

$$CaC_2 + N_2 \xrightarrow{\text{heat}} CaCN_2 + C$$

The calcium carbide is obtained by heating calcium carbonate to produce calcium oxide, which is in turn reacted with coke at 2200°C to form fused calcium carbide.

Use of this process is limited to just one or two locations in the world. Production from the only North American plant, at Niagara Falls, Ontario, is intended for nonagricultural purposes and only small amounts are used in the formulation of specialty fertilizers. However, calcium cyanamide is used more widely in European countries such as West Germany.

Direct Oxidation of Nitrogen. The oxidation of nitrogen has been taking place since the earth's atmosphere, as we know it now, evolved. With every flash of lightning some nitrogen is fixed. It was Cavendish in 1766 who first achieved this combination in a laboratory by passing electric sparks through a mixture of the two gases. The reactions are

$$N_2 + O_2 \longrightarrow 2NO$$

$$2NO + O_2 \longrightarrow 2NO_2$$

When the nitrogen dioxide is dissolved in water, nitric acid is formed:

$$3NO_2 + H_2O \longrightarrow 2HNO_3 + NO$$

The fixation of nitrogen by this method is appropriately termed the *arc process*.

Several attempts have been made in the United States to manufacture nitrates by this process, but none has been sufficiently successful to permit continued operations for more than a few years, for it has a large electrical requirement and it cannot compete with the less expensive ammonia fixation. A modification of the arc process was developed by Birkeland and Eyde in Sweden, where, because of cheap hydroelectric power, it has been successful. By this method nitrogen and oxygen are passed through an arc which is expanded in an electromagnet to increase the contact. The gas mixture leaving the furnace contains about 1.3 to 1.7% nitric oxide.

Ammonia Derivatives. Anhydrous ammonia is the starting point for at least 99% of all the nitrogen present in commercial fertilizers. Ammonium nitrate and urea are two solid nitrogen fertilizers that can be made from ammonia without additional raw materials. Other nitrogenous fertilizers manufactured from ammonia and various other raw materials include ammonium chloride, ammonium sulfate, calcium cyanamide, calcium nitrate, oxamide, and sodium nitrate.

Ammonia Oxidation. Nitric acid used in the manufacture of ammonium nitrate and other nitrate-containing fertilizer materials is made by oxidizing ammonia in the presence of a platinum catalyst at 800°C. The method now in use is known as the low-pressure process because it takes place under pressures of only 50 to 100 lb/in.[2]. It was developed by Ostwald and Brauer and first operated about 1908 in Germany. Conversion of ammonia to nitric acid proceeds in two steps: (1) oxidation and (2) absorption. The chemical reactions involved in these two steps are:

Oxidation:

1. $4NH_3 + 5O_2 \longrightarrow 4NO + 6H_2O$

2. $2NO + O_2 \longrightarrow 2NO_2$

Absorption:

3. $3NO_2 + H_2O \longrightarrow 2HNO_3 + NO$

4. $2NO + O_2 \longrightarrow 2NO_2$

5. Reactions 3 and 4 repeat through the absorber.

The usual preference for the oxidation catalyst is platinum containing between 2 and 10% rhodium.

Urea. Urea, or carbamide, as it is sometimes called, is a nonionic nitrogen material used industrially in the manufacture of plastics, in fertilizers, and as a protein supplement in the feed of ruminant animals. Its preparation is a bit more complicated than that of many fertilizer salts, which essentially require only the neutralization of an acid with ammonia. All commercial production of urea is achieved by reacting ammonia and carbon dioxide under very high pressure of about 2000 lb/in.[2] in the presence of a suitable catalyst. Since both reactants are produced at an ammonia plant, urea is a well-integrated commodity for manufacture in association with a synthetic ammonia plant. This reaction proceeds in two steps: (1) formation of ammonium carbamate and (2) dehydration of ammonium carbamate:

$$2NH_3 + CO_2 \longrightarrow NH_2COONH_4$$

$$NH_2COONH_4 \longrightarrow NH_2CONH_2 + H_2O$$

There are many variations of urea manufacturing processes, with most of the differences occurring in the methods employed to recover, separate, and recycle unreacted ammonia and carbon dioxide. The various processes are usually described as being either once through, partial recycle, or total recycle. Conversion and eventual recovery of ammonia nitrogen as urea nitrogen is about 30% for the once-through process, 50 to 80% for partial recycle processes, and up to 98% for total recycle.

An aqueous solution containing about 75% urea is the end product of the various urea synthesis processes. This solution can be used to manufacture urea-ammonium nitrate fertilizer solutions or it can be processed further to make solid urea or granular compound fertilizers.

A simplified flowsheet in Figure 10-2 illustrates the major features of the spherodizer granulation process which is used at many of the new urea manufacturing plants in the United States and Canada.

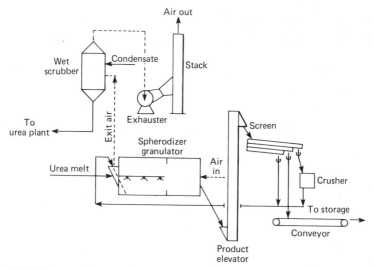

FIGURE 10-2. Simplified flowsheet for the C & I Girdler Spherodizer process for production of granular urea (and ammonium nitrate). (International Fertilizer Development Center and United Nations Industrial Development Organization, *Fertilizer Manual*, Reference Manual IFDC-R-1. Muscle Shoals, Ala.: IFDC, 1979.)

The desirable physical properties of granular urea were mentioned in Chapter 5. Consumption of urea as a nitrogen fertilizer is increasing rapidly because of its use in bulk blends, nitrogen solutions, and ease of handling for direct application.

Other Nitrogen Fertilizer Materials. Most of the nitrogen fertilizers discussed in Chapter 5 are made by different processes involving the use of the compounds just discussed—ammonia, nitric acid, and urea. Salts such as ammonium phosphate, ammonium sulfate, ammonium nitrate, and their various combinations are manufactured by neutralizing their acids with ammonia. Nitric phosphates are made by acidulating phosphate rock with nitric acid or mixtures of nitric acid with phosphoric or sulfuric acid and ammoniating. Many nitrogen solutions are made by dissolving ammonia gas in water and adding ammonium nitrate and/or urea.

The nitrogen content of the various solid and liquid nitrogen-fertilizer materials was discussed in Chapter 5.

Phosphate Fertilizer Materials

Today nearly all of the industrial and agricultural phosphates originate from phosphate rock. During the nineteenth century and the early part of the twentieth century, bones and guano were important sources of fertilizer phosphorus. The latter is the excreta and remains of seafowl and contains about 9% P_2O_5, most of which is water soluble, and up to 13% nitrogen. The once-rich deposits of guano along the western coasts of Lower California, South America, and Africa are now exhausted. Neither of these formerly prominent materials is now of any consequence in providing fertilizer phosphorus.

Phosphate Rock. The basic phosphorus compound in all commercially important deposits of phosphate rock is apatite. It may be a fluoro-, chloro-, or carbonatoapatite, as indicated in Chapter 6, though most occurs as fluoroapatite.

TABLE 10-3. World Phosphate Reserve Base (Million Metric Tons)

Country	Number of Deposits	Reserves *	Reserve Base †
North America			
United States	130	1,300	5,700
Canada	1	—	20
Mexico	2	120	140
Total *	133	1,400	5,900
South America			
Brazil	11	—	400
Colombia	1	—	15
Peru	1	—	120
Venezuela	1	12	12
Total	14	12	550
Europe			
Finland	2	—	45
Turkey	1	—	12
USSR	60	6,500	8,000
Total **	63	6,500	8,100
Africa			
Algeria	1	250	250
Egypt	5	—	260
Morocco	11	2,100	20,000
Western Sahara	1	850	850
Senegal	2	170	170
South Africa, Republic of	1	1,800	1,800
Togo	1	50	50
Tunisia	7	70	220
Other	2	—	12
Total **	31	5,300	24,000
Asia			
China	10	170	1,000
Israel	3	—	70
Jordan	3	530	530
Syria	2	170	180
Other	1	—	34
Total **	19	870	1,800
Oceania			
Australia	6	340	850
Nauru	1	23	23
Total **	7	360	870
World total **	267	14,000	41,000

* Cost less than $30 per metric ton. Costs include capital, operating expenses, taxes, royalties (if applicable), and miscellaneous costs; a 15% rate of return on investment is also included. Costs and resources are as of January 1981.

† Pending establishment of criteria for the reserve base, classification of data is based on a judgmental appraisal of current knowledge and assumptions. Cost $100 per ton or less.

** Data may not add to totals shown because of independent rounding.

Source: Stowasser, *Phosphate Rock.* Mineral Commodity Profiles. Pittsburgh, Pa.: U.S. Dept. of the Interior, Bureau of Mines, 1983.

The larger deposits are of sedimentary origin, laid down in beds in the ocean and then elevated to land masses. The phosphate is usually in the form of small pellets cemented by $CaCO_3$. It may be loose pebbles or hard rock. There is considerable speculation as to the amount of phosphate rock reserves. Identified world resources of this indispensable ore are given in Table 10-3.

More details on the occurrence of phosphate rock reserves and identified resources in the United States are provided in Table 10-4. Most of the mining of these deposits occurs in Florida and North Carolina, followed somewhat distantly by the western states. Of the 53.6 million metric tons of phosphate rock produced in 1981, 46.3 million metric tons were extracted in Florida and North Carolina, with an additional 1.3 and 6.0 million metric tons mined in Tennessee and the western states, respectively. The Florida mines currently account for about 30% of the world's phosphate rock production.

Table 10-5 brings in a basic concept in that as the price per recoverable ton increases, phosphate reserves increase. The amount of phosphate rock in the world has not changed. However, the reserves that may be economically exploited have increased tremendously. In fact, as price per ton increases, industry can afford to mine rock that at a lower price could not be made available in the marketplace. There is no shortage of phosphate rock in the world. All that is needed for it to be made available is time and money.

   Mining of Phosphate Rock. The working of phosphate rock is accomplished by both strip- and shaft-mining techniques. Strip mining is employed in Florida, Tennessee, North Carolina, and in some of the western deposits. Shaft mining is also practiced in some of the western deposits but not in any of the eastern producing areas.

TABLE 10-4.   Phosphate Rock Reserves and Reserve Base Estimates in the United States (Million Metric Tons)

State	Reserves *	Reserve Base †
Florida	550	2600
North Carolina	410	1300
Tennessee	20	30
Idaho	60	230
Utah	220	830
Wyoming	—	690
Montana	—	3
Total **	1,300	5,700

* Cost less than $30 per metric ton. Costs include capital, operating expenses, taxes, royalties (if applicable), miscellaneous costs, and a 15% rate of return on investment. Costs and resources are as of January 1981.

† Pending establishment criteria for the reserve base, classification of data is based on a judgmental appraisal of current knowledge and assumptions. Cost $100 per ton or less.

** Data do not add to totals shown because of independent rounding.

Source: Stowasser, *Phosphate Rock*. Mineral Commodity Profiles, Pittsburgh, Pa.: U.S. Department of the Interior, Bureau of Mines, 1983.

TABLE 10-5.   World Phosphate Reserves as Related to Price per Recoverable Ton (Millions of Tons)

Area	$8	$12	$20
North America	1,836	5,350	16,340
South America	53	290	930
Europe	829	2,050	4,100
Africa	1,770	8,430	20,500
Asia	335	1,186	4,600
Oceania	120	750	1,300
World	4,943	18,036	47,770

Source: Turbeville, "The phosphate rock situation." *Phosphate–Sulphur Symposium,* Tarpon Springs, Fla. (January 22–23, 1974).

Rock from most of the deposits has to be treated to separate the phosphate-containing fraction from the inert material. In some cases the ore must also be crushed. The phosphate-containing fraction is then separated from the waste material and concentrated by a complex system of washers, screens, classifiers, table agglomeration, and flotation. When dry, the concentrated ores are suitable for the manufacture of processed phosphate fertilizers. Figure 10-3 is a pictorial summary of what takes place during the various stages of beneficiation of phosphate mined in Florida.

Phosphate rock is a major commodity in world commerce. The world trade pattern from the principal producing areas of North America, North Africa, the USSR, and the Pacific Islands to consuming regions of the world is depicted in Figure 10-4. Morocco, the United States, the USSR, and the Pacific Islands of Nauru, Ocean, and Christmas were, in this order, the principal export sources of phosphate rock. These exports were destined primarily for western Europe, eastern Europe, Japan, Canada, and South America.

Treatment of Phosphate Rock. As pointed out in Chapter 6, the apatite bond in phosphate rock must be broken if the contained phosphate is to be rendered easily available to plants. This can be done either by heat or acid treatment. For various reasons, sulfuric acid is generally used for acidulation, followed by nitric and hydrochloric acids. Phosphoric acid itself may be used, but it is usually a secondary product. The prominence of sulfuric acid in the various methods for converting high-grade phosphate rock into finished phosphate fertilizer materials is apparent in Figure 10-5.

Defluorinated Phosphates. When phosphate rock is heated in a gas- or oil-fired chamber to around 1500 to 1600°C, the fluorine is driven off and the remaining calcium phosphate is of greater plant availability than the phosphorus in the original rock. This defluorination is brought about both by calcination and fusing. *Calcination* refers to the heating of rock with silica and steam to temperatures below the melting point of the mix (see Chapter 6). *Fusion* refers to the process in which the mix is heated to a temperature above its melting point so that the furnace charge will run together, or fuse, to form a glassy product.

Regardless of whether fusion or calcining is employed, the reaction believed to take place during the defluorination of rock with heat may be represented as follows:

$$Ca_{10}(PO_4)_6 \cdot F_2 + xSiO_2 + H_2O \longrightarrow 3Ca_3(PO_4)_2 + CaO \cdot xSiO_2 + 2HF$$

The phosphorus remaining is α-tricalcium phosphate, and the fluorine escapes as a gas. Several of the materials discussed in Chapter 6 resulted from one of these two processes, and their manufacture is based on the reaction just described.

Elemental Phosphorus Manufacture. If phosphate rock is heated above 1400°C in the presence of silica and carbon in a reducing atmosphere, elemental phosphorus is formed. The reaction is complex, but it can be expressed by the following generalized equation:

$$Ca_3(PO_4)_2 + 3SiO_2 + 5C \longrightarrow 3CaSiO_3 + P_2 + 5CO$$

The recovered elemental phosphorus is stored in the liquid state until ready for use. The U.S. Department of Defense uses elemental phosphorus in incendiary shells, grenades, and other items of warfare, but most of the industrial needs center around a high-purity phosphoric acid made from elemental phosphorus. This element is burned to form phosphorus pentoxide, which in turn is reacted with water to form phosphoric acid as indicated in the following equations:

$$2P + {}^5/_2O_2 \longrightarrow P_2O_5$$

$$P_2O_5 + 3H_2O \longrightarrow 2H_3PO_4$$

By far the greatest tonnage of fertilizer phosphate materials is manufactured by treating phosphate rock with acid rather than with heat.

Ordinary Superphosphate. Ordinary superphosphate is manufactured by the simple expedient of mixing gravimetrically equal parts of sulfuric acid and rock phosphate. The reaction, represented by the following simplified equation, gives off a considerable quantity of heat:

$$[Ca_3(PO_4)_2]_3CaF_2 + 7H_2SO_4 \longrightarrow 3Ca(H_2PO_4)_2 + 7CaSO_4 + 2HF$$

Three points are illustrated by this equation. First, the phosphate originally present as apatite is converted to water-soluble monocalcium phosphate. Second, one of the products of the reaction is gypsum, which is intimately mixed in with the monocalcium phosphate. Third, the reaction releases toxic hydrofluoric acid gas, which is usually recovered as a valuable by-product.

In one process, known as a *batch mix*, weighed quantities of rock phosphate and sulfuric acid of a certain concentration are combined. The ingredients are mixed and allowed to react for about 1 minute, after which the slurry is dumped into a compartment called a den. Here the phosphate may remain for about 15 minutes after the den is filled. It is then removed, stirred, and stored. However, removal may not take place for 24 hours. The acidulated phosphate in the den sets up into a hard block, and removal is accomplished by means of various mechanical excavators which are usually equipped with revolving knives. These knives cut into the block, and the disintegrated superphosphate is stored until needed.

The batch-type process is used to some extent in the United States, but a continuous process of rock acidulation in which the phosphate rock and the acid are added to a mixer is now the major method. Ingenious metering and weighing devices are required for the successful operation of this phase of the process. The mixture is agitated for two or three minutes, and is then discharged onto an endless slat conveyor on which it solidifies. The slat conveyor moves the block of hardened superphosphate toward a revolving cutter, which disintegrates the material. It is then transferred to a storage bin.

Regardless of the mixing process employed, the manufacture of superphosphate is a simple operation. The product is a very satisfactory fertilizer material which contains calcium, sulfur, and 7 to 9% available phosphorus (14 to 21% P_2O_5), the lowest content of any of the important sources of fertilizer phosphorus.

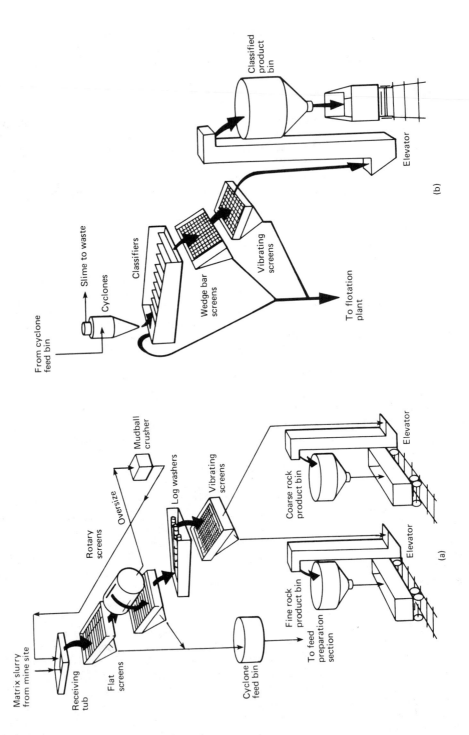

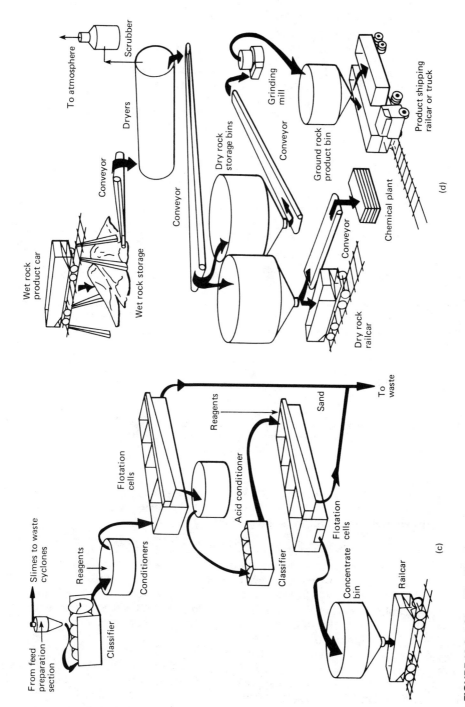

FIGURE 10-3. Simplified flowchart showing stages in beneficiation of mined Florida phosphate rock: (a) washing; (b) feed preparation; (c) flotation; (d) wet rock storage, drying, and shipping. [Hoppe, *Eng. Min. J.*, **177**(9):81 (1976).]

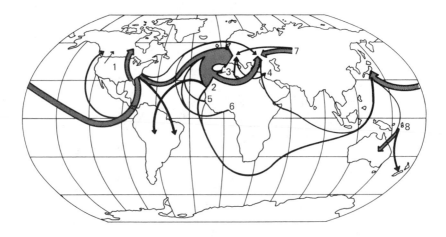

Exporting sources	Destination of exports	Quantity (thousand metric tons)	Exporting sources	Destination of exports	Quantity (thousand metric tons)
1. United States	Canada	3,200	5. Senegal	Western Europe	826
	Western Europe	3,525		Asia	215
	Asia	1,486			
	Eastern Europe	254	6. Togo	Western Europe	1,393
	South America	481		Eastern Europe	712
2. Morocco	Western Europe	10,181	7. USSR	Eastern Europe	4,067
	Eastern Europe	2,848		Western Europe	948
	South America	1,103			
	Asia	859	8. Pacific Islands	Australia	1,767
				New Zealand	853
3. Algeria and Tunisia	Western Europe	794		Indonesia, Republic	231
	Eastern Europe	807		of Korea, Malaysia,	
				Singapore, and Japan	
4. Israel and Jordan	Western Europe	1,431			
	Asia	1,962			
	Eastern Europe	1,438			

FIGURE 10-4. International trade in phosphate rock, 1981. Includes only principal exporting sources and destinations given in list. Numbers preceding exporting sources in list correspond to those on map. (Courtesy of The International Fertilizer Industry Association Ltd., Paris, France.)

Wet-Process Phosphoric Acid. Wet-process phosphoric acid is manufactured by extracting phosphate rock with sulfuric acid. The principal reaction taking place is represented by the following equation:

$$Ca_{10}(PO_4)_6F_2 + 10H_2SO_4 + 20H_2O \longrightarrow 10CaSO_4 \cdot 2H_2O + 6H_3PO_4 + 2HF$$

This reaction is carried out for about 8 hours in a digestion system. The reaction itself is essentially complete within a matter of a few minutes, but the additional time is needed to ensure the formation of gypsum crystals of a size adequate to permit more rapid filtration. The slurry is ultimately filtered and the acid concentrated to the desired strength by heating. A simplified representation of this process appears in Figure 10-6.

Phosphoric acid to be used in the manufacture of fluid fertilizers such as ammonium polyphosphate solution may require partial purification to remove troublesome impurities, including excessive magnesium, the compound $(FeAl)_3KH_{14}(PO_4)_8 \cdot 4H_2O$, and organic matter. It was mentioned briefly in Chapter 6 that wet-process acid utilized for production of fluid fertilizers must also be concentrated to superphosphoric acid, with P_2O_5 concentrations ranging from 68 to 72% and with 40 to 50% of the phosphorus as polyphosphates.

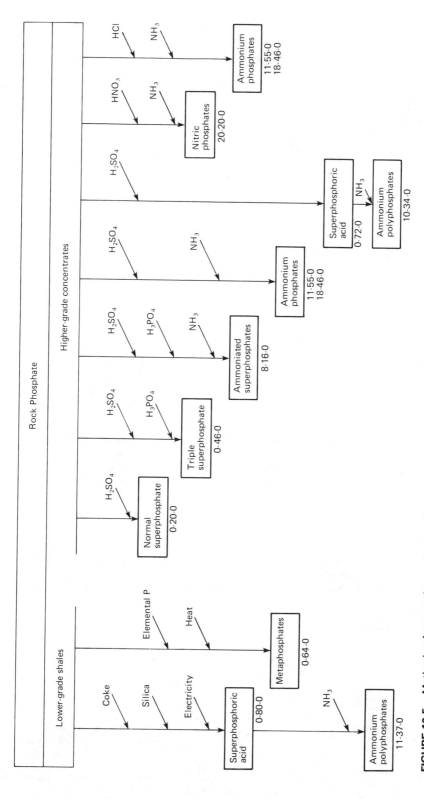

FIGURE 10-5. Methods of converting rock phosphate to phosphate fertilizers. (Nielsen and Janke, *Proc. Western Canada Phosphate Symp.*, pp. 140–158. Edmonton, Alberta: Alberta Soil and Feed Testing Laboratory, 1980.)

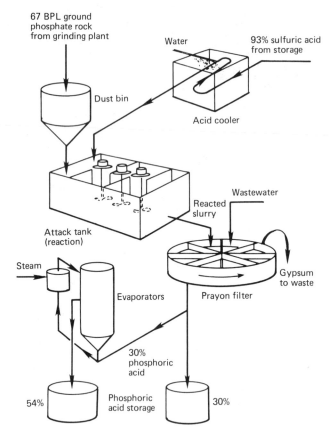

FIGURE 10-6. Major process stages in the manufacture of wet-process phosphoric acid. [Hoppe, *Eng. Min. J.*, **177**(9):81 (1976).]

Other Phosphatic Fertilizers. Wet-process phosphoric acid is the source of phosphorus used for the manufacture of triple superphosphate, ammonium phosphates, and liquid fertilizers (Figure 10-7). Essential features of the production of granular triple superphosphate and ammonium phosphates are shown in Figures 10-8 and 10-9, respectively.

Sulfuric Acid. Sulfuric acid is literally the workhorse of the fertilizer industry. More than 60% of the total consumption of this industrial acid is accounted for by the fertilizer industry alone, its largest single user.

Some of the reasons for its general use include its ready production from elemental sulfur, which is a relatively inexpensive raw material in abundant supply.

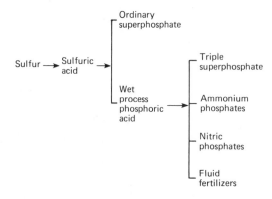

FIGURE 10-7. Phosphate fertilizer derivatives of sulfuric acid and wet-process phosphoric acid. (Bixby, in F. E. Khasawneh, E. C. Sample, and E. J. Kamprath, Eds., *The Role of Phosphorus in Agriculture*, Chapter 5.Madison, Wis.: American Society of Agronomy, 1980.)

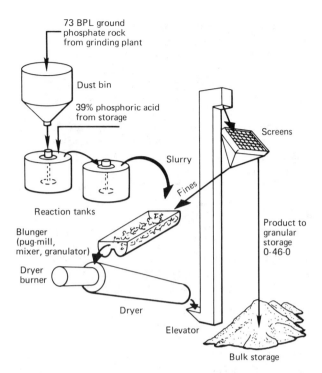

FIGURE 10-8. Simplified flowchart for granular triple superphosphate manufacture. [Hoppe, *Eng. Min. J.*, **177**(9):81 (1976).]

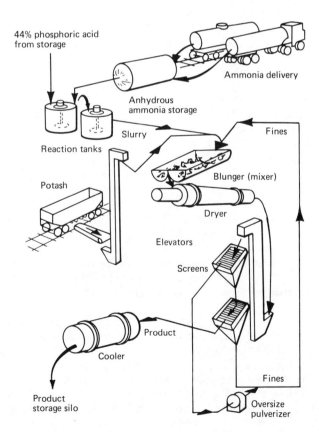

FIGURE 10-9. Simplified flowchart illustrating production of granular ammonium phosphate. [Hoppe, *Eng. Min. J.*, **177**(9):81 (1976).]

When sulfuric acid is reacted with phosphate rock, the resulting calcium sulfate is insoluble, thus permitting easy removal by filtration. This allows for good separation of the desired product, wet-process phosphoric acid.

Sulfuric acid is manufactured either from elemental sulfur or from the sulfur dioxide collected in the roasting of metal sulfides known as pyrites. In either case the sulfur dioxide formed is oxidized to sulfur trioxide and reacted with water to form sulfuric acid.

There are two principal processes in the manufacture of sulfuric acid. One is termed the *contact process*, the other the *lead-chamber process*. Although some lead-chamber plants still produce acid, all of the newer installations are contact plants.

In the lead-chamber process sulfur is burned in an oven to produce sulfur dioxide:

$$S + O_2 \longrightarrow SO_2$$

The reaction is exothermic. The fumes are pulled by a fan-induced draft into a series of large lead tanks, in which the sulfur dioxide is catalytically oxidized to sulfur trioxide by NO_2. This NO_2, known to the industry as niter gas, is recovered and used over and over again; otherwise, the process would not be economically possible. The generalized reactions are illustrated by the following equations:

$$2HNO_3 + H_2O + 2SO_2 \longrightarrow 2H_2SO_4 + NO + NO_2$$
$$SO_2 + H_2O + NO_2 \longrightarrow H_2SO_4 + NO$$

As the gases are swept along, the nitric oxide is reconverted to niter gas and recovered in concentrated sulfuric acid in a tank at the end of the system. The sulfuric acid produced is collected in and withdrawn from the various lead tanks or chambers in the system. It is then stored or used immediately, depending on the prevailing situation.

In the contact process (Figure 10-10) a mixture of sulfur dioxide and air is passed through iron tubes containing a finely divided catalyst, usually platinum. In the presence of this catalyst and at 400°C the sulfur dioxide is rapidly converted to sulfur trioxide.

$$SO_2 + \tfrac{1}{2}O_2 \longrightarrow SO_3 + 22{,}600 \text{ cal}$$

The sulfur trioxide is then passed into 98% sulfuric acid, the reason being that sulfur trioxide is not readily soluble in water. The concentration of the acid is maintained at 98% by the constant addition of water as the sulfur trioxide is passed into it.

This process is nearly 100% efficient and the heat it evolves is used to generate steam, which is used for other purposes.

Sources of Sulfur for Sulfuric Acid. Both elemental sulfur or pyrites are suitable sources of sulfur for the contact and lead-chamber processes. As indicated in Figure 10-10, sulfur dioxide, the intermediate raw material, is obtained by burning elemental sulfur or by roasting of pyrites.

The usual sources of elemental sulfur are dome sulfur, sour gas, and volcanic sulfur. Pyrites are common metal sulfide ores treated in metallurgical operations.

Dome Sulfur. Elemental sulfur is the principal source of sulfur dioxide in most countries, even in many in which there are substantial deposits of pyrites. Elemental sulfur is produced commercially in several ways. Deposits of the element are found along the Gulf coast of Mexico, Louisiana, and Texas. These deposits occur in domelike formations, in conjunction with gypsum, several hundred feet below the surface of the earth.

An ingenious engineer named Herman Frasch devised a scheme by which this

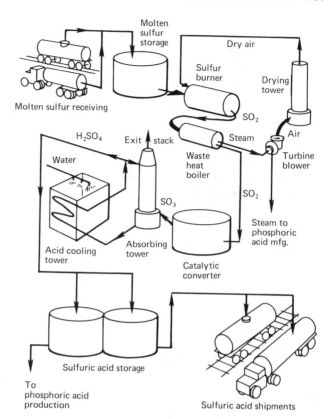

FIGURE 10-10. Simplified flowchart showing steps in the manufacture of sulfuric acid by the contact process. [Hoppe, *Eng. Min. J.*, **177**(9):81 (1976).]

sulfur could be brought to the surface. It consists of drilling into the sulfur–gypsum formation and inserting three concentric pipes leading from the deposit to the surface of the earth. Superheated water and compressed air are forced down through two of the pipes into the formation. As the temperature of the superheated water is higher than that of the melting point of sulfur, the sulfur melts. The compressed air forces the molten sulfur to the surface of the ground through the third pipe. It is essentially 99.5% pure as mined. It is either poured into molds, solidified, and stored in this state until ready for use or shipped in molten form. About 90% of all Frasch sulfur is handled in molten form.

In 1981, Frasch sulfur accounted for 53% of total U.S. sulfur production. Production of Frasch sulfur has averaged about 6.7 million tons annually during the period 1971–1981.

Sour Gas. Elemental sulfur is also recovered from natural gas containing hydrogen sulfide, or sour gas. Because of its toxicity and corrosive effects, hydrogen sulfide must be removed before the natural gas can be used commercially. Some natural gas fields such as those close to the Rocky Mountains in Alberta may contain over 50% hydrogen sulfide.

Approximately 8 million metric tons of sulfur were recovered from U.S. and Canadian sour gas in 1981. The Western Overthrust area of Wyoming and Utah is expected to become more important since it contains over one-third of the estimated 100 million metric tons of elemental sulfur in U.S. sour gas reserves. Large quantities of sulfur are also recovered from sour gas in France, the Middle East, West Germany, and the USSR.

Hydrogen sulfide is normally removed by scrubbing the natural gas with amine solutions, usually monoethanolamine. Separation of hydrogen sulfide from the amine solution is accomplished by steam stripping and the amine is recovered by washing the stripped gas with water.

Modified versions of the Claus process are generally used for converting the separated hydrogen sulfide into elemental sulfur. The chemistry of this conversion may be considered to take place in two steps. First, the concentrated hydrogen sulfide gas is burned in a combustion chamber in such a manner that one-third of its volume is converted to sulfur dioxide as follows:

$$2H_2S + 3O_2 \longrightarrow 2SO_2 + 2H_2O$$

The cooled combustion gases are mixed with the remaining two-thirds of the hydrogen sulfide and then passed through a catalyst-packed convertor so that the following reaction takes place:

$$2H_2S + SO_2 \longrightarrow 3S + 2H_2O$$

Following condensation of the vapor to liquid sulfur, the recovered sulfur is handled in the same way as that mined by the Frasch method.

Volcanic Sulfur. Local deposits of elemental sulfur throughout the world are associated with volcanic activity. The largest occurs in Sicily, where sulfur is mined by pick-and-shovel methods and marketed. In fact, before the discovery of the sulfur domes on the Gulf Coast of the United States, the Italians and Sicilians monopolized the world market. Today, however, Sicilian sulfur accounts for only a small fraction of the total world production of elemental sulfur.

Pyrites. Pyrites are the sulfides of heavy metals such as iron, lead, copper, and zinc. In recovering the metal from these ores, the sulfur is driven off by roasting or, more properly, by burning. The general reaction can be illustrated by the equation for the roasting of an iron pyrite:

$$2FeS_2 + {}^7\!/_2O_2 \longrightarrow 2SO_2 + Fe_2O_3 + heat$$

Roasting pyrites for their sulfur values is generally not economical. Acid produced by this method can compete in price with acid made from elemental sulfur only when the metal recovered from the pyrite can be profitably marketed. The sulfur dioxide produced from pyrites is converted to sulfuric acid, as previously described.

Many of these smelting operations are located at places inaccessible to the market and because of shipping costs, the SO_2 value has not contributed to the supply of sulfuric acid. Legislation aimed at reducing environmental pollution is forcing the smelting industry to recover SO_2, much of which was released into the atmosphere.

Another possible future source of sulfur is the SO_2 which is released by the coal- and oil-fired steam generating plants. The SO_2 emissions from the burning of fossil fuels and other activities in the United States in 1980 were estimated to be equivalent to about 13 million metric tons of elemental sulfur. Total annual emissions of sulfur oxides in Canada are judged to be equivalent to approximately 3.5 million metric tons of elemental sulfur. These emissions represent a significant potential additional supply of elemental sulfur if economical recovery processes can be found.

Nitric Acid Acidulation. Processes for this method of treating phosphate rock use nitric acid, either alone or in combination with sulfuric or phosphoric acid. Four basic processes involve acidulation with (1) nitric and phosphoric acid, (2) nitric and sulfuric acid, (3) nitric acid with potassium sulfate added to the slurry, and (4) nitric acid in which the slurry is ammoniated and carbonated with carbon dioxide. The first two are probably best suited for commercial application, although some European manufacturers have had considerable success with modifications of the fourth process as indicated in Chapter 6.

That nitric phosphates have not been widely accepted in the United States stems in part from two inherent drawbacks in most of the nitric phosphate processes. The first is the additional cost of obtaining a product in which more than 50 or 60% of the contained phosphorus is water soluble. The second is the limited number of ratios and grades that can be manufactured by a given nitric phosphate process. Not only is the number of grades limited, but it is a difficult and costly process to change from one grade to another. Under present economic conditions an ammonia–phosphoric acid plant combination is capable of producing highly water-soluble phosphates in a large number of grades and ratios to meet specific crop requirements.

The known reserves of elemental sulfur, sour gas sulfur, pyrites, and sulfur from fossil fuels are finite, although very large. The supply of elemental nitrogen, although not infinite, is for all intents and purposes so limitless that its shortage should not be considered. In the final analysis the economics of production will govern the part that nitric acid will take in the acidulation of phosphate rock.

Potassium Fertilizers

Like phosphates, potassium fertilizers are obtained from deposits found several hundred to several thousand feet below the earth's surface. Like phosphates, too, the potassium ores must be beneficiated to produce high-grade potassium fertilizers; but unlike phosphate rock, potassium salts do not require treatment with heat or strong acids to render the contained potassium available to plants, for they are water soluble.

The word *potash*, which is the trade term commonly applied to potassium-containing fertilizers, was derived from *pot ashes*. During the early days wood and other plant residues were burned in pots to obtain the salts, largely for the manufacture of soap. The ashes containing the salts were extracted with water and the solution was then evaporated. The residue consisted of a mixture of potassium carbonate and other salts. The production of these salts from wood ashes was one of the first important chemical enterprises in the early colonial days in the United States. The first patent was granted in 1790 for the preparation of potassium salts.

Potash Deposits. Potash deposits occur as beds of solid salts at varying depths in the earth's surface and also as brines in dying lakes and seas. World potash resources are immense as is evident in Table 10-6. Estimated reserves deemed economical to mine under current economic conditions are nearly 12 billion metric tons. They are sufficient to meet requirements for many centuries to come.

In addition to the total global potassium resources approaching 137 billion metric tons of K_2O, potassium can also be recovered in virtually unlimited quantities from seawater. Potassium resources are potentially minable ores which, because of cost or other constraints, are uneconomic at current prices but might be in the future. Recovery of potassium from seawater at this time is even less feasible.

Potassium reserves in Canada and the USSR represent more than half of the world's reserves and about 80% of the world's resources. There are extensive deposits also in East and West Germany and Israel. Mining of the Canadian deposits began in 1959 and by 1982 production capability reached slightly more than 25% of the world's potassium capacity of over 35 million metric tons of K_2O.

In the United States the largest known deposits occur in the Permian salt basin, which includes southeastern New Mexico, northwestern Texas, and some of the western part of Oklahoma. The center of production in this area is around Carlsbad, New Mexico. Mining of extensive potassium salt deposits in central Utah, which

TABLE 10-6. World Potassium Resources (Million Metric Tons of K₂O)

Location	Reserves *	Other Resources	Total Resources
North America			
Canada	9,000	58,000	67,000
United States	180	5,260	5,440
Total	9,180	63,260	72,440
South America			
Chile	9	9	18
Peru	0	9	9
Brazil	45	225	270
Total	54	243	297
Europe			
France	35	145	180
Germany, Democratic Republic	270	4,260	4,530
Germany, Federal Republic	180	3,080	3,260
Italy	9	27	36
Spain	27	154	181
USSR	1,800	43,500	45,300
United Kingdom	45	225	270
Total	2,366	51,391	53,757
Asia			
Israel and Jordan	218	870	1,088
China, People's Republic of	9	9	18
Laos	18	27	45
Thailand	55	9,015	9,070
Total	300	9,921	10,221
Africa			
Congo	0	180	180
Total	0	180	180
Oceania			
Australia	0	18	18
Total	11,900	125,013	136,913

* At average 1976 domestic mine price.
Source: U.S. Bureau of Mines, *Potash.* Mineral Commodity Profiles MCP-11. Pittsburgh, Pa., 1978. (Converted from short to metric tons.)

are located at depths of more than 2000 ft, was begun in 1964. Deposits are present in the Williston basin of North Dakota.

Important sources of lake brines in the United States are Searles Lake, California, the Great Salt Lake, and the Salduro Marsh in Utah. The last two are remnants of former Lake Bonneville. Figure 7-21 shows the location of the main potassium-producing areas in North America and also indicates the mining method employed.

The potassium-bearing minerals found in these deposits, together with their approximate potassium concentrations, are listed in Table 10-7. Sylvite is the most important and it usually occurs mixed with sodium chloride. Other minerals processed for potassium fertilizer production are carnallite, langbeinite, niter, and kainite.

Mining Potassium Salts. The method used to mine potash ores depends on the nature of the deposit. Shaft mining of the solid ore is employed in some areas, solution mining in others, and for the surface lake brines the method varies. A brief summary of these mining operations follows.

Solid Ore Mining. The problem of mining potassium-containing ores is complicated by the great depths at which the salts occur. In general, it is difficult to mine profitably at depths greater than 4000 ft. However, the depth at which

**TABLE 10-7.   Potassium Minerals
Present in Potash Deposits**

Mineral	Composition	Percent K_2O
Chlorides		
Sylvinite	$KCl \cdot NaCl$ mixture	approx. 28.0
Sylvite	KCl	63.1
Carnallite	$KCl \cdot MgCl_2 \cdot 6H_2O$	17.0
Kainite	$KCl \cdot MgSO_4 \cdot 3H_2O$	18.9
Hanksite	$KCl \cdot 9Na_2SO_4 \cdot 2Na_2CO_3$	3.0
Sulfates		
Polyhalite	$K_2SO_4 \cdot MgSO_4 \cdot 2CaSO_4 \cdot 2H_2O$	15.5
Langbeinite	$K_2SO_4 \cdot 2MgSO_4$	22.6
Leonite	$K_2SO_4 \cdot MgSO_4 \cdot 4H_2O$	25.5
Schoenite	$K_2SO_4 \cdot MgSO_4 \cdot 6H_2O$	23.3
Krugite	$K_2SO_4 \cdot MgSO_4 \cdot 4CaSO_4 \cdot 2H_2O$	10.7
Glaserite	$3K_2SO_4 \cdot Na_2SO_4$	42.6
Syngenite	$K_2SO_4 \cdot CaSO_4 \cdot H_2O$	28.8
Aphthitalite	$(K, Na)_2SO_4$	29.8 *
Kalinite	$K_2SO_4 \cdot Al_2(SO_4)_3 \cdot 24H_2O$	9.9
Alunite	$K_2 \cdot Al_6(OH)_{12} \cdot (SO_4)_4$	11.4
Nitrates		
Niter	KNO_3	46.5

* Assuming equimolecular proportions of potassium and sodium.
Source: International Fertilizer Development Center and United Nations Industrial Development Organization, *Fertilizer Manual.* Reference Manual IFDC-R-1. Muscle Shoals, Ala.: IFDC, 1979.

the German and Canadian deposits are mined is considerably greater than that in the Carlsbad area in the United States.

In the room-and-pillar mining pattern, rectangular rooms are mined out and pillars of ore are left for support. The first stage takes about 50 to 60% of the ore. The second stage, *pillar robbing*, may result in removal of about 90% of the ore.

The first mining in the Carlsbad area was begun in 1931, and coal-mining methods common at that time were used in the initial effort. Later the operation was entirely mechanized, and all underground equipment was electrically powered. Continuous mining machines, which remove the ore continuously from the potassium vein, are largely used now. The loosened ore is loaded onto shuttle cars or continuous belts and hauled to the foot of an elevator shaft, where it is crushed to a maximum size of about 6 in. before being carried to the surface for further processing.

The development of continuous mining machines ranks as one of the major advances in potash mining. This equipment actually cuts the ore directly from the mine face and does away with undercutting, drilling, and blasting operations. One of the two most widely used types of these machines is pictured in Figure 10–11. This particular one has two rotors, each powered by a 500-hp electrical motor, that break up the potash as the vehicle moves forward into the ore bed at a rate of about 1 ft/min. It cuts a section about 17 ft wide and 11 ft high. The other popular Marietta machine has four rotors and cuts an area 8 ft high by 26 ft wide.

The long-room mining technique as practiced in Saskatchewan involves cutting rooms 67 ft wide and 4000 ft long. A pillar 50 to 60 ft wide is left and another room of the same dimension is cut. Then a second pillar 180 ft wide is left and the same series repeated. Use of this method is dependent on geological conditions of the area being mined. It is employed when the overlying strata are relatively stable with little shifting above the mine.

Where the strata are unstable, the stress-relief mining procedure is used. In

FIGURE 10-11. Goodman continuous-mining machine used for removal of potassium ore in Saskatchewan. [Courtesy of the Potash Corporation of Saskatchewan (1982).]

stress-relief mining, three to five parallel tunnels (called entries) are cut in sequence, with the center entries, which are the ones to be mined, cut last. The outer entries take the majority of the stresses and pressures from the surrounding area and will fail, with the ceiling falling and the floor heaving, as is intended. This relieves the pressure on the center entries, and they are thus suitable for mining.

   Solution Mining. Considerable attention has been given to the possibility of drilling wells down to the ore bed. The principle of this method is based on pumping hot solution down to the bed, dissolving the potassium salts, and returning the potassium-laden brine to the surface for refining. This technique is used at one location each in the United States and Canada (see Figure 7-21). It is advantageous when the ore body is irregular in shape or when the depth is too great for conventional mining (e.g., at depths below 3500 ft in the Saskatchewan deposits).

   Brines. At Searles Lake wells are driven to a few feet short of the bottom of the deposit and the brine is pumped a distance of several miles to the processing plant. Wells may have a life of several years before the composition of the brine becomes unsatisfactory.

   At Great Salt Lake in Utah there is a complex of almost 30,000 acres of specially constructed ponds for processing mineral-rich brines. The sun furnishes 90% of the energy in evaporation and time required to convert to brines to harvestible salts is 2 years. Potassium sulfate, sodium sulfate, magnesium chloride, and common salt are some of the products.

   At Salduro Marsh brine is pumped from a network of more than 50 miles of canals, 3 ft wide and 14 ft deep.

Cement Kiln Potassium. In the production of cement some of the potassium-bearing ingredients are given off in flue dusts and gases. These dusts and fumes are collected in a precipitator, and the product is sold as a fertilizer for its potassium and calcium content. Such materials contain about 30% calcium and 5% potassium. Limited quantities of cement kiln dusts are collected in Maryland and California.

Refining of Potassium Ores. As mentioned in Chapter 7, potassium chloride is, in the matter of tonnage, the most important of the potash fertilizers. In the following sections some of the details of the production of this salt are covered and a brief description of the processing of the other potash fertilizers is given.

Potassium Chloride. Recovery of potassium chloride from sylvinite ore is made by the mineral flotation process or by solution of KCl, followed by recrystallization. Flotation is by far the most widely used. Recovery from brines at Searles Lake is by fractional crystallization.

Flotation. Sylvinite ore is a mixture of interlocked crystals of potassium chloride and sodium chloride plus small quantities of clay and other impurities. To make an acceptable product, the clay and salt must be removed and the potassium chloride upgraded and sized to meet commercial requirements. The sequence of events in the milling process used to upgrade potassium chloride ore is highlighted in the flow sheet in Figure 10-12. It is apparent that flotation has a vital role in the beneficiation. A simplified description of the milling procedure follows:

1. The ore is ground to separate the crystals and disperse the clay slime.
2. The ground ore is suspended and agitated in a saturated NaCl-KCl brine.
3. The slurry is deslimed to remove the clay. This is important, for clay will increase the requirement for the flotation reagents.
4. The deslimed slurry is conditioned with aliphatic amine acetate salts to film the potassium chloride particles selectively. The sodium chloride particles are not filmed.
5. The conditioned slurry then passes to rougher flotation cells into which air is drawn by agitation. The air bubbles attach to the filmed potassium chloride particles and float them to the surface. The froth is mechanically skimmed off by paddles and passes into cleaner flotation cells for further purification.
6. The froth and brine are centrifuged and the potassium chloride is dried in rotary driers.
7. The dried potassium is moved to the screening area for sizing into the four principal products—granular, coarse, standard, and special standard.
8. Portions of the fine particle grades—special standard and standard—are sent to the compactor where they are passed between rollers under pressure to form dense sheets. These sheets are then broken up and screened to produce granular and coarse-sized products which are in high demand for formulation of bulk blends with other fertilizer ingredients.

At Salduro Marsh the brine is concentrated by solar evaporation in ponds covering 11,000 acres. The potassium chloride is separated from the sodium chloride by a flotation process.

Heavy Media Separation. Because of its coarse crystalline structure, Saskatchewan ore does not need to be crushed as fine as other ores. After screening, the coarse fraction is blended with a mixture of brine and magnetite. Potassium

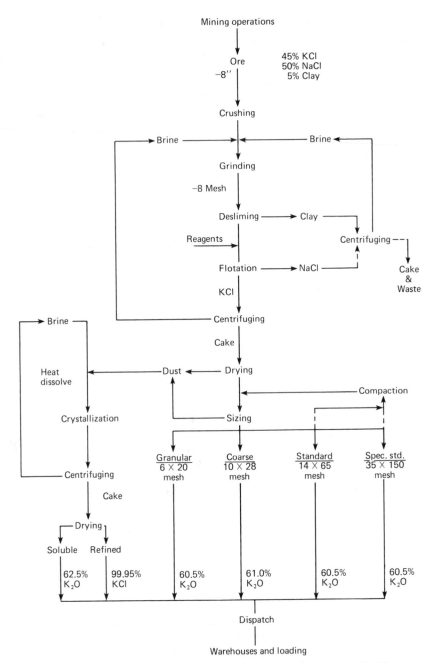

FIGURE 10-12. Milling process flowsheet for refining potassium chloride ore. (Courtesy of Cory Limited, Potash Corporation of Saskatchewan.)

chloride is separated by pumping through hydroclones which act as centrifugal separators. The KCl is then debrined, dried, crushed, and screened to size. This technique preserves large potassium chloride crystals which are already large enough to qualify for the granular and coarse grades without further processing.

Crystallization. The difference in the temperature–solubility relationships of potassium and sodium chlorides is the usual basis of this method of recovery. The solubility of potassium chloride increases rapidly with a rise in temperature, whereas sodium chloride solubility varies only slightly over a wide temperature range.

Cool brine saturated with both salts is heated and passed over the ore counter-current to the flow of the ore. Potassium chloride plus small amounts of sodium chloride are dissolved. The slurry is clarified and pumped through vacuum crystallizers to crystallize the potassium chloride, which is filtered out and dried.

At Searles Lake, where product recovery is accomplished by evaporation and fractional crystallization, the brine is concentrated in huge evaporators under vacuum. Twenty-three separate crystallization steps are required to separate potassium chloride and ten other basic chemical products. Among the by-products are sodium chloride, lithium carbonate, soda ash, salt cake, bromine, borax, boric acid, and pyrobor.

Potassium Sulfate. Potassium sulfate is a white salt which contains 41.5 to 44.2% potassium (50.0 to 53.2% K_2O). It is produced commercially by a number of processes.

Langbeinite Process. Production from langbeinite ($K_2SO_4 \cdot 2MgSO_4$) proceeds according to the following equation:

$$K_2SO_4 \cdot 2MgSO_4 + 4KCl \longrightarrow 3K_2SO_4 + 2MgCl_2$$

Trona Process. Burkite ($Na_2CO_3 \cdot 2Na_2SO_4$) is reacted with potassium chloride to form glaserite ($Na_2SO_4 \cdot 3K_2SO_4$) and reacted with potassium chloride brine to give potassium sulfate.

Hargreaves Process. Potassium sulfate and hydrochloric acid are made directly from sulfur and potassium chloride. Sulfur dioxide from a sulfur burner is mixed with water vapor and air and passed over heated beds of potassium chloride to give potassium sulfate and its by-product hydrochloric acid.

Mannheim Process. Although the reactions pass through two stages, the overall equation is the following:

$$2KCl + H_2SO_4 \longrightarrow K_2SO_4 + 2HCl$$

Glaserite Process. Commercial production of potassium sulfate by a two-stage process, involving the formation of glaserite as an intermediate, is scheduled to begin in late 1984 in Saskatchewan. In the first stage, as shown in the following equation, sodium sulfate and potassium chloride react to form glaserite:

$$4Na_2SO_4 + 6KCl \longrightarrow Na_2SO_4 \cdot 3K_2SO_4 + 6NaCl$$

Glaserite is then reacted with potassium chloride in the second stage to form potassium sulfate as expressed in the equation below:

$$Na_2SO_4 \cdot 3K_2SO_4 + 2KCl \longrightarrow 4K_2SO_4 + 2NaCl$$

Potassium Magnesium Sulfate. As sold in the United States, this material contains about 18% potassium (22% K_2O), 10.8% magnesium (18.0% MgO), 22% sulfur, and a maximum of 2.5% chlorine.

Langbeinite is mixed with sodium and potassium chlorides. The slower rate of dissolution of langbeinite is used as the basis for purification in which the ore is crushed and a countercurrent washing process removes the chloride salts.

Potassium Nitrate. There are several means of producing potassium nitrate. One recently patented method reacts anhydrous liquid nitrogen pentoxide with potassium chloride. Liquid chlorine is a by-product. A double decomposition reaction has been used for many years but the cost is comparatively high:

$$NaNO_3 + KCl \longrightarrow KNO_3 + NaCl$$

Methods have been proposed in which nitric acid and potassium chloride are the basis of new commercial production:

$$6KCl + 12HNO_3 \longrightarrow 6KNO_3 + 3Cl_2 + 6NO_2 + 6H_2O$$

The chlorine is liquefied and recovered. The nitrogen dioxide is converted to nitric acid and reused. Two grades of potassium nitrate, industrial and agricultural, are crystallized and filtered out.

Potassium Polyphosphates. Commercial processes have been developed to make materials with a wide range in water solubility and concentration. Some of the products are:

	Liquids	*Solids*
Orthophosphates	0-30-11	9-48-16
	0-20-20 (clear)	5-46-30
	0-25-20 (clear)	0-47-31
Polyphosphates	0-26-26 (clear)	0-50-40

General reactions are:

muriate of potash + sulfuric acid $\longrightarrow$ potassium bisulfate + HCl

sulfuric acid + rock phosphate + potassium bisulfate $\longrightarrow$
$$KH_2PO_4 + H_3PO_4 + CaSO_4$$

The H_3PO_4 reacts with KH_2PO_4 to produce polyphosphates. Solubility and concentration are varied in part by adding potassium or removing phosphorus. These materials have a low salt index, high analysis, varying solubility, and no chlorine.

Other Sources. Waste in the manufacture of tobacco products, consisting largely of the stems and ribs of tobacco leaves which are ground and sold for use in the fertilizer industry, contains 4 to 8% potassium and 2 to 4% nitrogen. This product also serves as a good conditioner for mixed goods.

Kelp, a giant seaweed, occurs in large beds extending off of the lower coast from Lower California to the Alaskan peninsula. The plants grow rapidly and the beds are accessible. It appears that kelp could furnish a considerable amount of potassium annually. The ash, which contains as much as 25% potassium, is also an important source of iodine. During World War I limited use was made of California kelp as a source of potassium.

A vast, untapped source of potassium is seawater, a cubic mile of which contains the equivalent of about 1.6 million tons of potassium. Considerable research has been conducted on the extraction of potassium from this source and, although little progress has been made, Norway has reported some success.

Mixed Fertilizers

The mixed-fertilizer industry is extensive in the United States and sales of mixed fertilizers date back more than 100 years. Numerous fertilizer mixtures have been

TABLE 10-8. Percentage of Plant Nutrients Consumed in the United States Applied in Mixtures (Selected Years from 1959 to 1982)

Year	Nitrogen (N)	Phosphorus (P_2O_5)	Potassium (K_2O)
1959–1960	37	79	87
1964–1965	31	80	81
1974–1975	24	82	58
1979–1980	22	84	48
1981–1982	21	88	46

Source: Hargett and Berry, "1980 fertilizer summary data." *TVA Bull. Y-165* (1981); and Crop Reporting Board, USDA, "Commercial fertilizers consumption for year ended June 30, 1982." Sp.Cr. 7 (11-82), Washington, D.C. (1982).

developed to meet a wide range of crop needs and soil conditions. Often, however, more grades were formulated than were actually needed. In 1980–1981 about 150 grades made up 76% of the fertilizer consumed in the United States.

Although the tonnage of plant nutrients supplied in mixtures has been increasing over the years, it can be seen in Table 10-8 that the percentage of total nitrogen and potassium applied in mixtures has declined, especially the former nutrient. This shift is due to increased demand for the direct application of nitrogen and potassium materials with resultant savings to farmers.

Figure 10-13 shows that while U.S. consumption of nutrients both in mixtures and materials has continued to expand, use of materials for direct application has, since the early 1970s, increased more rapidly. Reasons for the greater use of materials can be traced to the introduction of soil-fertility evaluation techniques, such as soil testing, which made it possible to determine which nutrient(s) was actually lacking. In addition, research results show that it is often best to apply corrective applications of the deficient nutrient(s) rather than to add several nutrients present in a mixture. As a result, use of materials supplying a specific nutrient(s) was encouraged more than the mixtures providing nutrients, which in some situations might be unneeded.

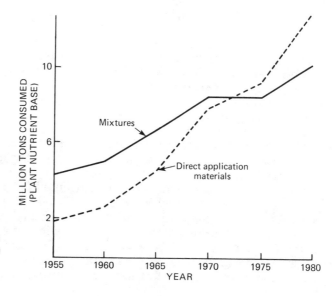

FIGURE 10-13. Consumption in the United States of total nutrients (N + P_2O_5 + K_2O) in mixtures and direct application materials for selected years 1955 to 1980. [Hargett and Berry, 1980 "Fertilizer summary data." *TVA/OACD-81/18, Bull. Y-165* (1981).]

The manufacture of mixed fertilizers constitutes a major agricultural industry. Before considering the properties of these fertilizers, a few of the more important terms in common use should be explained.

Terminology

A *fertilizer* is any substance containing one or more recognized plant nutrient(s) which is used for its plant nutrient content. Unprocessed animal and vegetable manures, marl, lime, limestone, wood ashes, and other products are exempt from this definition.

A *fertilizer material* or carrier is a fertilizer which either:

1. Contains important quantities of no more than one of the primary plant nutrients: nitrogen (N), phosphoric acid (P_2O_5), and potash (K_2O), or
2. Has 85% or more of its plant nutrient content present in the form of a single chemical compound, or
3. Is derived from a plant or animal residue or by-product or natural material deposit which has been processed in such a way that its content of plant nutrients has not been materially changed except by purification and concentration.

A *mixed fertilizer* is a fertilizer containing any combination or mixture of fertilizer materials.

A *complete fertilizer* contains the three major plant-nutrient elements—nitrogen, phosphorus, and potassium.

The *fertilizer grade* refers to the minimum guarantee of the plant-nutrient content in terms of total nitrogen, available phosphorus pentoxide, and soluble potassium oxide (6-24-24, for example).

The *fertilizer ratio* refers to the relative percentages of nitrogen, phosphorus pentoxide, and potassium oxide (a 6-24-24 grade has a 1-4-4 ratio).

The *fertilizer formula* is an expression of the quantity and analysis of the materials in a mixed fertilizer.

A *filler* is *make-weight* material added to a mixed fertilizer or fertilizer material to make up the difference between the weight of the added ingredients required to supply the plant nutrients in a ton of a given analysis and 2000 lb.

An *acid-forming fertilizer* is one capable of increasing the acidity of the soil, which is derived principally from the nitrification of ammonium salts by soil bacteria.

A *basic fertilizer* is capable of decreasing the acidity of the soil.

A *non-acid-forming* or neutral fertilizer is one that is guaranteed to leave neither an acidic nor a basic residue in the soil.

Dry bulk blending is the process of mechanically mixing solid fertilizer materials.

A *bulk fertilizer* is a fertilizer distributed in a nonpackaged form.

Clear liquid fertilizer is one in which the N-P-K and other materials are completely dissolved.

Suspension fertilizer is one in which some of the fertilizer materials are suspended as fine particles.

Fluid fertilizer is clear or suspension liquid fertilizer.

Compound fertilizer is a term often used in Europe and has about the same meaning as *mixed* in the United States.

A *specialty fertilizer* is a fertilizer distributed for nonfarm use.

The *Association of American Plant Food Control Officials* is an organization of officers and their deputies charged by law with regulating the sale of fertilizers and of research workers employed by state, dominion, or federal agencies engaged in the investigation of fertilizers. Its object is to promote uniform and effective legislation, definitions, and rulings, and to enforce the laws relating to the control of sale and distribution of fertilizers and fertilizer materials.

Acid, Neutral, and Basic Fertilizers

The acidifying effect of mixed fertilizers is largely the result of contained ammonium nitrogen, as explained in Chapter 5. Limestone may be added to mixed fertilizer in amounts sufficient to neutralize the expected acidity, and even greater amounts may be added to leave a basic residue.

Neutralizing Agents. Dolomitic limestone is commonly used as a neutralizer, for it is also a source of magnesium. In addition, the dolomite is less likely than calcite to cause reversion of monocalcium phosphate. Caution must be used in adding lime to mixed fertilizer, for large excesses may result in reversion of the soluble form of phosphate to the insoluble or less available forms as shown by the following equation:

$$Ca(H_2PO_4)_2 + CaCO_3 \longrightarrow 2CaHPO_4 + H_2CO_3$$

With dolomite, the water-soluble phosphorus decreases more rapidly in ammoniated than in nonammoniated mixtures because of the formation of magnesium ammonium phosphate.

Excess lime may cause loss of ammonia gas by reaction with the NH_4^+ ion as

$$NH_4^+ + OH^- \longrightarrow NH_4OH \longrightarrow NH_3 \uparrow + H_2O$$

With an inadequate liming program in humid areas the continued use of acid-forming fertilizers will increase soil acidity. Lime added in the fertilizer helps to maintain the content of soil bases. However, with higher-analysis mixed fertilizers, there is less room for lime; and with the gradual trend toward higher-analysis fertilizers, the proportion of acid-forming fertilizer is increasing.

Acid-forming mixed fertilizers may be manufactured for use on alkaline soils. Ammonium sulfate is one of the principal sources of nitrogen for this purpose, and elemental sulfur may also be included as an additional source of acidity.

High-Analysis Fertilizers

The cost of a ton of mixed fertilizer is influenced largely by two factors:

1. The cost of the plant nutrients in the materials used to make up the mixed fertilizer.
2. The fixed costs, which include manufacturing, transporting, and distributing.

For example, for equivalent amounts of nutrients only two-thirds as much fertilizer product would be required for a 15-15-15 as for a 10-10-10. Obviously, the fixed costs per pound of nutrients decrease with an increase in the nutrient content. Because of these lower fixed costs and in spite of somewhat greater costs of the materials in higher-analysis fertilizer, the cost of nutrients in the higher-analysis mixed fertilizer tends to be reduced.

Beyond a certain point, however, the extra cost of the higher-analysis materials required may offset any saving. For example, for normal superphosphate (16 to 22% P_2O_5 or 7 to 9.5% phosphorus) versus concentrated superphosphate (44 to 52% P_2O_5 or 19 to 23% phosphorus) the manufacturing costs of concentrated superphosphate at the plant are generally greater than those of ordinary superphosphate. Hence the savings that may be effected by the use of concentrated superphosphate must result from lower unit costs of manufacturing, packing, and distributing the higher-analysis mixtures.

Of course, the greater the transportation distance, the greater the advantage of higher-analysis fertilizers, for transportation costs are rising rapidly. Transportation and handling costs have been conservatively estimated to represent 25 to 34% of the cost of fertilizers to farmers.

More Concentrated Materials. The content of plant nutrients in materials from which mixed fertilizers are being manufactured has increased rapidly. Natural organics, which contain 6 to 8% nitrogen and were popular around the turn of the century, have been almost entirely replaced by higher-analysis materials, including urea (46% nitrogen). Anhydrous ammonia (82% nitrogen) is also being added to mixed fertilizers, and concentrated superphosphate and ammonium phosphates are replacing the low-analysis superphosphates. Potassium sources have shifted from kainite and manure salts to potassium chloride (60 to 62% K_2O or 50 to 52% potassium). As a consequence of the more concentrated materials, the nutrient content of mixed fertilizers is rising steadily (Table 10-9).

TABLE 10-9.   **Average Analysis of Mixed Fertilizers in the United States**

Year	Total N	Available P (P₂O₅)	Soluble K (K₂O)	Total
		Nutrient Content (%)		
1954–1955	5.24	5.19 (11.86)	8.96 (10.80)	19.39 (27.90)
1959–1960	6.50	5.70 (12.99)	10.00 (12.06)	22.20 (31.55)
1963–1964	7.68	6.57 (14.94)	10.50 (12.65)	24.75 (35.27)
1972–1973	10.32	8.3 (19.02)	10.6 (12.71)	29.2 (42.05)
1981–1982	11.23	8.9 (20.39)	10.3 (12.38)	30.4 (43.99)

Effect of Concentration on Cost of Nutrients. An example of the effect of concentration on the cost to the farmer is shown in Figure 10-14. On the basis of what was demonstrated in this figure, a considerable saving would be expected by going from a 10-10-10, still the leading grade in eight states in the United States in 1982, to a 15-15-15.

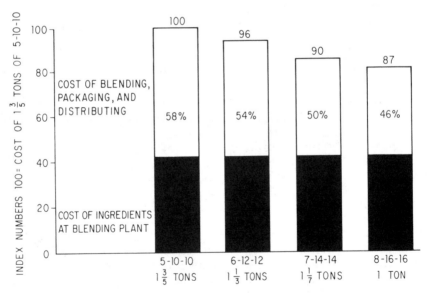

FIGURE 10-14.   Relative farm cost of plant nutrients in fertilizers of different concentrations. Maine. (Maurice H. Lockwood, "High-analysis mixed fertilizers," in K. D. Jacob, Ed., *Fertilizer Technology and Resources in the United States.* Vol. III. of *Agronomy.* Copyright 1953 by Academic Press. Inc.)

Make-Weight Materials. In mixed fertilizer there is often a difference between the weight of the materials required to furnish the nutrients in a ton of a given analysis and 2000 lb. This difference is made up with materials, known as *make-weight* or *filler*, to supply nutrients other than nitrogen, phosphorus, or potassium.

The example in Table 10-10 shows a 10-10-10 in which the weight of the materials containing the primary nutrients would be 1666 lb. An additional 334 lb of lime would be added per ton to make the grade. If no extra material were used when diammonium phosphate replaced the superphosphate, the analysis of the mixture would be 15-15-15. It is thus possible to avoid using those materials in mixed fertilizers that bear the make-weight stigma.

There are several kinds of make-weight materials and limestone is the principal one.

Growers Demand Higher-Analysis Goods. The advantages of higher-analysis fertilizers from the standpoint of the grower are listed:

1. Lower cost per unit of plant food.
2. Lower transportation cost.
3. Less storage space required.
4. Less labor in handling.
5. Increased speed of application in the field because of fewer stops.

Some growers may consider it worthwhile to buy high-analysis fertilizer even if the cost per unit of plant food is somewhat higher, because labor, storage, transportation, and/or time required for application may be critical factors.

Agronomic Factors Involved. Fertilizers must be applied according to their plant-nutrient content and not by the number of pounds per acre. For example, if a farmer has been applying 10-10-10 at the rate of 600 lb/A, a 15-15-15 would be applied at the rate of 400 lb/A to have the same amount of nitrogen, phosphorus, and potassium per acre. This may seem to be a simple point, but many of the unfortunate experiences that have resulted from the use of higher-analysis goods have been caused by failure to reduce the rate of application accordingly. This, in turn, is usually due to the effect of improper placement of the larger amount of fertilizer with respect to the seed or the plant.

TABLE 10-10. Formulation of a Mixed Granulated Fertilizer

	Analysis	10-4.4-8.3 (10-10-10) (lb)	15-6.6-12.5 (15-15-15) (lb)
Nitrogen solution 440	44% N	320	254
Sulfate of ammonia	21% N	320	667
Phosphoric acid	23.6% P	160	280
Superphosphate	8.7% P	588	—
Diammonium	18% N	—	333
phosphate	20.3% P	—	—
Muriate of potash	50.2% K	338	506
Limestone	—	334	—
Total *		2000	2000

* Data do not add to totals shown because of moisture in the ingredients.
Source: R. C. Smith and B. Makower, "Advances in manufacture of mixed fertilizer," Chapter 11. Reprinted from *Fertilizer Technology and Usage,* published and copyrighted by Soil Science Society of America, 1963, by permission of copyright owners.

The other side of the picture, however, is that many crops are inadequately fertilized. The application of higher-analysis fertilizers at the same rate once used for low-analysis goods is, of course, an advantage in that greater quantities of plant nutrients are applied.

Higher-analysis fertilizers generally contain less of the secondary plant-nutrient elements, calcium, magnesium, and sulfur. Although there is usually much less need for the addition of make-weight material such as dolomitic limestone, the problem is not serious. The calcium and magnesium requirements can be met by broadcast dressings of limestone or by commercial fertilizer sources of these two nutrients, which were described in Chapter 8. The cost of these treatments may offset to some extent the savings realized from the use of high-analysis fertilizers.

The sulfur supply is important in many areas of the United States, and the continued use of fertilizers that lack this element will result in a reduction in crop yield. Sulfur deficiencies have been observed on numerous crops, as indicated earlier in Chapter 3, and fertilizer manufacturers are now including sulfur sources in mixed, granular, and blended goods. The grower may also add sulfur in the form of gypsum or elemental sulfur or he may choose a nitrogen–sulfur material. Sulfur tends to acidify the soil, but when applied in amounts needed to supply sulfur as a plant nutrient the acidifying effect is minor.

High-analysis fertilizers are likely to contain fewer micronutrients than the lower-analysis goods. However, numerous examples of micronutrient deficiencies appear even when low-analysis materials are used. The most effective insurance against the incidence of these deficiencies is the addition of micronutrients in the form of their respective carriers when the deficiencies are known to exist.

Manufacture of Mixed Fertilizers

Many changes in the materials and in the composition of mixed fertilizers are taking place. Materials with widely different properties are mixed together and physical and chemical changes occur. As few as two or as many as 10 different materials may be combined in the manufacture of a complete fertilizer, and, with the trend toward ammonium phosphates and higher-analysis goods, improvement will continue as a direct result of efforts to market a more efficient product.

The present-day fertilizer manufacturer is confronted with a multiplicity of problems of formulation, processing, control of chemical reactions, and the physical condition of the mixed goods.

The various processes used in the manufacture of mixed fertilizers are:

1. Dry mixing of nongranular or pulverized materials.
2. Granulation of dry-mixed materials by processes in which chemical reactions are not an essential part of the process.
3. Granulation of dry materials with the addition of materials that react chemically, usually ammonia or solutions containing ammonia (ammoniation) and often sulfuric or phosphoric acid.
4. Slurry granulation in which the materials to be granulated are in the form of a slurry, usually derived from reaction of sulfuric, nitric, or phosphoric acid with ammonia, phosphate rock, or some combination of these materials. In some process modifications, solid materials may be added to the slurry during granulation; in other processes, all incoming materials are incorporated in the slurry.
5. Melt granulation in which all or a part of the mixture to be granulated is in the form of a hot, fluid melt, usually containing less than about 2% water, which solidifies on cooling.
6. Blending or bulk blending consisting of mechanical dry mixing of granular materials. The materials may be either straight or compound fertilizers. The mixture may be marketed in bags or bulk, usually the latter.

7. Fluid mixing is of two types.
   a. Liquids, in which all or nearly all of the ingredients are in solution, sometimes called "fertilizer solutions" or "clear liquid fertilizers."
   b. Suspensions, which are fluid mixtures containing solids, usually suspended in a saturated solution of fertilizer materials.

Processes 1, 2, and 6 are often referred to as mechanical processes, contrasted with those in which chemical reactions are an essential part of the process. Important aspects of the mixing techniques are discussed next.

Chemical Reactions. In addition to the reactions that occur in the chemical mixing of fertilizers, a number of reactions will occur in mechanically mixed fertilizers. These reactions in dry, mechanically mixed fertilizers are often detrimental to the physical properties of the products.

The extent and rate of these reactions are greatest in granulation processes, where high temperatures and moist conditions accelerate them. The particle size of the fertilizer materials also influences chemical reactions. The magnitude of chemical reactions is least in bulk blends, but even so they may occur, thereby lowering product quality.

Mixtures are usually prepared well in advance and are allowed to cure in storage piles at the factory. Temperature and moisture of the fertilizer mixtures are reduced before they are placed in the curing bin.

The principal categories of reactions that occur in dry mixtures follow.

Double Decomposition. Double decomposition sets in between two compounds, without a common ion, in the presence of moisture. Common examples are the following:

$$CaH_4(PO_4)_2 + (NH_4)_2SO_4 \longrightarrow CaSO_4 + 2NH_4H_2PO_4$$

$$NH_4NO_3 + KCl \longrightarrow NH_4Cl + KNO_3$$

$$(NH_4)_2SO_4 + 2KCl \longrightarrow 2NH_4Cl + K_2SO_4$$

Certain of these reactions may be relatively slow, and the first may require two months or more. It is evident that new compounds, with properties entirely different from those of the original materials, may be formed.

Neutralization. The most common reactions in this group, which are those that take place during ammoniation of superphosphate, are described in a subsequent section. In addition, there are other neutralization reactions involving the free acid in superphosphate:

$$2H_3PO_4 + CaCO_3 \longrightarrow Ca(H_2PO_4)_2 + H_2CO_3$$

$$H_3PO_4 + (NH_4)_2HPO_4 \longrightarrow 2NH_4H_2PO_4$$

Hydration. Hydration is an important reaction in that anhydrous forms of certain salts tie up free water and bring about chemical drying in the curing bin. It takes place most rapidly at temperatures of 50°C or lower:

$$CaSO_4 + 2H_2O \longrightarrow CaSO_4 \cdot 2H_2O$$

$$CaHPO_4 + 2H_2O \longrightarrow CaHPO_4 \cdot 2H_2O$$

$$MgNH_4PO_4 + 6H_2O \longrightarrow MgNH_4PO_4 \cdot 6H_2O$$

$$(NH_4)_2SO_4 + CaSO_4 + H_2O \longrightarrow (NH_4)_2SO_4 \cdot CaSO_4 \cdot H_2O$$

$$K_2SO_4 + CaSO_4 + H_2O \longrightarrow K_2SO_4 \cdot CaSO_4 \cdot H_2O$$

Decomposition. Under some conditions certain compounds may decompose in fertilizer mixtures:

$$CO(NH_2)_2 + H_2O \longrightarrow 2NH_3 + CO_2$$

The decomposition of urea is accelerated by higher temperatures, particularly in the presence of such compounds as monocalcium, dicalcium, or monoammonium phosphate. Breakdown of diammonium phosphate is also favored by high temperatures in storage:

$$(NH_4)_2HPO_4 \longrightarrow NH_4H_2PO_4 + NH_3$$

If the mixture contains calcium sulfate, the liberated ammonia will not be lost but ammonium sulfate and dicalcium phosphate will be formed:

$$NH_4H_2PO_4 + CaSO_4 + NH_3 \longrightarrow CaHPO_4 + (NH_4)_2SO_4$$

Ammonia released from urea can also be conserved by reactions with superphosphate and monoammonium phosphate:

$$Ca(H_2PO_4)_2 + NH_3 \longrightarrow CaHPO_4 + NH_4H_2PO_4$$

$$NH_4H_2PO_4 + NH_3 \longrightarrow \qquad (NH_4)_2HPO_4$$

Many of the reactions above are exothermic and cause heating within storage piles. The higher temperatures can result in rapid decomposition of nitrates with formation of toxic nitrogen oxides. Also, in acidic mixtures containing nitrates and organic matter, the rapid heating can lead to fires.

Urea forms adducts with fertilizer material such as monocalcium phosphate, ammonium chloride, gypsum, and phosphoric acid. The adduct it forms with monocalcium phosphate, with coincident release of water which will likely cause caking, is as follows:

$$Ca(H_2PO_4)_2 \cdot H_2O + 4CO(NH_2)_2 \longrightarrow Ca(H_2PO_4)_2 \cdot 4CO(NH_2)_2 + H_2O$$

Ammoniation. The majority of nitrogen going into mixed fertilizers in the United States is added through ammoniation.

The chemistry of the reactions that take place during the ammoniation of ordinary and concentrated superphosphates is somewhat complex. For the purposes of this discussion, these reactions are summarized by the following four equations:

$$Ca(H_2PO_4)_2 \cdot H_2O + NH_3 \longrightarrow NH_4H_2PO_4 + CaHPO_4 + H_2O \qquad (1)$$

$$2NH_3 + 2CaHPO_4 + CaSO_4 \longrightarrow Ca_3(PO_4)_2 + (NH_4)_2SO_4 \qquad (2)$$

$$NH_4H_2PO_4 + NH_3 \longrightarrow (NH_4)_2HPO_4 \qquad (3)$$

$$3CaHPO_4 + 2NH_3 \longrightarrow Ca_3(PO_4)_2 + (NH_4)_2HPO_4 \qquad (4)$$

Mixed Fertilizers. Reaction (1) marks the initial stages of the ammoniation of both ordinary and concentrated superphosphates. If the ammoniation is continued, there is, with ordinary superphosphate, a conversion of the phosphorus to the unavailable tricalcium phosphate and an attendant formation of ammonium sulfate, as indicated by equation (2). Equations (3) and (4) represent the reactions that occur when the ammoniation of concentrated superphosphate is continued.

The ammoniation of phosphoric acid in the formation of mono- and diammonium phosphates was covered in Chapter 6. Calcium metaphosphate and fused tricalcium phosphate do not combine with ammonia. As already mentioned, however, when calcium metaphosphate is hydrolyzed with sulfuric acid, it can be partially ammoniated.

Effect of Degree of Ammoniation on Phosphate Solubility. The ammoniation of ordinary superphosphate to much greater than 3% by weight of ammonia causes a marked reduction in the content of water-soluble phosphorus in the fertilizer, as shown by the data in Table 10-11. If precautions are taken to introduce the correct amount of ammonia, reaction (2) will be minimized.

Although there is a continuous linear decrease in the water-soluble phosphorus content of ordinary superphosphate with increasing ammoniation, this is not true of concentrated superphosphate. Regardless of the degree of ammoniation, the water-soluble phosphorus does not decrease below about 60%. The formation of di- and tricalcium phosphates during ammoniation is known as *reversion*, a term frequently used in the trade to refer to a decrease in phosphate availability.

Chlorine-Free Fertilizer. A European fertilizer manufacturer developed a process for making a chlorine-free N-P-K fertilizer based on ion exchange. Phosphate rock is digested by nitric acid. The solution of calcium, phosphorus and nitrogen is then pumped to the potassium-loaded ion exchanger where the calcium is absorbed. The resulting KCl solution is pumped to the ion exchanger, the potassium and calcium exchange places, and the chlorine is removed in the form of calcium chloride. Many high-analysis grades are made in this way.

Mixing Procedure. To ammoniate mixed fertilizers a batch of superphosphate, muriate of potash, and conditioner is placed in the mixing unit and sprayed with the ammoniating solution. The reactions of ammonia with superphosphates and acids are exothermic. If this heat is not dissipated fairly rapidly, additional reversion is favored. The ammoniated material must also be cooled to prevent overheating in the stored pile. To reduce the heat content of the finished product, it is frequently passed through a rotary cooler in which it is tumbled in a countercurrent stream of air.

A continuous ammoniation process is used in most plants. Figure 10-15 is a sketch of a conventional ammoniation-granulation plant. In this type of plant the ammoniating solution is introduced at a constant rate with the other materials. Thus there is no loss of time in recharging the granulator.

Materials Used in Manufacture. The number of materials used in the manufacture of fertilizers has decreased as a consequence of the need for regularly available, more uniform ingredients in more complex processes.

Factors influencing the eligibility of a material have been listed by Smith and Makower (1963): (1) nutrient content; (2) cost per pound of nutrient; (3) chemical form, such as nitrate or ammoniacal; (4) absence of toxic compounds; (5) availa-

TABLE 10-11. Composition of Ammoniated Superphosphate

Compound	Percentage of Composition with Percentage of Ammonia Added as Indicated						
	0.0	1.0	2.0	3.0	4.0	5.0	6.0
$CaH_4(PO_4)_2$	25.0	14.3	3.5	0.0	0.0	0.0	0.0
$CaHPO_4$	4.5	9.5	12.0	11.2	5.0	0.0	0.0
$Ca_3(PO_4)_2$	0.0	0.0	0.0	8.2	17.5	25.7	30.6
$NH_4H_2PO_4$	0.0	8.9	14.3	15.5	12.5	9.1	6.0
$(NH_4)_2SO_4$	0.0	0.0	0.0	2.7	8.5	14.3	20.0
$CaSO_4 \cdot 2H_2O$	62.0	62.0	62.0	58.3	51.0	42.0	32.0
Inerts	3.0	3.0	3.0	3.0	3.0	3.0	3.0
Total	94.5	97.7	94.8	98.9	97.5	94.1	91.6

Source: F.W. Parker, *Commer. Fert.*, **42**:28 (1931).

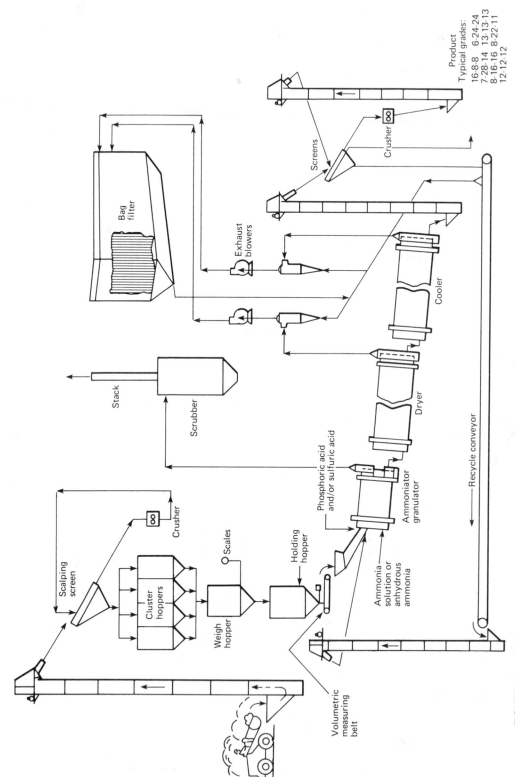

FIGURE 10-15. Diagram of a TVA-type ammoniation-granulation plant for N-P-K mixtures. (International Fertilizer Development Center and United Nations Industrial Development Organization, *Fertilizer Manual*, Reference Manual IFDC-R-1. Muscle Shoals, Ala.: IFDC, 1979.)

Scalping screen

Crusher

Cluster hoppers

Scales

Weigh hopper

Holding hopper

Volumetric measuring belt

Ammonia solution or anhydrous ammonia

Phosphoric acid and/or sulfuric acid

Ammoniator granulator

Dryer

Cooler

Recycle conveyor

Exhaust blowers

Bag filter

Scrubber

Stack

Screens

Crusher

Product
Typical grades:
16-8-8   6-24-24
7-28-14 13-13-13
8-16-16 8-22-11
12-12-12

bility; (6) content of other nutrients; (7) moisture content; (8) hygroscopicity; (9) reactions with other ingredients; (10) heat of reaction; (11) particle size; and (12) effect on physical properties.

Suppliers must continually be aware of those characteristics of materials needed to produce satisfactory fertilizer mixtures, as well as ones that enable plants to be operated most efficiently. A case in point is the introduction of "coarse"-grade potassium chloride to replace the finer conventional "standard"-grade potassium chloride. The larger particle size facilitated granulation by improving the desired agglomeration and by reducing the amount of product that had to be recycled in the process. This development increased production rates in many plants with undersized equipment.

Mixing Processes

Nongranulated Fertilizers. All of the early fertilizer plants in the United States produced nongranulated fertilizers but few are operating now.

Ammoniation is practiced because of the economy of the nitrogen source and the physical condition of the product. Standard (fine) solid materials are combined with concentrated superphosphate and sulfate of ammonia and nitrogen solutions. Muriate of potash is the customary source of potassium.

There are two kinds of mixing machines in general—the stationary and the rotary drum. The stationary mixer has a vertical cylinder fitted with a baffle system and the mixing is done by gravity.

The rotary drum, which is a horizontal cylinder equipped with a series of baffles, permits mixing and ammoniation to be done simultaneously. The materials are introduced into the drum, the ports are closed, and the solution is sprayed into the system as the materials are mixed. With high ammoniation a rotary cooler is necessary before the fertilizer is taken to bin storage. Two to four weeks of storage is generally adequate to permit completion of the reactions.

A conditioner added during mixing to serve as a parting agent for the particles tends to reduce crystal knitting during changes in moisture and temperature under pressure in storage.

Granulation. This step is performed at nearly all mixed fertilizer manufacturing plants. The rotary drum or tube is the most popular piece of equipment for granulating mixtures comprised of all dry materials. Other types of granulators include (1) pugmills, which are horizontal or inclined U-shaped troughs equipped with rotating blades or pins on one or two shafts to agglomerate the material and move it through the trough; (2) rotating pans, either horizontal or inclined, some of which may have mixing blades; (3) blungers; and (4) pear-shaped rotating mixers similar to concrete mixers. Roll compactors or extrusion machines which are used primarily to produce cylindrical pellets of either potassium chloride or ammonium sulfate may also be used to granulate compound fertilizers.

Granulation of Dry Materials. Granulation is essentially an extension of nongranulated fertilizer production in which particles of reasonably uniform size and stability are manufactured. Figure 10-16 shows the steps in steam granulation of all dry materials in a rotary drum. Incoming materials are screened to remove lumps and then weighed into feed hoppers. The mixture is fed continuously at a controlled rate into the granulator. Steam is supplied under the bed of material at the feed end, and water is sprayed on the bed at two or more points along the axis of the rotary drum. Granulation is controlled by the amounts of steam and water added. The granular product is dried and then sized. Oversize granules are crushed and the fines are recycled into the process. The final product contains about 1% moisture and the prime cause of caking is eliminated by drying down to this level.

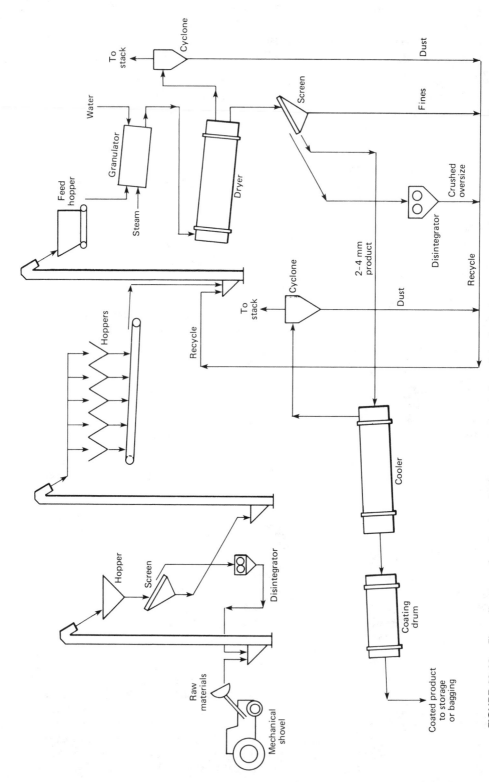

FIGURE 10-16. Flow diagram of main steps in the steam granulation of dry fertilizer materials. (International Fertilizer Development Center and United Nations Industrial Development Organization, *Fertilizer Manual*, Reference Manual IFDC-R-1. Muscle Shoals, Ala.: IFDC, 1979.)

The principal dry materials used for formulation of mixtures by this process are ammonium nitrate or urea, triple superphosphate, or powdered monoammonium phosphate.

Granulation by Ammoniation. Granulation often occurs during ammoniation and it can be controlled by addition of water or steam or by modifying the formulation to provide sufficient chemical heat. About two-thirds of all U.S. granulation plants use the TVA continuous ammoniator-granulator (Figure 10-15).

Granulation of Slurries. Slurry granulation processes are those in which all or most of the materials entering the granulation process are in the form of a slurry. The slurry is usually prepared by the reaction of nitric, phosphoric, or sulfuric acid or mixtures of these acids with ammonia and, occasionally, phosphate rock. Solid materials such as potassium salts may be added to the slurry before granulation or mixed with it in the granulator to produce a variety of complete fertilizers. Part of the reactions may occur in the granulator. The granulator is usually either a rotary drum or some type of pugmill or blunger.

The advantages of the slurry system are these:

1. Its reaction is rapid and easy to control.
2. There is no need for curing the final product.
3. It uses the more economical phosphate rock.
4. It uses the less expensive nitrogen raw materials, nitric acid and ammonia.

These materials tend to offset the additional equipment that would be needed to handle the slurry and the higher cost of removing the water. Nitric phosphates may also be produced with the slurry process.

Granulation of Melts. Dryers are usually the largest and most expensive unit in a granulation plant. Processes not requiring a dryer are advantageous because of savings in capital costs plus fuel costs for drying. Melt processes where dryers have been eliminated are now in commercial use in England and the United States.

The principle of melt processes is that reactions of certain combinations of sulfuric acid and phosphoric acid and/or nitric acid with ammonia will generate sufficient heat to evaporate all of the water, thus forming an anhydrous melt. TVA's pipe-reactor and the more recent pipe-cross reactor process are examples of melt processes. In the latter process sulfuric and phosphoric acid react with ammonia to form a melt that is sprayed into a rotary-drum granulator (Figure 10-17). This process is now operating in many plants in the United States and elsewhere in the world. It is being used to make a variety of compound and complete fertilizers such as 12-48-0, 10-40-0, 6-24-24, and 13-13-13.

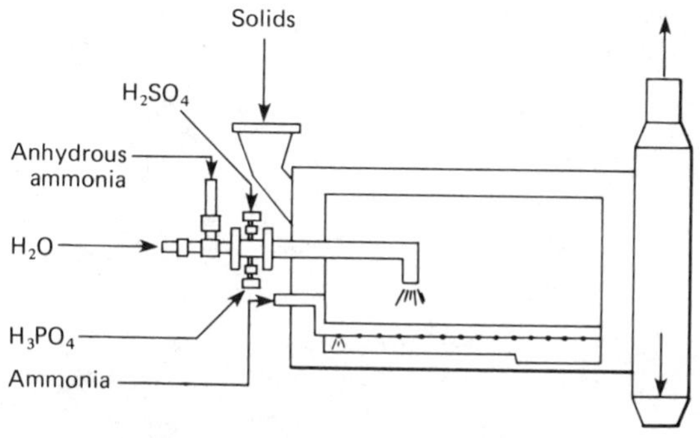

FIGURE 10-17.   Sketch of a TVA pipe-cross reactor used in conjunction with a rotary ammoniator-granulator. (Achorn et al., in *Situation 80, TVA Fert. Conf. Bull. Y-157.* Muscle Shoals, Ala.: NFDC–TVA, 1981.)

Nearly all mixed fertilizers in England are granulated. In the United States the trend is toward high-analysis granulated fertilizer.

Granulated products offer many advantages to the manufacturer and the farmer. When properly stored, granulated fertilizers maintain good physical and handling properties for some time. Caking is reduced to a minimum, and the manufacturer will have fewer operations to perform between mixing and bagging. Segregation is almost eliminated, and the grower is assured of a material that can be uniformly applied in the field. In addition, there is less dust to contribute to the physical discomfort of the individuals who work with the fertilizer. However, with the exception of the water-soluble phosphates discussed in Chapter 6, little agronomic advantage from granulation has been noted.

Storage. In off seasons the freshly mixed fertiizer is transferred to storage bins in which it is allowed to cool and cure. The fertilizer may be left in the bins for several months or even longer. If the fertilizer sets up or cakes, it may have to be crushed and screened before it is acceptable for use. During peak demand periods in the late winter and early spring, however, fertilizers may not be stored and will instead go directly into the marketplace.

Bagging. Packages vary in kind and size. Plastic bags keep the fertilizer in a virtually moistureproof condition. This is quite important, particularly with some fertilizers, and under conditions of high humidity. In addition, the appearance of the product is quite attractive. Paper bags with polyethelyene liners are used for much of the remaining bagged-goods market. Although at various times in the past much heavier bags of 200-, 100-, and 80-lb capacities were used, lighter 50-lb paper or plastic bags which are much easier to handle are now widely used in North America.

The tightness of the pack depends on the fertilizer. For a material that is likely to cake, a loose pack is preferred, for each time the bag is moved the contents are loosened. On the other hand, a tight pack is preferred for granulated mixtures or for materials that do not cake.

There is considerable variation in the cost of bagging fertilizers, but it typically runs from $20 to $25 per ton. Cost of the bags represents about one-half of this expense. Popularity of this form of marketing mixed fertilizers has diminished rapidly since the 1960s (Figure 10-18). Handling of all classes of dry fertilizers

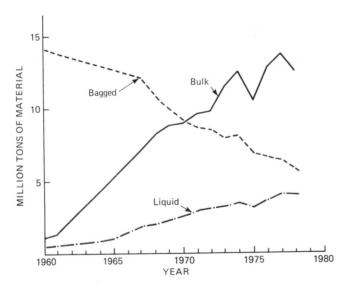

FIGURE 10-18. Consumption in the United States of classes of mixed fertilizer between 1960 and 1980. (Bridges, "Fertilizer trends 1979." *Bull Y-150*, Muscle Shoals, Ala.: NFDC–TVA, 1980.)

has undergone a similar rapid decline. For example, farmers in the United States in 1954 received 89% of their fertilizers in bags. In 1974 and 1979 only 23 and 15%, respectively, of their fertilizers were handled in bags. Large returnable bags of 0.5- and 1.0-ton capacities have been developed for machine handling of all types of dry fertilizer materials, including mixtures.

Physical Condition. A major problem faced by the fertilizer manufacturer is the physical condition of the product. In the manufacture of nongranulated fertilizer this problem generally becomes more serious as the nutrient content of mixtures is increased. Nothing discourages a grower more quickly than to have lumpy or caked fertilizer. Such materials are spoken of as "tombstones," which may also well describe the situation of the fertilizer manufacturer who continues to put them out. The grower demands a product that will flow easily through the distributing equipment and can be uniformly applied in the field. The principal factors that determine the drillability of fertilizer are its moisture content and its state of subdivision.

Hygroscopicity. The absorption of water from the air is called hygroscopicity. Materials such as calcium nitrate, ammonium nitrate, sodium nitrate, and urea absorb this moisture at fairly low humidity and temperature. On the other hand, superphosphate and potassium sulfate take up water only at a very high relative humidity. Mixtures of certain salts such as ammonium nitrate and urea are more hygroscopic than either material alone, and this complicates the problem. As a consequence, the fertilizer materials which make up a given formula must be selected with considerable care. Many plants limit the use of ammonium nitrate or urea to reduce moisture absorption.

Caking. The moisture in fertilizers dissolves some of the more soluble compounds to form a saturated salt solution. As the moisture content of the system decreases or the temperature drops, the dissolved salts crystallize and the crystals knit together. Crystal knitting may also occur when a new compound is formed by chemical reaction or when crystals flow together under pressure. Crystal phase-change inversions in ammonium nitrate caused by temperature cycling through 32°C (90°F) during storage can also result in serious caking problems. Although finely divided particles may cake slightly as the result of cohesion or adhesion, the main source of caking appears to be crystal knitting.

Under ideal curing conditions most of this knitting will take place before shipping. However, a state of equilibrium is seldom attained under conditions of manufacture of nongranulated fertilizer, and again temperature and moisture changes may take place in storage. During the rush season, when it is necessary to ship directly from production, there is a good chance of caking if the material is allowed to stand for any length of time.

Conditioners. Conditioning materials are added to nongranular and granular fertilizers to improve their physical condition or to decrease caking. Their actual purpose is to reduce crystal knitting during changes in moisture and temperature when the fertilizer is under pressure. Most anticaking agents do not function as moisture barriers or waterproofing agents.

In nongranulated fertilizers the conditioners serve as separators between the particles. Ground cocoa shells or ground corncobs may be used in amounts of about 100 lb/ton, or 5%. A basic material such as lime will react with any free acids to form hydrated salts.

Solid conditioning dusts used for coating granular fertilizers include the following materials, which have low bulk densities: diatomaceous earth (kieselguhr), kaolin clays, talc, and chalk. The proportions applied usually vary between 1 and

4% by weight. Modern granulation techniques with the accompanying reduction in surface area, drying, and cooling have greatly decreased the need for such conditioners.

In some cases "internal" conditioners are added to fertilizer materials before granulation. These substances acting internally, usually as hardeners or crystal modifiers, will improve physical properties and associated storageability. For example, the current practice in urea prill or granule production is to include 0.2 to 0.5% by weight of formaldehyde or urea-formaldehyde in the urea melt. Urea granules or prills containing these additives are hard and resistant to caking and they do not require solid conditioning dusts. Internal conditioners such as magnesium nitrate or a combination of ammonium sulfate and diammonium phosphate can be added at the rate of about 1% by weight to stabilize ammonium nitrate. They inhibit or modify crystal phase changes in ammonium nitrate.

Adherence of solid conditioning dusts can often be improved by spraying the fertilizer with a small amount of oil either in advance of or following application of the conditioner. Fairly viscous oils of high paraffin content are preferred and they are usually applied at the rate of only 0.2 to 0.5% by weight. Oils are also sometimes used to control caking and dust problems of the higher-purity ammonium phosphates.

Coated Fertilizers. An interesting development is the covering or encapsulating of granules with water-resistant or impermeable coatings. The purpose is to give a *metered* supply of nutrients, reduce injury from seed fertilizer contact, avoid excessive uptake of nutrients, reduce leaching losses, and improve physical condition. Researchers at the University of Wisconsin showed that encapsulation of fertilizer with polyethylene film will control rate of release of water-soluble fertilizer. The availability of fertilizer in the capsules is regulated by the number of pinholes. It is postulated that moisture entered the capsule through the pinholes primarily as a vapor and dissolved the salt. The saturated solution then flowed out by gravity.

With uncoated fertilizer the plant can take advantage of an early and rather high content of nutrients, but in later growth stages the supply is sharply reduced. This is in contrast to the even supply provided by coated fertilizers. Research results show reduced leaching losses of potassium and nitrates and less seedling injury from coated fertilizer. Actual agronomic value is still to be determined. Coated fertilizers have helped to maintain vigor in ornamentals and heavy applications can be safely made.

Various coating substances such as plastics, resins, waxes, paraffin, elemental sulfur, and asphaltic compounds are under investigation. Temperature and thickness of the coating have a marked influence on release, and raising the temperature from 10 to 20°C almost doubles the rate. However, pH has little effect and the release is not dependent on microbial action.

Sulfur as a coating substance was discussed in Chapter 5. The coating of urea and muriate of potash with sulfur has the advantage of being economical as well as of supplying a necessary element. The sulfur is sprayed on as a melt. Sulfur-coated urea is the most useful coated fertilizer at present.

Segregation. Nongranulated mixed fertilizers are composed of particles that vary considerably in size, shape, and density. The tendency is to segregate during such operations as bagging, transporting, or pouring into the distributor. This segregation results in fertilizers that are not of uniform chemical composition. In addition to the agronomic implications, it is more difficult to obtain representative samples for chemical control work.

Bulk Blending

Bulk blending is the mechanical mixing of two or more dry granular fertilizer materials to produce mixtures providing nitrogen, phosphorus, potassium, and other nutrients such as sulfur and magnesium, as well as micronutrients. The blend is usually a prescription mixture ordered on a custom basis by farmers. Bulk blends are normally spread soon after formulation; however, some may be bagged.

The desirable qualities of a top-quality bulk blend include the following:

1. Granular and free flowing.
2. Components are present in the right plant nutrient proportions and are mixed to a high degree of homogeneity.
3. High degree of nonsegregation during handling and spreading.
4. Analysis of any reasonable sample reflects the guaranteed plant nutrient concentrations.
5. Nondusty and reasonably nonhygroscopic.

This mode of fertilizer marketing originated in the United States in the early 1950s.

There has been a dramatic rise in popularity of bulk blending in the United States since 1960 (Figure 10-18). In 1967, 35% of all fertilizers used in the United States were supplied as bulk blends. This figure rose to 52% in 1979. In the early 1980s there were between 5000 and 5300 bulk-blending operations in the United States, up considerably from the estimated 4140 plants in 1968. Large numbers of bulk-blending facilities were built from the mid-1960s to the end of the 1970s in Canada.

Advantages of Bulk Blending. The main advantages of blending are:

1. The facility used to produce N-P-K blends and mixtures with other elements usually has a relatively low investment cost per unit of plant food.
2. The material cost for blends is usually low, particularly at times when there is an oversupply of granular products.
3. Plant operation is relatively easy and requires a minimum of trained personnel.
4. Maintenance cost is low. Maintenance of the plant is easy and can be performed with unskilled labor.
5. It is possible to supply farmers with a prescription mixture that closely fits their plant-food requirements.

All of these factors have contributed to savings to the farmer which can be substantial in comparison to purchases of N-P-K granulated fertilizers. Additional savings are made when costs of transporting and spreading the material are considered. Many of the costs between the plant and the farm are fixed costs per ton of material. Hence a higher-analysis mixture, characteristic of bulk blends, leads to a lowered cost per ton of nutrients spread on the field. TVA workers made a study of the effect of content of plant nutrients on the cost per unit for several ratios (Figure 10-19). For example, for a 1-1-1 ratio a total content of 54 units of $N + P_2O_5 + K_2O$, or an 18-18-18, was most economical. This would be 18-7.9-15 in terms of N-P-K. Although cost comparisons will depend on the assumptions made, and bases may change rapidly, the *economics* of bulk blending are attractive to the grower.

Materials. In the first attempt at bulk blending, muriate of potash was

dumped on top of rock phosphate in the spreader, and ammonium sulfate was subsequently blended with the mixture. The choice of materials has now moved to mainly monoammonium and diammonium phosphate, triple superphosphate, potassium chloride, ammonium nitrate, urea, and ammonium sulfate. Limited

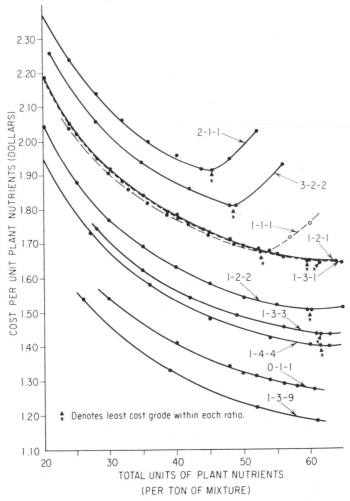

FIGURE 10-19.   Effect of concentration of plant nutrients on cost per unit (20 lb). [Douglas et al., *Commer. Fert.*, **101**(5):23 (1960).]

quantities of other products, such as an ammonium phosphate sulfate, granular normal superphosphate, nitric phosphates, granular elemental sulfur, and so on, are also used.

Ingredients used in the formulation of bulk blends must be chemically compatible. Only a few combinations of the customary materials are unsuitable (Figure 10-20). Blending of urea and ammonium nitrate must be avoided completely because of the high degree of wetting that occurs when the two are brought together. Urea will react with unammoniated normal or triple superphosphate to release water, which results in stickiness and caking. However, not all superphosphates are affected at the same rate, and with some the reaction is very slow.

A slow reaction can also take place between diammonium phosphate and unammoniated superphosphate. This causes caking and a reduction in water solubility of the contained phosphates. It is unimportant in bulk-spreading operations, but bagging of such a combination is not recommended. Ammoniation of superphosphate or use of a conditioner in the bagged blends are practical measures that can be taken to largely eliminate these difficulties.

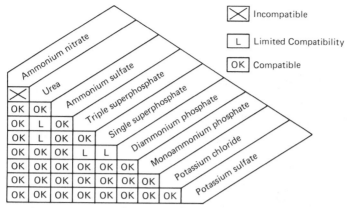

FIGURE 10-20. Chemical compatibility of blend materials. [Hoffmeister, *Proc. TVA Fert. Bulk Blending Conf.* (August 1–2, 1973).]

Bulk blending works best with materials that are well granulated, closely sized, and sufficiently dry and strong so that they do not cake or deteriorate. Particle sizes of the products are now generally well standardized at -6 to $+16$ mesh. However, size distribution can vary widely within this range. The shape of particles varies from round to irregular. Additionally, the apparent specific gravity of the particles in the materials can differ widely.

Bulk-Blending Plant Equipment. Examples of two bulk-blending plants are shown in Figures 10-21 and 10-22. These are ground storage buildings, which are used most frequently. Essential features of bulk plants are:

1. Material handling system to convey the product from railway cars and bulk transport trucks to storage bins.
2. Bins for storage of raw materials. Six bins, each with a capacity of 100 to 150 tons, are common.
3. Front-end loader or scoop truck to move the product from storage.
4. Screen to remove lumps and foreign matter.
5. Weigh scale.
6. Mixer.
7. Conveyor to load out blended fertilizer.

Some of the large modern plants, such as the one illustrated in Figure 10-22, are automated. In others equipped with a computer, an office employee provides information on the N-P-K requirements to the computer, and the computer will calculate the least-cost formulation, operate the mixer, and print out the desired setting for the fertilizer spreader, spreader speed, and so on.

Capacities of mixers in older bulk-blending plants were normally between 1 and 4 tons. The capacity of a plant equipped with a 1-ton rotary mixer is about 20 tons of blend per hour. In most new facilities the capacity of mixers is at least 5 tons, and some are as large as 10 tons. Output from a 5-ton mixer is about 50 tons of mixtures per hour.

Disadvantages of Bulk Blending. Blending of dry fertilizers is not without some problems. Some of the disadvantages of this practice are:

1. Tendencies to segregate during handling.
2. Difficulties in mixing small quantities of micronutrients uniformly without a binding agent.

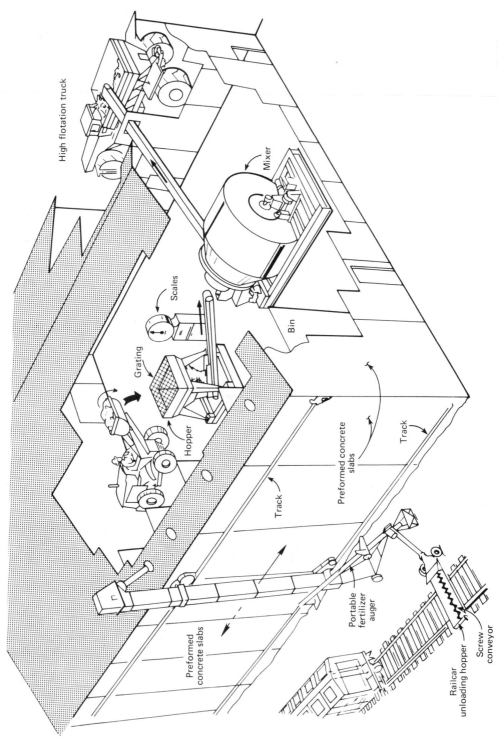

FIGURE 10-21. Flat-type bulk-blending plant with concrete-type mixer. (Achorn and Kimbrough, *in TVA Fert. Conf. 1976. Bull Y-106.* Muscle Shoals, Ala.: NFDC–TVA, 1976.)

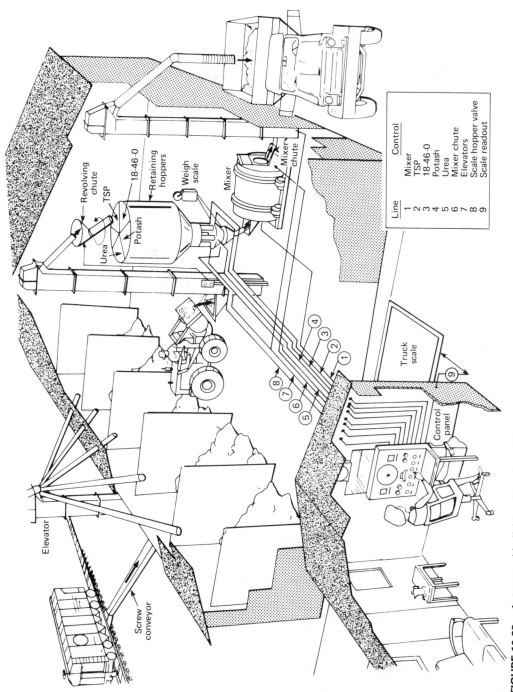

FIGURE 10-22. Automated bulk-blending plant with remote operating board. (Achorn and Kimbrough, in *TVA Fert. Conf. 1976. Bull. Y-106.* Muscle Shoals, Ala.: NFDC–TVA, 1976.)

3. Uneven distribution of fertilizer in the field due to segregation caused by poor application procedures and equipment.
4. Difficulties in obtaining uniform application of granular herbicides and insecticides included in blends.
5. Difficulties in meeting specific guaranteed nutrient analysis when segregation is high.

Segregation. One of the problems of bulk blends has been that of segregation, or the separating out of certain components of the blend during handling after mixing. This problem is due chiefly to particle-size differences among blend ingredients. Other properties of the blended materials, such as particle density and shape, are relatively insignificant in comparison with the effects of particle size.

In the United States, the unofficial, but generally accepted, particle-size limits for granular materials are 1.0 to 3.3 mm in diameter, which corresponds to −6 to +16 Tyler mesh (Figure 10-23). Somewhat larger granules in the −5 to +9 Tyler mesh size range are preferred in Europe.

There is a basic difference between the particle-size control required in production of satisfactory materials for bulk blending and that needed in production for nonblending uses such as direct application or bagging. For the latter, it usually is sufficient to control only the size range, whereas for blending purposes, attention must also be given to distribution of sizes within a specific size range.

The importance of particle-size and particle-size range in formulating an acceptable 28-28-0 bulk blend comprised of urea and diammonium phosphate is illustrated in Figures 10-24 and 10-25. Figure 10-24 shows that these two ingredients are well within the size-range limits of −6 to +16 mesh for granular fertilizers. However, they would segregate badly when blended and subsequently

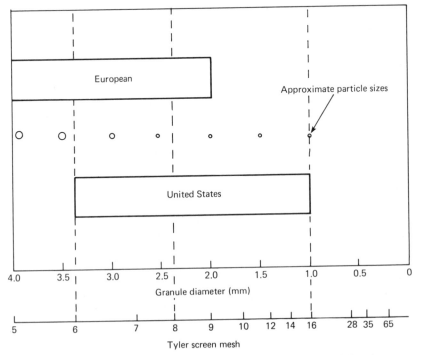

FIGURE 10-23. Size ranges of particles in granular fertilizers (particles drawn to approximately one-half scale) (Hoffmeister, "Particle-size control requirements for granular fertilizers." *Reprint Z-126.* Muscle Shoals, Ala.: NFDC–TVA, 1981.)

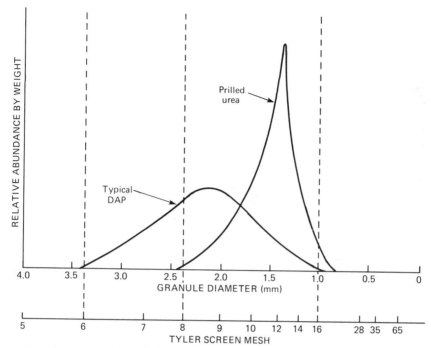

FIGURE 10-24. Particle-size distribution in a 1:1 blend of prilled urea with typical granular diammonium phosphate. (Hoffmeister, "Particle-size control requirements for granular fertilizers." *Reprint Z-126.* Muscle Shoals, Ala.: NFDC–TVA, 1981.)

handled, because of the obvious differences in size distribution. All parts of the blend larger than about +9 mesh are deficient in urea, while all portions finer than +9 mesh are deficient in diammonium phosphate.

Contrasted with the poor match of ingredients discussed above, Figure 10-25 shows a nearly perfect pairing of blend components. No significant segregation would occur in this formulation, which is based on the same source of diammonium phosphate but with urea produced in the TVA pan granulator.

Hoffmeister of the Tennessee Valley Authority recommends that materials intended for blending should be essentially 100% within the −6 to +16 mesh range. Additionally, the ingredients should contain at least 25%, but no more than 45%, of plus 8 mesh particles.

The Canadian Fertilizer Institute has used Hoffmeister's guidelines to develop the *Size Guide Number* (SGN) system for assisting in selection of compatible materials for blending. The SGN is defined as the medium size (in millimeters to three significant digits) determined by the 50% cumulative point on a curve of particle-size distribution plotted percent cumulative by mass versus millimeter diameter, multiplied by 100 and rounded to the nearest increment of "5." The closer the SGN of raw materials utilized in a blend, the lower the probability of segregation in the blended product.

Introduction of pesticides or micronutrients into the blend merely aggravates the problems of segregation and uniform distribution, again because of particle size. Careful granulation may be the answer. Addition of oil, H_2O, or liquid mixed fertilizer helps the micronutrient carriers to adhere to the particles.

Bulk Application of Blended and Manufactured Goods. Blending generally implies bulk handling and spreading of fertilizers. Although by tradition manufactured goods were handled in bags, most of these materials are now being

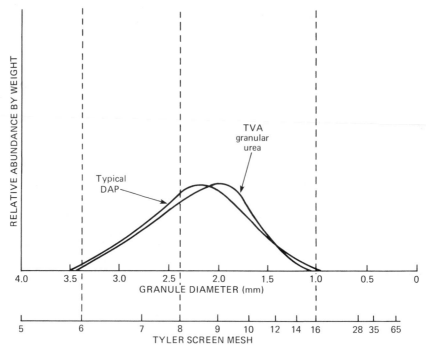

FIGURE 10-25. Particle-size distribution in a 1:1 blend of TVA granular urea with typical granular diammonium phosphate. (Hoffmeister, "Particle-size control requirements for granular fertilizers." *Reprint Z-126*. Muscle Shoals, Ala.: NFDC–TVA, 1981.)

handled in bulk. The growth of shipments of both solid and fluid fertilizers in bulk for retail markets in the United States is as follows:

Year	Solids (1000 tons)		Fluids (1000 tons)	
	Mixtures	Materials	Mixtures	Materials
1953–1954	469	1,360	28	560
1960–1961	1,357	2,490	531	2,244
1971–1972	9,800	7,700	3040	7,800
1980–1981	18,451	14,058	4883	13,796

Uniformity of Application. Quality control in bulk blending of solid fertilizers is important through to the final step of field application. Uneven application can result in lower crop yields and loss in returns to the farmer. Poor spreader adjustment and maintenance and unskilled driving are major potential sources of nonuniformity. Segregation in spreading, like it is in other handling, is chiefly the result of particle-size mismatching. Therefore, attention to size matching will greatly benefit the spreading operation.

Excessively high spinner speeds on spinning disk or fan-type spreaders and excessively wide swaths will lead to segregation and erratic application. High spinner speeds will also cause shattering of soft granular products, with resultant formation of fines that segregate from the remainder of the blend. Use of low spinner speeds and provision for considerable overlap through narrower swaths are advisable if a blend is comprised of poorly matched ingredients.

Research in Illinois and Missouri has shown that uneven spreading did not affect

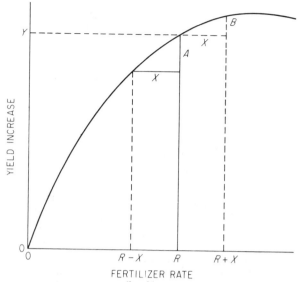

FIGURE 10-26. Thoretical effect of uneven spreading on yield as shown by decreases or increases in fertilizer rates (R). [Jensen and Pesek, *SSSA Proc.,* **26**:170 (1962).]

crop yields to a measurable degree. Theoretically, however, when crop yields are high, Jensen and Pesek postulate that uneven distribution of plant nutrients will have an adverse effect on yields. This is explained by the diminishing response from higher rates (Figure 10-26). The loss in yield caused by underfertilization (A) is larger than that gained by overfertilization (B). This holds only for the higher rates of fertilizer near the top of the yield response curve.

To meet the demand for small applications at planting, some blenders sell a part of their product in bags. Others mix standard grades on a registered basis.

Fluid Mixed Fertilizers

N-P-K and N-P grades of fluid mixed fertilizers are used in starter solutions and foliar sprays, as water-soluble fertilizers for use in irrigation water, and for direct application to the soil. This discussion deals largely with the first and last of these methods.

Plant-nutrient consumption as fluid fertilizers in the United States during the 1970s and early 1980s has grown at the rate of about 7% per year. This increase greatly exceeds the 4% annual gain in total nutrient consumption occurring during the same period.

Of the 23 million tons of plant nutrients ($N + P_2O_5 + K_2O$) used in 1980 in the United States, about 3.8 million tons or 17% were supplied as fluid sources. Suspensions currently represent about 40% of the fluids marketed in the United States.

In 1980, there were about 3200 fluid mix plants in the United States. The average throughput per plant is about 3700 tons of fertilizer per year and they generally service customers within a radius of 15 to 20 miles.

The majority of reasons given in Chapter 5 for rapid growth in the use of non-pressure nitrogen solutions are also applicable to fluid mixtures. Other advantages of clear liquid fertilizers include:

1. Energy requirements are lower and no evaporation or drying is required.
2. There are no problems of dust or fume in manufacture or use of liquids.

3. Problems of physical properties such as hygroscopicity and caking are absent.
4. Less labor is required for storage, handling, and loading for shipment than for solids.
5. Liquid fertilizers are adaptable to foliar application.
6. Equipment for producing liquid mixed fertilizers is simple and inexpensive provided that suitable materials are available from manufacturers of primary materials.

Contrasted with these desirable characteristics are the following disadvantages:

1. Materials for clear liquids must be water soluble; thus the choice is limited. Some constituents may be more expensive or supplies may be limited.
2. Liquid mixed fertilizers are less concentrated than solids; therefore, shipping costs are increased.
3. Low temperatures may cause crystals to form and "salt out." Concentration of nutrients must be reduced under cold conditions to guard against settling out and resultant plugging of equipment. Lower concentrations add further to transportation costs.
4. Specialized equipment is needed for storage and transport of liquids. Tanks, barrels, rail and road tankers, ships, barges, and pipelines are all used. This equipment is not necessarily more expensive than for solids, but it may not always be immediately available.
5. Expensive high-capacity flotation-type application equipment is often used, particularly in North America.

Manufacture. Fluid mixed fertilizers, either clear liquid or suspension, are usually made in two types of processes, generally known as hot-mix and cold-mix plants.

Hot-Mix. The hot-mix method has been widely used and is based on the neutralization of either wet-process (54% P_2O_5) or superphosphoric acid (69 to 72% P_2O_5) to make such base solutions as 10-15-0 (10-34-0), 11-16-0 (11-37-0), and 12-17-0 (12-40-0). Nitrogen solution is normally used as an additional source of nitrogen and finely divided potassium chloride is the usual source of potassium.

A diagram of a typical hot-mix plant is shown in Figure 10-27. The mixing equipment is relatively simple and a batch system is often employed. Mix tanks usually have a capacity of 4.5 to 18 metric tons (5 to 20 tons). Considerable heat is generated and the liquid is recirculated through a cooler to prevent excessive boiling in the mix tank.

Cold-Mix. In contrast to hot-mix plants, cold-mix plants are simple and storage tanks, meters, and a simple mix tank make up the equipment list. The base solution, 10-34-0 or 11-37-0, urea-ammonium nitrate solution, and potash are mixed together. Heat is not generally evolved. Clear liquids or suspensions can be made. The principal components of a typical liquid cold-mix plant are evident in Figure 10-28. The cold-mix process corresponds to bulk blending of dry granular fertilizer materials.

Suspensions. A suspension fertilizer is so concentrated that small crystals of fertilizer salts form. To prevent these crystals from settling out, a suspending agent, usually 2% attapulgite clay, is used. The main reason for interest in suspension fertilizers is that usually materials of lower purity and cost can be used than in the clear liquids. Also, suspensions are popular because high potash grades such as 3-4-25 (3-10-30) and 5-6-25 (5-15-30) can be made. The average plant-nutrient concentration of suspensions can exceed 45 units, whereas the average concentration for clear liquids is about 28 units.

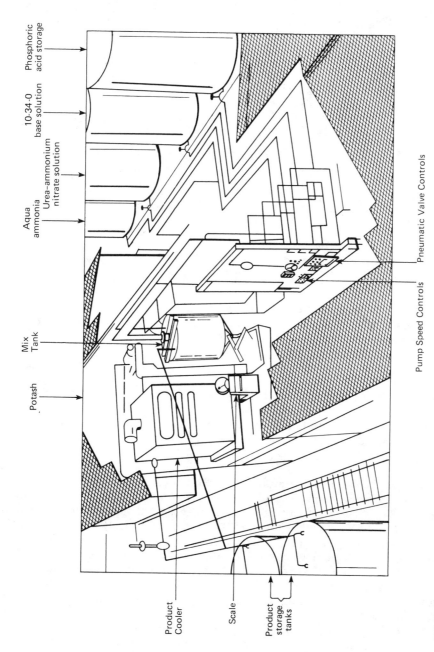

FIGURE 10-27. Major components of a liquid hot-mix plant. (*Farm Chemicals Handbook 1982*, p. B-41. Willoughby, Ohio: Meister Publishing Company, 1982.)

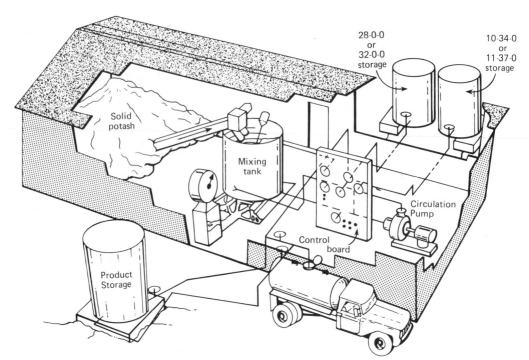

FIGURE 10-28. Main features of a liquid cold-mix plant. (*Farm Chemicals Handbook 1982*, p. B-41. Willoughby, Ohio: Meister Publishing Company, 1982.

Suspensions must remain fluid enough to be mixed, pumped, and applied to the soil without major alterations in the application equipment. They must also remain homogeneous during application. Suspensions may be stored for several weeks or months by using periodic agitation, but the general practice is to limit storage since crystals may increase to such a size as to cause difficulties in removal and application.

The increasingly greater proportion of the fluid fertilizer market being satisfied by suspensions is due to several reasons. In addition to those already mentioned, other attractive aspects include higher analysis, ease of including additives, elimination of salt-out problems, more economical, and more flexibility in grades produced.

Materials. The major problem in fluid mixed fertilizers is the cost of the phosphorus. Superphosphoric acid from the wet process, 29 to 31% phosphorus (68 to 72% P_2O_5), of which about 50% is in the polyphosphate form, or furnace process acid, 33 to 35% phosphorus (76 to 80% P_2O_5), of which 60 to 85% is in the polyphosphate form, may be used. Merchant-grade orthophosphoric acid, 23% phosphorus (52 to 54% P_2O_5), is lower in price. However, because of shipping distance the delivered cost of fluid fertilizer made from the superphosphoric acid is often no more than that from the wet process. Generally, merchant-grade wet process orthophosphoric acid is becoming more popular because impurities are being removed and it is now a relatively clear solution.

One of the difficulties in the production of clear solutions has been the incorporation of potassium. Potassium chloride is the most economical source, but it has low solubility. Potassium hydroxide or potassium carbonate can be used, for they react with phosphoric acid to form potassium phosphate. TVA reports that a 6-7.9-15 (6-18-18) can be made with potassium hydroxide but only a 3-4-7.5 (3-9-

9) with potassium chloride. Unfortunately, only the chemical-grade potassium compounds, which are currently quite expensive, give high solubility.

Pesticides (herbicides and insecticides) are being added successfully to fluid fertilizers for broadcast application. Such additions save trips over the field. Compatability charts are available in order to avoid undesirable chemical reactions or precipitation. In case of doubt, applicators are often advised to mix up a sample in a glass jar and observe it.

Micronutrients in various forms may be added readily, particularly with suspensions. Addition of chelating agents favors solubility. A polyphosphate such as 11-37-0 gives higher micronutrient solubility by sequestration. Sulfur may be added in several forms including ammonium thiosulfate and ammonium bisulfite.

Precipitation or salting out may occur in clear solutions at low temperatures. Careful attention must be given to the selection of materials for clear solutions to avoid salting out during storage or application. The general aim is to keep the salting-out temperature below 32°F. Suspensions do not salt out but under cool temperatures viscosity may increase making application difficult.

Grades. A wide variety of grades are formulated. The 10 leading grades of liquids and suspensions used in the United States in 1980 were as follows:

Clear Liquid Grades	Suspension Grades
20-10-0	3-10-30
4-10-10	4-12-24
7-21-7	3- 9 -27
20- 6 -12	6-18-18
7-22-5	13-13-13
5-10-10	3- 9 -18
6-18-6	10-30-0
4- 8 -12	5-10-15
16-20-0	20- 5 -10
18-18-0	5-15-30

Source: AAPFCO Fertilizer Plant Survey (1980).

Interest is growing in N-P materials, particularly in the West, on soils very high in potassium where potassium is not needed. Where potassium is needed in sizable amounts, broadcast application of potash materials may be used in addition to the fluid fertilizer.

Application. Equipment to apply clear and suspension fluid fertilizers is constantly being improved. Most of the latest changes in application equipment have been high-flotation applicators, which were referred to in the section on non-pressure nitrogen solutions in Chapter 5. Also, there have been significant developments in equipment for preplant deep banding of fluid fertilizers during primary tillage operations. Examples of flotation and deep banding applicators are shown in Figures 10-29 and 10-30, respectively.

Fluid fertilizers are applied in three main ways. They are often applied as starter fertilizers either in the seed row or banded close to it. Broadcasting of fluids is common in situations where soil fertility levels are high and where it is feasible and convenient to combine the application of fertilizer with pesticides. In dry regions where fertilizers close to the surface often lose their effectiveness, deep banding or injection of fluid fertilizers, usually combined with a tillage operation, has become popular.

FIGURE 10-29. Large modern flotation-type applicator for surface banding of fluid fertilizers. Note the streams of liquid fertilizer flowing from the nozzles in the boom at the back of truck. [Courtesy of National Fertilizer Solutions Association.]

Surface banding, which was first started in the early 1970s, is gaining wide acceptance, especially for fertilization of established forage crops and for topdressing winter wheat and fall rye. In some instances high-flotation applicators have been equipped to spot-spray weeds at the same time fluid fertilizer is being dribble-banded.

Factors Affecting Uniformity of Application. Care must be exercised to apply fluid fertilizers uniformly just as in the case of dry fertilizer products. The main factors that can influence uniformity of application include pressure, height, and attitude of nozzles and orifices; nozzle tips and orifices; nozzle and orifice spacing; pump characteristics; ground speed; specific gravity and viscosity of the fluid; uniformity of the fluid formulation; agitation; and swath marking. A helpful review of the effect of these factors appears in Follett et al. (1981).

FIGURE 10-30. Equipment for deep placement of fluid fertilizers during tillage. [Courtesy of Green Drop, Calgary, Alberta (1982).]

Effectiveness. Assuming the same type of placement and water solubility of phosphorus, responses to fluid and solid fertilizers are similar. When the water solubility of phosphorus is important, liquids are superior to solids that contain a high amount of water-insoluble phosphorus. In some instances, when zinc in the soil is deficient, the polyphosphates may have a sequestering effect and make the zinc more available to the plant.

Cost of nutrients in fluid fertilizer is somewhat higher than that in bulk blends but convenience is an important factor. Lower labor costs and the fact that liquids can be moved by gravity, air pressure, or pumps are distinct advantages over dry materials. Other important considerations include ease of adding sulfur and other elements such as micronutrients; uniformity of application; and combining their application with other programs such as pest control and tillage.

Agronomic Service

An advisory service including plant and soil analysis is now being provided by the fertilizer industry in many areas of North America. In many situations, plants are designed to serve a limited area. The managers of these plants, from whom the growers can obtain the exact ratio and amount of fertilizer called for by their soil-test recommendation sheets, can meet their customers on a more personal basis. Many farmers in the United States and Canada rely on fertilizer dealers and other representatives of the fertilizer industry for the latest information on fertilizer use and various other aspects of crop production.

The philosophy governing the use of fertilizer, which is to build and/or maintain soil fertility, should eventually eliminate inadequate plant nutrition as a factor limiting crop production. Fertilizer dealers have a prominent role in soil fertility management since they supply the plant nutrients required by their grower customers. Custom blends ordered from the dealer can also be custom-applied with no physical effort whatever on the part of the customer.

Fertilizer–Pesticide Mixtures

Fertilizer–pesticide mixtures combine two operations into one and save labor, time, and energy. Also, the uniformity of application and efficacy of soil-incorporated pesticides are often improved through the application of such combinations. Hence one of the important developments in fertilizers in recent years involves their use as pesticide carriers (herbicides, insecticides, and fungicides). This has brought on technical chemistry problems, application questions, and legal requirements. The incorporation of insecticides in fertilizers is not new, for in 1904 a patent was issued in France for such mixtures. Numerous patents have been issued since that time in the United States as well as in foreign countries. A problem is the change of approved pesticides over the years.

Formulation. Several types of commercial pesticides have been developed. These include sprayable concentrates, granular materials, dusts, solutions, and minor formulations.

Blends of granular pesticides with dry fertilizers are not normally recommended. Such blends, because of differences in particle size and density, do not maintain uniform composition, and as a result application in the field is usually uneven. More acceptable approaches include spraying the liquid pesticide materials on dry fertilizer, spraying onto a carrier such as vermiculite and mixing with fertilizer, or premixing prior to granulation or blending. Instructions furnished by the pesticide manufacturer must be followed carefully.

Because there are variations in pH, concentration and type of certain salts, water content, and perhaps other properties of fluid fertilizers, a given pesticide may react differently depending on the fertilizer materials used. Reactions may affect physical characteristics as well as performance of the pesticide. Although many pesticides are suitable for blending with sprayable fluid fertilizers, the manufacturer's label should always be consulted to be certain that the particular pesticide is approved for use in blended fluid fertilizers.

Crop protection chemicals may not always mix evenly throughout a sprayable fluid or the components may separate too quickly to produce a satisfactory blend. A simple compatability test using small quantities of the components to be mixed should be conducted. Mixing the appropriate quantities in a glass jar and observing the results is a desirable alternative to the loss in operating time and extra labor to clean tanks and clogged nozzles, as well as loss of unusable fertilizer and chemicals.

Insecticides. At one time the chlorinated hydrocarbons and the organophosphates were all recommended for use in combination with fluid fertilizers. Certain organophosphates and carbamates are approved for application with mixed fluid fertilizers. Labels sometimes restrict such use to clear liquids based on either ammonium polyphosphate or ammonium orthophosphate.

Insecticide stability when mixed with a fertilizer and stored for any length of time has been a problem. There are many substances known as compatability agents which will improve the compatability and stability of fluid fertilizer–herbicide and/or insecticide mixtures.

Use of Fertilizer–Insecticide Mixtures. Such mixtures are satisfactorily used on turf in both solid and fluid forms. Certain organophosphate and carbamates have limited use in fluid fertilizers banded beside the row for insects such as corn rootworm. Overall there is limited use of fertilizer–insecticide mixtures.

Problems. Distribution of a given amount of insecticide in the field is a complicating factor, for fertilizers vary greatly in analysis and in the amount required for various crops.

A multiplicity of problems is connected with insect control. The type of insect, the intensity of infestation, and the stage in the life cycle at which the treatment would be most effective must all be recognized in selecting the insecticide, the amount to apply, and the time and method of application.

Residual effects must be considered, since insecticides vary in their persistence in soil. For example, the organophosphate insecticides break down into harmless substances in plants and soils more rapidly than the chlorinated hydrocarbon type of insecticides. An additional concern is that of imparting undesirable flavors to subsequent crops. Benzene hexachloride, for example, has been found to give a distinctly musty taste to peanuts, Irish potatoes, and other food crops. Because of this quality such mixtures must bear labels that specify how, at what rate, and on what crops they may be applied.

Herbicides. The use of fertilizer–herbicide mixtures is widespread and is commonly called "weed and feed." Much work has been done concerning the compatibility of herbicides and fluid fertilizers. More herbicides are physically compatible in nitrogen solutions than in either suspensions or liquid mixed fertilizers based on either ammonium polyphosphate or ammonium orthophosphate. Although some separation of suspension–herbicide combinations can occur 2 hours after the last agitation, effectiveness of the herbicide or the fertilizer is not affected by time.

Use of Fertilizer–Herbicide Mixtures. Fluid or dry fertilizer–herbicide mixtures may be applied preplant. Some herbicides require soil incorporation but this is compatible with fertilizer usage. Herbicide–nitrogen solution mixtures may be applied preplant, immediately after planting, or after plant emergence as directed sprays to avoid some chemicals burning crop foliage. However, shallow or surface incorporation of nitrogen and other plant nutrients, particularly relatively immobile ones like phosphorus, may not be the best method of application. As will be discussed in Chapter 13, deeper placement of plant nutrients is often more effective, especially in dry regions and dry seasons.

The practice of dry bulk impregnation has grown steadily since the late 1960s in many areas of the United States and Canada. Advantages of dry bulk impregnation to farmers include:

One less trip over the field, saving energy and time and reducing compaction where it is a problem.
The grower is spared from storing, handling, and mixing the herbicide, as well as from container disposal.
The fertilizer–herbicide blend may either be custom-applied or applied by the farmers with their own or rented equipment.
Fertilizer granules carrying the herbicide are little affected by surface debris.
The efficacy of some herbicides may be increased.

The liquid/solid ratio is important in the formulation of acceptable fertilizer–herbicide blends. Ideally, enough herbicide in liquid form should be used to obtain uniform coverage of the dry fertilizer products. Too much liquid will cause the fertilizer mix to be excessively moist, causing clogging and/or calibration problems. At the other extreme, coverage of the fertilizer particles is incomplete, resulting in uneven distribution of the herbicide in the field and poor weed control. In most cases the lowest rates of dry fertilizer for proper impregnation are from 160 to 200 lb/A. At the other end of the scale, fertilizer rates should not exceed 750 lb/A.

Few, if any, herbicides should be combined with straight ammonium nitrate or its blends since their organic constituents may sensitize ammonium nitrate, making it potentially very hazardous. Sutan + and Eradicane and combinations should not be applied to single or triple superphosphate. Literature describing uses of Treflan specifies that it not be impregnated in straight-coated ammonium nitrate, clay coated urea, or straight limestone. However, blends containing mixtures of these may be impregnated with Treflan.

There is much need for additional research on fertilizer–pesticide mixtures. New pesticide chemicals, new fertilizers, and regulations all render this field a challenging one. Students are advised to consult manufacturers' product labels as ultimate sources of information on approved combinations and practices.

Fertilizer Control and Regulation

Laws and regulations governing the sale of fertilizers are necessary because of the opportunity to defraud. This has to do with both the quantity of nutrients and quality of carriers present in the fertilizer. Regulatory laws, which apply to mixed fertilizers and to fertilizer materials, protect the farmer and the reliable fertilizer manufacturer by keeping goods of questionable value off the market.

The fertilizer laws in the various states are similar in many respects, and this is helpful to fertilizer companies operating across state lines. When adjoining states have different requirements for labeling and guarantees, manufacturing problems are increased.

The Association of American Plant Food Control Officials has been directing considerable attention to a uniform state fertilizer bill. Although this model is under continual revision, essentially all of the states have adopted it.

Guarantee. The total nitrogen (N), available phosphorus pentoxide (P_2O_5), and the soluble potash (K_2O) must be guaranteed in terms of the percentage of each of these constituents present. For example, in a fertilizer grade such as 6-24-24 the nutrients are guaranteed as 6% total nitrogen, 24% available P_2O_5, and 24% soluble K_2O. The elements are listed in that order for mixed fertilizers in all states. In some states it is necessary to guarantee the percentage of water-insoluble nitrogen present, particularly in tobacco fertilizers. Fertilizer material must bear the guarantee of the primary nutrient it is carrying: for example, urea, 46% nitrogen.

The oxide expressions for phosphorus and potassium are actually inaccurate and confusing, for they are based on early practices with which chemists determined the elements by ignition and weighed the oxides. These methods have long since been discarded. Proponents of the change to the elemental expressions for phosphorus and potassium feel that the understanding of fertilizer formulation will be enhanced by the change. In 1963 several scientific societies, including the American Society of Agronomy, the Soil Science Society of America, and the Crop Science Society of America, initiated the practice in their publications of giving the elemental analysis of fertilizers, with the oxide in parentheses. Many states now use the elemental as well as the oxide expressions in extension and research publications. Some fertilizer companies print both the oxide and elemental guarantees for phosphorus and potassium on their bags or labels.

Countries such as Norway, New Zealand, and South Africa use the elemental basis. There is not a strong effort, however, to make this change in either the United States or Canada.

Uniform State Fertilizer Bill. There is some variability among states in the guarantees required for nutrients other than nitrogen, phosphorus, and potassium. Because of this lack of uniformity and the growing market for fertilizers containing micronutrients, the Association of Plant Food Control Officials, in cooperation with representatives of many other organizations, worked out the following proposed regulation for adoption under state fertilizer laws:

Other plant nutrients, when mentioned in any form or manner shall be registered and shall be guaranteed. Guarantees shall be made on the elemental basis. Sources of the elements guaranteed and proof of availability shall be provided the _____ upon request. The minimum percentages which will be accepted for registration are as follows:

Element	Percent
Calcium (Ca)	1.00
Magnesium (Mg)	0.50
Sulfur (S)	1.00
Boron (B)	0.02
Chlorine (Cl)	0.10
Cobalt (Co)	0.0005
Copper (Cu)	0.05
Iron (Fe)	0.10
Manganese (Mn)	0.05
Molybdenum (Mo)	0.0005
Sodium (Na)	0.10
Zinc (Zn)	0.05

Guarantees or claims for the above listed plant nutrients are the only ones which will be accepted. Proposed labels and directions for use of the fertilizer shall be furnished with the application for registration upon request. Any of the above listed elements which are guaranteed shall appear in the order listed immediately following guarantees for the primary nutrients of nitrogen, phosphorus and potassium.

A warning or caution statement is required on the label for any product which contains 0.1% or more of boron in a water soluble form. This statement shall carry the word "WARNING" or "CAUTION" conspicuously displayed, shall state the crop(s) for which the fertilizer is to be used, and state that the use of the fertilizer on any other than those recommended may result in serious injury to the crop(s).

Products containing 0.001% or more of molybdenum also require a warning statement on the label. This shall include the word *Warning* or *Caution* and the statement that the application of fertilizers containing molybdenum may result in forage crops containing levels of molybdenum which are toxic to ruminant animals.

The majority of the states have now adopted this regulation. A maximum guarantee of not more than 3% is required for the chlorine content of tobacco fertilizers in some states.

The Fertilizers Act for regulation and control of fertilizers in Canada is administered by one agency, the Feed and Fertilizer Division of Agriculture Canada. In many respects this Act is similar to the Uniform State Fertilizer Bill.

Labeling Information. The following information is usually required on the label of fertilizers distributed in bags or other containers:

1. Net weight of fertilizer.
2. Brand and grade.
3. Guaranteed analysis.
4. Name and address of registrant or licensee.

For fertilizers sold in bulk, the same information must be shown on the shipping bill or on a statement accompanying the shipment.

It is extremely important that farmers become well acquainted with the meaning of these data, particularly the N-P_2O_5-K_2O guarantee. It must be realized that not just fertilizer is being bought but that the plant-nutrient elements contained in the fertilizer are the items of purchase.

Inspection and Enforcement of Fertilizer Laws. Inspection and analysis are usually carried out by the state department of agriculture or the state agricultural experiment station. The expense of this service is borne by the tonnage tax on the fertilizer. Each manufacturer is required to register each grade or material annually with the regulatory organization in his state. From time to time during the year the fertilizers are sampled by inspectors who are trained for that particular job, and the samples are sent to the control laboratory for analysis. The findings are then checked with the registration on file, and if they fall outside the allowable tolerance, as specified in the law, a penalty is assessed.

The official methods of sampling and analysis are prescribed by the Association of Official Agricultural Chemists. This uniformity is essential to the fertilizer companies operating in several states.

In the manufacture of granulated and nongranulated fertilizers inspection of the product in the bin or in bags is reasonably simple. However, bulk blending presents a different situation. Some blenders mix standard grades in advance and register them. Others mix to the specific order of the final purchaser and load the mixtures into truck or spreader for immediate transport to the field. The physical task of getting an inspector on hand to sample a percentage of the loads is difficult, but it is being accomplished.

It is required by law that a bulletin bearing the guaranteed and found analysis of all fertilizer brands inspected be published each year. This sort of publicity is quite effective in minimizing fraud. If a company persists in the manufacture of fertilizers that regularly fall below the guarantee, this fact soon becomes known.

Generally, the amounts of nutrients found in most fertilizers are somewhat above the guarantee. Most companies overformulate to decrease the chance of a deficiency in the analysis.

Fertilizers purchased by mail order from outside a state are generally not subject to state inspection. Extravagant claims are sometimes made in local newspapers by these outside companies, but the control agencies have no authority to curb this activity.

Uniformity of Recommended Grades and Ratios

Over the years many grades and ratios have been developed. It is often difficult to determine how a given grade originated, but as a rule it was developed for a specific purpose. To manufacturers a large number of grades means a considerable expense in terms of money, time, storage bins, and labor if these grades are to be stocked. To the farmer a long grade list means a confusing array from which to make a choice. To the agronomist these long lists usually represent a needless duplication of grades or unsatisfactory and pointless analyses.

The variation between certain grades is often too small to be detected when it is considered that the fertilizers may be applied at the rate of 200 to 500 lb/A. However, once they have become established in a state and the industry has built up a demand, both farmers and industry dislike dropping them. In some instances industry markets a grade a little different in analysis from existing ones in order to have a distinctive grade.

To complicate matters further, adjoining states may have different recommended grades and ratios, which makes a change difficult if shipments are being made into several states. In view of this situation, various organizations and states over the country have been striving for more uniformity. For example, certain groups of states in the East are working toward the development of standard recommendations across state lines, and regional soil test work groups have made real strides in ironing out some of the differences.

In bulk blending the materials are often mixed together to meet the results shown by soil tests with no regard to grade or ratio. University recommendations now are largely in terms of pounds of N, P_2O_5, and K_2O per acre. When it is considered that extra nitrogen, phosphorus, and potassium may be applied separately as materials in accordance with recommendations or known needs, a minimum number of grades is really needed.

It is recognized, of course, that special fertilizers may be needed for some crops. This may require the addition of micronutrients to some of the foregoing ratios under certain conditions.

In Canada, the Feed and Fertilizer Division of Agriculture Canada is responsible for similar inspection and analysis of fertilizers. The majority of samples are analyzed at its central laboratory in Ottawa, Ontario, with some testing also done at its regional laboratory in Calgary, Alberta.

Summary

1. Most fertilizer nitrogen is produced from ammonia which is synthesized from hydrogen gas and atmospheric nitrogen. Ammonium sulfate and Chilean nitrate of soda are minor sources of nitrogen. The former is a by-product of coking of coal, metallurgical operations, and the production of caprolactam for the manufacture of nylon. In Germany some nitrogen fertilizer is still produced by the cyanamide process.

2. Ammonia manufacture is a very energy intensive process, requiring from 33 to 40 billion BTU/ton. Natural gas is the most widely used feedstock for ammonia production, with about 35,000 to 40,000 ft^3 of it consumed per ton of ammonia. Approximately 60% of the natural gas requirement is used to provide hydrogen, while the remaining 40% is used as fuel for heating and operating equipment.

3. Most of the other synthetic nitrogen fertilizers are manufactured from ammonia. Ammonia as such, or in aqueous solution, is used as a fertilizer. It can be oxidized to nitric acid from which various nitrate fertilizers are then manufactured. Ammonia is reacted with carbon dioxide to produce urea, which in turn is reacted with formaldehyde to synthesize slowly available urea-formaldehyde materials. The basic manufacturing processes in the production of the different nitrogen fertilizers were discussed.

4. For all intents and purposes rock phosphate is the original source of phosphatic fertilizers. The phosphorus compound in the rock is apatite, which is a highly insoluble material. This apatite structure is usually destroyed by treatment with either acid or heat in order to render the phosphorus more available.

5. Sulfuric acid is the most frequently used treatment for solubilizing phosphate rock. Other acids such as phosphoric and nitric are also used. Heat treatment of phosphate rock is usually done in gas- or oil-fired chambers. The various processes for treating phosphate rock and their reactions were discussed.

6. Sulfuric acid is the workhorse of the fertilizer industry. It is manufactured by burning elemental sulfur and converting the sulfur dioxide to sulfur trioxide. It can also be made from the sulfur dioxide that results from the burning of pyrites. The reactions and processes in the production of sulfur and sulfuric acid were covered.

7. The sources of potassium fertilizers are deposits of salts containing this element and the brines of dying lakes or seas. The salt deposits are situated in layers or beds several hundred to several thousand feet below the earth's surface and are mined by shaft-mining processes. Mining of these deposits by hot solution pumped into the deposits has also been perfected. Mining of brine is relatively simple. The brine is pumped to a centralized location for refining and the potassium salts are recovered. The techniques of the recovery of potassium salts from deposits and brines were presented.

8. The principal fertilizer-grade potassium salts are potassium chloride, potassium sulfate, potassium-magnesium sulfate, and potassium nitrate. Potassium nitrate, however, is not found to any extent in salt deposits and is made by reacting potassium chloride with nitric acid.

9. The fixed costs, which include manufacturing, bagging (in some instances), transporting, and distribution, and the cost of the plant nutrients influence the cost of a ton of mixed fertilizer. Although the cost of the nutrients may be somewhat greater, the lower fixed costs make higher-analysis fertilizers more economical. The grower appreciates the reduced labor of handling as well as the lower cost.

10. Higher-analysis fertilizers contain less of the micro- and secondary nutrients. Careful attention must be directed toward identifying deficiencies and toward adding supplements as needed.

11. A multiplicity of reactions takes place in dry fertilizers, including double decomposition, neutralization, hydration, decomposition, and ammoniation. The rate at which these reactions proceed is influenced by moisture, temperature, and particle size, all factors over which manufacturers must exercise close control.

12. The ammoniation–granulation process is widely used to produce high-analysis dry fertilizer mixtures. Basic features of this process were described.

13. Almost all dry fertilizers are granulated. Granulation of fertilizers reduces caking to a minimum, eliminates segregation, and results in a product that can be uniformly applied on the field.

14. Physical condition of the fertilizer, which is of major importance, is influenced by moisture content and state of subdivision. The moisture dissolves certain salts, new compounds form, and crystal knitting occurs. Conditioners such as ground cocoa shells are used to separate the particles in nongranulated fertilizers. Either internal conditioning substances or solid conditioning dusts are used to control caking and dust problems when they occur in granular materials.

15. Bulk blending or the mechanical mixing of solid, granular fertilizers is popular in the United States and Canada. The blend is usually a prescription mixture containing various combinations of nitrogen, phosphorus, potassium, other nutrients such as sulfur and magnesium, as well as micronutrients. The main components of bulk-blending facilities are described.

16. Uniformity of particle-size distribution in fertilizer products used in bulk blends is of major importance in preventing segregation and in producing consistent application in the field. Also, application equipment must be properly adjusted, maintained, and operated.

17. Fluid mixed fertilizers make up about 17% of the tonnage of mixed fertilizers in the United States. Neutralization of phosphoric acid with ammonia is the basic process and potassium is added as KCl. Factors such as dependability of supply and cost of phosphorus, economics, and convenience have all influenced acceptance of fluid mixed fertilizers. Suspensions in which fertilizer salts are suspended in their own saturated solutions are means of creating lower-cost, higher-analysis fluid fertilizers.

18. Fertilizer–pesticide mixtures combine two operations into one and save time, labor, and energy. Also, in some instances, uniformity of application and efficacy of pesticides can be improved. The multiplicity of the chemical reactions of organic pesticides with the fertilizer materials constitutes a challenge to the manufacturer. Care must be used to avoid undesirable physical reactions such as precipitates or the rendering of the pesticide ineffective. New chemicals and regulations make the practice a continually changing one.

19. Fertilizer–insecticide mixtures are being used in turf fertilizer and to a limited extent in liquid and solid starter fertilizer. Fertilizer–herbicide mixtures, or "weed and feed," are growing rapidly. The herbicide may be applied preplant in fluid fertilizer or in impregnated granular solids, and in nitrogen solutions just after planting or as a directed spray.

20. The control and regulation of fertilizer production is essential to protect both the grower and the reliable manufacturer. Fertilizers are sampled and analyzed systematically by the control agency. These results are reported to the manufacturer and publicized annually.

21. Guarantees for the nutrients nitrogen, phosphorus, and potassium are provided in terms of nitrogen (N), phosphorus pentoxide (P_2O_5), and potassium oxide (K_2O), respectively. All other nutrients are guaranteed on a elemental basis.

Questions

1. What are the three basic processes for the fixation of elemental nitrogen used in fertilizer manufacture?
2. Illustrate by appropriate equations the reactions of each of these processes.
3. Which of these processes is currently the most important?
4. What is the significance of natural gas in fertilizer production?
5. What are the three main steps in the manufacture of ammonia?
6. What is steam reforming?
7. List hydrogen feedstocks which are suitable for steam reforming. Identify the most important ones.
8. Does the price of natural gas contribute significantly to fertilizer manufacturing costs?
9. Why is ammonia such an important material in the fertilizer industry?
10. How is urea manufactured?
11. Why is urea a well-integrated commodity for production in association with a snythetic ammonia plant?
12. How is the oxidation of ammonia to nitric acid brought about in fertilizer manufacture?
13. How is defluorinated phosphate rock manufactured?
14. Which acid is commonly termed the *workhorse of the fertilizer industry*?
15. What are the three principal sources of the sulfuric acid used in fertilizer manufacture?
16. Are there adequate phosphate rock reserves in the world?
17. Does the United States have significant phosphate rock deposits? If so, where are they located?
18. What is *wet-process* phosphoric acid?
19. How is elemental phosphorus manufactured?
20. How does calcination of phosphate rock differ from fusion?
21. What is Frasch sulfur?
22. The Claus process is used to remove what unwanted constituent(s) of natural gas?
23. How is nitrogen obtained from coal?
24. What are sylvinite and sylvite?
25. Where are the principal deposits of potassium salts in North America?
26. What two processes are commonly employed in mining these salts?
27. What are the essential features of the long-room and stress-relief methods for mining potassium?
28. In what other form besides solid salt deposits do commercial potash sources occur?
29. What must be done to potassium-bearing ore before it is ready for use in the fertilizer industry?
30. Evaluate the practice of including lime in fertilizers.
31. Assume that a 10-10-10 fertilizer sells for $150 per ton. What would the difference be in the cost of a ton of the 10-10-10 and an equivalent amount of nutrients supplied in 15-15-15 priced at $190 per ton? What costs does this difference represent? What other benefits would farmers derive from the use of 15-15-15? Why have the quantities of fillers in fertilizers been decreasing?
32. What are some of the materials being used to formulate higher-analysis fertilizers? What are the basic costs that determine whether

a higher-analysis fertilizer will be more or less economical than a lower-analysis fertilizer?

33. What deficiencies may occur in crops grown with a higher-analysis fertilizer? Why?

34. Obtain prices on a 16-8-8 and a 12-12-12. How much is the nitrogen costing per pound in the 16-8-8? How does this compare with the price of nitrogen in ammonium nitrate, anhydrous ammonia, nitrogen solutions, urea, and sodium nitrate?

35. Why does the water solubility of phosphorus tend to decrease with continued ammoniation of normal superphosphate but not with concentrated superphosphate? Give equations.

36. What is granulation?

37. What are the main steps in (a) steam granulation and (b) granulation by ammoniation?

38. Why is there a need for drying in the granulation process? What are the advantages of granulated over nongranulated fertilizer?

39. What are melts? What costly step(s) can be eliminated by granulation of melts?

40. Are there commercial processes in which melt granulation is used? If so, describe one such process.

41. What functions do conditioners serve? On what principle are coated fertilizers based?

42. What reactions may occur during curing? What effect will such reactions have on physical condition? Why may caking be more of a problem in higher-analysis fertilizers than in lower-analysis goods?

43. What has been the trend in marketing of fertilizers in bags?

44. What are bulk blends? What is their level of acceptance?

45. Explain the difference in unit cost of plant nutrients in bulk blends and in granulated fertilizer.

46. List the major advantages of bulk blending.

47. What are the main components of a bulk-blending facility?

48. What are the problems in uniform spreading of bulk blends? Will this affect crop yields?

49. What particle-size guidelines should be followed to obtain acceptable high-quality bulk blends?

50. What is the significance of the Size Guide Number?

51. Identify points in the bulk-blend marketing system where the quality of blended product can be seriously reduced.

52. List the advantages and disadvantages of fluid mixed fertilizers. What are fertilizer suspensions?

53. (a) The maximum amount of chlorine that can safely be applied to tobacco is 40 lb/A. At planting you apply 1000 lb of a fertilizer containing 3% chlorine. Suppose then you want to make an additional N-K sidedressing of 15 lb of nitrogen and 35 lb of potassium oxide per acre. How much muriate and sulfate would you mix per acre to keep within the allowable chlorine limit and at the same time keep the cost of potassium at a minimum? The price of muriate is $150 per ton and the price of sulfate of potash $250 a ton. (b) What source of nitrogen would you use to sidedress corn if you could buy anhydrous ammonia at $280 per ton, nitrogen solution at $160 per ton, and urea at $240 per ton?

54. What are some of the latest developments in the application of fluid fertilizers? Which one(s) will in your opinion become the most widely accepted? Give reasons for your answer.

55. Describe the principal components of (a) hot-mix and (b) cold-mix fluid fertilizer plants.
56. What is bulk impregnation? State some of the obvious advantages of this technique?
57. Why are fertilizer–herbicide mixtures gaining in popularity? What precautions must be taken in mixing?
58. What nutrients must be guaranteed in the area where you live? Is it permissible to guarantee other nutrients? Which ones?
59. What information must be provided on (a) fertilizer labels and (b) on shipping bills or statements accompanying shipments of bulk fertilizers?

Selected References

Achorn, F. P., "Bulk blending of granular fertilizers." *Semin. ANDA*, São Paulo, Brazil, May 1970. Muscle Shoals, Ala.: Tennessee Valley Authority, 1970.

Achorn, F. P., "Application of fertilizer." *Meet. Far West Fert. Assoc.*, December 12–14, 1979. Muscle Shoals, Ala.: Tennessee Valley Authority, 1979.

Achorn, F. P., "Dry bulk impregnation: uniform application is essential with dry bulk impregnated fertilizer." *Fert. Prog.*, **13**(1):34 (1982).

Achorn, F. P., and T. R. Cox, "Production, marketing and use of solid, solution and suspension fertilizers," in R. A. Olson, T. J. Army, J. J. Hanway, and V. J. Kilmer, Eds., *Fertilizer Technology and Use*, 2nd ed., p. 381. Madison, Wis.: Soil Science Society of America, 1971.

Achorn, F. P., and N. L. Hargett, "NFSA and TVA survey of suspensions in 1971." Muscle Shoals, Ala.: Tennessee Valley Authority, 1972.

Achorn, F. P., and H. L. Kimbrough, "Uniform application of granular fertilizers." *TVA Fert. Conf. 1975. Bull. Y-96*. Muscle Shoals, Ala.: National Fertilizer Development Center–Tennessee Valley Authority, 1975.

Achorn, F. P., and H. A. Kittams, "Avoiding problems when producing suspensions from MAP or DAP." *Solutions*, **26**(4):42 (1982).

Achorn, F. P., and E. B. Wright, Jr., "Fertilizer spreading patterns and how to correct." *Bull. Y-68*. Muscle Shoals, Ala.: Tennessee Valley Authority, 1973.

Achorn, F. P., D. G. Salladay, and J. L. Greenhill, "Effects of high costs on fertilizer production." *Situation 80, TVA Fert. Conf. Bull. Y-157*, pp. 5–12. Muscle Shoals, Ala.: National Fertilizer Development Center–Tennessee Valley Authority, 1980.

Association of American Plant Food Control Officials, "Uniform State Fertilizer Bill (Tentative 1981)." *Official Publ. 35* (1982).

Bauer, H. J., and F. P. Achorn, "What's new in fertilizer bulk blending." *Situation 81, TVA Fert. Conf. Bull. Y-170. TVA/OACD-82/5*, pp. 41–44 (1981).

Bond, B. J., "The thrust of TVA's fertilizer program in the 1980's." *Situation 81, TVA Fert. Conf. Bull. Y-170. TVA/OACD-82/5*, pp. 5–10 (1981).

Bridges, J. D., "Fertilizer trends 1979." *Bull. Y-150*. Muscle Shoals, Ala.: National Fertilizer Development Center–Tennessee Valley Authority, 1980.

Davis, C. H., "New developments in U.S. chemical fertilizer technology." *180th Nat. Meet. Am. Chem. Soc.*, Las Vegas, Nev. (August 24–29, 1980).

Energy, Mines, and Resources Canada, "Phosphate rock, an imported mineral commodity." *Miner. Bull. MR 193*. Ottawa, Ontario: Canadian Government Publishing Centre, 1981.

Energy, Mines, and Resources Canada, "Potash, a proposed strategy." *Miner. Bull. MR 194*. Ottawa, Ontario: Canadian Government Publishing Centre, 1982.

Engineering and Technology Committee, "SGN: a system of materials identification." Ottawa, Ontario: The Canadian Fertilizer Institute, 1982.

Farm Chemicals, "Savings at production end will help ammonia dealers." *Farm Chem.*, **145**(6):54 (1982).

Fasullo, P. A., "United States outlook for sulphur recovery from hydrocarbons," in Raw Materials Committee II, *Sulphur and Gas & Ammonia*, pp. 21–33. Vancouver, B.C.: International Fertilizer Industry Association Ltd., September 8–10, 1982.

The Fertilizer Institute, *Fertilizer Reference Manual*. Washington, D.C.: The Fertilizer Institute, 1982.

"Fertilizers," in R. E. Kirk and D. F. Othmer, Eds., *Encyclopedia of Chemical Technology*, 2nd ed., Vol. 9, pp. 25–150. New York: Wiley-Interscience, 1966.

Follett, R. H., L. S. Murphy, and R. L. Donahue, *Fertilizers and Soil Amendments*. Englewood Cliffs, N.J.: Prentice-Hall, 1981.

Getsinger, J. G., G. A. Slappey, and M. F. Broder, "What's new in fluid fertilizers." *Situation 81, TVA Fert. Conf. Bull. Y-170. TVA/OACD-82/5*, pp. 25–40 (1981).

Gribbins, M. F., "Conversion of ammonia to fertilizer materials," in K. D. Jacob, Ed., *Fertilizer Technology and Resources in the United States*, pp. 63–84. Vol. 3 of *Agronomy*. New York: Academic Press, 1953.

Hackett, H., "Dry bulk impregnation, dealer experiences in the plant and field." *Fert. Prog.*, **13**(1):29 (1982).

Hargett, N. L., and J. T. Berry, "1980 fertilizer summary data." *Bull. Y-165. TVA/OACD-81/18* (1981).

Hoffmeister, G., "Quality control in a bulk blending plant." *Proc. TVA Fert. Bulk Blending Conf.*, August 1–2, 1973, *Reprint Z-49* (1973).

Hoffmeister, G., "Particle-size control requirements for granular fertilizers." *Conf. Canadian Fert. Inst.*, Ottawa, Ontario, March 3, 1981, *Reprint Z-126*. Muscle Shoals, Ala.: National Fertilizer Development Center–Tennessee Valley Authority, 1981.

Hoppe, R. W., "Phosphates are vital to agriculture—and Florida mines for one-third the world." *Eng. Min. J.*, **177**(5):79 (1976).

Hoppe, R. W., "From matrix to fertilizers: Florida's phosphate industry girds to produce over 50 million tpy." *Eng. Min. J.*, **177**(9):81 (1976).

International Fertilizer Development Center and United Nations Industrial Development Organization, *Fertilizer Manual*, Reference Manual IFDC-R-1. Muscle Shoals, Ala.: IFDC, 1979.

Jacob, K. D., Ed., *Fertilizer Technology and Resources in the United States*. Vol. 2 of *Agronomy*. New York: Academic Press, 1953.

Janzen, K., "Dry bulk impregnation, in use over 10 years." *Fert. Prog.*, **13**(1):28 (1982).

Jensen, D., and J. Pesek, "Inefficiency of fertilizer use resulting from nonuniform spatial distribution: I." *SSSA Proc.*, **26**:170 (1962).

Kilmer, V. J., S. E. Younts, and L. B. Nelson, Eds., *The Role of Potassium in Agriculture*. Madison, Wis.: American Society of Agronomy, Crop Science Society of America, and Soil Science Society of America, 1968.

Minister of Supplies and Services, "Office Consolidation, Fertilizers Act and Fertilizers Regulations." *Cat. YX75-F-9-1979*. ISBN 662-50859-9 (1979).

Murphy, L. S., L. J. Meyer, O. G. Russ, and C. W. Swallow, "Evaluating fluid fertilizer–herbicide combinations." *Fert. Solut.*, **17**(4):26 (1973).

National Fertilizer Solutions Association, *NFSA Additives Handbook*. Peoria, Ill.: National Fertilizer Solutions Association, 1979.

Nelson, W. L., and G. L. Terman, "Nature, behavior, and use of multinutrient (mixed) fertilizers," in *Fertilizer Technology and Usage*, pp. 379–427. Madison, Wis.: Soil Science Society of America, 1963.

Nielsen, K. F., and W. E. Janke, "World reserves of rock phosphate and the manufacture of phosphate fertilizers." *Proc. Western Canada Phosphate Symp.*, pp. 140–158. Edmonton, Alberta: Alberta Soil and Feed Testing Laboratory, 1980.

Olson, R. A., T. J. Army, J. J. Hanway, and V. J. Kilmer, Eds., *Fertilizer Technology and Use*, 2nd ed. Madison, Wis.: Soil Science Society of America, 1971.

Sargent, A., and G. Hoffmeister, "Quality control in bulk blending plant segregation in holding bins." *TVA Fert. Conf. 1975. Bull. Y-96*. Muscle Shoals, Ala.: National Fertilizer Development Center–Tennessee Valley Authority, 1975.

Sauchelli, V., Ed., *Fertilizer Nitrogen. Its Chemistry and Technology.* ACS Monogr. 161. New York: Reinhold, 1964.

Slack, A. V., "Liquid fertilizers," in V. Sauchelli, Ed., *Chemistry and Technology of Fertilizers*, pp. 513–537. ACS Monogr. 148. New York: Reinhold, 1960.

Stowasser, W. F., *Phosphate.* Mineral Commodity Profiles. Pittsburgh, Pa.: U.S. Dept. of the Interior, Bureau of Mines, 1979.

Stowasser, W. F., "The United States Phosphate Rock Industry Resources and Supply/Demand Outlook," in Raw Materials Committee I, *Phosphate Rock and Potash*, pp. 32–46. Vancouver, B.C.: International Fertilizer Industry Association Ltd., September 8–10, 1982.

Striplin, M. M., D. McKnight, and T. P. Hignett, "Compound fertilizers from rock phosphate, nitric and sulfuric acid, and ammonia." *Ind. Eng. Chem.,* **44:**236 (1952).

Tennessee Valley Authority, *Proc. TVA Bulk Blending Conf.* (1973).

Tuller, W. N., Ed., *The Sulphur Data Book,* New York: McGraw-Hill, 1954.

Turbeville, W. J., Jr., "The phosphate rock situation." *Phosphate–Sulphur Symp.,* Tarpon Springs, Fla. (1974).

United States Department of Agriculture, *1980 Fertilizer Situation.* FS-10. Washington, D.C.: Economics, Statistics, and Cooperatives Service, 1979.

Chapter 11

Soil Acidity and Liming

LIMING, as the term applies to agriculture, is the addition to the soil of any calcium or calcium- and magnesium-containing compound that is capable of reducing acidity. Lime correctly refers only to calcium oxide (CaO), but the term almost universally includes such materials as calcium hydroxide, calcium carbonate, calcium-magnesium carbonate, and calcium silicate slags. This chapter is devoted to a discussion of the theory of soil acidity, the practice of liming, and to some of the materials employed for this purpose.

What Is Acidity?

An acid is a substance that tends to give up protons (hydrogen ions) to some other substance. Conversely, a base is any substance that tends to accept protons. This concept of acids and bases was developed by Brönsted and Lowry and is generally accepted as describing the behavior of materials that yield or gain protons in all liquid media, including water. The earlier theory of Arrhenius described the behavior of compounds in water, and an acid was accordingly defined as any substance yielding hydrogen ions (protons) when dissolved in water. A base was defined as any substance yielding hydroxyl ions when dissolved in water. The second definition is a special case covered, of course, by the broader concept of an acid and base defined by the Brönsted–Lowry theory. Soil acids are aqueous systems, and therefore the terms hydrogen and hydroxyl ions rather than proton, proton donor, and proton acceptor are employed in this book.

Active and Potential Acidity.
The characteristics of acid solutions are based on the activity of the hydrogen ion (H^+). An acid when mixed with water dissociates or ionizes into hydrogen ions and the accompanying anions, as represented by the hypothetical acid HA:

$$HA \overset{H_2O}{\rightleftharpoons} H^+ + A^-$$

The H^+ ions to the right indicate *active* acidity, and the more the reaction tends toward that direction, the greater the activity of the H^+ and the stronger the acid is said to be. The HA on the left side of the equation is the *potential* acidity.

Strong and Weak Acids.
Acids are arbitrarily classified according to the extent to which they dissociate in water. If the dissociation is great, the acid is said to be strong; well-known examples are nitric, sulfuric, and hydrochloric. Acids that dissociate to only a slight extent, examples of which are acetic, carbonic, and boric, are termed weak acids.

Expressions of Acid Concentration. The total acidity of a solution is the sum of the concentrations of the active and potential acidities. As an example, suppose that a solution of an acid is 0.099 M with respect to its active acidity and 0.001 M with respect to its potential acidity. The total acid concentration is 0.100 M and would certainly be classed as strong. The foregoing is illustrative of most strong acids in dilute solution. The activity of the H^+ is so nearly equal to the concentration of the total acidity that an expression of the latter is for all practical purposes an automatic expression of the former. In other words, there is little need for a separate designation of the active and total acidity of dilute solutions of strong acids.

With weak acids, however, a knowledge of the total concentration is of no value in predicting the activity of the H^+ ions unless the ionization constant is known. Many weak acids are dissociated to less than 1%. Assume that the total concentration of a weak acid HA is 0.1 M and that it is 1% dissociated. This means that the activity of the H^+ in such a solutions is $0.1 \times 0.01 = 0.001$ M. Obviously, a measure of the total acidity gives no indication whatever of the active acidity.

The pH Concept. There are means of determining the H^+ ion activity of solutions, and this activity can be expressed as is the case with the strong acids. But with extremely weak acids this mode of expression becomes inconvenient and the H^+ ion activity is generally stated in terms of pH values. This connotation was developed a number of years ago by Sorenson, a Swedish chemist, and has now been almost universally adopted for describing the H^+ ion activity of very dilute acid solutions. The pH is defined as the logarithm of the reciprocal of the H^+ ion activity, or symbolically,

$$pH = \log \frac{1}{A_{H^+}} = -\log A_{H^+}$$

where A_{H^+} is the hydrogen ion activity in moles per liter. A solution with an H^+ ion activity of 0.001 M will have a pH of 3.0; one with an H^+ ion activity of 0.0001 M, 4.00, and so on. The pH connotation is an integral part of the terminology employed in soil fertility and should be thoroughly understood.

Neutralization. Quite important in any consideration of acids and bases is neutralization, the reaction of an acid with a base to form a salt and water. This reaction is represented by the following equation, in which HA and BOH are the hypothetical acid and base:

$$HA + BOH \rightarrow BA + HOH$$

If a given quantity of acid is titrated with a base and the pH of the solution is determined at intervals during the titration, a curve is obtained by plotting pH values against the amounts of base added. Titration curves so obtained for strong and weak acids differ markedly, as can be seen in Figure 11-1. Neutralization of soil acidity is brought about by liming.

$$CH_3COOH \rightleftharpoons H^+ + CH_3COO^-$$
$$CH_3COONa \rightleftharpoons Na^+ + CH_3COO^-$$

Buffer Mixtures. Buffers or buffer systems are compounds that can maintain the pH of a solution within a narrow range when small amounts of acid or base are added. The term *buffering*, then, defines the resistance to a change in pH. An example of a commonly known buffer system is acetic acid and sodium acetate. The salt is essentially dissociated and gives rise to a large concentration of acetate ions, which, by mass action, suppress the dissociation of the acetic acid because it tends to form the undissociated acetic acid. If an acid such as hydrochloric is added to this system in small quantities, the excess of acetate ions forms molecular acetic acid, with the net result that there is little change in the active acidity of the solution. Conversely, if sodium hydroxide is added, the hydroxyl ions are neutralized by the hydrogen ions forming water. Because of the large supply of undissociated acetic acid, the equilibrium shifts to the right, replacing the hydrogen ions removed by the preceding neutralization. The net effect is, again, only a slight change in the active acidity of the solution, hence only a small change in the pH. The resistance offered by this system to changes in pH, even though acids or bases may be added, is termed buffering. This brief review of weak acids and buffering has been made only because soils behave in many respects like buffered weak acids. In the soil, however, humus and aluminosilicate clays, with their ability to retain aluminum and other cations, act as the buffer medium.

Soil Acidity

Soil acidity has been considered seriously by soil scientists for more than fifty years, and numerous concepts have been advanced to explain its observed patterns and behavior. In fact, these explanations have almost completed a cycle, a point made very clearly by Hans Jenny of California in an excellent review paper, "Reflections on the Soil Acidity Merry-Go-Round." This article is included in the list of references at the end of the chapter and is recommended reading for those interested in the development of ideas relating to this important soil property.

Nature of Soil Acidity. According to generally held views, acidity in soils has several sources: humus or organic matter, aluminosilicate clays, hydrous oxides of iron and aluminum, exchangeable aluminum, soluble salts, and carbon dioxide.

Humus. As mentioned in Chapter 4, soil organic matter or humus contains reactive carboxylic and phenolic groups that behave as weak acids. They will dissociate, releasing H^+ ions, depending on the dissociation constant of the acid formed and pH conditions. The heterogeneity of soil organic matter is such that it varies from place to place, and its contribution to soil acidity will vary accordingly.

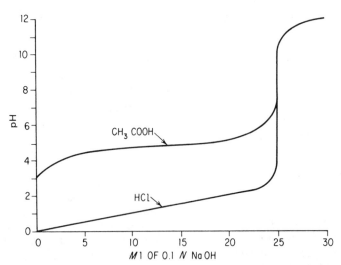

FIGURE 11-1. Titration of 0.10 N CH_3COOH and 0.10 N HCl with 0.10 N NaOH.

However, it is a significant factor, particularly in peat and muck soils and in mineral soils containing large amounts of organic matter.

Aluminosilicate Clay Minerals. These minerals are the two- and three-layer clays typified by kaolinite and montmorillonite. The charge on these clays originates from the isomorphous substitution in the crystal lattice of a cation of lower valence for a cation of higher valence. Charges on clays may also originate from the dissociation of hydrogen ions from hydroxyl groups or from bound water of constitution, both of which are structural components of the crystal lattice.

It was mentioned in Chapters 4 and 6 that the total charge on soil colloidal materials can be separated into two categories, one of which is called the permanent charge and the other termed the pH-dependent or variable charge. The mechanisms of charge generation are illustrated by the following reactions:

On sesquioxides (pH dependent):

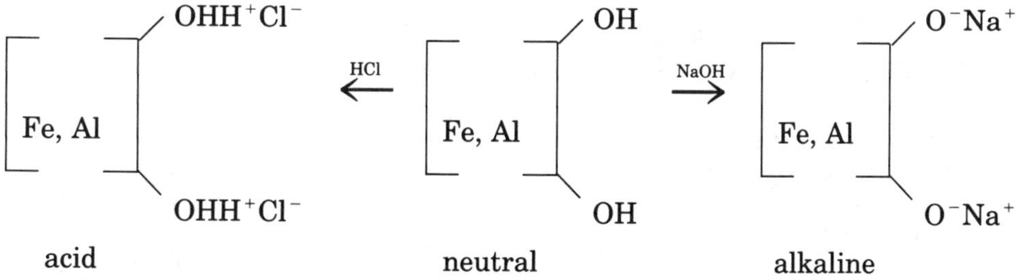

$$
\left[\text{Fe, Al} \Big\langle \begin{array}{c} \text{OHH}^+\text{Cl}^- \\ \text{OHH}^+\text{Cl}^- \end{array} \right] \quad \xleftarrow{\text{HCl}} \quad \left[\text{Fe, Al} \Big\langle \begin{array}{c} \text{OH} \\ \text{OH} \end{array} \right] \quad \xrightarrow{\text{NaOH}} \quad \left[\text{Fe, Al} \Big\langle \begin{array}{c} \text{O}^-\text{Na}^+ \\ \text{O}^-\text{Na}^+ \end{array} \right]
$$

acid   neutral   alkaline

On clay minerals (pH dependent):

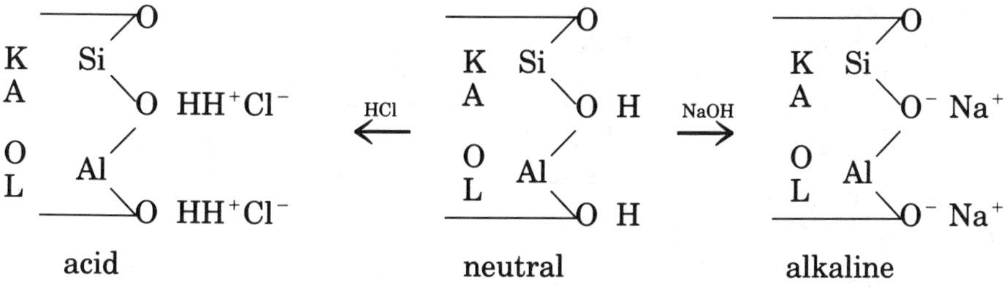

acid   neutral   alkaline

On organic matter (pH dependent):

$$
\text{R}-\text{COOH} \xrightarrow{\text{NaOH}} \text{R}-\text{COO}^- \;\; \text{Na}^+
$$
acid   alkaline

$$
\text{R}-\text{OH} \xrightarrow{\text{NaOH}} \text{R}-\text{O}^- \;\; \text{Na}^+
$$
acid   alkaline

On clay minerals (permanent due to isomorphous substitution):

Mg^{2+} replacing Al^{3+} in octahedral layer
Al^{3+} replacing Si^{4+} in tetrahedral layer

Most soil scientists recognize that acid soil development is due to the presence of exchangeable Al^{3+} combined with one or more of the following factors: depletion of basic cations such as calcium, magnesium, potassium, and sodium by leaching and crop removal; decomposition of organic residues; application of fertilizers, particularly ammoniacal nitrogen sources; and atmospheric acidity (e.g., acid rain).

Aluminum and Iron Polymers. The Al^{3+} ions displaced from clay minerals by cations are hydrolyzed to monomeric and polymeric hydroxyaluminum complexes. Hydrolysis of the monomeric forms is illustrated by the following stepwise reactions, which in each case liberate H^+ and lower pH unless there is a source of OH^- to neutralize the released H^+ ions. Each successive step occurs at a higher pH.

$$Al(H_2O)_6^{3+} + H_2O \rightleftharpoons Al(OH)(H_2O)_5^{2+} + H_3O^+$$

$$Al(OH)(H_2O)_5^{2+} + H_2O \rightleftharpoons Al(OH)_2(H_2O)_4^+ + H_3O^+$$

$$Al(OH)_2(H_2O)_4^+ + H_2O \rightleftharpoons Al(OH)_3(H_2O)_3^0 + H_3O^+$$

$$Al(OH)_3(H_2O)_3^0 + H_2O \rightleftharpoons Al(OH)_4(H_2O)_2^- + H_3O^+$$

These aluminum hydrolysis products can be readsorbed by the clay minerals causing further hydrolysis. H^+ ions resulting from the hydrolysis of aluminum and iron compounds react with and dissolve or decompose soil minerals. Further, the addition of salts, such as those contained in fertilizers, to sesquioxide-coated interlayered minerals increases the hydrolysis of nonexchangeable iron and aluminum resulting in an increase in the hydrogen-ion concentration of the soil solution, and hence a lower pH.

If a base is added to this soil, the H^+ ions will be neutralized first. When more of the base is added, the aluminum hydrolyzes with the production of H^+ ions in amounts equivalent to the aluminum present. At pH values below 4.7 the Al^{3+} ion is predominant, while $Al(OH)_2^+$ is the principal species between pH 4.7 and 6.5. Between pH 6.5 and 8.0, $Al(OH)_3^0$ is the major species and $Al(OH)_4^-$ is most prevalent at pH values above 8.0. The relationship between pH and the occurrence of these soluble aluminum species is shown in Figure 11-2. It should be noted that insoluble $Al(OH)_3$ will be precipitated throughout this pH range whenever its solubility product is exceeded.

Because of the amphoteric nature of aluminum in soil it can function as either

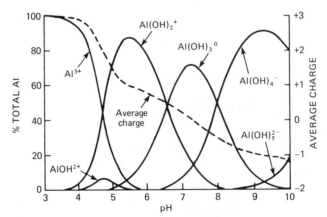

FIGURE 11-2. Relationship between pH and the distribution and average charge of soluble aluminum species. [Marion et al., *Soil Sci.*, **121**:76 (1976).]

an acid or a base, as illustrated in the following equations:

Soil aluminum as a base:

$$Al(OH)_3 + H^+Cl^- \rightleftharpoons Al(OH)_2^+ \; Cl^- + H_2O$$

$$Al(OH)_2^+ \; Cl^- + H^+Cl^- \rightleftharpoons Al(OH)^{2+} \; 2Cl^- + H_2O$$

$$Al(OH)_2^{2+} \; 2Cl^- + H^+Cl^- \rightleftharpoons Al^{3+} \; 3Cl^- + H_2O$$

$$Al(OH)_3 + 3H^+3Cl^- \rightleftharpoons Al^{3+} \; 3Cl^- + 3H_2O$$

Soil aluminum as an acid:

$$Al^{3+} \; 3X^- + NaOH \rightleftharpoons Al(OH)^{2+} \; 2X^- + Na^+X^-$$

$$Al(OH)^{2+}2X^- + NaOH \rightleftharpoons Al(OH)_2^+ \; X^- + Na^+X^-$$

$$Al(OH)_2^+ \; X^- + NaOH \rightleftharpoons Al(OH)_3 + Na^+X^-$$

Soil aluminum as an anion:

$$Al(OH)_3 + NaOH \rightleftharpoons Al(OH)_4^- \; Na^+$$

The monomeric hexaquoaluminum [$Al(H_2O)_6^{3+}$] ion is exchangeable. However, because of its trivalent charge it will be retained strongly by many soil colloids.

The hydroxy aluminum ions described in the previous equations tend to polymerize rapidly to form large, multicharged units. Such polymerization is favored in the presence of soil colloid surfaces. The mechanism for their formation is the sharing of hydroxyl groups by adjacent aluminum ions as represented in the following equations:

$$\begin{bmatrix} & OH & \\ (H_2O)Al & & \\ & H_2O & \end{bmatrix}^{2+} + \begin{bmatrix} H_2O & & \\ & Al(H_2O)_4 & \\ HO & & \end{bmatrix}^{2+} \longrightarrow$$

$$\begin{bmatrix} & OH & \\ (H_2O)_4Al & & Al(H_2O)_4 \\ & OH & \end{bmatrix}^{4+} + 2H_2O$$

They often age, releasing H^+,

$$\begin{bmatrix} & OH & \\ (H_2O)_4Al & & Al(H_2O)_4 \\ & OH & \end{bmatrix}^{4+} + 2H_2O \underset{slow}{\rightleftharpoons}$$

$$\begin{bmatrix} & O & \\ (H_2O)_4Al & & Al(H_2O)_4 \\ & O & \end{bmatrix}^{2+} + 2H_3O^+$$

or polymerize further and become firmly bound to colloid surfaces. These multicharged polymers are positively charged and are essentially nonexchangeable.

Cation exchange capacities of the soil colloids can be affected by the formation of these positively charged polymers on their surfaces. At high pH values the polymeric positive charges become less and can even become negative as a result of the dissociation of protons. As a result, cation exchange capacities of the soil colloid–polymer complex will increase. Decreasing pH in soils with large amounts of these aluminum polymers closely associated with soil colloids will lower soil cation exchange capacities by increasing the positive charge on the polymers.

Iron hydrolysis, as shown in the following equation, is similar to that of aluminum. Although this reaction is more acidic than the corresponding hydrolysis of aluminum, the acidity that develops is buffered by the reactions of aluminum. Thus iron hydrolysis will have little effect on soil pH until most of the soil's substantial reserves of aluminum have reacted:

$$Fe(H_2O)_6{}^{3+} + H_2O \rightleftharpoons Fe(OH)(H_2O)_5{}^{2+} + H_3O^+$$

These aluminum and iron polymers may occur in amorphous or crystalline colloidal form as coatings on clay and other mineral particles. They are also held between the lattices of expanding soil minerals, preventing collapse of these lattices as water is removed during drying or freezing. The occurrence of interlayer groups of aluminum and iron polymers and their involvement in potassium availability were discussed in Chapter 7.

Soluble Salts. The presence of salts—acid, neutral, or basic—in the soil solution is accounted for by mineral weathering, organic-matter decomposition, or their addition as fertilizer compounds. The cations of these salts will displace adsorbed aluminum and cause an increase in the acidity of the soil solution, which is easily measured in soil pH determinations. Divalent cations usually have a greater effect on lowering soil pH than monovalent metal cations.

The addition of fertilizer in a band will result in a high soluble-salt concentration in and immediately surrounding this band. This, in turn, will bring aluminum, iron, and manganese into solution, resulting, through hydrolysis, in a lowering of the soil pH. With high rates of band-applied fertilizer, this effect on soil pH in the band can be appreciable. In soils where the pH is marginal, further acidification may be detrimental to plant growth.

Carbon Dioxide. In soils near neutrality or those containing appreciable quantities of carbonates or bicarbonates the pH is influenced to a great extent by the partial pressure of the carbon dioxide in the soil atmosphere. The pH of a soil containing free calcium carbonate and in equilibrium with carbon dioxide at the pressure of normal aboveground air is 8.5. If the carbon dioxide pressure of the soil atmosphere in such a system increases to 0.02 atm, the pH will drop to about 7.5.

Decomposition of organic residues will result in formation of carbon dioxide, which combines with water to form carbonic acid. Dissociation of this weak acid into H^+ and $HCO_3{}^-$ will thus provide another source of H^+ for soil acidification. Root activity and metabolism may also serve as sources of carbon dioxide.

Factors Affecting Soil Acidity. Soil acidity in crop production systems is influenced by factors such as (1) use of commercial fertilizers, especially ammoniacal sources which produce H^+ ions during nitrification; (2) crop removal of basic cations, including calcium, magnesium, potassium, and sodium in exchange for H^+; (3) leaching of these cations being replaced first by H^+ and subsequently by Al^{3+}; and (4) decomposition of organic residues.

Acidity and Basicity of Fertilizers. Some fertilizer materials leave an acid residue in the soil, others a basic residue, and still others seemingly have no influence on the soil pH. Results of numerous experiments have shown that among the plant nutrients nitrogen, phosphorus, and potassium, carriers of nitrogen have the greatest effect on both the soil pH and the loss of cations by leaching. The acidity resulting from ammoniacal nitrogen sources is discussed in Chapter 5.

Phosphoric acid released from dissolving phosphate fertilizers such as triple superphosphate (monocalcium phosphate) and monoammonium phosphate can temporarily acidify localized zones at the site of application. The former material will lower pH to as low as 1.5, while the latter will decrease pH to approximately 3.5. Their acidity, however, is rapidly neutralized, but the acidic reaction products may remain to influence soil properties.

Diammonium phosphate will initially raise pH of the soil affected by fertilizer treatment to about 8. Although this change is short-lived in very acid soils, the resulting basic phosphate compounds may persist for some time. Acidity produced by the nitrification of the ammonium component of diammonium phosphate will generally offset this initial basic effect. Considerable acidic effect also results from the nitrification of monoammonium phosphate.

A method for determining the acidity or basicity of fertilizers was developed by Pierre in 1933. His method is based on the assumption that:

1. The acid-forming effect of fertilizer is caused by all of the contained sulfur and chlorine, one-third of the phosphorus, and one-half of the nitrogen.
2. The presence of calcium, magnesium, potassium, and sodium in fertilizer will increase the lime status, hence the pH of fertilized soil.

He assumed that half of the applied fertilizer nitrogen was taken up in nitrate form and accompanied by equivalent amounts of basic cations, such as K^+, Ca^{2+}, Mg^{2+}, or Na^+. Uptake of the other half of nitrate was believed to be associated with H^+ as the counterion or exchanged for HCO_3^- from plant roots. Although not apparently considered by Pierre, significant amounts of NH_4^+ can be utilized directly by plants, with resultant acidification of soil close to roots.

According to Pierre's calculations, 1.8 lb of chemically pure calcium carbonate would be required to neutralize the acidity resulting from the addition of each pound of ammoniacal nitrogen. Sources of nitrogen such as sodium nitrate or calcium nitrate would leave a basic residue due to the accompanying basic cation. This method, modified for estimating the equivalent acidity or basicity of complete fertilizers, was adopted by the Association of Official Agricultural Chemists (AOAC) and is currently recognized as the official procedure.

Pierre has demonstrated that plant growth reduces the potential or theoretical amount of acidity produced by nitrification of ammoniacal fertilizer materials. This reduction is caused by unequal absorption of inorganic cations and anions by the crop or crop part that is removed by harvesting.

Another researcher, Andrews, has also studied the acidic and basic effects of fertilizers on soils. He makes no allowances for the crop effects referred to above. Thus his lime equivalent values for neutralization of the acidity formed by fertilizer materials are higher than those of Pierre. They both maintain that each pound of fertilizer nitrogen as NH_3 will require 3.57 lb of CP calcium carbonate to neutralize the acidity if converted to the nitrate form. Also every pound of nitrogen leached from the soil as the nitrate (NO_3) takes with it 3.57 lb of $CaCO_3$ or its equivalent in basic cations. He has accordingly calculated the amounts of limestone he maintains will be required to neutralize the acidity formed by the various sources of nitrogen. These values, along with those determined by the Pierre method, are shown in Table 11-1. The figures currently listed for the acidity or basicity of mixed fertilizers and straight goods are determined by the official AOAC procedure, which is that of Pierre.

Regardless of which method, whether that of Pierre or Andrews, more closely estimates the acidifying effects of the different forms of fertilizer nitrogen, the fact remains that some materials are acid forming and others are not. The importance

TABLE 11-1.   Equivalent Acidity and Basicity of Nitrogenous Fertilizer Materials According to Andrews and Pierre *

Material	Percent Nitrogen	Pure Lime Necessary to Make Lime Salts †		Pounds of Pure Lime ** For Neutralization	
		Per lb of Nitrogen	Per 100 lb of Material	Per lb of Nitrogen	Per 100 lb of Material
Inorganic sources of nitrogen					
Sulfate of ammonia	20.5	7.14	146	5.35	110
Ammo-phos A	11.0	6.77	74	5.00	55
Anhydrous ammonia	82.2	3.57	293	1.80	148
Calcium nitrate	15.0	0.42	6	1.35B	20B
Calnitro	16.0	0.66	11	1.31B	21B
Calnitro	20.5	1.77	36	0	0
Crude nitrogen solution	44.4	2.98	132	1.20	53
Nitrate of soda	16.0	0.00	0	1.80B	29B
Potassium nitrate	13.0	0.00	0	2.00B	26B
Manufactured organic nitrogen					
Cyanamid	22.0	1.18B	26B	2.85B	63B
Urea	46.6	3.57	166	1.80	84
Urea-ammonia liquor	45.5	3.57	162	1.80	82
Natural organic nitrogen					
Cocoa shell meal	2.7	2.37	6	0.60B	2B
Castor pomace	4.8	2.67	13	0.90	4
Cottonseed meal	6.7	3.17	21	1.40	9
Dried blood	13.0	3.52	46	1.75	23
Fish scrap	9.2	2.67	25	0.90	8
Fish scrap	8.9	1.78	16	0.01	0
Guano, Peruvian	13.8	2.72	38	0.95	13
Guano, white	9.7	2.22	21	0.45	4
Milorganite	7.0	3.47	24	1.70	12
Tankage, animal	9.1	1.92	17	0.15	1
Tankage, garbage	2.5	0.93B	2B	2.70B	7B
Tankage, high grade	8.4	2.52	21	0.75	6
Tankage, low grade	4.3	5.43B	23B	7.20B	31B
Tankage, packing house	6.0	0.12	1	1.65B	10B
Tankage, process	7.4	3.32	25	1.55	12
Tobacco stems	1.4	16.03B	22B	17.80B	25B
Tobacco stems	2.8	2.53B	7B	4.30B	12B
Sources of potash					
Manure salts	0	0	0	0	0
Muriate of potash	0	0	0	0	0
Potassium nitrate	13.0	0	0	2.00B	26B
Sulfate of potash	0	0	0	0	0
Sulfate of potash – magnesia	0	0	0	0	0
Sources of phosphorus					
Ammo-phos A	11.0	6.77	74	5.00	55
Precipitated bone	0	0	0	0	29B
Superphosphate	0	0	0	0	0
Triple superphosphate	0	0	0	0	0

* B, lime in excess of that required to make neutral salts or neutral fertilizers.

† Data to make lime salts from organic sources of nitrogen were obtained by adding 1.77 lb per pound of nitrogen to data for neutral fertilizers.

** Official method for neutralizing fertilizers.

Source: Andrews, *The Response of Crops and Soils to Fertilizers and Manures,* 2nd ed. Copyright 1954 by W. B. Andrews.

of this property fades into insignificance in a well-run farming enterprise. In such an operation the maintenance of proper soil pH in an adequate liming program is mandatory, and in such an operation the factors determining the choice of fertilizer nitrogen are its applied cost, market availability, and ease of application.

Removal of Basic Cations. Loss of basic cations from soils by cropping and leaching is the reason that they become acidic. Crop uptake of such cations, as will be explained later, can either reduce or increase the soil acidity produced by nitrification of ammonium from all sources, including fertilizers, crop and animal wastes, or in organic matter. A priori reasoning would lead to the conclusion that nitrate sources carrying a basic cation should be less acid forming than ammoniacal sources. Further, ammoniacal sources that carry an acidic anion such as SO_4^{2-}, which is not absorbed as rapidly as the NO_3^- ion, would be more acidic than a material such as ammonia or urea. Experimental evidence has shown this to be the case.

Pierre and his associates reviewed the factors that contribute to the acidity or basicity of fertilizer nitrogen materials. Greenhouse studies showed that in uncropped soil the acidity developed by ammonium nitrate was about equal to the theoretical value, provided that soluble salts were removed from the soil before the pH measurements were made.

The studies of Pierre and his co-workers showed that deviations in acidity from the theoretical values were related to crop types and parts of the crop harvested. Their results showed that these variations were quantitatively explained by differences in the numbers of chemical equivalents of nitrogen and excess bases taken up by the plant or plant part, excess bases (EB) being defined as total cations $(Ca^{2+}, Mg^{2+}, K^+,$ and $Na^+)$ minus total anions $(Cl^-, SO_4^{2-}, NO_3^-,$ and $H_2PO_4^-)$.

Plants with an EB/N ratio below 1.0 decrease the acidity formed by nitrification, whereas those with a ratio above this value increase it. This ratio was determined in a total of 149 samples taken at different stages of growth of 26 crop species grown under widely different conditions. Only a few crops—buckwheat, tobacco, and spinach—had values slightly above 1.0. Cereal and grass crops had average ratios of 0.43 and 0.47, respectively, meaning that only 43 and 47%, respectively, of the nitrogen they received was acid forming. The EB/N ratio of different crop parts also varied greatly.

Westerman at Oklahoma State University calculated the effect of various aspects of a winter wheat production system on the acidity produced by the application of ammoniacal nitrogen. His values are presented in Table 11-2. Theoretical potential acidity from nitrogen fertilization was reduced 77 and 97%, respectively, due to grazing and grain removal. Straw removal without grain or forage harvests increased the theoretical acidity resulting from nitrogen fertilization by 120%. Theoretical potential acidity was reduced 87, 64, and 70% by combinations of practices listed in order; grazing plus grain removal, grain plus straw removal, and grazing plus both grain and straw removal.

Volatilization of Nitrogen. Volatilization of NH_3 will reduce nitrification and the resultant development of soil acidity. Denitrification losses of N_2, NO, or N_2O will leave a basic residue due to the formation of OH^- ions equivalent to the amount of NO_3^- reduced.

Acidulation with Sulfur. Sulfuric acid formed from elemental sulfur and its compounds will acidify soils. This topic was discussed at considerable length in Chapter 8, as was the occurrence of acid precipitation, which is a significant acidifying agent in some areas.

TABLE 11-2.   Effect of Winter Wheat Production on the Theoretical Potential Acidity Resulting from the Application of Ammoniacal Nitrogen

Fertilizer and Crop Removal System	lb/A			Meq/100 g			100% ECCE Lime Required for Neutralization (lb/A)	Theoretical Potential Acidity (%)
	Yield of Dry Matter	N Removed	Excess Base * Removed	N	EB	EB/N		
200 lb N—fallow	—	—	—	—	—	—	360	100
200 lb N + grazing †	2000	80	65	286	66	0.23	82	23
200 lb N + grain	3600	72	10	143	5	0.03	11	3
200 lb N + straw	4000	28	118	50	60	1.20	432	120
200 lb N + grazing + grain	5600	152	75	194	25	0.13	47	13
200 lb N + grain + straw	7600	100	128	94	33	0.36	130	36
200 lb N + grazing + grain + straw	9600	180	193	134	40	0.30	108	30

* Expressed as $CaCO_3$.
† Calculations for grazing are for removal and do not include basic cations and nitrogen returned to the soil in manure.
Source: Westerman, *Solutions*, **25**(3):64 (1981).

The Soil As a Buffer

The foregoing discussion suggests that soil acidity is influenced by numerous factors, not the least of which is the presence of aluminum ions on the exchange complex and their hydrolysis in the soil solution to variously charged polymers with simultaneous release of H^+ ions. It was also pointed out that the soil behaves like a buffered weak acid and that it will resist sharp changes in pH accordingly. This buffering mechanism is explained in the following somewhat oversimplified discussion.

For the purpose of this discussion, it can be assumed that the clay micelle is a large acid radical which behaves as a weak acid when saturated predominantly with Al^{3+} ions. The adsorbed aluminum ions will maintain an equilibrium with aluminum ions in the soil solution, the hydrolysis of which gives rise to H^+ ions in the solution as indicated in the following equation:

$$Al^{3+} + H_2O \longrightarrow AlOH^{2-} + H^+$$

If these H^+ ions are then neutralized by the addition of small amounts of a base and the aluminum ions in solution are precipitated as $Al(OH)_3$, the equilibrium of the system, hence the pH, will tend to be maintained by the movement of adsorbed aluminum ions to the soil solution. These aluminum ions in turn hydrolyze, producing H^+ ions, and the pH tends to remain as it was before the addition of the base.

As more base is added to the system, however, the above-given reaction continues with more of the adsorbed aluminum being neutralized and replaced on the soil colloid with the cation of the added base. As a result, there is a gradual increase in the pH of the system rather than the abrupt change in pH that is characteristic of the neutralization of unbuffered systems or of strong acids such as HCl.

The reverse of the reaction described above also occurs. As acid is continually added to a neutral soil, OH^- ions in the soil solution are neutralized. Gradually, as these OH^- ions are consumed by the added H^+ ions, the $Al(OH)_3$ dissolves, enters the soil solution, and gradually replaces the basic cations held on the soil clay. As this change takes place, there is a continual but slow decrease in soil pH as the Al^{3+} replaces the adsorbed basic cations.

The total amount of clay and organic matter in a soil and the nature of the clay minerals will determine the extent to which soils are buffered. Soils containing large amounts of mineral clay and organic matter are said to be highly buffered and require larger amounts of added lime to increase the pH by any given number of units than do soils with a lower buffer capacity. Sandy soils with small amounts of clay and organic matter are poorly buffered and require only small amounts of lime to effect a given change in pH. As a general rule, soils containing large amounts of the 1:1 type clays (Ultisols and Oxisols) are generally less strongly buffered than soils in which the predominant clay minerals are of the 2:1 type (Alfisols and Mollisols).

Determination of Active and Potential Acidity in Soils

Active Acidity. Active soil acidity can be determined in several ways. Indicator dyes are frequently employed, particularly in rapid soil test methods. Unless they are handled by a skilled operator, however, considerable error can result. Currently, the most accurate and most widely used method employs a pH meter, or glass electrode potentiometer, as it is sometimes called.

Soil pH_w is measured by placing a suspension of soil and distilled water in contact with the glass electrode of a pH meter and reading the result on the dial. In soil-water suspensions a phenomenon known as the suspension effect is sometimes observed. If the soil-water suspension is stirred and a reading is taken, the needle will drift to a higher pH_w value as the suspension settles. As soil colloids behave as weak acids, in which the unionized phase is made up of solids containing surface-adsorbed acidity, the presence of a solid phase may be expected to give a lower pH_w value when in intimate contact with the electrode. This is exactly what is shown, for the pH_w of the suspension is always lower than the pH_w of the supernatant liquid.

Not all workers agree that the suspension effect is produced by the mechanism just suggested. Some feel that it is caused by a junction potential resulting from the salt bridge used with the electrodes attached to the pH meter. The successful measurement of pH by a glass electrode potentiometer requires that the transference of K^+ and Cl^- ions from the salt bridge take place at the same rate. If the salt bridge were in contact with the clay, there would be a more rapid diffusion of K^+ than Cl^-, for the K^+ would be attracted to the negatively charged clay. This would give rise to a potential difference above that resulting from the H^+ ion activity. The thicker the suspension around the electrode and the salt bridge, the greater the junction potential. To get around this difficulty the glass electrode may be placed so that it will come in contact with the soil suspension and the salt bridge will remain in the supernatant liquid.

The presence of soluble salts influences soil pH, and to obtain a true estimate of the acidity of the soil these salts must be removed before a pH determination. This is usually done by leaching with distilled water and then measuring the pH of the salt-free soil, a time-consuming operation when routine determinations on many samples are required.

Another approach used in correcting the acidifying effects of salts is to add a salt solution to the soil instead of water. At first this may seem to be an absurd approach—eliminating the effect of salt by adding a salt solution! The trouble with salt, however, is that it can never be determined just how much is present in the soil and whether the pH measured reflects the true acidity of the soil. Therefore, a 0.01 M KCl or $CaCl_2$ solution is added to the soil and the pH_s is measured. Such pH values, of course, are lower than those made with water, but the salt content of the soil solution is negligible contrasted with that of the added salt solution. Therefore the effect of the former will be of negligible effect in comparison to the effect of the latter on the pH_s of suspension. This means that differences in soil pH caused by differences in the salt concentration of the soil solution will have no effect on the pH_s measured in the added soil-salt solution suspension, which is then a more precise estimate of the acidity status of the soil than that measured in a soil-water suspension. Of course, the soil could be leached with distilled water, but it is simpler to measure the pH_s of the soil in the added salt solution.

Workers in Great Britain, who used this method, have calculated a lime potential for soils expressed by the formula

$$pH - \tfrac{1}{2}p(Ca + Mg)$$

It has been shown that this value for a particular soil is fairly constant over a wide range of soil-salt levels.

Soil pH is a useful indicator of the presence of exchangeable aluminum and hydrogen ions. Exchangeable hydrogen is normally present in measurable quantities at pH values only below 4, while exchangeable aluminum customarily occurs in significant amounts at pH levels lower than about 5.5. Multicharged aluminum polymers are dominant in the pH range 5.5 to 7.0.

Potential Acidity. Soil pH is an excellent single indicator of general soil conditions. However, as a measure of active acidity in a medium that behaves as a weak acid, it gives no indication of the amount of lime to be applied. Potential acidity must then be considered. Some method of relating a change in soil pH to the addition of a known amount of acid or base is necessary and this is termed a lime-requirement determination.

The lime requirement of a soil is related not only to the soil pH but also to its buffer or cation exchange capacity. As indicated previously, some soils are more highly buffered than others, and a lime requirement determined for one soil will in all probability not be the same as that determined for another. The buffer or exchange capacity is related to the amount of clay and organic matter present: the larger the amount, the greater the buffer capacity. Hence soils classed as clays, peats, and mucks have higher buffer capacities and, if acid, will have a high lime requirement. Coarse-textured soils with little or no organic matter will have a low buffer capacity and, even if acid, will have a low lime requirement.

The implications of the statements made in the preceding paragraph are of tremendous practical importance. The indiscriminate use of lime on coarse-textured soils could lead to excessively alkaline conditions and to serious consequences, such as deficiencies of iron, manganese, and other microelements. Conversely, the application to an acid clay soil of the amount of lime detrimental to crops growing on sandy soils may well be insufficient to raise the pH the desired amount. Adequate liming recommendations are based on a knowledge of the buffer capacity of a soil. Since the pH value alone is no criterion of the amount of lime that should be added to a soil, the indiscriminate use of pH test kits by persons unfamiliar with the rudiments of soil chemistry is undesirable.

Determining the Lime Requirement of Soils. Many experiment stations and soil-testing laboratories have determined the lime requirement of the major soil series and types in the areas they serve. Once this has been done, a knowledge of the pH and the soil type will make possible an immediate liming recommendation. If the lime requirement has not been determined and a recommendation must be made, the texture and organic matter content must also be taken into consideration.

The lime requirement of a soil can be determined by several different methods. One, which is not used much because it is time consuming, is discussed here to illustrate the mechanics of the determination. It is a simple technique which illustrates the meaning of buffer capacity and lime requirement. It can also be used to determine the amount of acid or sulfur needed to lower soil pH.

Such data are obtained by adding to a series of small beakers or flasks a known quantity of soil. To each beaker is then added a given amount of acid or base. The base usually employed is calcium hydroxide and the acid is hydrochloric. Water is added to equalize the volume of liquid in all beakers, and the samples are allowed to equilibrate. pH determinations are made, and the values obtained are plotted against the milliequivalents of acid or base added. A buffer curve is then constructed. From these data it is simple to determine the amount of lime to be added. The data obtained are illustrated in Figure 11-3.

In an example of how these data may be used the pH of the soil to which no acid or base was added is 5.7. Suppose that we wish to add sufficient lime to raise the pH to 6.2. The milliequivalents of added base or acid corresponding to the untreated soil, of course, are zero. Commencing at the pH value of 6.2 on the Y-axis, draw a line parallel to the X-axis until the line crosses the lime requirement curve. At this point of intersection drop a line to the X-axis parallel to the Y-axis. This perpendicular strikes the X-axis at 0.067 meq of base. Since the weight of soil added to each beaker was 10.0 g, 0.067 meq. of a base must be added to each 10 g of the soil in question to change the pH from 5.7 to 6.2. If it were decided to use finely divided calcite, the amount required would be 0.067×0.05, or 0.00335 g. Calculated on an acre basis, the amount of calcite required would be 670 lb, assuming that the weight of an acre furrow slice of this soil is 2 million lb.

An even more time-consuming method involves adding known amounts of $CaCO_3$ to measured quantities of soil and incubating the mixtures for several months to allow the reaction to go to completion. pH values of the samples are then measured and a buffer curve constructed as in the previous case, relating soil pH at the end of the incubation period to amount of $CaCO_3$ added to the sample. An example of the results of one such study is shown in Figure 11-4.

The methods in common use are based on the change in pH of a buffered solution when a sample of soil is added to a given amount of solution. When a sample of acid soil is added to a measured quantity of the solution, the pH is depressed in proportion to the original soil pH and its buffer capacity. By calibrating pH changes of the buffered solution which accompany the addition of known amounts of acid, the amount of lime required to bring the soil to some prescribed pH can be calculated.

Numerous buffer methods have been developed over the years. The Shoemaker, McLean, and Pratt (SMP) single-buffer (SB) method for rapid measurement of lime

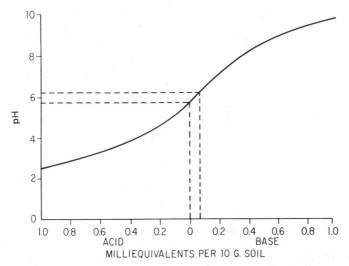

FIGURE 11-3. Lime requirement curve for a Maumee sandy loam.

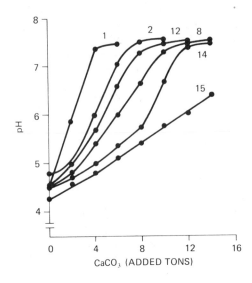

FIGURE 11-4. Titration curves for representative soils from Ohio after incubation with $CaCO_3$ for 17 months. [Shoemaker et al., *SSSA Proc.,* **25**:274 (1961).]

requirement of acid soils has been widely adopted by U.S. soil-testing laboratories. Their buffer solution is a dilute mixture of triethanolamine, paranitrophenol, potassium chromate, and calcium acetate.

Uniformly acceptable results have been obtained with the SMP method for soils with wide differences in lime requirement. It is especially well suited for soils possessing the following properties: lime requirement > 4 meq per 100 g (> 4000 lb of lime per acre), pH < 5.8, organic-matter concentration of < 10%, and appreciable quantities of soluble (extractable) aluminum. The SMP-SB method, however, is known to have shortcomings when used on soils with low lime requirements. Lime requirements based on this method will frequently result in overliming of many low-base-status soils.

Many comparisons have been made of the reliability of the various buffer methods. A comprehensive study was conducted by McLean and his associates at Ohio State University. A group of 54 soils from 12 states of the United States representing a wide range in lime requirements was used to evaluate (1) the Yuan double-buffer method; (2) the regular SMP-SB procedure; (3) a modified SMP method employing the double-buffer feature of Yuan at several pH pairings and soil/buffer ratios; and (4) the Mehlich single-buffer method, based on exchangeable acidity in a group of soils, primarily Ultisols, of North Carolina. Although the SMP-SB procedure was improved considerably by modifying it to include double-buffer, quick-test, and mathematical adjustment features, it was concluded that the regular SMP-SB method was still the most satisfactory compromise between simplicity of measurement and acceptable accuracy of results for soils of a wide range in lime requirement. A recommended modification to the standard SMP-SB technique included a narrower soil/solution ratio and shorter shaking time plus an adjustment for incomplete reaction with the soil during the shorter "quick test" reaction interval.

In tropical areas the lime requirement may be based simply on the amounts needed to lower the solubility of aluminum. For example, in Colombia, it is recommended that 1 ton of liming material be applied for each milliequivalent of exchangeable Al^{3+} per 100 g of soil. The following relationship is used in Brazil for determining the amount of liming material needed to neutralize exchangeable aluminum.

$$\text{meq CaCO}_3/100 \text{ g} = 2 \times \text{meq exchangeable Al}/100 \text{ g}$$

Kamprath in North Carolina has reported that the amounts of liming material needed to reduce aluminum solubility range from 1.5 to 3.3 tons/meq of aluminum per 100 g of soil. Other researchers have found that in Hydric Dystrandept and Andept soils 1000 lb of lime is needed to neutralize 1 meq of aluminum.

Soil pH for Crop Production

For many years the optimum pH of soils for crop production was considered to be between 6.5 and 7.0. Work carried out over the past several years, however, has cast doubt on the validity of this assumption. Researchers in North Carolina have proposed that liming of the Oxisols and Utisols of the warm, humid southern United States to pH values greater than 6.0 or 6.2 may be not only unnecessary but harmful. It is the opinion of this group, and other scientists as well, that lime sufficient to neturalize the exchangeable aluminum is all that need be added to the red and yellow soils of the humid, warm areas of the United States. Work by scientists in tropical countries has generally confirmed this. Liming of most soils that are high in the hydrous oxides of aluminum and iron sufficient only to neutralize most of the exchangeable aluminum will bring the pH to about 5.6 or 5.7 and the exchangeable aluminum to less than 10% of the effective CEC. Liming is known to be unnecessary for satisfactory growth of crops on some Hydric Dystrandepts and Andepts, even though soil pH may be lower than 5.5. Extractable aluminum levels in such soils may be low in relation to the amounts of calcium present.

The North Carolina workers have presented data which show that the liming of Ultisols and Oxisols to pH values approaching 7.0 does result in deleterious effects. Kamprath reported that liming red and yellow soils to pH 7.0:

1. Reduced water percolation.
2. Reduced the growth of legumes and nonlegumes.
3. Reduced plant uptake of phosphorus.
4. Reduced the micronutrient uptake by plants.

In northern Alberta and northeastern British Columbia, where lime must be transported long distances, and with the exception for alfalfa, it is most economical to add just enough lime to reduce 0.02 M $CaCl_2$-soluble aluminum below the 1 ppm toxic level for sensitive crops.

Workers in the North Central States, notably McLean and his associates at Ohio, believe that for the soils of the Alfisol and Mollisol types liming to pH 6.5 to 6.8 will give the most desirable results. McLean points out that if all nutrients are supplied in adequate amounts and no elements are present in toxic concentrations, adequate plant growth can be obtained at pH values considerably lower than 6.5. Because these conditions do not prevail in most soils, liming to pH values of 6.5 to 6.8 is preferred on the Alfisols and Mollisols in the midwestern United States. These workers state that in most of the midwestern United States there appears to be no generally adverse effect from liming to a nearly neutral pH value where crops depend for most of their nutrients on the various release mechanisms of the soil.

The chemical and physical properties of Oxisols and Ultisols in the southern United States are largely controlled by hydroxyaluminum and hydroxyiron (the multicharged polymers referred to earlier) coatings on the clay fractions, whereas in Mollisols these coatings are much less common. This helps to explain why the pH best for crop production in the southern United States is lower than the ideal pH in the midwestern and other areas of the United States. It must be kept in mind that this holds only for general field crops such as corn, small grains, soybeans, alfalfa, clovers, and similar crops. Crops such as tobacco, potatoes, and others which may be affected by diseases or micronutrient deficiencies associated with high or low pH values would be excepted and their individual pH requirements should be met.

Liming Materials

The materials commonly used for the liming of soils are the oxides, hydroxides, carbonates, and silicates of calcium or calcium and magnesium. The presence of these elements alone does not qualify a material as a liming compound. In addition to these cations, the accompanying anion *must* be one that will reduce the activity of the hydrogen and hence aluminum in the soil solution.

The reaction mechanisms of liming materials with acid soils are complex. For most of the common liming materials to be described later these reactions begin with the neutralization of H^+ ions in the soil solution by either OH^- or SiO_3^{2-} ions furnished by the liming material. The basic reaction of a liming material when added to the soil can be illustrated with the case for calcium carbonate. In water, $CaCO_3$ behaves as follows:

$$CaCO_3 + H_2O \longrightarrow Ca^{2+} + HCO_3^- + OH^-$$

$$H^+ \text{(soil soln.)} + OH^- \longrightarrow H_2O$$

The rate of the reaction above, and thus of the solution of $CaCO_3$, is directly related to the rate at which the OH^- ions are removed from solution. As long as sufficient H^+ ions are in the soil solution, Ca^{2+} and HCO_3^- ions will continue to go into solution. When the H^+ ion concentration is lowered, however, solution of the Ca^{2+} and HCO_3^- ions is reduced.

In acid soils, the concentration of the H^+ ions in solution is related to the hydrolysis of Al^{3+} or hydroxyaluminum or hydroxy-Fe^{3+} ions. Their hydrolysis in turn is influenced by the amount of clay and organic matter in the system. The continued removal of H^+ from the soil solution will ultimately result in the precipitation of the aluminum and iron ions and their replacement on the adsorption sites with calcium and/or magnesium and other basic cations. When the exchangeable aluminum and the hydroxyaluminum and hydroxy-Fe^{3+} have been precipitated as $Al(OH)_3$ and $Fe(OH)_3$, the soil acidity that remains arises from the other sources mentioned earlier in this chapter.

As the neutralization of soil solution H^+ by a material is necessary for it to be classed as a liming agent, gypsum ($CaSO_4 \cdot 2H_2O$) and other neutral salts cannot qualify as such. In fact, the addition of neutral salts will actually lower soil pH. Their addition, especially as in band placement, results in replacement of adsorbed Al^{3+} in a localized soil zone, sometimes with a significant lowering of the pH in this region.

Calcium Oxide. This is the only material to which the term *lime* may be correctly applied. Calcium oxide (CaO), known also as unslaked lime, burned lime, or quicklime, is a white powder, quite disagreeable to handle. It is manufactured by roasting calcitic limestone in an oven or furnace. The carbon dioxide is driven off, leaving calcium oxide. The purity of the burned lime depends on the purity of the raw material. This product is shipped in paper bags because of its powdery nature and its caustic properties. When added to the soil, it reacts almost immediately.

When unusually rapid results are required, either this material or calcium hydroxide should be selected. Complete mixing of calcium oxide with the soil may be difficult, however, for immediately after application absorbed water causes the material to form flakes or granules. These granules may harden because of the formation on their surfaces of calcium carbonate, and in this condition they may remain in the soil for long periods of time. Only by very thorough mixing with the soil at application time can this caking be prevented.

On a pound-for-pound basis, calcium oxide is the most effective of all the liming materials commonly employed, for the pure material has a neutralizing value or calcium carbonate equivalent (CCE) of 179%, compared with pure calcium carbonate. The full significance of this last statement will become apparent when the neutralizing value of the various materials is discussed.

Calcium Hydroxide. Calcium hydroxide [$Ca(OH)_2$] is frequently referred to as slaked lime, hydrated lime, or builders' lime. Like calcium oxide, it is a white, powdery substance, difficult and unpleasant to handle. Neutralization is rapidly effected, as it is with calcium oxide.

Slaked lime is prepared by hydrating calcium oxide. Much heat is generated, and on completion of the reaction the material is dried and packaged in paper bags. The purity of the commercial product varies, but the chemically pure compound has a neutralizing value of 136, making it pound-for-pound the second most efficient of the commonly used liming materials.

Calcium and Calcium-Magnesium Carbonates. The carbonates of calcium and magnesium occur widely in nature and in a number of different forms.

Crystalline calcium carbonate ($CaCO_3$) is termed calcite or calcitic limestone. Crystalline calcium-magnesium carbonate [$CaMg(CO_3)_2$] is known as dolomite when the calcium carbonate and magnesium carbonate occur in equimolecular proportions. In other proportions they are said to be dolomitic limestones. Metamorphism of these high-grade limestones produces marble. Deposits of high-grade limestone are widespread in the United States and Canada.

Limestone is most often mined by open-pit methods. First the overburden of soil and undesirable rock is removed, after which explosives are used to blast the exposed limestone. Blasting breaks out the rock, generally in sizes that can be accommodated by the quarrying and crushing equipment. The loosened material is crushed to sizes of 1 in. or less, and the limestone is ready for grinding and pulverizing. The pulverized material is classified by passing it through one or more screens of some specified size. This is required particularly of agricultural-grade limestone, specifications for which are quite rigid. Having been thus processed, it is generally stored in the open in piles and may be shipped either in bulk or in bags.

The quality of crystalline limestones depends on the degree of impurities they contain, such as clay. The neutralizing values usually range from 65 to 70% to a little more than 100%. The neutralizing value of chemically pure calcium carbonate has been established arbitrarily at 100%, and, theoretically, chemically pure dolomite may have a neutralizing value of nearly 109%. As a general rule, however, the neutralizing value or CCE of most agricultural limestones is between 90 and 98% because of impurities.

Marl. Marls are soft, unconsolidated deposits of calcium carbonate. They are frequently mixed with earth and usually quite moist. Marl occurs in many states in the eastern part of the United States and is easily mined. The deposits are generally thin, though the layers have been known to range up to 30 ft in thickness. Marl is recovered by dragline or power shovel after the overburden has been removed. The fresh material is stockpiled and allowed to dry before being applied to the land.

Marls are almost always low in magnesium. Their value as liming materials depends on the amount of clay they contain. Their neutralizing value usually lies between 70 and 90%, and their reaction with the soil is the same as that of calcite.

Slags. Several types of material are classed as slags, three of which are important agriculturally.

Blast-Furnace Slag. Blast-furnace slag is a by-product of the manufacture of pig iron. In the reduction of iron the calcium carbonate in the charge loses its carbon dioxide, and calcium oxide then combines with the molten silica to form a slag that is tapped off and either air-cooled or quenched with water. The cooled product is ground, screened, and shipped in open cars or trucks.

As a liming material, slag behaves essentially as calcium silicate. Metasilicic acid, which is formed when the slag is added to acid soils, is weakly dissociated, and the pH of the soil is raised. The neutralizing value of blast-furnace slags ranges from about 75 to 90%. These slags usually contain appreciable amounts of magnesium. Results of field tests indicate that, when applied on the basis of equivalent amounts of calcium and magnesium, they are just as effective in producing crops as ground limestones.

Basic Slag. A second type of slag is known as basic or Thomas slag and was discussed in Chapter 6. This slag is a by-product of the basic open-hearth method of making steel from pig iron, which, in turn, is produced from high-phosphorus iron ores. The impurities in the iron, including silica and phosphorus, are fluxed with lime and slagged off. The slag is cooled, finely ground, and usually marketed in 80- or 100-lb bags. In addition to its phosphorus content, basic slag has a neutralizing value of about 60 to 70%. It is generally applied for its phosphorus content rather than for its value as a liming material, but because of its neutralizing value it is a good material to use on acid soils. Its selection in relation to other sources of phosphorus is determined by economic factors.

Electric-Furnace Slag. A third type of slag results from the electric-furnace reduction of phosphate rock in the preparation of elemental phosphorus. The slag is formed when the silica and calcium oxide fuse, and the product is thought to be largely calcium silicate. It is drawn off and quenched with water. The slag is a waste product which is marketed at a low price and usually only within a limited radius of the point of production. It contains 0.9 to 2.3% P_2O_5 and is not ground. The neutralizing value ranges from 65 to 80%. Its reaction with the soil is similar to that indicated for blast-furnace slag.

Electric-furnace slags are also produced in the manufacture of pig iron and steel. They are comprised mainly of calcium and silica and may have neutralizing values as high as 89%. The results from a long-term growth-chamber experiment presented in Table 11-3 indicate that they are effective liming agents.

TABLE 11-3. Response of Red Clover to Kimberley Electric Furnace Iron Slag and Other Liming Materials

	Yield (g) of Oven-Dry Tissue per Pot *					
	Alouette Silt Loam			Pitt Silty Clay		
Liming Treatment	30% Base Satn.	60% Base Satn.	Means	30% Base Satn.	60% Base Satn.	Means
Agricultural lime	1.55a	1.51a	1.53a	4.29a	4.49a	4.39a
Slag (broadcast)	2.03	2.28	2.16	4.42a	4.54a	4.48a
Slag (incorporated)	2.48	3.19	2.84	4.52a	5.02a	4.77a
Dolomitic limestone	2.25	2.66	2.45	4.04a	4.94a	4.49a
Magnesium carbonate	1.58a	1.13b	1.35	4.14a	3.28	3.71
Marl	1.58a	1.66a	1.62a	4.29a	4.64a	4.47a
Control (no liming)	1.10	1.10b	1.10	2.30	2.30	2.30
Mean	1.80	1.93	1.86	4.00	4.17	4.09

* Averaged over 10 harvests and five replicates. Any two means, within a given set of means, that have the same letter are not significantly different ($P < 0.05$) by the Duncan multiple range test.
Source: Beaton et al., *Can. J. Plant Sci.,* **48:**455 (1968).

Miscellaneous Liming Materials. Other materials that are used as liming agents in localized areas close to their source include fly ash from coal-burning power generating plants, sludge from industrial water treatment plants, Cotrell lime or flue dust from cement manufacturing, sugar lime, pulp mill lime, carbide lime, acetylene lime, packinghouse lime, and so on. These by-products contain varying amounts of calcium and magnesium compounds and other materials.

Neutralizing Value or Calcium Carbonate Equivalent of Liming Materials. Liming materials differ markedly in their ability to neutralize acids. The value of limestone for this purpose depends on the quantity of acid that a unit weight of the material will neutralize. This property, in turn, is related to the molecular composition of the liming material and its purity; in other words, its freedom from inert contaminants such as clay. Pure calcium carbonate is the standard against which other liming materials are measured, and its neutralizing value is considered to be 100%. Calcium carbonate equivalent (CCE) is defined as the acid-neutralizing capacity of an agricultural liming material expressed as a weight percentage of calcium carbonate.

The molecular constitution is the determining factor in the neutralizing value of chemically pure liming materials. Consider the reactions illustrated by the following equations:

$$CaCO_3 + 2HCl \longrightarrow CaCl_2 + H_2O + CO_2 \tag{1}$$

$$MgCO_3 + 2HCl \longrightarrow MgCl_2 + H_2O + CO_2 \tag{2}$$

In each of these equations the molecular proportions are the same; that is, one molecule of each of these carbonates will neutralize two molecules of acid. However, the molecular weight of calcium carbonate is 100, whereas that of magnesium carbonate ($MgCO_3$) is only 84. In other words, 84 g of magnesium carbonate will neutralize the same amount of acid as 100 g of calcium carbonate. How much more effective then is 100 g of magnesium carbonate than the same quantity of calcium carbonate in neutralizing an acid? This is demonstrated quite easily by the following simple proportion:

$$\frac{84}{100} = \frac{100}{x}$$

$$x = 119$$

Therefore, magnesium carbonate on a weight basis will neutralize 1.19 times as much acid as the same weight of calcium carbonate; hence its neutralizing value or CCE in relation to $CaCO_3 = 100$ is $1.19/1 \times 100$, or 119%. The same procedure is used to calculate the neutralizing value of other liming materials. The neutralizing values for several compounds are shown in Table 11-4. Their use makes

TABLE 11-4.   Neutralizing Value (CCE) of the Pure Forms of Some Commonly Used Liming Materials

Material	Neutralizing Value (%)
CaO	179
Ca(OH)$_2$	136
CaMg(CO$_3$)$_2$	109
CaCO$_3$	100
CaSiO$_3$	86

possible the simplest and most straightforward comparison of one liming material with another in regard to neutralizing properties. These values are one guarantee which may be made for commercial liming agents. There are other methods of expressing the value of limestones, and these are discussed next.

Calcium Magnesium. The composition of liming materials is sometimes expressed in terms of the elemental content of calcium and magnesium. Chemically pure calcite, for example, contains 40% calcium and chemically pure magnesium carbonate contains (24/84)100, or 28.6% magnesium. Obviously, to convert the percentage of calcium to its calcium carbonate equivalent it is necessary to multiply by a factor of 2.5; and to convert the percentage of magnesium to the percentage of magnesium carbonate, it is necessary to multiply by a factor of 84/24, or 3.5. If a limestone carries the guarantee in terms of the element, it is a simple matter to calculate the percentage in terms of the carbonate to which the elemental percentage is equivalent.

Calcium and Magnesium Oxide Content. The quality of a limestone may also be expressed in terms of its calcium or magnesium oxide equivalent. As an example, pure calcite is 100% calcium carbonate and contains 40% calcium. Suppose that we wished to express these quantities, not in terms of either of the constituents, but rather in terms of the oxide (CaO). Calcium oxide has a molecular weight of 56, which means of course that 16 g of oxygen is combined with 40 g of calcium. Therefore, if the calcium present in calcium carbonate were expressed as the oxide, it would contain (56/100)100, or 56% calcium oxide equivalent. Thus to convert the percentage of calcium to percentage of calcium oxide, we need only to multiply the calcium by 56/40, or 1.4; and to convert the percentage of calcium carbonate to the percentage of calcium oxide, we have but to multiply the percentage of calcium carbonate by 56/100, or 0.56. Similar figures may be derived for the magnesium-containing limestones.

Total Carbonates. Another expression of the quality of limestones is that of total carbonates. This is a summation of the percentages of the carbonates contained in a given liming material. For example, assume that a limestone contains 78% calcium and 12% magnesium carbonate. The total carbonate content would be 90%.

Conversion Factors for the Various Methods of Expression. The conversion figures for a few transformations, such as the percentage of calcium to that of calcium oxide, have been given. It is on occasion desirable to convert the percentage of magnesium oxide or magnesium carbonate to the calcium carbonate equivalent. This is illustrated in the following example. Assume that a limestone contains this guarantee:

35% CaO

15% MgO

Assume further that we wish to express this analysis in terms of the calcium carbonate equivalent. The conversion may be obtained if it is remembered that 56 g of calcium oxide is equivalent to 100 g of calcium carbonate. The percentage of calcium oxide is multiplied by 100/56, and the calcium carbonate equivalent in this sample is 35% × 1.785, or 62.5%.

How is the conversion of magnesium oxide to calcium carbonate to be handled? A glance at their molecular composition shows that 1 mole of each will neutralize the same quantity of acid. A mole, or 1 gram molecular weight, of the magnesium oxide is 40, whereas that of calcium carbonate is 100. In other words 40 g of

TABLE 11-5.   Limestone Conversion Factors

Percent		Percent		Factor
Ca	to	CaO	multiply by	1.40
Ca	to	Ca(OH)$_2$	multiply by	1.85
Ca	to	CaCO$_3$	multiply by	2.50
Mg	to	MgO	multiply by	1.67
Mg	to	Mg(OH)$_2$	multiply by	2.42
Mg	to	MgCO$_3$	multiply by	3.50
Mg	to	Ca	multiply by	1.67
Mg	to	CaCO$_3$	multiply by	4.17
MgO	to	CaCO$_3$	multiply by	2.50
MgCO$_3$	to	CaCO$_3$	multiply by	1.19

magnesium oxide will neutralize the same amount of acid as 100 g of calcium carbonate. Therefore, all that remains is to multiply the 15% magnesium oxide in the sample by the factor 100/40, or 2.5. The result, 37.5%, is added to the 62.5% and the total neutralizing value, or calcium carbonate equivalent, for the limestone in question is 100%.

Conversion factors which make possible an expression of the value of a limestone may be determined in any way desired, provided that the content of one of the constituents is given. A few of these factors are listed in Table 11-5. The student should be able to derive them as well as others not shown.

Fineness of Limestone

Molecular constitution and freedom from inert impurities are not the only properties that limit the effectiveness of agricultural limestones. The degree of fineness is equally important, because the speed with which the various materials will react is dependent on the surface in contact with the soil. Materials such as calcium oxide and calcium hydroxide are by nature powdery, so that no problem of fineness is involved, but the crystalline limestones are an entirely different matter.

When a given quantity of crushed limestone is thoroughly incorporated with the soil, its reaction depends upon the size of the individual particles. If they are coarse, the reaction will be slight, but if they are fine, the reaction will be extensive. This is strikingly illustrated by the data in Table 11-6. It is obvious that compared to the coarser fractions, the finely ground limestone reacts rapidly with soil. These results show very clearly a steady rise in the effectiveness of the coarser fractions relative to the <100 fraction with increased exposure period in soil. Also, these relative efficiencies based on change in soil pH are influenced by the magnitude of the pH change used for making the comparisons. The coarser fractions show higher efficiencies relative to the <100 fraction the lower the reference pH.

Amounts of a particular limestone fraction needed to produce a given rise in pH 2 years after application are shown graphically in Figure 11-5. Much less of the finer fractions than of the coarser fractions is needed to achieve a certain pH, particularly at lower reference pH values.

The data shown in Table 11-6 and Figure 11-5 indicate that finely divided agricultural liming materials will produce the most rapid upward adjustments in soil pH. However, the cost of limestone increases with the fineness of grinding. What is needed is a material that requires a minimum of grinding, yet contains enough fine material to effect a pH change rapidly. As a result, agricultural limestones contain both coarse and fine materials. Many states require that 75 to 100% of the limestone pass an 8- to 10-mesh screen and that 20 to 80% pass anywhere from an 8- to 100-mesh screen. In this way there is fairly good distribution of both the coarse and fine particles.

TABLE 11-6. **Relative Efficiency of Various Dolomitic Limestone Fractions As Affected by Reference pH and Time of Equilibration in the Withee Silt Loam Under Field Conditions at Marshfield, Wisconsin**

Fraction (Mesh Size)	pH 5.5				pH 6.0			pH 6.5	
	1 mo.	1 yr	2 yr	3 yr	1 yr	2 yr	3 yr	2 yr	3 yr
	Where < 100 fraction at 3 years' equilibration for each pH level = 100								
8–20	3	9	27	—	—	21	54	13	24
20–40	5	29	77	100	11	50	80	28	55
40–60	8	53	83	100	16	72	100	53	73
60–100	14	67	100	100	28	85	100	67	92
< 100	40	67	100	100	48	93	100	90	100
	Where < 100 fraction in each column = 100								
8–20	6	14	27	—	—	22	54	15	24
20–40	11	44	77	100	24	54	80	31	55
40–60	21	79	83	100	33	78	100	59	73
60–100	36	100	100	100	59	91	100	74	92
< 100	100	100	100	100	100	100	100	100	100

Source: Love et al., *Trans. 7th Int. Congr. Soil Sci.*, IV.37: 293 (1960).

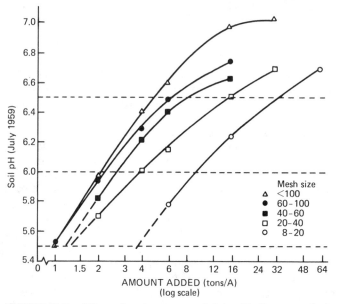

FIGURE 11-5. Effect of various rates of dolomitic limestone fractions on the pH of Withee silt loam after 2 years of equilibration under field conditions. [Love et al., *Trans. 7th Int. Congr. Soil Sci.*, IV.**37**:293 (1960).]

Rapid Methods for Evaluating Agricultural Limestones

The best criterion for evaluating agricultural limestones is their performance in field soils. Their reaction with soils in small containers in the laboratory is also an effective method of evaluation. Both methods, however, require too much time, for limestone takes 2 to 6 months to react with the soil. Its reactivity is determined not only by its purity and particle-size distribution but also by its hardness and magnesium content. The rapid evaluation of large numbers of limestone samples is desirable from the standpoint of the commercial utilization of these materials, for some knowledge of their reactivity is needed for the development of sound promotional and sales programs and for the protection of the purchaser.

Several laboratory tests have been developed. Workers in Maryland reported on limestone samples digested in a boiling solution of sodium EDTA. The efficiency of the limestones was then calculated from the results of similar analyses made on pure calcite. These efficiency values were compared with the effectiveness of these same materials in changing the pH of soil samples. Their results are shown in Table 11-7. It is obvious that with most of them the chemical method gave a good estimate of the effectiveness of the liming materials.

Another method, developed by Shaw and Robinson, is based on the dissolution of limestones in ammonium chloride. Comparison of results by this method with results from incubating limestones with soil showed that the ammonium chloride treatment gave a very good estimate of the agricultural value of the limestones tested.

Using the method just described, the reactivity of different size separates of a calcite and a dolomite was determined. Results are shown graphically in Figure 11-6 and two points are brought out. The coarser the size separate, the less reactive the material. This is true for both dolomite and calcite. The second point is that dolomite is less reactive than calcite for each of the size separates.

The reactivity of limestones is related in part to the fineness with which they are ground. The greater the surface of the limestone exposed, the faster its reaction with the soil. However, apparent surface, calculated on the basis of the size of solid, smooth particles, may not represent total surface. Limestone particles may be spongy, as found by workers at the U.S. Department of Agriculture, who used a technique of surface adsorption of an inert gas. They found that the total surface measured by this gas adsorption technique was much greater than simple surface measurements based on particle size determinations. They also found that their method produced results that were closely related to the reactivity of limestone measured by an ammonium chloride technique.

TABLE 11-7.   Relative Efficiency of Limestone in Producing pH Changes in a Soil Suspension

Commercial Limestone Samples	Relative Efficiency	Change in pH * (%)
Pure calcite passing 200 mesh	100	100
1	100	97
2	101	95
3	92	93
4	97	90
5	84	84
6	78	82
7	79	79
8	78	77
9	74	74
10	73	73
11	71	71
12	64	64
13	59	61
14	41	58
15	36	45
16	19	8

* Total change in pH studied was from 3.90 to 7.00.
Source: El Gibaly et al., *SSSA Proc.,* **19**:301 (1955).

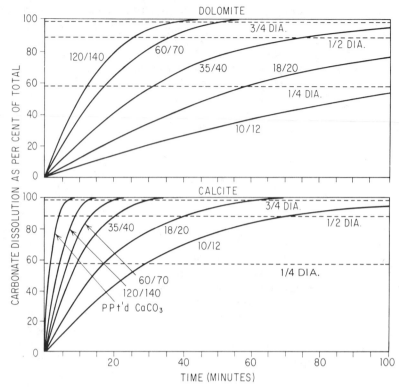

FIGURE 11-6. Reaction rates of separates of calcite and dolomite of different sizes in boiling 5 N NH$_4$Cl solution. [Shaw et al., *Soil Sci.*, **87**:262 (1959). Reprinted with permission of The Williams & Wilkins Co., Baltimore.]

Selection of a Liming Material

The quality of liming materials is determined by such properties as neutralizing value, magnesium content, degree of fineness, reactivity, and moisture content. Degree of fineness is widely recognized as a major factor in the selection of liming materials. In the United States, some states utilize a system in which particles remaining on a screen with 8 meshes to the inch are rated as totally ineffective and are given a fineness rating of 0. Particles passing through an 8-mesh sieve but collecting on a 60-mesh sieve are classified as 50% effective, while those passing through the 60-mesh sieve are considered to be 100% effective. The effectiveness factors used in Canada are 0 for particles retained on a No. 10 Tyler sieve, 40% for those which pass a No. 10 sieve but are retained on a No. 60 sieve, and 100% for those which pass a No. 60 sieve. These effectiveness ratings may not accurately represent the actual solubility of a material, but they do permit practical evaluations of the suitability of different materials.

The effective calcium carbonate (ECC) rating of a limestone or liming agent is the product of its calcium carbonate equivalent (purity) and the fineness factor. The latter is the sum of the products of the percentage of material in each of the three size fractions, specified earlier, multiplied by the appropriate effectiveness factor. Examples of two lime sources with different effective calcium carbonate contents are given in Table 11-8.

The magnesium content of limestone should also be considered. Many soils are deficient in this element, and the use of dolomitic lime is to be encouraged.

For limestones that are ground to meet local specifications and which contain roughly the same amount of magnesium, an excellent criterion of selection is the

TABLE 11-8.   Fineness Effects on Effective Calcium Carbonate (ECC) Content of Two Lime Sources

	Solid Agricultural Lime	Suspendable Lime
Percent $CaCO_3$ equivalent	90	98
Sieve analysis		
% of 8-mesh sieve	2	0
% on 60-mesh sieve	21	2
% passing 60 mesh sieve	77	98
Calculated fineness factor		
% on 8-mesh × 0% effectiveness	$2 \times 0 = 0$	$0 \times 0 = 0$
% on 60-mesh × 50% effectiveness	$21 \times 50\% = 11.5$	$2 \times 50\% = 1$
% through 60-mesh × 100% effectiveness	$77 \times 100\% = 77$	$98 \times 100\% = 98$
Fineness factor (%)	88.5	99
Percent ECC = purity × fineness factor	$90 \times 88.5\% = 79.6$	$98 \times 99 = 97.0$

Source: Murphy et al., in *Situation 1978, TVA Fert. Conf. Bull. Y-131,* Muscle Shoals, Ala.: National Fertilizer Development Center–Tennessee Valley Authority, 1978.

cost per unit of neutralizing value applied to the land, the calculation of which has already been covered. As indicated in Chapter 15, the returns per dollar spent on lime are phenomenal.

As a general rule, for the same degree of fineness, the material that costs the least per unit of neutralizing value applied to the land should be selected. Assume that there are available a calcitic limestone, CCE = 95%, and a dolomitic limestone, CCE = 105%, each with the same fineness or mechanical analysis. Assume also that they both cost $12 per ton applied to the land. Based on the neutralizing value, the first will cost 105/95 × 12 or $13.26 per ton, compared with the dolomite at $12 per ton. In addition, the dolomite supplies magnesium, which is a nutrient not overly abundant in many humid-region soils.

Whenever possible, some measure of the reactivity of the limestone, based on one of the rapid chemical methods, should be obtained. When not possible, the purchaser should be guided by the fineness of the material, its neutralizing value, its magnesium content, and the cost per ton applied to the land.

Use of Lime in Agriculture

Application of lime to acid soils in many areas of the United States and Canada produces striking increases in plant growth. The areas in the United States in which the need for liming is greatest are the humid regions of the East, South, and middle western and far western states. Soils in Canada requiring lime are confined mainly to areas of high precipitation in the Atlantic and central provinces plus the western province of British Columbia.

It is noteworthy that increasing soil acidity is a relatively new soil management problem in the inland Pacific Northwest region, which includes eastern Washington and northern Idaho. Soil acidity is also a recent problem in dryland agriculture of the Canadian prairie provinces and northeastern British Columbia, where a total of over 5.8 million acres of cultivated land have pH values of 6.0 or less.

Lime is seldom needed in those areas where rainfall is low and leaching is minimal, such as parts of the Great Plains states and the arid, irrigated saline and saline-alkali soils of the Southwest, intermountain, and far western states. This is also true for the majority of soils in the Prairie Provinces of Canada. However, the presence of lime can be helpful in the management of sodic soils.

When crop responses are obtained from the application of materials carrying the major plant nutrients, nitrogen, phosphorus, and potassium, it is assumed, and usually correctly, that the response was the direct result of overcoming a deficiency of one of these nutrient elements. Responses from the application of lime, however, may not always be attributed to the plant-nutrient value of the calcium or magnesium.

Direct Benefits. Aluminum toxicity is probably the most important growth-limiting factor in many acid soils, particularly those having pH levels below 5.0 to 5.5. Excess aluminum interferes with cell division in plant roots; fixes phosphorus in less available forms in soils and in or on plant roots; decreases root respiration; interferes with certain enzymes governing the deposition of polysaccharides in cell walls; increases cell wall rigidity by cross-linking with pectins; and interferes with the uptake, transport, and use of several elements (calcium, magnesium, and phosphorus) and water by plants. Thus it is not surprising that perhaps the greatest single direct benefit of liming acid soils is the reduction in the activity or solubility of aluminum and manganese. Both of these ions in anything other than very low concentrations are toxic to most plants. The poor growth of crops that is observed on acid soils is due largely to the large concentration of these two ions in the soil solution. When lime is added to acid soils, the activity of the aluminum and manganese is reduced and they are removed from solution. Table 11-9 illustrates the beneficial effect of liming an acid soil on growth of barley. It is readily apparent that the lime treatment raised soil pH while greatly reducing levels of extractable aluminum.

Not only are these ions toxic to plants, the presence of increasing amounts of aluminum in the soil solution also decreases the uptake of calcium and magnesium by plants. The restrictive influence of aluminum on calcium uptake by plants is demonstrated in Figures 8-16 and 11-7.

At pH 4.5 or less another benefit is the removal of H^+ toxicity, which damages root membranes and also is detrimental to the growth of many beneficial bacteria.

It has been found that different crops and even different varieties of the same crop will differ widely in their susceptibility to aluminum toxicity. Foy and his co-workers at the USDA found that different varieties of crops such as soybeans, wheat, and barley show wide ranges in their tolerance to high concentrations of aluminum in the soil solution. An example of this differential effect is shown in Figure 11-8.

Crops also vary in their tolerance to excessive amounts of manganese. For example, rapeseed or canola is very sensitive to manganese toxicity, while barley is more tolerant.

TABLE 11-9. Effects of Lime on Hudson Barley and on the pH and Level of Extractable Aluminum in Tatum Surface Soil

$CaCO_3$ Added (ppm)	Yield of Barley Tops *,† (g/pot)	Soil Properties *	
		pH	KCl-Extractable Al (meq/100 g)
0	0.29 e	4.1	5.75
375	0.91 d	4.3	4.81
750	2.72 c	4.5	4.33
1,500	4.29 b	4.8	2.75
3,000	5.07 a	5.5	0.37

* Averages of three replications.
† Any two yields having a letter in common are not significantly different at the 5% level by the Duncan multiple range test.

Source: Foy et al., *Agron. J.,* **57**:413 (1965).

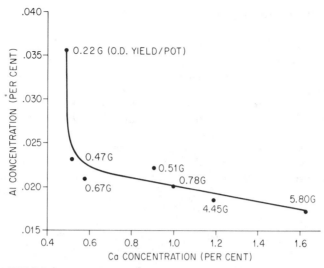

FIGURE 11-7. Relationship between aluminum and calcium concentration in cotton tops from nonleached subsoil. [Soileau et al., *SSSA Proc.*, **33**:919 (1969).]

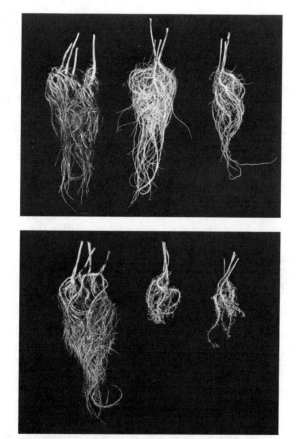

FIGURE 11-8. Differential effects of aluminum on root growth of Perry (top) and Chief (bottom) soybean varieties grown in $^1/_5$ Steinberg solution containing 2 ppm. calcium. *Left to right*: 0, 8, 12 ppm aluminum added. [Foy et al., *Agron. J.*, **61**:505 (1969).]

Indirect Benefits. Although the importance of calcium and magnesium in plant nutrition is not questioned, the scope of benefits derived from the application of lime where needed is much broader than would be expected from a simple response to the addition of a deficient nutrient element. Many of the indirect benefits have been covered in this text in conjunction with other topics. They are summarized here, however, and a few additional advantages are listed.

Effect on Phosphorus Availability. The relationship between the available soil phosphorus and the soil pH has been covered. At low pH values and on soils high in aluminum and iron, phosphates are rendered less available because of their reaction with these compounds. The addition of a liming agent to these soils will inactivate the iron and aluminum, thus increasing the level of plant-available phosphorus.

If the soil pH is greatly increased by the addition of excessive amounts of lime, phosphate availability will again be decreased because of precipitation as calcium or magnesium phosphates. A liming program should be planned so that the pH can be kept between 5.5 and 6.8 to 7.0 if maximum benefit is to be derived from the applied phosphate. As a usual practice, soils are not limed much above 6.5 because of the possibility of decreasing the availability of certain microelements, as discussed in a previous section.

Micronutrient Availability. The effect of soil pH on micronutrient availability was covered in Chapter 9. With the exception of molybdenum, the availability of the microelements increases with a decrease in pH. This can be detrimental because of the toxic nature of many of the elements in anything other than minute concentrations. The solubility of aluminum, iron, and manganese increases with increasing acidity. In addition to toxic effects, their presence may interfere with the absorption of calcium, magnesium, and other basic cations. The addition of adequate lime causes their inactivation, and soil at a pH value of 5.6 to 6.0 is usually most satisfactory from the standpoint of minimum toxicity and adequate availability of these elements. The effect of high pH on the availability of manganese and iron has been discussed.

Molybdenum nutrition of crops is improved by liming (see page 379) and deficiencies are infrequent in those soils limed to pH values in excess of 7.0. Because of the effect on availability of other micronutrients, liming to this value or above is not normally recommended for most crops in humid areas.

Nitrification. Most of the organisms responsible for the conversion of ammonia to nitrates require large amounts of active calcium. As a result, nitrification is enhanced by liming to a pH of 5.5 to 6.5. Decomposition of plant residues and breakdown of soil organic matter are also more rapid in this pH range than in more acidic soils. The effect of liming on both mineralization of organic nitrogen and nitrification is shown in Table 11-10. Application of lime just prior to incubation almost doubled the mineralization of organic nitrogen. However, lime added 1 or 2 years before sampling had little or no effect on release of mineral nitrogen in two of the soils. Adding lime at the start of the incubation improved nitrification, but the earlier field applications of lime had an even greater effect on nitrification in this laboratory experiment.

Nitrogen Fixation. The process of nitrogen fixation, both symbiotic and nonsymbiotic, is favored by adequate liming. With legumes, as was noted in Chapter 5, activity of some rhizobia species such as *R. meliloti* is greatly restricted by soil pH levels below 6.0 (Figure 5-2). Thus liming will increase the growth of legume plants because of the greater amount of nitrogen fixed and thereby return larger quantities of organic matter and nitrogen to the soil. With the nonsymbiotic, ni-

TABLE 11-10. Mineralization of Organic Nitrogen and Nitrification in Three Acid Soils Incubated Four Weeks With and Without Lime *

Soil	Treatment	Organic N Mineralized (ppm)	Percent Nitrification
Site 1 (pH 5.5, 0.20% soil N)	No lime	36a	8a
	Limed at start of incubation	61b	66b
	Limed 2 yr before in the field	33a	94c
Site 2 (pH 5.4, 0.13% soil N)	No lime	40a	7a
	Limed at start of incubation	72b	64b
	Limed 1 yr before in the field	44a	93c
Site 3 (pH 5.7, 0.83% soil N)	No lime	90a	28a
	Limed at start of incubation	177c	83b
	Limed 1 yr before in the field	134b	94c

* For each site, values not followed by the same letter are significantly different at $P = 0.05$.
Source: Nyborg and Hoyt, *Can. J. Soil Sci.,* **58:**331 (1978).

trogen-fixing organisms the greater fixation of atmospheric nitrogen taking place in adequately limed soils makes possible the more rapid conversion to humus of carbonaceous crop residues, such as those of corn and small grains. The increased level of soil nitrogen means a higher content of stable organic matter and a general increase in the fertility status of the land.

Soil Physical Condition. The structure of fine-textured soils may be improved by liming. This is largely the result of an increase in the organic matter content and to the flocculation of calcium-saturated colloids. Favorable effects of lime on soil structure include reduced soil crusting, better emergence of small-seeded crops, and lower power requirements for tillage operations. However, the overliming of Oxisols and Ultisols (liming in excess of that needed to neutralize the exchangeable aluminum) can result in the deterioration of soil structure with the consequent decrease in water percolation through such soils. This was mentioned in an earlier section of this chapter.

The presence of fine calcium carbonate particles is also known to improve the physical conditions of sodic soils. It appears that increased electrolyte concentration due to calcium carbonate dissolution is responsible for preventing clay dispersion and losses in hydraulic conductivity of such soils.

Disease. Correction of soil acidity by liming may have a significant role in the control of certain plant pathogens. An example of the beneficial effect of lime on reducing the harmful effects of clubroot is given in Table 11-11. Clubroot is a disease of cole crops which reduces yields and causes the infected roots to enlarge and become distorted. Lime does not directly affect the clubroot organism, but at soil pH greater than 7.0, germination of clubroot spores is inhibited.

On the other hand, liming will increase the incidence of diseases such as scab in root crops. Severity of take-all infection in wheat with resultant reductions in yield is known to be increased by liming soils to near neutral pH.

Application of Liming Materials

Placement. Since lime moves slowly in soils, its initial beneficial effects occur only in the immediate vicinity of application. Thus surface applications of lime without some degree of mixing in the soil are not immediately effective in correcting subsoil acidity. In several studies it was observed that 10 to 14 years were required for surface-applied lime to raise soil pH at a depth of 15 cm.

TABLE 11-11. Effect of Liming on the Harmful Effects of Clubroot Disease in Cauliflower

Lime (tons/A)	Lime Applied in 1978			Lime Applied in 1979		
	Yield Percent Marketable	Clubroot Rating *	pH at Harvest	Yield Percent Marketable	Clubroot Rating *	pH at Harvest
0	48	3.3	5.6	28	3.8	5.7
2.5	73	1.8	6.6	39	3.8	6.4
5.0	81	1.1	6.9	63	3.4	6.7
10.0	86	0.2	7.1	74	2.5	7.2

* Clubroot rating = $\dfrac{\text{sum of (number of roots at a rating} \times \text{rating)}}{\text{total number of roots}}$

Rating: 0, no visible clubroot; 1, less than 10 galls on the lateral roots; 2, more than 10 galls on the lateral roots, taproot free of clubroot; 3, galls on taproot; 4, severe clubbing on all roots.
Source: Waring, Proc. 22nd Annu. Lower Mainland Hort. Improvement Assoc. Growers' Short Course, Lower Mainland Horticultural Improvement Association and British Columbia Ministry of Agriculture, pp. 95–96 (1980).

Results from a long-term experiment on a Wooster silt loam soil in Ohio demonstrate that deep movement of cations can be significant (see Table 11-12). It is evident that maintenance of surface soil at pH 6.0, 6.5, and 7.2 reduced acidity deeper in the root zone. Similar effects have been observed at other locations including a 6-year experiment in Maine. Keeping surface soils at the proper pH over a period of years is a practical way of at least partially overcoming the problem of subsoil acidity.

Neutralization of subsoil acidity through deep incorporation of surface-applied liming agents is possible with tillage equipment now available. Investigations conducted in Alabama on the effect of depth of incorporation of surface applied lime on growth of cotton and corn showed that the amount and depth of cotton rooting was encouraged by mixing lime to depths of 45 cm. (Figure 11-9). However, the seed cotton yields were not increased by mixing lime any deeper than 15 cm at one location (Figure 11-10). At the two other sites there was no yield advantage for lime incorporation below the 30-cm depth. Corn grain and stover yields were unaffected by depth of lime incorporation in one soil, while deep mixing was beneficial at the second site. It is likely in some situations that optimum amounts and distribution of precipitation during the growing season will tend to compensate for shallow root systems.

An example from Brazil is shown in Figure 11-11, where mixing lime even deeper, to depths of 60 cm, markedly increased yields of corn.

With no-till cropping systems the surface inch or two becomes quite acid in a few years, primarily because of the surface application of nitrogen fertilizers (Table 11-13). Soil management to control this acid-producing effect is an important part

TABLE 11-12. Effect of Continued Liming on the Surface and Subsoil pH of Wooster Silt Loam Soil in Ohio

Soil Depth (in.)	pH at Various Depths Resulting from Surface Application of Limestone				
0– 7	4.9	5.5	6.0	6.5	7.2
7–14	4.9	5.2	5.9	6.7	7.2
14–21	4.7	4.8	5.2	5.4	6.5

Source: Ohio Agronomy Guide 1983–84. Columbus, Ohio: Cooperative Extension Service, Ohio State Univ., 1983.

FIGURE 11-9. Amount and depth of cotton rooting as affected by depth of lime incorporation. From left to right: unlimed; 0–15 cm (0–6 inches) limed; 0–45 cm (0–18 inches) limed. [Doss et al., *Agron. J.,* **71**:541 (1979).]

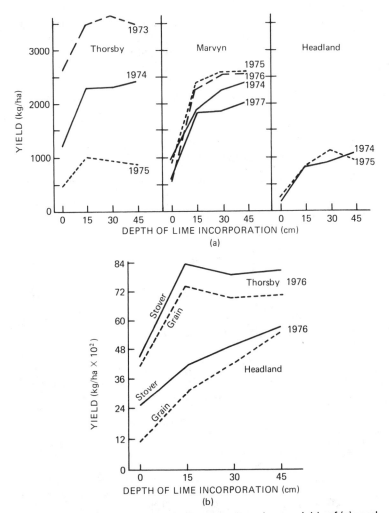

FIGURE 11-10. Effect of depth of lime incorporation on yields of (a) seed cotton and (b) corn grain and stover. Treatments imposed in 1973 at Thorsby and in 1974 at Marvyn and Headland. [Doss et al., *Agron. J.,* **71**:541 (1979).]

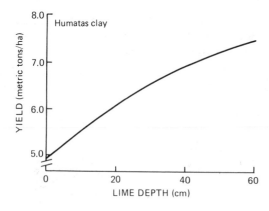

FIGURE 11-11. Effect of depth of lime incorporation on corn yield. [Bouldin, *Cornell Int. Agr. Bull. 74*, Ithaca, N.Y. Cornell Univ. (1979).]

of continuous no-tillage row crop production systems. Fortunately, this increased acidification is concentrated in the soil surface, where it can be readily corrected by surface liming.

Researchers in Oregon have investigated segmental or partial mixing of lime with soil as a means of reducing liming rates. They obtained maximum yields of wheat dry matter under growth-chamber conditions when lime was confined to 30% of the soil rather than complete mixing (Table 11-14).

One of the factors contributing to low yields on acid soils is more difficult weed control. Figure 11-12 shows the effect of pH in a Maury silt loam soil on the

TABLE 11-13.   Soil pH After 7 Years of Continuous Corn Grown on Maury Silt Loam Soil in Kentucky Affected by Tillage Methods, Nitrogen Fertilization, and Liming

Nitrogen Treatment	Soil Depth (cm)	Conventional Tillage		No-Tillage	
		Limed	Unlimed	Limed	Unlimed
High N rate	0–5	5.3	4.9	5.5	4.3
(336 kg/ha)	5–15	5.9	5.1	5.3	4.8
	15–30	6.0	5.5	5.8	5.5
Moderate N rate	0–5	5.9	5.2	5.9	4.8
(168 kg/ha)	5–15	6.3	5.6	5.9	5.5
	15–30	6.2	5.7	6.0	5.9

Source: Blevins et al., *Agron. J.,* **40**:322 (1978).

TABLE 11-14.   Effect of Mixing Lime with Varying Soil Volumes on the Growth of Nugaines Winter Wheat in the Growth Chamber (g/box *)

Rate of Lime (tons/ha)	Percent of Soil Limed				
	0	10	30	60	100
0	1.27	—	—	—	—
2.24	—	1.95	2.42	2.03	1.83
6.72	—	1.85	3.10	3.02	2.60
11.20	—	1.72	3.50	2.90	2.60
SEM 0.153					

* Percent × rate interaction significant at the 0.01 level.
Source: Kauffman and Gardner, *Agron. J.,* **70**:331 (1978).

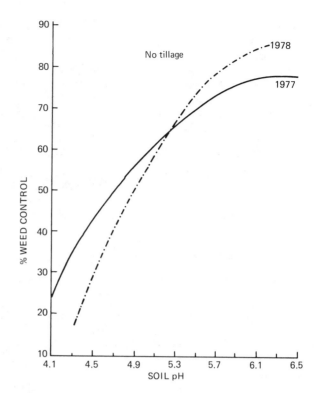

FIGURE 11-12. Effect of pH of the Maury silt loam on weed control in corn. [Kells et al., *Agronomy Notes* 12(2), Univ. of Kentucky (1979).]

effectiveness of broadcast triazine herbicides. Poor weed control is related to the increased rate of degradation and absorption of simazine at the lower soil pH values.

Equipment

Bulk application by the supplier who hauls lime to the farm is most efficient. The spinner truck spreader, which throws the lime in a semicircle from the rear of the truck, is often used. Uniform spreading is more difficult with this equipment than with the kind that drops the lime from a covered hopper or conveyor.

Suspending liming agents of various types in water or fertilizer solution, frequently referred to as "fluid lime," is a relatively new approach to lime application. Mixing equipment used for conventional suspension fertilizers can be readily utilized to produce fluid lime. Liming materials with particle sizes ranging from -20 mesh to -325 mesh are satisfactory. However, -100-mesh materials are perhaps most commonly used. Formulations composed of 50% lime and 50% water are most popular. Some of the salient features of fluid lime are:

1. An excellent distribution pattern can be obtained with no dust.
2. The finely divided lime reacts rapidly with soil.
3. Only a small amount of liming material is applied at any one time, (e.g., 500 to 1000 lb of material per acre).
4. Very-low-pH soils can be corrected quickly.
5. Regular annual applications help to maintain pH.
6. Cost of lime applied by this method is usually two to four times more expensive than dry additions, so economics must be considered.
7. Fluid fertilizer dealers are supplying fluid lime in areas where conventional lime has been difficult to obtain.

Urea-ammonium nitrate (UAN) solutions can be used for suspending the lime component. Ammonia volatilization is not a problem as long as CaO is absent. Incorporation soon after application of the suspension will eliminate potential ammonia losses following urea hydrolysis.

Nitrogen–potassium–lime suspensions are being used successfully. The feasibility of including phosphorus sources in lime suspensions has yet to be established. Limited studies conducted by Winter at Kansas State University indicate that lime–herbicide suspensions can be used safely for grain sorghum and soybean production.

Regardless of the method employed, care should be taken to ensure uniform application. An examination of the distribution pattern at the start of the operation is helpful in correcting nonuniform spreading and providing proper lapping. Nonuniform distribution can result in excesses and deficiencies in different parts of the same field and corresponding nonuniform crop growth.

Factors Determining the Selection of a Liming Program

It is obvious that a number of factors will determine the selection of a liming program.

1. Lime requirement of the crop to be grown.
2. Texture and organic matter content of the soil as well as the pH.
3. Time and frequency of liming.
4. Nature and cost of the liming material.
5. Depth of tillage.

Lime Requirement of the Crop. Plants differ widely in their response to added lime. The nature of this response, as previously pointed out, is not always known, but it is a matter of common observation that certain plants will grow well in acid soils, whereas others will not. In considering the liming program for a given soil, the type of crop to be grown ranks first in importance. Plants such as blueberries, cranberries, azaleas, and camellias do best on soils that are distinctly acid, whereas plants such as sweet clover, alfalfa, and sugar beets make their best growth on neutral to slightly alkaline soils. On Oxisols and Ultisols crops such as corn and wheat grow well at pH values of 5.5 to 5.8. On Mollisols and similar soils that are not as highly weathered as the Oxisols and Ultisols, best growth may be obtained in the pH range of 6.0 to 6.4. It should be noted that this evidence on Mollisols is mainly based on yields where legumes in the rotation are the source of nitrogen for corn.

Texture and Organic Matter Content of the Soil. The importance of this point has already been covered. If a soil is coarse-textured and has a low organic matter content, the quantity of lime to be applied will certainly be less than required to effect the same pH change in a fine-textured soil or one high in organic matter. The overliming of coarse-textured soils is not uncommon, but a knowledge of the basic chemistry of soils can prevent it. In areas in which a soil-testing service is available advice should be sought. In the absence of such an organization a soil-test kit in the hands of an experienced person will suffice.

Time and Frequency of Liming Applications. For rotations that include leguminous crops lime should be applied 3 to 6 months before the time of seeding. This is particularly important on very acid soils. Liming just a few days before seeding alfalfa under such conditions may produce disappointing results, for the lime may not have adequate time to react with the soil. If clover is to follow fall-seeded wheat, the lime is best applied when the wheat is planted. The caustic forms of lime [CaO and Ca(OH)$_2$] should be spread well before planting to prevent injury to germinating seeds. With bulk spreading there is a greater tendency to add the lime whenever convenient. For example, during the summer and fall after hay harvest or on pastures is a good time to get the truck out on the fields and it also helps to distribute the peak demands on lime dealers.

The frequency of application generally depends on the texture of the soil, nitrogen applied, crop removal, and the amount of lime applied. On sandy soils frequent light applications are preferable, whereas with fine-textured soils larger amounts may be applied less often. The type of limestone added will also determine to a certain extent the frequency of application. The finely divided materials react more quickly, but their effect is maintained over a shorter period than that of materials containing appreciable amounts of coarse particles.

The most satisfactory means of determining reliming needs is by soil tests. Samples should be taken every 3 to 5 years—for sandy soils somewhat more frequently.

Depth of Tillage. More farmers are increasing the depth of tillage from the customary 6 in. to 7 to 10 in. Lime recommendations are presently made on the basis of a 6-in. furrow slice. When land is tilled to a depth of 10 in., the lime recommendations should be increased by at least 50%.

Summary

1. An acid is a substance that tends to give up protons (hydrogen ions) to some other substance. A base is any substance that tends to accept protons.

2. Acids range all the way from strong to weak. Those that are strong are those that dissociate to a great degree, whereas those that are weak dissociate only slightly. The strength of acids and the methods of expressing it were discussed. The pH concept, which is a convenient means of describing the activity of weak acids, was explained. Neutralization of acids was covered and illustrated with appropriate equations and the principle of buffering and buffered systems was explained.

3. In many ways soils are similar to buffered weak acids. This and other current ideas on the fundamental nature of soil acidity were discussed. Soil acidity is affected by the nature and amount of humus and clay colloids present, the amount of hydrous oxides of aluminum and iron, especially aluminum, the content of soluble salts in the soil, and the level of carbon dioxide in the soil atmosphere. The importance of aluminum in soil acidity was stressed.

4. Soil acidity in crop production systems is influenced by the use of commercial fertilizers, especially ammoniacal sources, which produce H$^+$ ions during nitrification; crop removal and leaching of basic cations; and decomposition of organic residues.

5. Some fertilizer materials leave an acid residue in the soil, others a basic residue, and still others have no influence on soil pH. The effect of various fertilizer materials on soil pH and the mechanisms responsible for their action were covered.

6. Plant growth reduces the potential or theoretical amount of acidity produced by nitrification of ammoniacal fertilizer materials. Reasons for this variance were discussed.

7. Soil pH, or active acidity, is registered by a pH meter on a sample of soil and water. Potential acidity is determined by adding known amounts of acid and base to a series of soil samples and measuring the pH change. The curve plotted from such data is termed a buffer curve and is used in calculating the lime requirement of field soils. Lime requirements can also be measured using various strongly buffered solutions.

8. Soil pH determinations are helpful for indicating the presence of exchangeable aluminum and hydrogen ions. Exchangeable hydrogen is normally present in measurable quantities at pH values below 4, while exchangeable aluminum usually occurs in significant amounts at pH levels lower than about 5.5. Multicharged aluminum polymers are dominant in the pH range 5.5 to 7.0.

9. Several materials, including dolomitic and calcitic limestones, burnt lime, hydrated lime, marl, and slags, are employed commercially in the liming of soils. Their properties and reactions in soil were discussed.

10. The neutralizing value, or CCE, of limestones is a measure of their effectiveness in neutralizing soil acidity. Several methods of determining this property were described.

11. Lime is one of the most important of the production inputs in the farming system. Its effect on phosphate and microelement availability, nitrification, nitrogen fixation, soil structure, and disease influences crop production in many ways.

12. Movement and placement of lime were discussed. New developments in liming methods such as fluid lime were described.

13. The selection of a liming program to be followed is determined by the lime requirements of the crop, the pH, texture, and organic matter content of the soil, the liming material to be used, and the time and frequency of the lime applications.

Questions

1. What is an acid?

2. How are acids neutralized?

3. What ions are the principal sources of soil acidity?

4. Can pH measurements be used to identify the principal sources of soil acidity? What are the general relationships?

5. For what reasons do soils become acid?

6. What, if any, influence do aluminum and iron polymers have on soil acidity? Do they influence other soil properties? Which ones?

7. Is soil pH affected by fertilizer applications?

8. Are there differences in crop uptake of basic cations?

9. The term *agricultural* lime usually refers to what material?

10. Chemically speaking, lime refers to what compound?

11. Distinguish between active acidity and potential acidity. Which of these two forms is measured when a pH determination is made?

12. What is meant by the terms *buffer* and *buffer capacity*?

13. What soil properties determine its buffer capacity?

14. How is the buffer capacity of a soil related to the lime requirement of that soil?

15. Define the term *lime requirement.*

16. How is the lime requirement of a soil determined?

17. Calcium chloride ($CaCl_2$) contains 53% calcium. Express this as an equivalent percentage of calcium oxide. Can calcium chloride be used as a liming agent? Why?

18. What three types of slag can be used as effective liming materials?

19. What are marls? What determines their value as liming materials?

20. Define *neutralizing value* or *calcium carbonate equivalent* as it refers to liming materials.

21. What is the calcium carbonate equivalent of sodium carbonate?

22. You analyze a limestone and find that it has a neutralizing value of 85%. How many tons of this limestone would be equivalent to 3 tons of chemically pure calcium carbonate?
23. In addition to its purity and neutralizing value, what other property of crystalline limestones is important with respect to their value as agricultural liming materials?
24. What are the principal direct benefits of adding lime to soil?
25. What are several indirect benefits of adding lime to soil?
26. A solution has a pH value of 6.5. To what hydrogen ion activity does this correspond?
27. Does lime move quickly in soil?
28. Are benefits derived from deep mixing of lime in soil?
29. Can long-term liming of surface soil influence subsoil acidity?
30. What are several factors that will determine the frequency and rate of liming?
31. You have two fields, A and B, that need liming. The characteristics of the soils in each of these fields are the following:

| | Field | |
Soil Property	A	B
Organic matter (%)	0.8	3.1
Clay (%)	10.0	38.1
Sand (%)	74.0	51.2
Type of clay	1:1	1:2
pH	5.2	5.2

You have lost the liming recommendations sent to you by the soil laboratory but you do recall that 3 tons/A were recommended for field B. Because the pH is the same in both fields, you apply 3 tons to field A as well. Have you acted wisely? Why?

32. Solution A has a pH of 3.0. Solution B has a pH of 6.0. The active acidity of solution A is how many times greater than that of solution B?
33. What is fluid lime? What are its major advantages and disadvantages?
34. Assume that you have available three dolomitic limestones of equal neutralizing value but of the following mechanical analyses:

| | Limestone | | |
	A	B	C
Coarser than 8 mesh	20	5	0
Coarser than 60 mesh	70	30	20
Coarser than 100 mesh	95	60	50

(a) Which limestone would you not buy? (b) Which one would you select for the quickest results? (c) Under what circumstances could you afford to buy limestone B?

35. In what parts of the United States and Canada is liming needed and a very important practice? Where is it relatively unimportant? In your own particular area, is liming a needed practice? Why?
36. Define the term *effective CEC*. How and why does it differ from the CEC as determined by the ammonium acetate method?

37. With what type of soil would the ammonium acetate method give a fairly good approximation of the effective CEC of soil? On what types of soil would it not give a good estimate? Would the estimate by high or low? Why? What method would give a better estimate of the CEC of such soils?

Selected References

Abruna-Rodriguez, F., J. Vicente-Chandler, R. W. Pearson, and S. Silva, "Crop response to soil acidity factors in Ultisols and Oxisols: I. Tobacco." *SSSA Proc.*, **34:**629 (1970).

Adams, F., and R. W. Pearson, "Neutralizing soil acidity under Bermudagrass sod." *SSSA Proc.*, **33:**737 (1969).

Adams F., and R. W. Pearson, "Differential response of cotton and peanuts to subsoil acidity." *Agron. J.*, **62:**9 (1970).

Andrew, C. S., and D. O. Norris, "Comparative responses to calcium of five tropical and four temperate pasture legume species." *Australian J. Agr. Res.*, **12:**40 (1961).

Assoc. Am. Pl. Fd. Cont. Offic., Model Agricultural Liming Materials Bill (tentative, 1970).

Barrows, H. L., A. W. Taylor, and E. C. Simpson, "Interaction of limestone particle size and phosphorus on the control of soil acidity." *SSSA Proc.*, **32:**64 (1968).

Beaton, J. D., R. C. Speer, and J. T. Harapiak, "Response of red clover to Kimberley electric furnace iron slag and other liming materials." *Can. J. Plant Sci.*, **48:**455 (1968).

Bertsch, P. M., and M. M. Alley, "Conventional and suspension limestone influence on soil chemical properties and corn and soybean yields." *Agron. J.*, **73:**1075 (1981).

Blevins, R. L., L. W. Murdock, and G. W. Thomas, "Effect of lime application on no-tillage and conventionally tilled corn." *Agron. J.*, **70:**322 (1978).

Bohn, H., B. McNeal, and G. O'Connor, *Soil Chemistry*. New York: Wiley, 1979.

Bouldin, D. R., "The influence of subsoil acidity on crop yield potential." *Cornell Int. Agr. Bull. 74*, Cornell Univ., Ithaca, N.Y. (1979).

Brown, I. C., "A rapid method of determining exchangeable hydrogen and total exchangeable bases of soils." *Soil Sci.*, **56:**353 (1943).

Buckman, H. O., and N. C. Brady, *The Nature and Properties of Soils*, 7th ed. New York: Macmillan, 1969.

Coleman, N. T., E. J. Kamprath, and S. B. Weed, "Liming." *Adv. Agron.*, **10:**475 (1958).

Collins, J. B., E. P. Whiteside, and C. E. Cress, "Seasonal variability of pH and lime requirements in several southern Michigan soils when measured in different ways." *SSSA Proc.*, **34:**56 (1970).

Doss, B. D., W. T. Dumas, and Z. F. Lund, "Depth of lime incorporation for correction of subsoil acidity." *Agron. J.*, **71:**541 (1979).

El Gibaly, H., and J. H. Axley, "A chemical method for the rating of agricultural limestones as soil amendments." *SSSA Proc.*, **19:**301 (1955).

Eno, C. F., "The relationship of soil reaction to the activities of soil microorganisms. A review." *Soil Crop Sci. Soc. Florida, Proc.*, **17:**34 (1957).

Evans, C. E., and E. J. Kamprath, "Lime response as related to percent Al saturation, solution Al, and organic matter content." *SSSA Proc.*, **34:**893 (1970).

Fisher, T. R., "Crop yields in relation to soil pH as modified by liming acid soils." *Missouri Agr. Sta. Res. Bull.*, 947 (1969).

Fleming, A. L., and C. D. Foy, "Root structure reflects differential aluminum tolerance in wheat varieties." *Agron. J.*, **60:**172 (1968).

Foy, C. D., "Effects of aluminum on plant growth," in E. W. Carson, Ed., *The Plant Root and Its Environment*, Chapter 20. Charlottesville, Va.: University Press of Virginia, 1974.

Foy, C. D., A. L. Fleming, and W. H. Armiger, "Aluminum tolerance of soybean varieties in relation to calcium nutrition." *Agron. J.*, **61:**505 (1969).

Foy, C. D., W. H. Armiger, L. W. Briggle, and D. A. Reid, "Aluminum tolerance of wheat and barley varieties in acid soils." *Agron. J.*, **57**:413 (1965).

Foy, C. D., A. L. Fleming, G. R. Burns, and W. H. Armiger, "Characterization of differential aluminum tolerance among varieties of wheat and barley." *SSSA Proc.*, **31**:513 (1967).

Heddelson, M. R., E. O. McLean, and N. Holowaychuck, "Aluminum in soils: IV. The role of aluminum in soil acidity." *SSSA Proc.*, **24**:91 (1960).

Hoyt, P. B., "Liming research for dryland agriculture in western Canada." *Better Crops Plant Food*, **66**:13 (Fall 1982).

Hunter, A. S., H. Kinney, C. W. Whittaker, J. H. Axley, M. Peech, and J. E. Steckel, "Reproducibility of ratings and correlations with chemical and physical characteristics of materials (lime)." *Agron. J.*, **55**:351 (1963).

Hutchinson, F. E., and A. S. Hunter, "Exchangeable aluminum levels in two soils as related to lime treatment and growth of six crop species." *Agron. J.*, **62**:702 (1970).

Jenny, H., "Reflections on the soil acidity merry-go-round." *SSSA Proc.*, **25**:428 (1961).

Jolley, V. D., and W. H. Pierre, "Profile accumulation of fertilizer-derived nitrate and total nitrogen recovery in two long-term nitrogen-rate experiments with corn." *Agron. J.*, **41**:373 (1977).

Kamprath, E. J., "Exchangeable aluminum as a criterion for liming leached mineral soils." *SSSA Proc.*, **34**:252 (1970).

Kamprath, E. J., "Potential detrimental effects from liming highly weathered soils to neutrality." *Soil Crop Sci. Soc. Florida, Proc.*, **31**:200 (1971).

Kauffman, M. D., and E. H. Gardner, "Segmental liming of soil and its effect on the growth of wheat." *Agron. J.*, **70**:331 (1978).

Lee, C. R., "Influence of aluminum on plant growth and tuber yield of potatoes." *Agron. J.*, **63**:363 (1971).

Lee, C. R., "Influence of aluminum on plant growth and mineral nutrition of potatoes." *Agron. J.*, **63**:604 (1971).

Lin, C., and N. T. Coleman, "The measurement of exchangeable Al in soils and clays." *SSSA Proc.*, **24**:444 (1960).

Lindsay, W. L., M. Peech, and J. S. Clark, "Determination of aluminum ion activity in soil extracts." *SSSA Proc.*, **23**:266 (1959).

Long, F. L., and C. D. Foy, "Plant varieties as indicators of aluminum toxicity in the A_2 horizon of a Norfolk soil." *Agron. J.*, **62**:679 (1970).

Love, J. R., R. B. Corey, and C. C. Olsen, "Effect of particle size and rate of application of dolomitic limestone on soil pH and growth of alfalfa." *Trans. 7th Int. Congr. Soil Sci.*, **37**:293 (1960).

Love, K. S., and C. W. Whittaker, "Surface area and reactivity of typical limestones." *J. Agr. Food Chem.*, **2**:1268 (1954).

McLean, E. O., and E. J. Kamprath, "Letters to the Editor." *SSSA Proc.*, **34**:363 (1970).

McLean, E. O., S. W. Dumford, and F. Coronel, "A comparison of several methods of determining lime requirements of soils." *SSSA Proc.*, **30**:26 (1966).

McLean, E. O., D. J. Eckert, G. Y. Reddy, and J. F. Trierweiler, "An improved SMP soil lime requirement method incorporating double-buffer and quick-test features." *Soil Sci. Soc. Am. J.*, **42**:311 (1978).

Mahler, R. L., F. E. Koehler, and A. R. Halvorson, "Implications of gradual acidification of farmland in semi-arid regions of the Pacific Northwest." *Proc. 32nd Annu. Northwest Fert. Conf.*, pp. 3–14, Billings, Mont. (1981).

Marion, G. M., D. M. Hendricks, G. R. Dutt, and W. H. Fuller, "Aluminum and silica solubility in soils." Soil Sci., **121**:76 (1976).

Moschler, W. W., G. D. Jones, and G. W. Thomas, "Lime and soil acidity effects on alfalfa growth in a red-yellow podzolic soil." *SSSA Proc.*, **24**:507 (1960).

Moschler, W. W., D. C. Martens, C. I. Rich, and G. M. Shear, "Comparative lime effects on continuous no-tillage and conventionally-tilled corn." *SSSA Proc.*, **65**:781 (1973).

Motto, H. L., and S. W. Melsted, "Efficiency of various particle size fractions of limestone." *SSSA Proc.*, **24**:488 (1960).

Munson, R. D., "Potassium, calcium, and magnesium in the tropics and sub-tropics." *Tech. Bull. IFDC-T-23.* Muscle Shoals, Ala.: International Fertilizer Development Center, 1982.

Murphy, L. S., K. T. Winter, and D. E. Kissel, "Agronomic response and comparative economics of fluid lime." *1980 Annu. Meet., Am. Chem. Soc.,* Las Vegas, Nev. (August 24–29, 1980).

Murphy, L. S., K. Winter, and R. E. Lamond, "The potential impact of lime suspensions." *Situation 1978, TVA Fert. Conf. Bull. Y-131,* pp. 59–68. Muscle Shoals, Ala.: National Fertilizer Development Center–Tennessee Valley Authority, 1978.

National Fertilizer Development Center. *Nat. Conf. Agr. Limestone,* Muscle Shoals, Ala. (1980).

Nyborg, M., and P. B. Hoyt, "Effects of soil acidity and liming on mineralization of soil nitrogen." *Can. J. Soil Sci.,* **58:**331 (1978).

Nye, P., D. Craig, N. T. Coleman, and J. L. Ragland, "Ion exchange equilibria involving aluminum." *SSSA Proc.,* **25:**14 (1961).

Ohio Agronomy Guide 1983–84. Columbus, Ohio: Cooperative Extension Service, Ohio State Univ., 1983.

Olson, R. A., et al., Eds., *Fertilizer Technology and Use,* 2nd ed. Madison, Wis.: Soil Science Society of America, 1971.

Pearson, R. W., and F. Adams, Eds., *Soil Acidity and Liming.* Madison, Wis.: American Society of Agronomy, 1967.

Pierre, W. H., and W. L. Banwart, "Excess-base and excess-base/nitrogen ratio of various crop species and parts of plants." *Agron. J.,* **65:**91 (1973).

Pierre, W. H., J. Meisinger, and J. R. Birchett, "Cation–anion balance in crops as a factor in determining the effect of nitrogen fertilizers on soil acidity." *Agron. J.,* **62:**106 (1970).

Pionke, H. B., R. B. Corey, and E. E. Schulte, "Contributions of soil factors to lime requirement and lime requirement tests." *SSSA Proc.,* **32:**113 (1968).

Plucknett, D. L., and G. D. Sherman, "Extractable aluminum in some Hawaiian soils." *SSSA Proc.,* **27:**39 (1963).

Pratt, P. F., "Phosphorus and aluminum interactions in the acidification of soils." *SSSA Proc.,* **25:**467 (1961).

Pratt, P. F., and F. L. Bair, "Buffer methods for estimating lime and sulfur applications for pH control of soils." *Soil Sci.,* **93:**329 (1962).

Reid, D. A., et al., "Differential aluminum tolerance of winter barley varieties and selections in associated greenhouse and field experiments." *Agron. J.,* **61:**218 (1969).

Rixon, A. J., and G. D. Sherman, "Effects of heavy lime applications to volcanic ash soils in the humid tropics." *Soil Sci.,* **94:**19 (1962).

Rysler, G. J., G. R. Gist, and G. W. Volk, "Equivalent amounts of liming materials." Mimeo., Agron. Dept., Ohio State Univ.

Schollenberger, C. J., and C. W. Whittaker, "The ammonium chloride-liming reaction." *J. Assoc. Offic. Agr. Chem.,* **36:**1130 (1953).

Schollenberger, C. J., and C. W. Whittaker, "A comparison of methods for evaluating activities of agricultural limestones." *Soil Sci.,* **93:**161 (1962).

Shainberg, I., and M. Gal, "The effect of lime on the response of soils to sodic conditions." *J. Soil Sci.,* **33:**489 (1982).

Shaw, W. M., "Rate of reaction of limestone with soils." *Tennessee Univ. Agr. Exp. Sta. Bull. 319* (1960).

Shaw, W. M., and B. Robinson, "Chemical evaluation of neutralizing efficiency of agricultural limestone." *Soil Sci.,* **87:**262 (1959).

Shoemaker, H. E., E. O. McLean, and P. F. Pratt, "Buffer methods for determining lime requirement of soils with appreciable amounts of extractable aluminum." *SSSA Proc.,* **25:**274 (1961).

Soileau, J. M., O. P. Engelstad, and J. B. Martin, Jr., "Cotton growth in an acid fragipan subsoil: II. Effects of soluble calcium, magnesium, and aluminum on roots and tops." *SSSA Proc.,* **33:**919 (1969).

Thorup, J. T., "pH effect on root growth and water uptake by plants." *Agron. J.,* **61:**225 (1969).

Tisdale, S. L., "The use of sulphur compounds in irrigated arid-land agriculture." *Sulphur Inst. J.*, **6**:2 (1970).

Turner, R. C., and W. E. Nichol, "A study of the lime potential: 1. Conditions for the lime potential to be independent of salt concentration in aqueous suspensions of negatively charged clays." *Soil Sci.*, **93**:374 (1962).

Turner, R. C., and W. E. Nichol, "A study of the lime potential: 2. Relation between lime potential and per cent base saturation of negatively charged clays in aqueous salt suspensions." *Soil Sci.*, **94**:58 (1962).

Walsh, L. M., and J. D. Beaton, Eds., *Soil Testing and Plant Analysis*, rev. ed. Madison, Wis.: Soil Science Society of America, 1973.

Webber, M. D., P. B. Hoyt, M. Nyborg, and D. Corneau, "A comparison of lime requirement methods for acid Canadian soils." *Can. J. Soil Sci.*, **57**:361 (1977).

Westerman, R. L., "Factors affecting soil acidity." *Solutions*, **25**(3):64 (1981).

White, R. P., "Effects of lime upon soil and plant manganese levels in an acid soil." *SSSA Proc.*, **34**:625 (1970).

Whittaker, C. W., C. J. Erickson, K. S. Love, and D. M. Carroll, "Liming qualities of three cement kiln flue dusts and a limestone in a greenhouse comparison." *Agron. J.*, **51**:280 (1959).

Whittaker, C. W., J. H. Axley, M. Peech, J. E. Steckel, E. O. McLean, and A. S. Hunter, "Relationship of ratings and calcium carbonate content of limestones to their reactivity in the soil." *Agron. J.*, **55**:355 (1963).

Wolf, B., "Evaluation of calcined magnesite as a source of magnesium for plants." *Agron. J.*, **55**:261 (1963).

Woodruff, C. M., "Determination of the exchangeable hydrogen and lime requirement of the soil by means of the glass electrode and a buffered solution." *SSSA Proc.*, **12**:141 (1948).

Yuan, T. L., "Some relationships among hydrogen, aluminum and pH in solution and soil systems." *Soil Sci.*, **95**:155 (1963).

Chapter 12

Soil Fertility Evaluation

HISTORICALLY, crop production has been based on the use of plant nutrients already in the soil. Although the addition of plant nutrients is increasing, most crops continue to be grown on the basis of mining the soil for some or most nutrients. Soils, of course, vary greatly in how long they can be cropped without yield reduction before a given nutrient must be added. Diagnostic techniques, including identification of deficiency symptoms as well as soil and plant tests, are helpful in determining when additions are needed.

The selection of the proper rate of plant nutrients is influenced by a knowledge of the nutrient requirement of the crop and the nutrient-supplying power of the soil on which the crop is to be grown. When the soil does not furnish adequate quantities of the elements necessary for normal development of plants, it is essential that the required amounts be supplied. This necessitates finding a method that will permit the determination of those deficient elements. Obviously, looking at a given soil will tell little about its nutrient-supplying power. Red, gray, or black soils may all be deficient in nitrogen, phosphorus, potassium, or other nutrients.

Diagnosis of the needs of plants is comparable in many ways to diagnosis of human ills. The medical doctor observes the patient, obtains all the information possible with his questions, and then makes the appropriate tests, all of which are helpful in diagnosing the case. Similarly, the grower or agricultural worker observes the plants, obtains information on past management, and may make tests on the soil or the plant. The success of his diagnosis depends on his understanding of the fundamentals of plant and soil science and on a correct interpretation of the facts at hand.

Diagnostic measurements of the ailing plant or soil are often classed as troubleshooting. They can and are being used for this purpose, but *a more important application is in preventive measures*; for by the time a plant has shown deficiency symptoms a considerable reduction in the potential yield will already have occurred and the grower will have lost money. By the time potassium deficiency symptoms appear in potatoes, yield reduction may be as much as 50%.

Approaches Employed

The problem of predicting plant-nutrient needs has been under study for many years. In 1813, Sir Humphry Davy stated that if a soil is unproductive the cause of sterility can be determined by chemical analysis. This has not always been the case, but much work has been done on soil analysis and other techniques, and a gradual improvement in methods of predicting the fertility status of the soil has become evident.

In contrast to chemical soil analysis, which depends on chemical reagents for the measurement of available plant nutrients, biological methods make use of plants as extracting agents to achieve the same result. Generally speaking, biological soil tests are of two general types—those employing higher plants and those employing lower plants, such as bacteria and fungi. The criterion of treatment effect will vary with the method and may be expressed in a number of ways, including yield in bushels of grain, amount of a given nutrient extracted, quality, disease resistance, standability, or diameter of a mycelial growth.

In a consideration of the merits of chemical and biological tests it should be understood that to be of value as a basis for making lime and fertilizer recommendations the results must be correlated with crop responses in the field. With no prior knowledge of the relationship of the test results to crop response, the tests themselves are of little practical use.

Several techniques are commonly employed to assess the fertility status of a soil:

1. Nutrient-deficiency symptoms of plants.
2. Analyses of tissue from plants growing on the soil.
3. Biological tests in which the growth of either higher plants or certain microorganisms is used as a measure of soil fertility.
4. Chemical soil tests.

Nutrient-Deficiency Symptoms of Plants

Many of the methods for evaluating soil fertility are based on observations of or measurements on growing plants. These methods have considerable merit because the plants act as integrators of all growth factors (Figure 12-1) and are the products in which the grower is interested.

An abnormal appearance of the growing plant may be caused by a deficiency of one or more nutrient elements. If a plant is lacking in a particular element, more or less characteristic symptoms may appear. This visual method of evaluating soil fertility is unique in that it requires no expensive or elaborate equipment and can be used as a supplement to other diagnostic techniques.

Occurrence of Symptoms. Nutrient-deficiency symptoms may be classified as follows:

1. Complete crop failure at seedling stage.
2. Severe stunting of plants.
3. Specific leaf symptoms appearing at varying times during the season.
4. Internal abnormalities, such as clogged conductive tissues.
5. Delayed or abnormal maturity.
6. Obvious yield differences, with or without leaf symptoms.
7. Poor quality of crops, including unseen chemical composition differences, as in protein, oil, or starch content and in keeping or storage quality.
8. Yield differences detected only by careful experimental work.

In addition, nutrient deficiencies have a marked effect on extent and type of root growth (Figure 12-2). The underground portion of the plant has not received much attention because of the difficulty of making observations. However, when we consider that the roots are the main avenue of entry for nutrients, the importance of this aspect of plant development looms large.

Deficiency of an element does not directly produce symptoms. Rather, the normal plant processes are thrown out of balance, with the result that there is an accumulation of certain intermediate organic compounds and a shortage of others. This leads to the abnormal conditions recognized as symptoms and has a definite re-

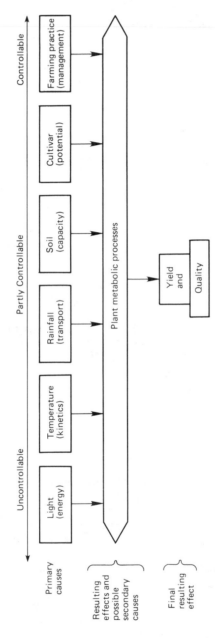

FIGURE 12-1. Schematic representation of the interrelationships between crop yield and quality, metabolic process, and external and genetic factors. [Beaufils, *Soil Sci. Bull. 1*, Univ. of Natal, Pietermaritzburg, South Africa (1973).]

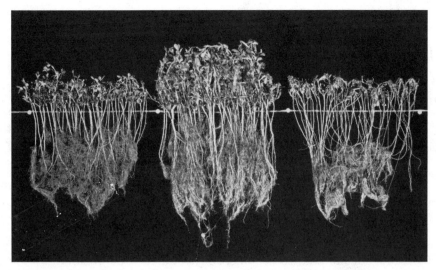

FIGURE 12-2. Omitting phosphorus *(left)* or potassium *(right)* reduced the growth of alfalfa roots as well as tops in the spring after seeding. There is evidence of alfalfa heaving above the groundline marked by the string. This soil tested low in phosphorus and potassium. (Courtesy of the Potash and Phosphate Institute, Atlanta, Ga.)

lation to shortages of elements. For example, diamine putrescine forms in some potassium-deficient plants and causes characteristic symptoms. Actually, a plant containing adequate potassium will show symptoms when injected with this compound.

Each symptom must be related to some function of the element in question. However, a given element may have several functions to perform, and this makes it difficult to explain the physiological reason for a particular deficiency symptom. For example, when nitrogen is deficient, the leaves of most plants tend to become pale green or light yellow. When the quantity of nitrogen is limiting, chlorophyll production is reduced, and the yellow pigments, carotene and xanthophyll, show through. Any one of a number of nutrient deficiencies, however, may produce pale green or yellow leaves, and the difficulty must be further related to a particular leaf pattern or location on the plant.

Deficiencies are actually relative, and a deficiency of one element implies adequate or excessive quantities of another. For example, manganese deficiency may be induced by adding large quantities of iron, provided the manganese supply is close to the critical point. This was discussed in Chapter 9. In addition, a sufficient supply under one condition may become deficient as other elements become more abundant. At a low level of nitrogen supply, the corn plant may not require much phosphorus, but with an adequate level of nitrogen the same phosphorus supply may become critical. In other words, once the first limiting factor is eliminated, the second limiting factor will appear.

Precautions. In the field it is often difficult to distinguish among the deficiency symptoms. It is not infrequent that disease or insect damage will resemble certain minor element deficiencies. An example is the confusion of leaf hopper damage with boron deficiency in alfalfa. It has been observed that boron deficiency is accompanied by a red coloration of the leaves near the growing point when the plant is well supplied with potassium. On the other hand, when the potassium content is low, yellowing of the alfalfa leaves occurs.

A symptom is a secondary effect and may be the result of more than one cause. For example, accumulated sugar in corn may combine with flavones to form anthocyanins (purple, red, and yellow pigments). Sugar accumulation may be based on several factors, such as an insufficient supply of phosphorus, cool nights and warm days, insect damage to the roots, nitrogen deficiency, or transverse creasing of the leaves. Insufficient phosphorus restricts the phosphorylation of sugars which is necessary for movement of sugars into cells. A shortage of two or more nutrients will also complicate the situation.

Nutrient-deficiency symptoms as a means of evaluating soil fertility represent an excellent example of closing the door after the horse has left the barn. These symptoms appear only after the supply of an element is so low that the plant can no longer function properly. In such cases it would have been profitable to have applied fertilizer long before the symptoms appeared.

If the symptom is observed early, it might be corrected during the growing season. This may be true with nitrogen, potassium, and certain micronutrients. Of course, the principal objective is to get the limiting nutrient into the plant as quickly as possible. With some elements and under some conditions this may be accomplished with foliar applications, or through irrigation water; otherwise, side-dressings must be used. Usually, the yield is reduced below the quantity that would have been obtained if adequate nutrients had been available at the beginning. However, if the trouble is properly diagnosed, the deficiency can be fully corrected the following year.

The points just discussed in relation to deficiency symptoms are raised so that the diagnostician may be aware of some of the pitfalls. It should be emphasized that the wise use of deficiency symptoms in conjunction with other methods of diagnosis, such as plant or soil analyses, can do much to promote proper fertilization. For a detailed description the reader is referred to Sprague (1964).

Hidden Hunger

Hidden hunger refers to a situation in which a crop needs more of a given element, yet has shown no deficiency symptoms (Figure 12-3). The content of an element is above the deficiency symptom zone but still considerably below that needed to permit the most profitable crop performance. Because agriculture has changed from a way of life to a business, the grower will make a determined effort to avoid obvious deficiency in his crops. However, he may not add quantities sufficient to obtain the most profitable yield or maximum economic yield. With most nutrients on most crops, significant responses can be obtained even though no recognizable symptoms have appeared.

In the beginning stages of use of a plant nutrient in an area, deficiency symptoms point toward first recognition of trouble. However, as use of the nutrient increases

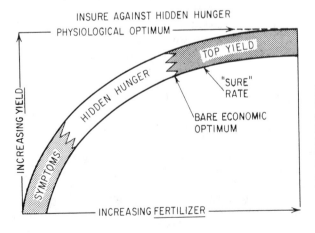

FIGURE 12-3. *Hidden hunger* is a term used to describe a plant that shows no obvious symptoms, yet the nutrient content is not sufficient to give the top profitable yield. Fertilization with the "sure" rate rather than the bare economic optimum for an average year helps to obtain the top profitable yield. (Courtesy of the Potash & Phosphate Institute, Atlanta, Ga.)

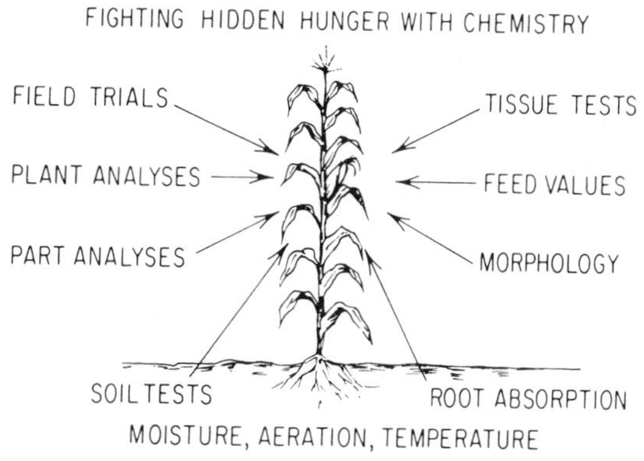

FIGHTING HIDDEN HUNGER WITH CHEMISTRY

FIELD TRIALS ⟋   ⟍ TISSUE TESTS

PLANT ANALYSES →   ← FEED VALUES

PART ANALYSES ⟋   ⟍ MORPHOLOGY

SOIL TESTS   ROOT ABSORPTION

MOISTURE, AERATION, TEMPERATURE

FIGURE 12-4. Detecting hidden hunger in crops is an increasing problem as yield goals rise and higher profits are sought. In this zone, with no symptoms to guide us, we must turn to more diagnostic chemistry to evaluate needs more accurately. Many diagnostic tools are available. (Courtesy of the Potash & Phosphate Institute, Atlanta, Ga.)

and higher yields are desired, deficiency symptoms are of less value and can be classified as a problem of the marginal farmer.

The question then is how best to eliminate hidden hunger (Figure 12-4). Testing of plants and of soils is helpful for planning or modifying plant-nutrient programs to avoid this problem in subsequent crops. In both approaches careful consideration must be given to past management practices.

Seasonal Effects

Nutrient shortages in the soil may be caused by or intensified by abnormal weather conditions. Nutrients may be present in sufficient quantities when conditions are ideal, but in drought, excessive moisture, or unusual temperature the plant may not be able to obtain an adequate supply. For example, under cooler temperatures less nitrogen, phosphorus, and potassium are taken up by tomatoes.

Likewise, moisture stress influences nutrient uptake. As moisture stress increased, the concentrations of N-P-K in corn leaves decreased (Table 12-1). Application of these nutrients reduced the effects of moisture stress, but concentrations were still below the optimum in stress years.

There are instances, not too uncommon, of nutrient deficiency symptoms appearing during early growth. Such deficiencies may disappear as the growing

TABLE 12-1. Influence of Applied Nitrogen, Phosphorus, and Potassium and Moisture Stress on Percent Nitrogen, Phosphorus, and Potassium in Corn Leaves

Nutrients Applied (kg/ha)			N-P-K Concentration	
N	P	K	No Stress Days	Maximum Stress
			% N	
0	78	47	2.0	1.5
179	78	47	2.9	2.2
			% P	
179	0	47	0.26	0.12
179	78	47	0.32	0.18
			% K	
179	39	0	1.1	0.7
179	39	93	1.6	1.2

Source: Voss, *Proc. 22nd Annu. Fert. Agr. Chem. Dealers' Conf.,* Iowa State Univ. (1970).

season progresses, or there may be no measurable yield benefit from supplemental additions of the nutrient(s) in question. It has frequently been observed that fertilizer phosphorus will improve early growth of crops but at harvest, there is no measurable yield response. Such occurrences are probably related to some of these seasonal effects or to penetration of roots into areas of the soil having higher fertility levels. Disappearance of deficiency symptoms could also be due to slowing of the growth rate, and thus a lessening in demand for nutrients.

Homeowners buy fire insurance for their homes, hoping never to collect. It is worthwhile to consider the insurance feature and add enough nutrients to meet seasonal variations. To eliminate plant nutrients as a limiting factor the nutrient content of the plant must be raised to the sure or insurance level rather than to the bare economic optimum (adequate in some years). Fertilizing to this level helps to take advantage of a good season and leaves nutrients in the soil for the succeeding crop.

Plant Analyses

Two general types of plant analysis have been used. One is the tissue test which is customarily made on fresh tissue in the field. The other is the total analysis performed in the laboratory with precise analytical techniques.

Plant analyses are based on the premise that the amount of a given element in a plant is an indication of the supply of that particular nutrient and as such is directly related to the quantity in the soil. Since a shortage of an element will limit growth, other elements may accumulate in the cell sap and show high tests, regardless of supply. For example, if corn is low in nitrate, the phosphorus test may show high. This is no indication, however, that if adequate nitrogen were supplied to the corn the supply of phosphorus would be adequate.

Tissue Tests. Rapid tests for the determination of nutrient elements in the plant sap of fresh tissue have found an important place in the diagnosis of the needs of growing plants. In these tests the sap from ruptured cells is tested for unassimilated nitrogen, phosphorus, and potassium.

They are semiquantitative tests intended mainly for verifying or predicting deficiencies of nitrogen, phosphorus, or potassium. The results are read as very low, low, medium, or high. Through the proper application of tissue testing it is possible to anticipate or forecast certain production problems—while still in the field. Although not as widely used, similar on-the-spot tissue tests have been developed for other nutrients, including sulfur, magnesium, manganese and zinc.

The plant roots absorb the nutrients from the soil, and these nutrients are transported to other parts of the plant where they are needed. The concentration of the nutrients in the cell sap is usually a good indication of how well the plant is supplied *at the time of testing*. The cell sap of the conducting tissues might be compared with the conveyor belts in a factory. If the factory is to operate at full capacity, all the belts bringing in raw materials must be running on schedule. If one raw material is short, its belt will run empty, the other raw materials will pile up, and production will be drastically reduced. An alert factory superintendent will make sure that there are no shortages. Similarly, an alert farmer or agricultural worker will make certain that no nutrient is limiting crop growth and that the supplies of nutrients are in the proper balance.

General Methods. The release in 1926 of Purdue Agricultural Experiment Station Bulletin 298, "Testing Corn Stalks Chemically to Aid in Determining Their Plant Food Needs," by G. N. Hoffer, did much to lay the groundwork for tissue testing. Since that time numerous developments have taken place.

In one test the plant parts may be chopped up and extracted with reagents. The intensity of color developed is compared with standards and used as a measure of the supply of the nutrient in question. This is the system on which the Purdue Soil and Plant Test Kit, the Spurway Kit, and others were based.

In another more rapid test plant sap is transferred to filter paper by squeezing the plant tissue with pliers. The tests for nitrogen, phosphorus, and potassium, which are then made with various reagents, are simple and easy to perform. Semiquantitative values for the nitrogen, phosphorus, and potassium status of a plant can be obtained in about a minute. Researchers at Pennsylvania State University and the University of Illinois developed the methods.

Tissue tests are gaining in popularity because of ease of handling and the small amount of equipment needed. Because a number of tests can be made in a few minutes, it is unpardonable to guess at the nutritional status of a plant when some of the guesswork can be avoided. These methods have the advantage over those of the laboratory. The lab tests require more time for an answer, and as a result there is a tendency to let the diagnosis go with a guess rather than send samples to the laboratory.

Plant Parts to Be Tested. It is essential to test that part of the plant which will give the best indication of the nutritional status. Considerable work is still needed on this point, but certain principles are fairly well established.

As the supply of nitrogen decreases, the upper part of the plant, in which maximum utilization of plant nutrients is in progress, will show a low test for nitrates first. In the case of phosphorus and potassium the reverse is true, and the lower part of the plant will become deficient first. In general, the conductive tissue of the latest mature leaf is used for testing, while immature leaves at the top of the plant are avoided. For certain plants the specific part recommended for tissue testing is given in Figure 12-5. If the goal is top profit yields the tests should be high at all stages of growth.

The best part to use for testing is generally that showing the greatest range of levels as the nutrient goes from deficient to adequate levels (Figure 12-6).

Time of Testing. The stage of maturity is of considerable importance in tissue testing. The average farm crop grows for a period of 100 to 150 days or longer, and its nutritional status will change during that period. Plants testing high in nutrients when small might test lower later on. However, if deficiencies were expected and the plants were tested early, there would be an opportunity to correct the difficulty.

In general, the most critical stage of growth for tissue testing is at the time of bloom or from bloom to early fruiting stage. During this period the utilization of nutrients is at its maximum, and low levels of nutrients are more likely to be detected. In corn the leaf opposite and just below the uppermost ear at silking is sampled. Although there are indications that corn yields can be increased by additions of nitrogen at silking, little is usually done about the current crop because of the special equipment required for such application.

Forage crops lend themselves well to tissue tests and can be analyzed after considerable growth has been made in the spring. If deficiencies are found, top-dressings of the required nutrients are effective in later growth.

Time of day has an influence on the nitrate level in plants, for nitrates are usually higher in the morning than in the afternoon if the supply is short. They accumulate at night and are utilized during the day as carbohydrates are synthesized. Therefore tests should not be made early in the morning or late in the afternoon.

Sampling Chart			
Plant ▼	Test ▼	Part to sample ▼	(To avoid hidden hunger) Minimum level ▼
Corn			
Under 15 in.	NO_3	Midrib, basal leaf	High
	PO_4	Midrib, basal leaf	Medium
	K	Midrib, basal leaf	High
15 in. to ear showing	NO_3	Base of stalk	High
	PO_4	Midrib, first mature leaf*	Medium
	K	Midrib, first mature leaf*	High
Ear to very early Dent	NO_3	Base of stalk	High
	PO_4	Midrib, leaf below ear	Medium
	K	Midrib, leaf below ear	Medium
Soybeans			
Early growth to midseason	NO_3	Not tested	
	PO_4	Pulvinus (swollen base of petiole), first mature leaf*	High
	K	Petiole, first mature leaf	High
Midseason to good pod development	PO_4	Pulvinus, first mature leaf	Medium
	K	Petiole, first mature leaf	Medium
Cotton			
To early bloom	NO_3	Petiole, basal leaf*	High
	PO_4	Petiole, basal leaf*	High
	K	Petiole, basal leaf*	High
Boll setting to 2/3 maturity	NO_3	Petiole, first mature leaf*	High
	PO_4	Petiole, first mature leaf*	High
	K	Petiole, first mature leaf*	High
2/3 maturity to maturity	NO_3	Petiole, first mature leaf*	Medium
	PO_4	Petiole, first mature leaf*	Medium
	K	Petiole, first mature leaf*	Medium
Alfalfa			
Before 1st cutting	PO_4	Middle 1/3 of stem	High
	K	Middle 1/3 of stem	High
Before other cuttings	PO_4	Middle 1/3 of stem	Medium
	K	Middle 1/3 of stem	Medium
Small Grains Shoot stage to milk stage	NO_3	Lower stem	High
	PO_4	Lower stem	Medium
	K	Lower stem	Medium

*First Mature Leaf—Avoid the immature leaves at the top of the plant. Take the most recently fully matured leaf near the top of the plant.

FIGURE 12-5. Part of plant used for tissue tests. [Wickstrom et al., *Better Crops Plant Food.* **47**(3):18 (1964).]

A few key points are listed:

1. It is ideal to follow the uptake of nutrients through the season by testing five or six times. Nutrient levels should be higher in the early season when the plant is not in stress.
2. The plants' greatest need for nutrients generally comes at the time when they are preparing to make seed (i.e., flowering stage). If the field is to be checked only once a season to determine the adequacy of the fertilization program, this will be the time.
3. Comparison of plants in a field is helpful. Test plants from deficient areas and compare with plants from normal areas.
4. Plants vary. Test 10 to 15 plants and average the results.

Use. Tissue tests and plant analyses are made for the following reasons:

1. To aid in determining the nutrient-supplying power of the soil. They are employed in conjunction with soil tests and management history.
2. To help identify deficiency symptoms and, even more important, to determine nutrient shortages days or weeks before they appear.
3. To aid in determining the effect of fertility treatment on the nutrient supply in the plant. This is helpful in measuring the effect of additional fertilizers even though yield responses are not available. In some cases added plant nutrients may not be assimilated because of improper placement, dry weather, leaching, fixation, or poor aeration.
4. To study the relationship between the nutrient status of the plant and crop performance.
5. To survey large areas.
6. To interest more people in sound soil testing programs.

Interpretation. The plant diagnostician must be well acquainted with the physiology of the plant with which he or she is working. Some of the more important factors that should be considered before making a decision are these:

1. General performance and vigor of the plant.
2. Level of other nutrients in the plant.
3. Incidence of insects or disease.
4. Soil condition, such as poor aeration.

FIGURE 12-6. Selection of sugarbeet leaves for analysis. A leaf stalk from any one of the recently matured, fully expanded leaves marked *A* may be included in the sample. The small leaves in the center or the old leaves should be avoided. (Courtesy of Albert Ulrich, and the Potash & Phosphate Institute, Atlanta, Ga.)

5. Soil moisture.
6. Climatic conditions.
7. Time of day.
8. Yield goal.

If a plant appears to be discolored or stunted and gives a high test for nitrogen, phosphorus, and potassium, it is not necessarily evidence that these nutrients are present in adequate amounts. It suggests, however, that some other factor is limiting growth to that level. Only after this condition has been corrected can tissue tests be expected to reveal which of the major plant nutrients may be a limiting factor in growth.

Generally, low to medium tests for nitrogen, phosphorus, or potassium in the early part of the growing season mean that a plant will yield considerably less than optimum. At blooming time a test of medium to high is adequate in most crops. Much more work is needed on calibration at very high yield levels however.

Total Analysis. Total analysis is performed on the whole plant or on plant parts. Precise analytical techniques are used for measurement of the various elements after the plant material is dried, ground, and ashed. Simultaneous analysis of up to 22 elements is possible by inductively coupled plasma optical emission spectrometry. This method is at least as sensitive as older but less automated techniques such as flame atomic absorption spectrometry. The spectrograph can also be used for the simultaneous determination of several elements. Electron microprobes and microscopes also are being developed. Analytical results from these various instruments are often compiled and reported by computer.

With total analysis many elements, such as nitrogen, phosphorus, potassium, calcium, magnesium, sulfur, manganese, zinc, boron, copper, iron, molybdenum, cobalt, silicon, and aluminum, can be determined. As in tissue tests, the plant part selected is of first importance. Similar to tissue tests, recently matured material is preferable (Figure 12-6).

Interpretation. It has been suggested that in some crops the relationship of potassium content in the lower leaves to potassium content in the upper leaves is an indication of deficiency or sufficiency. If the potassium content of the lower leaves is below that of the upper leaves, the plant is deficient. However, if the potassium content of the lower leaves is equal to or greater than that of the upper leaves the plant is not deficient.

For some purposes plant tissue tests on green material are thought to be more valuable than total analysis. For example, if the nutrient supply had just been exhausted, the difficulty would be more likely to be found by use of the tissue tests on the cell sap. Both tissue tests and total analysis, however, have been employed to considerable advantage in following the nutrient status of the plants through the growing season.

Critical Nutrient Levels and Ranges. The concept of critical nutrient concentration (CNC) as a basis for diagnosing nutritional problems is well established. It is usually designated at a single point within the bend of the curve relating nutrient concentration to yield. As depicted in Figure 12-7, the single CNC points are customarily located in that portion of the curve where the plant-nutrient concentration changes from deficient to adequate. Common definitions of CNC include:

Concentration that is just deficient for maximum growth.
Concentration that is just adequate for maximum growth.
Point where growth is 10% less than the maximum (too risky in modern agriculture).

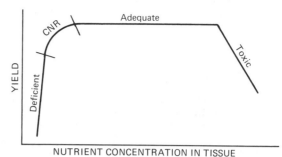

FIGURE 12-7. Relationship between nutrient concentration in plant tissue and crop yield, showing the proposed critical nutrient range. (CNR). [Dow and Roberts, *Agron: J.*, **74**:401 (1982).]

Concentration where plant growth just begins to decrease.
Lowest amount of nutrient in the plant accompanying the highest yield.

The most meaningful CNC definition for efficient growers is the level of a nutrient below which crop yield, quality, or performance is unsatisfactory. For example, in corn about 3% nitrogen, 0.3% phosphorus, and 2% potassium in the leaf opposite and below the uppermost ear at silking time are considered critical points. However, it is difficult to choose a specific concentration because the level of other nutrients in the plant may affect the critical point for a particular element. Weather conditions also affect concentrations.

For crops such as sugar beets or malting barley where excessive concentrations of nitrogen seriously affect quality, the CNC for nitrogen is a maximum rather than a minimum.

It is difficult to experimentally determine a specific CNC as usually there is considerable variation in the points plotted in the transition zone between deficient and adequate nutrient concentration. Consequently, it is more realistic to use the concept of critical nutrient range (CNR), which is defined as that range of nutrient concentration at a specified growth stage above which the crop is amply supplied and below which the crop is deficient. Figure 12-8 is a generalized illustration of CNR and how it varies during the growing season.

Critical nutrient ranges have been developed for most of the essential elements in many crops. In Georgia the concentration of each element is reported as less than, greater than, or within the sufficiency range.

        Increase in Yield with Increase in Nutrient Content. Up to a given point, increasing the amount of a plant nutrient supplied to a crop, for example, nitrogen, will increase the elemental content of the plant as well as the yield. An example is shown in Figure 12-9 in which the applied nitrogen increased the percentage of nitrogen in the corn leaf in direct proportion to the increase in yield.

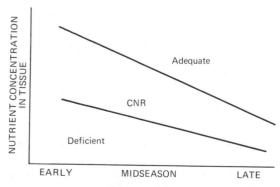

FIGURE 12-8. Generalized interpretive guide based on the concept of critical nutrient range (CNR) for tissue sampled at different times through the season. [Dow and Roberts *Agron. J.,* **74**:401 (1982).]

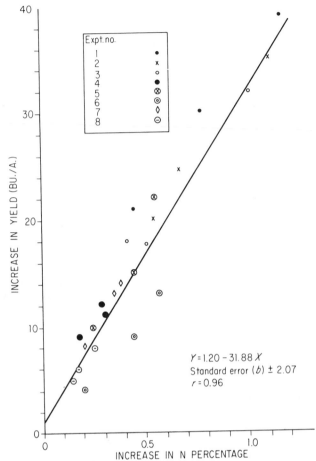

$Y = 1.20 - 31.88\,X$
Standard error $(b) \pm 2.07$
$r = 0.96$

FIGURE 12-9. With nitrogen applied on corn, the yield increase was directly proportional to the increase in percentage of nitrogen in the corn leaf. [Hanway, *Better Crops Plant Food*, **46**(3):50. (May-June 1962).]

The relationship between corn grain yield and percent potassium in the leaf is shown in Figure 12-10. It is suggested that rather than referring to a critical level, it would be better to refer to a critical zone which in this instance would appear to be above 2% potassium.

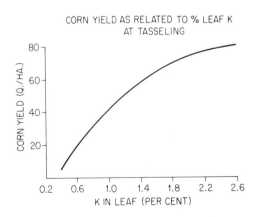

FIGURE 12-10. Corn grain yields increased with increasing levels of potassium in the leaf sampled at silking. [Loué. *Fertilite*, **20** (November–December 1963).]

Balance of Nutrients.   One of the problems in the interpretation of plant analyses is that of balance among nutrients. Ratios of nutrients in plant tissue are frequently used to study mineral balances in crops. Commonly used ratios which often reflect some known nutritional antagonism are N/S, K/Mg, K/Ca, Ca + Mg/K, N/P, and so on. Correlation of these ratios with crop yields has been only partially successful because they have been misunderstood or misinterpreted.

Sumner at the University of Georgia has explored the meaning of such nutrient ratios in order to define and understand their relationship with crop yields. The terms "insufficient" and "excessive" which he uses frequently were defined at the beginning of Chapter 3. The meaning of "optimal" is given later in this discussion. He points out that when a nutrient ratio has an optimal value, any yield is possible, the actual level being determined by the factors contributing to yield. When a ratio is too low, a response to the element in the numerator will be obtained if it is limiting. If the element in the denominator is excessive, a yield response may or may not occur depending on the level of other yield factors. When the ratio is too high, the reverse is true. These conclusions are supported by the following examples based on the assumption of an optimum range for the N/S ratio in a particular plant part within which the crop yield is maximized.

When the N/S ratio is in this optimum range or balanced it will be identified by a horizontal arrow ($\longrightarrow$). Ratios above the optimum will be recognized by an upward vertical arrow ($\uparrow$), and those below it will be assigned a downward vertical arrow ($\downarrow$).

In situations with N/S = $\longrightarrow$ or in its optimal range, three possibilities exist:

$$\frac{N \rightarrow}{S \rightarrow} \qquad \text{or} \qquad \frac{N \uparrow}{S \uparrow} \qquad \text{or} \qquad \frac{N \downarrow}{S \downarrow}$$

| Both numerator and denominator optimal | Both numerator and denominator excessive | Both numerator and denominator insufficient |

It is not possible from the ratio alone to determine which of the situations above represent what is actually taking place in the plant. All that can be said is that the two nutrients are in relative balance.

Where the N/S ratio is either above or below the optimal range, two possibilities exist in each case:

$$\frac{N}{S} = \uparrow \qquad \frac{N \rightarrow}{S \downarrow} \qquad \text{or} \qquad \frac{N \uparrow}{S \rightarrow}$$

$$\qquad\qquad \text{S insufficiency} \qquad \text{N excess}$$

$$\frac{N}{S} = \downarrow \qquad \frac{N \rightarrow}{S \uparrow} \qquad \text{or} \qquad \frac{N \downarrow}{S \rightarrow}$$

$$\qquad\qquad \text{S excess} \qquad\quad \text{N insufficiency}$$

With N/S above the optimal range, a response to sulfur will be obtained only if sulfur is lacking. If nitrogen is excessive and sulfur normal, additional sulfur may not necessarily be beneficial. The same is true with respect to nitrogen when the N/S ratio is below the optimal range. This analysis demonstrates why, when a ratio has a given value outside the optimum range, a yield response is not always obtained.

Consideration of more than one ratio at a time improves the chances of making a correct diagnosis. Use of many ratios is an integral part of the Diagnosis and Recommendation Integrated System (DRIS), which will be described in the next section.

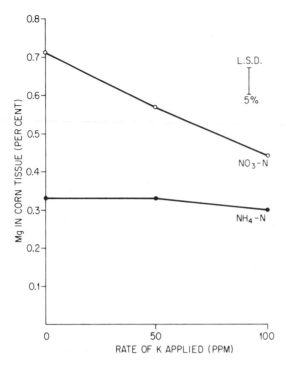

FIGURE 12-11. NH$_4$ ion has a greater effect than potassium in decreasing magnesium in corn (Fincastle silt loam). [Claassen and Wilcox, *Better Crops Plant Food*, **57**(4):10. (1973–1974).]

It has been shown that plants under uniform environmental conditions tend to take in a constant number of cations, including ammonium, on an equivalent basis. Similarly, the sum of the anions generally remains constant. If, for example, the potassium in the plant were increased, calcium and magnesium would tend to decrease, and vice versa.

Low concentrations of magnesium can seriously lower the quality of fodder consumed by cattle. The effect of the NH$_4^+$ ion in depressing magnesium uptake is illustrated on corn 37 days old (Figure 12-11). Similar restrictive action of excesses of potassium and aluminum on magnesium availability was covered in Chapter 8. These examples serve to illustrate the problem of nutrient balance in using the actual quantity of a given element as an indication of adequacy or deficiency.

Another problem occurs when a plant is low in nitrogen, and phosphorus and potassium accumulate to show high values. With nitrogen additions, phosphorus and potassium may drop drastically. This in part may be a depressing effect and in part dilution.

Diagnosis and Recommendation Integrated System (DRIS). Plant analysis interpretations based on the CNR and sufficiency range concepts have limitations. Stage of growth greatly influences nutrient concentrations and unless the crop sample is taken at the proper time, the analytical results will be of little value. Also, considerable skill on the part of the diagnostician is needed to interpret the crop analysis results in terms of the overall production conditions.

DRIS is a new approach to interpreting leaf or plant analysis which was developed by Beaufils at the University of Natal, South Africa. It is a comprehensive system which identifies all the nutritional factors limiting crop production and in so doing increases the chances of obtaining high crop yields by improving fertilizer recommendations. Index values which measure how far particular nutrients in the leaf or plant are from the optimum are used in the calibration to classify yield factors in order of limiting importance. Provision of all the details of the DRIS approach is beyond the scope of this book. However, some of its essential features follow.

To develop a DRIS for a given crop, the following requirements must be met whenever possible:

1. All factors suspected of having an effect on crop yield must be defined.
2. The relationship between these factors and yield must be described.
3. Calibrated norms must be established.
4. Recommendations suited to particular sets of conditions and based on correct and judicious use of these norms must be continually refined.

Establishment of DRIS Norms. A survey type of approach is first employed in accumulating the basic data required to establish a data bank from which norms are determined. In this phase a large number of sites where a crop is growing are selected at random in order to represent the whole production area of a county, state or district. At each site, plant and soil samples are taken for all essential element analyses. Other parameters likely to be related directly or indirectly to yield are also recorded. In addition, details of soil treatments (fertilizers, herbicides, etc.), climatic conditions (rainfall, etc.), cultural practices and any other relevant types of information are recorded and stored in a computer for ready access.

Second, the entire population of observation is divided into two subpopulations (high and low yielders) on the basis of vigor, quality and yield. Each element in the plant is expressed in as many ways as possible. For example, the percentage of N in the dry matter or ratios N/P, N/K, or products N-P, N-K, and so on, may be used. The mean of each type of expression for each subpopulation is calculated. Each form of expression which significantly discriminates between the high- and low-yielding subpopulations is retained as a useful diagnostic parameter. The mean values for each of these forms of expression then constitute the diagnostic norms. Using N-P-K in corn leaves, as an example, the significant forms of expression were found to be N/P, N/K, and K/P.

Determination of Relative NPK Requirement. An example follows on how to diagnose the relative NPK requirements of corn using the DRIS chart illustrated in Figure 12-12. The chart is constructed of three axes for N/P, N/K, and K/P, respectively, with the mean value for the subpopulation of high yielders (> 160 bu/A) located at the point of intersection for each form of expression. These values are N/P = 10.04, N/K = 1.49, and K/P = 6.74. This point of intersection

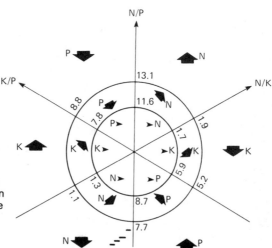

FIGURE 12-12. DRIS chart for obtaining the qualitative order of requirements for N-P-K in corn. Means of significant expressions (value at origin in chart) are: N/P = 10.04, N/K = 1.49, and K/P = 6.74. [Sumner, *Solutions,* **22**(5):68 (1978).]

of the three axes therefore represents the composition for which one is striving and at which one should achieve the highest yield permitted by limiting factors other than N, P, and K. The concentric circles can be considered as confidence limits, the inner being set at the mean ± 15% and the outer at the mean ± 30% for each expression.

A qualitative reading of this chart can be done by using arrows in the following conventional manner: Horizontally for values within the inner circle of the chart (between 1.30 and 1.71 for N/K, 8.73 and 11.55 for N/P and 5.86 and 7.75 for K/P) corresponding to a nutritionally balanced situation; diagonally [↘↗]for values between the two circles representing a tendency to imbalance (e.g., between 1.15 and 1.30 or 1.71 and 1.94 for N/K) and vertically [↑↓] for values found beyond the outer circle (e.g., beyond 1.15 or 1.94 for N/K) representing nutrient imbalance.

The way in which this chart is used will be illustrated by means of an example. N, P, and K concentrations for corn leaves are presented in Table 12-2 from which the expressions N/P, N/K, and K/P are calculated. Because an excess of one plant nutrient corresponds to a shortage of another, by convention only insufficiencies are recorded for the purpose of diagnosis, which is done stepwise for each function. Identical diagnoses are obtained by considering either excesses or insufficiencies or both. Using the data in the first line of Table 12-2 as an example, one finds that the value of the function N/P (13.33) lies in the zone of phosphorus insufficiency, giving: (1) N P↓ K; while the value of N/K (1.27) lies between the two circles, adding a tendency to nitrogen insufficiency: (2) N↘ P↓ K; while that of K/P (10.48) lies in the zone of phosphorus insufficiency, giving: (3) N↘ P↓ ↓ K.

Once the three forms of expression have been read, the remaining element is assigned a horizontal arrow. The final reading then becomes: (4) N↘ P↓ ↓ K→, which gives the order of N-P-K requirements of the crop in terms of limiting importance on yield: P > N > K.

Calculation of DRIS Indices. The arrow notation used thus far can be quantified by calculating DRIS indices. The equations used for calculating such indices will not be given here, but they appear in Sumner's technical papers. When using these indices, keep in mind that the most negative index is the one most required, while the most positive one is least needed.

Using DRIS Indices to Diagnose NPK Requirements. An example of using DRIS indices in diagnosing the N-P-K requirements of corn is presented in Table 12-2. These data are taken from a field experiment in which yield responses

TABLE 12-2. Use of DRIS Norms in Diagnosing N-P-K Requirements of Corn Based on the Selection of a Plot in an N-P-K Factorial Experiment, Diagnosing the Requirement, and Satisfying It by Selecting the Plot in Which the Required Element Is Applied (Leaf Sample Taken at Tasseling)

Treatment With N-P_2O_5-K_2O (lb/A)	Leaf Composition (%)			Forms of Expression			Chart Reading			DRIS Indices			Corn Yield * (%)
	N	P	K	N/P	N/K	K/P	N	P	K	N	P	K	
0-0-0	2.80	0.21	2.20	13.33	1.27	10.48	↘	↓↓	→	7	−22	15	28
0-50-0	3.20	0.28	1.00	11.43	3.20	3.57	→	→	↓↓	31	13	−44	49
0-50-60	2.93	0.28	2.60	10.46	1.13	9.29	↘	↓	→	−6	−9	15	55
0-100-60	2.60	0.26	2.44	10.00	1.07	9.38	↓	↓	→	−9	−8	17	60
100-100-60	3.16	0.33	2.45	9.58	1.29	7.42	↘	→	→	−5	−1	6	75
200-100-60	3.40	0.34	2.40	10.00	1.42	7.06	→	→	→	−1	−1	2	100

* Percentage of the highest value.
Source: Sumner, *Solutions,* **22**(5):68 (1978).

TABLE 12-3.   Effect of Age of Crop Sampled on the DRIS Diagnosis of N-P-K Requirements of Corn

Age of Crop at Sampling (days)	Leaf Composition (%)			Forms of Expression			Chart Reading			DRIS Indices		
	N	P	K	N/P	N/K	K/P	N	P	K	N	P	K
30	4.6	0.30	3.4	15.33	1.35	11.33	→	↓ ↓	→	16	− 32	16
60	3.9	0.26	2.4	15.00	1.63	9.23	→	↓ ↓	→	19	− 25	6
80	3.4	0.24	1.9	14.17	1.79	7.92	→	↓ ↘	↘	19	− 18	− 1
110	3.0	0.20	1.8	15.00	1.67	9.00	→	↓ ↓	↘	20	− 24	4

Source: Sumner, *Solutions,* **22**(5):68 (1978).

to nitrogen, phosphorus, and potassium were obtained. The indices will be used to show that they are capable of diagnosing the yield responses observed in the field. Starting with the control plot 0-0-0, one diagnoses that phosphorus is the most limiting nutrient. In order to see the response to phosphorus, one selects the treatment in which phosphorus was applied (e.g., 0-50-0). One finds that phosphorus application increased the yield and that potassium is now the most required nutrient.

The potassium requirement is satisfied by selecting treatment 0-50-60. When this is done the yield has been increased by the potassium treatment and phosphorus is again the most limiting nutrient. When phosphorus is applied in treatment 0-100-60, the yield increases further and nitrogen becomes the most limiting factor. Addition of nitrogen in treatments 100-100-60 and 200-100-60 results in the yield increasing to the maximum. At this stage, the diagnostic criteria indicate a relatively well-balanced N-P-K nutritional status, which means that factors other than N-P-K are limiting yield.

The same diagnosis is made by both the qualitative (arrow) and quantitative (index) procedures. One can see that requirements for phosphorus (treatment 0-50-60) and for nitrogen (treatment 100-100-60) were correctly predicted by the DRIS system when the nitrogen and phosphorus levels in the leaf were above the critical values noted earlier in the discussion on critical nutrient levels. In many cases, the DRIS system is capable of diagnosing requirements that would not be obvious when using the critical or sufficiency level approach.

In order to show that a diagnosis using DRIS norms can be made over a range of plant age, data from a corn experiment in which leaf samples were taken at various times will be used (Table 12-3).

Irrespective of the age of the crop when leaf samples were taken, the DRIS approach diagnoses that phosphorus is more required than potassium, which is more required than nitrogen (e.g., P > K > N). Using the critical value norms given earlier, the diagnosis would only have been possible after the 80-day stage. Thus the DRIS approach results in greater flexibiity in diagnosis.

It has also been shown that the DRIS method of determining N-P-K requirements of corn is little affected by other factors, such as leaf position on the plant or cultivar.

Although this discussion of the DRIS approach has dealt only with the nitrogen, phosphorus, and potassium requirements of corn, it has been successfully applied to other crops, including sugarcane, rubber, soybeans, potatoes, wheat, sunflowers, alfalfa, and ryegrass. Also, norms for calcium and magnesium in corn have been established and additional norms for other essential nutrients are expected in the future.

In summary, the DRIS system has a number of distinct advantages over the classical critical level approach in making diagnoses for fertilizer recommendation purposes:

1. The importance of nutritional balance is taken into account in deriving the norms and making diagnoses. This is particularly valuable at high yield levels where balance is often critical in determining yield.
2. The norms for the elemental content in leaf tissues can be universally applied to the particular crop, regardless of where it is grown.
3. Diagnoses can be made over a wide range in stages of crop development, irrespective of cultivar.
4. The nutrients limiting yield either through excess or insufficiency can be readily identified and arranged in order of their limiting importance on yield.

Time of Sampling. The percentage of certain plant nutrients may drop rapidly from early to late stages of growth. This tendency is portrayed in Figure 12-8 and a large decline in the nitrogen, phosphorus, and potassium concentrations in corn is shown in Table 12-3. Hence stage of growth for sampling must be carefully selected and identified.

Surveys. Analysis of plant samples from many fields gives a general indication of the levels of nutrients. To permit interpretation these levels, of course, must be compared with critical levels observed in controlled plots. This method has been particularly useful in obtaining preliminary information on elements such as zinc, boron, cobalt, and copper.

Caution should be exercised in extrapolating plant-nutrient concentration data from controlled growth-chamber and greenhouse experiments to crops grown under field conditions. The short growing periods, environmental conditions, interactions, and so on prevailing in these kinds of experiments will often result in nutrient concentrations that are unrepresentative of those occurring in naturally grown crops.

Routine Use. Quantitative plant analyses are employed extensively in research to obtain another measure of the effect of treatment. However, crops on a commercial scale, such as sugarcane and pineapples, are analyzed periodically in many areas. Public and commercial analytical services are often available for tree crops. Numerous public and private organizations maintain a plant analysis service on agronomic and horticultural crops. California workers have established safe levels and deficiency ranges for nitrogen, phosphorus, and potassium in a long list of crops for use by growers (*Western Fertilizer Handbook*).

Crop variety has an effect on plant analyses in some instances. When one looks at an average of many fields the effect may not be apparent but when one is evaluating a given situation the influence may be significant.

Plant analysis is another helpful tool in evaluating the nutrient status of the plant. It must be considered along with soil testing and crop-management practices in diagnosing problems. Its use in crop logging is a case in point.

Crop Logging. An excellent example of the use of plant analyses in crop production operations is the crop logging carried out for sugarcane in Hawaii. The crop log, which is a graphic record of the progress of the crop, contains a series of chemical and physical measurements. These measurements indicate the general condition of the plants and suggest changes in management that are necessary to produce maximum yields.

A critical nutrient concentration approach is used in the crop log system and nutrient concentrations in leaf sheaths 3, 4, 5, and 6 are utilized for diagnosis of calcium, magnesium, sulfur, and micronutrient deficiencies. It uses complex indices based on tissue nutrient concentration, sheath moisture, and other factors to diagnose nitrogen, phosphorus, and potassium deficiencies.

During the growing season plant tissue is sampled every 35 days and analyzed for nitrogen, sugar, moisture, and weight of the young sheath tissue. Analyses are made for phosphorus and potassium at critical times, and adjustments in management practices are introduced as needed (Figure 12-13). Knowledge of the percentage of moisture makes it possible to regulate irrigation, particularly during

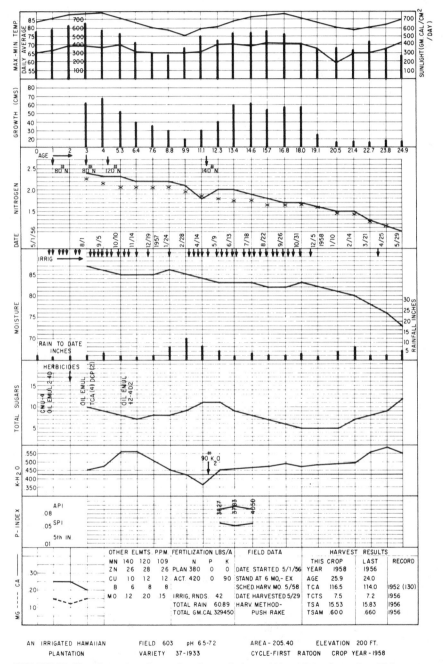

FIGURE 12-13. Completed crop log for an irrigated Hawaiian plantation. This approach has been valuable in a complete diagnostic approach. (Clements, in Walter Reuther, *Plant Analysis and Fertilizer Problems,* p. 132. Copyright 1960 by the American Institute of Biological Sciences.)

the ripening period. It has been found, for example, that the moisture content should drop gradually to around 73% when the crop is ripe, and irrigation is regulated accordingly. Many plantations in Hawaii are using crop logging, for it has been found that yields gradually increase under this system of record keeping. One important factor is that, periodically, key personnel are able to take a close look at the fields, and this examination is helpful from the standpoint of observing signs of trouble.

The diagnostic accuracy of the conventional crop log approach, a locally calibrated crop log system, and DRIS were compared in 28 past fertilizer experiments on the island of Hawaii. Both the locally calibrated crop log and the DRIS systems gave slightly more accurate diagnoses of nutrient deficiencies than the conventional crop log approach.

Isotopic Dilution Techniques.   The radiochemical analyses of plants grown on soils which have been treated with fertilizers containing elements such as radioactive phosphorus may be used to calculate the phosphorus supply of the original soil. Two isotopic dilution equations have been developed by researchers in the United States and Europe. Equations for the two approaches are given below. These equations, based on the concept that a plant will absorb a nutrient from two sources in direct proportion to the amount available from each source, reduce essentially to identities in which $A = Y$.

Fried and Dean: $$A = \frac{B(1 - y)}{y}$$
(or A value)

Larsen: $$Y = \frac{x(C_0 - C)}{C}$$
(or L value)

where A or Y = available phosphorus in the soil
$\quad$ B or x = amount of phosphorus applied to soil
$\quad\quad\quad$ y = fraction of phosphorus taken by the plant from B
$\quad\quad\quad$ $C_0 = k \times$ total phosphorus in the plant
$\quad\quad\quad$ $C = k \times$ phosphorus taken by the plant from x
$\quad\quad\quad$ k = proportionality constant

For example, if 50 lb/A of phosphorus were applied and 20% of the phosphorus in the plant came from the fertilizer, the A or Y values would be 200 lb/A. Both equations are applicable to studies with other labeled nutrients.

Potential sources of error in both the A- and L-value techniques are related to the assumptions that (1) the amount of nutrient absorbed from the soil is independent of the rate of fertilizer application, and (2) the utilization percentage of the fertilizer is the same for all rates of application. These assumptions are not always correct because uptake of soil phosphorus as well as fertilizer phosphorus are known to vary with increased rate of application. When the applied phosphorus or other nutrient stimulate root development, soil supplies of nutrients become more accessible.

Biological Tests

Use of the growing plant understandably has much appeal in the study of fertilizer requirements, and much attention has been devoted to this method for measuring the fertility status of soils.

Field Tests.   The field-plot method is one of the oldest and best-known of the biological tests. The series of treatments selected depends on the particular question the experimenter wishes to have answered. The treatments are then randomly assigned to an area of land, known as a replication, which is representative of the conditions. Several such replications are used to obtain more reliable results and to account for variations in soil and management.

These experiments are helpful in the formulation of general recommendations. When large numbers of tests are conducted on soils that are well characterized, recommendations based on such studies can be extrapolated to other soils with similar characteristics. Field tests are expensive and time consuming, and one is unable to control climatic conditions and other limiting factors. They are valuable tools, however, and are widely used by experiment stations, although they are not well adapted for use in determining the nutrient status of large numbers of soils. Rather, they are used in conjunction with laboratory and greenhouse studies as a final proving ground and in the calibration of soil and plant tests.

Strip Tests on Farmers' Fields. Strips of fields are being treated with fertilizer by extension, industry, and growers alike to check on recommendations based on soil or plant tests. The results of these tests must be interpreted with caution if they are unreplicated and use of two or three replicates is recommended. Repetition of strip tests at several locations is also helpful.

Laboratory and Greenhouse Tests. Simpler and more rapid biological techniques which still involve higher plants but utilize small quantities of soil have been developed. These methods have met with wide acceptance on the European continent and have frequently been employed in the United States.

Mitscherlich Pot Culture. In this method oats are grown to maturity in pots containing 6 lb of soil. A total of 10 pots is used. The yields of the N-P and N-K treatments are expressed as a percentage of the yield from the complete N-P-K treatment. For example, if the N-P-K treatment yielded 80 g and the N-K treatment, 60 g, the yield would be 75%. With these percentage yields, the plant-nutrient reserve in the unfertilized soil can be read in pounds per acre from yield tables prepared by Mitscherlich, and from these same tables predictions of the percentage increases in yield expected from the addition of given amounts of nutrients can be obtained. The growth equations derived from these studies were discussed in Chapter 2.

Neubauer Seedling Method. The Neubauer technique is based on the uptake of nutrients by a large number of plants grown on a small amount of soil. The roots thoroughly penetrate the soil, exhausting the available nutrient supply within a short time. The nutrients removed are usually determined quantitatively by chemical analysis of the entire plant. In some procedures, however, the tops and the roots are harvested and analyzed separately. Tables have been set up to give the minimum values for satisfactory yields of various crops. The Neubauer method has been used for the availability of several nutrients including phosphorus, potassium, calcium, micronutrients, or fertilizer materials.

Short-Term Method. The short-term method helps to bridge the gap between chemical extraction and greenhouse pot methods. Plants deficient in the element under study are grown in sand contained in a cardboard carton with the bottom removed. A dense mat of roots is formed at the bottom in 2 to 3 weeks. The roots are then placed in contact with soil, or soil plus fertilizer, contained in a second carton. An uptake time of about 1 week is allowed, after which the plants are analyzed. Uptake is generally well correlated with uptake in conventional pot tests that require several weeks.

Microbiological Methods. Winogradsky was one of the first to observe that in the absence of mineral elements certain microorganisms exhibited a behavior similar to that of higher plants. It was shown that the growth of *Azotobacter* served to indicate the limiting mineral nutrients in the soil, especially calcium, phosphorus, and potassium, with greater sensitivity than chemical methods. Since that time several techniques which employ different types of microorganisms have been developed and a few are described. In comparison with methods that utilize higher plants, microbiological methods are rapid, simple, and require little space.

Sackett and Stewart Technique. The Sackett and Stewart technique is based on Winogradsky's work and was used to study the phosphorus and potassium status of Colorado soils. A culture is prepared of each soil, phosphorus is added to one portion, potassium to another, and both elements to a third portion. The cultures are then inoculated with *Azotobacter* and incubated for 72 hours. The soil is rated from very deficient to not deficient in the respective elements, depending on the amount of colony growth.

Aspergillus Niger. To determine phosphorus and potassium small amounts of soil are incubated for a period of four days in flasks containing the appropriate nutrient solutions. The weight of the mycelial pad or the amount of potassium absorbed by these pads is used as a measure of the nutrient deficiency. Mehlich devised a more refined technique in which the mycelial pad is also analyzed for potassium. An example of the criteria used is tabulated.

Weight of Four Pads (g)	Potassium Absorbed by Aspergillus niger per 100 Grams of Soil (mg)	Degree of Potassium Deficiency
< 1.4	< 12.5	Very deficient
1.4–2.0	12.5–16.6	Moderate to slight deficiency
> 2.0	> 16.6	Not deficient

Mehlich's *Cunninghamella*-Plaque Method for Phosphorus. The soil is mixed with the nutrient solution, a paste is made, spread uniformly in the well of a specially constructed clay dish, inoculated on the surface in the center of the paste, and allowed to incubate for $4\frac{1}{2}$ days. The diameter of the mycelial growth on the dish is used to estimate the amount of phosphorus present.

Soil Testing

A soil test is a chemical method for estimating the nutrient-supplying power of a soil. Although biological methods for evaluating soil fertility have certain advantages, most of these tests have the disadvantage of being time consuming, hence not well adapted for use with large numbers of samples. A chemical soil test, on the other hand, is much more rapid and has the added advantage over deficiency symptoms and plant analyses in that one may determine the needs of the soil *before* the crop is planted.

A soil test measures a part of the total nutrient supply in the soil. The values are of little use in themselves. To employ such a measurement in predicting nutrient needs of crops the test must be calibrated against nutrient rate experiments in the field and in the greenhouse, especially the former.

Chemical tests are used extensively in the United States as well as in other countries. Almost all states offer soil-testing services by state-operated agencies.

California and Illinois are two states where these public services are unavailable. Lack of a state operated soil testing laboratory in California is probably due in part to the very wide variety of soils and crops in the state. Large numbers of soil tests have been made; in 1980 1.5 million samples were reported to have been analyzed in state-operated agencies. About the same number of samples were processed by the many independent private soil-testing laboratories in the United States.

Many other countries, including Ireland, Canada, Denmark, the Netherlands, Scotland, England, Germany, France, and India, have well-developed soil-testing programs. Both publicly and privately operated soil testing services are provided in Canada. In addition, many soil samples from Canada are processed by several of the larger commercial laboratories in the United States.

Objectives of Soil Tests. Information gained from soil testing is used in many ways.

1. *To build and/or maintain fertility status of a given field.* An attempt is made to extract some portion of the nutrients for calibration with the capacity of the plant for taking nutrients from the soil. When one considers the wide range in soil characteristics and the many crops grown, it is not surprising that much effort has been expended over the last century to improve soil-testing methods.
2. *To predict the probability of obtaining a profitable response to lime and fertilizer.* Although response to applied nutrients will not always be obtained on low-testing soils because of other limiting factors, the probability of a response is greater than on high-testing soils.
3. *To provide a basis for recommendations on the amount of lime and fertilizer to apply.* These basic relations are obtained by careful laboratory, greenhouse, and field studies.
4. *To evaluate the fertility status of soils on a county, soil area, or state-wide basis by the use of soil-test summaries.* Such summaries are helpful in developing plans for research and educational work.

Expressed simply, the objective of soil testing is to obtain a value that will help to predict the amount of nutrients needed to supplement the supply in the soil. For example, with a high test value the soil will not require as much supplementing as it will at a low test value (Figure 12-14). It must be kept in mind that soil fertility is only one of the many factors affecting plant growth.

Sampling the Soil. One of the most important aspects of soil testing is the matter of obtaining a soil sample that is representative of the area. Usually, a composite sample of only 1 pint of soil (about a pound) is taken from a field. In a 10-acre field there are about 20 million pounds of surface soil, and with this extremely small sample there is a considerable opportunity for error. Hence it is quite important that the sampling instructions outlined by agricultural agencies be followed carefully, for a poor sample is worse than none at all. If the sample does not represent the field, it is impossible for a good recommendation to be made. The error in sampling a field is generally greater than the error in laboratory analyses. The variability which may occur in a field is shown in Figure 12-15.

Soils are normally heterogeneous and variability is frequently observed by research scientists and farmers. Wide variability can occur in even "uniform" fields. Generally, soil variability has not received sufficient attention despite the fact that it is extremely important in sampling procedures and soil-test interpretation. Intensive soil sampling has been shown by researchers at Utah and Washington State Universities to be the most efficient way to evaluate variability and also to be the best guide for fertilizing soils according to actual needs.

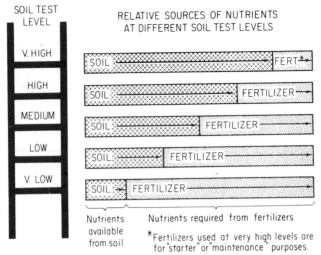

FIGURE 12-14. As the soil tests higher in a plant nutrient, the amount needed from fertilizers becomes less. The purpose of soil testing is to determine the levels of nutrients. (Courtesy of the Potash & Phosphate Institute, Atlanta, Ga.)

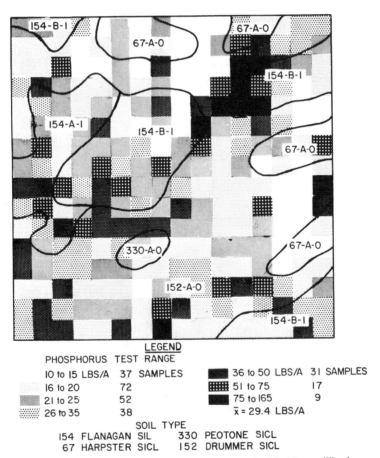

FIGURE 12-15. Variability of soil phosphorus in this 40-acre Illinois field helps to illustrate the problem in accurately sampling a soil. (Peck and Melsted, "Field sampling for soil testing" in Leo M. Walsh and J. D. Beaton, Eds., *Soil Testing and Plant Analysis*, p 67. Madison, Wis. Soil Science Society of America, 1971.)

Lack of uniformity in the nutrient status of soil can occur on a small scale (Figure 12-16) as well as on a field scale. In this example the total plot area of 0.73 acre was divided into plots 11.3 ft × 50 ft. It is apparent that soil phosphorus and potassium levels varied considerably from plot to plot. Truly diligent researchers will utilize only sites that have acceptable uniformity of soil fertility properties.

Tools. There are two important requirements of a sampling tool: first that a uniform slice be taken from the surface to the depth of insertion of the tool and, second, that the same volume of soil be obtained from each area. Soil tubes in general meet these two requirements very well. An additional advantage is that the volume taken is small enough that fifteen to twenty cores can be placed in a pint container. These tubes have 5 to 15 in. cut away on one side, except at the cutting head which has 1 in. of solid tube of a smaller bore. These tubes work well under most conditions, except in dry or gravelly ground. Other tools used are trowels, augers, spades, or power-driven samples. Hydraulically operated sampling equipment is especially helpful for obtaining deep samples and in taking large numbers of samples.

Areas to Sample. The size of the area from which one sample may be taken varies greatly but usually ranges from 5 to 20 acres or more. Areas that vary in appearance, slope, drainage, soil types, or past treatment should be sampled separately (Figure 12-17), and small areas that cannot be treated separately by lime and fertilizer applications might well be omitted from the sample. On the other hand, with the trend toward higher and higher production and more uniform crop growth, many growers are directing attention to these small spots in their fields and treating them as needed. The idea is to have every acre yielding as well as every other acre. Hence separate samples from localized areas of poor crop growth are helpful.

Number of Spots to Sample for the Composites. Each soil sample is a composite consisting of the soil from cores taken at several places in the field.

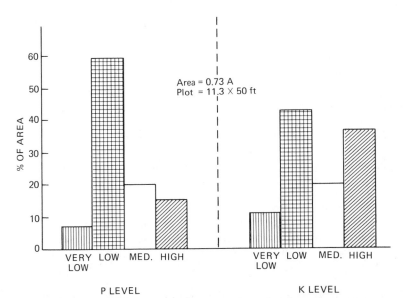

FIGURE 12-16. Variability in available soil phosphorus and potassium at a potato research site in the Columbia Basin of Washington. [Kunkel, *Proc. 17th Annu. Pacific Northwest Fert. Conf.*, pp. 29–47, (1966).]

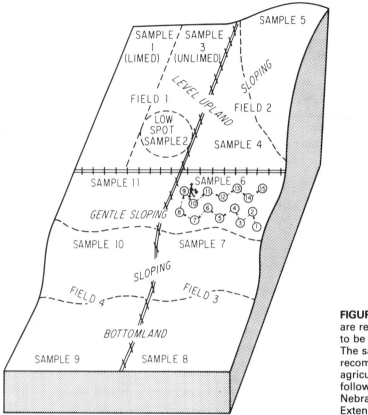

FIGURE 12-17. Samples that are representative of the field to be fertilized are important. The sampling pattern recommended by the various agricultural agencies should be followed. (Courtesy of the Nebraska Agricultural Extension Service.)

The purpose of this procedure is to minimize the influence of any local nonuniformity in the soil. For example, in fields in which lime or fertilizer applications have been made in the last 2 or 3 years the plant nutrients may be incompletely mixed with the soil. In addition, there may be spots in which the fertilizer or lime was spilled or dumped, or on which plant refuse was burned, or where crop residue from harvesting was concentrated. A sample taken entirely from such an area would be completely misleading. Consequently, most recommendations call for taking borings at 15 to 20 locations over the field for each composite sample (Figure 12-17). If more than a pint is obtained, the soil is mixed well and a subsample of the mixture is taken for the analysis.

As larger amounts of fertilizer are applied in the row, careful attention must be given to sampling between visible rows. The tendency toward heavier broadcast applications will increase the difficulty of obtaining representative samples unless the soil has been plowed and worked at least twice. The need for special sampling procedures for fields where fertilizers have been deep banded has received some consideration.

To obtain good representative samples about 25 sampling sites are required for each average-sized field. For small fields a minimum of eight places should be sampled and more intensive sampling is encouraged. In very large fields at least one site should be sampled for each 5 to 7.5 acres. These guidelines are for situations in which there is no extreme variability within the field.

Depth of Sampling. For cultivated crops samples are ordinarily taken to a plow depth of 6 to 9 in. In some areas, however, growers are chiseling or plowing as deep as 12 in., and this fact should be taken into consideration by sampling to this depth. The operations of land preparation tend to mix previous lime and fertilizer applications with the whole plow layer. When lime and fertilizer are broadcast on the surface for established pastures and lawns, a sample from the upper 2 in. is most satisfactory. With no-till or minimum till it is best to take a sample from the surface 2 in. and another sample from the 2- to 8-in. layer. It is recognized that plants obtain nutrients from below the plow layer, but little information is available for interpreting the analyses on such subsoil samples. Workers in some states have characterized the subsoils of major soil series in respect to levels of phosphorus and potassium and adjust their recommendations accordingly. Obviously the fertility level of the subsoil will not change appreciably over a 5- to 10-year period.

An exception is sampling 2 to 6 ft in some of the lower rainfall areas in the United States in order to measure nitrate and moisture in the profile. These amounts are then considered in the recommendations. Sampling to 2 ft is encouraged in the Canadian Prairie Provinces for measurement of nitrate, sulfate, and soil salinity.

Time of Sampling. Ideally, samples should be taken just prior to seeding or when the crop is growing. However, these times are largely impractical because of constraints in taking samples, obtaining test results, and supplying the needed lime and fertilizer.

Consequently, samples are customarily taken any time soil conditions permit. For spring-planted crops, sampling in the fall after harvest is often practiced. In drier regions where nitrate levels are used to assess the nitrogen status of soil, sampling in the fall to diagnose the needs of annual spring-seeded crops is often postponed until the surface soil temperature drops to 5°C.

On some soils the potassium level is lower during the summer. This has been found in West Virginia and in Illinois. In Illinois it is recommended that, depending on the soil type, 30 to 60 lb be subtracted from the soil test for samples taken before May 1 and after September 30.

Most recommendations call for testing each field about every 3 years, with more frequent testing on the lighter soils. In most instances this is often enough to check on the lime level in the soil and to determine whether the fertilization program is adequate for the crop rotation. For instance, if the phosphorus level is decreasing, the rate of application can be increased. If it has risen to a satisfactory level, application may be reduced to maintenance rates.

Copies of the soil-test results from previous samplings are often maintained in the office of the agricultural leader or the fertilizer dealer. When the grower decides what crop to grow for a given year, he or she may check back with these people to get the recommendation.

Analyzing the Soils. Any chemical soil test should be designed to permit an estimate of the amount of plant nutrients contributed by the cation exchange fraction of the soil, by the fraction retaining the phosphorus, and under some conditions by the decomposition of organic matter. The major plant-nutrient cations available for use by plants are held in exchangeable form. Among those anions that are of major importance in soils the phosphates are retained most strongly, the sulfates less strongly, and the nitrates not at all. The method of retention of many of the micronutrients has not been established, but some of them are held in the exchangeable form.

Several kinds of extracting solutions have been employed in an effort to correlate soil test results with plant growth. However, when it is considered that the soil sample will be in contact with the extracting solution for only a few minutes, whereas the plant will absorb from the soil during the entire growing season, the complexity of the task becomes apparent.

Organic matter contributes to plant nutrition. Certain fractions retain cations in exchangeable form; other fractions are decomposed or mineralized by microorganisms, and nitrogen, phosphorus, and sulfur as well as other nutrients are released.

Soil acidity is an important characteristic and is a good index to a number of conditions. It gives an indication of the base saturation and a lead to possible toxicity or deficiency of certain elements, particularly micronutrients.

Cations. The principle underlying the determination of alkaline earth cations is the replacement of all or a proportionate amount of the cation from the exchange complex. Cations such as H^+, Ba^{2+}, NH_4^+ (if NH_4^+ is not to be determined), or sodium may be used, but ammonium acetate is a common extractant for potassium, calcium, and magnesium. Generally, the soils are dried before extracting for chemical analysis. However, evidence has been obtained that on some soils potassium uptake by plants is better correlated with the exchangeable potassium determined in undried rather than air-dried samples because of the release or fixation that occurs in the drying process. Hence, in 1964 Iowa State University adopted procedures to test undried samples.

Mention was made in Chapter 7 of the importance in some soils of nonexchangeable potassium present in feldspars and micas in the gradual replenishment of solution and exchangeable potassium. Sands of the Great Plains and the sandy soils along the Atlantic coastal plain are examples of such soils. Acidification by additions of sulfuric or nitric acids is used to extract from these soils the exchangeable and a readily available portion of the nonexchangeable potassium.

The percentage base saturation refers to the percentage of exchange capacity made up of exchangeable bases that include ammonia but not hydrogen and aluminum. Rather than the actual amounts, some laboratories are basing recommendations on percentage base saturation. The concept is based on the principle that the availability of a given cation to a plant is influenced by the concentration of other cations present.

It has been found that at a given level of potassium in the soil, a sandy soil will have a greater proportion of the exchangeable potassium in solution than will a heavier soil. Hence, this may be part of the reason why percentage base saturation may be useful in helping to predict needs. In Ohio the optimum potassium test for corn has been related to the cation exchange capacity (CEC) using the following equation:

$$\text{optimum potassium soil test (lb K/A)} = 220 + (5 \times \text{CEC})$$

Hence, on a soil with a CEC of 15, the optimum potassium soil test would be 295. However, on soybeans the optimum test has been found to be 60 lb higher.

The foregoing approach used in Ohio is known as the basic cation saturation ratio (BCSR) in contrast to the sufficiency level of available nutrients (SLAN) concept. McLean indicates considerable evidence on the suitability of simply using selected sufficiency levels of cations, say 300 lb of exchangeable potassium or 150 lb of magnesium. The soil has mechanisms that serve as buffers or ionic reservoirs that remove nutrient ions from and return them to the soil solution (Figure 12-18). The relative amounts of each type of cation exchange bond are quite different from soil to soil.

Phosphorus. Extracting solutions, ranging from water, alkalies, and weak acids all the way to relatively strong acids containing ammonium fluoride, have been used for the extraction of phosphorus. In a study conducted by the Soil Test Work Group on a wide range of soils all over the United States the Bray No. 1 method, which employs 0.025 N HCl + 0.03 N NH_4F, gave especially good correlation with A values in the greenhouse and with crop responses. Olsen's

Chemical equilibrations with soil buffer

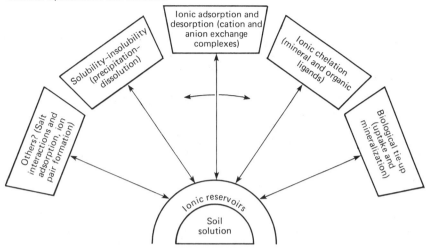

FIGURE 12-18. Mechanisms regulating reactivity and plant availability of ions in soil. [McLean, *Commun. Soil Sci. Plant Anal.*, **13**(6):411 (1982).]

method of employing 0.5 N $NaHCO_3$ has been satisfactory on alkaline soils. Strong acids cannot be used in alkaline soils because the tricalcium phosphate, which is not available to crops, is dissolved out. Contrasted with this is the possiblity that weak acids are neutralized in basic and calcareous soils, thereby losing their ability to extract phosphorus.

Mehlich's 0.05 N HCl + 0.025 N H_2SO_4 and the Morgan sodium acetate extractants have been popular in some areas. Another extractant developed by Mehlich is being used in some states.

Once the nutrients are extracted quantitative equipment such as the flame photometer, photoelectric colorimeters, atomic absorption, auto analyzers, inductively coupled plasma optical emission spectrometer, and so on, are generally used to measure the amounts in the extract.

Micronutrients. A number of laboratories make soil tests for micronutrients, mostly on a special basis. Several extractants are used but chelating agents such as DTPA appear most promising for zinc, copper, manganese, and iron. The pH of the extraction can be controlled and chemical attack on lime and mineral fractions is minimized. Analyses for boron are usually made by the hot-water method. Two major problems stand out with micronutrient soil testing. One is interpretation and the other is laboratory control.

Soltanpour and Schwab at Colorado State University developed an NH_4HCO_3-DTPA soil test capable of simultaneously extracting phosphorus, potassium, zinc, iron, copper, and manganese from soils. They have since demonstrated that with minor changes it is possible to use inductively coupled plasma spectrometry for the simultaneous determination of all six of these nutrients in this extract.

Organic Matter and Nitrogen. A knowledge of the organic matter content is helpful in estimating the cation exchange capacity and the nitrogen-supplying power of the soil. Organic matter is usually determined by wet combustion methods in which the soil is subjected to treatment with sulfuric acid and potassium dichromate. Certain of the less resistant fractions, known as *easily oxidizable organic matter*, are oxidized.

Key methods for measurement of nitrate nitrogen are phenoldisulfonic acid and the nitrate electrode. In some laboratories nitrification tests are being made. The soil is incubated under optimum moisture and temperature conditions for two weeks, at the end of which time the nitrates are leached out and determined.

Consistent results from nitrate-nitrogen tests are complicated by the dependency of nitrogen availability on decomposition of organic matter and amount of leaching. Environmental conditions such as moisture and temperature affect decomposition, hence seasonal variations may be great. Satisfactory predictive values are obtained in soils in the dry areas of the Great Plains in the United States and Canada. Past cropping practices are widely used in predicting needs.

The USDA scientists Stanford and Smith developed a useful procedure for estimating nitrogen mineralization potential. It involves a long-term incubation of soil for 30 weeks at 35°C. Mineral nitrogen (NO_3^-, NO_2^-, and NH_4^+) is leached from the soil by means of $0.01\ M\ CaCl_2$ and a minus-N nutrient solution at intervals of 2, 4, 8, 12, 16, 22, and 30 weeks.

Sulfur. Determination of the sulfur needs by soil tests is complicated by the various forms and methods by which sulfur is held in the soil. The organic matter contains sulfur and many factors influence retention of sulfate on the inorganic soil fraction. A number of laboratories now determine sulfur with water, $Ca(H_2PO_4)_2$, or $CaCl_2 \cdot 2H_2O$ as the most common extractants. A turbidimetric measurement of $BaSO_4$ is usually used.

In general, plant tests for assessing sulfur needs of crops have been somewhat more successful than soil tests. Absolute levels of total sulfur vary with species, but for some crops, especially alfalfa, levels have been set which fairly well describe the need for sulfur. For many crops, however, such levels have not been determined but the N/S ratio (% total N/% total S) has been used as an indication of the need for sulfur. N/S ratios of 14 to 16:1 are generally considered to be satisfactory, but ratios in excess of 17:1 suggest the need for sulfur fertilization. Shortcomings in the use of such ratios for assessing the nutrient status of plants were reviewed earlier in this chapter.

Soil Acidity and Lime Requirement. This subject was discussed in Chapter 11. The soil acidity determination is accomplished by the use of the pH meter which is standard equipment in most laboratories. Color indicator dyes are employed in field kits but must be used by relatively experienced operators if correct results are to be obtained. These dyes are helpful in pointing out gross needs, and if an acid soil is found a lime requirement test in the laboratory is essential.

The soil pH, along with a consideration of the organic matter and amount and type of clay, is the basis for many lime recommendations. This method, however, is subject to a good bit of estimation and human error. Many laboratories use a method involving a buffer in which hydrogen is replaced from the exchange complex and the depression in pH is read and related to lime requirement.

Labor-Saving Equipment. Soil testing on a service basis involves the handling of large numbers of samples. Special laborsaving equipment has been developed in order that the volume of work may be dealt with rapidly and accurately. For example, extraction racks may hold 12 bottles and can be manipulated as a unit; extraction solutions may be added to 12 samples at a time from specially built dispensers, and electrically powered automatic pipettes dispensing 1 to 5 ml at each stroke may be used. A few seconds saved in each of several manipulations is extremely important. In addition, as the operation is made easier and more routine, the laboratory technician is less subject to fatigue, hence less subject to errors. Many laboratories have modern equipment, including computers, that translates the instrument readings to punch cards or provides a printout of the results.

Detailed analytical procedures are used in research work. More rapid but carefully performed tests are used for the routine soil samples. In many soil-testing laboratories, however, the accuracy of the routine soil tests approaches that of the detailed procedure. The use of well-trained personnel who understand the chemical reactions and the operation of special equipment contributes much toward more reliable results.

Physical Properties of Soils. Rapid estimates of soil structure have been rather difficult to make. The soil tube described in this chapter is an effective tool for visual examination of the soil profile to a depth of 18 in. Shaving the exposed side of the core has helped to detect compact layers, and a suspension of precipitated chalk in water dropped on the cut surface is also useful. Soil layers in good tilth will absorb both lime and water. Compact layers will absorb the water but not the lime, and the intensity of the lime spot increases with compactness.

Small penetrometers are available which are suited for use in the field to investigate soil structural problems. They measure the resistance of soil to penetration by a small plunger.

Physical condition of the soil becomes increasingly important at higher yield levels. In the future more effort will probably be diverted to the examination and characterization of the physical properties of the root zone as affected by management.

Central Laboratories and County Laboratories. Most states use a central or regional laboratory system, but a few maintain laboratories in counties in addition to a central control laboratory. There are advantages to both systems.

Analyses are probably more accurately performed in central laboratories because of more skilled technicians, better equipment, and better control. It is also somewhat easier to make uniform recommendations; and if changes are made, these changes can be put into effect quickly. The lime and fertilizer recommendations may be made at the laboratory, or the results may be sent out to regional representatives or county agents for handling. The latter procedure reduces the criticism that individuals unfamiliar with local conditions are making recommendations. Many laboratories are now using the computer to make recommendations. The program is set up to consider soil tests along with many management factors and the recommendations are printed directly on a sheet. The regional representatives, public or private, may choose to make adjustments based on their knowledge of the managerial capacity of the farmer, the productivity of the soil, and other factors.

In Canada there is one central soil-testing facility in each province. In most cases they are operated under the jurisdiction of the respective provincial governments. Several of these laboratories function in close association with the provincial university.

Private Laboratories. Private laboratories are important in all areas of the country. Some which are connected with the fertilizer industry provide tests as a service to customers. Others offer soil testing as a part of a wide range of analytical services. In any event, there is a marked trend toward the use of private laboratories in some areas.

Most important in private laboratories, as in those operated by state agencies, is the correct interpretation of results. Therefore, these laboratories often have access to check samples from the state agencies as well as tables showing the interpretation of the results. In some states a system of approved laboratories has been developed.

Growers may be suspicious of recommendations made on tests in a laboratory supervised by a fertilizer company. In some cases this suspicion may be justified, but the rapid trend toward service by the industry, the employment of trained personnel, and the competitive need for a program that will be profitable to the grower should lead to reliable recommendations.

Use of private laboratories is also significant in Canada. Most aspects of these laboratories discussed above also apply to the Canadian situation.

Calibrating Soil Tests. Although the chemical analysis of soils presents some difficulties, perhaps the greatest problem in a testing program is the calibration of the tests. It is essential that the results of soil tests be calibrated against crop responses from applications of the nutrients in question. This information is obtained from field and greenhouse fertility experiments conducted over a wide range of soils. Yield responses from rates of applied nutrients can then be related to the quantity of available nutrients in the soil.

Much additional work in calibration is needed in the United States as well as in other countries, particularly at maximum yield levels. In certain European countries, such as the Netherlands and Denmark, large numbers of experiments have been used in calibration studies. With the increased interest in fertilizers in the developing countries, soil testing is receiving more emphasis. Unfortunately, in these countries as in the early days in the United States, few calibration data are available.

Many of the testing laboratories in the United States classify the fertility level of soils as very low, low, medium, high, or very high, based on the results of chemical tests. Some, however, report results in terms of pounds per acre or ppm of phosphorus and potassium. In general, the very low to very high classification seems to be somewhat more easily understood by the grower, but crops differ in their requirements and what is low for potatoes may be high for small grain; what is low for a clay loam may be high for a sandy loam. In any case, it is important that the grower know the meaning of the results reported.

Some states use a fertility index such as the one in service in North Carolina shown in Table 12-4. This is the relative sufficiency expressed as a percentage of the amount adequate for top yields. It is possible to convert the percentage figures to pounds per acre.

A simplified approach for developing countries simply plots percentage yield and the soil test value. A critical level is established at about 75%.

The probability of a response to fertilization as related to soil-test results has been emphasized. As already stated soil fertility is only one of the factors influencing plant growth, but in general there will be a greater chance of obtaining a response from a given element if the soil test is low in that element. This concept is presented diagrammatically in Figure 12-19. For example, more than 85% of the fields testing very low may give a profitable increase; in the low range 60 to 85% may give increases, whereas in the very high range less than 15% would respond. These values are arbitrary, but they illustrate the idea of expectation of

TABLE 12-4.   Fertility Index As Related to Soil Test, Responses, and Recommendations

Soil Test Rating	Crop Response	Fertility Index	Recommendations
Very low	Highly probable	0–10	Crop requirement + substantial buildup
Low	Probable	10–25	Crop requirement + buildup
Medium	Possible	25–50	Crop requirement + modest buildup
High	Unlikely	50–100	Maintenance
Very high	Highly unlikely	100 +	Vegetable crops—some maintenance

Source: D. W. Eaddy, North Carolina Dept. of Agriculture, personal communication.

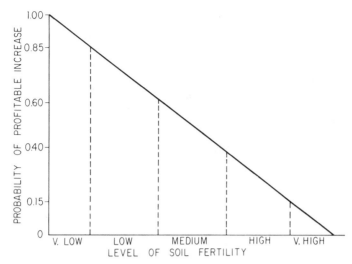

FIGURE 12-19. There is a greater probability of obtaining a profitable response from fertilization on soils testing low in an element than from soils testing high in that element. [Fitts, *Better Crops Plant Food,* **39**(3):17 (1955).]

response. The better informed farmers who are using good management practices will be more likely to obtain a response in the high range.

The variable yield response of barley to applied nitrogen and phosphorus in different experiments related to soil levels of these two nutrients is well illustrated in Figure 12-20. It is obvious that the gains from applied fertilizer nutrients were greatest at low soil-test levels. Some minor yield benefits also occurred at high soil-test levels.

The concept of probability of response is quite important. Probability varies because of the many factors affecting response to added nutrients. The information below indicates a greater probability of response at a given soil test level when the fertilizer is placed beside the row than when broadcast.

	Probability of Response (%) to P and K	
	Broadcast *	Beside the Row †
Very low	95–100	—
Low	70–95	95–100
Medium	40–70	65–95
High	10–40	30–65
Very high	0–10	10–30

* Purdue University
† University of Minnesota

Probabilities of barley responding to application of potassium increased greatly at the low levels of exchangeable potassium in soils of central Alberta (Table 12-5). Also, the magnitude of the response was much greater at low levels of available soil potassium.

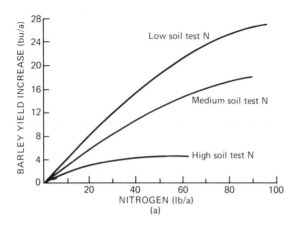

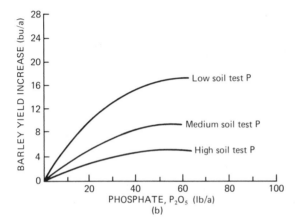

FIGURE 12-20. Expected barley yield increases from (a) nitrogen and (b) phosphorus at different soil test levels in central Alberta. [McLelland, "Barley Production in Alberta." *Agdex 114/20-1.,* Alberta Agriculture, (1982).]

TABLE 12-5. Yield Response of Barley to Added Potassium Fertilizer at Different Levels of Exchangeable Potassium

Exchangeable K (lb/A) in Soil 0–6 in. in Depth	*Percent of Test Sites Giving Yield Response*	*Percent Increase in Yield from K Fertilizer**
< 50	100	> 1000
51–100	75	242
101–150	66	47
151–200	24	30
201–250	18	34
> 250	3	11

* Percent yield increase calculated for those field sites that gave yield increases from potassium fertilizer.
Source: Walker, *Better Crops Plant Food,* **62**:13 (Summer 1978).

According to North Dakota experience, the chance that a crop will respond to an application of fertilizer when grown on fields with different soil test categories can be expressed as follows:

Soil Test Category	Percent of Fields Where a Yield Response Is Expected *
Low	90
Medium	40–60
High	10–20
Very high	Less than 5

* North Dakota State University

An example of the calibration of the soil test for zinc with crop response is shown in Figure 12-21. Eighty-six percent of the soils testing below 0.55 ppm zinc responded to zinc sulfate, whereas 77% above this level did not respond.

In a North Central regional experiment on corn, responses to potassium were calibrated with soil-test values. As the exchangeable potassium increased, the amount of potassium needed to obtain the most profitable return was reduced.

The calibration of soil tests is complicated by the fact that many factors other than fertility level influence the response obtained. Temperature, water, soil properties, stand, cultural practices, and pests are more readily controlled in the greenhouse than in the field. However, it is essential to conduct field trials after the preliminary greenhouse tests have been completed.

Variety may have a marked effect on response to applied fertilizer as shown below for paddy rice in the Philippines.

Rice	Yield (tons/ha)	
	With No Potassium	With 300 kg/ha K_2O
Native variety	1.7	1.9
Improved variety	1.4	4.8

A major problem in soil tests is that they are sometimes calibrated at a yield level too low to be useful to progressive farmers. Many of the calibration data are out of date because of the development of new technology to control more of the limiting factors which results in higher yields. The reason for the emphasis on higher yields in calibration trials is simple. Higher yields use more nutrients, the opportunity for responses is greater, and results are more meaningful for the progressive grower.

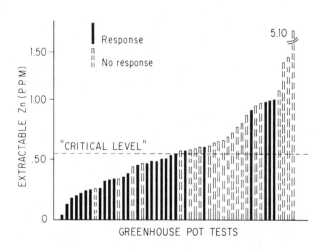

FIGURE 12-21. Response of sweet corn to zinc as related to dithizone-extractable zinc in 55 California soils. Each solid line represents a soil that responded; each dashed line shows no response. [Brown et al., *Calif. Agr.* **15**:15 (1961).]

The Changing Philosophy of Interpretation of Soil Tests. Much progress has been made in soil testing in measuring the available nutrients in soils. However, the big problem is interpreting the results in terms of fertilizer needed. The degree of accuracy depends on several factors including knowledge of the soil, the yield level expected, level of management, and weather. When a soil test is first developed in an area we simply want to know the critical level and if a particular nutrient should be added. As knowledge is gained in a given area the soil-test level may be divided into three or more categories and the fertilization rate adjusted to the soil fertility level.

The percentage yield concept is based on the idea that the expected yields, as a percentage of the maxiumum yield, are predicted from the phosphorus and potassium analyses of the soil (Figure 12-22). Sufficient amounts are added to bring yields to 95% or more of the maximum. While this concept fits a number of broad situations, interactions may cause considerable deviation. When this concept was developed it was indicated that it applied only when the same pattern and rates of planting were used and with about the same soil and seasonal conditions. For example, in Indiana on the same soil it was found that in dry years potassium per acre increased corn yields 39 bushels, in average years 8 bushels and in wet years 48 bushels.* In Kentucky the following information was obtained:

Corn Plants per Acre	Response to 100 lb P_2O_5 (bu/A)	Response to 200 lb K_2O (bu/A)
15,700	2	21
24,500	22	39

Hence, considering the weather effects and improved practices of progressive farmers, the importance of individual attention to each situation looms large.

Recommendations for Different General Yield Levels. Soil test interpretation involves an economic evaluation of the relation between the soil test value and the fertilizer response. However, the potential response may vary due to several factors among which are the soil, weather, and management ability of the farmer (Figure 12-23).

Some laboratories may vary recommendations with expected yield level (Table 12-6). The amount of nitrogen recommended depends on previous crop and yield goal.

Many laboratories make one recommendation assuming best production practices for the region and the area representative may make adjustments as necessary. However, some base them on average production practice and this tends to discourage progressive farmers from using the laboratory.

As technology and management practices improve or with increased economic incentives, yield potential increases and recommendations increase. *For the commercial grower the goal is to maintain the plant nutrients at a level for sustained top profits per acre, which means that nutrients should not be a limiting factor at any stage, from plant emergence to maturity.* This approaches the "sure" rate (Figure 12-3).

Types of Recommendations. In general, there are four courses of action if the soil is low in phosphorus or potassium.

1. *Buildup or basic treatment.* Corrective applications are added to bring the soil up to the desired level. A rule of thumb is that about 9 lb of P_2O_5 will raise the P_1

* S. A. Barber, Purdue University, personal communication.

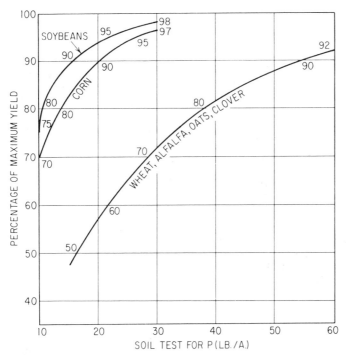

FIGURE 12-22. Crops vary in need for phosphorus. For example, with a 20-lb phosphorus test the percentage of the maximum yield is 57, 90, and 94% for wheat, corn, and soybeans, respectively. [Bray, *Better Crops Plant Food,* **45**(3):18 (1961).]

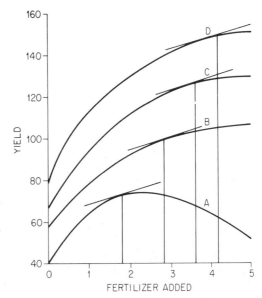

FIGURE 12-23. The yield response to fertilizer depends on the potential yield level with A being the poorest and D the greatest potential. (Barber, "The changing philosophy of soil test interpretations," in L. M. Walsh and J. D. Beaton, Eds., *Soil Testing and Plant Analysis*, p. 203. Madison, Wis.: Soil Science Society of America, 1971.)

test 1 lb and about 4 lb of K_2O will raise the potassium 1 lb. However, in Michigan the amount of P_2O_5 was found to vary from 4.5 to 11.5 lb and K_2O to vary from 2 to 4.5 lb, depending on soil texture. On calcareous soils in the dry regions of the western United States and Canada as much as 53% of applied fertilizer phosphorus will appear as a change in $NaHCO_3$ available phosphorus (Figure 12-24).

TABLE 12-6. **Phosphorus and Nitrogen Recommended for Corn**

Soil Test-Bray P_1 (lb/P/A)	Yield Goals (bu/A)		
	80	120	160
	Annual Application of P_2O_5 (lb/A)		
5	55	70	85
15	45	60	75
25	35	50	65
30–60	30	45	60
75	20	30	45
90	20	20	30
	Annual Application of Nitrogen (lb/A)		
Continuous corn	40	115	200

Source: Ohio Agronomy Guide 1983–1984. Columbus, Ohio: Cooperative Extension Service, Ohio State Univ., 1983.

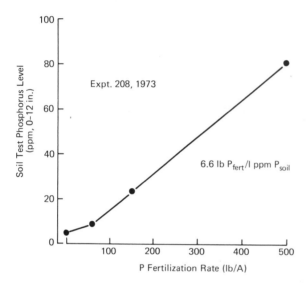

FIGURE 12-24. Effect of phosphorus fertilization on sodium bicarbonate extractable phosphorus levels in a calcareous Portneuf silt loam soil, Kimberly, Idaho. [Westermann, *Proc. 28th Annu. Northwest Fert. Conf.*, pp. 141–146 (1977).]

The soil is retested in 2 to 3 years to see if additional corrective applications are needed. Maintenance amounts are then added to replace losses by crop removal, erosion, leaching, and fixation.

2. *Annual application.* Phosphorus and potassium are added to each crop in the rotation. This may be in amounts to effect a gradual buildup or merely to maintain the soil test. This approach demands well-calibrated tests and there is considerably more chance of error. When capital is limited, the fertilizer-using area is new, or the land is being rented, the maintenance program is probably the preferred method. Yield will not be so high and profit per acre will be lower, but returns per dollar spent will be higher than in the buildup method.

3. *Rotation.* This method is perhaps the most widely used and a gradual buildup approach may be used. In a rotation involving soybeans, such as corn–soybeans, many farmers just fertilize the corn and supposedly add enough to take care of the soybeans. The favorable effect of applying high rates of potassium to corn on the following soybean crops is readily apparent in Figure 12-25. Additionally, these data show that high levels of available soil potassium are necessary for high yields of soybeans.

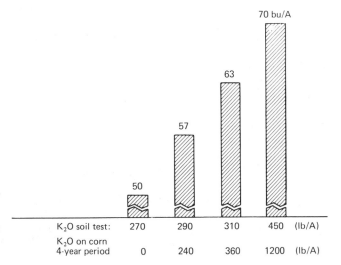

70 bu/A

63

57

50

FIGURE 12-25. Carryover potassium and applied the previous 4 years on corn increases soil test and soybean yield. [Welch, *Better Crops Plant Food*, **58**(4):(1974–1975).]

K$_2$O soil test:	270	290	310	450 (lb/A)
K$_2$O on corn 4-year period	0	240	360	1200 (lb/A)

It is sometimes difficult to impress on farmers and recommenders the high amounts of nutrients removed in their ever-increasing yields. Removals in crop, other losses and some buildup in the soil must be evaluated in the recommendations. As shown below, nutrient removal in just the grain of corn and soybeans is substantial.

	Nutrient (lb/A)				
	N	P$_2$O$_5$	K$_2$O	Mg	S
200 bu corn grain	150	87	57	18	15
60 bu soybean grain	240 *	48	84	17	12
Total	390	135	141	35	27

* Legumes get much of their nitrogen from the air.

In rotational fertilization certain points must be considered when deciding how to recommend the fertilizer among the crops in the rotation:

a. Apply the broadcast plowdown before most responsive and profitable crops

b. Apply row phosphorus and perhaps potassium for corn, particularly in northern areas.

c. Forage crops remove large amounts of potassium. Annual applications are essential to maintain yields.

d. Soybeans may respond more to a high soil fertility level than to direct application of fertilizer. However, on soils medium or less in fertility apply fertilizer directly for soybeans.

Double-cropping or two crops in one year is practiced in some areas. This may be small grain–soybeans or small grain silage-soybeans. In such instances sufficient phosphorus and potassium is recommended to be applied for both crops before the small grain is planted. Again the removals, losses, and any buildup must be considered in making the recommendations.

4. *Replacement system.* As soils are built up to what is considered to be an adequate level in phosphorus and potassium, amounts may be recommended to replace removals on the basis of expected yield. For example, if 60 bushels of soybeans were expected and at 0.75 lb of P$_2$O$_5$ and 1.4 lb of K$_2$O per bushel, about

48 lb of P_2O_5 and 84 lb of K_2O would be suggested. With an 8-ton yield of alfalfa and a removal of 15 lb of P_2O_5 and 60 lb of K_2O per ton, 120 lb of P_2O_5 and 480 lb of K_2O would be suggested.

There are several aspects which must be considered in such a system:

a. On soils with a high supplying power only 50% of removal might be recommended.
b. What is an adequate level in the soil?
c. Will the farmer wish to increase the level as yield potential increases?
d. The content of phosphorus, potassium, and other elements in grain may vary considerably and even more so in forage.
e. Will adding just what is removed maintain soil fertility level? This will depend on fixation and release in the soil and certain losses, but generally it will not.
f. In some soils the amount added over removal may need to be increased 10 to 25%.
g. If 100 lb of P_2O_5 or K_2O is added, can it be expected that the crop is 100% efficient in absorbing it over a period of years?

In view of these points, if this system is used, the soil must be monitored periodically to determine if the soil fertility level is decreasing or increasing.

Nitrogen. Recommendations depend on many factors, including the amount of nitrate and moisture in the soil profile in areas of low precipitation, previous crop, yield goal, and nitrogen management on the previous crop.

For dryland wheat production in Idaho, the following credits and adjustments in estimated nitrogen fertilizer requirements are made according to yield goal and the previous crop:

Previous Crop	Estimated Nitrogen * (lb/A) for a Potential Yield (bu/A) of:			
	40	60	80	100
Grain (residue returned)	50–70	70–90	90–110	110–130
Grain (residue removed)— peas, lentils, fallow	20–30	30–40	50– 65	65– 80
Alfalfa or green-manure crop	0–15	15–30	30– 50	55– 75

* University of Idaho.

Nitrogen credits of 5 lb/ton of manure, 1 lb/bu of soybeans grown, and so on, may also be subtracted from the normal recommendations.

The strong dependency of crop responses to fertilizer nitrogen on moisture supplies during the growing season was emphasized in Chapter 5.

A system has been developed at Washington State University for estimating the amount of nitrogen fertilizer that potential yields of winter wheat will need. Potential yields of winter wheat in eastern Washington depend on three factors:

1. Soil moisture content.
2. Expected precipitation after sampling for soil moisture.
3. Fertility level, especially nitrogen, that will be available to the crop. The Washington State University system carefully coordinates the two most limiting factors—available moisture and available nitrogen.

Results of their research indicate that it takes 4 in. of precipitation to grow the crop up to the time it sets seed. For each remaining inch of moisture available over this amount, it is expected that 7 bu/A of wheat will be produced. The moisture that will be available is calculated by determining the moisture content of soil samples taken to a depth of 6 ft and adding to it the rainfall that can be expected after the sample has been taken.

If, for example, the actual and expected moisture is 20 in., the predicted yield potential would be 112 bu/A. It has also been determined that 2.7 lb of nitrogen is required for each bushel of wheat produced. Using the potential yield of 112 bushels based on the available moisture in the example used, it would take 302 lb of nitrogen per acre.

To determine the amount of nitrogen that will have to be applied to the soil, the soil samples are analyzed for NO_3-N. The total nitrogen available is calculated by adding the NO_3-N content and the amount of nitrogen that is expected to be released from the decay of organic matter. This amount is then subtracted from 302 lb, and the remainder is the amount of nitrogen that will have to be applied in the form of fertilizer.

For example, if the soil contains 80 lb NO_3-N and the amount expected to be released from organic matter is 60 lb, the soil already has 140 lb of nitrogen available. That leaves 162 lb of nitrogen fertilizer to be applied.

Nitrogen recommended for corn following a heavy small grain crop would be increased because of the straw while recommendations on small grain following a heavily fertilized corn crop would be reduced. With continuous corn, if the desired yield level had not been reached, more nitrogen would be applied than if it had been reached. Nitrogen use efficiency decreases as yields increase. For example, at 100 bushels, 1 lb of nitrogen/bu might be adequate but at 200 bushels, 2 to 3 lb/bu might be needed.

As potential yield increases, nitrogen needs increase considerably because crop needs for this nutrient are high. There will be a corresponding rise in requirements for phosphorus and potassium and other nutrients.

Responses on High Testing Soils. Many factors affect crop response to a nutrient—temperature, moisture, time, placement, tillage, yield goal, crop, and so on. Hence even on a soil testing very high in phosphorus in such states as Georgia or Wisconsin, a starter fertilizer is recommended for corn because it is planted early under cool, wet conditions. Similarly, spring-planted small grain in New York responds to phosphorus even on a very high phosphorus soil. Vegetable crops often respond to phosphorus and/or potassium on soils very high in these elements. Extent of root systems, rapidity of uptake need, and root respiration are all part of the picture.

Prescription Method. The prescription method for making recommendations is based on the idea that plants can secure certain percentages of nitrogen, phosphorus, and potassium contained in the soil, manure, and fertilizer. When the approximate number of pounds per acre required to produce a given yield is known, the amount of supplemental manure and/or fertilizer is calculated. The principle behind this method is to formulate fertilizer recommendations fitted to needs, which are determined by the rotation followed, crop management, soil analysis, and immediate crop to be grown.

Several different methods have been used. An example of one approach is given in Table 12-7. The estimated percentages of nitrogen, phosphorus, and potassium available to a crop such as corn in any one year are listed. Similar calculations have been made for other crops, including small grain, legumes, potatoes, tomatoes, and sugar beets.

It is recognized that this method has limitations. The problem of developing soil tests which will give an accurate measure of the amounts of nutrients available has been discussed. Tests vary somewhat among the states, and a given prescription must be related to the soil-test method being used. The percentages of elements

TABLE 12-7.   Estimated Percentages of Nitrogen, Phosphorus, and Potassium in Soil, Manure, and Fertilizer Available to a Crop Such As Corn During One Season

Sources	Percentages Obtained During One Season		
	Nitrogen	Phosphorus	Potassium
Soil (available)	40	40	40
Manure (total)	30	30	50
Fertilizer (available)	60	30	50

Source: K. C. Berger, *Farm Chem.*, **117**:47–50 (1954).

available from manure and fertilizer are approximations for they change with crop, soil, and climatic conditions.

In spite of these limitations, such methods serve to point out the nutrient needs of crops and that higher yields will demand greater quantities. This emphasizes the attention that is being given to eliminating as many of the limiting factors in crop production as possible.

Lime.   This subject has been covered in detail in Chapter 11. It is mentioned here only to emphasize that the first step in a buildup fertility program in acid soils is *adequate* lime application. Applications often do not raise the soil pH to the desired level. Testing after 3 years may reveal the need for reapplication.

Recommendations Made by Electronic Computers.   The computer is fast becoming the farmer's electronic hired hand. Many farmers are already depending on computerized farm management and dairy production records. The ability of computers to integrate data such as soil-test results as well as other growth factors is being put to use in developing effective crop production management practices. Economic aspects are usually included in designing these strategies.

Many handwritten recommendations consider only the soil test results. On the computer many other factors, such as soil type, yield potential, crop sequence and moisture, may be considered. Many samples can be processed in a minute, yet the farmer gets more information than ever before possible. The recommendations are printed on the soil-test report. For satisfactory computer-made recommendations it is helpful to adjust them at the dealer or agent level so that they correspond to the managerial ability of farmers.

Importance of Soil Tests to the Farmer and the Lime and Fertilizer Industries.   It should be recognized that soil testing is not an infallible guide to crop production. The problems of representative samples, accurate analyses, correct interpretation, and environmental factors which influence crop responses all enter in. However, the soil test helps to reduce the guesswork in fertilizer practices. The soil test may be used to monitor the soil periodically to determine general soil fertility levels. Rates of lime and fertilizer are then applied to supply adequate quantities to the current crop or crops.

The lime and fertilizer industries can be, and in many cases are being, helpful in state or private soil-testing programs. Agronomists, salesmen, and dealers can encourage their customers to have soils tested or can take samples for their customers, and dealers can make a point of stocking the recommended fertilizers. If growers apply what their crops need, they are much more likely to be able to pay their bills in the fall. On the other hand, if they apply phosphorus and potassium when lime or nitrogen happens to be limiting, their bills may go unpaid. Occasionally, a fertilizer dealer will refuse to sell fertilizer to a customer until the soil has been properly limed.

The difficulty in getting growers to take samples is great, and most dealers or private laboratories provide a soil sampling service.

Industry classifies the two most important functions of soil testing as providing a basis for more profitable recommendations and opening the door to customer counseling. Farmers have been hearing about soil tests and fertilizer for years, and many farmers are convinced of their value, but they want to know how much to apply. This takes the soil test out of the category of an educational tool and into that of a profit-guaranteeing service.

More and more fertilizer dealers recognize the importance of soil tests, as well as other diagnostic techniques, in helping to predict the plant nutrients that growers will need on their fields. Fertilizer dealers sell their service along with products in a personalized approach to sound development of lime and fertilizer use.

Fertilizer recommendations based on soil analysis results will differ. Although the soil test is not an infallible guide as has been pointed out, it should still be used whenever possible. A soil test is the place to start when designing a fertility program for maximum-profit crop yields.

Soil-Test Summaries

The purpose of soil testing is to give the individual growers dependable information regarding the fertilizer and lime needs of their fields. However, not all growers will have their soils tested, and neither will all growers taking soil samples include all the fields on their farms. Better use of lime and fertilizer may be attained on most of these untested fields by drawing on the information gained from the fields tested. This should lead to the best average recommendation.

To accomplish this purpose, summaries of the results of soil tests have been made in most states by county or on the basis of soil association, type of farming, or crop. A summary map will serve to alert agricultural workers and industry personnel to the sectional problems of the state. This map can be helpful in the overall direction of a program and in indicating where more intensive educational efforts can be made.

It should be stressed that although the content of lime, phosphorus, or potassium in the soils of a given county or area may, on the average, be high the soils in individual fields may test all the way from very low to very high. This serves to direct attention to the need for specific soil samples from each field for the most intelligent use of plant nutrients.

The soil phosphorus levels as related to certain crops in North Carolina are listed in Table 12-8. The large percentages of samples in the high and very high ranges for tobacco and potatoes are related closely to the use of fertilizer. Tobacco, for

TABLE 12-8. Phosphorus Level Related to the Crop to Be Grown in North Carolina (1981)

Crop	Number Tested	Percent of Samples Testing				
		VL	L	M	H	VH
Legume–grass pasture	3,499	19.6	21.9	21.1	21.1	16.3
Soybeans	24,232	3.2	8.1	17.8	31.0	39.9
Small grains	16,204	2.1	6.8	18.1	31.4	41.6
Corn	44,270	2.9	7.5	17.2	29.9	42.5
Tobacco (FC and Bur.)	17,169	2.1	4.9	8.2	17.0	67.8
Potatoes, Irish	1,055	1.5	2.3	6.3	16.7	73.2

Source: D. W. Eaddy, North Carolina Dept. of Agriculture, personal communication.

TABLE 12-9. Summary of Potassium Tests (Percent) for Field Crops by Soil Management Group, Michigan

Soil Management Group	Pounds of Potassium per Acre					
	< 110	110–159	160–209	210–249	250–299	> 300
Organic	12.8	13.2	16.7	10.5	7.5	39.4
Clay loam or loam	11.1	19.7	26.0	14.8	11.1	17.2
Sandy loam	22.0	23.6	24.1	11.1	8.3	11.0
Loamy sand	43.4	23.8	16.9	6.3	3.9	5.8

Source: V. W. Meints, Michigan State Univ., personal communication.

example, has been receiving the equivalent of 50 to 60 lb/A of phosphorus annually from which not more than 10 lb/A is removed. High rates are also applied to cotton and potatoes. Peanuts not listed here receive little phosphorus directly but are often grown in rotation with cotton.

Another summary by a soil management group is shown in Table 12-9. Such summaries can be readily made when soil testing data and site information are punched on cards.

It is interesting that in organic soils 39% test more than 300 lb/A of available potassium. On the other hand, 43% of the loamy sands contained less than 110 lb/A. Although there will be considerable variation within each group, these data are helpful to agricultural workers. Other summaries may be prepared by soil areas in the state, such as in Iowa.

A summary by general subsoil fertility groups has been made in some states. An example, for Wisconsin, is shown in Figure 13-25. Subsoil fertility will affect response to fertilizer, and a knowledge of this fertility will contribute to a satisfactory interpretation of soil tests.

Results of soil test summaries can be viewed in different ways. For example, an article on them may have a title such as:

Sixty-six percent of fields test high in phosphorus and/or potassium

The implication may be that high is synonymous with *excessive*. It should be synonymous with *desirable*. Consider a different headline for the same situation:

One out of every three fields tests low to medium in phosphorus and/or potassium

The implication is that one-third of the acreage is being underfertilized for top yields, particularly under stress conditions and top management. If one-third of the people in North America were nutrient deficient in vitamin C, protein, and so on, it would be viewed as a national calamity. *An eventual goal should be to have every planted acre testing high in its content of plant nutrients.* Anything less means a loss in yield potential.

Periodic Summaries. Several states make periodic summaries. Some of these show that phosphorus and/or potassium levels are gradually increasing as the result of higher rates of fertilizer in the past decade. There is some concern that levels may be too high on a few soils and fertilizer practices may need to be reevaluated. This may be true for a few soils and these should be studied carefully. However, the following should be kept in mind:

1. Yields are higher.
2. Yield will continue to increase.
3. As other elements become limiting, they should be added.
4. High levels help plants to endure stress periods.

5. Nutrients are in soil and can be used in case application cannot be made—for example, because of adverse weather, fertilizer shortages, or lack of funds.

A point might be mentioned in regard to the accuracy of a summary from routine soil samples. Studies in Wisconsin, North Carolina, Kentucky, Alabama, and Indiana indicate that such a summary agrees quite well with systematic samples.

Summary

1. The selection of adequate lime and fertilization practices depends on the requirement of the crops, yield goal, weather, and the soil characteristics and necessitates finding a method that will reveal the deficiencies in the soil.

2. Although diagnostic approaches are used in troubleshooting, they are more important as preventive measures.

3. Deficiency symptoms are helpful guides in new fertilizer-using areas. In high-use areas they are a sign of mediocre farming practices and are often difficult to interpret because of their complication with many other problems.

4. Hidden hunger is insidious, but careful plant and soil tests will help to avoid it.

5. The plant integrates all factors in the environment into itself, and tests can be highly revealing. Quick tissue tests in the field on growing plants are useful, but careful interpretation is essential.

6. As a nutrient is added, the percentage of that nutrient in the plant usually increases. It is important to identify the point at which there is no further economic yield increase.

7. Critical nutrient range rather than a specific critical nutrient concentration seems more meaningful in the evaluation of plant nutrient status.

8. Caution must be exercised in the interpretation of plant nutrient ratios.

9. The Diagnosis and Recommendation Integrated System (DRIS) is a greatly improved approach to plant analysis.

10. Plant analyses are of real value in making surveys of incipient micronutrient problems in a given area.

11. Balance among nutrients in the plant may be just as important as actual amounts. The relationships among Ca, Mg, K, and NH_4, Mn and Fe, and Zn and P are examples.

12. There are many biological short-term methods in which both higher and lower plants are used for determining nutrient needs. Eventually, all of these methods must be related to field responses for calibration.

13. The principle of soil testing is to obtain a value that will help to predict the amount of nutrients needed to supplement the supply in the soil. Soil tests are of little use in themselves. They must be calibrated against nutrient rate experiments in the field and in the greenhouse. Also, soil fertility is only one of many factors influencing crop production.

14. Soil is very heterogeneous or nonuniform. This characteristic must be adequately taken into account in soil sampling.

15. The physical properties of soils become more and more important as the top profitable yields are approached, but much more work is needed to identify favorable and unfavorable physical conditions.

16. Soil tests can be classified on the basis of probability of response. For example, in a low test for phosphorus there would be a high probability of response to this element, although in some cases it might not be obtained because of other limiting factors.

17. Most of the calibration experiments need to be rerun using the new technology in crop production, to get maximum yields for the soil and environment and to provide greater opportunity for response. These results will have more meaning for the commercial grower.

18. Two or more levels of recommendation or provision for higher yields, made by some laboratories, help to provide benefits from soil testing for the leading growers. The goal is to *maintain soil fertility at a level for top profit yields*.

19. In general, there are four approaches in recommendations: (a) *buildup* with heavy broadcast rates plus maintenance; (b) *annual application* with modest additions to each crop in the rotation; (c) *rotational fertilization*, a combination of the first two; (d) *replacement* to replace nutrients in crops removed.

20. Prescription and debit and credit methods, in which the contributions of the soil, crop residues, fertilizer, and manure are related to crop needs, are useful. Although the values obviously are influenced by many factors, the methods serve to point out broad needs and possible adjustments.

21. The lime and fertilizer industries can utilize soil and plant tests in a complete service program to provide better identification of the nutrient needs of farmers' fields.

22. Summaries of soil tests help to determine the most needed fertilizer ratios and corrective applications. It is possible to develop more realistic general fertilizer recommendations for use by the growers, the vast majority of whom still do not have their soils tested.

23. Phosphorus and potassium levels are being built up in some soils. Excessive buildup must be studied carefully. However, increased fertility levels encourage higher yields and aid the plant in periods of stress. Nutrient balance must be continually evaluated.

Questions

1. For what reason may a plant develop an unusual red or purple color? What factors encourage this change in color? Distinguish between nitrogen- and potassium-deficiency symptoms in corn.
2. What factors must be taken into consideration in interpreting tissue tests?
3. What are critical nutrient concentrations and critical nutrient ranges?
4. What is the Diagnosis and Recommendation Integrated System (DRIS)? What are its advantages?
5. Is it possible to misinterpret plant nutrient ratios in diagnosing nutritional problems?
6. What is the difference between tissue analysis and plant analysis?
7. Why cannot just any part or growth stage of crops be used in most systems of plant analysis? Are there approaches that eliminate such restrictions?
8. How does cation balance influence the interpretation of plant analysis for a given cation?
9. Why must soil tests be calibrated with crop response? How would you set up a series of experiments to determine the calibration of the phosphorus test for corn soils in your state?
10. Explain why a response to phosphorus would be more generally expected in the northern United States than in the South.

11. Corn plants treated with phosphorus at rates of 20, 40, and 60 lb/A absorb 20, 25, and 30 lb/A of phosphorus, respectively. Using the extrapolation technique to obtain the *A* value, what would be the content of available phosphorus in the soil?

12. Why can the growth of *Cunninghamella* be used as an indication of the amount of phosphorus a soil will supply for higher plants?

13. From the standpoints of the grower and the agencies making the tests, what are the greatest problems in the soil-testing program in the area in which you live? What can be done to remedy the situation?

14. Would you apply a given nutrient if there were a 50% chance of obtaining a response? A 25% chance? Why or why not?

15. Of what value to a county agent or district agriculturist is a summary by crops of the soil-test results? How can a summary be prepared?

16. Name some advantages and disadvantages of deficiency symptoms, tissue tests, and soil tests for detecting plant-nutrient needs.

17. What is crop logging?

18. Ten percent of a grower's field is black lowland soil and the remainder is light-colored upland soil. How should the field be sampled? How frequently should it be resampled?

19. What complicates the securing of good correlation of soil tests for nitrogen with response to nitrogen in the field?

20. Why does probability of response to a nutrient vary at a given soil test level?

21. Can soil variability be a problem in the conduct of soil fertility research?

22. Explain the percentage maximum yield concept. Under what conditions might it not be accurate?

23. What are four general types of recommendations? List advantages and disadvantages of each in your area.

24. If you were a fertilizer dealer what kind of a crop production service would you have for your customer? (Remember that it costs money.)

25. Are soil fertility levels increasing in your area? What are the advantages? Disadvantages?

26. Is deep soil sampling a useful practice? State reasons for your answer.

Selected References

Barber, S. A., "The changing philosophy of soil test interpretations," in L. M. Walsh and J. D. Beaton, Eds., *Soil Testing and Plant Analysis*, p. 201. Madison, Wis.: Soil Science Society of America, 1971.

Barrett, W. B., C. F. Engle, and R. M. Smith, "Factors influencing the levels of exchangeable potassium in Gilpin and Cookport soils." *West Virginia Bull.* 622T (1973).

Bartholomew, W. V., "Soil nitrogen." International Soil Fertility Evaluation and Improvement Program, *Tech. Bull. 6*, North Carolina State Univ., Raleigh, N.C. (1972).

Beaufils, E. R., "Diagnosis and recommendation integrated system (DRIS)." *Soil Sci. Bull. 1.*, Univ. of Natal, South Africa (1973).

Berger, J. C., "Soil tests and fertilizer prescriptions." *Farm Chem.*, 117:47 (1954).

Bray, R. H., "You can predict fertilizer need with soil tests." *Better Crops Plant Food*, 45(3):18 (1961).

Brown, A. L., and B. A. Krantz, "Zinc deficiency diagnosis through soil analysis." *Calif. Agr.*, 15:15 (1961).

Cate, R. B., and L. A. Nelson, "A rapid method for correlation of soil test analysis with plant response data." International Soil Testing Program, *Tech. Bull. 1*, North Carolina State Univ., Raleigh, N.C. (1965).

Claassen, M., and G. E. Wilcox, "Another factor affecting your magnesium level." *Better Crops Plant Food*, **57**(4):10 (1973–1974).

Clements, H. F., "Crop logging of sugar cane in Hawaii," in H. Reuther, Ed., *Plant Analyses and Fertilizer Problems*, p. 131. Washington, D.C.: American Institute of Biological Sciences, 1960.

Cope, J. T., "Fertilizer recommendations and the computer programs key." *Auburn Univ. Agr. Exp. Sta. Cir. 172* (1972).

Dahnke, W. C., "NDSU-Soil test recommendation tables," in *North Dakota 1982 Crop Production Guide*, pp. 60–64. North Dakota State Univ., Fargo, N.D. (1981).

Diagnostic Techniques for Soils and Crops. Washington, D.C.: Potash & Phosphate Institute, [formerly American Potash Institute], 1948.

Dow, A. I., and D. W. James, "Intensive soil sampling: uniformly high soil fertility for uniformly high crop production." in *Proc. 23rd Annu. Northwest Fert. Conf.*, pp. 91–99. Boise, Idaho, (July 18–20, 1972).

Dow, A. I., and S. Roberts, "Proposal: critical nutrient ranges for crop diagnosis." *Agron. J.*, **74**:40 (1982).

Dumenil, L., "N and P composition of corn leaves and corn yields in relation to critical levels and nutrient balance." *SSSA Proc.*, **25**:295 (1961).

Engle, C. F., F. E. Koehler, K. J. Morrison, and A. R. Halvorson, "Fertilizer guide: dryland wheat nitrogen needs." *FG-34* (Reprint), Washington State Univ., Pullman, Wash. (1975).

Fitts, J. W., "Using soil tests to predict a probable response from fertilizer application." *Better Crops Plant Food*, **39**(3):17 (1955).

Fitts, J. W., and J. J. Hanway, "Prescribing soil and crop nutrient needs," in R. A. Olson, T. J. Army, J. J. Hanway, and V. J. Kilmer, Eds., *Fertilizer Technology and Use*, 2nd ed., p. 57. Madison, Wis.: Soil Science Society of America, 1971.

Hanway, J. J., "Plant analysis guide for corn needs." *Better Crops Plant Food*, **46**(3):50 (1962).

Hanway, J. J., "Test undried soil samples." *Better Crops Plant Food*, **48**(5):1 (1964).

Hauser, G. F., "Guide to the calibration of soil tests for fertilizer recommendations." *FAO Soils Bull. 18* (1973).

Hipp, B. W., and G. W. Thomas, "Method for predicting potassium uptake by grain sorghum." *Agron. J.*, **60**:467 (1968).

Hoffer, G. N., "Testing corn stalks chemically to aid in determining their plant food needs." *Purdue Agr. Exp. Sta. Bull. 298* (1926).

Hoffer, G. N., "Soil aeration and crop response to fertilizers." *Better Crops Plant Food*, **31**(12):6 (1947).

James, D. W., and A. I. Dow, "Intensive soil sampling: costs versus returns." *Proc. 23rd Annu. Northwest Fert. Conf.*, pp. 101–108. Boise, Idaho, (July 18–20, 1972).

Johnson, J. W., and W. Wallingford, "Soybeans respond to potash." *Better Crops Plant Food*, **63**:3 (Winter 1979–1980).

Jones, C. A., and J. E. Bowen, "Comparative DRIS and crop log diagnosis of sugarcane tissue analyses." *Agron. J.*, **73**:941 (1981).

Jones, J. B., "Soil testing—changing role and increasing need." *Commun. Soil Sci. Plant Anal.*, 4(4):241 (1973).

Kemmler, G., "Response of high yielding paddy varieties to potassium: experimental results from various rice growing countries." *Int. Symp. Soil Fert. Eval.*, p. 391. New Delhi: Indian Society of Soil Science, 1971.

Kunkel, R., "Requirements for maximum potato yields." *Proc. 17th Annu. Pacific Northwest Fert. Conf.*, pp. 29–47. Wenatchee, Wash. (July 12–14, 1964).

Loúe, A., "A contribution to the study of inorganic nutrition in maize with special attention to potassium." *Fertilite*, **20** (November–December 1963).

McDole, R. E., J. P. Jones, and R. W. Harder, "North Idaho fertilizer guide: wheat." *Curr. Inf. Ser. 453*, University of Idaho, Moscow, Idaho (1978).

McLean, E. O., "Chemical equilibrations with soil buffer systems as bases for future soil testing programs." *Commun. Soil Sci. Plant Anal.*, **13**(6):411 (1982).

McLean, E. O., "Approaches to soil testing: double equilibration with soil buffer system by assessing nutrient sufficiency versus basic cation saturation." *Proc. Wisconsin Fert., Aglime, Pest Manag. Conf.*, Madison, Wis. (1982).

Mehlich, A., "Uniformity of expressing soil test results—a case for calculations in a volume basis." *Commun. Soil Sci. Plant Anal.*, **3**(5):417 (1972).

Mengel, K., "Potassium availability and its effect on crop production," in *Japanese Potash Symposium*, p. 141. Tokyo: Yokendo Press, Kali Kenkyu Kai, 1971.

Ohio Agronomy Guide 1983–84. Columbus, Ohio: Cooperative Extension Service, Ohio State Univ., 1983.

Ohlrogge, A. J., "The Purdue soil and plant tissue tests." *Purdue Univ. Agr. Exp. Sta. Res. Bull. 635* (rev. 1962).

Parker, F. W., W. L. Nelson, E. Winters, and I. E. Miles, "The broad interpretation and application of soil test information." *Agron. J.*, **43**:105 (1951).

Plank, C. O., *Plant Analysis Handbook for Georgia.* Athens, Ga.: Univ. of Georgia, 1979.

Potash and Phosphate Institute, *Soil Testing in High Yield Agriculture.* Atlanta, Ga., 1982.

Potash Institute of North America, "Fight hidden hunger with chemistry." *Better Crops Plant Food*, **48**(3):1 (1964).

Reed, J. F., and W. L. Nelson, *The Diagnostic Approach.* Atlanta, Ga.: Potash and Phosphate Institute, 1984.

Reuther, W., Ed., *Plant Analysis and Fertilizer Problems.* Washington, D.C.: American Institute for Biological Sciences, 1960.

Soil Test Work Group, "Soil tests compared with field, greenhouse and laboratory results." *North Carolina Tech. Bull. 121* (1956).

Soltanpour, P. N., S. M. Workman, and A. P. Schwab, "Use of inductively-coupled plasma spectrometry for the simultaneous determination of macro- and micronutrients in NH_4HCO_3-DTPA extracts of soils." *Soil Sci. Soc. Am. J.*, **43**:75 (1979).

Sparks, D. L., "Chemistry of soil potassium in Atlantic coastal plain soils: a review." *Commun. Soil Sci. Plant Anal.*, **11**:435 (1980).

Sprague, H. B., Ed., *Hunger Signs in Crops.* New York: David McKay, 1964.

Stanford, G., "Assessment of soil N availability," in F. J. Stevenson, Ed., *Nitrogen in Agricultural Soil*, Chapter 17. Agronomy Series 22. Madison, Wis.: American Society of Agronomy, 1982.

Stanford, G., and S. J. Smith, "Nitrogen mineralization potentials of soils." *Soil Sci. Soc. Am. J.*, **36**:465 (1972).

Sullivan, L. J., "What motivates U.S. farmers in use of fertilizer." *Phosphorus Agr.*, **61**:19 (1973).

Sumner, M. E., "Interpretation of nutrient ratios in plant tissue." *Commun. Soil Sci. Plant Anal.*, **9**(4):335 (1978).

Sumner, M. E., "A new approach for predicting nutrient needs for increased crop yields." *Solutions*, **22**(5):68 (1978).

Sumner, M. E., "The diagnosis and recommendation integrated system (DRIS)." *Soil/Plant Analysts' Semin.*, Counc. Soil Test. Plant Anal., Anaheim, Calif., (December 1, 1982).

Sumner, M. E., and F. C. Boswell, "Alleviating nutrient stress," in G. F. Arkin and H. M. Taylor, Eds., *Modifying the Root Environment to Reduce Crop Stress*, Chapter 4. ASAE Monogr. 4. St. Joseph, Mo.: American Society of Agricultural Engineers, 1981.

Syltie, P. W., S. W. Melsted, and W. M. Walker, "Rapid tissue tests as indicators of yield, plant composition, and soil fertility for corn and soybeans." *Commun. Soil Sci. Plant Anal.*, **3**(1):37 (1972).

Terman, G. L., D. R. Bouldin, and J. R. Webb, "Evaluation of fertilizers by biological methods." *Adv. Agron.*, **14**:265 (1962).

Terman, G. L., F. E. Khasawneh, S. E. Allen, and O. P. Engelstad, "Yield–nutrient absorption relationships as affected by environmental growth factors." *Agron. J.*, **68**:107 (1976).

Tyler, K. B., F. W. Fullmer, and O. A. Lorenz, "Plant and soil analysis for potatoes in California." *Trans. 7th Int. Congr. Soil Sci.*, 130 (1960).

Voss, R. D., "P—most limiting nutrient for corn in Iowa." *Proc. 22nd Annu. Fert. Agr. Chem. Dealers Conf.*, **141**:1, Iowa State Univ., Ames, Iowa (1970).

Walker, D. W., "Potassium fertilization in central Alberta." *Better Crops Plant Food*, **62**:13 (Summer 1978).

Walsh, L. M., and J. D. Beaton, Eds., *Soil Testing and Plant Analysis*. Madison, Wis.: Soil Science Society of America, 1973.

Welch, L. F., "High K soils sometimes need more K." *Better Crops Plant Food*, **58**(4):3 (1974–1975).

White, R. P., and E. C. Doll, "Phosphorus and potassium fertilizers affect soil test levels." *Michigan State Res. Rep. 127* (1971).

Wickstrom, G. A., N. D. Morgan, and A. N. Plant, "Ask the plant in the field." *Better Crops Plant Food*, **47**(3):18 (1964).

Wilcox, G. E., and R. Coffman, "Simplified plant evaluation of potassium status." *Better Crops Plant Food*, **56**(1):8 (1972).

Chapter 13

Fundamentals of Fertilizer Application

COMMERCIAL fertilizers are used in modern agriculture to correct known plant-nutrient deficiences; to provide high levels of nutrition, which aid plants in withstanding stress conditions; to maintain optimum soil fertility conditions; and to improve crop quality. Adequate fertilization programs supply the amounts of plant nutrients needed to sustain maximum net returns. In essence, fertilizers are used to make certain that soil fertility is not a limiting factor in crop production.

The major factors influencing the selection of the rate and placement of fertilizer are the crop characteristics; soil characteristics; climate, especially moisture supply; yield goal; and the cost of the fertilizer in relation to the sale price of the crop. The economics of lime and fertilizer use are dealt with in Chapter 15.

Crop Characteristics

Nutrient Utilization. The approximate amounts of nitrogen, phosphorus, potassium, magnesium, and sulfur in certain crops are shown in Table 13-1. The values do not include the quantities contained in the roots. Although uptake will vary considerably, depending on a number of factors, these data indicate the comparative uptake among crops. These factors include: (1) variety or hybrid, (2) moisture availability, (3) temperature, (4) soil type, (5) nutrient levels and their balance in the soil, (6) final plant population, (7) tillage practices, and (8) pest control. Method of harvest of the crop, of course, determines the amounts of nutrients actually removed from the field. Harvest of the complete crop will result in the greatest removal of nutrients. However, in the case of nitrogen and phosphorus as much as 65 to 75% of total uptake can be accounted for in just the harvest of grain. Much smaller proportions of potassium and magnesium removal are accounted for in the grain.

The percentage of total uptake during various growth stages for corn, soybeans, and sorghum is shown in Table 13-2. During the first two growth stages all three crops take up a higher percentage of their total potassium than of nitrogen or phosphorus. This illustrates the importance of an adequate supply of potassium early in the life of the plant. In the last two growth periods the percentage uptake of nitrogen and phosphorus is greater than that for potassium. An important key is to have enough nutrients in the right proportions in the soil to supply crop needs during the entire growing season.

TABLE 13-1. **Approximate Utilization of Nutrients by Selected Crops**

Plant	Yield per Acre	Nitrogen (lb)	P_2O_5 (lb)	K_2O (lb)	Magnesium (lb)	Sulfur (lb)
Alfalfa	10 tons	600	120	600	53	51
Orchard grass	6 tons	300	100	375	25	35
Coastal Bermuda	10 tons	500	140	420	50	50
Clover-grass	6 tons	300	90	360	30	30
Corn						
Grain	200 bu	150	87	57	18	15
Stover	8000 lb	116	27	209	47	18
Sorghum						
Grain	8000 lb	120	60	30	14	22
Stover	8000 lb	130	30	170	30	16
Corn silage	32 tons	266	114	266	65	33
Cotton						
1500 lb lint, 2250 lb seed		94	38	44	11	7
Stalks, leaves, and burrs		86	25	82	24	23
Oats						
Grain	100 bu	80	25	20	5	8
Straw		35	15	125	15	11
Peanuts						
Nuts	4000 lb	140	22	35	5	10
Vines	5000 lb	100	17	150	20	11
Potatoes, Irish						
Tubers	500 cwt	173	73	280	14	15
Vines	300 cwt	96	17	265	36	7
Potatoes, sweet						
Roots	300 cwt	73	34	169	8	—
Vines		83	30	144	10	—
Rice						
Grain	7000 lb	77	46	28	8	5
Straw	7000 lb	35	14	120	6	7
Soybeans						
Grain	60 bu	240	45	84	17	12
Straw	7000 lb	84	16	58	10	13
Tobacco, flue-cured						
Leaves	3000 lb	85	15	155	15	12
Stalks	3600 lb	41	11	102	9	7
Tobacco, burley						
Leaves	4000 lb	173	17	189	21	24
Stalks	3600 lb	115	19	131	11	21
Tomatoes						
Fruit	30 tons	100	23	216	8	21
Vines	4400 lb	80	25	120	20	20
Wheat						
Grain	80 bu	92	44	27	12	5
Straw	6000 lb	42	10	135	12	15
Barley						
Grain	100 bu	110	40	35	8	10
Straw		40	15	115	9	10
Sugar beets						
Roots	30 tons	125	15	250	27	10
Tops	16 tons	130	25	300	53	35
Sugarcane						
Stalks	100 tons	160	90	335	40	54
Tops and trash		200	66	275	60	32

Source: Better Crops Plant Food, **63:**4 (Spring 1979).

Root Characteristics. The growth and appearance of the aboveground portion of plants vary considerably from one species to another and even within species, depending on environment. It might be expected that root systems would also vary greatly in rapidity and extent of development. Since the roots are the principal organs through which plant nutrients are absorbed, understanding of the characteristic rooting habits and relative activity should be helpful in developing fertilization practices. Root systems are usually classified as fibrous or tap. The fact that crops may be annuals, biennials, or perennials is also important.

The ability of crops to exploit the soil for nutrients and water is dependent on the morphological and physiological characteristics of roots. Root radius, root length, root surface/shoot weight ratio, and root hair density are the main morphological features. The presence of mycorrhizae is another important characteristic which is often overlooked. Morphological root properties such as root radius, root surface per unit of shoot weight, and the distribution of roots in the soil profile are important for phosphorus and potassium uptake by corn. Additionally, it has been found that uptake of these two nutrients depends on physiological root characteristics.

Species Differences. A knowledge of the early rooting habits of various crop plants is helpful in determining the most satisfactory method of placing the fertilizer. If a vigorous taproot is produced early, applications may best be placed directly under the seed. If many lateral roots are formed early, side placement may be best. If the rooting habits during the period of rapid growth are known, it should be possible to determine the most effective placement of fertilizer to be used by the plant during this period of growth. Only limited data are available, however, for rooting habits are difficult to study.

A technique has been employed in which small quantities of radioactive phosphorus (^{32}P) are injected at various places in the soil with respect to the plant. The plant is analyzed for ^{32}P, and thus it is possible to study rate of root development

TABLE 13-2.   Percentage of the Total Nutrient Requirement Taken Up at Different Growth Stages

| | Corn Growth Periods (days) | | | | |
	0–25	26–50	51–75	76–100	101–115
N	8	35	31	20	6
P	4	27	36	25	8
K	9	44	31	14	2

| | Soybean Growth Periods (days) | | | | |
	0–40	41–80	81–100	101–120	121–140
N	3	46	3	24	24
P	2	41	7	25	24
K	3	53	3	21	20

| | Sorghum Growth Periods (days) | | | | |
	0–20	21–40	41–60	61–85	86–95
N	5	33	32	15	15
P	3	23	34	26	14
K	7	40	33	15	5

Source: Basic data on soybeans and sorghum from North Carolina and Kansas, respectively. Corn, soybean, and sorghum data appeared in *Better Crops Plant Food,* **56**(2)(1971), and **57**(4)(1973).

and activity within given soil zones. A diagram of the extent of root development of several crops 2 and 3 weeks after planting is shown in Figure 13-1. At 2 weeks the corn root system is more extensive than that of tobacco and cotton. The root development of tobacco and cotton would suggest that at least for early absorption of nutrients the presence of the nutrients under the plants is important. At 3 weeks corn has the most extensive root system and cotton has the most restricted. These differences tend to persist as long as 3 months after planting.

Carrots show considerable root activity at 33 in. (Figure 13-2) and much more than onions, peppers, and snap beans. The reduction in activity at 13 in. marks the beginning of the mottled compact subsoil in the Blount soil. There was a marked effect of soil, and activity was much less at 33 in. in peat.

Many observations have shown that corn depends heavily on phosphorus near the roots early in the season. From the knee-high stage on, however, corn develops a very extensive root system and has a great capacity for utilizing the nutrients distributed through a large soil zone. The root system of corn, and to a lesser extent that of soybeans, exploits the soil rather thoroughly in contrast to potatoes (Figure 13-3). At later stages corn and soybeans utilize only small quantities of fertilizer phosphorus as compared with potatoes. Potatoes have a limited root system, often

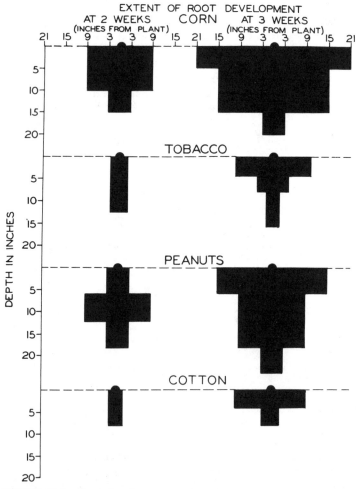

FIGURE 13-1. Root development at 2 and 3 weeks of corn, tobacco, peanuts, and cotton. [*North Carolina Agr. Expt. Sta. Tech. Bull. 101* (1953).]

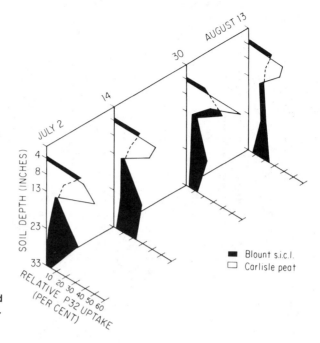

FIGURE 13-2. Root activity of carrots on Blount silty clay loam and Carlisle peat. [Hammes et al., *Agron. J.*, **55**:329 (1963).]

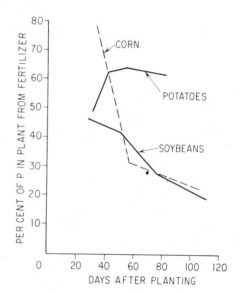

FIGURE 13-3. Percentage of phosphorus in corn, soybeans, and potatoes derived from the fertilizer. [Krantz et al., *Soil Sci.*, **68**:171 (1949). Reprinted with permission of The Williams & Wilkins Co., Baltimore.]

being confined by the hilled row, and the roots penetrate only a small volume of soil. Traditionally, the effective rooting depth of potato plants has been believed to be 20 to 24 in., but Idaho research suggests that it may be only 11 to 15 in. below level ground.

In some areas potatoes are grown during a cooler time of year than corn and have a shorter growing season. These factors plus the differences in rooting habits help to explain the greater dependence on the fertilizer and less on the soil phosphorus.

By far the greatest early root growth of corn, and for that matter most crops, is in the topsoil (Figure 13-4). Both root surface per plant and root density in the

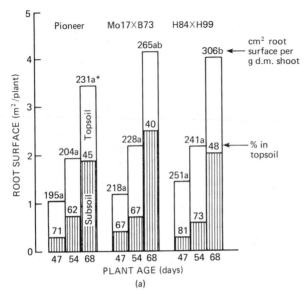

*Different letters indicate significant differences at
particular plant age.

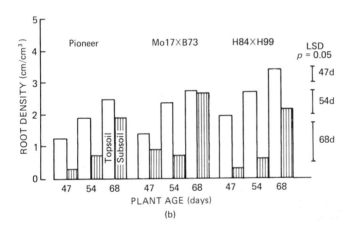

FIGURE 13-4. Root surface per plant (a) and root density (b) in topsoil and subsoil for three corn genotypes at three harvest dates. [Schenk and Barber, *Plant Soil*, **54**:65 (1980).]

subsoil, however, increased considerably with plant age and by the end of the last harvest they were equal to or greater than those in the topsoil.

On a sandy soil in Nebraska, corn roots reached a depth of more than 8 ft and completely extracted readily available soil moisture down to 6 ft. Corn root weights of 3000 lb/A or more have been found, and there is as much variation in root growth as in top growth.

Small grains such as wheat have a fairly extensive root system that compares favorably with corn (Figure 13-5). The early response of small grains to phosphorus placed near the seed even on soils well supplied with this element is characteristic. Little information is available about the rooting habits of crops such as Ladino clover, but the root system is generally considered to be fairly shallow. This is contrasted with alfalfa, the roots of which may penetrate 25 ft if soil conditions are favorable. Depths of 8 to 10 ft are common even on quite compact soils. One

WEIGHT OF ROOTS PER PLANT

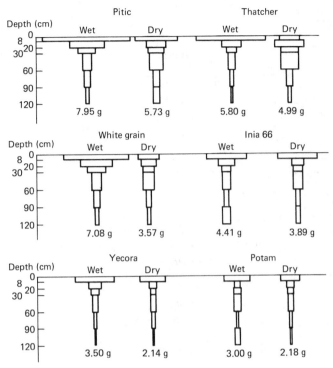

FIGURE 13-5. Weight of roots of wheat cultivars grown at high and low moisture levels. [Hurd, *Agron. J.,* **60**:201 (1968).]

of the advantages of deep-taprooted crops such as alfalfa and sweet clover is that they tend to loosen compact subsoils by root penetration and subsequent decomposition. Also, such legumes in pastures help to provide more feed during moisture shortages in the summer than do the more shallow-rooted grasses.

There is a tendency for root systems of the same species not to interpenetrate (Figure 13-6). This suggests an antagonism or toxic effect. With the trend to closer rows and thicker planting of crops the characteristic root pattern is altered and there may be a tendency for deeper rooting if soil conditions permit.

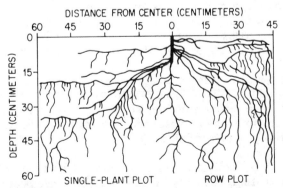

FIGURE 13-6. Contrasting rooting patterns of soybeans in single-plant plots and row plots. [Raper and Barber, *Agron. J.,* **62**:581 (1970).]

TABLE 13-3. **Effect of Tillage Treatment on the Corn Root-Weight Distribution (mg/100 cm³ of Soil) with Depth**

Depth (cm)	Tillage Treatment				
	Conventional	Conventional, No Residues	Chisel	Rototill	No-Till
0–5	29	49	26	69	137
5–10	38	52	104	136	100
10–15	96	218	110	137	74
15–30	85	111	68	88	73
30–45	73	61	47	52	28
45–60	61	59	35	52	37

Source: Barber, *Agron. J.,* **63:**724 (1971).

Tillage system also affects root-weight distribution with depth (Table 13-3). When the soil was plowed annually (conventional) corn roots developed more extensively than with no-till. Rototill and chisel were in between. Also, when residues were removed there was greater root growth in the surface 15 cm. Thus it appears that residue decomposition products were inhibiting root growth.

Varietal Differences. Differences in extent of root development among varieties of corn and wheat are evident in Figures 13-4 and 13-5, respectively. Numerous other crops, including soybeans and sweet potatoes, have exhibited differences in root development related primarily to deeper rooting.

Feeding Power. As discussed in Chapter 4, differences exist in cation exchange capacity of roots among various plants. The exchange capacity of the roots of dicotyledonous plants is much higher than that of monocotyledonous plants.

Nonlegumes simply have a lower requirement of divalent cations and take up more monovalent cations to meet needs. The relative absorption of cations and anions by the root is related to the release of H^+ or HCO_3^- by the root. Mention was made in Chapters 5 and 6 of the localized acidity that develops from the release of H^+ from the root in response to absorption of NH_4^+ ions. The opposite effect on soil reaction produced by the release of HCO_3^- and/or OH^- ions following plant uptake of NO_3^- was also referred to earlier. Such changes in pH of the rhizosphere affect the solubility of many plant nutrients, including phosphorus, boron, copper, iron, manganese, and zinc. Calcium phosphates may accumulate on the root surface. Soils vary greatly as to the effect of the pH change on nutrient availability and salt accumulation around the root.

Plants vary in extent of roots, and the proximity of the absorbing plant roots to the soil surfaces increases nutrient uptake. Even then it has been estimated that roots occupy one-half of 1% of the volume of the topsoil and much less in the subsoil.

Mycorrhizal fungi are often associated with the roots of plants grown under conditions of low soil fertility. These organisms increase the ability of plants to absorb nutrients such as phosphorus, potassium, copper, and zinc. Under field conditions addition of nitrogen, phosphorus, and complete fertilizer will reduce the presence and activity of mycorrhiza.

Soil Effects. Soil characteristics have a pronounced influence on the depth of root penetration, and soils with compact B horizons are highly restrictive. Effects of soil series in Illinois on corn root development are shown below. A number of factors were operating, including bulk density, organic-matter content, acidity and plant nutrient content, old root channels, and oxygen supply.

	Flanagan	Muscatine	Clarence	Wartrace	Cisne
Corn root weights (lb/A)	1846	2008	1758	3136	2647
Depth of penetration (in.)	60	66	38	48	60
Water-holding capacity to rooting depth (in.)	12.8	17.4	6.4	17.1	17.3

The yield of a crop is often directly related to the availability of stored water in the soil. This amount is related to soil characteristics. The following yields of corn were reported:

	Corn Yield (bu/A)		
Available H_2O in Profile (in.)	Lafayette, Indiana	Urbana, Illinois	Ames, Iowa
4	79	85	88
8	121	128	129
12	130	136	135

Basically, the soil is a rooting medium and a storehouse for nutrients and water. Hence it is essential that the roots fully exploit the soil in order that the plants not only obtain nutrients but also root deeply enough to obtain water to help the plant better withstand periods of water stress.

In drought-prone areas in the Mississippi Delta where the soil is below pH 5.5, annual subsoiling to 18 in. and liming is suggested for soybeans. A crop such as soybeans has only limited ability to penetrate even moderately compacted soil layers, and both soybeans and corn have responded to in-row subsoiling in the South. An additional subsoil barrier may be the acidity and the accompanying high degree of aluminum saturation. The harmful effects of aluminum on roots and on uptake of water and nutrients were referred to in Chapters 7 and 11.

Soil compaction will become an increasing problem on some soils as other limiting factors are removed, higher yields sought, and heavier equipment used. However, in many instances in the past, attempts to loosen plowpans or heavy subsoils have not been entirely successful. The operation is most effective when the subsoil is dry so that shattering of the soil occurs. However, in most cases there is a rapid resealing of the subsoil. One cultivation with a disk or other such implement may almost eliminate any effect of subsoiling. A practice called vertical mulching has received some attention. Chopped plant residues blown into the slit behind the subsoiler serve to keep the channel open and improve water uptake.

Sodic claypan or Solonetzic soils, formed as a result of salts high in sodium at or near the soil surface, occupy 17 to 20 million acres of land in the northern Great Plains of the United States. There are a further 20 million acres of these tough, impermeable soils in the Prairie Provinces in Canada. Crop production on such soils may be only about one-third of that possible on sodium-free soils in the same area. Deep plowing to depths of 16 to 24 in. and the subsequent mixing of surface soil, hardpan layer, and lime-salt layer in approximately equal proportions is a satisfactory corrective measure in some Solonetzic soils.

Plant-Nutrient Effects. Adequate fertilization of surface soil and proper management are important in encouraging deeper rooting. Proper nutrition encourages not only greater top growth but also a more vigorous and extensive root system.

It is well known that crops develop masses of roots in the zone of fertilizer placement. The proliferation of roots resulting from contact with localized zones of high nutrient concentration in infertile soil is illustrated in Figure 13-7. This stimulation of root development is related to the buildup of high concentrations of nitrogen and phosphorus in the cells that hastens division and elongation. This favors branching and is accompanied by an increase in growth-regulator auxins.

The effects of localized supply of phosphate, nitrate, ammonium, and potassium on root form of barley were shown in Figure 3-3. Exposure of parts of the main seminal roots (axes) to either phosphate, nitrate, or ammonium had little effect on the extension of seminal axes but greatly enhanced the initiation and extension of both first- and second-order lateral roots. When roots were locally enriched with nitrogen, phosphate, or potassium, the affected plants appeared to compensate for negligible absorption by the remainder of the root system in zones of low plant-nutrient concentration. The source of nitrogen and phosphorus for encouraging root proliferation in any part of the root system must be externally in the soil solution rather than internally.

Plants absorb nutrients only from those areas in the soil in which roots are active. It is also well known that plants cannot absorb nutrients from a dry zone. Hence root systems modified by shallow applications of fertilizer may be less effective in time of drought. It appears that the bulk of the less mobile nutrients should be distributed through the plow layer. In general, fertilizer should be placed in that portion of the root zone where stimulation of root growth is wanted.

Another aspect is that since oxygen is consumed when nutrients are absorbed, intense absorption from a fertilizer band may result in a temporary deficiency of oxygen. Of course, oxygen deficiency or anaerobic conditions may prevail in wet seasons and in poorly drained soils.

Hence it is quite apparent that soil chemistry, fertilizer chemistry, and plant physiology are all important in the fertilizer band. The anions, such as NO_3^-, $H_2PO_4^-$, and Cl^- and cations such as calcium, magnesium, NH_4^+ and potassium are competing with one another for entrance into the plant. New compounds are being formed and some may actually be toxic to the plant roots. Much is yet to be learned about these interactions.

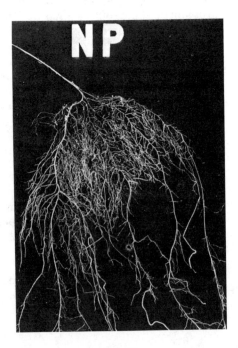

FIGURE 13-7. Development of a single soybean root growing through a nitrogen- and phosphorus-fertilized zone. [Wilkinson and Ohlrogge, *Agron. J.*, **54**:288 (1962).]

An example of the effect of proper management on growth of corn roots in a Cisne silt loam is shown in Figure 13-8. The Cisne has a claypan and is low in native fertility. The use of adequate plant nutrients with proper cropping systems, including legumes, was effective in the development of a much deeper root system. With top management, including irrigation, yields of over 200 bu/A have been obtained at this location.

Root weight of Coastal Bermuda was increased from 6765 lb with no nitrogen to 8098 lb with 1600 lb of nitrogen. Eighty-five to ninety percent of the roots were in the surface 24 in. on a fine sandy loam.

The influence of plant nutrition on resistance to winter killing is important. By the addition of adequate plant nutrients to the soil, alfalfa is now being grown on many soils in which growth was once thought to be impossible. Similar effects of plant nutrients on extending root growth of wheat have been observed (Figure 13-9).

Winter killing is generally due to the following conditions:

1. *Heaving.* Plants are lifted from the soil and roots are broken by alternate freezing and thawing.
2. *Smothering.* Ice-sheet formation causes internal accumulation of toxic products of aerobic and anaerobic respiration.
3. *Physiological drought.* A frozen soil is like a dry soil.
4. *Direct effect of low temperature.* As plant tissues freeze and ice forms, the cells rupture and dry out.
5. *Indirect effect of low temperature.* Low-temperature plant pathogens protected by snow cover.

The effect of fertilizer treatments on the winter killing of alfalfa is given in Table 13-4. High levels of lime, phosphorus, and potassium increased sugars, soluble proteins, and retention of water, all of which are directly associated with less winter killing.

Soil Characteristics

A most important soil factor in the determination of rate and placement of fertilizers is the crop to be grown and the amount of soil nutrients that will be needed by the crop during the growing season. Soil and tissue tests, history of management, and observations of plant growth are of great help in evaluating the nutrient-supplying power of the soil. This was covered in Chapter 12. The effects of fixation capacity, certain physical characteristics that influence water movement through

TABLE 13-4.   Effect of Plant Nutrients on Winterkilling and Composition of Alfalfa

Treatment				Starch plus Sugar (%) (Nov. 26)	Soluble Protein (%) (Nov. 26)	Moisture Retained on Drying (%) at 50°C
Lime (tons)	Phosphorus (lb)	Potassium (lb)	Winterkill (%)			
0	0	0	90	14.68	10.36	2.70
5	0	0	50	15.53	16.20	3.09
5	132	250	> 20	—	—	3.82
10	132	250	> 20	19.74	15.37	4.18
5	264	0	50	17.90	16.55	3.85
5	0	500	> 20	18.32	14.99	3.46
5	264	50	> 20	19.74	17.10	4.45

Source: Wang et al., *Agron. J.*, **45**:381 (1954).

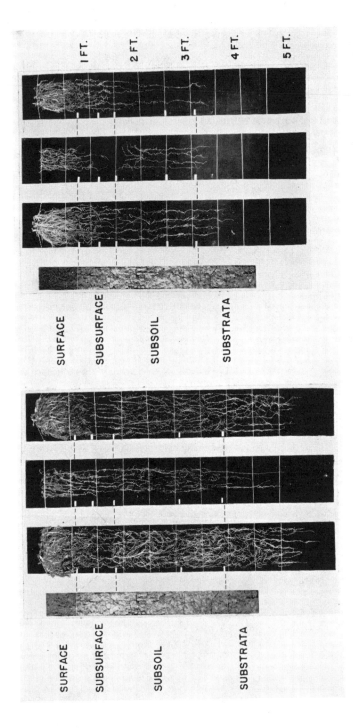

FIGURE 13-8. Soil treatment affects root growth. A rotation including corn, small grains, and legumes was followed. The corn roots on the left, however, were grown in soil receiving adequate lime, phosphorus, and potassium. Those on the right were grown in soil receiving no fertilizer or lime. (Fehrenbacher et al., *Soil Sci.,* **17**:281. Copyright 1954 by The Williams & Wilkins Co., Baltimore.)

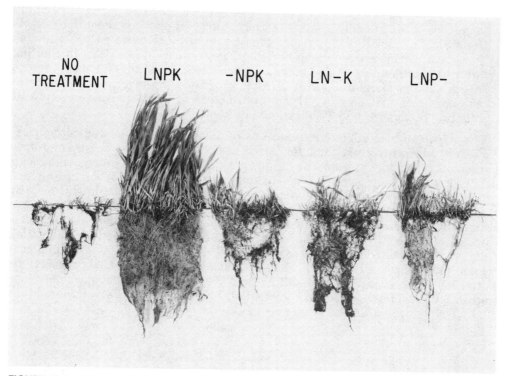

FIGURE 13-9.   Balanced fertility aids winter survival of wheat. Early spring vigor means more stooling and more yield. (Courtesy of the Brownstown Agronomy Research Center, Illinois.)

the soil, and climatic factors on placement of the fertilizer are discussed in appropriate sections of this chapter.

Fertilizer Placement

An important item in the efficient use of fertilizers is that of placement in relation to the plant. Determining the proper zone in the soil in which to apply the fertilizer ranks in importance with choosing the correct amount of plant nutrients. Fertilizer placement is important for at least three reasons:

1. *Efficient use of nutrients from plant emergence to maturity.* A fast start and continued nutrition are essential for sustained maximum profit. Merely applying fertilizer does not ensure that it will be taken up by the plant. It is usually important to place some of the fertilizer where it will be intercepted by the roots of the young plant and to place the bulk of the nutrients deeper in the soil, where they will be more likely to be in a moist zone the greater part of the year.

2. *Prevention of salt injury to the seedling.* Soluble nitrogen, phosphorus, potassium, or other salts close to the seed may be harmful. An important rule is that there should be some fertilizer-free soil between the seed and the fertilizer band, especially for small seeded and/or sensitive crops such as flax, rapeseed/canola, peas, sunflowers, and so on. An important exception is that of low rates of fertilizer drilled in directly with the seed of some crops, such as the small grains.

3. *Convenience to the grower.* With a premium on labor saving, speed, and timeliness, placement methods assume additional importance. Growers want much

of their fertilizer applied in advance of planting when time is less critical and a much less amount of effectively placed fertilizer at planting time. Under some conditions none will be applied at planting in order to avoid delays. In many areas, corn yields are reduced a bushel or more for each day of delay in planting after the optimum date. The statement has been made that a farmer's time is worth $200 or more an hour at corn-planting time. However, profit must also be considered and there is a trend toward banding fertilizer beside the row at planting because of a yield advantage in some situations.

The importance of adequate quantities of plant nutrients is well recognized. However, full return from their increased use is sometimes limited by improper placement, particularly from the standpoint of fertilizer salt injury, which can delay germination and even be the cause of reductions in stand. When large quantities of fertilizer are applied, maximum returns may not be obtained if most of the nutrients are concentrated near the seed row. The varying effects of fertilizer materials on increasing the salt concentration or electrical conductivity in a calcareous, alluvial soil are obvious in Figure 13-10. These elevated conductivities were inversely related to the clay content of soil. Consequently, potential problems related to fertilizer salts are likely to be greatest in coarse-textured soils. High initial NH_4^+-N concentrations resulting from the addition of ammoniacal nitrogen sources will also increase osmotic suction and favor the temporary accumulation of NO_2^--N, which is toxic to plants.

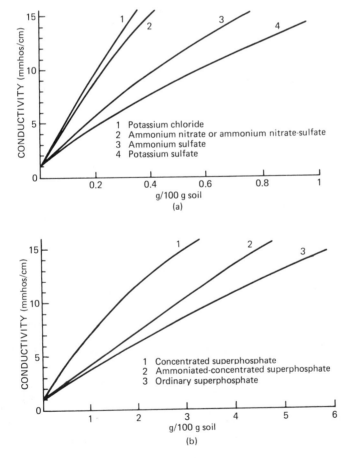

FIGURE 13-10. Effect of (a) potassium and nitrogen, and (b) phosphate fertilizers on the conductivity of saturation extracts of Norwood silt loam. [Chapin et al., *Soil Sci. Soc. Am. J.,* **28**:90 (1964).]

It should be stressed that as the soil fertility level increases the benefit from fertilizer applications at planting is generally decreased. The important point is that adequate amounts of nutrients be available to the crop.

Methods of Placement. There are many variations of broadcasting and banding, the two main methods of fertilizer application: (1) surface broadcast; (2) disked; (3) plowdown; (4) seed placed; (5) banding, including all combinations of distances below and to the side of the seed; (6) plow-sole placed; (7) deep placement; (8) layering; (9) strip placement; and so on, as well as many combinations of the various specific methods. Many of the popular application methods are illustrated in Figure 13-11.

1. *Broadcast.* The fertilizer is applied uniformly over the field before planting the crop and it is incorporated by tilling or cultivating. Where there is no opportunity for incorporation such as on perennial forage crops and no-till cropping systems, fertilizer materials may be broadcast on the surface. However, some form of placement is preferred for application of phosphorus and potassium during seeding of no-till annual crops.
2. *Sideband.* The fertilizer is applied in bands to one or both sides of the seed or plant. Special equipment places the bands of fertilizer 2 to 3 in. to the side and 1 to 2 in. below the seed or transplant (Figure 13-12).It is essential that careful attention be directed to the proper adjustment of equipment and that the placement be checked frequently to make certain that the distributor is not out of adjustment. Disastrous results have been reported because of poor adjustment of equipment.
3. *"With seed."* These applications cover a number of methods. The fertilizer may be run down the same spout with the seed, as in a small grain drill, or it may be placed as a pop-up with the seed. As higher rates of higher-analysis fertilizers are used, the emergence of seedlings is delayed and in some cases yields are reduced (Figure 13-13). The fertilizer may also be placed at the bottom of a rather deep furrow. The soil is then bedded back on the row before planting. Air seeders can be used to pneumatically apply seed and fertilizer simultaneously in a range of row widths. They are relatively new to North America and more will be said about them in Chapter 14.
4. *Top-dressed or side-dressed.* The fertilizer is applied to the crop after emergence. The term *top-dressed* usually refers to broadcast applications on crops such as small grain or forage. The term *side-dressed* refers to fertilizer placed beside the rows of a crop such as corn or cotton.

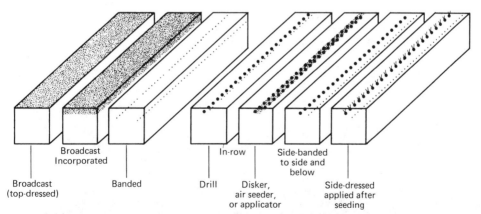

FIGURE 13-11.   Cross section of soil profile showing fertilizer placement. [Robertson, "Fertilizer placement." *Agdex 542-5.* Alberta Agriculture (August 1982).]

FIGURE 13-12. Band application of fertilizer to the side and somewhat below the seed helps to avoid injury to the plant and permits more efficient use of fertilizer. (Courtesy of the Potash & Phosphate Institute, Atlanta, Ga.)

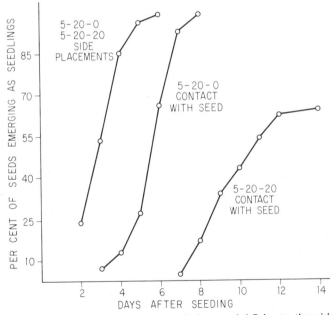

FIGURE 13-13. Placement 1.5 in. below and 1.5 in. to the side resulted in faster emergence of wheat seedlings under greenhouse conditions in contrast to placement in contact with the seed. [Lawton et al., *Agron. J.,* **52**:326 (1960).]

5. *Other banding methods.* These include the injection of anhydrous and aqua ammonia, deep placement during tillage and cultivation, banding with air seeders and pneumatic fertilizer applicators, and shallow placement with seed drills. Some no-till drills have the capability of deep banding fertilizers below the seed row.

Results of many experiments might be cited to show the effect of different methods of fertilizer placement on crop stands and yields. The correct placement of fertilizers on vegetable crops is of particular importance. Large quantities of fertilizers are applied, but any retardation in the rapidity of growth is undesirable. These two factors combine to make sidebands quite beneficial. For example, with potatoes a summary of results from several states in which 1000 to 2000 lb/A of fertilizer was used showed that when the fertilizer was mixed in the row with the seed a yield of 304 bu/A was obtained. When the fertilizer was placed 2 in. to the side and 2 in. below, a yield of 348 bushels was obtained.

Movement of Fertilizer

Soluble salts are dissolved in the soil solution surrounding the zone of fertilizer application, which thus becomes quite concentrated. The rate and distance of movement of the salts from point of application depends on the nature of the salts, application rates, the character of the soil, and climatic conditions.

Phosphorus movement from the point of placement is generally limited, for the phosphate ion is only slightly mobile in the soil.

The nitrogen salts move up and down in the soil solution, depending on direction of water movement. Of the two general types of nitrogen salts, nitrate moves more readily, for it does not attach itself to soil particles. On the other hand, ammoniacal nitrogen is adsorbed by the soil colloids. As it is converted to the nitrate, it, of course, becomes mobile.

The potassium ion is positively charged. It tends to attach itself to the colloidal complex and is restricted in movement. The anionic make-up of ammoniacal and potassium salts as well as the cation replaced must be considered in any evaluation of possible salt effects. If the accompanying anion is sulfate or chloride and the potassium replaces magnesium on the exchange complex, there is still a soluble salt present. On the other hand, if calcium is replaced, a much less soluble salt is formed if the added anion is sulfate.

As the soils dry out, the concentration of the soil solution is increased, soil water moves upward by capillary movement, and the salts move with it. In some instances they may be deposited on the surface just above the fertilizer band. This may be seen as a white or a light brown deposit coming from dispersed organic matter. Rain immediately after planting followed by a long dry period is conducive to solution of the salts and upward salt movement. With considerable rain the soluble salts tend to move downward again. It would be expected that on soils having a relatively low water-holding capacity the increase in concentration of the soil solution would be greater than on heavier soils having a larger water-holding capacity.

In view of possible movement, the concentration of fertilizer in proximity to the seed is generally hazardous. Excessive concentration of soluble salts in contact with the roots or the germinating seeds causes injurious effects through plasmolysis, restriction of moisture availability, or actual toxicity. The term "fertilizer burn" is often used. The plant will lose water and dry out just as effectively as if it had been placed in an oven.

Certain of the nitrogen-bearing compounds contribute more to damage of germinating seeds than appears to be explained by the osmotic effect. Evidence is that free ammonia is a toxic factor and can move freely through the cell wall, whereas NH_4^+ cannot. Urea, diammonium phosphate, ammonium carbonate, and ammonium hydroxide may cause more damage than materials such as monoammonium phosphate, ammonium sulfate, and ammonium nitrate. This was discussed in Chapter 5. Placement to the side and below the seed is an effective method of avoiding the problem.

Salt Index

Fertilizers increase the salt concentration of the soil solution. The salt index of a fertilizer is a measure of this phenomenon and is determined by placing the material under study in the soil and measuring the osmotic pressure of the soil solution. Osmotic pressure is expressed in atmospheres. Salt index is actually the ratio of the increase in osmotic pressure produced by the material in question to that produced by the same weight of sodium nitrate, based on a relative value of 100.

Fertilizer salts differ greatly in their effect on the concentration of the soil solution. Mixed fertilizers of the same grade may also vary widely in salt index, depending on the carriers from which they are formulated. It would be well to emphasize that the higher-analysis fertilizers will generally have a lower salt index per unit of plant nutrients than water-soluble, lower-analysis fertilizers because they are usually made up of higher-analysis materials. For example, to furnish 50 lb of nitrogen, 250 lb of ammonium sulfate would be required, whereas with urea 110 lb would be required. Hence the higher-analysis fertilizers have less of a tendency to produce salt injury than equal amounts of plant nutrients in the lower-analysis fertilizers.

The salt index per unit, 20 lb, of plant nutrient for several materials is shown in Table 13-5. Nitrogen and potassium salts have much higher salt indices and

TABLE 13-5. Salt Index per Unit of Plant Nutrients Supplied for Representative Materials

Material	Analysis *	Salt Index per Unit of Plant Nutrients
Nitrogen carriers		
Anhydrous ammonia	82.2	0.572
Ammonium nitrate	35.0	2.990
Ammonium sulfate	21.2	3.253
Monammonium phosphate	12.2	2.453
Diammonium phosphate	21.2	1.614
Nitrogen solution 2A	40.6	1.930
Potassium nitrate	13.8	5.336
Sodium nitrate	16.5	6.060
Urea	46.6	1.618
Phosphorus carriers		
Superphosphate	20.0	0.390
Superphosphate	48.0	0.210
Monoammonium phosphate	51.7	0.485
Diammonium phosphate	53.8	0.637
Potassium carriers		
Manure salts	20.0	5.636
Potassium chloride	50.0	2.189
Potassium chloride	60.0	1.936
Potassium nitrate	46.6	1.580
Potassium sulfate	54.0	0.853
Potassium magnesium sulfate	21.9	1.971

* By analysis is meant the percentage of nitrogen in nitrogen carriers, of P_2O_5 in phosphorus carriers, and of K_2O in potassium carriers.

Source: Rader et al., *Soil Sci.*, **55**:201. Copyright 1943 by The Williams & Wilkins Company.

**TABLE 13-6. Comparative Salt Index of
5-4.4-8.3 and 10-8.8-16.6**

| | 5-4.4-8.3 (5-10-10) | | 10-8.8-16.6 (10-20-20) | |
	Pounds	Salt Index	Pounds	Salt Index
Ammoniating solution	148	5.79	—	—
Diammonium phosphate	—	—	373	6.37
Urea	—	—	260	9.71
Ammonium sulfate	195	6.51	—	—
Treble superphosphate	—	—	417	2.10
Superphosphate	1000	3.90	—	—
Muriate of potash (41% K)	400	21.89	—	—
Muriate of potash (50% K)	—	—	667	38.72
Conditioner	100	—	110	—
Filler	157	—	180	—
	2000	38.09	2000	56.90

are much more detrimental to germination than phosphorus salts when placed close to or in contact with the seed. From this information the relative salt index of a mixed fertilizer can be calculated if the formulation of the fertilizer is known. The comparison between the 5-4.4-8.3 (5-10-10) and 10-8.8-16.6 (10-20-20) is shown in Table 13-6. In fertilizers of these formulations the salt index for equal quantities of plant nutrients is 38.09 and 28.45, respectively. This then is another advantage of higher-analysis fertilizers.

In considering the placement of fertilizers at planting it is important that the differences in salt effects among fertilizers be kept in mind. Obviously, if higher amounts per acre of higher-analysis fertilizers are to be used, the problem becomes increasingly important.

General Considerations

Band Applications. Early stimulation of the seedlings is usually advantageous, and it is desirable to have N-P-K near the plant roots.

The early growth of the plant top is essentially all leaves. In a crop such as corn leaf growth is completed in about 60 days. Since the photosynthesis is carried on in the leaves, the number of leaves produced in this period will influence the grain produced in the next 45 days. It is important to have a small amount of nutrients near the very young plants to promote early growth and the formation of large healthy leaves (Figure 13-14). Also, the various plant parts such as the ear bud in corn are differentiated very early in the growth cycle and lack of stress from any source is desirable. Alberta investigators report that the yield potential of barley is determined by the four-leaf stage. Environmental factors through the remainder of the growing season will determine the degree to which this potential is expressed.

Starter response to N-P-K is often independent of fertility level. Under cool temperatures the early available nutrient supplies may be inadequate because of slow mineralization of nitrogen, phosphorus, sulfur, and so on, from the soil organic matter; restricted release of plant nutrients from soil minerals; reduced diffusion of phosphorus and potassium; or limited absorption of phosphorus, potassium, and other nutrients by the plant. Localized applications of fertilizer at planting are commonly referred to as starter or planting fertilizers.

The advantage of early stimulation depends on the crop and seasonal conditions. Some of the factors that might be considered are the following.

FIGURE 13-14. A readily available supply of nutrients near the young plant helps to ensure rapid early growth and the formation of large leaves essential in photosynthesis. (Courtesy of the Potash & Phosphate Institute, Atlanta, Ga.)

Resistance to Pests. Under adverse conditions a fast-growing young plant is usually more likely to resist insect and disease attacks.

Competition with Weeds. Weed control is facilitated when the young plants start off rapidly. Herbicides are more likely to be effective and/or cultivation can be started before the weeds become established, and the number of cultivations may be reduced. If one cultivation is eliminated, this saving should be considered in calculating cost of fertilization. Vigorous early growth of uncultivated crops such as seedling forage plants is also important in reducing weed competition.

Early Maturity. Particularly with vegetables, an early crop is in most cases of paramount importance. A delay of only 3 or 4 days may make the difference between a good price on an early market and a break-even situation.

Maintenance. Phosphorus might be applied in a band at planting in sufficient amounts for maintenance of optimum levels of available soil phosphorus for a crop such as corn. This will permit more efficient use as well as saving a trip over the field. It is of interest that for high-yield corn production the University of Georgia recommends a sideband application of fertilizer regardless of the soil test level.

Broadcast Applications. Broadcast applications usually imply large amounts of nutrients for soil buildup. However, this approach may be used for maintenance. Phosphorus and potassium as well as some secondary or micronutrients are usually disked or plowed down, although plowing may be preferable because the nutrients can be placed deeper. With the trend to reduced tillage and less plowing more nutrients remain near the surface. This point will be considered in the next section.

Depending on the form of nitrogen used it may be plowed down, injected into the soil, or placed on the surface. There are several points in favor of broadcasting the major amounts of nutrients.

1. Application of large amounts of fertilizer is accomplished without danger of injuring the plant.
2. If tilled into the soil, distribution of nutrients throughout the plow layer encourages deeper rooting. Placement of nutrients in a band near the seed tends to encourage concentration of roots in the fertilizer zone and less complete exploration of the soil for water and nutrients takes place. The exception would be with small amounts of nutrients.
3. Labor is saved during planting. The fertilizer marketing season is spread out through fall, winter, or early spring applications.
4. As a consequence of item 3, the application of adequate amounts of nutrients is made easier.
5. This method is the only practical means of applying maintenance fertilizer to established stands of forage.

Uniformity of Spreading. Although the importance of uniformity of application of blends of dry fertilizer products was discussed in Chapter 10, it will be stressed again that uniform and accurate spreading of dry and fluid fertilizers, lime, and pesticides is essential for effective utilization by the crop. The effects of uneven application of recommended rates of N-P-K fertilizer on yield of crops are shown in Table 13-7.

The influence of uneven spreading patterns on yield is affected by soil fertility levels. As might be expected, fertilizer rates well below those needed for low-fertility soils will result in significant yield losses. There may also be yield reductions from overfertilization on soils medium or higher in fertility. Nitrogen on small grain would be an example.

Placement of Nutrients

Phosphorus. Since this element tends to move very little, the phosphorus should generally be placed in the zone of root development. Obviously, surface applications after the crop is planted will not be in the zone of root activity and will be of little value to row crops in the year of application. With subsequent tillage it will be better distributed for later crops.

An exception to the inefficiency of surface application is with forage-crop fertilization. Top-dressed phosphorus for maintenance purposes is an efficient method of placement. Some of the phosphorus is absorbed by crowns of the plant as well

TABLE 13-7.   Effects of Uneven Application of Recommended Rates of N-P-K Fertilizer on Crop Yields at Two Locations in Virginia

	Yield *			
	Capron	Orange		
Spread Pattern	Barley (kg/ha)	Barley, Forage † (ton/ha)	Soybeans (kg/ha)	Corn (kg/ha)
No fertilizer	592 b	1.41 b	1278 a	2059 c
Uniform	2809 a	2.38 a	1345 a	8060 a
Skewed	2540 a	2.35 a	1264 a	7571 ab
Single pyramid	2454 a	2.29 a	1318 a	6993 b
Double pyramid	2793 a	2.33 a	1197 a	7740 ab

* Values in columns followed by a letter in common are not significantly different at the 5% level.
† Silage yields. All other data in this table are grain yields.
Source: Lutz et al., *Agron J.*, **67**:526 (1975).

as by very shallow roots. In addition, such applications come in contact with less soil than applications disked in, and there is less opportunity for fixation. Injection can overcome the difficulty of placing phosphorus in the soil for a growing sod crop. Surface banding is also possible.

Zone fertilization by banding P_2O_5 or K_2O on the surface and plowing down was found to be more effective than banded beside the row or broadcast plowdown on a low phosphorus soil (Table 13-8). Bands of P_2O_5 on surface and plowed down were superior at all rates. Similar results were obtained with potassium. Less fixation and more concentrated zones of nutrients are possible reasons. This would vary with the soil.

The question of localized applications in the row versus broadcast application is one of considerable importance. Fixation as well as proximity to the plant must be considered. Progressive fixation of added phosphorus by certain fractions in the soil tends to diminish its utilization by crops. The mechanism of phosphorus fixation was discussed in Chapter 6.

When all of the phosphorus is either banded or broadcast, the relative efficiency is related to both the phosphorus status of the soil and the rate of application. As the rate of phosphorus increases, in most cases, broadcast applications become adequate and sometimes even superior to banding (Figure 13-15). When the phosphorus application is split between band and broadcast, it is noted that at no point will band alone reach the maximum yield, thus the advantage of building up the general soil phosphorus level.

University of Nebraska data show that differences between seed-placed phosphorus and broadcast phosphorus for wheat declined markedly with increasing levels of available soil phosphorus. Research conducted at North Dakota State University also indicated that at high soil test phosphorus levels the differences between broadcast and deep placement applications of nitrogen and phosphorus diminished.

When surface soil dries out or becomes very hot, phosphorus in the upper part of the root zone can become unavailable. Placement of nutrients at depths of 4 to 7 in. for dryland crops in the Great Plains of the United States and the Prairie Provinces of Canada can improve their availability because of greater soil moisture at lower depths. This method of application has also increased fertilizer efficiency for irrigated crops in the Great Plains region. More information is needed on responses to deep placement in humid areas, particularly as related to soil-test levels.

The presence of ammoniacal nitrogen in the deep placement zone is likely a positive factor in phosphorus absorption. Reasons for this were discussed in Chapter 6 and are reviewed again later in this chapter.

The placement of water-soluble phosphorus in bands tends to reduce contact with the soil and should result in less fixation than broadcast application. With broadcasting and thorough mixing, the phosphorus comes into intimate contact with a large amount of soil. Results of radioactive phosphorus studies indicate

TABLE 13-8. Effect of the Method of Phosphorus Application on Corn Yields (bu/A, Five-Year Average)

	P_2O_5 (lb/A)		
	30	60	120
Banded near row	115	115	115
Broadcast—plowdown	118	121	122
Narrow bands on surface 30 in. apart and plowed down	128	132	133

Source: S. A. Barber, Purdue Univ., personal communication.

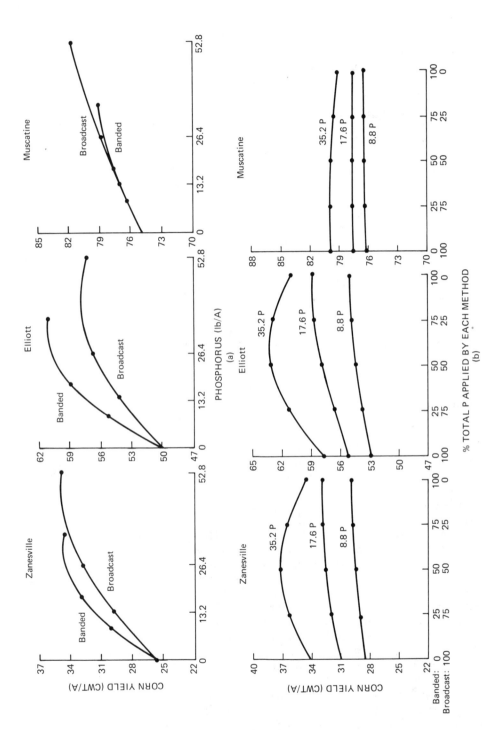

FIGURE 13-15. Corn yields at (a) various rates of phosphorus applied either banded or broadcast and (b) various percentages of the total phosphorus banded and broadcast (numbers on curves are total pounds of phosphorus per acre).[Welch et al., *Agron. J.*, **58**:283 (1966).]

that with the more soluble materials such as monocalcium phosphate there is a greater uptake from band placement. With the less soluble sources such as dicalcium or tricalcium phosphate there is greater uptake when the materials are mixed in the soil. The latter method allows more root–fertilizer contact, and this is apparently important with the less soluble forms. With water-soluble forms granules are more effective, whereas with the less soluble forms finely divided materials are most satisfactory. This strong interdependence of water solubility and particle size on effectiveness of phosphorus sources was discussed in Chapter 6.

Crops during any one season are generally able to recover only a small fraction of the added fertilizer phosphorus, usually less than 25%. This is in marked contrast to the recovery of applied nitrogen and potassium, which may be 50 to 75%. Even with band placement, fertilizer phosphorus is rather inefficiently used. On a Bladen silt loam fairly low in phosphorus, and with a band application of 11 lb/A of phosphorus, corn used only 15.2%, soybeans 10.4%, and potatoes 27.2% of the fertilizer phosphorus. Broadcast application would probably have resulted in still lower utilization.

Phosphorus placement for small grains is often more critical than for row crops and perennials. Limited root systems, shorter growing seasons, and cooler temperature enhance the response to banded over broadcast phosphorus, especially when available phosphorus levels in soil are low. Traditionally, the method of applying phosphorus in many of the small-grain producing areas is to band it at low rates with the seed at planting. However, when high rates of phosphorus are used in dry and/or coarse-textured soils, banding away from the seed at planting appears to be superior to banding with the seed. It is evident in Figure 13-16 that at rates greater than 60 kg/ha of P_2O_5, the highest yields of wheat were obtained when phosphorus was banded below the seed rather than placing it with the seed or banding it to the side and below the seed row. Similar results are shown for barley in Figure 13-16. The loss in efficiency of high rates of seed-placed phosphorus supplied as either monoammonium- or diammonium phosphate is probably due to NH_4^+ toxicity and to the effects of fertilizer salts in general.

Placing the phosphorus directly under the drill row for forage plants is also superior to broadcast applications or to placement one in. to the side (Figure 13-17). This practice is called band seeding and has been particularly successful in helping to ensure stands. The seeds are dropped behind the disks of the drill and directly over the fertilizer. The beneficial effect of the placement of phosphorus directly under tomato seed instead of 1.5 in. to the side is shown in Figure 13-18. With such seedlings the initial roots grow directly down. Workers in Michigan have shown that phosphorus should be placed directly under the onion sets.

It is generally true that the most efficient use of limited quantities of phosphorus at planting and the highest return from each dollar spent will be obtained by band or localized applications. However, if one considers the profit per acre and has sufficient capital, a different decision may be reached for some crops and under some conditions. On soils low in phosphorus or other nutrients it is usually difficult to produce top yields by row applications. There is some advantage in building up soil fertility, hence productivity, in a long-term fertilizer program. Some research suggests that the soil be built up to a medium to high level in phosphorus by broadcast applications. Maintenance can then be taken care of by periodic broadcast applications and/or by applications at planting on the most responsive crops in the rotation.

The beneficial effect of building up soil phosphorus levels on the value of four successive irrigated crops in Idaho is shown in Table 13-9. When phosphorus is applied only once in such a cropping sequence net returns will be even more attractive if crops such as potatoes, having a high phosphorus requirement, are grown as soon as possible after the phosphorus fertilization. These results demonstrate that high phosphorus fertilization rates may be profitable over several

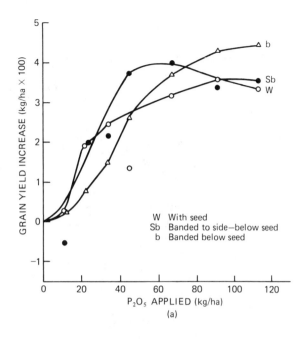

W With seed
Sb Banded to side—below seed
b Banded below seed

(a)

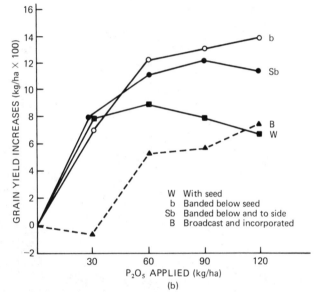

W With seed
b Banded below seed
Sb Banded below and to side
B Broadcast and incorporated

(b)

FIGURE 13-16. Effect of phosphorus placement on yields of (a) wheat (average check 2525) in Saskatchewan and (b) barley (average check 1695) in Manitoba. (Bailey et al., *Proc. Western Canada Phosphate Symp.*, pp. 200–229. Edmonton, Alberta: Alberta Soil and Feed Testing Laboratory, 1980.)

TABLE 13-9. Crop Values from Phosphorus Fertilization over a Four-Year Cropping Period Under Irrigation in Idaho

| P Applied (lb/A), Fall, 1972 * | Gross Crop Value ($/A) | | | | | P Fertilizer Costs ($/A) | 4-Year Net Returns ($/A) |
	1973 Sugar-beets	1974 Spring Wheat	1975 Potatoes	1976 Silage Corn	Total		
0	577	240	1214	232	2263	0	—
60	661	267	1204	250	2382	24	95
150	655	291	1277	239	2462	60	139
500	660	306	1512	252	2730	200	267

* Initial STPL 5.6 ppm phosphorus (0–12 in.).
Source: Westermann, *Proc. 28th Annu. Northwest Fert. Conf.*, pp. 141–146 (1977).

FIGURE 13-17.   Forage plants on the left resulted from broadcasting both seed and fertilizer. Plants on the right resulted from drilling seed and banding fertilizer 1 in. below the seed. Both plots were planted September 17 and photographed March 31. [Courtesy of Wagner et al., *Natl. Fert. Rev.,* **29**:13 (1954).]

crops. When high-acre-value crops are to be grown on soils low in phosphorus, it is good insurance to increase the soil level of this element.

In some cropping situations a combination of soil building additions of phosphorus plus annual band applications will be more effective than either treatment alone. Additional yield responses from banded phosphorus may be explained by the high levels of available phosphorus helping to alleviate crop stress caused by

FIGURE 13-18.   At 5 weeks of age 20 lb/A of phosphorus 2 in. directly under the tomato seed *(center)* produced much better growth than 1.5 in. to the side and the same depth *(right)*. (Courtesy of G. E. Wilcox, Purdue Univ.)

unfavorable climatic and soil conditions. USDA researchers in North Dakota found that current-year banding plus residual effects of banding in previous years increased spring wheat yields 10% over those obtained from 6-year residual phosphorus responses. They attributed this beneficial effect of banding to growth stimulation of spring-planted wheat in the cool springs that prevail in the northern Great Plains.

Nitrogen.   In contrast to phosphorus, the nitrate salts are mobile and move vertically or horizontally within the soil as the water moves. While ammoniacal nitrogen is held to the clay and organic matter fractions, a mobile salt is produced as soon as it is converted to the nitrate. In fine-textured soils the movement of nitrogen is restricted. A comparison of soil nitrogen losses from sodium nitrate and ammonium sulfate applied to a Norfolk sandy loam and Cecil clay was made in Alabama. In the Norfolk series 88% of the nitrate and 11.5% of the ammoniacal nitrogen was lost. In the Cecil series comparable losses were 57.8% and 8.1%, respectively. Another example of nitrogen movement in soils was given in Chapter 5.

Under conditions in which large quantities of a wide C/N ratio of organic matter are present in the soil, some of the nitrogen may be temporarily absorbed by the microorganisms and rendered unavailable to the growing plant. Under such conditions fall applications hold fairly well in the soil. The nitrogen may be absorbed by the microorganisms decomposing the organic matter, or the reducing conditions brought about by the actively decomposing organic matter may discourage nitrification of the ammoniacal nitrogen. One characteristic of anhydrous ammonia placed deep in certain soils is that the lack of aeration or decomposing residues delays nitrification and some NH_4-N may stay in the soil a good part of the season. There is increasing evidence of fixation of ammonium by the clay minerals, at least temporarily. All of these aspects of nitrogen behavior in soils were covered in Chapter 5.

Small amounts of nitrogen are important in starting plants off, but because of its mobility, particularly of the nitrate salts, the fertilizer must be kept well away from the plant roots. On sandy soils extreme care must be taken to avoid damage to the plants when considerable nitrogen is used in the fertilizer.

The addition of ammoniacal nitrogen to the fertilizer at planting has beneficial effects on absorption of phosphorus by the plant (see Chapter 6). Intimate association of nitrogen and phosphorus in the band is essential. Controversy exists over the exact mechanisms responsible for this interaction and over its relative importance under field conditions. Although some researchers have credited this interaction as being a major cause of the effectiveness of dual deep placement of nitrogen–phosphorus, others believe that more favorable positional availability is the principal factor.

It is usually undesirable to apply all nitrogen in the row at planting because of possible injury to the crop. A more rational procedure is to apply the bulk either before planting or as a top- or sidedressing after the crop is growing. Water movement carries the dissolved nitrate salts down to the plant roots. If ammoniacal nitrogen is used, it must be nitrified before it moves down in appreciable quantities. An exception is on soils of very low exchange capacity in which some of the ammoniacal nitrogen may move down directly.

More efficient use of nitrogen for row crops such as corn can be obtained by sidedressing part of it when the corn is knee- to waist-high, or even at tassel. This is particularly applicable on coarse-textured soils but also applies on those of medium and fine texture. An important guideline is that there should be ample nitrogen in corn plants until the black layer forms in kernels. However, with limited irrigation on medium to fine-textured soils all nitrogen should be applied preplant according to studies in Kansas.

Production from permanent tame grassland and native range in the dry regions of the western United States and Canada is limited primarily by moisture and nitrogen availability. Low rates of nitrogen application, (i.e., less than 150 lb/A of nitrogen), are generally ineffective, probably because of the need to satisfy the substantial needs of the "nitrogen pool" in grassland soils of the semiarid northern Great Plains. Researchers at Montana State University and with Agriculture Canada in Saskatchewan have shown that high rates of nitrogen will greatly increase yields.

Several items related to the Saskatchewan research reported in Table 13-10 are of interest. When NO_3-N levels in the treated soils were measured, it was observed that the low fertilizer rates had virtually no effect. However, at the higher rates NO_3^--N increased dramatically and most of it was concentrated at depths of 60 to 90 cm. Considerable amounts also occurred in the 30 to 60-cm zone. This deeper positioning of NO_3^--N helped to reduce the year-to-year variations in yield caused mainly by limitations in moisture supply.

The Montana investigators showed that the total amount of nitrogen fertilizer needed could be reduced by applying it to strips rather than broadcasting over the entire area. Both downward and lateral movement of nitrogen from the fertilized strips combined with root extension into the areas of high nitrogen concentration helped compensate for the restricted spreading patterns and lower nitrogen rates. Similar favorable results from strip or dribble application of fertilizer on forages have occurred in Kansas and Texas.

Potassium. Potassium salts are much less mobile than the nitrates but more mobile than phosphorus. Concern is sometimes expressed, however, over leaching losses of potassium. Although some losses on sandy soils may occur, losses from most soils are negligible. Studies on several Illinois silt loams over a period of years has shown small annual losses of 2 to 5 lb/A of potassium.

The lack of movement of potassium is attributed to the fact that as a cation it

TABLE 13-10.   Effect of Fertilization with High Rates of Nitrogen on Production from Permanent Tame Grassland and Native Range

Grass	Initial Application Rate of Nitrogen (kg/ha)	Cumulative Yield over 9 Years (metric ton/ha)
Crested wheatgrass	0	5.57
	150	6.25
	330	9.03
	445	12.31
	560	14.84
	740	15.43
Russian wild ryegrass	0	3.25
	150	4.07
	330	5.17
	445	8.73
	560	10.87
	740	11.49
Native rangeland	0	2.20
	150	2.80
	330	4.78
	445	7.57
	560	9.06
	740	9.56

Source: Leyshon and Kilcher, *Proc. 1976 Soil Fert. Workshop,* Soil Manag. 510, Publ. 244, Univ. of Saskatchewan, Saskatoon (1976).

is adsorbed on the base exchange complex in the soil. Hence amount and type of clay and amount of organic matter will influence movement.

Potassium is loosely held in organic soils. They often flood and the potassium may be moved out of the plow layer or root zone. As much as 60% of the available potassium has been found to be lost from the 0- to 6-in. layer between fall and spring soil samplings. Potassium will tend to move less in a properly limed soil than in an acid soil. Mention was made in Chapter 7 of the effect that liming can have on lowering potassium mobility as a result of the collapse of silica layers in expanded clays and the subsequent trapping and fixing of potassium within the affected clays. Also, there is more opportunity in limed soils for potassium to be held on the cation exchange complex rather than in the soil solution.

Because of solubility, potassium salts cannot be placed in contact with the seed in any great quantity. Although the tendency for movement is much less than that of nitrate salts, fertilizers high in potassium should for most crops be placed in a band to the side and below the seed or transplant. There are exceptions such as for barley and other small grains which will respond to low rates of from 15 to 30 lb/A of K_2O placed with the seed.

Starter responses from potassium similar to those from nitrogen and phosphorus are recognized on many crops and soils. It is generally accepted that potassium should be added to help promote early growth and large leaves except on soils testing high in this element. However, barley responds to potassium fertilizer at planting in North Dakota even on soils testing as high as 600 to 800 lb exchangeable. Similar benefits have occurred with the use of supplemental potassium for barley and winter wheat on high-potassium soils in Montana. The trend to plant crops earlier, under cooler, wetter conditions and other conditions of stress improves the probability of responses to potassium, even on high-testing soils.

Even in so-called normal years, seed may be placed in cold, wet ground. Although soil conditions improve later in the growing season, starter fertilizer may still increase yield, as is apparent in Table 13-11. Only 20 lb/A of K_2O banded 2 in. to

TABLE 13-11.   Starter Potassium Overcomes Cold, Wet Soil in a "Normal" Spring to Produce More Corn on a Kansas Soil Testing Very High in Available Potassium (> 700 ppm)

Nitrogen (lb/A)	Yield (bu/A) with:		
	No K₂O	20 lb/A Banded K₂O	Increase from Starter
0	72	80	8
75	128	137	9
150	167	182	15
225	166	182	16
300	167	185	18

Source: H. Sunderman, 1980 Report of Progress 382. Colby Branch Exp. Sta. Kansas State Univ.

the side of the corn row produced an additional eight to 18 bu/A depending on the nitrogen rate. This type of response is unlikely every year and it should be noted that the results here are only for one year. Nevertheless, this example shows the value of using more potassium on this soil already well supplied with available potassium.

Broadcast potassium is usually less efficient than banded. In Illinois it was found that the relative efficiency ranged from 0.33 to 0.88 on soils low to medium in potassium. As the soil-test level increases, there is generally less difference, however. Thus the importance of placement decreases as higher rates of fertilizer are used over the years.

Workers in Wisconsin showed that alfalfa recovered more of the potassium added to the surface of a silt loam than when applied at various depths. Recovery of the potassium was 41, 29, 19, 16, 10, 15, and 11% from the surface and soil depths of 7.5, 22.5, 37.5, 52.5, 67.5, and 82.5 cm, respectively.

Conservation Tillage. This includes practices all the way from no-till to chiseling to moldboard plowing. With reduced tillage treatments, much of the applied phosphorus and potassium, along with the phosphorus and potassium brought up by the roots and recycled into the crop residue, tends to become concentrated in the upper 2 to 4 in. of soil. If slope and soil permit, plowing every 4 or 5 years to distribute these nutrients more uniformly throughout the root zone may be desirable. Wherever feasible, soils low in fertility should be brought up to medium or higher before initiating no-till.

Broadcast applications of phosphorus and potassium are effective under many conditions, particularly in the more humid areas. With the surface residues there is more moisture near the surface and roots grow in this area. However, under low-fertility conditions and/or cooler and drier areas surface-applied phosphorus and potassium may not be sufficiently available. This is illustrated for potassium in Table 13-12 based on results from Wisconsin. Note the lower potassium in corn leaves at silking with no-plowing. Increased application of potassium increased the potassium level in the leaves under both tillage systems and reduced the yield loss from not plowing. The same principle holds for phosphorus.

Another key principle is illustrated in Figure 13-19. Yield increases from row fertilizer (40 lb/A of N, 40 lb/A of P_2O_5, 40 lb/A of K_2O) 2 in. below and 2 in. to the side were greater on the unplowed till-planted system than on the plowed system. With conservation tillage there are greater amounts of surface residues. This leads to cooler and wetter conditions at planting and lower availability of nutrients in the soil.

TABLE 13-12. Potassium Fertilization Helps Compensate for Losses in Corn Yields Due to Reduced Tillage of a Wisconsin Soil Medium in Available Soil Potassium

K_2O Applied Annually 1973–1976 (lb/A)	Yield Loss (bu/A) from Not Plowing (Plowed-Unplowed)	% K in Ear Leaf Tissue	
		Plowed	Unplowed
0	37	0.73	0.59
80	26	1.40	1.04
160	13	1.71	1.42

Source: Schulte et al., in Soils, Fertilizer and Agricultural Pesticides Short Course, Minneapolis, Minn. (December 12–13, 1978).

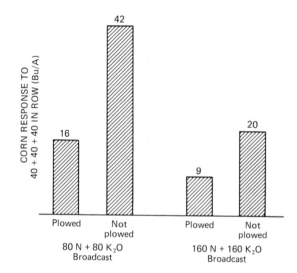

FIGURE 13-19. Use of row fertilizer greatly reduced the yield difference between plowed and unplowed treatments on the Plano silt loam at Arlington, Wisconsin. [Schulte, *Better Crops Plant Food,* **63**:25 (Fall 1979).]

Time of Application

The time at which to apply a fertilizer depends on the soil, climate, nutrients, and crop. With respect to the soil factor, soils differ greatly in the speed with which water will move through them. They also vary greatly in their capacity to fix plant nutrients.

Climate is important in any consideration of fertilizer application. The amount of rainfall between the time of application and the time of utilization by the plant will influence the efficiency of a material. Temperature also affects the availability of certain elements; for example, release of nitrogen, phosphorus, and sulfur from organic matter. It also affects nitrification and the absorption of phosphorus and potassium by the plants.

The nature of the crop itself will determine the need for split applications. Split-nitrogen applications for corn will improve nitrogen efficiency and they are also recommended for winter wheat production in warm, humid regions. When nitrogen is needed on perennial grass crops in areas of high precipitation, it is best to apply it in two to four supplements during the year. In many areas the effect of a nitrogen application will last for about 2 months or less.

Year-Round Application. This is the process by which farmers and/or fertilizer dealers plan crop fertilization to take advantage of opportunities that exist throughout the year. The bulk of the fertilizer goes on in the spring in many areas. However, with the trend to plant earlier, the increasing use of fertilizer, and the decline in transportation capacity to deliver the fertilizer to farmers on time, there is a real need and opportunity to apply any time during the year when soil, weather, and crop conditions permit.

The summer offers opportunity for applications on forages or after small grain is harvested. Considerable interest has been shown by farmers, agronomists, and the fertilizer industry in fall or winter fertilization in preparation for row crops the following year and for topdressing of forages. It is economically advantageous to the fertilizer manufacturer and to those engaged in bulk spreading if the period of fertilizer application and use could be extended. The farmer, however, must have a satisfactory answer to the question: "How will it benefit me?"

There are a number of factors that point to more fall and winter spreading:

1. Farmers want to plant earlier and any operation such as fertilization which can be moved to an earlier time is beneficial. Timeliness is the main key in crop production.
2. Fertilizer usage is increasing and transportation systems are less able to meet requirements during peak demand periods. However, warehousing facilities in heavy-use areas are increasing.
3. Farmers are more apt to get the fertilizer they need. In a wet spring they are inclined to plant without all the fertilizer that is required.
4. Custom application by the dealer is increasing, so spreading can begin as soon as a field is harvested. The farmer's labor is not involved.
5. Favorable pricing and payment incentives may make it economically attractive to purchase and apply fertilizer in the fall and winter.
6. Soil compaction is becoming of increasing importance and soils are likely to be drier in the fall than in early spring.

The heavy, relatively level soils which often remain wet in spring and delay broadcasting fertilizers by truck or pull-type spreaders are just the soils in which fall plowdown of fertilizer is agronomically feasible. Phosphorus or potassium applied in the fall will be held safely by the clay and organic matter. Application of nitrogen requires special precautions to be mentioned later.

Sandy soils, which drain more rapidly, are usually less of a problem in the spring because they will support fertilizer spreaders ahead of time to plow and plant. Also, it is not advisable to apply nitrogen and/or potassium fertilizers in the fall on sandy or organic soils because of the danger of leaching.

Winter application of phosphorus and potassium is feasible in the warmer areas. In colder areas fertilizers can be applied up to a 4 to 5% slope with a good cover of residues present. With a thin covering of snow the fertilizer pellets melt through and react with the soil.

Phosphorus. In theory, phosphorus should not be applied very far in advance of seeding because soluble forms revert to less available forms. The magnitude varies greatly with the fixing capacity of the soil. In actual practice the time of application is adjusted to labor available and other field operations.

Phosphorus can be applied in the fall for a spring-planted crop without danger of loss by leaching. On soils of low to moderate fixing capacity broadcasting in the fall is one of the most effective methods. On soils medium to high in phosphorus, the time and method become of less importance (Figures 13-15 and 13-16). On such soils proportionately larger applications every 2 to 4 years may be recommended. However, a vast majority of soils in the world are low in phosphorus and the preferred time is at planting with placement in bands near the seed or transplant. Low rates can be placed directly with the seed of many crops, but rates for fertilizer-salt-sensitive crops should not normally exceed about 20 lb/A of P_2O_5.

Preplant deep banding is also effective for crops such as corn and small grains in the drier and/or low-phosphorus soil areas.

Heavy initial applications at planting of forage crops are sometimes suggested. Generally, the yields are high in the beginning but may gradually decrease when compared with annual application. This is the result of fixation as well as crop removal.

Nitrogen. In contrast to phosphorus, the possibility of nitrogen losses must be considered in selecting the time at which it may be applied. Theoretically, it would be most desirable to add nitrogen as close as possible to the peak requirement of the crop. In addition to time and amount needed by plants, climate and soil type influence the time of application.

The amount and distribution of rainfall is important. The map in Figure 13-20 shows the relation of annual water surplus to geographical area. To obtain the annual surplus, the potential evapotranspiration (after Thornthwaite) is subtracted from the annual precipitation. A 4-in. waterholding capacity is assumed. The greater the surplus, the greater is the possibility of loss of nitrogen through leaching if the crop is not growing vigorously or if the land is not protected by a plant cover. In addition, conditions conducive to denitrification are likely to occur when soils become waterlogged due to the accumulation of excessive amounts of water.

Another aspect is temperature. Farther south the temperatures are more nearly optimum for nitrification during a greater portion of the year. Ammoniacal nitrogen applied before planting would thus be more subject to nitrification and leaching. Nitrification takes place at lower temperatures than was once assumed, for slow nitrification has been detected at temperatures close to freezing. Nitrogen stabilizers would be of help with fall applications. The effects of temperature and inhibitory substances on nitrification were covered in Chapter 5.

In actual practice ammoniacal forms of nitrogen are generally recommended in the fall in the north central United States when the temperature drops to 45 to 50°F at 4 in. except on sandy or organic soil. Research in Michigan and other corn-growing states has shown that spring applications of ammonium nitrogen are more efficient than fall, 105 to 110% on fine- and medium-textured soils and 110 to 130% on coarse-textured soils.

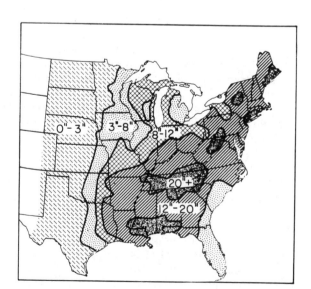

FIGURE 13-20. Average annual water surplus in inches. The surplus is the amount precipitation exceeds evapotranspiration. [Nelson et al., *SSSA Proc.* **19**:492 (1955).]

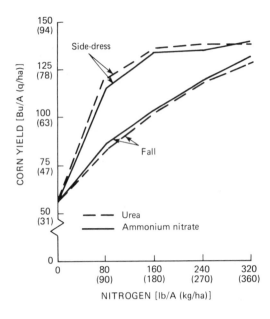

FIGURE 13-21. Corn yields related to source and rate of nitrogen and time of application, DeKalb, Illinois, 1969–1972. [Welch, "Nitrogen use and behavior in crop production." *Bull. 761.*, Univ. of Illinois at Urbana-Champaign (1979).]

The effect of time of nitrogen application on corn yields in Illinois is shown in Figure 13-21. Sidedressing after crop emergence was clearly superior to the fall application. Not shown in this figure are the results of spring preplant incorporation, which were similar to those for sidedressing. Differences between the two sources of nitrogen were small. As yields increase and nitrate accumulates in the profile and in the residues, response to nitrogen decreases and time of application becomes of lesser importance. Also, at higher than optimum rates, time of application would not be so important. The grower must weigh all of these factors in relation to convenience, supply and cost of product, timeliness, efficiency of use, and ease of earlier applications. Knowledge of the amount of nitrates in the soil profile is essential.

Delaying most of the nitrogen application until after the 2-ft-high growth stage of corn may result in decreased utilization of the added nitrogen. It is interesting to note, however, that nitrogen applications at the pretasseling stage, when deficiency was rather marked, were effective in increasing yields. In Indiana, applying 60 lb/A of nitrogen at tasseling increased corn yields by 10 to 15 bu/A.

On fall-planted small grain on heavy soils in cool climates or in drier areas such as the Great Plains, application of all or most of the nitrogen in the fall has possibilities under many conditions. Where necessary, nitrification inhibitors can be used to stabilize ammoniacal sources. In warmer humid regions yields will be somewhat below those obtained by top-dressing nitrogen in late winter because of leaching or gaseous losses. However, there are several important advantages to fall applications on small grains. In late winter the ground may be too wet for machinery to be operated, it may be difficult to convince the farmer of the importance of early applications, and there is usually a saving in labor. On fall-planted small grain a delay in the late winter or early spring nitrogen application reduces the yields.

Recommendations for optimal levels of fertilizer nitrogen for many crops, especially those in dry regions of the western United States and Canada, are conditioned by the amount of NO_3-N in the soil profile. It is widely assumed that soil NO_3-N substitutes for fertilizer nitrogen on an equal basis. Haby and co-workers in Montana determined the substitution of soil NO_3-N for fertilizer nitrogen in dryland winter wheat experiments. They showed that soil NO_3-N in the upper 1.2

m of soil sampled before fall seeding was only about one-third as efficient as fertilizer nitrogen. Reasons advanced for this variance were differences in time and spatial distribution of soil NO_3-N. Both sources of NO_3-N are apparently equally effective when present in the zones of maximum root activity, which are normally in the upper parts of the profile. Nitrate nitrogen from spring-applied fertilizer will initially be close to the soil surface when plant demands are high, whereas soil NO_3-N at deeper depths will not be within early reach of plants.

Long-term residual benefits of nitrogen do not usually receive as much recognition as do nutrients such as phosphorus and potassium which are less mobile and not as subject to loss by biological transformations. Evidence of the residual effects of high rates of nitrogen is illustrated in Figure 13-22. The residual response from 150 lb of nitrogen applied in 1960 and 1961 was about equal to the yield realized from an application of 50 lb of nitrogen in 1962.

Much of the nitrogen is applied preplant and measures for stabilizing ammoniacal sources may be utilized. One approach is the use of nitrification inhibitors, which was described in Chapter 5. Other techniques that show promise with ammoniacal sources entail some form of deep localized placement. This was also covered in Chapter 5.

Potassium. Potassium is commonly applied before or at planting. This time of application is usually more efficient than side-dressing, for opportunity is provided to incorporate the element in the soil. Because potassium is a cation, it does not move down into the soil very rapidly; hence side-dressed potassium would be less likely to move to the root zone of this year's row crop. Fall application of potassium is even more feasible than for either phosporus or nitrogen, for there is usually less loss in efficiency. Some crops—for example, peanuts, flax, and rapeseed/canola, which are taprooted plants—may respond better to applications made the year before than to direct applications. This is believed to be the result in part of movement of some of the potassium into the lower soil zones and to its more uniform distribution throughout the absorbing zone of the roots.

Under some cropping practices fertilizers high in potassium may be broadcast once or twice in the rotation. Fall plowdown is a practical approach. These applications are usually made before the more responsive crops, such as corn and legumes. However, in a corn–soybean rotation, potassium may be applied after

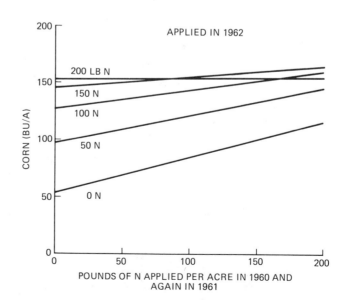

FIGURE 13-22. Residual effects of 1960–1961 applied nitrogen on corn were quite marked in 1962 on this black prairie soil in Indiana. (Courtesy of S. A. Barber, Purdue Univ.)

corn because it will be plowed down before planting soybeans. Less tillage is normally practiced following soybeans and before planting corn because of potential problems of soil erosion. Potassium may be included in the starter or planting fertilizer. When rates of 30 to 50 lb/A of K_2O are being used, all of it may be most conveniently applied in the starter fertilizer.

Maintenance application on forage crops can be made at almost any time. Fall applications are generally desirable, for the potassium will have had time to move down into the root zone. On hay crops an application made after the first cutting is desirable, for the first crop will have had the benefit of any potassium made available over the winter. There is some evidence that increased yields and efficiency are obtained by top-dressing alfalfa after each cutting. However, the grower may weigh the inconvenience against the value of the yield increase and prefer one large application.

Fertilization of the Rotation

In many respects the problem of distribution of fertilizer in the rotation is essentially one of time of application. Results of only a few comprehensive experiments are available in which comparisons have been made between splitting the application of fertilizer among all the crops in the rotation, applying all at one time, or treating only a few specific crops. Any added effectiveness from small frequent applications must be balanced against the extra time and cost of making them. Actually, there are many times at which to fertilize in the rotation (Figure 13-23).

The effect of place in the rotation of the application of 53 lb of phosphorus in a corn, oats, alfalfa-brome rotation on a calcareous Ida silt loam in Iowa is significant. Splitting the phosphorus between oats and second-year forage instead of applying it all to oats reduced oat yields by 7 bu but increased corn yields by 9 bu/A. Splitting the phosphorus between oats and corn instead of applying it all to oats decreased oat yields by 8 bushels and forage yields 0.5 ton but increased corn yields 19 bushels. Splitting the application was desirable on this soil to obtain the greatest efficiency of the applied phosphorus. It is interesting to note that 106 lb of phosphorus on oats gave the highest yield of all crops, regardless of the placement in

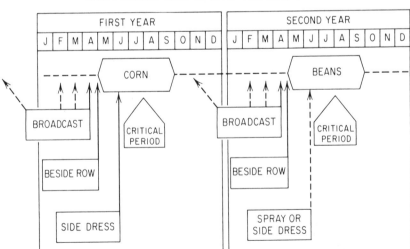

FIGURE 13-23. There are many times when fertilizer can be applied in a rotation. An example of a corn–soybean rotation is shown.

the rotation. *This illustrates the point that as higher rates of fertilizer are used and soil fertility increases the problem of time of fertilizer application becomes less important.*

Corn–soybeans is a common rotation and sufficient quantities of phosphorus and potassium may be applied to corn to take care of nutrient requirements of both crops. However, soybeans respond to these nutrients to about the same extent percentage-wise. In Iowa, 0-60-60 on a low-fertility soil for each crop increased yields of soybeans 5.5 bushels and corn 14 bushels. Formerly, soybeans were considered as scavengers or "eating at the second table." The high amounts of nutrients removed just in the grain of high-yielding crops must be recognized, as was pointed out in Chapter 12.

Forage crops, such as alfalfa, were formerly considered to be luxury consumers of potassium and recommendations were made that the element be applied to other crops in the rotation. However, greater yields are accompanied by increased amounts of potassium in the plant. More frequent cutting implies that younger plants are utilized and that they are higher in potassium. The amounts of nutrients in eight tons by cuttings is shown in Table 13-13. The percentage of potassium in the four cuttings ranged from 2.25 to 2.50.

In tropical soils with insufficient water, nitrogen, and phosphate, the potassium released by soil minerals may be sufficient for low yields of rice. However, with the use of irrigation, nitrogen, phosphorus, and new high-yielding rice varieties, the response to potassium increases, as illustrated by the following table, the response to potassium being 6, 12, and 33%, respectively (Mengel, 1971).

Nutrient (kg/ha)			Yield of Paddy Rice (tons/ha) in:		
N	P	K	First Year	Second Year	Third Year
120	80	0	4.98	4.45	4.02
120	80	80	5.29	4.97	5.35

High Acre-Value Crops. In crops such as vegetables, fruits, and tobacco the acre value of the produce is quite high. Potatoes or tomatoes, for example, may be worth $1500/per acre or more, and the cost of the fertilizer is a relatively small item. In such situations large quantities of fertilizer may be applied as an insurance measure, with the thought that in a good growing season nutrients must not be a limiting factor. This is particularly true of phosphorus and potassium. Rates of 100 to 200 lb/A of phosphorus may be applied to potatoes, whereas the crop may remove less than 75 lb of this nutrient.

TABLE 13-13.   Amounts of Plant Nutrients in Four Cuttings of Alfalfa (lb/A)

	1st Cut	2nd Cut	3rd Cut	4th Cut	Total
N	136	111	93	75	415
P_2O_5	31	24	22	17	94
K_2O	124	107	98	72	401
Ca	50	41	36	24	151
Mg	13	9	7	7	36
S	6	8	7	5	26
Yield (T/A)	2.35	2.10	2.03	1.52	8

Source: Flannery, *Better Crops Plant Food,* **57**(2):1 (1973).

Cover Crops. Wherever intensive vegetable growing is practiced there is a need for a well-developed system of crop management that will supply organic matter. Green manure crops, such as small grains or crimson clover, are sometimes grown. In some areas intensive fertilization of these crops is practiced not only to produce large quantities of organic material but also to provide a source of nutrients for the next crop. The nutrients, in a sense, are stored in the growing crops and released on decomposition.

There are some advantages to this procedure. Large quantities of fertilizer are usually applied to truck crops, and there is a possibility of fertilizer injury. The cover crops will release the nutrients gradually for distribution throughout the root zone. Results of numerous studies with radioactive phosphorus have shown that this element in a green manure crop may be more readily available than equivalent quantities supplied as superphosphate. If vegetable crops are being planted under cool temperatures, it is usually necessary to use starter fertilizer in the row, for under such conditions the green manure may decompose too slowly.

Carryover Effects

Applications of nutrients to help to set the stage for maximum profits will result in a certain portion of these nutrients being left in the soil. The amounts remaining will depend on the amounts added, the yield, the method of harvesting, and the soil effects. On the average, a nonlegume crop will remove about one-half to three-fourths of the nitrogen and variable amounts of phosphorus and potassium, depending on the crop and method of harvest. What is left, however, is not all carryover because of leaching losses, fixation, and surface erosion.

Examples. Rotational fertilization or soil buildup programs are based on carryover effects, examples of which were cited earlier in the chapter. Another good example of carryover nitrogen is shown in Figure 13-22. Examples of carryover nitrogen equal to one-fourth of the preceding year's application are quite common. A general rule of thumb is that soybeans leave in the soil about 1 lb of nitrogen for each bushel of beans produced.

The carryover effect of phosphorus is well known. An example can be cited on corn in Iowa, where 17.6 lb/A of phosphorus increased yields by 18 bushels in the first year and with no further application increased yields 15 bushels in the second, for a total 33-bushel gain.

An example of residual effect of potassium on soybeans is shown in Figure 13-24. Actually, response by soybeans to carryover fertilizer has been generally quite good provided enough is applied to the previous crop.

Beneficial carryover effects can also occur with other nutrients. For example, USDA researchers at Pendleton, Oregon, studied the response from sulfur applied as gypsum at seeding of winter wheat in the first year of a wheat–pea rotation lasting 8 years. In their study on a marginally sulfur-deficient soil in an 18-in. rainfall area, no yield response to sulfur occurred in the first wheat crop when optimum rates of 40 to 60 lb of nitrogen were applied. However, one of four second-year wheat crops, all three third-year wheat crops, and both of the two fourth-year wheat crops showed a yield response to the initial application of sulfur at optimum nitrogen fertilization. An example of the sulfur response obtained in the fourth-year wheat crop is given in Table 13-14.

Significance. As fertilizer is applied in increasing quantities, it becomes apparent that increased attention must be given to the value of the carryover. In many cases the cost of fertilization is charged to the crop treated. However, carryover fertilizer is like money in the bank and is part of fertilizer economics.

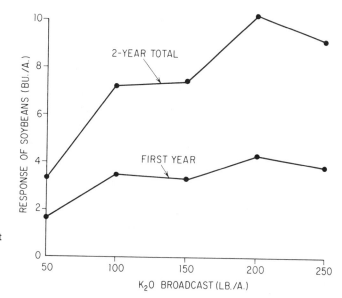

FIGURE 13-24. Response of soybeans to residual potassium on a soil low in that element. Note the marked response in the second year. [Miller et al., *Soybean Dig.* 21(3):6 (1961).]

TABLE 13-14. Yield Response (bu/A) of the Fourth Crop of Soft White Winter Wheat* to Sulfur Applied at the Seeding of the First Wheat Crop in a Wheat–Pea Rotation, Athena Silt Loam, Weston, Oregon

	Nitrogen (lb/A)			Sulfur (lb/A) †			
Initial	2nd Crop	3rd Crop	4th Crop	0	15	30	60
0	50	50	50	30.9 d	36.3 bc	39.8 ab	35.6 bc
40	50	50	50	30.1 d	37.1 abc	37.7 abc	37.6 abc
80	50	50	50	30.8 d	33.4 cd	39.2 ab	41.6 a
160	50	50	50	30.9 d	34.5 cd	37.1 abc	40.0 ab

*The fourth wheat crop was the seventh crop grown in a wheat–pea rotation following sulfur application. Yields are derived from 13 field experiments conducted over an 8-year period.

† Means followed by the same letter or letters are not significantly different at the 0.05 level of probability.

Source: Ramig and Rasmussen, *Proc. 23rd Annu. Northwest Fert. Conf.,* pp. 125–137, Boise, Idaho (July 17–20, 1972).

Micronutrients

Approaches. Two approaches are commonly followed in supplying micronutrients to crop plants:

1. Specific nutrients to take care of certain needs are applied in areas known to be severely deficient in one or more of the micronutrients or to crops known to have especially high requirements. The micronutrient may be added to a mixed fertilizer and, as in boron on alfalfa, in which 0-4.4-25 + B (0-10-30 + B) may be recommended generally over an entire state. In other instances the material may be applied separately such as a broadcast application of zinc or copper.

2. In a second approach small amounts of a micronutrient mixture are added to fertilizers intended for general use. The objective here is to add the material as insurance in quantities that will not harm the more sensitive crops but that will take care of the needs of many crops under certain soil conditions. Justification is offered as follows: (a) it is impossible to determine needs on each

field; (b) the absence of visual symptoms does not preclude hidden hunger; (c) it is better to anticipate needs rather than to wait and lose money; and (d) low-cost insurance. If a crop is suffering from a severe deficiency of one micronutrient, this approach is not likely to correct the problem.

Placement. Considerable specific information on the use of micronutrients was provided in Chapter 9. Micronutrients are applied in small amounts and placement methods are directed toward the greatest efficiency.

Identification of Needs. Some of the progress made in the development of diagnostic tools for identifying micronutrient deficiencies was summarized in Chapter 12. Soil analysis methods are available, but generally, satisfactory calibration of these tests is lacking except possibly for boron and zinc. This is not surprising in view of the small amounts of elements involved and the marked effect of a multitude of interacting environmental, soil, and plant factors on crop responses. Plant analyses are usually more reliable than soil tests.

In diagnosing a plant nutrient problem, the following steps should be considered:

1. Deficiency symptoms must be recognized.
2. Soil types or locations must be observed.
3. The crop response probability list must be checked.
4. A complete soil test must be made; pH is especially important.
5. An analysis must be made on the foliage.
6. Yield goal must be considered.

The micronutrient content of the seed of certain crops may be an indication of need, as determined with molybdenum on soybeans in Indiana. Yield increases from molybdenum occurred when the soybean seeds on the untreated plots contained 1.6 ppm or less of this element.

Use. There are hazards in the indiscriminate use of micronutrients. On an acid soil the addition of manganese would not as a rule be beneficial and might be detrimental. The following is an example of a response to copper but a depressing effect from additional micronutrients on wheat on a Hyde silt loam in North Carolina:

No micronutrients	15 bu/A
5 lb/A of Cu	53 bu/A
5 lb/A each of Cu, Fe, Mn, Zn	36 bu/A

The interaction of copper and lime on a Hyde soil containing 20.4% organic matter and having a pH of 4.9 has also been studied. Wheat yields were increased by copper and lime, but the response was greater to copper at the low lime levels.

It has been pointed out that the grower may add micronutrients even though experiments do not show much improvement in yield. The cost is not great and only a small increase will be required, although this increase may not be significant at 19:1 odds. However, the grower might be interested even if the odds were only 1:1.

In Chapter 10 the regulation of plant nutrients in addition to N-P-K was discussed. The amounts are considered to be too low in some states. Michigan, for example, has higher minima: 1% manganese, 0.125% boron, 0.1% iron, 0.5% copper, 0.5% zinc, and 0.04% molybdenum.

Farmers may want to try an all-inclusive micronutrient mix on part of their land. If a response is obtained, further tests must be carried out to identify the responsible nutrient. Such a trial mixture might be applied at per acre rates of

0.5 lb of manganese, 2 lb of zinc, 1 lb of copper, 0.5 lb of boron, and 0.2 lb of molybdenum on soil with a pH of more than 6.0. This mixture should be added to N-P-K fertilizer and placed in bands about 2 in. away from the seed to prevent fertilizer injury. Boron should not be used in the mix for crops such as beans or small grain, which are easily injured. Rates should be doubled for peats or mucks.

As pointed out in Chapters 8 and 9, the addition of certain acid-forming sulfur-containing fertilizers in a band near the seed or transplant may also serve to correct micronutrient deficiencies that are induced by high pH values. Elemental sulfur, ammonium polysulfide, ammonium thiosulfate, and H_2SO_4 have been found to be effective under such conditions.

Banding of nitrogen sources that have a strong acidifying action will markedly increase levels of available manganese in soil. The effect of potassium chloride and other neutral chloride-containing salts on encouraging the release of available manganese in acid soils was noted in Chapter 9. Also, banding of some types of phosphorus fertilizers may correct marginal manganese deficiencies.

Trends. Micronutrient deficiencies are increasing and can be expected to continue. Higher yields are being obtained and are putting a greater drain on all nutrients. Interaction among major and micronutrients will assume greater importance. As an example, it was mentioned in Chapter 9 that phosphorus fertilization of soils marginally or very deficient in zinc can result in crop production problems caused by excessive uptake of phosphorus. Normal potato plants tend to have a P/Zn ratio of less than 400, and in deficient plants the ratio is generally greater than 400, all of which points to the fact that rapid advances in diagnostic techniques and a new philosophy of the use of micro- and macroelements is needed.

Utilization of Nutrients from the Subsoil

The utilization of nutrients from the subsoil has long been under investigation, but little information is available. Soil structure, aeration, pH, drainage, and root distribution are factors that are important with respect to physiological or positional availability but are beyond the scope of this chapter. Two general aspects of the problem are mentioned, however.

1. Utilization of the native nutrients from the subsoil.
2. Addition of nutrients to the subsoil.

Native Nutrients. The subsoils of most soils in the humid regions are generally acid and low in fertility. In the lower-rainfall areas, however, the subsoil may be well supplied with certain nutrients. Barber at Purdue University estimates that corn takes less than 1% of its phosphorus needs and less than 10% of its potassium requirements from a low-fertility subsoil.

On certain soils, particularly those derived from loess, the B and C horizons may be fairly high in phosphorus. The amount of phosphorus the plants can derive from these lower depths under field conditions has not been accurately determined. Results of greenhouse studies indicate that the phosphorus content of the B and C horizons of certain soils may be more available than that of the surface horizons. It has been suggested that the continued growing of either alfalfa or sweet clover, both deep-rooted crops, will increase the available phosphorus in the surface by upward transfer from the subsoil. This improvement is brought about as the organic residues are returned to the surface soil and decomposed. The surface horizons of forest soils are commonly higher in nutrients than the subsoil horizons because of this upward transfer.

Certain soils, particularly loess or alluvial, may be uniformly high in potassium throughout the profile, and there is evidence that subsoil potassium can be utilized by plants. In some areas difficulty is experienced in correlating soil test results for potassium in the surface soil with potassium response. When the content of potassium in the subsoil is taken into consideration, the relation between exchangeable potassium and crop response may be considerably improved.

Some states have made a systematic analysis of the subsoils of major series. An example from Wisconsin is shown in Figure 13-25. Such data could be helpful in making more accurate fertilizer recommendations.

The absorption of micronutrients from the subsoil by crops has been considered by some to be one of the advantages of crop rotation. These elements are supposedly concentrated in the surface horizons as the organic residues from certain crops are returned. This may be a reasonable assumption, but at present there is only limited experimental evidence to support this view.

Added Nutrients. Deeper application of nutrients may sometimes be desirable. As discussed earlier in this chapter and in Chapter 3, application of nitrogen and phosphorus and perhaps other nutrients to a deficient soil zone will greatly enhance root development in the treated zone. Lime added to an acid subsoil (refer to Chapter 11) will not only supply calcium and magnesium but will also reduce the quantities of aluminum, iron, and manganese in solution.

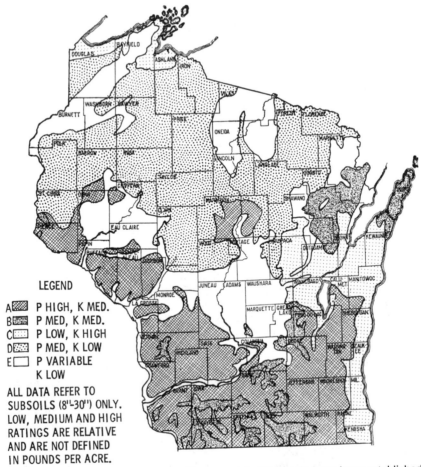

LEGEND

A P HIGH, K MED.
B P MED, K MED.
C P LOW, K HIGH
D P MED, K LOW
E P VARIABLE
    K LOW

ALL DATA REFER TO
SUBSOILS (8'-30') ONLY.
LOW, MEDIUM AND HIGH
RATINGS ARE RELATIVE
AND ARE NOT DEFINED
IN POUNDS PER ACRE.

FIGURE 13-25. General subsoil fertility groups in Wisconsin have been established. (Courtesy of M. T. Beatty and R. B. Corey, Univ. of Wisconsin.)

TABLE 13-15. Effect of Subsoiling and Deep Incorporation of Phosphorus and Potassium on the Yield of Barley Grain

Treatment	Grain Yield (tons/ha at 85% dry matter)				
	1974	1975	1976	1977	Mean
None	4.89	2.30	3.43	2.90	3.38
Subsoiled alone	5.23	3.79	4.46	3.53	4.25
Subsoiled + P and K	6.21	4.69	4.51	4.34	4.94
P and K to topsoil	4.53	1.90	3.77	3.20	3.35
SEM of a difference ±	1.075	0.240	0.261	0.338	0.302

Source: McEwan and Johnston, *J. Agr. Sci. (Camb.)*, **92**:695 (1979).

Much of the early work on subsoil fertilization involved the placement of fertilizer at a depth of 18 to 24 in. behind a subsoiler. Results of many field tests have failed to show consistent benefits, perhaps because of the wide spacing between fertilizer bands.

Some experiments conducted in the United States, however, showed that subsoiling alone increased crop yields by 14%, whereas subsoil incorporation of fertilizers produced yield increases of 24%. There was considerable variation in the beneficial effect of these treatments with season and crop. These yield increases are in reasonable agreement with those reported for research carried out at the Rothamsted Experimental Station in England (Table 13-15). As shown here, subsoiling to a depth of 23 to 46 cm increased the 4-year mean yield of barley by 24% and subsoil incorporation of phosphorus and potassium increased mean yields of barley an additional 20%.

Profile modification, a unique approach, received considerable attention in North Carolina. It was postulated that root penetration may be impeded for a number of reasons which include cementation of soil particles, compaction, poor aeration, soil acidity, toxicity, or nutrient deficiency. Accordingly, field trials were conducted on Norfolk soils which were plowed to a depth of 24 to 36 in. A portion of the lime and phosphorus was broadcast before plowing and the remainder after plowing. Results are reported to have been favorable.

In this situation the principal benefit may be related to the more efficient use of annual precipitation. More water should enter the soil and the roots should penetrate more deeply. If the plant is utilizing water effectively from a layer 24 to 36 in. thick, in contrast to one only 12 in. thick, there will be a greater reserve against drought. Under some conditions, turning up heavy clay subsoil material may cause the surface soil to seal off more rapidly and to decrease water intake.

Plowing a Palouse-area wheat soil to a depth of 36 in. has been beneficial. This soil has a dense clay subsoil but is permeable to 18 in. Summers are relatively dry in the Palouse area, but winter rainfall recharges the subsoil water. Hence with deep plowing the crops can absorb water from a 36-in. rather than an 18-in. soil layer. In the southeastern United States excellent yield responses of corn and soybeans have been obtained to fertilizer placed directly under the row with a chisel.

In southern Illinois a low-organic-matter low-fertility claypan soil was tilled 9, 18, 27, and 36 in. deep and four rates of lime, phosphorus, and potassium were applied. Root penetration and growth during moisture stress were increased, but corn yields were not increased by deeper tillage and fertilization. Moisture penetration was reduced, probably due to the low-organic-matter clay soil material being brought to the surface.

In Michigan a soil with a sand lens at 20 in. was tilled 26 in. deep. The sand lens interfered with the downward movement of water and roots. Seven years after deep tillage, corn yields were 128 bushels as compared to 87 bushels for 10-in. depth tillage.

Production problems associated with Solonetzic soils characterized by dense, sodic claypans were mentioned briefly earlier in the chapter. Plowing 24 in. deep with the addition of sulfur or gypsum increases water and root penetration. In some trials wheat yields were increased from 19 bu/A with a 9-in. plowing depth to 34 bu/A with a 27-in. plowing depth.

In somewhat the reverse of deep tillage, Michigan workers developed the idea of installing an asphalt barrier at 24 to 30 in. in very sandy soils to help hold the water. This approach has been quite effective, but it may be an uneconomical practice even for high-value crops.

From this discussion it is apparent that the effects of physical manipulation of the soil profile varies according to the soil. Much more needs to be learned. In the meantime, deeper plowing or chiseling to break up plowpans and improved management practices to encourage deeper rooting are of help.

Fluid Fertilizers

Response to surface or plowdown application of fluid fertilizers does not differ from that of solid fertilizers if placement is the same. Certain other aspects will be considered here.

Fertigation. This is the application of fertilizer in irrigation water in either open or closed systems. The open systems include lined and unlined open ditches and gated pipes that are used for furrow and flood irrigation methods. Sprinkler, spitter, trickle, drop, and dual-wall tubing systems are the main types of open systems.

Nitrogen and sulfur are the principal nutrients applied by fertigation. Phosphorus fertigation has been less common because of concerns over the formation in high-calcium and high-magnesium waters of precipitates of uncertain solubility. Application of sulfur in soluble forms through sprinkler irrigation systems is effective. Potassium and highly soluble forms of iron and zinc have also been successfully applied by fertigation.

Application of anhydrous ammonia or other fertilizer materials such as non-pressure urea-ammonium nitrate solutions containing free ammonia to irrigation waters high in calcium, magnesium, and bicarbonate may result in precipitation of calcite and/or magnesite. These precipitates will cause scaling and plugging problems in equipment. Their formation can be prevented or corrected by the addition of sulfuric acid.

Anhydrous ammonia and aqua ammonia are not usually applied in closed irrigation systems because they tend to volatilize at the discharge point, thus resulting in high nitrogen losses.

There are several advantages of fertigation. One, nutrients, especially nitrogen, can be applied close to the time of greatest plant need. With corn, for example, the plant uses most of its nitrogen from the rapid growth stage to the milk stage. Applying some of the nitrogen at this time is more efficient on sandy soils and just as efficient on heavier soils. Second, one or more field operations are eliminated. Labor is saved, for it takes little effort to meter fluid fertilizer from a tank or nurse wagon into the water, allowing it to do the work. With rank-growing, long-season crops such as sugarcane, it is difficult to get through the field to distribute fertilizer. Application in the irrigation water is one answer. Finally, midseason deficiencies in crops can be corrected by fertigation.

The question of uniformity of application of nutrients in irrigation water is sometimes raised. This should not be a problem with skilled irrigation management and properly designed irrigation systems since the dissolved nutrients accompany the water wherever it goes. However, unsatisfactory distribution of nutrients can occur under some conditions and with low rates of fertilization. Under row irrigation a large proportion of the nutrients may be deposited near the inlet. With sprinkler irrigation the nutrients, of course, fall with the water.

To prevent nutrients from being leached beyond the reach of roots or from accumulating near the surface, inaccessible to the crop, they should not be introduced at the initiation of irrigation. Best results are obtained when the fertilizer materials are supplied toward the middle of the irrigation period and their application terminated shortly before completion of the irrigation.

Clear liquid mixed fertilizers can be applied, but on annual crops in most instances sufficient phosphorus and potassium are best applied at planting time. Farmers must consider the economics and advantages discussed above before reaching a decision to use fertigation.

Foliar Applications. Certain of the fertilizer nutrients soluble in water may be applied directly to the aerial portion of plants. The nutrients must penetrate the cuticle of the leaf or the stomata and then enter the cells. This method provides for more rapid utilization of nutrients and permits the correction of observed deficiencies in less time than would be required by soil treatments. However, the response is often only temporary.

When problems of soil fixation of nutrients exist, foliar application constitutes the most effective means of fertilizer placement. Workers in the USSR have emphasized the importance of foliar feeding in the Arctic regions, where permafrost retards nutrient release from plant residues, root growth, and nutrient uptake. Eventually, the practice may be helpful to early-planted crops in the temperate zone, and it can be combined with regular disease- and insect-spray programs, with a corresponding saving in labor.

So far the most important use of foliar sprays has been in the application of micronutrients. The greatest difficulty in supplying nitrogen, phosphorus, and potassium in foliar sprays is in the application of adequate amounts without severely burning the leaves and without an unduly large volume of solution or number of spraying operations. Nutrient concentrations of generally less than 1 to 2% are employed to avoid injury to foliage. Nevertheless, foliar sprays are excellent supplements to soil applications.

Foliar fertilization can be accomplished by means of overhead sprinkler systems and by application through equipment customarily used for spraying pesticides. Ground-spray equipment used for foliar feeding is usually of the high-pressure, low-volume type, designed for uniform spraying of foliage and for keeping water volume to a minimum. The nutrient spray may be applied through single- or multiple-nozzle hand guns; multiple-nozzle booms; or by multiple-nozzle oscillating or stationary cyclone-type orchard sprayers. Droplet size must be carefully controlled since it will affect crop response.

Foliar sprays in which urea supplies the nitrogen have been successful on apples, citrus, and a number of other crops. Spraying produces more rapid absorption of nitrogen than soil applications. Foliar applications, however, on small grains and wheat are no more efficient than soil additions, but late applications may increase the protein content of grain. It has been found that greater absorption of urea nitrogen by tobacco occurs at night or when epidermal hairs are injured but that no difference exists between upper and lower leaf surfaces. Pineapples receive as much as 75% of their nitrogen fertilizer needs as urea sprays and about 50% of their supplemental phosphorus and potassium by foliar fertilization.

In Alaska potato yields were increased from weekly potassium sprays. Potassium-containing sprays have also been used on other crops, among which are apples, celery, and pineapple. As with phosphorus, however, the problem of adding sufficient amounts becomes critical.

Micronutrients lend themselves more readily to spray applications because of the small amounts required, and several have been supplied in this way. Zinc is a notable example, and leaf and dormant spray applications have been found to be many times more efficient than soil applications for fruit trees such as citrus or peach. On some soils difficulty has been experienced in obtaining much uptake from soil applications because of fixation of the zinc. Manganese and iron are other examples. Spray applications of 1 to 2 lb/A of manganese a few weeks after planting soybeans are commonly recommended on problem soils. Pineapples in Hawaii are regularly sprayed with ferrous sulfate because soil applications of iron are ineffective.

Efforts to correct iron chlorosis have not always been successful and more than one application may be needed on some crops. Chlorosis is a common problem on soybeans grown on high-pH soils under low-rainfall conditions. Randall in Minnesota found that spraying with an iron chelate was effective about 80% of the time. Magnesium has also been applied as a foliar spray on certain fruit trees and celery. Boron sprays on fruit trees have also been quite successful.

Investigations have been conducted at Iowa State University to determine if foliar fertilization helped maintain photosynthetic activity of leaves late in the growing season in order to supply adequate carbohydrate for roots and seeds. A fertilizer solution with an N/P/K/S ratio of 10:1:3:0.5 prepared from urea, potassium polyphosphate, and potassium sulfate was used for these studies. Significant yield increases were obtained from two to four sprayings at several locations. In a large number of experiments in subsequent years, however, foliar fertilization has not consistently increased soybean yields.

Foliar applications of phosphorus are used less than nitrogen largely because most phosphorus compounds are damaging to leaves when sprayed on in quantities large enough to make the application beneficial. Research at Iowa State University has showed that ammonium tripolyphosphate, ammonium tetrapolyphosphate, and phosphoryl triamide could be applied successfully to corn and soybeans. Corn tolerated sprays of tri- and tetrapolyphosphate solutions containing up to 1.3% phosphorus. Soybeans were damaged when the phosphorus concentration in these solutions was over 1.1%. The maximum concentration of phosphorus tolerated as orthophosphate was 0.5% for corn and 0.4% for soybeans.

Inclusion of sucrose in foliar sprays applied to soybean leaves will reduce the severity of damage to the leaves from urea and, to a lesser extent, from phosphorus in the form of orthophosphoric acid. Various environmental factors, including temperature, humidity, and light intensity, also affect the rate of absorption and translocation of nutrients applied to the foliage.

To be most effective two or three spray applications repeated at short intervals may be needed, particularly if the deficiency has caused severe stunting. Care must be taken to identify the nutrient needed, or additional problems may develop. The microelements are usually required in only very small amounts, and too much of one element may be detrimental.

Much more needs to be learned about foliar application, and its value in supplementing standard soil fertility programs must be resolved. The approach to both problems should be the addition of plant nutrients, as revealed by soil and plant tests and crop performance, and the cost of the practice in relation to the increased value of the crop.

Starter Solutions. The application of a nutrient solution around the roots of vegetable crops has been widely used for transplants such as tomatoes and peppers. The principal response is from phosphorus, and such starter solutions are usually high in this element. Starter solutions, however, have generally proved to be of little economic value for flue-cured tobacco.

It has been observed that plants recover more rapidly from the shock of transplanting when starter solutions are used. Increased yields have also been obtained, and in several instances the crops matured at an earlier date. The reason for the observed responses is probably that the transplants have a limited root system and consequently a limited capacity for the absorption of water and nutrients. The addition of a dilute solution of these plant nutrients at the time of planting should make it possible for the plant to absorb these needed elements more readily. Starter solutions, however, must have a low salt concentration if injury from plasmolysis is to be avoided.

Summary

1. Although nutrient uptake by crops cannot be used as an accurate guide for fertilizer recommendations, it does indicate differences among crops and gives an idea of the rate at which the reserves or storehouse nutrients in the soil are being depleted.

2. Root development helps to give an idea of the most effective placement of fertilizer. For example, potato roots are much less extensive than corn; hence potatoes can utilize nutrients closer to the plant more effectively.

3. Varietal differences in root characteristics may also result in large variances in feeding power of plants.

4. Soil characteristics influence depth of rooting and yields may be directly proportional to the water available to the roots.

5. Proper fertilizer placement is important in the efficient use of nutrients from emergence to maturity, prevention of salt injury, enhancement of deeper rooting to compensate for dry conditions at the soil surface, and convenience to the growers.

6. Movement of some fertilizer salts in soils is appreciable. Nitrates move most freely, but ammoniacal nitrogen is adsorbed by the soil colloids and moves very little until converted to the nitrate. Potassium is also adsorbed and moves little except in sandy soils. Phosphorus movement is generally limited but can be appreciable on sandy irrigated soils, in the presence of large quantities of organic residues, and when heavy batch applications are made.

7. The more concentrated materials have a lower salt index per unit of plant nutrient and when placed close to the roots have less salt effect on young plants.

8. Band application at planting is important in providing a rapid start in physiological processes and large healthy leaves. Under cool conditions nitrogen, phosphorus, potassium, and zinc are generally less available to the young plants, and band placement will enhance their adsorption.

9. Broadcast application is a means of applying large quantities of nutrients that cannot be conveniently added at planting. The nutrients are often mixed through the plow layer, which helps to provide a continuing source of nutrients later in the growing season. They are more likely to be in a moist zone for a greater part of the year. Because elements such as phosphorus and potassium move to the roots by diffusion through water films, this point is important.

10. With reduced tillage nutrients tend to accumulate in the top 1 to 2 in. of soil. This may necessitate higher application rates, more banding, and plowing every 4 or 5 years, if conditions permit.

11. Because of the limited mobility of phosphorus, it should, when used at low rates, be placed in the zone of root development. Band applications are generally the most efficient when soil test phosphorus is low and when low rates are applied. For high yields of many crops, however, it is essential to build up the soil level. Forages are an exception in that they can use top-dressed phosphorus by absorption through the crowns.

12. In no-till crop production where planting is done in killed sod or plant residues, surface application of phosphorus and potassium are generally available to the crop in humid regions. The soil is moist under the residues and the roots develop near the surface. If the soil is low in these elements, building the soil fertility of the plow layer is desirable before beginning no-till. In drier regions surface applications may not be as effective because of less moisture under the residues.

13. Small amounts of nitrogen in the planting fertilizer encourage absorption of phosphorus. Because nitrogen is mobile, or becomes mobile after nitrification, application of the major portion before planting or side-dress applications after planting are both effective. In general, the nearer the time of application to peak nitrogen demand, the more efficient the utilization. The amount and distribution of rainfall must be considered in connection with soil texture.

14. Leaching losses of potassium are insignificant except on sandy or organic soils under heavy rainfall. Hence band applications at planting and broadcast applications before planting or at some point in the rotation are effective. Starter responses from potassium similar to those from nitrogen and phosphorus are obtained on low-potassium soils.

15. In determining the method and time of application, convenience to the grower must be considered along with efficiency and safety.

16. Care must be exercised to spread fertilizers uniformly since yield losses can result from uneven application.

17. As higher rates of fertilizer are used in conjunction with top management, more attention is given to fertilization for the entire rotation. Bulk applications of phosphorus and potassium may be applied once or twice in a 4-year rotation in addition to starter applications at planting for the more responsive crops. Nitrogen is applied annually to nonlegume crops.

18. Year-around fertilization implies fertilization anytime the soil, crop, or weather permits and is becoming a must with increased volume of fertilizer, declining transportation facilities, and a need to save labor. As soil fertility levels increase, the point and time of application of phosphorus and potassium declines in importance. The point is to apply adequate quantities for maximum profit yield.

19. Application of nutrients for the most profitable yield will result in a portion remaining in the soil. In many cases the cost of fertilization is charged to the crop treated. However, if a critical evaluation of fertilizer use is made the carryover value must be considered.

20. Two approaches are followed in the application of micronutrients: (a) addition to take care of specific needs; (b) addition to the fertilizer of a small amount of a mixture of micronutrients for general use. The latter is a method of insurance. Development of suitable diagnostic tools for the recognition of needs and to help predict responses is a major problem. Plant analysis is a good guide.

21. The theory of subsoil fertilization is the promotion of deeper rooting, greater water penetration, and more efficient use of water. Field results from fertilization with a subsoiler have been variable.

22. With problems of soil fixation of nutrients, foliar application may constitute the most effective placement, particularly for certain micronutrients. There are indications that materials such as ammonium tripolyphosphate, ammonium tetrapolyphosphate, and phosphoryl triamide can be used to foliarly apply significant amounts of phosphorus to corn and soybeans.

23. Fertigation, application of plant nutrients in irrigation water, is an effective method of applying nitrogen. Other nutrients applied less frequently by this means include phosphorus, sulfur, potassium, zinc, and iron.

Questions

1. Why can phosphorus materials be placed close to the seed or plant? Why is it usually important that phosphorus be close to the seed or young plant? How do you account for the marked response of legumes to band seeding?

2. Why is root growth stimulated in response to plant nutrients on an infertile soil?

3. What kinds of roots seem to be stimulated the most by localized concentrations of nutrients?

4. What root characteristics influence the ability of crops to exploit soil for moisture and nutrients?

5. In what part of the root zone does most of the early root growth take place?

6. Are there differences in extent of root development among crop varieties?

7. What soil conditions might affect depth of crop rooting?

8. Under what soil texture conditions would ammoniacal nitrogen be more likely to move? Why? What soil environmental conditions would favor rapid transformation to nitrate forms?

9. What crops in your area are being underfertilized with phosphorus or potassium? Overfertilized?

10. What materials may be used in a low-analysis fertilizer such as 5-4.4-8.3 (5-10-10) to give it a lower salt index per pound of nutrients than a 10-8.8-16.6 (10-20-20)?

11. An experiment is being conducted on a sandy soil to determine the effects of nitrogen and potassium on snap-bean production. The fertilizer is placed in bands 2 in. to each side and 2 in. below the seed. In addition to the phosphorus, nitrogen and potassium are applied in quantities to furnish 50 lb/A of nitrogen (ammoniacal) and 60 lb/A of potassium. On all plots in which the complete fertilizer was applied the stand was poor; when nitrogen was omitted, the stand was poor, but when potassium was omitted the stand was good. Explain just what happened. What would you do to avoid this trouble?

12. Explain specifically why crops are more likely to experience salt injury on a sandy soil than on a silt loam. Why does potassium not move appreciably in a silt loam?

13. Why might the nature of the root system of the crop being grown affect the decision to build up the fertility level of the soil versus applying fertilizer in the row? How would the economic status of the farmer affect the decision?

14. Explain how band and broadcast applications complement each other in encouraging efficient crop production.

15. You are planning to apply phosphorus broadcast. You had the choice of broadcasting and plowing down, broadcasting and disking in after plowing, or subsurface application in a broad band. Which procedure would be most desirable? Explain fully.

16. What is meant by *soil building* and *maintenance applications of fertilizer?*
17. Under what relative levels of available soil-test phosphorus and potassium would you approve of making broadcast maintenance applications of these nutrients?
18. Under what conditions is surface broadcast phosphorus and potassium taken up by the plant? Explain.
19. What cropping systems exist in your area in which it might be desirable to apply all the phosphorus and potassium to one crop in the rotation?
20. Under what specific conditions in your area do you believe that all the nitrogen could be applied before planting? Under what conditions should none be applied before planting?
21. Why does ammoniacal nitrogen applied with phosphorus cause more phosphorus to be absorbed by the plant?
22. Under what conditions would you advocate fall fertilization in your area?
23. What are the possibilities for summer, fall, winter, and spring application of fertilizer in your area? Why is there a need to spread the fertilizer season?
24. Calculate the removal of phosphorus and potassium in a corn–corn–soybean–wheat–alfalfa rotation and in a corn–soybean–wheat–alfalfa–alfalfa rotation. Assume yields given in Table 13-1 and that corn, soybeans, and wheat are harvested for grain only.
25. What is meant by carryover fertilizer? Why is there an appreciable amount in a properly fertilized rotation?
26. Are there residual benefits from NO_3-N in soils?
27. Do crops benefit equally from soil NO_3-N and that derived from fertilizer? Explain any differences if they exist.
28. Give the pros and cons of the two approaches used in applying micronutrients.
29. What is fertigation, and what are its advantages and drawbacks?
30. What is foliar fertilization? Discuss any limitations.
31. What is dual deep placement of fertilizers? What are its advantages and disadvantages?
32. Is the distribution of soil fertility in the root zone modified by tillage?

Selected References

Achorn, F. P., and T. R. Cox, "Production, marketing, and use of solid, solution and suspension fertilizers," in R. A. Olsen, T. J. Army, J. J. Hanway, and V. J. Kilmer, Eds., *Fertilizer Technology and Use*, 2nd ed., Chapter 12. Madison, Wis.: Soil Science Society of America, 1971.

Alessi, J., and J. F. Power, "Effects of banded and residual phosphorus on dryland spring wheat yield in the Northern Plains." *Soil Sci. Soc. Am. J.*, **44**:792 (1980).

Alston, A. M., "Effects of depth of fertilizer placement on wheat grown under three water regimes." *Australian J. Agr. Res.*, **27**:1 (1976).

Anderson, C. K., L. R. Stone, and L. S. Murphy, "Corn yield as influenced by in-season applications of N with limited irrigation." *Agron. J.*, **74**:396 (1982).

Bailey, L. D., H. Ukrainetz, and D. R. Walker, "Effect of phosphorus placement on crop uptake and yield." *Proc. Western Canada Phosphate Symp.*, pp. 200–229. Edmonton, Alberta: Alberta Soil and Feed Testing Laboratory, 1980.

Barber, S. A., "Effect of tillage practice on corn root distribution and morphology." *Agron. J.*, **63**:724 (1971).

Barber, S. A., "The influence of the plant root system in the evaluation of soil fertility." *Int. Symp. Soil Fertil. Eval. Proc.* **1**:249. New Delhi: Indian Society of Soil Science (1971).

Barel, D., and C. A. Black, "Effect of neutralization and addition of urea, sucrose, and various glycols on phosphorus absorption and leaf damage from foliar-applied phosphate." *Plant Soil*, **52**:515 (1979).

Barel, D., and C. A. Black, "Foliar application of P: I. Screening of various inorganic and organic P compounds." *Agron. J.*, **71**:15 (1979).

Barel, D., and C. A. Black, "Foliar application of P: II. Yield responses of corn and soybeans sprayed with various condensed phosphates and P-N compounds in greenhouse and field experiments." *Agron. J.*, **71**:21 (1979).

Belcher, C. R., and J. L. Ragland, "Phosphorus absorption by sod planted corn from surface applied phosphorus." *Agron. J.*, **63**:754 (1972).

Boawn, L. C., and G. E. Leggett, "Phosphorus and zinc concentrations in Russet Burbank potato tissues in relation to development of zinc deficiency symptoms." *SSSA Proc.*, **28**:229 (1964).

Chapin, J. S., F. L. Fisher, and A. G. Caldwell, "Effect of fertilizers on the conductivity of saturated soil extracts." *Soil Sci. Soc. Am. J.*, **28**:90 (1964).

Choriki, R. T., D. E. Ryerson, and A. L. Dubbs, "Evaluation of nitrogen use and methods of application on mixed prairie vegetation in Montana in relation to forage yield, change in composition of vegetation, residual nitrogen, nitrate poisoning and beef gain per acre." *20th Northwest Fert. Conf.*, pp. 182–191, Spokane, Wash. (July 8–10, 1969).

Cochran, V. L., R. L. Warner, and R. I. Papendick, "Effect of N depth and application rate on yield, protein content, and quality of winter wheat." *Agron. J.*, **70**:964 (1978).

Cook, R. L., and J. F. Davis, "The residual effect of fertilizer." *Adv. Agron.*, **9**:205 (1957).

De Wit, C. T., "A physical theory on placement of fertilizers." *Staatsdrukkerij-Uitgeverijbedrijf*, The Netherlands (1953).

Drew, M. C., "Comparison of the effects of a localized supply of phosphate, nitrate, ammonium and potassium on the growth of the seminal root system, and the shoot in barley." *New Phytol.*, **75**:479 (1975).

Dumenil, L., J. Pesek, J. R. Webb, and J. J. Hanway, "Phosphorus and potassium fertilizer for corn: how to apply." *Iowa Farm Sci.*, **19**(10):11 (1965).

Fehrenbacher, J. B., and H. J. Snider, "Corn root penetration in Muscatine, Elliot and Cisne soils." *Soil Sci.*, **77**:281 (1954).

Fehrenbacher, J. B., P. R. Johnson, R. T. Odell, and P. E. Johnson, "Root penetration and development of some farm crops as related to soil physical and chemical properties." *Trans. 7th Int. Congr. Soil Sci.*, **3**:243 (1960).

Fitts, J. W., and W. V. Bartholomew, "Modifying the soil profile for deeper root penetration." *Better Crops Plant Food*, **44**(5):52 (1960).

Flannery, R., "Alfalfa absorbs much plant food." *Better Crops Plant Food*, **57**(2):1 (1973).

Follett, R. H., L. S. Murphy, and R. L. Donahue, *Fertilizers and Soil Amendments*. Englewood Cliffs, N.J.: Prentice-Hall, 1981.

Foth, H. D., K. L. Kinra, and J. N. Pratt, "Corn root development." *Michigan State Univ. Agr. Exp. Sta. Q. Bull.*, **43**(1):2 (1960).

Fox, R. L., and R. C. Lipps, "Distribution and activity of roots in relation to soil properties." *Trans. 7th Int. Congr. Soil Sci.*, **3**:260 (1960).

Garcia, R., and J. J. Hanway, "Foliar fertilization of soybeans during seed-filling period." *Agron. J.*, **68**:653 (1976).

Gardner, B. R., and R. L. Roth, "Applying nitrogen in irrigation waters," in R. D. Hauck, J. D. Beaton, C. A. I. Goring, R. G. Hoeft, G. W. Randall, and D. A. Russel, Eds., *Nitrogen in Crop Production*, Chapter 26B. Madison, Wis.: American Society of Agronomy, Crop Science Society of America, and Soil Science Society of America (in press).

Gardner, E. H., "Nitrogen fertilizer rates for Columbia Plateau dryland wheat." *Proc. Pendleton–Walla Walla Fert. Dealers Conf.*, pp. 1–3, Pendleton, Oreg. (1980).

Grunes, D. L., "Effect of N on the availability of soil and fertilizer P to plants." *Adv. Agron.*, **11**:369 (1959).

Haby, V. A., C. Simons, M. S. Stauber, R. Lund, and P. O. Kresge, "Relative efficiency of applied nitrogen and soil nitrate for winter wheat production." *Agron. J.*, **75**:49 (1983).

Hall, N. S., W. V. Chandler, C. H. M. van Bavel, P. H. Reed, and J. H. Anderson, "A tracer technique to measure growth and activity of plant root systems." *North Carolina Agr. Exp. Sta. Tech. Bull. 101* (1953).

Ham, G. E., W. D. Poole, and G. W. Randall, Soybean yield improvement with foliar fertilization. *Proc. Natl. Fert. Solut. Assoc. Roundup*, Kansas City, Mo. (1978).

Hammes, J. K., and J. F. Bartz, "Root distribution and development of vegetable crops as measured by radioactive P injection technique." *Agron. J.*, **55**:329 (1963).

Hansen, C. M., L. S. Robertson, H. J. Retzer, and H. M. Brown, "Grain drill design from an agronomic standpoint." *Trans. ASAE*, 5(1):8 (1962).

Holt, E. C., and F. L. Fisher, "Root development of coastal Bermuda grass with high N fertilization." *Agron. J.*, **52**:593 (1960).

Hurd, E. A., "Growth of roots of seven varieties of spring wheat at high and low moisture levels." *Agron. J.*, **60**:201 (1968).

Hurd, E. A., and E. D. Spratt, "Root patterns in crops as related to water and nutrient uptake," in U. S. Gupta, Ed., *Physiological Aspects of Dryland Farming*, pp. 167–235. New Delhi: Oxford & IBH, 1975.

Illinois Agronomy Handbook, Dept. of Agron., Univ. of Illinois (1983–1984).

Kissel, D. E., "Topdress or preplant applications of N can boost winter wheat yields." *Solutions*, 26(3):76 (1982).

Krantz, B. A., W. L. Nelson, C. D. Welch, and N. S. Hall, "A comparison of phosphorus utilization by crops." *Soil Sci.*, **68**:11 (1949).

Lambeth, V., "Desirable soil nutrient level for commercial vegetables." *Better Crops Plant Food*, **42**:14 (1958).

Lavy, T. L., and S. A. Barber, "A relationship between the yield response of soybeans to molybdenum applications and the molybdenum content of the seed produced." *Agron. J.*, **55**:154 (1963).

Lawton, K., and J. F. Davis, "Influence of fertilizer analysis and placement on emergence, growth and nutrient absorption by wheat seedlings in the greenhouse." *Agron. J.*, **52**:326 (1960).

Leyshon, A. J. and M. R. Kilcher, "High nitrogen fertilizer applications on grasses and long term residual effects." *Proc. 1976 Soil Fert. Workshop*, Soil Manag. 510, Publ. 244, pp. 114–120, Univ. of Saskatchewan (1976).

Linford, K., and R. E. McDole, "Survey to investigate potato rooting in southern Idaho soils." *Proc. 28th Annu. Fert. Conf. Northwest*, Twin Falls, Idaho (July 12–14, 1977).

Longnecker, D., and F. G. Merkle, "Influence of placement of lime compounds on root development and soil characteristics." *Soil Sci.*, **72**:71 (1952).

Lucas, R. E., "Micronutrients for vegetables and field crops." *Michigan State Univ. Ext. Bull. E-486* (1973).

Lucas, R. E., "Organic soils (Histosols) formation, distribution, physical and chemical properties, and management for crop production." *Michigan State Univ. Res. Rep.* (1982).

Lutrick, M. C., "The downward movement of K in Eustis loamy fine sand." *Soil Crop Sci. Soc. Florida, Proc.*, **18**:198 (1958).

Lutz, J. A., Jr., G. D. Jones, G. W. Hawkins, and T. B. Hutcheson, Jr., "Effects of uneven spreading of fertilizer on crop yields." *Agron. J.*, **67**:526 (1975).

McEwen, J., and A. E. Johnston, "The effects of subsoiling and deep incorporation of P and K fertilizers on the yield and nutrient uptake of barley, potatoes, wheat, and sugar beet grown in rotation." *J. Agric. Sci. (Camb.)*, **92**:695 (1979).

Martens, D. C., G. W. Hawkins, and G. D. McCart, "Field response of corn to $ZnSO_4$ and Zn-EDTA placed with the seed." *Agron. J.*, **65**:135 (1973).

Mengel, K., "Potassium availability and its effect on crop production," in *Japanese Potash Symposium*. Tokyo: Yokendo Press, Kali Kenkyu Kai, 1971.

Miller, M. H., "Effects of nitrogen on phosphorus absorption by plants," in E. W. Carson, Ed., *The Plant Root and Its Environment*, Chapter 21. Charlottesville, Va.: University Press of Virginia, 1974.

Miller, R. J., J. T. Pesek, and J. J. Hanway, "Soybean yield responses to fertilizer." *Soybean Dig.*, **21**(3):6 (1961).

Munson, R. D., and W. L. Nelson, "Movement of applied K in soils." *Agr. Food Chem.*, **11**(3):193 (1963).

Murphy, L. S., "Significant contributions of the '70s—starting point for the '80s." *Solutions*, **24**(6):84 (1980).

Murphy, L. S., R. S. Rauschkolb, and S. J. Locascio, "Phosphorus application in irrigation water: problems and possibilities." *Solutions*, **23**(6):70 (1979).

Nelson, L. B., and R. E. Uhland, "Factors that influence loss of fall applied fertilizers and probable importance in different sections of the United States." *Soil Sci. Soc. Am., Proc.*, **19**:492 (1955).

Nielsen, N. E., and S. A. Barber, "Differences among genotypes of corn in the kinetics of P uptake." *Agron. J.*, **70**:695 (1978).

Olson, R. A., D. H. Sander, and A. F. Dreier, "Soil analyses—Are they needed for nursery data interpretation?" *Proc. Int. Winter Wheat Conf.*, p. 179. Ankara, Turkey: USDA and Univ. of Nebraska, 1972.

Olson, R. A., A. F. Dreier, C. Thompson, K. Frank, and P. H. Grabouski, "Using fertilizer nitrogen effectively on grain crops." *Nebraska Agr. Exp. Sta. Bull. SB479* (1964).

Ozanne, P. G., and A. Petch, "The application of nutrients by foliar sprays to increase seed yields." in A. R. Ferguson, R. L. Bieleski, and I. B. Ferguson, Eds., *Proc. 8th Int. Colloq., Plant Anal. Fert. Prob.*, pp. 361–366, Auckland, New Zealand (August 28–September 1, 1978).

Parks, W. L., and W. M. Walker, "Effect of soil potassium, fertilizer potassium and method of fertilizer placement upon corn yields." *SSSA Proc.*, **33**:427 (1969).

Peterson, G. A., D. H. Sander, P. H. Grabouski, and M. L. Hooker, "A new look at row and broadcast phosphate recommendations for winter wheat." *Agron. J.*, **73**:13 (1981).

Peterson, L. A., and D. Smith, "Recovery of K_2SO_4 by alfalfa after placement at different depths in a low fertility soil." *Agron. J.*, **65**:769 (1973).

Phillips, R. E., G. W. Thomas, and R. L. Blevins, "No-tillage research." Research reports and reviews, Univ. of Kentucky (1978).

Pumphrey, F. V., F. E. Koehler, R. R. Allmaras, and S. Roberts, "Method and rate of applying zinc sulfate for corn on zinc-deficient soil in western Nebraska." *Agron. J.*, **55**:235 (1963).

Rader, L. F., L. M. White, and C. W. Whittaker, "The salt index—a measure of the effects of fertilizers on the concentration of the soil solution." *Soil Sci.*, **55**:201 (1943).

Ramig, R. E., and P. E. Rasmussen, "Residual availability of sulfur to wheat grown in a wheat-pea rotation in northeastern Oregon." *Proc. 23rd Annu. Northwest Fert. Conf.*, pp. 125–137, Boise, Idaho (July 17–20, 1972).

Randall, G. W., "Efficiency of nitrogen use as related to application methods," in R. E. Hauck, J. D. Beaton, C. A. I. Goring, R. G. Hoeft, G. W. Randall, and D. A. Russel, Eds., *Nitrogen in Crop Production*, Chapter 37. Madison, Wis.: American Society of Agronomy, Crop Science Society of America, and Soil Science Society of America (in press).

Raper, C. D., Jr., and S. A. Barber, "Rooting systems of soybeans: I. Differences in root morphology among varieties." *Agron. J.*, **62**:581 (1970).

Richards, G. E., Ed., *Band Application of Phosphatic Fertilizers*. Little Rock, Ark.: Olin Corporation, 1977.

Riley, D., and S. A. Barber, "Effect of ammonium and nitrate fertilization on P uptake as related to root-induced pH changes at the root-soil interface." *Soil Sci. Soc. Am. Proc.*, **35**:301 (1971).

Robertson, J. A., "Fertilizer placement." *Agdex 542-5*, Alberta Agriculture (August 1982).

Runge, E. C. A., "How weather and soil moisture affect corn yield." *Get Your Answer from Us.* Univ. of Illinois Agronomy Field Day (1973).

Sandoval, F. M., J. J. Bond, and G. A. Reichman, "Deep plowing and chemical amendment effect on a sodic claypan soil." *Trans. ASAE*, **15**(4):68 (1972).

Schenk, M. K., and S. A. Barber, "Phosphate uptake by corn as affected by soil characteristics and root morphology." *Soil Sci. Soc. Am. J.*, **43**:880 (1979).

Schenk, M. K., and S. A. Barber, "Root characteristics of corn genotypes as related to P uptake." *Agron. J.*, **71**:921 (1979).

Schenk, M. K., and S. A. Barber, "Potassium and phosphorus uptake by corn genotypes grown in the field as influenced by root characteristics." *Plant Soil*, **54**:65 (1980).

Schulte, E. E., "Buildup soil K levels before shifting to minimum tillage." *Better Crops Plant Food*, **63**:25 (Fall 1979).

Schulte, E. E., L. M. Walsh, J. Moncrief, and A. E. Peterson, "Fertilizer needs under conservation tillage." Soils, Fertilizer and Agricultural Pesticides Short Course, Minneapolis, Minn. (December 12–13, 1978).

Shickluna, J. C., R. E. Lucas, and J. F. Davis, "The movement of potassium in organic soils." *Proc. 4th Int. Peat Congr. I–IV*, p. 131 (1972).

Smith, E. S., D. D. Wolf, and M. Lentner, "Straighten up and spread right." *Crops Soils*, pp. 9–12 (August–September 1981).

Terman, G. L., D. R. Bouldin, and J. R. Webb, "Evaluation of fertilizer by biological methods." *Adv. Agron.*, **14**:265 (1962).

Tidmore, J. W., and J. T. Williamson, "Experiments with commercial nitrogenous fertilizers." *Alabama Exp. Sta. Bull. 238* (1937).

Triplett, G. B., Jr., and D. M. VanDoren, Jr., "Nitrogen, phosphorus and potassium fertilization of non-tilled maize." *Agron. J.*, **61**:637 (1969).

Vavra, J. P., "The effect of deep tillage and subsoil fertilization on growth of corn." *Illinois Fert. Conf. Proc.*, p. 60 (1966).

Vitosh, M. L., R. E. Lucas, and R. J. Black, "Effect of nitrogen fertilizer on corn yield." *Ext. Bull. E-802*, Michigan State Univ. (1979).

Wagner, R. E., and W. C. Hulburt, "Better forage stands." *Natl. Fert. Rev.*, **29**:13 (1954).

Wang, L. C., O. J. Attoe, and E. Truog, "Effects of lime and fertility level on the chemical composition and winter survival of alfalfa." *Agron. J.*, **45**:381 (1954).

Welch, L. F., "Nitrogen use and behavior in crop production." *Bull. 761*, Univ. of Illinois at Urbana-Champaign (1979).

Welch, L. F., P. E. Johnson, G. E. McKibben, L. V. Boone, and J. W. Pendleton, "Relative efficiency of broadcast versus banded potassium for corn." *Agron. J.*, **58**:618 (1966).

Welch, L. F., D. L. Mulvaney, L. V. Boone, G. E. McKibben, and J. W. Pendleton, "Relative efficiency of broadcast versus banded phosphorus for corn." *Agron. J.*, **58**:283 (1966).

Westermann, D. T., "Phosphorus fertilization economics." *Proc. 28th Annu. Northwest Fert. Conf.*, pp. 141–146 (1977).

Western Fertilizer Handbook, 6th ed. Danville, Ill.: Interstate Printers and Publishers, 1975.

Wilkinson, S. R., and A. J. Ohlrogge, "Fertilizer nutrient uptake as related to root development in the fertilizer band. Influence of N and P fertilizer on endogenous auxin content of soybean roots." *Trans. 7th Int. Congr. Soil Sci.*, **3**:234 (1960).

Wilkinson, S. R., and A. J. Ohlrogge, "Principles of nutrient uptake from fertilizer bands: V. Mechanisms responsible for intensive root development in fertilized zones." *Agron. J.*, **54**:288 (1962).

Wittwer, S. H., M. J. Bukovac, and H. B. Tukey, "Advances in foliar feedings of plant nutrients," in *Fertilizer Technology and Usage*, pp. 429–455. Madison, Wis.: Soil Science Society of America, 1963.

Wolf, D. D., and E. S. Smith, "Seed and fertilizer patterns with spinning and oscillating spreaders." *Agron. J.*, **70**:1019 (1978).

Younts, S. E., "Response of wheat to rates, dates of application, and sources of copper and to other micronutrients." *Agron. J.*, **56**:266 (1964).

Younts, S. E., and R. P. Patterson, "Copper–lime interactions in field experiments with wheat. Yield and chemical composition data." *Agron. J.*, **56**:229 (1964).

Zubriski, J. C., E. H. Vasey, and E. B. Norum, "Influence of nitrogen and potassium fertilizers and dates of seeding on yield and quality of malting barley." *Agron. J.*, **62**:216 (1970).

Chapter 14

Cropping Systems and Soil Management

THE principal objective of any cropping practice or soil management program is sustained profitable production. The Soil Conservation Service, begun in 1935 in the United States, has focused much attention on this aspect of agriculture. Soil conservation is essentially good soil management and embraces more than just the prevention of soil losses. Soil erosion is a *symptom* of poor soil management, whether it be inadequate plant nutrients or improper cropping systems. Erosion is a symptom, not a primary cause of soil destruction. The primary cause of soil destruction by erosion is impoverishment of nutrients, especially N.

Yield Trends

Crop yields in the United States have gradually increased since 1900 but most of the improvement has occurred since the 1950s (Figure 1-1). In 1982 the average yields in the United States of the three major crops, corn, soybeans, and wheat, were 114.8, 32.2, and 35.6 bu/A, respectively. These are record highs for corn and wheat and the soybean yield equals the previous high attained in 1979. Improved management, including varieties, spacing and population, pest control, tillage, and fertilization, all contribute to this trend to higher yields. The series of technological, cultural, and management practices adopted between 1930 and 1979 by Minnesota farmers for corn production are compiled in Table 14-1. Contributions of the various changes to increased corn yields are identified in Table 14-2.

A partial explanation for the slow increase in yields prior to the 1950s is provided in Figure 14-1. According to estimates from Ohio, the productivity of soil decreased 40% in the 60-year period from 1870 to 1930. This corresponds in general to estimates made on the productivity of Iowa soils over the period 1890 to 1950 and is closely related to fertility levels. Soil organic matter, hence nitrogen supply, has decreased. The removal of such elements as phosphorus, potassium, calcium, magnesium, and sulfur has generally been greater than the amounts returned to the soil in the form of manure and commercial fertilizers.

The actual increase in crop yields in Ohio of about 15% during the period 1870 to 1930 was the result of such compensating factors as improved varieties, insect and disease control, cultural practices, machinery, drainage, and increased fertilization and liming. With these developments it was estimated that crop yields should have increased 40 to 60% if soil fertility had been maintained. These data from Ohio illustrate the fallacy of using yield trends alone over a period of years as a measure of soil fertility. Yields of crops in Ohio have tripled in the period 1930–1982. Similar dramatic increases occurred in the United States as a whole.

TABLE 14-1. Changes in Corn Production Practices in Minnesota by Decades for the 50-Year Period 1930–1979

Practices	1930s	1940s	1950s	1960s	1970s	1979
			Mean Values by Decade			
Area planted to hybrids (%)	8	82	96	98	100	100
Seeding method						
% check rowed	90	70	47		1	NA
% hill dropped			24	55	10	NA
% drilled			29	45	89	NA
Seeding rate (plants/ha)	30,740	33,345	38,780	39,270	46,440	49,780
Fertility						
Manure (tons/ha)	18	17	14	13	9	8
Commercial						
Area fertilized (%)	NA *	NA	14.0	67.0	93.0	95.0
Total kg/ha (N + P_2O_5 + K_2O)	4.5	50.5	68.5	86.4	236.9	229.0
Nitrogen (kg/ha)	NA	NA	16.8	33.7	97.7	112
Phosphate (kg/ha)	NA	NA	24.2	15.8	55.0	51.5
Potassium (kg/ha)	NA	NA	9.9	14.8	32.0	33.4
Green manure (sweet clover)						
(ha × 10^6)	0.16	0.30	0.20	0.08	—	—
Pest control						
Cultivation (no. of times)	3.0	3.0	2.8	2.6	1.6	1.5
Herbicide, area treated (%)	—	T †	12	58	89	93
Insecticide, area treated (%)	—	—	4	29	32	33
Tillage						
Fall plowed (%)	21	NA	70	66	62	72
Spring plowed (%)	79	NA	30	26	26	10
Chisel or minimum tilled (%)	—	—	—	8	12	18
Row spacing (cm)	107	107	102	96	91	90
Seeding date, May	21	21	20	19	13	11 **
Insects						
European corn borer		Spread	————Economic problem levels————			
Western corn rootworm			Spread	Economic problem levels		
Yield (kg/ha)	2,012	2,460	3,150	4,275	5,338	6,287

* NA, data not available.
† T, trace.
** Average of last 5 years.
Source: Cardwell, *Agron. J.,* **74**:984 (1982).

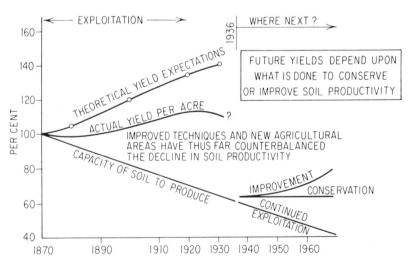

FIGURE 14-1. Improved practices in Ohio since 1870 should have resulted in yields 40 to 60% higher per acre in 1930, but the aggregate yield increased less than 15%. Improved practices only slightly more than counterbalanced the decline in the ability of the soil to produce. Yields can be increased only if proper soil-management programs are adopted. [*Ohio State Agr. Ext. Serv. Bull. 175* (1936).]

TABLE 14-2. Sources of Corn Yield Increases with Changing Production Practices in Minnesota from 1930 to 1979

Cultural Practice or Yield Limiting Factor	Unit	Rate or Magnitude of Change	1930 to 1979 *	Contribution to 1979 Yield	
				kg/ha	% Net Gain
Pre-1930 yield levels	kg/ha	—	—	2012	—
Hybrids					
Double cross	kg/ha	371	100% of hectarage	371	9
Three-way crosses	kg/ha	163	17% of hectarage	28	1
Single crosses	kg/ha	314	75% of hectarage	235	6
Genetic gain	kg/ha/yr	36.5	50 years	1825	43
Fertilizer N	kg/ha/kg N/ha	18.9	112 kg N/ha × 93% of corn area	2003	47
Plant population	kg/ha/100 plants/ha above 30,740	47.4	19,130 plants/ha	905	21
Herbicides	kg/ha	1,048	93% of hectarage	975	23
Row spacing	kg/ha/cm	10.2	− 17 cm	173	4
Planting date	kg/ha/day	36.4	10 days	364	8
Drilling vs. hill drop	kg/ha	408	79%	322	8
Fall plowing	kg/ha	440	51%	224	5
Rotations					
Soybeans	kg/ha/kg N/ha	18.9	25.6 kg N/ha	484	11
Alfalfa/clovers	kg/ha/kg N/ha	18.9	− 7.2 kg N/ha	− 136	− 3
Sweet clover	kg/ha/kg N/ha	18.9	− 16.8 kg N/ha	− 318	− 7
Interference effect	kg/ha	874	− 33% of hectarage	− 291	− 7
Manure	kg/ha/kg N/ha	18.9	− 6.7 metric tons/ha × 0.5% N = 33.5 kg N/ha × 50 yr = 1.35%	− 633	− 15
Organic matter	%/year	− 0.027	1.35% × 22.4 kg N/% O.M. = 302 kg N/ha	− 571	− 13
Insects					
Corn borer	kg/ha/borer/plant	−185	1.2 borer/plant	− 220	− 5
Corn rootworm	kg/ha	−446	32.5% of hectarage	− 145	− 3
Soil erosion	kg/ha/cm	−69	− 0.1 cm/yr × 50 yrs	− 345	− 8
Unidentified negative factors	—	—	—	− 975	− 23
Net gain	—	—	—	4275	—
1977–1979 yield level	—	—	—	6287	—

* See Table 14-1 for actual levels in 1930s and 1979.
Source: Cardwell, *Agron. J.,* **74**:984 (1982).

The marked increase in yields has been due to rapid advances in technology of which fertilizer is only one part. The USDA estimated in 1964 that elimination of nitrogen and phosphorus fertilizer would decrease corn yields in Illinois by 37%, grapefruit in Florida by 94%, and alfalfa in Arizona by 34%. The values would be greater now. Also, if fertilizer were not used on corn in Iowa, 29% more land would be needed. This would mean less suited land would go into production with greater opportunity of erosion. Much more dramatic results could be reported in developing countries. In Mexico, wheat yields increased from 775 kg/ha in 1943 to 3600 kg/ha in 1978–1980 (80% irrigated).

Although in dryland regions water is more of a limiting factor than fertility, serious losses in soil fertility and productivity have often occurred as a result of cropping systems involving frequent summer fallowing. Improved management, especially in stress years, is just as important in those regions as it is in more humid ones. For example, in the nine Great Plains states annual rainfall was very low during 1930–1939 and 1950–1956, 20.4 and 20.7 in., respectively. However, wheat yields differed markedly, 10.6 bu/A in the 1930s and 15.2 bu/A in 1952–1956. Much of the drop in production capacity of soils in the low precipitation areas is more related to wind and water erosion and deep percolation of mobile nutrients during fallowing than it is to actual depletion of fertility by cropping. Contrasted with this are the soils of the Southeast, which are inherently low in fertility and have been farmed intensively for 100 to 200 years.

The continued increase in yields on experiment station plots and by top farmers is important, for they serve as a guide as to what can be done. In those states and with those crops on which experimental yields have continued to increase, the average state yields, although considerably less than those of experiment stations, have continued to rise. This emphasizes the importance of continued intensive research to overcome the effect of those factors that might be limiting crop yields.

Aim of Crop and Soil Management

The aim of all cropping systems should be sustained maximum profit from the farming operation. In evaluating a crop and soil management system for its effects on sustained high production there are several factors that must be kept in mind, including:

1. Organic matter and soil tilth.
2. Plant-nutrient supply.
3. Incidence of weeds, insects, and diseases.
4. Water intake and soil erosion.

Soils differ greatly in their characteristics, hence in their management requirements. For example, one soil, a silty clay loam high in organic matter, may have excellent tilth and could endure management practices that decrease tilth for some little time before problems are encountered. Another soil, a silt loam low in organic matter, might be in poor tilth, and the same management practices would cause trouble immediately. It is important that soils be evaluated with respect to their management requirements. Experimental data should be available to show the effects of management practices on different soils.

Figure 14-2 shows that erosion must not be taken for granted and it is imperative that crop production be carried on in such a way as to minimize the destructive effect of water and wind erosion. More than one-third of cropland in the United States is subject to erosion severe enough to shorten significantly the productive life of soils. It should be noted in Table 14-2 that erosion detracted from the progress made in achieving higher corn yields in Minnesota.

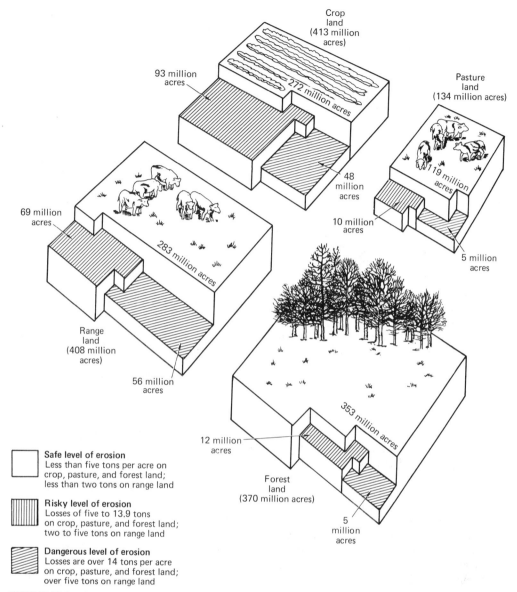

Crop
land
(413 million
acres)

93 million
acres

272 million acres

Pasture
land
(134 million acres)

119 million
acres

48
million
acres

10 million
acres

5 million
acres

69 million
acres

283 million acres

Range
land
(408 million
acres)

56 million
acres

353 million acres

12 million
acres

Forest
land
(370 million acres)

5
million
acres

□ Safe level of erosion
Less than five tons per acre on
crop, pasture, and forest land;
less than two tons on range land

▥ Risky level of erosion
Losses of five to 13.9 tons
on crop, pasture, and forest land;
two to five tons on range land

▨ Dangerous level of erosion
Losses are over 14 tons per acre
on crop, pasture, and forest land;
over five tons on range land

FIGURE 14-2. How serious is erosion? Shown here are the results of the National Resource Inventory conducted by the Soil Conservation Service in 1977. [Fraedrich, *Farm Futures,* **8**(9):28 (1980).]

Organic Matter in the Soil

Much emphasis was once placed on the organic matter content of a soil as an indication of its productivity. With the increasing use of nitrogen it is not necessary or even wise to rely on nitrogen release from soil organic matter for high yields of crops such as corn or wheat. At one time summer-fallowed fields on the Canadian prairies were considered self-sufficient in nitrogen, but more than 20% of them now require fertilizer nitrogen. Bradfield (1963) sums up the subject as follows:

> For most farmers, the only economical way to get more organic matter in their soil is to grow more organic matter on their own farms. Larger crops will mean

more roots, more stalks and stubble, more feed for livestock, and hence more manure to return to the soil. The cheapest way to grow these larger crops is by more liberal fertilization and by use of good soil-building rotations in which the soil is so handled that maximum efficiency is obtained from the fertilizers. This will require the best seed, the best adapted cultivation practices, and the most efficient use of all organic residues. Organic farming with chemical fertilizers will result in even higher yields per acre and even more organic matter in our more productive soils.

Should Organic Matter be Maintained or Increased?

To answer this question it is necessary to review some of the functions of organic matter:

1. It acts as a storehouse for nutrients—nitrogen, phosphorus, sulfur, boron, zinc, and so on.
2. It increases exchange capacity.
3. It provides energy for microorganism activity.
4. It releases carbon dioxide.
5. It increases water-holding capacity.
6. It stabilizes structure and improves tilth.
7. It provides surface protection and thus reduces crusting and increases infiltration.
8. It reduces the effects of compaction.
9. It buffers the soil against rapid changes in acidity, alkalinity, and salinity.

It is interesting to note that most of these functions, with the exception of surface and compaction protection, depend on decomposition. Hence the production of large quantities of residues, and their subsequent decay, is necessary to good crop and soil management. It is significant that one of the plant-nutrient problems of the Arctic is the resistance to decay of organic matter under low temperatures, and organic matter accumulates even on gravel ridges. In contrast, in the subtropics and tropics, although much organic matter is produced, it decays very rapidly.

Maintenance of organic matter for the sake of maintenance alone is not a practical approach to farming. It is more realistic to use a management system that will give sustained top profitable production. The greatest source of soil organic matter is the residue contributed by current crops. Consequently, the selection of the cropping system and method of handling the residues are equally important. Proper management and fertilization will produce high yields, the principal point in which farmers are interested. A by-product of these high yields is the organic residue. Therefore, those soils that are being managed to produce large yields are being improved at the same time.

Workers at Michigan State University state: "The practices we perform to grow our top yields do the best job in conserving and building our soils."

Effect of Cropping Systems. Cropping systems affect the organic matter in the soil in several ways:

1. Tillage of the soil produces greater aeration, thus stimulating more microbial activity, and increases the rate of disappearance of soil organic carbon (Table 14-3). This decline is reported to be most rapid during the first 10 years after the initiation of cultivation and then it continues at a gradually diminishing rate for several decades. Eventually, an apparent equilibrium is reached after about 40 years under conditions prevailing in the northern Great Plains. A cropping system with a large proportion of row crops thus encourages a much more rapid loss of organic matter than a cropping system with a large proportion of close-growing or sod crops. Tillage in some instances has increased the yields

TABLE 14-3. Carbon Loss from Virgin Grassland Due to Cropping Under Stubble Mulch and Conventional Management Systems on a Variety of Soils in the Northern Great Plains

Soil Depth (cm)	Montana/North Dakota/Wyoming *	Grant County, North Dakota †	
		Stubble Mulch	Conventional
0–15.2	41	27	38
15.2–30.5	20	7	14

* Two field stations in each state, sampled in 1950 after about 40 years of cropping.
† Thirty-six farm fields sampled in 1979 after about 70 years of cropping.
Source: Bauer and Black, *Soil Sci. Soc. Am. J.,* **45**:1166 (1981).

of nonlegumes on soils containing a good supply of organic matter but to which insufficient nitrogen was applied. This is probably related in part to the effect of increased aeration on organic matter decomposition and subsequent release of nitrogen. Too, new surfaces are continually exposed, and there is greater opportunity for wetting and drying of the soil. Of course, all of this causes gradual depletion of the soil nitrogen.

2. Excessive tillage of the soil tends to facilitate erosion, thus resulting in the physical loss of organic matter and other components. Primary tillage at the beginning of a cropping or fallow season, however, will often improve soil structure, porosity, and roughness. This increases water penetration and the soil's resistance to erosion.

3. Cropping systems differ in the amount of plant residues they contribute. Corn for grain may add 3 to 7 tons/A of stover. Roots may furnish another 1 to 2 tons. Corn for silage results in removal of the stover and grain. A grass–legume sod will produce about the same amount of residues, but most of it will be removed in hay and grazing. With a good yield of small grain, 3 tons of residue may be returned if the straw is left on the field, but only about 0.5 ton may be returned if the straw is removed. In peanuts the tops, nuts, and many of the roots are removed, a system that will encourage much more rapid loss of organic matter. In a livestock program, in which the grain and plant parts are consumed by the animal, only a fraction of the organic matter will be returned to the land, and then the manure will be applied to those fields closest to the barn.

4. Cropping systems vary in nitrogen content of the plant residues, and the accumulation of soil organic matter is in part related to this content. If residues low in nitrogen are turned under, much of the carbon will be evolved as carbon dioxide in decomposition before the ratio approaches 10 or 12:1.

Corn well fertilized with nitrogen will probably contain at least 1.0% in the stover, although 0.75% is considered more typical. In contrast, corn receiving small amounts of nitrogen may have less than 0.5% in the stover. Because of this and the much larger amount of organic matter produced, the stover from the well-fertilized corn will be more effective in maintaining organic matter than that from poorly fertilized corn. Most crop residues containing around 1.5% nitrogen do not need additional amounts for rapid breakdown and conversion into humus. This need for adequate nitrogen in residue decomposition was discussed in Chapter 5.

There has been much discussion of the effect of adding extra nitrogen to accelerate residue decomposition and encourage formation of soil organic matter. The data in Table 14-4 show that increasing the nitrogen supply to decomposing wheat

TABLE 14-4. **Effect of Adding Ammonium Nitrate–Sulfate on the C/N Ratio and Nitrogen and Carbon Concentrations in Decomposing Wheat Straw Incubated for Nine Weeks in the Absence of Soil**

Treatment	Time	C/N Ratio	N (%)	C (%)
Straw only	Initially	107	0.38	40.4
Straw with added N		30	0.38	40.4
Straw only	After 9 weeks	116	0.35	40.8
Straw with added N		76	0.55	41.7

Source: Cochran et al., *Soil Sci. Soc. Am. J.,* **44:**978 (1980).

straw greatly reduced the C/N ratio while substantially increasing the nitrogen percentage and marginally raising the carbon concentration.

Under some conditions a lack of sulfur can retard the decomposition of organic residues. This was discussed in Chapter 8, where it was pointed out that the addition of fertilizer sulfur is necessary under some conditions to bring about decomposition of the added organic material.

Soil Organic Matter. Much work has been done in many areas and on many different types of soil to determine the effect of cropping systems on soil organic matter. In general, in experiments started on virgin soils it has been difficult to maintain the soil organic matter even with the best cropping systems. On soils in which the organic matter had been depleted before experimentation began, cropping systems with a high proportion of sod or close-growing crops and a minimum of cultivated crops are likely to result in an increase in the organic matter and nitrogen content.

In areas of high mean annual temperatures, such as in the southern United States, decomposition continues over a considerable part of the year. It is difficult to increase the organic matter content under such conditions. In contrast, in the northern United States the soil organic matter content, if depleted, can be increased more readily by certain cropping systems, particularly ones that feature less tillage and the return of more crop residues. This is well illustrated in the example in Table 14-5, which shows that soil organic matter and total nitrogen increased in systems that reduced the frequency of fallowing and increased the return of organic residues.

TABLE 14-5. **Effect of Cropping Systems and Manure on Nitrogen and Organic Matter of Soils After Thirty-seven Years of Cropping**

Rotation	Organic Matter (%)		Total Nitrogen (%)	
	Control	Manure	Control	Manure
Fallow–wheat	3.7	4.1	0.19	0.28
Fallow–wheat–wheat	4.9	5.5	0.26	0.30
Fallow–wheat–wheat–wheat	4.7	5.5	0.25	0.28
Wheat continuous	7.2	7.6	0.36	0.38
Average	5.1	5.6	0.27	0.31
Alfalfa fallow–wheat–wheat–wheat	5.8	—	0.28	—
Grass fallow–wheat–wheat–wheat	6.3	—	0.31	—

Source: Ridley and Hedlin, *Can. J. Soil Sci.,* **48:**315 (1968).

Many factors determine whether the soil organic matter is increased or decreased by cropping systems. *The key point is to keep large amounts of crop residues (stover and roots) passing through the soil. Continued good management, including adequate fertilization, helps bring this about.*

Effect of Added Plant Nutrients. The amount of lime and fertilizer in the cropping system affects not only the yield of the harvested crop and its composition but also the amount of crop residues produced. This increased quantity of crop residues brought about by greater amounts of plant nutrients is important in organic matter maintenance. In addition, higher yields mean more extensive root systems which distribute organic matter deeper in the soil.

Nitrogen. The effect of nitrogen on corn grain and stover yields is illustrated by the data shown in Table 14-6. The higher rates of nitrogen not only raised grain yields but also increased stover yields 50%. It is noteworthy that grain yields increased relatively faster than stover.

The general relation that exists between additions of nitrogen and loss of soil nitrogen is shown in Figure 14-3. This indicates that the yearly losses of soil nitrogen, hence soil organic matter, decrease with larger additions of nitrogen. It would appear that if nitrogen additions were equal to or slightly greater than crop removals the losses in the soil would be reduced to a minimum.

TABLE 14-6. Effect of Nitrogen on Corn Grain and Stover Yields

Nitrogen (lb/A)	Grain (bu/A)	Stover (lb/A)
0	22.8	3614
40	52.1	3356
80	81.7	5078
120	100.5	5459
160	102.3	5328

Source: Krantz and Chandler, *North Carolina Agr. Exp. Sta. Bull 366* (rev. 1954). Adequate phosphorus and potassium were applied.

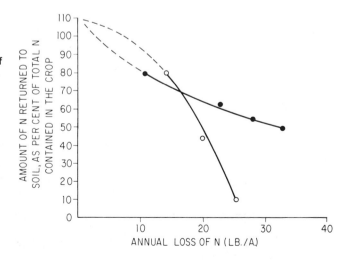

FIGURE 14-3. Yearly losses of nitrogen decrease as nitrogen, expressed as a percentage of that element contained in the crop, is returned to the soil in increasing amounts. Solid circles, values for soils with more than 4500 lb/A of total nitrogen; open circles, values for less than 2500 lb/A of total nitrogen. (Melsted, "New concepts of management of corn belt soils," in A. G. Norman, Ed., *Advances in Agronomy,* Vol. VI. Copyright 1954 by Academic Press, Inc.)

On a Cisne silt loam in southern Illinois over a twelve-year period the addition of nitrogen to a corn–soybean–wheat (legume, grass) rotation already receiving lime, phosphorus, and potassium increased the nitrogen content of the soil 0.014% and the carbon 0.15%.

Phosphorus. The effect of phosphorus in increasing the grain and straw yields of spring wheat is considerable. It should be kept in mind that with soils low in phosphorus, adequate fertilization can increase the amounts of plant residues turned back to the soil. Such residues should be helpful in maintaining or even increasing soil organic matter.

Other Nutrients. The effects of nitrogen and phosphorus are cited as examples. Yield responses to other nutrients such as potassium, lime, sulfur, or micronutrients will also increase the amount of residues. Figure 14-4 shows both the individual effect of sulfur and the influence of a nitrogen–sulfur interaction on increasing grain and straw yields of winter wheat.

Effects of Added Organic Matter on Availability of Nutrients in the Soil. The effects of organic matter on availability of nutrients already in the soil will be mentioned briefly. The large quantity of carbon dioxide evolved during organic matter decomposition is thought to be important to the release of certain nutrients, inorganic phosphorus in particular. The carbon dioxide dissolves in water and forms carbonic acid. The result is a decrease in soil pH. This effect would be of greater importance on neutral or alkaline soils. Under such conditions the temporary reduction in pH would increase the rate of release of other elements such as boron, copper, zinc, manganese, and iron as well as phosphorus.

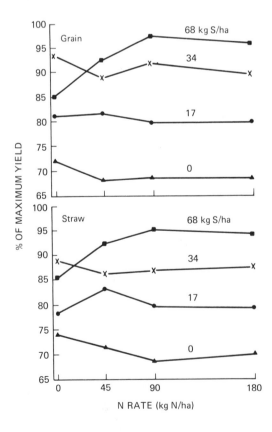

FIGURE 14-4. Grain and straw yield response to residual sulfur in the fourth wheat crop following fertilization (N rate refers to nitrogen applications to first wheat crop; 56 kg/ha of nitrogen applied to each of the second, third, and fourth wheat crops). [Ramig et al., *Agron. J.,* **67**:219 (1975).]

Certain of the intermediate products of organic matter decomposition are believed to form complexing or chelating ions. Phosphorus or certain of the micronutrients attach to these ions and are maintained in a weakly ionized state. The ions are retained against fixation by the soil but remain in a form that can be utilized by plants.

Surface Soil Versus Subsoil

Profitable crop production on eroded soils has been an important agricultural problem. The generally reduced crop yields of nonlegumes on subsoil is well known. On permeable soils this is largely the result of less organic matter and the subsequent lower release of nitrogen.

Research in Ohio on a permeable subsoil and on a tight subsoil high in clay from which the surface had been removed provides an interesting example. A corn, small-grain, and alfalfa rotation was used. On the permeable subsoil, with adequate lime, phosphorus, and potassium, corn yields were 95% of those grown on the topsoil. Added nitrogen did not produce a notable response because the alfalfa supplied this element.

On the tight subsoil there was considerable difficulty in establishing stands of corn and legume seedings, and yields of nonlegumes were generally low. Once a stand of alfalfa was obtained, however, yields of hay were reasonably good.

The results of this study illustrate that on permeable subsoils high yields of nonlegume crops can be had with added nitrogen or in rotation with alfalfa. In some years, however, moisture may be limiting because of reduced water entrance and a lower available waterholding capacity in the subsoil.

An additional aspect of this problem deals with the availability of zinc and sometimes phosphorus or sulfur in exposed subsoils. In subsoils exposed by leveling for irrigation, zinc deficiency may become acute on crops such as corn and beans. Zinc deficiency in these soils is accentuated by high pH, lime, low organic matter, and soil compaction. Deficiencies of sulfur in leveled soils seems to be largely due to low quantities of organic matter.

Legumes in the Rotation

Forage legumes were a mainstay of some rotations for many years. Their main purpose now is to supply large amounts of high-quality forage, whether hay or pasture. An additional and valuable benefit has been the nitrogen supplied to companion or succeeding crops. Further, legumes, particularly deep-rooted ones such as alfalfa and sweet clover, are beneficial on soils with poor physical condition.

Table 14-7 shows that barley grown for a 5-year period after legumes contained a total of 30 to 80 lb of nitrogen and yielded 60 to 70 bushels more in total than unfertilized (N) barley not following a legume. These values are somewhat conservative because in 1969 the "no-legume" plot was summer-fallowed, which accounts for the high yield and nitrogen uptake of barley in 1970. It can also be seen that the beneficial effects of legumes were most pronounced in the first few years after breaking, but continued to have some residual effects even after 5 years.

In spite of these advantages of legumes in a rotation, this practice may not always be attractive to growers. Farmers in some areas may not have a ready use or market for forage crops or the resulting high levels of soil nitrogen may be detrimental to crops such as flue-cured tobacco. Thorough extraction of soil moisture reserves might also be a disadvantage in the drier areas of western United States and Canada.

Many aspects of nitrogen fixation by *Rhizobia* and other symbiotic microorganisms were covered at the beginning of Chapter 5. However, several points related to this beneficial characteristic of legumes in rotations will be discussed in the next section.

TABLE 14-7. Yield and Nitrogen-Uptake of Barley Grown After Legumes

	Yield of Barley * (bu/A)			N Uptake of Barley (lb/A)		
	No Legume	Alfalfa †	Red Clover †	No Legume	Alfalfa	Red Clover
1970	66	41	70	59.4	44	68.2
1971	27	51	51	26.4	63.8	22
1972	26	50	40	26.4	55	41.8
1973	32	52	48	26.4	46.2	33
1974	27	35	37	19.8	28.6	24.2
1975	22	31	26	—	—	—
Total	200	260	272	158.4	237.6	189.2
Mean	34	43	45	—	—	—

* All plots received phosphorus, potassium, and sulfur.
† Grown in 1968 and 1969.
Source: Leitch, in *Alfalfa Production in the Peace River Region,* pp. C1–C5. Beaverlodge, Alberta: Alberta Agriculture and Agriculture Canada Research Station, 1976.

Nitrogen Fixation by Legumes. Elemental nitrogen occupies 78% of the air by volume. A group of bacteria (rhizobia), called nodule or symbiotic bacteria, utilizes this free nitrogen by attaching to the roots of legumes and causing nodules to form. This mutually beneficial relationship is known as symbiosis.

Nodule formation is accomplished by bacteria entering a single-celled root hair. The bacteria then rapidly increase in number, grow toward the base of the root hair, and pierce the cortex of the root. As a result of this penetration, marked cell proliferation takes place, and a nodule, which is a mass of root tissue containing millions of bacteria, is formed. Nodules should not be confused with certain nematode infections which are merely thickened roots.

Nodule bacteria use both the carbohydrates and minerals contained in the host plant and in turn fix atmospheric nitrogen. This nitrogen may be utilized by the host plant, it may be excreted from the nodule into the soil and be used by other plants growing nearby, or it may be released by decomposition of the nodules or legume residues after the legume plant dies or is plowed down.

Amounts of Nitrogen Fixed. The amount of nitrogen fixed by *Rhizobia* varies with the yield level, the effectiveness of inoculation, the nitrogen obtained from the soil, either from decomposition of organic matter or residual nitrogen, and environmental conditions. Optimum pH, water, aeration, and nutrient supply are essential. A high-yielding legume crop such as soybeans, alfalfa, or peanuts contains large amounts of nitrogen. In general about 50 to 80% of the total nitrogen in the plant is fixed by the nodule bacteria.

Welch estimates that in Illinois soils each bushel of soybeans removes 1.7 lb of nitrogen from the soil, and soybeans fix 45% or more of the total nitrogen in the plant.* However, on lighter-colored soils, soybeans might fix 80% or more. Other legumes would behave much the same way.

Legumes in combination with grasses for forage generally supply nitrogen for both crops. However, in the South, with a longer growing season, rotational grazing, and more complete utilization of forage, N-P-K is applied to grass–legume mixtures.

Legumes Versus Commercial Nitrogen. Formerly, one of the reasons for including legumes in a rotation was to supply nitrogen, but with the development of the synthetic nitrogen industry and the resulting availability of inexpensive nitrogen, agriculture is no longer dependent on legumes for this element.

* L. F. Welch, University of Illinois, personal communication.

The selection of the program that farmers should follow becomes a matter of economics; they should choose the program yielding the greatest net return on their investment.

The future cost of fertilizer nitrogen is uncertain. Mention was made in Chapter 10 of the impact of increasing costs of natural gas feedstock on nitrogen-fertilizer manufacturing costs. As a consequence, there is considerable interest in legumes as a possible alternative source of portions of the nitrogen requirements of non-legume crops. However, it is obvious on the basis of 1.25 lb of nitrogen needed per bushel of corn and a nitrogen cost of $1.00 per pound that profits could still be realized even if the nitrogen fertilizer input cost amounted to $1.25 per bushel.

In certain areas, particularly in some tropical countries, commercial nitrogen may not be available or the grower has no money to pay for it. Therefore, a well-planned cropping system which includes legumes is essential to help to supply the nitrogen needed for the growth of nonlegumes. The main drawback, however, is often the lack of adapted legumes.

When legumes are used for forage in a livestock system of farming, the problem is different. The legumes serve the dual purpose of feed for livestock and provide some of the nitrogen for the grain crops. In this system legumes are essential to furnish at least part of the feed. The alternative is to grow only grass and to make heavy applications of nitrogen. Feeding trials generally demonstrate the superiority of legumes over nitrogen-fertilized grass. Legumes are generally of superior quality, including higher protein, higher mineral concentrations, and beneficial biochemical differences.

Adequacy of Legume Nitrogen for Corn. It is generally recognized that in corn production fertilizer nitrogen must supplement the amount fixed by legumes. Good growth of a legume such as alfalfa will furnish sufficient nitrogen for average but not top yields of corn. Results from Ohio have shown corn after alfalfa to require the following quantities of additional nitrogen; about 50 lb/A the first year, 100 lb/A the second year, 150 lb/A the third year, and 200 lb/A the fourth year.

The quantity of nitrogen that will be produced is often uncertain. Legumes have been relied on to furnish nitrogen for crops in rotation but have fallen short of supplying adequate quantities. This may have been caused by stand failure of the legumes, improper inoculation, or inadequate fertility. When interpreting experimental results, it is sometimes difficult to determine whether plant nutrients have been present in sufficient quantities for adequate growth of the legume. A good legume plowed down may supply 100 lb/A of nitrogen. However, the farmer often overestimates the quality of the legume sod and only half or less of this amount may be furnished. Another factor to bear in mind is the effect of time of legume breaking on release of available nitrogen from the decomposing residue. Early breaking will allow more time for decomposition and accumulation of nitrogen.

Two-year average results from three locations in Iowa showed yields of corn of 83 and 91 bu after Madrid sweet clover and Ladino clover, respectively. However, nitrogen alone at the rate of 51 and 95 lb gave yields of 90 and 98 bu, respectively.

Rotations Versus Continuous Cropping

Continuous cropping, or monoculture, is not new. There are examples from all over the world—rice in the Far East, wheat in the subhumid areas of the United States, cotton in the South. Although monoculture was once considered a sign of poor farming, the greatly increased supply of chemical nitrogen in the 1950s prompted much interest in continuous corn on soils on which erosion was not a serious problem. Information obtained indicated that the value of different rotations should be reexamined under conditions in which crop yields would not be restricted by an inadequate supply of plant nutrients. There is nothing inherent in the corn plant that makes it hard on the soil.

Until about 1950 continuous corn plots were used as examples to demonstrate the undesirability of the system. Most of these plots did not receive adequate fertilization, particularly nitrogen, and were compared with rotations that included legumes. Hence continuous corn showed up poorly. Since that time more logical comparisons have been set up. In general under high-yield conditions, results show that continuous corn has a yield 15% lower than the yield of corn in rotations. A 7-year average from the Morrow plots at the University of Illinois showed 138 bu/A with continuous corn and 161 bu/A in a corn–soybean rotation. In some periods of limited moisture, continuous corn may be superior to corn after alfalfa. Alfalfa takes up water deep in the profile and the following corn crop may be short of water.

Continuous corn does not mean that the whole farm is in corn. Rather, corn may be grown in the more adapted fields and sod cropping confined to the other areas. For example, if a farm has both level and hilly land, the corn needs of the farmer and the cropping needs of the soils may be met by growing corn on the level land and leaving the hilly land in sod.

Calculations may be made to compare the returns from a rotation with those of continuous corn. The grower may accept a somewhat lower yield with continuous corn but still be ahead financially.

A crop may have a harmful effect on the following crop, be it another crop or the same crop. There is some evidence that substances released from roots or formed during the decomposition of residues are responsible for the toxicity. The comparison of continuous corn versus a corn–soybean rotation is an example. Seeding alfalfa following alfalfa is often unsatisfactory, for unknown reasons. Allelopathy is the term used to describe the antagonistic action of one plant on another. More on the subject of allelopathy as related to weed competition was given in Chapter 2.

Control of Disease, Weeds, and Insects. Monoculture may lead to difficulties with certain diseases, weeds, or insects. In most cases introduction of unrelated or nonsusceptible crops and/or initiation of other cropping practices will help control these problems. To curb diseases such as root rots in wheat and other cereals, cropping sequences, together with resistant varieties, clean seed, and field sanitation practices, are necessary. New information on the suppressive action of chloride on dryland root rot and take-all root rot in wheat and other cereals was referred to in Chapter 3. Legumes, other dicotyledons, and even cereals such as oats, barley, or corn are often suitable alternate crops in place of wheat when take-all occurs. However, in some instances this disease will be severe, even in wheat following alfalfa, soybeans, and grass crops.

A number of diseases other than root rots of cereals may be controlled by a rotation system, particularly when other measures are taken, including seed treatment, proper cultivation, and sanitation. Corn root rots have been reduced by rotations, and the severity of several seedling diseases has been reduced by rotations combined with field sanitation. Susceptible crops should be grown on the same field only once in every 3 to 4 years.

Crop rotation is an important approach for control of nematodes feeding on the roots of annual crops. In the southern United States, grass crops are commonly used in rotation to control root knot nematodes. Acceptable yields of irrigated cotton were obtained in Arizona following 2 or more years of root knot–resistant alfalfa. Two years of clean fallow also effectively controlled root knot. Few important bacterial or viral diseases are controlled by crop rotation.

Weeds. The role of crop rotation for weed control depends on the particular weed and the ability to control it with available methods. If all the weeds can be conveniently controlled in any crop a farmer wishes to grow, then crop rotation is not a vital part of a weed control program. However, there are situations where rotations are necessary for control of a troublesome weed.

Insects. Rotation was once a common practice for insect management, but its use declined with the development of cheap and effective organic insecticides in the 1950s. Interest in this practice has been renewed because of insect resistance to the chemicals and increased cost of imputs. Rotation can be helpful where the insects have few generations a year or where more than one season is needed for the development of a generation. The most striking example of a serious insect problem in a major crop is the northern corn rootworm. In Illinois and Iowa, a rotation of soybeans and corn has replaced the need for automatic use of soil insecticides for keeping this insect in check. Replacement of lespedeza by soybeans in rotation with rice in Arkansas initially eliminated problems with grape colaspis. Rotation is only partially successful in the curbing of cotton bollworm. Sorghum, when planted at the proper time, will protect cotton from worms, and the worms seldom harm the sorghum.

Effect on Soil Tilth. It is believed by some that with today's soil management practices crop rotations are no longer needed to preserve the physical structure of most soils. Current crop production practices provide good plant cover and return large amounts of crop residues to the soil. In addition, there is less tillage and the resultant detrimental effects of compaction and deterioration of soil structure. The important issue is not one of monoculture versus rotation but, rather, involves two factors: the amount of residues returned to the soil, and the nature of the soil tillage necessary in the rotation.

Rotation can greatly improve the soil structure and tilth of many medium- and fine-textured soils. Pasture grasses and legumes in rotation exert significant beneficial effects on the soil. When soils previously in sod are plowed, they crumble easily and readily shear into a desirably mellow seedbed. Plow draft is often reduced in fine-textured soils when less intensively tilled crops are grown in the rotation. Internal drainage can be improved so that ponding and time for the soil to conduct away excess water are reduced.

Corn in monoculture is unique since on many soils it maintains reasonably acceptable soil physical conditions. The compensating factors are the return of several tons of crop residues to the soil when corn is harvested for grain and the corn crop is well adapted to reduced tillage and the decreased damage from traffic on the soil.

More information is required concerning the soil conditions under which deep-rooted legumes are needed in the rotation. Additional knowledge is also required on the conditions of the soil under which continuous production of a highly fertilized row crop such as corn will result in increased yields over the years. In this competitive period it is not sufficient just to maintain yields.

Double cropping, such as small grain–soybeans or small grain–corn, triple cropping, or even four crops of rice in areas with long growing seasons and possibilities of irrigation are being considered to a greater extent. With four crops a year, 27 tons/ha of rice is a real possibility. This necessitates utilizing soil, solar, and water resources to the maximum. If adequate fertility and pest control are provided and varieties are improved, soil productivity should gradually increase. Thus more attention will be directed toward measuring yield per unit area per year.

Advantages of Both Systems

Rotations

1. Deep-rooted legumes may be grown periodically over all fields.
2. There is more continuous vegetative cover with less erosion and water loss.
3. Tilth of the soil may be superior.
4. Crops vary in feeding range of roots and nutrient requirements: deep-rooted versus shallow-rooted, strong feeder versus weak feeder, and nitrogen fixer versus nonlegume.
5. Weed and insect control are favored, although chemicals are becoming increasingly effective.
6. Disease control is favored. Changing the crop residues fosters competition among soil organisms and may help reduce the pathogens.
7. Broader distribution of labor and diversification of income are effected.

Continuous Cropping or Monoculture

1. Profit may be greater.
2. A soil may be especially adapted to one crop; for example, corn, rice, or forage.
3. The climate may favor one crop; for example, corn is more suitable than oats in the Corn Belt, wheat in the Great Plains.
4. Machinery and building costs are probably lower.
5. The grower may prefer a single crop and become a specialist. Few people can become well enough informed to do an expert job of growing a large number of crops and also produce livestock. Monoculture demands greater skills, including pest, erosion, and fertilization control.
6. The grower may not wish to be fully occupied with farming year round.

Other Fertility Effects of Cropping Systems

Concentration of Nutrients in Surface. Crop plants vary considerably in their content of primary, secondary, and micronutrients. In addition, crops may absorb nutrients from different soil zones, thus making the choice of cropping sequences important to plant nutrition. Deep-rooted crops absorb certain nutrients from the subsoil. As their residues decompose in the surface soil, shallow-rooted crops may benefit from the remaining nutrients.

On a soil close to the borderline in its content of a particular micronutrient it is entirely possible that the preceding crop could have a considerable effect on the supply of this element to the current crop.

For example, rye on a Collington loam contained 1.1 ppm of molydenum, and rye and vetch contained 11.1 ppm (Table 14-8). It appears that as the rye and vetch residues decompose in the surface soil a considerably greater amount of molybdenum will become available than after rye alone. It is possible that this effect may be one of the important benefits of a cropping system or rotation, particularly for the elements that are not contained in fertilizers.

Forests in the tropics and subtropics are an excellent example of a crop that transports nutrients to the surface. The fallen leaves, twigs, and stems decompose, but the nutrients remain to form the basis for the shifting cultivation in which the inhabitants clean, burn, and then farm their fields for 2 to 5 years. After this period the nutrients in the surface are almost entirely exhausted and the land is allowed to go back to trees for rejuvenation.

Effect on Phosphorus and Potassium in the Soil. The net effect of cropping practices on phosphorus and potassium levels depends on the removal of nutrients by the harvested portion of the crop, nutrients supplied by the soil, and supplemental fertilization. The variations in nutrient content of crops as well as

TABLE 14-8. Micronutrient Composition of Certain Crop Plants

	Parts per Million					
Crop Plants	Boron	Molybdenum	Copper	Manganese	Zinc	Cobalt
Grown on						
Sassafras loam						
Bean tops	50	1.6	20	50	112	0.64
Carrot tops	30	0.9	18	120	163	0.38
Rye	15	1.5	12	48	93	0.30
Rye and vetch	90	3.4	17	90	263	0.72
Rye grass	15	2.9	19	80	123	0.76
Grown on						
Collington loam						
Bean tops	75	2.2	19	40	551	0.24
Carrot tops	56	0.8	18	160	460	0.28
Rye	30	1.1	15	32	456	0.20
Rye and vetch	100	11.1	16	80	465	1.32
Rye grass	30	2.4	20	80	175	0.30

Source: Bear, *SSSA Proc.,* **13**:380 (1948).

those resulting from rotation fertilization were discussed in Chapter 13. Flue-cured tobacco may be used as an illustration:

	Pounds per Acre		
	N	P_2O_5	K_2O
5-10-15, 1500 lb/A added	75	150	225
13-0-44, 200 lb/A added	26	0	88
Leaves removed, 3000 lb/A	85	15	155
Net gain (lb/A)	16	135	158

In contrast, peanuts receive little or no fertilization, and the nutrient-element removal by a good crop (2 tons of nuts and 2.5 tons of hay) is about 240 lb of N, 40 lb of P_2O_5, and 185 lb K_2O/A. Considerable differences between soils on which tobacco and peanuts are grown might be expected. These differences tend to be minimized, however, because crops in rotation with tobacco receive small quantities of fertilizer, whereas those in rotation with peanuts are heavily fertilized.

Periodic soil tests serve to point out changes in fertility levels and a knowledge of removals and additions helps to explain trends. Unfortunately, less than 10% of fields may be tested, but as pointed out in an earlier chapter, soil-test summaries are effective in revealing the influence of cropping systems on the status of soil fertility. The use of these summaries in research and extension programs should be helpful in formulating an overall approach to the problem of cropping, as related to fertility requirements and maintenance.

Effect of Rotation on Soil and Water Losses

The erosion equation is $A = RLSKCP$. The soil loss per unit area-tons per acre annually (A) is the product of factors for rainfall (R), slope length (L), slope steepness (S), soil erodibility (K), cropping and management (C), and conservation practices (P).

In this discussion primary attention will be directed to cropping and management of which soil fertility is an important part. The higher the crop yield, the less the runoff. Also, the greater proportion of water entering the soil, the lower the runoff.

Erosion is a symptom, not a primary cause of soil destruction. The primary causes are impoverishment of nutrients, especially nitrogen, and inadequate plant population.

Some of the characteristics of cropping systems and/or fertility management related to soil losses are the following:

1. The denseness of cover or canopy. This affects the amount of protection from impact of rain and evaporation. The amount of transpiration of water, thus making room for more water, is another factor. Residues and stems reduce velocity of water and evaporation. Residues, when turned back to the soil, make it more permeable to water. Research has revealed the tremendous kinetic energy of the raindrop as it hits the soil with a bomblike blast and has helped to bring about a better understanding of the vital importance of plant cover in intercepting this force.
2. The proportion of time that the soil is in a cultivated crop versus the amount of time in a close-growing crop such as small grains or forage.
3. The time that the crop grows in relation to the distribution and intensity of rainfall. The period May to September would be most vulnerable.
4. The type and amount of root system.
5. The amount of residues returned. Points 4 and 5 will affect the soil structure.

The effect of the yield level of crops on soil loss has been well established by USDA researchers Wischmeir and Smith. They developed a series of curves showing the moderating influence of mulch and plant canopy on soil loss caused by rain drop impact.

Results from Missouri indicate that fertilizer applied to small grains reduced erosion:

Crop	Fertilizer	Soil Loss (tons/A)
Wheat after soybeans	No fertilizer	4.3
Wheat after soybeans	0-18-17	2.7
Oats after corn	No fertilizer	3.7
Oats after corn	0-18-33	1.8

These effects are largely the result of a denser cover and a more extensive root system of the growing crop when fertilizer was applied.

The effect of improved practices, including approximately tripling the rate of fertilizer and manure and increasing soil pH from 5.4 to 6.8 in a corn–wheat–forage–forage rotation in a small watershed in the hilly area of eastern Ohio, is shown in Table 14-9. Yields were increased about 50% and runoff and erosion

TABLE 14-9.   Effect of Management Level on Crop Yields, Runoff, and Erosion (1945–1968)

	Prevailing Practices	Improved Practices
Corn (metric ton/ha)	5.1	7.3
Wheat (metric ton/ha)	1.5	2.3
Hay (metric ton/ha)	4.3	7.8
Runoff, growing season (cm)	1.9	1.0
Peak runoff rate (cm/hr)	2.3	1.5
Erosion from corn (metric ton/ha/yr)	10.6	3.1

Source: Edwards et al., *SSSA Proc.,* **37**:927 (1973).

drastically reduced. Management practice did not cause strong differences among soil characteristics. The effects on runoff and erosion were due in part to better surface protection through a quicker cover in the spring, a denser cover throughout the season, and a more extensive root system of the growing crop. Contouring was used rather than straight rows across the slope.

Winter Cover Crops

Winter cover crops are planted in the fall and plowed down in the spring. These may be a nonlegume, a legume, or a combination grown together. There are several advantages to the latter practice. A greater amount of organic matter is produced, the nonlegume can benefit from the nitrogen fixation, and because the nonlegume is usually more easily established, a stand of at least one crop is assured.

Nitrogen Added. One important reason for using green-manure legume crops is that they supply additional nitrogen. The amount of nitrogen supplied depends on the amount of root and top growth turned under, so a large amount of plant growth, of course, is desirable.

When a nonlegume is turned under, only the nitrogen from the soil or that supplied in fertilizer is returned. Although there are certain special cases in which a nonleguminous crop is needed, for example, before flue-cured tobacco, a legume is most often preferred in order that the nitrogen supplied may help to justify the practice.

The effects of vetch and rye cover crops with varying levels of nitrogen on corn silage production were studied in New Jersey. The highest corn silage yield of 25.5 tons/A was obtained following rye that had received 180lb/A of nitrogen. It was only slightly lower at 24.3 tons/A following unfertilized vetch. The latter gave the highest net return per acre and the lowest production cost per ton.

Organic Matter Added. One of the benefits attributed to winter cover crops is the organic matter supplied to the soil. Organic matter is certainly a factor affecting the tilth of very fine sandy loams and fine-textured soils. Green manures will help maintain the soil organic matter or will sometimes even increase it. In North Carolina organic matter was increased by vetch cover crops plus high nitrogen on the corn on soils low in organic matter and was maintained on soils with relatively high contents of organic matter. Water infiltration was improved.

In rotations in which the crops return little residue, maintenance of soil productivity may be particularly difficult. The lengthening of the rotation to include crops such as well-fertilized small grains, corn, and green-manure crops could be beneficial. This has been found to be a necessity in vegetable-growing areas. The acreage of corn and sorghum silage is increasing in some areas. This leaves the soil with almost no surface residues. Oats or rye seeded immediately after harvest or seeded by airplane before harvest will help to protect the soil.

Supply of Nutrients. The phosphorus in green manure may be even more effective than fertilizer phosphorus, probably because of its gradual release during decomposition, localized placement, the presence of organic acids to maintain availability, and the formation of certain complex ions.

An interesting green-manure program was carried on by the Seabrook Farming Corporation of New Jersey. This concern, which grew large quantities of vegetables, found it necessary to obtain large quantities of organic matter from cover crops in order to produce maximum vegetable yields. Most vegetable crops return only small amounts of plant residue to the soil. Small grains such as barley or small grains plus crimson clover were thus effectively used. The cover crops were planted on time, and yields of 3 to 5 tons of dry matter per acre were plowed down. With the extremely high fertilization remaining from previous vegetable crops, the cover crops grew well and produced high quantities of organic matter containing as much as 100 lb of nitrogen, 30 lb of phosphorus, and 250 lb of potassium per acre.

Nitrogen was added at turning to ensure rapid decomposition of the organic matter, and more fertilizer was applied directly to the vegetable crops. The larger share of nutrients, however, came from the decomposing organic matter. This means that less fertilizer was needed for the vegetable crop and that there was less chance of fertilizer injury. In addition, the nutrients were distributed through the plow layer.

Protection of the Soil Against Erosion. Protection against erosion has been suggested as one of the most important reasons for winter cover crops. Benefits, however, should be related to the distribution of rain during the year. In the South the most intense rains occur in the summer months.

Critical studies of the problem indicate that the effect of cover crops on soil loss is generally rather small when winter cover crops are turned under in the usual manner in early spring. A mulch balk treatment in which rye residues were left on the surface between the rows was much more effective than plant residues turned under.

Surface residues from summer crops and weed growth, if left undisturbed, may provide more protection than cover crops seeded in the fall. The greater the percentage of the soil surface covered by mulch, the less the soil loss. Freshly prepared land is quite susceptible to erosion, and considerable time is required before the cover crop can provide enough protection to have much effect on curbing soil loss. For example, rye cover has been grown after corn but in comparison to heavy corn residues, it has not generally given much additional benefit.

Methods of handling cornstalks in the fall have a marked effect on soil loss. Shredding reduced soil losses to about half that obtained from cornstalks as left by corn harvesters (Figure 14-5.) The shredded cornstalks provide a cushion that

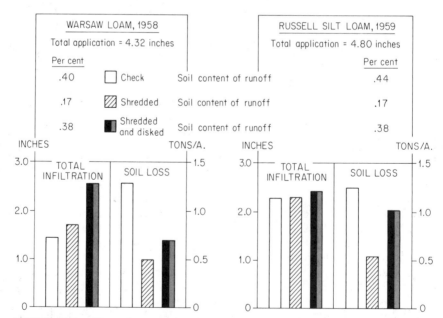

FIGURE 14-5. Effect of cornstalk residue management on infiltration and soil loss under simulated rainfall with an intensity of 2.4 in./hr. Warsaw, 4 to 4.5% slope; Russell, 3 to 3.5% slope. Check is cornstalks left by harvester. [Mannering et al., *SSSA Proc.,* **25**:506 (1961).]

TABLE 14-10.   Effect of the Method of Incorporation on the Loss of Nutrients (lb/A) in Runoff Water from a Silt Loam Soil Subjected to 5 Inches of Artificial Rainfall *

| | *Previous Crop* | | | |
| | *Fallow* | | *Bluegrass Sod* | |
Method of Incorporation	*Soluble P*	*Soluble* $(NH_4 + NO_3) - N$	*Soluble P*	*Soluble* $(NH_4 + NO_3) - N$
Check—no fertilizer	0.041	1.21	0.002	0.41
Plowed down	0.056	10.36	0.002	0.69
Disced in	0.281	28.59	0.011	0.81
Surface applied	0.627	26.11	0.071	2.96

* Fallow soil treated with 150 lb. ammonium nitrate-N and 50 lb. superphosphate-P per acre. Area previously in sod was treated with 200 lb. of N and 50 lb. of P per acre in liquid suspension form.
Source: Nelson, reprinted from *Fert. Solut.,* **17**:10 (May–June 1973).

protects the soil from the impact of raindrops. Disking shredded cornstalks is undesirable from the standpoint of soil conservation. Therefore, when land is not fall-plowed, shredded cornstalks represent a good possibility for surface cover.

For perennial crops, such as peaches and apples planted on steep slopes, continuous cover is helpful in reducing erosion. Since the trees and the cover crops occupy the land simultaneously, care must be taken, particularly in young orchards, to prevent competition for water and nitrogen. In some of the muck soils suitable for vegetables a strip of small grain or a row of trees at intervals helps to reduce losses from wind erosion.

On fallow land plowing down the fertilizer gave the least loss of nitrogen and phosphorus (Table 14-10). This was on a silt loam with 6% slope and the rainfall was applied at the rate of 2.5 in. in 1 hour 2 weeks apart. Hence, with no cover, the opportunity for loss of nutrients is great if they are not plowed down on such a slope.

Residual Effects. Decomposition of green-manure crops is rapid, but the residual effects are well recognized. The smallest residual effects generally are expected in areas in which the mean annual temperature is high and the soil is sandy. However, data obtained in the South by using oats as an indicator crop show surprising results. On Norfolk sandy loam one year after the last vetch crop, the residual effect produced 50 bushels of oats. On Norfolk loamy sand the residual effect was 32 bushels 2 years after the last vetch crop.

In Norfolk sandy loam the residual effect of the high rate of nitrogen on corn was considerable. This effect presumably came from the stover as it decomposed as well as from residual nitrogen in the soil. In Norfolk sandy loam the residual effect of the high rate of nitrogen was much less than that from vetch.

Grazing. Small grain or other crops can be grazed in late fall and winter when the amount of growth and soil conditions permit. Adequate fertility, either residual or added, is a necessity and extra nitrogen is needed. Grazing supplies tangible evidence of profit from cover crops.

Animal Manure

Animal manures are by-products of the livestock industry. In any successful business full use must be made of by-products, and animal manure is no exception. Much greater attention is being given to effective disposal of animal manures for the following reasons.

1. Public Law 92-500 gives the U.S. Environmental Protection Agency authority to make all water bodies clean again. All animal manures must be confined to the farm or feedlot. Leachings from the manure are not to pollute the water table.
2. Many domestic livestock are raised or fattened in confinement and disposal of the waste is part of an efficient business. Finding a market is often a challenge.

Collection and Disposal. The original method was to collect the manure or manure plus bedding and spread it on the fields. Liquid waste systems have since been developed. The animal manure is washed with water into pits, oxidation, ditches, or lagoons.

Methods for handling and storing manure will affect its final content of plant nutrients such as nitrogen, phosphorus, and potassium. Nitrogen losses from various systems are given in Table 14-11. Phosphorus and potassium losses are only 5 to 15% under all but the open lot and lagoon waste systems. In an open lot about 50% of these nutrients is lost. In a lagoon much of the phosphorus settles out and is lost from the liquid applied on the land.

The three principal methods used for field application of manure are:

1. Spreading of solid material when weather, soil, and crop permits.
2. Injecting the slurry of water and manure into the soil or spraying it on the surface.
3. Injecting the slurry into a sprinkle irrigation system.

Nitrogen loss is greatly affected by the method of application (Table 14-12). Immediate incorporation will minimize volatilization of the nitrogen. The effectiveness of liquid swine manure injected 8 in. deep has been improved by maintaining the nitrogen in ammonium form through the addition of nitrification inhibitors. In large feeding operations in which the animals are confined to feedlots, manure dried and bagged for the speciality turf and garden trade is a valuable by-product.

Composition. The composition of animal manure varies according to type and age of animal, feed consumed, bedding used, and waste management system. The dry matter, nitrogen, phosphorus, and potassium levels in manure from various classes of livestock which has been processed with the most common storage methods are shown in Tables 14-13 through 14-15. As might be expected,

TABLE 14-11. Effect of the Method of Handling and Storing on Nitrogen Losses from Animal Manure

Handling and Storing Method	Nitrogen Loss * (%)
Solid systems	
Daily scrape and haul	25
Manure pack	35
Open lot	55
Deep pit (poultry)	20
Liquid systems	
Anaerobic pit	25
Oxidation ditch	60
Lagoon	80

* Based on composition of waste applied to the land vs. composition of freshly excreted waste, adjusted for dilution effects of the various systems
Source: Sutton et al., *Purdue Univ. 1D-101* (1975).

TABLE 14-12.   Effect of the Method of Application of Manure on Volatilization Losses of Nitrogen

Method of Application	Type of Waste	Nitrogen Loss * (%)
Broadcast without cultivation	Solid	21
	Liquid	27
Broadcast with cultivation †	Solid	5
	Liquid	5
Knifing	Liquid	5
Irrigation	Liquid	30

* Percent of total nitrogen in waste applied which was lost within 4 days after application.
† Cultivation immediately after application.
Source: Sutton et al., *Purdue Univ., 1D-101* (1975).

dry matter is highest in the solid waste system. Nitrogen is highest when manure is handled as a liquid, and phosphorus and potassium are generally also higher in this system.

Manure in a Complete Fertility System. Manure is still a mainstay in the fertility program on many livestock farms. However, unless an unusually large quantity of legumes is produced on the farm and considerable commercial feed is consumed, thereby capitalizing on fertility from elsewhere, a livestock program in itself will tend to gradually deplete soil fertility.

Many comparisons have been made of the effects of manure on crop production with those obtained from the application of equivalent amounts of nitrogen, phosphorus, and potassium in commercial fertilizers. On a fine sandy loam at Rothamsted in England, commercial fertilizers used for over 100 years have been just as effective as manure for continuous wheat production. However, manure is often superior. In Colorado, 27 tons/A of manure increased corn yields an average

TABLE 14-13.   Approximate Dry Matter and Fertilizer Nutrient Composition and Value of Various Types of Animal Manure at Time Applied to the Land—Solid Handling Systems

Type of Livestock	Waste Handling System	Dry Matter (%)	Nutrient (lb/ton raw waste)				Value per Ton **
			Nitrogen		P_2O_5	K_2O	
			Available *	Total †			
Swine	Without bedding	18	6	10	9	8	$ 5.10
	With bedding	18	5	8	7	7	4.14
Beef cattle	Without bedding	15	4	11	7	10	4.26
	With bedding	50	8	21	18	26	10.44
Dairy cattle	Without bedding	18	4	9	4	10	3.36
	With bedding	21	5	9	4	10	3.60
Poultry	Without litter	45	26	33	48	34	24.72
	With litter	75	36	56	45	34	26.22
	Deep pit (compost)	76	44	68	64	45	35.16

* Primarily ammonium N, which is available to the plant during the growing season.
† Ammonium N plus organic N, which is slow releasing.
** Value per pound is 24 cents for available N, 30 cents for P_2O_5, and 12 cents for K_2O, as calculated by C. D. Spies, 1983.
Source: Sutton et al., *Purdue Univ. 1D-101* (1975).

TABLE 14-14. Approximate Dry Matter and Fertilizer Nutrient Composition and Value of Various Types of Animal Manure at Time Applied to the Land— Liquid Handling Systems *

| Type of Livestock | Waste Handling System | Dry Matter (%) | Nutrient (lb/ton raw waste) | | | | Value per 1000 Gallons ‡ |
| | | | Nitrogen | | P_2O_5 | K_2O | |
			Available †	Total **			
Swine	Liquid pit	4	20	36	27	19	$15.18
	Oxidation ditch	2.5	12	24	27	19	13.26
	Lagoon	1	3	4	2	0.4	1.80
Beef cattle	Liquid pit	11	24	40	27	34	17.94
	Oxidation ditch	3	16	28	18	29	12.72
	Lagoon	1	2	4	9	5	3.78
Dairy cattle	Liquid pit	8	12	24	18	29	11.76
	Lagoon	1	2.5	4	4	5	2.40
Poultry	Liquid pit	13	64	80	36	96	37.68

* Application conversion factors: 1000 gal = about 4 tons; 27,154 gal = 1 acre-in.
† Primarily ammonium N, which is available to the plant during the growing season.
** Ammonium N plus organic N, which is slow releasing.
‡ Value per pound is 24 cents for available N, 30 cents for P_2O_5, and 12 cents for K_2O, as calculated by C. D. Spies, 1983.
Source: Sutton et al., *Purdue Univ. 1D-101* (1975).

TABLE 14-15. Nutrient Value of Manure per Animal Unit (lb/1000 lb liveweight) per Year

| Handling and Disposal Method | Swine | | | Beef | | | Dairy | | | Broilers | | |
	N	P_2O_5	K_2O	N	P_2O_5	K_2O	N	P_2O_5	K_2O	N	P_2O_5	K_2O
Manure pack												
Broadcast	84	107	124	63	77	99	77	50	112	215	200	149
Broadcast and cultivation	102	107	124	77	77	99	91	50	112	263	200	149
Open lot												
Broadcast	58	61	80	44	45	64	51	30	59	—	—	—
Broadcast and cultivation	70	61	80	53	45	64	61	30	59	—	—	—
Manure pit												
Broadcast	95	111	119	69	82	95	87	54	107	—	—	—
Knifing	124	111	119	94	82	95	114	54	107	—	—	—
Irrigation	92	111	119	65	82	95	84	54	107	—	—	—
Lagoon												
Irrigation	24	25	89	18	18	71	23	14	80	—	—	—

Source: Sutton et al., *Purdue Univ. 1D-101* (1975).

of 20 bu/A over those obtained with either 220, 360, or 460 lb/A of nitrogen per year supplied by fertilizers.*

The favorable effect of manure on increasing the available moisture content of soils is shown in Table 14-16. Another interesting finding in this study was the capacity of manure to moderate soil acidification resulting from repeated applications of N-P-K fertilizer.

* S. R. Olsen, USDA, Ft. Collins, Colorado, personal communication.

TABLE 14-16. Laboratory Measurements of Field Capacity, Wilting Point, and Available Water of the Topsoil from the Nil and Barnyard Manure Treatments (Average of Six Years 1966–1971)

Soil	No Manure	Barnyard Manure
	Field Capacity (% moisture)	
Beryl FSL	19.4	22.6
Hazelmere L	21.6	24.3
Nampa CL	23.8	27.5
	Wilting Point (% moisture)	
Beryl FSL	4.8	5.8
Hazelmere L	7.4	8.8
Nampa CL	9.2	10.0
	Available Water (% moisture)	
Beryl FSL	14.6	16.7
Hazelmere L	14.1	15.6
Nampa CL	14.6	17.5

Source: Hoyt and Rice, *Can. J. Soil Sci.,* **57**:425 (1977).

Distribution of Manure by Grazing Animals. One aspect of the animal manure problem which has received little attention has been the question of distribution of manure by grazing animals. This presents a problem in the maintenance fertilization of pastures. A study was conducted in North Carolina over a period of years in which the pattern of distribution of manure was determined. One beef animal unit per acre was grazed, and the distribution of the manure over a given area in relation to time was observed. It was found that for elements such as nitrogen, which do not remain in effective concentrations for more than a year, about 10% of the grazing area would be effectively covered in one year. On the other hand, with elements such as phosphorus which are not leached or removed in large quantities, some effect might be obtained from a given application as much as 10 years later. The data showed that nearly all of the pasture area would have received deposits of manure in a 10-year period. Potassium is intermediate between nitrogen and phosphorus in retention in the soil, and manure-deposited potassium would be effective to some degree for at least 5 years. During this period about 60% of the area would have been covered.

Studies in Indiana revealed that with low stocking rates, animal excreta will essentially have no effect on soil fertility. On highly productive pastures with a high carrying capacity, excreta could have a beneficial effect on soil fertility over a period of time. Grain feeding on pastures had considerable effect on soil fertility and each increase of 4.5 tons of grain fed per acre resulted in an increase of 53 steer-days of grazing.

Benefits of Manure. Reasons for the favorable action of manure are unclear, but they probably include one or more of the following.

1. An additional supply of NH_4-N.
2. Greater movement and availability of phosphorus and micronutrients due to complexation, and so on.
3. Increased moisture retention.
4. Improved soil structure with corresponding increases in infiltration rate and decreases in soil bulk density.
5. Higher levels of carbon dioxide in the plant canopy, particularly in dense stands with restricted air circulation.

6. Increased buffering capacity against drastic changes in pH.
7. Complexation of Al^{3+} thereby reducing its toxicity.

Sewage

Disposal of processed sewage materials from treatment plants is of increasing concern because of population pressures, more stringent laws, and increases in energy costs. There are about 18,000 municipal sewage treatment plants in the United States. About 75% of the sewage handled by these plants is of human origin and the remaining 25% is from industrial sources.

The end products of all sewage treatment processes are sewage sludge and sewage effluent. The former is a general term used to describe the solids produced during the treatment of sewage. Sewage effluent is essentially clear water which contains low concentrations of plant nutrients and traces of organic matter. It may be chlorinated and may also be discharged into a stream or lake. Figure 14-6 illustrates the major stages and steps in wastewater treatment.

Sewage sludge is disposed of in a number of ways, including (1) recycled on cropland by approved methods; (2) incinerated with loss of organic matter and nitrogen and fuel is consumed; and (3) buried in land fill sites, where it will produce methane for many years.

Treatment. Sewage plants are designed to remove solids and organic matter from the sewage waste, the primary or raw sludge. This sludge is then decomposed by digestion in either anaerobic or aerobic environments. The liquid effluent separated from the raw sludge may be sprayed over rocks or gravel beds or aerated mechanically. Further treatment may include chlorination or adsorption on activated carbon.

Sludge Composition. Sludge is a heterogeneous material, varying in composition from one city to another and even from one day to the next in the same city. Before developing plans for land application of sludge, it is essential to obtain representative samples of the sludge over a period of time and determine its typical chemical analysis.

Table 14-17 summarizes an analysis reported by the University of Illinois to be representative of fresh, treated, and aerobically digested sewage sludge.

Application of Sludge to Agricultural Land. Use of sludge on agricultural land has some benefits; at the same time, there can be problems. Both aspects are reviewed briefly next.

Benefits. Sludge is a source of nitrogen, phosphorus, potassium, and micronutrients, including boron, copper, iron, manganese, and zinc. Additionally, it is a source of organic matter. Furthermore, this disposal procedure is a feasible, cost-effective alternative to more costly methods, such as burning or burying.

Problems. It is essential that appropriate application and soil management techniques be used to protect the environment and the health of human beings and animals. Because of the possibility of applying excessive nitrogen and subsequent movement of NO_3-N into surface water and groundwater, careful monitoring is necessary.

Care must be taken to avoid cadmium contamination of crops. Although there is no immediate human health problem, the U.S. Environmental Protection Agency has established recommendations concerning the maximum amount of cadmium that can be applied.

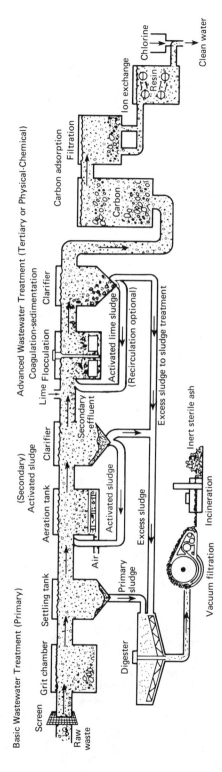

FIGURE 14-6. Stages in wastewater treatment. (Manufacturing Chemists Association, *Water Pollution Causes and Cures*. Washington, D.C., 1972.)

TABLE 14-17. Typical Analysis of Sewage Sludge

Component	Concentration on Dry Weight Basis
	percent
Organic carbon	50
N	
Ammonium	2
Organic	3
Total	5
P_2O_5	6.8
K_2O	0.5
Ca	3
Mg	1
S	0.9
	ppm
Fe	40,000
Zn	5,000
Cu	1,000
Mn	500
B	100
Cd	150
Pb	1,000
Ni	400

Source: "Utilization of sewage sludge on agricultural land." *Univ. Illinois Soil Manag. Conserv. Ser. SM-29* (1975).

There is a possibility of disease transmission due to the presence of bacteria, parasites, or viruses. Treatment of sludge by aerobic and anaerobic digestion, air drying, composting, and lime stabilization all help to reduce the pathogen content.

Contamination of crops by stable organic compounds such as polychlorinated biphenyls (PCBs) can be a problem. This can be minimized by incorporation and it is mandatory that sludges containing greater than 10 ppm PCBs be so treated.

Other difficulties include objectionable odors and impaired plant growth due to the antagonistic action of heavy metals such as copper, lead, nickel, and zinc. The former is eliminated by soil injection.

Application Rates. Soil testing and fertilizer recommendations are used in conjunction with sewage sludge characteristics to arrive at official application rates. The annual rate of sludge is based on the lowest tonnage that will satisfy nitrogen requirements of the crop or on the maximum quantity that can be used without exceeding permissible limits for cadmium.

Nitrogen. It was shown in Table 14-17 that sludges contain both inorganic and organic forms of nitrogen. Because of possible volatilization losses of up to 20 to 50% of the ammonium from surface application of sludge and only partial release, of the order of 20 to 30%, of nitrogen from the organic fraction, exact rates of nitrogen application are difficult to determine. However, if the sludge is incorporated into the soil very little ammonium will be lost. After the first year, when 20% of the organic nitrogen is mineralized, about 3% of the remaining nitrogen will be released annually during the next few years.

Nitrogen and phosphorus additions to soil from a single application of sludge can in some areas be as high as:

1. 800 lb/A (900 kg/ha) of total nitrogen.

2. 400 lb/A (450 kg/ha) of ammonium nitrogen.

3. 1025 lb/A (1150 kg/ha) of P_2O_5.

Use of fertilizer nitrogen and phosphorus on sludge-applied lands is generally not required for at least 2 years following sludge application. However, the possibility of having to correct imbalances of other nutrients such as potassium and sulfur should not be overlooked.

Cadmium. When cadmium is the main concern, sludge should be applied only to soils of pH 6.5 or above. For tobacco, leafy vegetables, and root crops, cadmium rates must be less than 0.44 lb/A. The maximum allowable for other crops is being reduced from 1.8 lb/A, and it must not exceed 0.44 lb/A after January 1, 1987.

The life of a sludge application site is determined by the cumulative addition of metals. Guidelines developed by universities, the USDA, and the EPA for buildup of cadmium and other heavy metals in soils are summarized in Table 14-18. Use of cation exchange capacity (CEC) as a controlling soil factor does not necessarily mean that all of these metals are retained on the exchange complex. Rather, CEC was chosen as a single soil property that can be easily measured and one that is positively related to soil components that may minimize plant availability of metals added to soils in sludges. On submarginal soils of little agricultural value, higher levels of cadmium may be permitted.

Purdue University has developed a worksheet for calculating application rates of sewage sludge on cropland. The following information is used in the worksheet.

1. Sludge composition data.
   a. Total nitrogen (N).
   b. Ammonium nitrogen (NH_4^+N).
   c. Nitrate nitrogen ($NO_3^- - N$).
   d. Phosphorus (P).
   e. Potassium (K).
   f. Lead (Pb).
   g. Zinc (Zn).
   h. Copper (Cu).
   i. Nickel (Ni).
   j. Cadmium (Cd).
   k. PCBs (if 10 ppm or greater, sludge must be incorporated into soil).

TABLE 14-18. Maximum Amount of Metal Suggested for Agricultural Soils Treated with Sewage Sludge

	Maximum Amount of Metal (lb/A) When Soil Cation Exchange Capacity (meq/100 g) Is:		
Metal	*< 5*	*5–15*	*> 15*
Lead	440	880	1760
Zinc	220	440	880
Copper	110	220	440
Nickel	110	220	440
Cadmium	4.4	8.8	17.6

Source: Sommers et al., *Purdue Univ. AY-240* (1980).

2. Soil and crop data.
   a. Available P and K.
   b. Fertilizer N, P, and K recommendations for crop grown.
   c. Soil pH and lime requirement to adjust soil to pH 6.5.
   d. Soil cation exchange capacity (CEC).

The maximum approved rates for single applications of several sewage sludge types on various classes of disposal sites in Alberta are listed in Table 14-19. These quantities range from 2.5 to 25 tons/ha or 1.1 to 11.2 tons/A.

Crop Response. Studies on crop responses to sludge have been carried out for over 50 years in states such as Wisconsin. In the past 25 years there has been renewed interest and much work in the future is anticipated because of the possible environmental impact.

Some regulatory agencies restrict the application of sludges to crops such as forages, oilseed crops, small grains, commercial sod, and trees. Unacceptable crops may include root crops, vegetables and fruit, tobacco, and dairy pastures. Direct grazing of sludge-treated forage lands is not usually recommended for a period of 3 years immediately following application. Wheat is preferable to barley. Oats are not recommended in the first two growing seasons following sludge treatment.

Response of crops to sewage sludge is at least equal to commercial fertilizer in the first year after application and may be somewhat greater in subsequent years because of residual effects of the added plant nutrients. Also, many of the favorable effects of the organic matter in sludges, probably very similar to the ones associated with manure, may be long lasting. Additionally, it may take one or more years for the applied nutrients to become effectively distributed in the root zone.

Sewage Effluent. This liquid may be a valuable water and nutrient resource for crops or a pollutant to land and waters. Studies have been under way for the past 25 years to study the effects on plant growth as related to soil characteristics.

Large quantities of water are involved since it is largely a disposal problem. Hence it is essential that the soil be (1) internally well drained and medium textured, having a pH of between 6.5 to 8.2, and (2) be supporting a dense stand of trees, shrubs, or grasses. The groundwater should be monitored periodically for nitrates.

Wastewater irrigation studies have been carried out in the United States and Europe on crops such as small grains, corn, soybeans, and potatoes. Use of effluent on trees has also been investigated in the United States and elsewhere.

Forage crops have been chosen in Canada because of their long growing season with resultant high seasonal evapotranspiration, their high nutrient uptake, and their capacity to stabilize the soil and prevent erosion. Because forages are not eaten directly by human beings, the transfer of human diseases is unlikely.

Yields of five forage crops irrigated with sewage effluent are shown in Table 14-20. Alfalfa was the most suitable forage crop when the system was operated for optimum utilization of the wastewater. The water did not supply sufficient nitrogen for a high production of the grasses.

The previous discussion indicates that there is much to be learned about the use of sewage sludge or effluent on agricultural crops. Continuing studies are called for. Society will benefit from wise application of this material and thus recycling of a valuable resource. A zero-risk philosophy is not feasible in a complicated environment.

TABLE 14-19. Maximum Addition of Sludge Solids and Sludge-Borne Nitrogen for Each Site Class and Sludge Type (Single Application)

Sludge Type	Solids Application Rate * (tons/ha dry weight basis)			Total Nitrogen Application Rate (kg/ha)			Available Nitrogen [(NH$_4$ + NO$_3$) – N] Application Rate for Surface Spreading † (kg/ha)		
	Class 1 Sites	Class 2 Sites	Class 3 Sites	Class 1 Sites	Class 2 Sites	Class 3 Sites	Class 1 Sites	Class 2 Sites	Class 3 Sites
Digested	25	20	10	900	700	400	450	350	200
Waste stabilization pond	10	8	5	800	600	300	400	300	150
Undigested **	5	4	2.5	600	500	200	300	250	100

* For surface application a maximum hydraulic loading rate of 100 m³/ha per day is imposed for sludges containing less than 5% solids. Allowable solids and nitrogen rates for such sludges may be achieved by making several incremental additions with the soil cultivated between each addition. There is no hydraulic loading rate restriction on subsurface injection.

† For subsurface injection the maximum available nitrogen application rate is 200 kg/ha on class 1 and 2 sites and 150 kg/ha on class 3 sites.

** Undigested sludge must have been stored for a minimum of 6 months, have a pH of greater than 12, or have a volatile solids content of less than 60% of the total weight of solids.

Source: McCoy et al., "Guidelines for the application of municipal wastewater sludges to agricultural lands." Alberta Environment (1982).

TABLE 14-20. Four-Year Annual Dry Matter Yield of Five Forage Species as a Function of Wastewater and Fertilizer Levels

Wastewater Irrigation (cm/yr)	Fertilizer N-P (kg/ha/yr)	Forage (metric tons/ha)					
		Alfalfa	Reed Canary	Brome	Altai Wildrye	Tall Wheat	Mean *
62.5	0–0	8.69	5.81	5.98	5.84	5.29	6.32d
	56–48	9.44	8.82	8.92	8.38	7.67	8.65c
125	0–0	10.20	9.35	9.95	9.59	7.91	9.40b
	56–48	10.05	11.48	11.11	11.73	9.82	10.84a
Mean *		9.59a	8.86b	8.99b	8.89b	7.67c	8.80

* Means in a row or column followed by the same letter do not differ significantly at the 5% level by Duncan's multiple range test.
Source: Bole and Bell, *J. Environ. Qual.*, **7**(2):222 (1978).

Tillage

Traditionally, soils in many areas of North America have been prepared for annual crops by moldboard plowing either in the fall or the spring, depending on the soil and the climate. Essentially all crop residues are turned under. This is followed with several secondary tillage operations, such as disking, harrowing, and field cultivating.

Interest has developed over the years in developing tillage practices to give greater protection to the soil against soil and water losses. The amount of surface residues and surface roughness both have an effect. Crop residue management has been developed to leave more of the harvest residues, leaves, and roots on or near the surface. Stubble mulching has been studied in the Great Plains with particular reference to wheat and other small grains since the 1930s and is used widely.

Conservation Tillage. This is a term used to describe any tillage system that reduces soil and/or water loss compared to unridged or clean tillage.

Advantages:

1. Higher crop yields, except possibly in level, fine-textured, poorly drained soils.
2. Less soil erosion by water and wind.
3. Improved infiltration and more efficient use of water.
4. Acreage of land that can safely be used for row crops is increased because crops can be grown on more sloping land.
5. Timing of planting and harvesting can be improved.
6. Lower machinery and fuel costs.

Disadvantages:

1. More pests, such as insects, diseases, weeds, and rodents.
2. Cooler soil temperatures in spring, resulting in slower germination and more stand problems. Ridge till planting helps.
3. Greater management ability is required.

General Types of Conservation Tillage

Chiseling. A chisel-type implement may till 8 to 15 in. deep with points 12 to 15 in. apart. A considerable amount of surface residues is left on the surface and the surface is rough.

Stubble Mulching. There are two types of tillage equipment used—those such as disks, chisel plows, and field cultivators which mix crop residues with the soil, and those such as sweeps or blades and rodweeders which cut beneath the surfaces without inverting the soil. Maximum moisture efficiency and reduced wind erosion are primary goals of stubble mulching.

No-till. A seed zone 2 in. wide or less is prepared in previously untilled ground. This is made by a fluted coulter running ahead of a planter unit with disk openers. Seeding by this method is successful in corn or soybean residues and in small grain stubble. A higher percentage of the residues are left on the surface. No-till has been most successful in the better drained soils.

Narrow hoe openers or narrow-angle double-disk openers are also used for minimum and zero-tillage seed placement. No-till drills may have pressures of up to 400 psi below the openers, while in conventional drills pressures of only 70 to 75 psi are developed.

Air seeders are now being used for no-till planting in some areas of the United States and Canada, especially in the dry western regions of both countries. Use of air to distribute seed to furrow openers was tried in Germany in the late 1950s and early 1960s. The concept was further developed in Australia in the 1970s for seeding cereals on large dryland farms. By the end of the 1970s air seeders represented 65% of the Australian seeder market.

A schematic diagram of a typical air-seeding unit appears in Figure 14-7. In operation, an air seeder is quite simple. Seed from the grain tank is metered into an airflow which carries seed to the main manifold. From this primary manifold, grain is distributed evenly to smaller manifolds and then to the plastic boot behind each cultivator sweep or chisel. One of the main advantages of air seeders is that they have only three moving parts.

Most air seeders can also be used for two types of fertilizer application. They can be used to apply fertilizer with the seed at planting time by metering fertilizer from one tank or compartment and grain from the other and applying both through

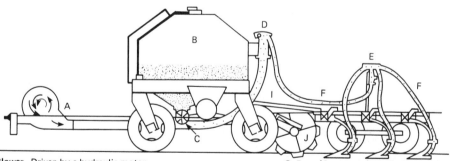

A. **Blower**—Driven by a hydraulic motor to be run from tractor hydraulic system or a separate hydraulic system.

B. **Grain Tank**—Hopper bottom. One fill-hole. Cleanout door.

C. **Metering Device**—Meters all types of grain accurately into the airstream.

D. **Primary Manifold**—Distributes grain accurately to secondary manifolds.

E. **Secondary Manifold**—Distributes grain accurately to each shank.

F. **Flexible Tubing**—Carries grain. This allows wings to be folded with no alterations to seeding system.

G. **Boot [Seed Tube]**—Plastic boot is located behind shank to protect it from rocks.

H. **Liquid Fertilizer Holder**—Holds liquid fertilizer tubes in place behind seed tube.

I. **Rolling Hitch**—Allows chisel plow to follow contour of ground for accurate depth control.

J. **Ground Drive Wheel**—Drives seed metering device and liquid fertilizer pump with no clutches. Wheel is lifted off ground when chisel plow is raised.

FIGURE 14-7. Sketch showing essential features of an air seeder. [United Grain Growers, *Grainews*, pp. 3, 8 (March 1980).]

the same seed boots. In addition, when equipped with chisel points or knives and alternate banding boots, they can also be used for fertilizer banding. Some units are equipped to band fluid fertilizers during the seeding operation.

Till Plant. This is a once-over tillage–planting operation. Planter units work on ridges made the previous year during cultivation or after harvest. The planter shoves the old stalk, root clumps, and clods into the area between the rows. This is perhaps most useful on fine-textured, poorly drained soils. Ridges are retained in the same position year after year, hence wheel tracks are in the same place.

Effects of Tillage. Effects vary greatly depending on soil, crop, and weather conditions. These effects are discussed in the following sections.

Residues on the Surface. The differences among the various systems are shown in Table 14-21. No-till leaves a higher proportion of residues on the surface.

Soil Loss. Some types of tillage increase soil losses (Figure 14–8), but no-till markedly reduces these losses.

Soil Temperature. Soil temperatures early in the growing season tend to be lower under conservation tillage than for conventional tillage, due to the

TABLE 14-21. Residue Cover Immediately After Planting, Corn After Corn

Tillage System	Percent Soil Cover (Average of Four Trials)
Conventional, fall plow	1.0
Till-plant	8.4
Disk twice	12.9
Chisel	19.0
Strip rotary	62.0
No-tillage	76.0

Source: Griffith et al., *J. Soil Water Conserv.*, **32**(1):20 (1977).

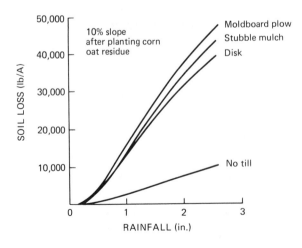

FIGURE 14-8. General relation existing between tillage practice and loss of soil by erosion. [Dickey et al., *Farm, Ranch Home Q.* Univ. of Nebraska **29**(1):18 (1982).]

insulating effect of the unincorporated crop residues on the surface. Soil temperatures in the top 6 in. of the root zone during May and June can be 4 to 5°C lower under zero till than under conventional tillage. Black and Siddoway found in Montana that daily mean soil temperature measured below the tillage depth (at 15.2 cm) was 2°C and 3°C warmer for the 7.6- and 12.5-cm tillage depths, respectively, than for no-till. This temperature difference persisted for nearly 50 days.

Decomposition of crop residues and soil organic matter with subsequent release of plant nutrients, including nitrogen, phosphorus, and sulfur, is restricted by low soil temperatures. Thus recycling of essential nutrients may be delayed. Further, low soil temperatures will retard root development and activity.

In tropical and semitropical regions the cooling effect of crop residues on soil temperatures is believed to be beneficial.

Water Relations. Method of tillage affects water loss. Increasing residues on the surface and greater surface roughness reduce runoff and soil erosion.

Water infiltration is greater with less tillage and/or more surface residues. Note the increased available soil water after no-till (Figure 14-9). Reduced tillage may result in faster movement of the water from the surface in wet springs if conventional tillage has caused considerable compaction. The bulk density of the soil after planting is usually higher with no-till than with conventional methods, however.

Chiseling to about 15 in., if soil conditions permit, will gradually create a deeper root zone with greater water intake and water-holding capacity. This may provide an extra inch or two of water at critical stages of growth.

Fertilization with Conservation Tillage. With moldboard plowing, nutrients broadcast on the surface are plowed down and mixed with the soil. Conservation tillage means less or no inversion of the soil. Hence, with broadcast applications, nutrients tend to stay on or near the surface. Limited success of conservation tillage may, in some cases, be related to low fertility in the root zone.

Phosphorus and Potassium. The tendency for phosphorus and potassium to accumulate under conservation tillage in the top few inches of soil in contrast to moldboard plowing was covered in Chapter 13. In a Minnesota study the potassium concentration in the ear leaf at silking was 2.2% with moldboard plowing and 1.9% with no-till. The desirability of building soil-test levels of phosphorus and potassium to medium or high before starting conservation tillage was also discussed previously. Under some economic conditions or in some terrain it may not be feasible.

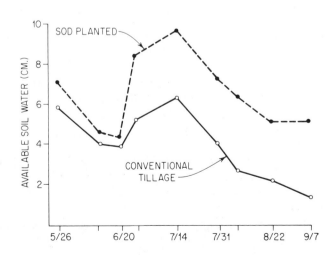

FIGURE 14-9. Available soil water in 0-to 60-cm profile as affected by tillage practice for corn with orchardgrass. [Bennett et al., *Agron. J.,* **65**:488 (1973).]

With chisel plowing more of the phosphorus and potassium is worked down into the soil over the years than with till plant or no-till. With no-till if the terrain and soil permit, periodic plowing every 4 or 5 years may be desirable for two reasons:

1. Distribution of fertility throughout the plow zone provides insurance against positional unavailability, especially in dry years, when root activity in the surface layer may be reduced.
2. Runoff and erosion of surface soils containing high amounts of plant nutrients are detrimental to the environment and a loss economically. As mentioned earlier, band applications at planting are often beneficial.

Results obtained for winter wheat production in eastern Washington indicate how fertilizer requirements may change following implementation of conservation tillage. Only nitrogen and sulfur are normally used on a regular basis for conventional cropping systems in this specific area. Conclusions drawn from this study are:

1. Maximum yields were obtained only when all four nutrients were applied.
2. Benefits of sulfur fertilization appeared to be even greater under conservation tillage than with conventional tillage.
3. Phosphorus fertilization becomes more important in no-till cropping.
4. Additions of potassium were beneficial, even though this nutrient is generally assumed to be in adequate supply in soils of this region.

Nitrogen. Research in several areas, including Kentucky and Maryland, indicates that higher rates of surface-applied nitrogen are required with no-till than with conventional tillage (e.g., with the moldboard plow).

There may be several reasons for this higher need with no-till.

1. Ammonia volatilization.
2. Fertilizer nitrogen is immobilized by microorganisms in the organic layer that accumulates on surface with no tillage.
3. More nitrogen is mineralized from the soil organic matter when plowed.
4. Higher yield potential with no-till in some areas.
5. More water movement through the soil with greater loss of nitrates.

Placement of nitrogen below the soil surface results in substantial increases in yield of corn compared to surface application.

Lime. With no-till the surface may become acid quite quickly because most of the nitrogen is applied to the soil surface. Soil samples from the 0- to- 2 in. layer are usually more acid than the lower depths. Hence it is important to keep close watch on the soil pH of the surface 2 in. Certain herbicides such as the triazines are inactivated by a low pH at the surface and weed control can break down. These points were discussed in detail in Chapter 11.

Factors Favoring No-Till. Two developments have contributed greatly to the feasibility of no-till cropping systems. They are:

1. More effective herbicides.
2. Planters that are effective in the presence of large quantities of surface residues.

Opportunities for Adoption of No-Till. This practice is destined to increase, and it is estimated that before the turn of the century, over 50% of cropland in the United States will be farmed by no-till systems. The attractive features of no-till are:

1. Row crop production on sloping lands is more practical. Less nutrients, soil, and water are lost.

2. Increased water use efficiency (WUE).
3. No-till corn production can readily follow soybeans.
4. Double cropping of row crops such as soybeans, sorghum, or corn planted immediately after wheat harvest is possible in many areas.
5. Seeding legumes and/or nonlegumes in rundown pastures can be accomplished by the same principle. Fertilizer may be placed beneath the seed or broadcast. Herbicides are used to kill or retard existing grasses and weeds. Again, this technique is useful on sloping lands that should not otherwise be tilled, as well as on areas that are more level.
6. Reductions in energy, time, and machinery costs.

Remote Sensing

Remote sensing is a term used to describe the collection and analysis of data on features of the earth's surface. The basic information is acquired through cameras and other imaging systems mounted in or on aircraft or space vehicles. Although remote sensing has progressed beyond the acquisition and analysis of aerial photographs, the field of aerial photography is still a basic component.

The scope of aerial photography is limited by the spectral sensitivities of photographic film emulsions, the spectral resolution of photographic filters, and the logistical problems associated with obtaining photographs via unmanned space vehicles. These limitations have been overcome through the development of various nonphotographic imaging systems installed on earth-orbiting satellites. Data acquired by means of the satellite imaging systems provide scientists with views of the earth not achievable with conventional aerial photography. Such satellite systems provide repetitive coverage of very large areas with a spatial resolution that allows each half hectare to be studied.

Through satellite imaging it is feasible to monitor land cover changes, both natural and those resulting from human activities, that would otherwise require thousands of aerial photographs if conventional methods were used. Interpretation of satellite images is still a relatively new science, and it will probably be some time before crop conditions can be assessed by simply studying a satellite photograph. However, satellite images are currently used for indicating the location and size of weather systems. Figure 14-10 is a satellite image of the Prairie Provinces of Canada, and part of the northern United States, showing the areas affected by a late summer frost.

Aerial photography and field surveys are still required for detailed studies of relatively small areas. In recent years, aerial photography has been used successfully as a research tool for studies of population densities in livestock and wildlife, forest inventories, movement of ocean currents, land-use patterns over large areas, physical and chemical characteristics of soils, plant diseases, insect infestations, crop vigor, and winter injury to perennial crops such as alfalfa and to fall-seeded small grains.

Infrared aerial crop photography is becoming accepted as a reliable means of monitoring crop production systems. Near-infrared film is used to take vertical photographs at approximately 8000 ft above sea level. The photographs will often cover as much as 1 square mile of land.

Healthy green plants reflect a large amount of infrared light, whereas plants under stress caused by such factors as drought, nutrient deficiency, disease, weeds, and chemical spray damage do not reflect infrared light. These differences can be clearly recorded on infrared film. Other crop management problems that can be detected by this technique include inadequate drainage, poor stand establishment, and uneven application of fertilizers and herbicides. A number of such problems are identified in Figure 14-11, which is a black-and-white reproduction of an aerial infrared photograph.

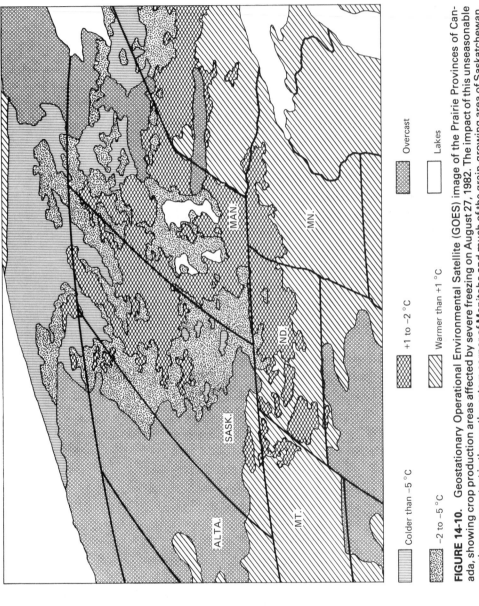

Colder than −5 °C   +1 to −2 °C   Overcast

−2 to −5 °C   Warmer than +1 °C   Lakes

FIGURE 14-10. Geostationary Operational Environmental Satellite (GOES) image of the Prairie Provinces of Canada, showing crop production areas affected by severe freezing on August 27, 1982. The impact of this unseasonable weather was greatest in the southwestern corner of Manitoba and much of the grain-growing area of Saskatchewan. The rest of the area either had milder temperatures or was protected from the front by cloud cover. [Canadian Wheat Board, *Grain Matters* (September 1982).]

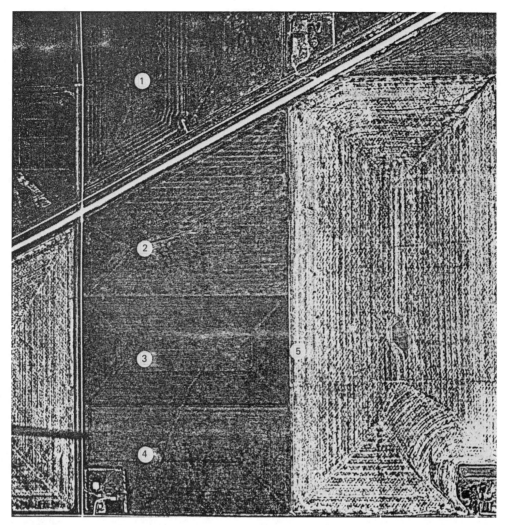

FIGURE 14-11. Black-and-white reproduction of an aerial infrared photograph showing the effects of improper fertilization and poor drainage on crop growth. The dark black areas correspond to a deep red color on the infrared photograph and they represent healthy plant growth. Gray portions of the reproduction equivalent to a pinkish color on the original are indicative of stressed plants. White areas in this figure are the counterpart of a greenish tint on the infrared photograph and they signify exposed soil showing through a seedling crop or sparse plant growth. 1, caused by uneven spreading of granular fertilizer; 2, liquid fertilizer misses; 3, nitrogen fertilizer not applied to center portion of field; 4, drainage problem in a flax field; 5, phosphorus fertilizer not applied to outer edges of field. [*The Furrow* (John Deere), p. 27 (January 1978).]

Yield Potential

It is important to determine the yield potential of the soils on an individual grower's farm. Yield potentials vary among soils. It may be 300 bu/A of corn on one soil and 150 bu/A on another; 100 bu/A of soybeans on one and 50 bu/A on another; 160 bu/A of wheat on one and 80 bu/A on another; 12 tons/A of alfalfa on one and 6 tons/A on another.

Setting up a High-Yield Demonstration Area. How much a high-yield effort succeeds or fails depends largely on how well the details of the management plan are carried out. Nothing should be left to chance.

Let us look at some suggestions for a high-yield area. Specific recommendations will vary with each field. It depends on the crop, yield goal, past management, yield history, willingness of the farmer to accept change, and other factors:

Yield goal. Set a high but realistic goal.

Soil. Should be the most productive on farm and well drained.

Tillage. Consider deep chiseling, till planting, or no-till if appropriate for the soil type.

Hybrid or variety. Use two or more with high yield potential developed for the region.

Row width. Consider narrow rows with uniform spacing within the row.

Planting date. Plant early. The specific date will depend on the crop and region.

Pest control. This must be perfect, with no compromises. Use a preventive rather than a corrective program.

Duration. Practice it for at least 5 years.

Rotation. Practice a rotation that will mutually benefit those crops in the rotation.

Population. Most stands in an area usually average much lower than required for a maximum-yield program. As fertility and management improve, increase population.

Fertility. Maintain soil pH between 6.2 and 7.0. Somewhat higher values are acceptable in western soils, lower in the South and the Southeast. It pays to build phosphorus and potassium soil tests to very high levels and maintain. Apply nitrogen for yield goal and *do not skimp*. Remember to add secondary and micronutrients if there is any chance for response. Add manure if available.

Soil testing. Sample annually and test for as many parameters as is practical.

Plant analysis. Take samples several times during the growing season. Watch closely for *all* essential nutrients.

Measure. Be sure to measure the yield and quality carefully—grain protein, and so on.

Figure. Know the cost effectiveness of the operations. Calculate production cost per unit of output—bushel, ton, and so on.

Evaluate. Determine which high-yield management practices can be adapted to the whole farm.

Only the farmer can judge what management practices are economical for him or her. The fertilizer dealer can certainly help farmers obtain high yields. The dealer can encourage customers to test on their own farms the economical practicality of the techniques used in maximum yield research.

One way is to work with a few of the better growers to help them set up high-yield areas. This can be done by working with a customer individually. Or it can be done by forming a high-yield club where facts are openly exchanged among farmers.

A good example of how a high-yield club operates is now at work in the Midwest, a group of 10 farmers organized around an interested dealer and a consulting agronomist. Such a club functions as follows:

Plots. Corn–soybeans are grown in rotation on 10 acres each, 20 acres total.

Farmer. He or she furnishes all material, labor, scouting; monitors the crop and keeps accurate records; and openly shares plans and results with other farmers in the club.

Dealer. He or she furnishes technology, coordination, and information; conducts several diagnostic clinics during the growing season; and assembles the group after harvest for exchange of results and planning ideas.

A successful club depends on each member having a strong commitment . . . to share . . . to have an open mind . . . to promote top management. It is *not* a contest. It is *not* competition among farmers. It is a *joint* learning experience. It emphasizes gradual improvement in management and yields over time.

Summary

1. Marked increases in yields of crops have taken place in the past 10 years as a result of improved management practices, including more adequate fertilization. Before that time yields increased slowly.

2. The aim of a crop and soil-management program should be to realize sustained maximum profit from the cropping program. Effects on tilth, water intake and erosion, plant nutrient supply, and pest control must be considered.

3. Cropping systems affect the amount of soil organic matter largely by the amounts of residue produced. As added plant nutrients increase yields, larger amounts of residues are obtained.

4. Organic matter serves as a storehouse for nutrients, increases exchange capacity, provides energy for microorganisms, releases carbon dioxide, improves tilth, increases moisture retention, and provides surface protection. All of these functions except the last depend on decomposition. Hence organic matter accumulation in the soil is not an end in itself. The important point is the production of high quantities of organic matter and its subsequent decay.

5. One of the problems in soil erosion is related to the lower supply of nutrients, nitrogen in particular. On permeable subsoils adequate fertilization and careful timing of tillage operations in regard to soil moisture content can go a long way toward producing satisfactory yields when combined with other good management practices.

6. The main function of legumes in a rotation, with the exception of legumes for grain, has been to furnish large quantities of high-quality forage. An additional benefit has been the nitrogen supplied on some soils. With the development of the commercial nitrogen industry growing legumes for nitrogen alone is not economical in most instances.

7. It is recognized that supplemental nitrogen must be applied for highly profitable production of first-year corn after a legume.

8. Legumes or a nonlegume–legume mixture in forages reduces fertilizer nitrogen requirement and generally results in superior animal performance.

9. The amount of nitrogen fixed by a legume is dependent on the amount of nitrogen the soil or residual fertilizer will supply. When the soil supplies a high amount of nitrogen the legume may fix less than 50% of its needs.

10. Where soil, solar, and water resources permit, two to four crops may be grown in a single year. Double cropping is common in the United States. In some tropical areas three or four crops may be possible.

11. Monoculture or continuous cropping to one crop is practiced in many parts of the world. Continuous corn may yield somewhat less than corn in rotation, but economics may favor the former when production costs and the value of other crops in the rotation are considered.

12. Sod crops, including legumes, in rotation improve tilth and water intake, while continuous row cropping may favor compacting and breakdown of soil structure. Much information is needed on long-time effects on high production.

13. Deep-rooted crops may bring nutrients to the surface. The net effect of cropping practices on fertility levels depends on removal of nutrients by the harvested portion, amount supplied by the soil, and the amount of supplemental fertilization.

14. A primary cause of *erosion is impoverishment of nutrients*. The effects of cropping systems on soil losses can be drastically altered by the denseness of the cover, which in turn can be increased by more adequate fertilization and thicker planting.

15. Winter cover crops will supply residues, and nitrogen if they are legumes. They are most useful in cropping systems in which few residues remain on the land over the winter. One of the uncertainties is the amount of growth produced. Early planting of these crops and late plowing to turn them under are helpful in increasing amount of growth.

16. Farm manure is a valuable by-product of the livestock industry. Because of losses by volatilization and leaching, generally not more than one third to one half of the value is realized. Manure alone on a farm will result in a gradual depletion of soil fertility, but it is a valuable supplement to a well-designed lime, fertilization, and soil-management program.

17. Disposal of human waste or sludge from municipal treatment plants is receiving increasing attention. The sludge will supply large amounts of most essential elements. The possible toxic effects of heavy metals is still to be answered.

18. Conservation tillage is a principle and not a practice. More residues are left on the soil surface. It is based on fewer trips over the field, results in better soil tilth, lower production costs, and generally higher yields, if other limiting factors are taken care of. This makes for more efficient use of fertilizers.

19. When maximum profit yield systems are practiced with conservation tillage, a deeper root zone is created, there is more water intake and greater water-holding capacity. An extra inch or two of water may be provided for critical growth stages.

20. No-till planting directly in sod or crop residues involves more herbicides but saves energy, water, and soil. It is particularly important on lighter soils and on sloping lands not suitable to other tillage because of erosion losses.

21. Establishing a maximum-yield system on a farm scale tests the system over a wide range of conditions and helps farmers determine what works for their particular conditions.

Questions

1. Why are yield trends over a period of years likely to be misleading as a measure of soil productivity? What might happen to yield trends if plant breeding studies ceased?
2. Why should a continued increase in ceiling yields in university and government research plots be conducive to increased state yields?
3. What is the aim of a crop and soil-management program?
4. Explain why additions of nitrogen equal to crop removal help to reduce loss of organic matter.
5. Under what soil condition may a corn–small grain (alfalfa) rotation be preferable to corn plus commercial nitrogen as well as other nutrients each year? Under what conditions may the latter cropping system be preferable?
6. What influences how much of its total nitrogen a legume will fix?
7. On what soils in your area could organic matter be increased? Under

what soil conditions in your area would additions of organic matter be beneficial other than for the nutrients supplied?

8. What nutrient may be the most likely one to give a marked yield response on corn grown on an eroded soil? Why?

9. Loss of surface soil is serious, but the seriousness varies considerably with the soil. In what soils in your area is the loss likely to be the most serious?

10. Why is it important that plant residues decompose? What functions do they serve in the undecomposed state?

11. List the advantages of rotations and monoculture.

12. What cropping systems have depleted phosphorus and potassium in soils in your area? Explain. In what cropping systems have these elements been increased? Explain.

13. In what ways may nitrogen, phosphorus, and potassium be lost other than by crop removal? In what other ways than fertilization may the supplies be increased?

14. Are there examples in your area in which the farmer is relying too heavily on a legume to furnish the nitrogen needs of a cropping system? Explain.

15. Explain the statement that a primary cause of erosion is depletion of plant nutrients.

16. In what cropping systems may winter cover crops fit? Why?

17. Why will fertility level on a given farm gradually decrease if manure is the only carrier of plant nutrients used? Explain the fertility distribution problem in a pasture.

18. What measures can be taken to prevent losses of plant nutrients from manure?

19. What are the most popular manure storage and application systems used in your area? Are they effective?

20. What is sewage sludge? Are there problems in its disposal? Explain.

21. What is sewage effluent? Are there problems in its disposal? Explain.

22. In which of the major nutrients is sludge from sewage plants usually low? Is sludge being used in your area? If so, where?

23. Is sewage effluent being used in your area? If so, where and what are its benefits?

24. What is conservation tillage? What are the advantages?

25. Would no-till fit in your area? If so, where? Why?

26. Is planting sometimes more difficult under no-till? Explain.

27. Can air seeders be used for planting operations in no-till cropping? Do they have other uses?

Selected References

Backtell, M. A., C. J. Willard, and G. T. Taylor, "Building fertility in exposed subsoil." *Ohio Agr. Exp. Sta, Res. Bull.* 782 (1956).

Bandel, V. A., "Fertilization technique for no-tillage corn." *Agr.-Chem. Age*, 14–15 (July 1981).

Barnes, G., "Crop rotation vs. monoculture. Insect control." *Crops Soils*, **32**(4):15 1980).

Bateman, H. P., and W. Bowers, "Planning a minimum tillage system for corn." *Illinois Agr. Ext. Cir. 846* (1962).

Bauer, A., and A. L. Black, "Soil carbon, nitrogen, and bulk density comparisons in two cropland tillage systems after 25 years and in virgin grassland." *Soil Sci. Soc. Am. J.*, **45**:1166 (1981).

Bear, F. E., "Variation in mineral composition of vegetables." *SSSA Proc.*, **13**:380 (1948).

Benacchio, S. S., G. O. Mott, D. A. Huber, and M. F. Baumgartner, "Residual effect of feeding grain to grazing steers upon the productivity of pasture." *Agron. J.*, **61:**271 (1969).

Bennett, O. L., E. L. Mathais, and P. E. Lundberg, "Crop responses to no-till management practices on hilly terrain." *Agron. J.*, **65:**488 (1973).

Black, A. L., and F. H. Siddoway, "Hard red and durum spring wheat response to seeding date and N-P fertilization on fallow." *Agron. J.*, **69:**885 (1977).

Black, A. L., and F. H. Siddoway, "Influence of tillage and wheat straw residue management on soil properties in the Great Plains." *J. Soil Water Conserv.*, **34:**220 (1979).

Blake, G. R., "Crop rotation vs. monoculture. Soil physical properties." *Crops Soils*, **32**(6):10 (1980).

Bole, J. B., and R. G. Bell, "Land application of municipal sewage wastewater: yield and chemical composition of forage crops." *J. Environ. Qual.*, **7:**222 (1978).

Bole, J. B., J. M. Carefoot, C. Chang, and M. Oosterveld, "Effect of wastewater irrigation and leaching percentage on soil and groundwater chemistry." *J. Environ. Qual.*, **10:**177 (1981).

Bradfield, R., in W. H. Garman, Ed., *The Fertilizer Handbook*, p. 120. Washington, D.C.: National Plant Food Institute, 1963.

Campbell, C. A., and V. O. Biederbeck, "Changes in the quality of soils of the prairies as a result of agricultural production." *Prairie Production Symp.: Soil Land Resourc.*, Advisory Committee to the Canadian Wheat Board (October 29–31, 1980).

Cardwell, V. B., "Fifty years of Minnesota corn production: sources of yield increases." *Agron. J.*, **74:**984 (1982).

Christensen, R. P., and R. O. Aines, "Economic efforts of acreage control programs in the 1950's." *ERS, USDA, Agr. Econ. Rep. 18*, p. 23 (1962).

Coble, H. D., "Crop rotation vs. monoculture. Weed control." *Crops Soils*, **32**(5):8 (1980).

Cooper, R. L., "Nitrogen response of soybeans with different crop residues in the absence of moisture and lodging stress." *Agron. Abstr.*, Am. Soc. Agron. (1972).

Copley, T. L., L. A. Forest, and W. G. Woltz, "Soil management of bright tobacco in lower Piedmont." *North Carolina Agr. Exp. Sta. Bull. 392* (1954).

Dickey, E. C., R. S. Moonrow, D. P. Shelton, and C. J. Kisling-Crouch, "Reducing soil erosion with oat residues." *Farm, Ranch Home Q.* (Univ. of Nebraska), **29**(1):18–19 (1982).

Doran, J. W., "Tilling changes soil." *Crops Soils*, **34**(9):10 (1982).

Edwards, W. M., J. L. McGuinness, D. M. Van Doren, Jr., G. F. Hall, and G. E. Kelley, "Effect of long-term management on physical and chemical properties of the Coshocton watershed soils." *SSSA Proc.*, **37:**927 (1973).

Engelstad, O. P., and W. D. Shrader, "The effect of surface soil thickness on corn yields: II. As determined by an experiment using normal surface soil and artificially exposed subsoil." *SSSA Proc.*, **25:**497 (1961).

Fenster, C. R., "Conservation tillage in the Northern Plains." *J. Soil Water Conserv.*, **32**(1):37–42 (1977).

Fenster, C. R., and T. M. McCalla, "Tillage practices in Western Nebraska with a wheat–sorghum–fallow rotation." *Nebraska Agr. Exp. Sta. SB 515* (1971).

Flannery, R. L., "Hairy vetch vs. rye cover crops for corn silage production using no-till." *Better Crops Plant Food*, **66:**22 (Summer/Fall 1981).

Fraedrich, R., "Are we farming our land too hard?" *Farm Futures*, **8**(9):28 (1980).

Griffith, D. R., J. V. Mannering, and W. C. Moldenhouer, "Conservation tillage in the eastern Corn Belt." *J. Soil Water Conserv.*, **32**(1):20–28 (1977).

Halvorson, A. D., and P. O. Kresge, "Flexcrop: a dryland cropping systems model." *USDA, ARS, Production Res. Rep. 180* (1982).

Hart, S. A., "Manure management." *Calif. Agr.*, **18**(12):5 (1964).

Hoyt, P. B., and W. A. Rice, "Effects of high rates of chemical fertilizer and barnyard manure on yields and moisture use of six successive barley crops grown on three gray luvisolic soils." *Can. J. Soil Sci.*, **57:**425 (1977).

Ibach, D. B., and J. R. Adams, "Crop yield response to fertilizer in the United States." *U.S. Dept. Agr. Sta. Bull. 431* (1968).

Kamprath, E. J., W. V. Chandler, and B. A. Krantz, "Winter cover crops—their effects on corn yields and soil properties." *North Carolina Agr. Exp. Sta. Tech. Bull. 129* (1958).

Koehler, F. E., and R. W. Meyer, "Fertilization for wheat production under a no-till system in Washington." *Proc. 27th Annu. Northeast Fert. Conf.*, pp. 75–76, (July 13–15, 1976).

Krantz, B. A., and W. V. Chandler, "Fertilize corn for higher yields." *North Carolina Agr. Exp. Sta. Bull. 366* (rev. 1954).

Larson, W. E., and W. C. Burrows. "Effect of amount of mulch on soil temperature and early growth of corn." *Agron. J.*, **54**:19 (1962).

Leitch, R., "The role of legumes in crop rotation," in *Alfalfa Production in the Peace River Region*, pp. C1–C5. Beaverlodge, Alberta: Alberta Agriculture and Agriculture Canada, Research Station, 1976.

Lucas, R. E., and M. L. Vitosh, "Soil organic matter dynamics." *Michigan State Univ. Res. Rep. 358* (1978).

Lutwick, G., and D. N. Graveland, "Effects of anaerobically digested sewage sludge on yield and elemental accumulation in alfalfa and soils near Edmonton, Alberta." Earth Sciences Division, Alberta Environment (1983).

McCalla, T. M., J. R. Peterson, and C. Lue-hing, "Properties of agricultural and municipal wastes," in L. F. Elliott and F. J. Stevenson, Eds., *Soils for Management of Organic Wastes and Waste Waters*, pp. 11–43. Madison, Wis.: Soil Science Society of America, American Society of Agronomy, and Crop Science Society of America, 1977.

McCormick, R. A., D. W. Nelson, D. M. Huber, and A. L. Sutton, "Improving the fertilizer value of liquid swine manure with a nitrification inhibitor." *Purdue Agr. Exp. Sta. Bull. 342* (1981).

McCoy, D., D. Spink, J. Fujikawa, H. Regier, and D. Graveland, "Guidelines for the application of municipal wastewater sludges to agriculture lands in Alberta." Alberta Environment (1982).

Mannering, J. V., "Conservation tillage to maintain soil productivity and improve water quality." *Agronomy Guide AY-222*, Purdue Univ. (1979).

Melsted, S. W., "Sewage sludges and effluents: effect on soils, plants and fertilizer markets." *Illinois Fert. Conf. Proc.* (1973).

Meyer, L. D., and J. V. Mannering, "Minimum tillage for corn: its effect on infiltration and erosion." *Agr. Eng.*, **42**(2):72 (1961).

Meyer, L. D., and J. V. Mannering. "Tillage and land modification for water erosion control," in *Tillage for Greater Crop Production*, ASAE (1967).

Moncrief, J. F., and E. E. Schulte, "The effect of tillage system on the N, P and K requirements of corn and on soil temperature, moisture and compaction." *Proc. 1980 Fert., Aglime Pest Manag. Conf.*, pp. 134–144 (1980).

Nelson, D. W., "Losses of fertilizer nutrients in surface runoff." *Fert. Solut.*, **17**:10 (1973).

No-Tillage Systems Proc. Sponsored by Ohio State Univ., OARDC and Chevron Chemical Co., Columbus, Ohio (1972).

Oveson, M. M., "Conservation of soil nitrogen in a wheat-summerfallow farming practice." *Agron. J.*, **58**:444 (1966).

Parker, D. T., and W. E. Larson, "Effect of tillage on corn nutrition." *Crops Soils*, **7**(4):15 (1965).

Peterson, J. B., "The relation of soil fertility to soil erosion." *J. Soil Water Conserv.*, **19**(1):15 (1964).

Peterson, J. R., T. M. McCalla, and G. E. Smith, "Human and animal wastes as fertilizers," in R. A. Olson, T. J. Army, J. J. Hanway, and V. J. Kilmer, Eds., *Fertilizer Technology and Use*, 2nd ed. Madison, Wis.: Soil Science Society of America, 1971.

Peterson, R. G., W. W. Woodhouse, Jr., and H. L. Lucas, "The distribution of excreta by freely grazing animals and its effect on pasture fertility: l. Excretal distribution." *Agron. J.*, **48**:440 (1956).

Phillips, J. A., "No tillage fertilization principles." *No-tillage Res. Conf.*, Univ. of Kentucky (1970).

Phillips, R. E., et al., "No-tillage agriculture." *Science*, **208**:1108–1113 (1980).

Pinck, L. A., F. E. Allison, and V. L. Gaddy, "The effect of green manure crops of varying carbon–nitrogen ratios upon nitrogen availability and soil organic matter content." *J. Am. Soc. Agron.*, **40**:237 (1948).

Potash & Phosphate Institute. *Maximum Economic Yield Manual—A Guide to Profitable Crop Production.* Atlanta, Ga. (1982).

Ramig, R. E., P. E. Rasmussen, R. R. Allmaras, and C. M. Smith, "Nitrogen–sulfur relations in soft white winter wheat: I. Yield response to fertilizer and residual sulfur." *Agron. J.*, **67**:219 (1975).

Ridley, A. O., and R. A. Hedlin, "Soil organic matter and crop yields as influenced by the frequency of summerfallowing." *Can. J. Soil Sci.*, **48**:315 (1968).

Rohweder, D. A., and R. Powell, "Grow legumes for green manure." *Wis. Ext. Fact Sheet A2477* (1973).

Salter, R. M., R. D. Lewis, and J. A. Slipher, "Our heritage—the soil." *Ohio State Agr. Ext. Serv. Bul. 175* (1936).

Sanchez, P. A., D. E. Bundy, J. H. Villachica, and J. J. Nicholaides, "Amazon Basin soils: management for continuous crop production." *Science*, **216**:821 (1982).

Scarseth, G. D., *Man and His Earth.* Ames: Iowa State University Press. 1962.

Schulte, E. E., and J. F. Moncrief, "Influence of tillage systems on N requirement of corn." *Proc. 1982 Fertilizer, Aglime Pest Manag. Conf.*, pp. 203–207 (1982).

Shrader, W. D., and R. D. Voss, "Crop rotation vs. monoculture. Soil fertility." *Crops Soils*, **32**(8):15 (1980).

Smith, J. H., and C. L. Douglas, "Wheat straw decomposition in the field." *SSSA Proc.*, **35**:269 (1971).

Sommers, L. E., D. W. Nelson, and C. D. Spies, "Use of sewage sludge in crop production." *Purdue Univ. AY-240* (1980).

Sommers, L. E., et al., "Effects of sewage sludge on the cadmium and zinc content of crops." *Counc. Agr. Sci. Technol. Rep. 83* (1980).

Sutton, A. L., J. V. Mannering, D. H. Bache, J. F. Marten, and D. D. Jones, "Utilization of animal waste as fertilizer." *Purdue Univ. 1D-101* (1975).

Triplett, G. B., Jr., and D. M. van Doren, Jr., "Chemicals make possible no tillage corn." *Ohio Farm Home Res.*, **48**(1):6 (1963).

Triplett, G. B., C. A. Osmond, and P. Sutton, "Fertilizer application methods for no-till corn." *Ohio Report*, OARDC, **57**(3):39 (1972).

van Doren, D. M., Jr., and G. B. Triplett, Jr., "Mulch and tillage relationships in corn culture." *SSSA Proc.*, **37**:766 (1973).

Walker, W. M., E. R. Swanson, and S. G. Carmer, "Have corn yields reached a plateau?" *Better Crops Plant Food*, **66**:28 (Spring 1982).

Walsh, L. M., and R. F. Hensler, "Manage manure for its value." *Wisconsin Ext. Circ. 550* (1971).

Walters, H. J., "Crop rotation vs. monoculture. Disease control." *Crops Soils*, **32**(7):7 (1980).

Welch, C. D., W. L. Nelson, and B. A. Krantz, "Effects of winter cover crops on soil properties and yields in a cotton–corn and in a cotton–peanut rotation." *SSSA Proc.*, **15**:29 (1951).

Whitaker, F. D., V. C. Jamison, and J. F. Thornton, "Runoff and erosion losses from Mexico silt loam in relation to fertilization and other management practices." *SSSA Proc.*, **25**:401 (1961).

Wischmeier, W. H., "Relation of field plot runoff to management and physical factors." *SSSA Proc.*, **30**:272 (1966).

Wischmeier, W. H., "Conservation tillage to control water erosion." *Conservation Tillage, Proc. Nat. Conf.*, Soil Conserv. Soc. Am. (1973).

Wischmeier, W. H., and D. D. Smith, "Predicting rainfall erosion losses—a guide to conservation planning." *USDA Agriculture Handbook 537* (1978).

Yoshida, S., F. T. Parao, and H. M. Beachell, "A maximum annual rice production trial in the tropics." *Int. Rice Com. Newslett.*, **21**(3):27 (1972).

Chapter 15

Economics of Plant-Nutrient Use

AGRICULTURE is an up-and-down industry profit-wise. It always pays to aim for the maximum profit per acre regardless of crop prices. This is the point at which the last dollar spent to produce the crop returns just a dollar. Such crop yields are higher than most people think and it means applying all good management practices to the limit with the latest technology.

Higher crop yields continue to offer the greatest opportunity for reducing per-bushel or per ton production costs. Modern growers have the attitude of any other good business managers. They would like to know if a given expenditure is the best they can make with their available funds and if their production costs are properly distributed.

Growers realize that they must spend money to make money. This is certainly true of expenditures for lime and fertilizer. The increasing demand for fertilizer in North America and in other parts of the world is evidence that growers recognize the returns that can be realized from added plant nutrients.

In spite of the rapid rise in the use of plant nutrients, reliable estimates show that much land is still underfertilized and that plant-nutrient additions could be profitably increased. Ohio applied more plant nutrients per acre in 1982 than any other state in the North Central region. Yet when one considers the yield potentials of row crops, small grain, hay, and pasture, the state is still at less than two-thirds of its potential even with present technology. As one moves West, South, or East, potentials are even greater. Lime use in the United States is about 35% of the need.

For the developing countries in Asia and the Far East, the Near East, Africa, and Latin America, N-P-K use in 1982 is estimated at about one-half of the 1992 demand. Economic development must give high priority to agriculture. Rostow says, "Put another way, the rate of increase in output in agriculture may set the limit within which modernization proceeds." Use of fertilizers is an index of the use of modern agricultural methods. The close correlation between average fertilizer use and the value index of crop production has been pointed out. As higher rates of plant nutrients are required, it becomes more important that the nutrients be applied so that they will be most efficiently utilized.

Attention must be given to water control, tillage, variety, date and rate of seeding, plant spacing, fertilizer placement, cultivation, weed, insect and disease control, and harvesting practices. Proper use of fertilizers complements the effects of other management practices, and vice versa.

Prices of Fertilizers Compared with Other Farm Prices

To obtain a given level of production, farmers can vary the inputs of land, fertilizer, labor, machinery, and so on. The actual use of each depends on relative costs and returns. It is interesting to compare the costs of certain farm inputs shown in Figure 15-1. In the period 1977–1979 fertilizer price increased slowly. On the other hand, machinery, wages, and real estate increased more rapidly. Hence farmers have tended to substitute agricultural chemicals, fertilizer, and lime, as well as machinery and other items, for labor and land (Figure 15-2). Although the price of fertilizers and lime will continue to rise, the price may not rise as fast as some of the other inputs.

Fertilizer Use and Farm Income

Use of fertilizers in a given year has tended to follow farm income of the year before. If growers have a good income in one year, they are more likely to increase their fertilizer use in the next. On the other hand, if their income was low, they reduce their expenditures for fertilizer. Agriculture is the only industry where data at least one year old are used in making decisions. Future decisions tend to be based on current cash prices of crops.

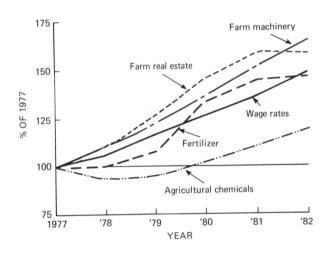

FIGURE 15-1. Prices of selected farm inputs, 1977–1982. 1981 preliminary, 1982 projected. Farm machinery includes tractors and self-propelled machinery. (*ERS, USDA, 1983 Handbook of Agricultural Charts; USDA Agriculture Handbook 619.*)

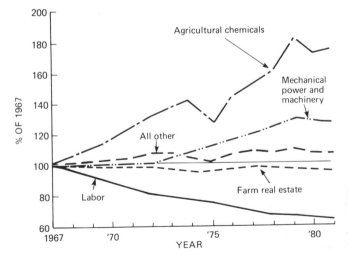

FIGURE 15-2. Use of selected farm inputs, 1977–1981. (*ERS, USDA, 1983 Handbook of Agricultural Charts; USDA Agriculture Handbook 619.*)

However, because of increased emphasis on research and education in fertilizer use, growers are spending an increasingly higher proportion of their income on plant nutrients. The price of produce and availability of markets are powerful influences on the introduction of improved practices by farmers the world over.

Maximum Yield Research and Maximum Economic Yield

Maximum yield research is the study on the effect of one or more variables and their interactions in a multidisciplinary system that strives for the highest yield possible for the soil and climate of the research site. Maximum yield is a constantly moving target because of continuing technological advances.

Maximum profit yield comes from an economic analysis of maximum yield research data. It is somewhat lower than the maximum yield (Figure 15-3). It is the point where the last increment of an input just pays for itself.

The maximum profit yields vary among soils. It might be 300 bushels/A of corn on one soil or 150 bushels on another; 100 bushels of soybeans on one or 50 bushels on another; 160 bushels of wheat on one or 80 bushels on another; or 12 tons of alfalfa on one and 6 tons on another. Regardless of the soil the maximum profit yield is much higher than is generally thought.

Researchers strive for higher and higher yields without regard for economics. They know that something not economically feasible today may become so tomorrow through technological advances spurred by a researcher who has moved the yield potential upward.

Wortman and Cummings challenge the modern researcher in their book, *To Feed The World: The Challenge and the Strategy* (1978):

> As part of their agricultural research, biological scientists in each country must continuously test the limits of available technology as well as their ability to put together new technological components to achieve highest yields. Maximum yields may not be economically practical, but agricultural scientists should not, for supposed economic reasons, fail to attempt to raise the limits on productivity imposed by technology.

Today's farmers want to produce maximum profit yields—to get the highest possible net return they can. Higher yields are the key to this goal. L. F. Welch, University of Illinois soil fertility researcher, says: "Even though wide differences in crop and fertilizer prices are used, the most profitable fertilizer rate turns out to be the one that results in yields being near the top of the response curve."

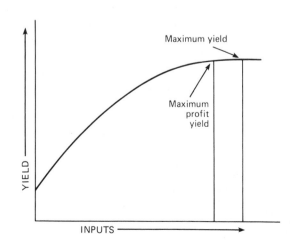

FIGURE 15-3. Maximum profit yield is slightly lower than maximum yield.

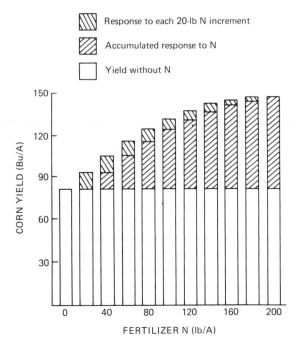

FIGURE 15-4. Diminishing returns in yield response of corn to fertilizer nitrogen. [Voss, Pesek, and Webb, *Iowa State Univ. Pm 651* (1975).]

Law of Diminishing Returns. When the soil is deficient in a nutrient needed for a desired crop yield, the first added increment of the nutrient will result in a large yield increase. The next added increment may also give an increase, but not as large proportionately as the first because of the "law of diminishing returns" (Figure 15-4). Consequently responses to fertilizer increments continue diminishing to the point where the last incremental yield value just equals cost of input. It is this application rate that gives maximum profit.

Urgency for Higher Yields. The urgency comes from two challenges.

Expanding the World Food Needs. The world population is expected to about double in the next 30 to 40 years. It was pointed out in Chapter 1 that the area of cultivated land will probably increase by only about 20% during this period. Productivity per unit area will have to increase substantially to satisfy expected food requirements.

Economic Necessity for the Farmer. Cost of production is expected to continue to increase. Prices of crops will increase but not as rapidly as cost of production. Break-even yields will continue to increase. This is illustrated in Table 15-1. For a production cost of $300 per acre with $3.00 corn, a 100-bushel yield is required just to break even ($300 ÷ 100 = $3.00).

TABLE 15-1.   Break-even Yields of Corn Increase with Higher Cost of Production

Cost of Production ($/A)	Break-even Yield of Corn * (bu/A)
300	100
400	133
500	167
600	200

* At $3.00/bu.

Yield Level and Unit Cost of Production

Practices that will increase yield per acre usually lower the cost of producing a bushel or ton of the crop, for it costs just as much to prepare the land, plant, and cultivate a low-yielding field as it does a high-yielding field. Land, buildings, machinery, labor, and seed will be essentially the same, whether the production is high or low. These and other costs are called *fixed* and must be paid regardless of yield. *Variable costs* are those that vary with the magnitude of the total yield, such as the quantity of fertilizer applied, pesticides, harvesting, and handling.

The effect of fertilizer on increasing yields and decreasing the cost per unit of production, in this case a quintal, is well illustrated in Table 15-2. Machinery, labor, management, interest, and other cash costs remained the same at all levels of fertilizer use. However, the operating costs, including fertilizer, lime, machinery operation, and seed and pest control, increased. The lowest cost per unit and the greatest return were obtained at the highest rate of fertilizer input. Although the cost of inputs and price of product will vary, this principle remains the same the world over.

The total cost of growing an acre of corn in the Midwest was estimated to be about $350 per acre in 1982, and this figure is climbing each year. The average yield in the United States in 1982 was 114 bu/A. This means that the farmer must receive $3.07 per bushel to break even.

Another example with corn is shown in Table 15-3. Increasing the yield increased the total cost of production per acre but decreased the cost per bushel and increased net profit. Many practices enter into increasing corn yields from 100 bu/A to 175 bu/A. It may take a period of years because much of this increase will come from gradually improving management practices. At the 175-bushel yield, not only were there more bushels, but the profit per bushel was greater ($2.75 − $2.18 = 57 cents/bu,57 cents × 175 bushels = $100/A net profit).

Level of Management. Two important keys in obtaining the most efficient use of the land and inputs are the weather and the ability of the farmer. In Chapter 16 the relationships between soil fertility and efficiency of water use will be discussed.

Time of planting, proper variety selection, tillage, plant spacing, pest control, and timely harvesting are some of the factors that can be controlled by farmers. Delaying planting soybeans one month may reduce soybean yields 10 to 20 bushels

TABLE 15-2.   Effect of the Rate of N-P$_2$O$_5$-K$_2$O (kg/ha) on Corn Yields and Profits

Corn Production Details	0-0-0	67-22-22	134-45-45	270-90-90
Corn yield (quintal/ha)	48	70	85	98
Gross income per hectare	$188	$274	$336	$390
Operating costs *	56	69	82	111
Other cash costs †	25	25	25	25
Fixed costs (machinery, labor, and management)	88	88	88	88
Interest **	82	82	82	82
Total costs	251	264	277	306
Cost per quintal	5.3	3.8	3.3	3.1
Return to land, labor, and management	− 63	+ 10	+ 59	+ 84

* Fertilizer, lime, machinery operation, seed, and pest control.
† Real estate taxes, land maintenance, and overhead.
** Interest costs of land.
Source: Sullivan, *Phosphorus Agr.,* **61**:19 (June 1973).

TABLE 15-3. **Effect of Increasing Yields on the Cost per Bushel of Corn and the Net Profit per Acre ($2.75/bu)**

Yield (bu/A)	Production Costs		Net Profit ($/A)
	$/A	$/bu	
100	331	3.31	− 56
125	343	2.74	1
150	359	2.39	54
175	383	2.18	100

Source: Adapted from Hinton, *Farm Economics Facts and Opinion,* Univ. of Illinois (January 1982).

in some areas. Delaying planting corn one day may cut yield 1 to 2 bushels a day after a given date. Ten percent of the corn or soybean crop may be left in the field. Fifty percent of the forage may be wasted in a pasture. When only the corn grain is harvested, about half of the total digestible nutrients may be left in the field in the form of stover.

In Ohio the average cost of production for corn farms grouped by yield into three categories was about the same, yet the profit ranged from $10 to $86 per acre. The same was true on cost of production for three groups of soybean farmers, yet the profit ranged from $20 to $90.* The soil or the weather might be blamed, but most of the difference was probably due to the management ability of the farmer.

With superior management, higher rates of fertilizer can and must be used. This general principle is illustrated in Figure 15-5, in which A is the most profitable rate with average management and B the most profitable rate with superior management. In Figure 15-6 it is seen that with a given rate of fertilizer, Q or R, a higher yield is obtained when all controllable growth factors are adequate.

Maximize Profits or Minimize Losses. Regardless of crop prices, it always pays to aim for the maximum profit yield. In the example in Table 15-3, if

* J. E. Beuerlein, Ohio State University, personal communication.

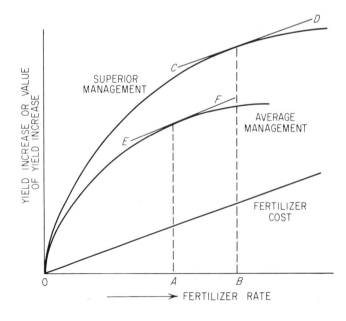

FIGURE 15-5. With superior management, higher rates of fertilizer can be profitably used than with average management. With average management, rate A is optimum, whereas with superior management, rate B is optimum. [Stritzel, *Iowa Farm Sci.,* **17**(12):14 (1963).]

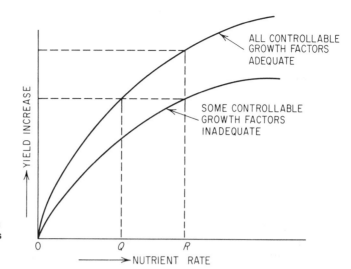

FIGURE 15-6. With a given rate of fertilizer (Q or R), a higher yield is obtained when all controllable growth factors are adequate. [Stritzel, *Iowa Farm Sci.,* **17**(12):14 (1963).]

the corn were $3.00 per bushel, at the 175-bushel yield the profit would be 82 cents/bu or $144 per acre. On the other hand, if the corn were $2.00 per bushel the loss at a 175-bushel yield would be only 18 cents per bushel or $32 per acre. At the 125-bushel yield, the loss would be 74 cents/bu or $93 per acre. Losses were minimized with the higher yield.

Hence, in this example, if farmers have the capital or can secure the credit, it would pay them to set their goal at 175 bu/A and apply the necessary inputs. Although on some soils the goal might be only 125 bu/A, it pays to aim for the highest appropriate yield. Maximum-yield research, experience, and consultation with progressive agricultural advisors can help establish these goals.

Effect of Added Nutrients on the Economics of Production. There are many examples of responses of crops to major nutrients, secondary nutrients, and micronutrients. The magnitude of the response will depend on many factors, including the content of a given nutrient in the soil.

Have low-cost nutrients in adequate supply in order to obtain the greatest return from nitrogen. With nonlegumes, nitrogen is the highest-cost nutrient per pound and/or the highest cost in terms of total amounts needed to produce maximum profit yields. It is important to have needed phosphorus, potassium, calcium, magnesium, sulfur, and micronutrients in place to get the greatest possible response from the added nitrogen.

Effect of Price of Crop or Fertilizer on Optimum Rate of Application. Price of crop or fertilizer does not affect optimum fertilizer rates appreciably. In a nitrogen trial on corn in Illinois, if the price of corn drops from $3.00 to $2.00, the optimum rate drops only 9 lb/A with 15-cent nitrogen (Table 15-4). This is closer than an applicator can be set. With $3.00 corn if the price of nitrogen is doubled from 15 cents to 30 cents per pound, the optimum rate drops only 15 lb/A.

Returns per Dollar Spent or Profit per Acre

Some farmers must consider the returns on each dollar spent for fertilizer, lime, or any other farming practice. They often have only a limited amount of cash or can obtain only a fixed amount of credit. They must decide whether to apply

TABLE 15-4. Effect of Crop and Fertilizer Prices on the Optimum Rate of Nitrogen

Corn Price ($/bu)	Optimum N Rate (lb/A)	
	15-cent N	30-cent N
4.00	192	180
3.50	190	178
3.00	187	172
2.50	183	166
2.00	178	154

Source: L. F. Welch, Univ. of Illinois, personal communication (1982).

nitrogen, phosphorus, potassium, and/or lime, in what quantities and on what fields, and whether to buy extra feed for their livestock or improved seed for their crops. In each operation they are interested in the return that can be obtained per dollar spent.

In general, as the rate of a particular nutrient is increased, the returns per dollar spent decrease. This decrease is the result of a reduced response for each successive incremental input of any given treatment. Eventually, the point is reached at which there is no further response to increasing amounts of a particular element, which is a corollary of the growth-response curves discussed in Chapter 2. Information on the effect of rate of nitrogen on net return per added dollar invested is given in Table 15-5. On the soil in this particular example, the extra 20 lb/A of nitrogen over 160 lb to give a total of 180 lb returned 12 cents (12%) over cost for each dollar spent on this 20-lb increment.

Credit agencies are very much interested in the returns the farmer realizes for each dollar spent on fertilizer or lime. As a general rule, $1 to $3 is the expected net return for each dollar spent. With this information available, the credit agencies would be in a more favorable position to extend credit to farmer clients. Certainly, compared with current rates, the purchase of fertilizer and lime with a 100 to 300% return would be an excellent investment for the farmer and a safe one for the credit agency.

Progressive growers recognize that although returns per dollar spent are important, the significant figure is the net return per acre. Several examples have been cited. With adequate cash or credit, the farmer must select the treatment

TABLE 15-5. Effect of the Rate of Nitrogen on the Net Return per Added Dollar Invested for Corn

Nitrogen Rate (lb/A)	Added Input (lb N/A)	Net Return per Added Dollar Invested
20	20	$7.25
40	20	5.75
60	20	5.00
80	20	3.87
100	20	2.38
120	20	1.63
140	20	0.88
160	20	0.50
180	20	0.12
200	20	−0.62

Source: P. R. Robbins and S. A. Barber, Purdue Univ., personal communication.

that will earn the greatest net return per acre. A low rate of fertilizer or another needed input may result in a high unit cost of production.

Higher Yields Can Service More Debt. Many farmers carry a high debt load from interest and principal payments on land purchases, machinery, and/or purchased inputs. Higher yields from improved management, including soil fertility, offer the opportunity to service a larger debt. If practices such as improved fertility and plant spacing increase net profit on corn, soybeans, or wheat by $25 per acre, this will service over $150 of debt at 16% interest. In today's agriculture, higher yields from increased soil fertility offer an excellent means of helping carry the heavy interest load or in paying off the principal.

Land Cost. When price of land rises rapidly, those who have bought land recently have a much higher land cost. For example;

$500/A land 10% down at 8% interest   = $36 A/yr/interest
($450 × 8% = $36)

$1200/A land 10% down at 12% interest = $130 A/yr/interest
($1080 × 12% = $130)

To realize the same net returns from the same land, the new buyer must increase the corn yield 38 bu/A immediately ($2.50 corn) or the soybean yields 16 bu/A immediately ($6.00 per bushel of soybeans). This vividly shows the economic urgency of higher yields.

Many Practices Cost Little or Nothing. As a grower aims for increased yields, much of the initial increase will come from improved practices, not just extra fertilizer. Many of these practices cost little or nothing. Some key examples follow.

Timeliness. This is the key in planting, tillage, equipment adjustment, pest control, observations, and harvesting.

Date of Planting. In Montana seeding barley April 6, May 6, and June 3 gave yields of 55, 42, and 33 bu/A, respectively. In Louisiana planting soybeans May 17 and June 15 gave yields of 50 and 33 bu/A. Delaying corn planting beyond the optimum time may cut yields of corn 1 to 2 bu/A per day.

Pest Control. As an example, wild oat removal at the one- to two-leaf stage gave no yield reduction of wheat, but removal at the three- to four-leaf stage gave an 18% yield reduction in Saskatchewan.

Variety Selection. The difference among varieties or hybrids in crop performance trials is well known. A superior variety of soybeans or wheat may yield 20 bu/A more than other varieties. A superior variety of corn may yield 50 bu/A more, and a superior variety of alfalfa or Coastal Bermuda, an extra 2 tons/A.

Plant Spacing Has Become of Increasing Importance. Plants per acre and row widths are prime examples. The average population of corn is around 20,000 plants per acre. An extra 5000 seeds would cost no more than $5.00 and could increase corn yields 15 to 20 bu/A in many instances.

Thirty-inch rows are common for soybeans. Drilling in 7-inch rows could increase yields 20% or more in many areas. Four-inch drill rows have been shown to be 10 to 20% superior to 8-inch drill rows for wheat.

TABLE 15-6. Yields of Continuous Peanuts
and Peanuts Following Corn

| | Peanut Yield (lb/A) | | |
Years	Continuous	Rotated	Increase
1965–1969	1400	1650	250
1970–1974	1780	2650	870
1975–1979	2220	3090	870

Source: J. T. Cope, *Highlights of Agriculture Research,* Vol. 28, No. 1, Alabama Agr. Exp. Sta. (1981).

Drills in place of planters for soybeans incur an extra cost per acre, but this extra cost can be amortized over acres and years.

    Rotation. Rotating crops is a valuable no-cost input to increase crop yields and profit. Rotation may not only reduce weed, disease, and insect problems, but also improve soil structure.

There is much information available. Table 15-6 shows the beneficial effect of rotating peanuts with corn and how the benefit tended to increase with time. Long-term trials in the Morrow plots at the University of Illinois showed continuous corn to yield 141 bu/A and corn after soybeans 165 bu/A.

Emphasis on practices costing little or nothing is included because:

These practices will improve the profit picture.
Higher yields will affect amounts of certain inputs needed.

What Is the Most Profitable Rate of Plant Nutrients?

Growers are asking this question. The answer is influenced by several factors:

1. Expected increase in yield from each increment.
2. Level of management.
3. Price of fertilizer and crops.
4. Additional harvesting and marketing costs.
5. Residual effects.
6. Buildup of soil fertility.

The most profitable rate of fertilization will change from year to year because of the variation in weather, prices of farm products, and price of fertilizer. Agronomists and economists are thus faced with a sizable task of calculation if the growers' questions are to be answered each year. They must work with expected yields, prices, and costs.

Crop production has always entailed considerable risk because of such factors as floods, droughts, insects, and diseases. There will always be risk in crop production, but the grower can reduce risks through superior management and increase the probability of increased yields and profits in the long term.

Expected Increase in Yield from Each Increment. This must be based on the best experimental evidence available and for the environmental conditions under which the question is raised. Many experiment stations have such data for some soil conditions in the state, but often these data are not for high yield levels. As an example, yield responses to nitrogen on corn over a number of years at eight experiment stations in Illinois indicated that economically optimal rates ranged from 100 lb/A to 240 lb/A.

An important factor in determining yield responses is the amount of nutrients present in the soil. The use of soil tests to monitor these levels helps to predict the need for lime and fertilizer. Many more calibration studies will be needed in the future at maximum yield levels to characterize crop responses with modern technology and the multiplicity of soil, management, and environmental conditions.

In one such example shown in Figure 15-7, the soil level of phosphorus and potassium affected the response of Coastal Bermuda grass to nitrogen. At low soil levels, the maximum profit rate was 200 lb of nitrogen and only $40 per acre. With high levels, however, the optimum rate was 310 lb, giving a profit of $79.55. This illustrates the importance of balanced soil fertility in obtaining maximum profit.

When too much fertilizer is added, the economic loss is not as great as when a crop is underfertilized by the same proportion (Table 15-7). The residual effect must be considered and helps to compensate for the interest on the extra fertilizer. Over a period of years it appears more profitable to use the optimum amount, even if the rate would be more than optimum in unfavorable years. This, of course, applies to nutrients that do not leach from the soil.

Price of Fertilizer in Relation to Value of Crop. Although the cost per pound of nutrients in a given fertilizer may fluctuate, these variations are much

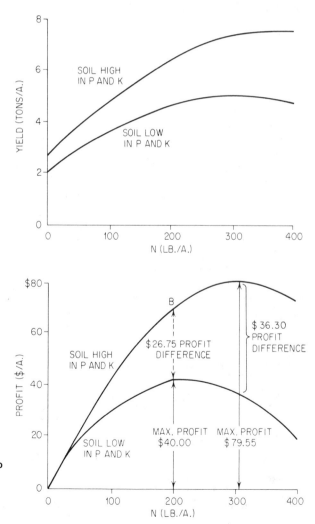

FIGURE 15-7. Level of phosphorus and potassium in the soil influences response of Coastal Bermudagrass to nitrogen. (L. F. Welch, cited by J. E. Engibous, *The Fertilizer Handbook.* Copyright 1972 by The Fertilizer Institute, Washington, D.C.)

TABLE 15-7.   Effect of Underfertilizing Versus Overfertilizing on the Net Return from Added Fertilizer

	Yield of Corn (bu/A)	Net Return from Fertilizer (bu/A)	Difference from Optimum (bu/A)
1/4 less	142	50.4	−5.8
Optimum	151	56.2	0
1/4 more	153	52.0	−4.2
None	79	—	—

Source: S. A. Barber, Purdue Univ., personal communication.

less than the fluctuations in crop prices. For some crops, however, government regulations have stabilized the floor price, and growers may use this figure as a basis for calculations. When crop prices have not been regulated, careful consideration of trends and outlook will be helpful in establishing a profitable rate of fertilization.

There are several methods that take into consideration both the cost of the nutrient and the price of the crop. In any case, the response of the crop in question to increasing rates of a nutrient must be known for the general soil condition. These data are obtained by research and are available from the agricultural experiment stations.

One such method is to use the crop–nutrient price ratio. This is shown for the corn/nitrogen price ratio in Table 15-8. If the yield potential is 160 bushels and the corn/nitrogen price ratio is 15:1, the most profitable nitrogen rate is 200 lb/A.

Residual Effects of Fertilizers. This has been treated in considerable detail in preceding chapters and is mentioned here because residual effects are definitely part of fertilizer economics. In general, the entire cost of moderate soil treatment with nitrogen, phosphorus, and potassium is charged to the crop being treated, whereas lime is charged off over a period of 5 to 10 years. With high rates of fertilizer, however, residual effects are striking.

At optimum fertilization rates it has been shown that on some soils about one-third of the nitrogen may be residual for next year's crop. Although the carryover of phosphorus varies greatly with the soil, it usually amounts to 40 to 60% on well-limed soils. The residual effects of potassium depend on the soil and how the crops are managed but vary from 25 to 60%. The lower figure would apply when hay, straw, or stover are removed from the land.

The increased yield of next year's crop due to residual effects may be sufficient to pay for much of the fertilizer application on this year's crop. On a dark-colored soil in Illinois, rates of potassium were applied on continuous corn over a 4-year period to give the rates shown in Figure 15-8. The soil test value for potassium

TABLE 15-8.   Most Profitable Nitrogen Rate (lb/A) for Corn Based on a Computer Model for Predicting Yield and Corn/Nitrogen Price Ratios

Corn/Nitrogen Price Ratio	Yield Potential (bu/A)							
	85	100	115	130	145	160	175	190
5/1	80	90	100	110	130	140	150	170
10/1	90	110	130	140	16ʹ	180	190	210
15/1	100	120	140	160	180	200	220	240
20/1	110	130	150	170	190	210	230	250
25/1	120	140	160	180	200	220	240	260

Source: Vitosh et al., *Michigan State Univ. Ext. Bull. E-802* (1979).

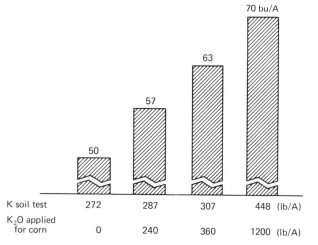

FIGURE 15-8. Potassium applied to continuous corn for 4 years increased the potassium soil test and the yield of soybeans the fifth year 20 bu/A. [Welch, *Better Crops Plant Food,* **63**(4):3 (1974).]

K soil test	272	287	307	448 (lb/A)
K_2O applied for corn	0	240	360	1200 (lb/A)

was increased from 272 lb/A to 448 lb/A. The average increase in corn yield for the four years was 25 bu/yr. The fifth-year soybeans were grown with no additional potassium and yields were increased 20 bu/A, from 50 to 70 bu/A.

Carryover fertility paid most of the fertilizer bill. With K_2O at 13 cents/lb, the cost of the 1200 lb/A for corn was $156. With soybeans at $6 a bushel, 20 bushels equals $120 and there was still much potassium left in the soil.

Cost of Building Soil Fertility. High crop yields are impossible with low levels of fertility. The fertility level of soils is a plant growth factor that is easily controlled, and up to a certain point increasing fertility level will be profitable. However, the initial cost of building soil fertility from low to high levels may discourage growers, if viewed as an annual rather than as a long-term investment.

Because of their immobile nature, phosphorus and potassium can be built up in soils. This buildup should be thought of as a capital investment to be amortized over a period of years.

The cost of building up soil phosphorus levels from 45 to 55 lb/A (P_1) is an example. Using a value of 9 lb. of P_2O_5 to raise the P_1 test 1 lb, 90 lb/A of P_2O_5 would be required. The initial cost is $24.30 per acre with P_2O_5 at $0.27 per pound.

The annual payment necessary to amortize the buildup cost with various interest rates and amortization periods is given in Table 15-9. Using a payoff period of 15

TABLE 15-9. Annual Payment Necessary to Amortize the $24.30 per Acre Initial Cost of Buildup Phosphorus with Various Interest Rates and Amortization Periods

Payoff Period (yr)	Annual Payment for Payoff at an Interest Rate of:		
	8%	*12%*	*16%*
1	$26.24	$27.22	$28.19
5	6.09	6.74	7.42
10	3.62	4.30	5.03
15	2.84	3.57	4.36
20	2.47	3.25	4.10

Source: Welch, *Better Crops Plant Food,* **66**:3 (Fall 1982).

years and an interest rate of 12%, the annual payment for payoff would be $3.57. A yield increase of 1.2 bushels of $3.00 corn or 0.6 bushel of $6.00 soybeans would pay for this cost.

Price per Pound of Nutrients

In a discussion of the economics of fertilizer use it is important that the cost per pound of nutrients be considered. Farmers, of course, are interested in the most economical source, but they may be accustomed to buying on the basis of cost per ton of fertilizer rather than on the cost per ton of plant nutrients.

In Materials. The variation that may occur in the cost of plant-nutrient elements in different carriers is shown in Table 15-10. The prices should be recognized as examples and will, of course, vary considerably from one area to another with year, time of year, and discounts. Local prices can be used for more exact comparisons. The wide variations in the cost per pound of nitrogen are significant. Of course, other factors, such as cost of application, content of secondary nutrients, and the effect on soil acidity must be taken into consideration.

With the advent of fluid fertilizers, such as anhydrous ammonia and nitrogen solutions, and bulk handling, different application equipment is needed. On small acreages custom application or rented equipment is generally advisable, and the extra charge for application must be considered. Usually, the charge is on a per acre basis.

The economy of ordinary and triple superphosphate will depend on location and the need for sulfur. With respect to potassium carriers, muriate is the most economical. Potassium from sulfate of potash or potassium nitrate is more expensive than that derived from muriate because of manufacturing costs, but these forms may be required for special-purpose fertilizers. Also, the added value of the sulfur and nitrogen must be calculated.

In Mixed Fertilizers. The grower is often faced with the choice of using all mixed fertilizers, mixed and straight materials, or all straight materials. A knowledge of the method of calculation of costs is important.

For example, a farmer has a choice of 12-24-24 or 6-24-24 plus ammonium nitrate or anhydrous ammonia.

6-24-24 costs $194/ton

12-24-24 costs $206/ton

Assuming that the cost of the phosphorus and potassium is the same in both mixtures, the additional nitrogen in 12-24-24 amounts to $12.00 or 10 cents/lb of nitrogen.

TABLE 15-10. Approximate Prices of Fertilizer Materials and Cost of Nutrients per Pound (for Illustration Only)

Material	Analysis	Price per Ton	Cost of N, P_2O_5, or K_2O (cents/lb)
Ammonium sulfate	20.5% N	$ 95	23.2
Ammonium nitrate	33.5% N	170	25.3
Urea	46% N	215	23.3
Anhydrous ammonia	82% N	220	13.4
Nitrogen solution	28% N	135	24.1
Superphosphate (triple)	44% P_2O_5	205	23.2
Muriate of potash	60% K_2O	150	12.5
Sulfate of potash	52% K_2O	250	24.0

Another calculation that farmers need to make relates to the economics of high-analysis fertilizer. In a comparison of 5-10-10 and 10-20-20, 2 tons of 5-10-10 are required to furnish the same amount of nutrients contained in 1 ton of 10-20-20. If 5-10-10 costs $100.00 and 10-20-20 costs $180.00 per ton, the 10-20-20 will be $20 cheaper than 2 tons of 5-10-10.

Other Factors. In addition to the actual cost of the material, farmers must consider the cost of transportation, storage, and labor used in applying the fertilizer. These costs may be difficult to evaluate, but if the actual price of the nutrients from one source is the same as another, growers will take the one requiring the least labor. The higher-analysis goods require less labor in handling. Time is also saved because fewer stops are made in applying the material.

Service. The term *service* usually refers to some form of agronomic assistance to the grower. It may include sampling soils; testing the soils in the company laboratory or sending them to another laboratory; fertilizer and lime recommendations; maps of farm showing requirements; planning rotations; and advice on other agronomic practices, including tillage, varieties, pest control, and harvesting. The agronomist or dealer may spend time with the growers in their fields during the summer to evaluate their programs and suggest improvements. Such specialized advice from qualified individuals can be particularly helpful to growers over a period of years.

Personalized service costs money and the grower must eventually pay for it. Actually, however, the investment per acre is not large. If the extra cost of the fertilizer is $10 per ton, the additional yield of corn needed to pay for it is calculated as follows (price of corn at $2 per bushel):

Rate of Application (lb/A)	Corn (bu/A)
200	0.5
400	1.0
600	1.5

A sound advisory service program could well increase yields 10 to 20 bu/A; therefore, such a service should enter into growers' decisions when they are ready to make their choice of fertilizer.

Lime and an Efficient Return from Nitrogen, Phosphorus, and Potassium

Lime supplies the least expensive nutrients. The returns are quite high when it is applied where needed (Table 15-11). In spite of a high return, however, lime is often neglected in the fertility programs in humid regions. This may be because:

Visual responses to lime are often not so spectacular as those obtained with such nutrients as nitrogen, phosphorus, or potassium, unless the soil is particularly acid.

Its effects last over a period of time and the returns are not all realized in the first year.

Lack of a concerted promotion and sales campaign by some lime companies.

At the Brownstown Soil Experiment Field in Illinois, the effect of lime in a corn–soybean–wheat–hay rotation was studied on an acid soil for many years. With

TABLE 15-11. **Effect of Changing Soybean Prices, Limestone Rate, and Yield Response on Net Return to Liming, $/A ***

Lime Needed (ton/A)	Annual Yield Increase							
	3 bu		6 bu		9 bu		12 bu	
	$6	$8	$6	$8	$6	$8	$6	$8
1	14	20	32	44	50	68	68	92
2	10	16	28	40	46	64	64	88
3	6	12	24	36	42	60	60	84
4	2	8	20	32	38	56	56	80

* Limestone cost amortized over 5 years at 10% interest assuming a total cost of $15/ton applied, with net return being rounded to the nearest dollar.

Source: Hoeft, *Nat. Conf. Agr. Limestone,* National Fertilizer Development Center, Muscle Shoals, Ala. (1980).

adequate N-P-K over a period of 8 years the unlimed treatment showed a net annual loss per acre of $13.64, whereas with lime and N-P-K the net increase was $19.75 per acre.

Lime is the first step in any sound soil-management program, and broadcast applications in accordance with soil and plant requirements are essential for greatest returns from fertilizer.

Animal Wastes

When the supply of commercial fertilizers is plentiful, the high cost of labor and equipment used in spreading animal manure makes it difficult to calculate its value. However, front-end loaders and large-capacity spreaders for solid wastes and improved techniques for distributing liquid manure make it less of a problem.

Benefits in addition to those from added primary, secondary, and micronutrients may be related to the organic components which might improve soil moisture relations, increase the downward movement of nutrients such as phosphorus, or perhaps raise the carbon dioxide level in crop canopies.

There is much variability in manure, depending on methods of storing and handling. However, it appears in general that with current prices of nutrients, labor, and equipment it is profitable for the grower to use livestock manure. Because it is largely an N-K fertilizer, best returns should be obtained by applying to nonlegumes. Hauling charges can be reduced by applying it on fields close to the barn or feedlot and using commercial fertilizer on the more distant fields.

In some farm management programs it is imperative that the manure be disposed of, and its value may be balanced against equipment and hauling costs. The general feeling, however, is that it is worth a little more than cost of handling. In areas or times when fertilizers are difficult or impossible to obtain, animal manures play an important role in supplying fertilizer needs.

Maintaining Total Production on Fewer Acres

In some areas and for some crops, emphasis is being placed on controlled production. This has the effect of preventing a glutted market and thus maintaining a reasonable price structure. The history of any crop has been that the yields per acre increase. This is due primarily to two factors: first, the less productive acres are abandoned or put in soil-improving crops, and second, more attention is devoted to proper practices on the acres under cultivation.

As yields are increased, fewer acres are required to produce a given amount of a crop. For example, if additional fertilizer and other improved practices increased production 20%, acreage could be reduced accordingly. Total profit would be greater because fixed costs would be saved on the reduced acreage.

Plant Nutrients as Part of Increasing Land Value

When buying land, the farmer may be faced with the possibility of choosing high-priced or low-priced property. There is usually the question of which is the better buy. The higher-priced land is generally in a better state of fertility and tilth and has better improvements in terms of buildings and drainage.

The lower-priced land may actually be a good buy, however, provided that it has not been severely eroded or has no other physical limitations. Such land, however, is usually infertile and may need considerable lime, nitrogen, phosphorus, and/or potassium. Adequate liming and heavy fertilization, as indicated by soil tests and combined with other good practices, can rapidly bring the land up to good production levels. Such expenditures to improve fertility may be considered as part of the cost of the land. If the problem is considered in this light, $100 per acre for liming and buildup fertility may not seem out of reason. Thus by proper management it is possible to increase land productivity and value. The cost can be amortized over a period of years, as indicated earlier.

Additional Benefits from Maximum Profit Yields

Greater Flexibility. Although farming is a business, it differs from other businesses in one distinct way. Yields are influenced greatly by the weather. The production package could be in place for a specific yield goal, but weather, pests, disease, and so on, may not allow the grower to reach that yield. On the favorable side the yield might be higher than expected, the price higher than expected, or both higher than expected. However, there are three distinctly negative possibilities that occur more often than "perfect" conditions and they show how yield goal affects flexibility. Yields may be lower than expected, price lower than expected, and yield and price both lower than expected.

Yield Drop. Figure 15-9 shows that a 120-bushel corn yield goal gives the farmer only about 8 bushels between profit and loss. A 160-bushel yield goal gives him 41 bushels between profit and loss. Higher yield management increases the farmer's chance for profit in the face of uncertainties of environment.

Price Drop. Figure 15-10 shows that the farmer can take only about a 23 cents/bu price drop at the 120-bushel yield before falling out of the profit zone, but about an 84 cents/bu drop at the 160-bushel yield. Higher yields help make farmers much more flexible in their marketing operations and reduce risk.

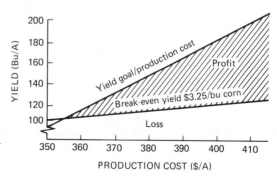

FIGURE 15-9. Higher yields increase profit potential ($3.25 per bushel corn). (Potash & Phosphate Institute, *Maximum Economic Yield Manual*, Atlanta, Ga., 1982.)

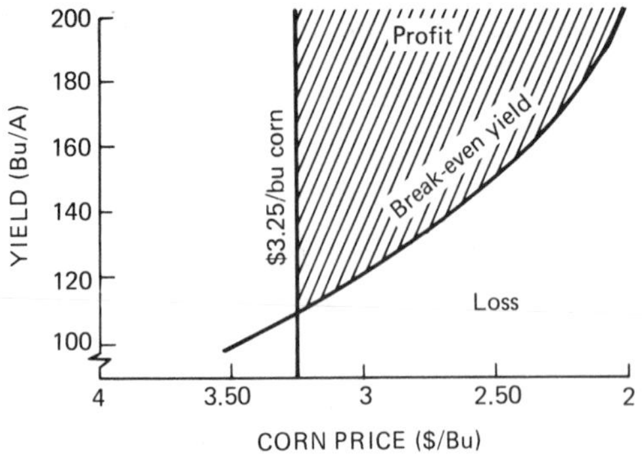

FIGURE 15-10. Higher yields mean lower break-even prices for corn. (Potash & Phosphate Institute, *Maximum Economic Yield Manual*, Atlanta, Ga., 1982.)

Increase in Energy Efficiency. Higher yields are an effective means of improving energy efficiency in agriculture. Higher yields require more input energy per acre, but the energy cost per bushel or ton is less. The reason for this is that some costs are the same regardless of yield level. For example, it takes just as much fuel to till a field yielding 40 bu/A of soybeans as one yielding 60 bu/A.

Reduction in Soil Erosion. Raindrops strike the soil with surprising force, dislodging particles and increasing soil erosion. However, growing crops, crop residues, and roots absorb the impact and slow water movement, decrease soil erosion, and increase water intake. The damaging effects of wind erosion are also reduced by the presence of crops and their residues.

Maximum profit yield systems match perfectly with soil conservation:

Crop canopy development is speeded.
Crop canopy density is increased.
More top and root residues are left.

On a Hickory silt loam with a 10% slope, management makes a difference.*

Poorly managed alfalfa (60% cover)   9.2 tons/A annual soil loss

Well managed alfalfa (95% cover)   1.1 tons/A annual soil loss

Conservation tillage practices such as no-till and chisel plowing leave more residues on the surface than moldboard plowing. However, with any given tillage practice, higher amounts of residues will decrease soil losses.

Increase in Soil Productivity. This is a long-term process and soil organic matter is increased over a long time period. Note the effects of increasing corn yields and soil organic matter in a model developed in Michigan on a loam soil:

Corn Yield (bu/A)	Soil Organic Matter (%)
50	1.7
100	3.1
150	4.0

In areas of higher temperatures and moisture it is difficult to increase organic matter. However, larger amounts of decomposing organic residues keep the soil in better physical condition. Infiltration of water is increased and supply for the

* R. Walker, University of Illinois, personal communication.

plant is improved. Thus plants are better able to withstand periods of drought. Also, less water runs off and erosion is reduced.

Reduction in Grain Moisture. More adequate fertilizer, particularly nitrogen and phosphorus, decreases the amount of water in the grain at harvest. This, of course, results in lower drying costs.

Energy efficiency in grain crops may be increased by allowing for field drying. For this to take place, it is imperative that the crop stand and not lodge. Another aspect of drying is that a plant such as corn stays green almost to the maturity of the grain. Water is lost from the plant and ear through the leaves by transpiration. If the plant dies from insufficient nutrients or pests, the water must be lost through the husks and drying is slow. Adequate plant nutrition makes for a more healthy plant which resists lodging and stays green.

Improvement in Crop Quality. Adequate plant nutrition improves crop quality, be it grain or forage. Increasing grain protein of hard red spring wheat with nitrogen additions will increase market price. In western Canada in 1980–1981 growers delivering wheat with a protein content over 13.5% received a premium of 13 cents/bu.

On a low-potassium soil, supplemental potassium not only increased soybean yields but decreased disease and mold in the seed (Table 15-12). Note the effects of potassium on this soil low in potassium on decreasing disease and decreasing dockage at the market place.

TABLE 15-12.   Effect of Adequate Potassium on Soybean Yield, Disease, and Dockage

K_2O (lb/A)	Yield (bu/ A)	Diseased and Moldy Beans (%)	Dockage (cents/ bu)	Value at $6 per Bushel ($/A)
0	38	31	54¢	$207.48
120	47	12	22¢	271.66

Source: M. Kroetz, Ohio State Univ., personal communication.

Maltsters pay a premium for plumper barley kernels. Note the effect of adequate nutrients on the percentage of plump kernels (Figure 15-11).

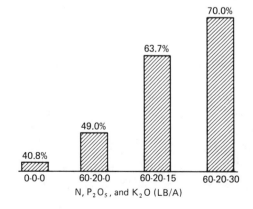

FIGURE 15-11.   N-P-K improves percent plump kernels of malting barley (Manitoba). [Vasey and Soper, *Better Crops Plant Food,* **50**(4):2 (1966–1967).]

70.0%

63.7%

49.0%

40.8%

0-0-0   60-20-0   60-20-15   60-20-30

N, P_2O_5, and K_2O (LB/A)

Summary

1. Maximum yield is a constantly moving target because of continual technological advances. Maximum-yield research is the study of one or more variables and their interactions in a multidisciplinary system that strives for the highest yield possible for the soil and the climate of the research site.

2. Maximum profit yield is based on maximum yield. It is the point where the last unit of an input just pays for itself and gives the greatest net return per acre or unit of land area.

3. Maximum profit yields continue to offer the greatest opportunity for improving net profit per acre and for reducing production costs per bushel or per ton. Adequate fertility is a major factor in obtaining higher yields. Although fertilizer use has increased in the past few years, it is still below optimum for maximum profit yields in most areas.

4. The urgency for higher yields is great. Cost of production is expected to increase in the coming years. Although prices of crops will increase, they will not increase as fast as cost of production. Hence higher and higher yields are a necessity for the farmer.

5. Fixed costs are those that remain about the same, regardless of yield. Hence practices that increase yields usually lower the cost of production per unit.

6. In a typical response curve each successive increment of a plant nutrient gives a smaller yield increase in accordance with the law of diminishing returns. Hence the return per dollar spent decreases, an important consideration to the grower with a limited amount of capital.

7. Progressive growers recognize that profit per acre is more important than return per dollar spent. Maximum profit from fertilizer application is obtained when the added return in yield just equals the cost of the last increment of fertilizer.

8. The most profitable rate of plant nutrients is related to expected increases in yield from each increment, weather, level of management, price of fertilizer, expected price of crop, additional harvesting and marketing costs, residual effects, and soil fertility level.

9. Management level is the degree to which all the factors affecting crop production are successfully controlled. At higher rates of fertilization, with expectation of higher yields, managerial ability becomes more and more important.

10. Regardless of crop prices it always pays to aim for the maximum profit yield. This will maximize profits or minimize losses.

11. Price of crop or fertilizer nutrients does not affect optimum fertilization appreciably.

12. Many practices cost little or nothing. Examples are timeliness, selection of variety, plant spacing, and rotations. Having these practices in place helps to increase returns from the higher cost inputs, such as fertilizer, land, labor, and machinery.

13. Residual effects of fertilizers and lime are an important part of fertilizer economics. With the increasing amounts of fertilizers being applied, the value of the residual fertilizer must be considered.

14. The cost of building soil fertility can be amortized over a period of years. The practices should be considered as long term rather than as an annual investment.

15. Price per pound of nutrients varies among sources of materials. Higher-analysis mixed fertilizers are generally the most reasonable in cost and in labor required for application.

16. Personalized agronomic service to the grower by the dealer will enter more and more into the grower's choice of a fertilizer dealer.

17. Priority in use of funds is of prime importance to the grower. It is generally more profitable to invest funds in fertilizer and lime applied in accordance with soil test level than in other parts of the farm business.

18. Lime usage is only about 35% adequate in the United States. Where needed, the returns from lime are much higher than those obtained from other plant nutrients.

Questions

1. What is maximum profit yield? How is it determined?

2. Why are higher crop yields urgent in the coming years?

3. Explain how maximum profit yields minimize losses in periods of low crop prices.

4. What are fixed costs? Variable costs? How do they affect unit cost of production?

5. When farmers have a limited amount of capital for fertilizer, are they going to be more interested in the rate that will give them the greatest return per dollar spent or the rate that will give them the greatest net profit? How would soil tests help in the decision? Would the situation be changed if a farmer had unlimited capital? Explain.

6. What are the banks in your area doing in regard to credit for lime and/or fertilizer?

7. How does use of irrigation make it easier to decide what rate of fertilizer to apply in semiarid and arid areas?

8. Discuss the factors that determine the most profitable rate of plant nutrients.

9. Why does level of management affect the return from a given level of fertilization?

10. What are some of the yield-improving practices costing little or nothing? How do they influence returns from high cost inputs?

11. Explain why there is considerable leeway in the amount of nutrients that can be added after a reasonable level of application is reached.

12. Explain how the following affect degree of carryover of fertilizer: soil, weather, application rate, yield increase, crop fertilized and harvested, and nutrient considered.

13. Why is it desirable to amortize the cost of building phosphorus and potassium levels or added lime over a period of years?

14. What nitrogen carrier would you choose to fertilize corn? Why? You are quoted the following prices: 3-12-12 at $120 per ton and 6-24-24 at $200 per ton. Which would you choose, and why?

15. Explain why funds invested in plant nutrients usually return more profit than investments in other phases of the farm business.

16. Obtain the average yield of corn, soybeans, or wheat in your area and the number of acres used for this production. What would a reasonable yield of corn per acre be if recommended practices were followed, and how many acres would be required to maintain present total production? What would you estimate the profits to be under each system?

17. What is holding back lime use in your area?

18. Is farm manure a valuable source of plant nutrients in your area? Why or why not?

19. How would you evaluate the residual nutrients in your area? Considering present crop production levels, are the soil test levels suggested in the 1970s adequate now?

20. Why do maximum profit yields increase energy efficiency? Reduce soil erosion? Increase crop quality?

21. How does higher-yield management give more flexibility and increase a farmer's chance for profit?

Selected References

Beaton, J. D., "Response to K: yield and economics," in *Potash for Agriculture Situation Analysis*, pp. 67–108. Atlanta, Ga.: Potash & Phosphate Institute, 1980.

Cooke, G. W., *Fertilizing for Maximum Yield*, 3rd ed. New York: Macmillan, 1982.

Hinton, R. A., "Consideration in crop selection and program participation in 1983." *Farm Economics Facts and Opinions*, Sec. 6, No. 4. Dept. of Agr. Economics, Univ. of Illinois (January 1982).

Hoeft, R. G., "Effects of agricultural liming materials on crop yield, quality and profits." *Nat. Conf. Agr. Limestone*, National Fertilizer Development Center, Muscle Shoals, Ala. (1980).

Marten, J. F., "1980 is history—plan now to make top yields in 1981." Potash & Phosphate Institute folder, Atlanta, Ga. (1980).

Marten, J. F., "Maximum economic yields." Potash & Phosphate Institute folder, Atlanta, Ga. (1981).

Petritz, D. C., "Higher alfalfa yields: costs, returns, considerations." *Better Crops Plant Food*, **66**: 34 (Fall 1982).

Potash & Phosphate Institute. *Soil Fertility Manual*. Atlanta, Ga., 1978.

Potash & Phosphate Institute. *Maximum Economic Yield Manual—A Guide to Profitable Crop Production*. Atlanta, Ga., 1982.

Stritzel, J. A., "You have a choice of fertilizer rates." *Iowa Farm Sci.*, **17**:14 (1963).

Sullivan, L. J., "What motivates U.S. farmers in use of fertilizer." *Phosphorus Agr.*, **61**:19 (June 1973).

Swanson, E. R., C. R. Taylor, and L. F. Welch, "Economically optional levels of nitrogen fertilizer for corn: an analysis based on experimental data, 1966–1971." *Illinois Agr. Econ.*, **13**(2):16 (1973).

Thornton, M., "Maximizing profit or reducing expenses." *Better Crops Plant Food*, **66**:20 (Fall 1982).

Vitosh, M. L., J. F. Davis, and B. D. Knezek, "Long-term effects of fertilizer, manure and plowing depth on corn." *Res. Rep. 198*, Michigan State Univ. (1972).

Vitosh, M. L., R. E. Lucas, and R. J. Black, "Effect of nitrogen fertilizer on corn yield." *Michigan State Univ. Ext. Bull. E-802* (1979).

Voss, R. D., J. T. Pesek, and J. R. Webb, "Economics of N fertilizer for corn." *Iowa State Univ. Pm 651* (1975).

Wagner, R. E., and W. K. Griffith, "Maximum economic yields are cost cutters." *Better Crops Plant Food*, **47**(3):34 (1964).

Walsh, L. M., and J. D. Beaton, Eds., *Soil Testing and Plant Analysis*, p. 204. Madison, Wis.: Soil Science Society of America, 1973.

Waters, W. K., "Corn and alfalfa returns and costs in 1981." Pennsylvania Ext. Serv., *Farm Management Reports* (1982).

Welch, L. F., "Nitrogen use and behavior in crop production." *Illinois Agr. Exp. Station Bull. 761* (1979).

Welch, L. F., "Economics favor building soil fertility." *Better Crops Plant Food*, **66**:3 (Fall 1982).

Wischmeier, W. H., and D. D. Smith, "Predicting rainfall erosion losses—a guide to conservation planning." *SEA, USDA Handbook 537* (1978).

Wortman, S., and R. W. Cummings, *To Feed the World: The Challenge and the Strategy*. Baltimore: The John Hopkins University Press, 1978.

Chapter 16

Fertilizers and Efficient Use of Water

WATER stress has often been a convenient scapegoat on which to blame any poorly growing crop, even though nutrient deficiency, pests, and other factors were fullfledged accomplices. There is much talk about good and poor seasons. Most of the time a good season in unirrigated regions is interpreted as one that has an ample amount of rain. Distribution of precipitation during the growing season can be just as important as the total amount received.

Agriculture so far generally has had first call on irrigation water. This may not continue indefinitely as pressures grow for increased industrial, recreational, and urban use. Agriculture will have to justify the water it uses, and agricultural needs will be balanced against other demands.

All of this means that water must be used as efficiently as possible. Finding ways of raising this efficiency is a prime challenge to agriculture. It is estimated that on the basis of what is now known, overall efficiency of water in irrigation farming is only about 50%. It is probably much less than that in unirrigated areas. In general, any growth factor that increases yield will improve the efficiency of water use. These factors include tillage, variety, plant spacing, pest control, time of planting, and plant nutrient supply.

The importance of the effect of fertilization on water requirement of plants has been recognized for many years. In 1913 Briggs and Shantz, USDA scientists, made this statement:

> Almost without exception the experiments herein cited show a reduction in the water requirement accompanying the use of fertilizer. In highly productive soils this reduction amounts to only a small percentage. In poor soils the water requirement may be reduced one-half or even two-thirds by the addition of fertilizers. Often the high water requirement is due to a deficiency of a single plant-food element. As the supply of such an element approaches exhaustion, the rate of growth as measured by the assimilation of carbon dioxide is greatly reduced but no corresponding change occurs in the transpiration. The result is inevitably a high water requirement.

About 50 years later another USDA scientist (Viets, 1962) said:

> Whether fertilizers increase consumptive use not at all or only slightly, all evidence indicates that water use efficiency, or dry matter produced per unit of water used, can be greatly increased if fertilizer increases yield. So fertilization for the adequate nutrition of all crops plays a major role in the efficient use and

conservation of water resources. Fertilizers may also increase root development within the soil so that soil water is used to higher tensions and at greater depths. This effect is important in dryland agriculture and even in farming in humid areas during periods of drought.

Hence it is quite clear that adequate fertility helps crops use water more efficiently and more of the crop is produced per inch of water. The improving of crop tolerance to low-rainfall conditions can be explained in part as follows:

Root exploration of the soil is increased. Adequate fertility favors expanded root growth and proliferation. When roots explore the soil a foot deeper, another inch or two of water will be obtained.

The major portion of the phosphorus and potassium moves to roots by diffusion through the water films around the soil particles. Under moisture stress the films are thin and path length of ion movement increases. Hence movement of phosphorus and potassium to the roots is reduced. Increasing the concentration of phosphorus and potassium in the soil and thus in the soil solution increases delivery to the roots.

Increased soil moisture tension (lower moisture) exerts a physiological effect on the roots. Elongation, turgidity, and the number of root hairs decrease with increasing tension. Mitochondria development slows, and *carrier concentration* and *phosphorylation* decrease. This all serves to reduce nutrient uptake.

Adequate fertility decreases water requirement. Transpiration is reduced. Potassium has been shown to aid in closing the stomata, thus reducing water loss.

Foliage canopy is increased and the soil is covered more quickly. This causes less evaporation directly from the soil to the air and most of the water is used by the plant. In a survey in Georgia, only 35% of the soybean fields had a full canopy in September. Sixty percent of the fields tested less than pH 6.0, 30% medium or less in phosphorus, and 79% medium or less in potassium.

Adequate fertility speeds maturity. This is particularly important for corn to help ensure pollination before summer drought hits. Similarly, small grains are adversely affected when growth is delayed so that summer drought occurs during and following heading.

The amounts of plant and root residues are increased. With any given tillage practice, a higher amount of residues will break the impact of raindrops, slow water movement, and increase water infiltration. The erosive effect of wind is also reduced by these residues.

Soil Moisture Level and Nutrient Absorption

Water is a key factor in all three mechanisms of nutrient uptake.

Interception. Roots intercept more nutrient ions when growing in a moist soil than when growing in a drier one because growth is more extensive. This is especially important for calcium and magnesium.

Mass Flow. Mass flow of soil water to supply the transpiration stream transports most of the nitrate, sulfate, calcium, and magnesium to roots.

Diffusion. Roots usually will not receive enough phosphorus and potassium by these first two methods. A third method, diffusion, is important. The plant absorbs nutrients adjacent to the roots and a concentration gradient is established. Nutrients then diffuse slowly from areas of higher concentration to areas of lower concentration but at distances no greater than $1/8$ to $1/4$ in. As this occurs through the water films, the rate of diffusion depends partly on the water content of the soil. With thicker water films or with a higher nutrient content in the soil the elements can diffuse more readily.

Considerable work has been done to establish the relationship between soil moisture levels and nutrient absorption by the plant. Nutrient absorption is affected directly by level of soil moisture as well as indirectly by the effect of water on the metabolic activity of the plant, soil aeration, and the salt concentration of the soil solution.

In Iowa as moisture stress increased, the concentration of nitrogen, phosphorus, and potassium in the corn leaves decreased. While application of nitrogen, phosphorus, and potassium increased the nutrient content of the plants, concentrations were still below the optimum range.

Potassium. The effect of decreasing soil H_2O (higher pF) on decreasing corn growth is shown in Figure 16-1. However, as percent potassium saturation is increased growth increased at all three moisture levels. Hence an adequate amount of nutrient, in this instance potassium, helps to overcome some of the stress associated with a low supply of water.

Summer rainfall (June, July, and August) is related to the response of corn to potassium. The lower the rainfall, the greater the response. This is related to at least two factors:

1. Most of the potassium absorbed must move to the plant root from adjacent areas. It passes through the water films, and the lower the water content the harder it is for the potassium to move. Hence at low soil moisture the potassium absorbed by the plant may originate in the immediate vicinity of the root. At high soil moisture it may originate at a greater distance from the root.
2. In some soils the subsoil contains less potassium than the surface. When the surface soil is exhausted of water in dry periods, the plant roots must feed in the subsoil where they cannot absorb as much potassium.

In a very wet year (about 25 in. seasonal rainfall) response was again large. This is related to restriction of aeration. Plant roots respire to get energy to absorb nutrients and this requires oxygen. Adequate added potassium helps to meet the needs of the plant even when root respiration is restricted.

Soybean yield response to phosphorus varies from year to year. This was found to vary most closely with amount of rainfall in the 12 weeks after planting over an 18-year period (Figure 16-2). The lower the rainfall, the greater the percentage response to phosphorus. The same relationship was found for potassium.

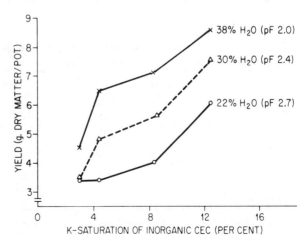

FIGURE 16-1. Dry-matter yield of corn after 3 weeks' growth in relation to potassium saturation and water supply. [Grimme et al., *Int. Symp. Soil Fertil. Eval. Proc.,* **1**:33. New Delhi: Indian Society of Soil Science (1971).]

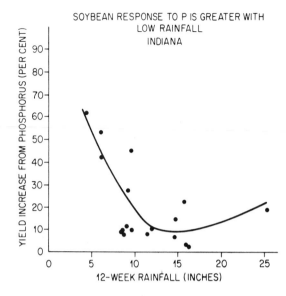

SOYBEAN RESPONSE TO P IS GREATER WITH LOW RAINFALL
INDIANA

FIGURE 16-2. The less the rainfall for 12 weeks after planting, the greater the percentage yield response of soybeans to phosphorus (Indiana). [Barber, *Better Crops Plant Food,* **55**(2):9 (1971).]

The effect of years of good rainfall and dry years on response of corn and soybeans to potassium is shown in Table 16-1. On a soil testing medium in potassium, there was little or no yield response or profit in years of good rainfall. In the stress years potassium gave a 48-bushel response on corn and 18-bushel response on soybeans with excellent profit.

In dry years the high rates of potash do not bring the corn or soybeans to the sufficiency potassium level in the leaves. The inability to take up adequate amounts of nutrients probably contributes to lower yields in dry years.

In Table 16-2 note the effect of increased fertility and peanut hulls on increasing corn yields in both dry and wet seasons. Even in a dry year added fertility and hulls increased yields to 157 bushels per acre. With good rainfall yields were 266 bushels per acre. Peanut hulls greatly increased potassium uptake. Moisture content was increased in the 0- to 15-cm depth by reducing evaporation and runoff and increasing infiltration.

Phosphorus. Black of the USDA found for spring wheat in Montana that 83% of the response to N-P fertilization in dry years was due to the effects of phosphorus. Yields were over 3.3 times greater in the wet years and phosphorus accounted for only 43% of the beneficial effects of N-P additions.

TABLE 16-1. **Effect of Potassium on Corn and Soybean Yields and Profits in Years of Good Rainfall and in Dry Years**

K_2O (lb/A)	Corn (bu/A)		Soybeans (bu/A)		K Soil Test, Initial 162 (lb/A)
	Good Year	Stress Year	Good Year	Stress Year	
0	163	81	56	30	129
50	163	113	59	42	152
100	167	121	60	48	196
200	163	129	58	48	236
Response	0	48	4	18	—
Profit ($/A)	0	87	18	104	—

Source: Johnson and Wallingford, *Crops and Soils,* **36**(6):15 (1983).

**TABLE 16-2.   Effect of Adequate Fertility on Corn
Yields (bu/A) in Good- and Poor-Rainfall Years**

Peanut Hulls Applied Annually (metric tons/ha)	1975 (Good Year)		1976 (Poor Year)	
	Unfertilized	100-100-100	Unfertilized	100-100-100
0	126	205	53	111
21	188	196	100	129
42	161	239	107	138
84	231	266	136	157
Average	176	226	99	134

Source: Lutz and Jones, *Agron. J.,* **70:**784 (1978).

In France there is a clear indication of the inverse relation between response of cereals to phosphorus and rainfall (Figure 16-3). The percent yield response of cereals to phosphorus is greater with low rainfall. In Syria added phosphorus improved the tolerance of cereal crops to acidity on a soil low in phosphorus.

Nitrogen. Although absorption of nitrogen is definitely reduced on dry soils, it is usually not reduced as much as that of phosphorus and potassium. Ammoniacal nitrogen does not move readily, but nitrate nitrogen is an anion and moves in and with the soil water. In heavy rains nitrates move downward in the soil profile and provide a storehouse of nitrogen for later use, provided that they do not move out of the root zone.

Another aspect is that under drought conditions, organic-matter decomposition, hence nitrogen release, is slowed.

Micronutrients. Much more information is needed about the effect of soil moisture levels on uptake of micronutrients. Temporary boron deficiency during periods of dry weather is quite common. Explanations include

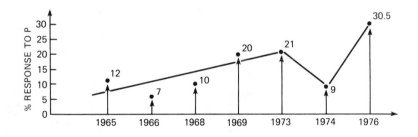

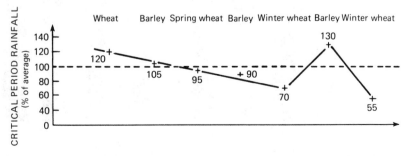

FIGURE 16-3.   Response of cereals to phosphorus is inversely related to the amount of rainfall. [Ignaze, *Phosphorus Agr.,* **70:**85 (June 1977).]

1. Much of the boron is in the organic matter, and under dry conditions organic-matter decomposition and release of boron are slowed down.

2. In some areas the lower soil horizons are lower in boron content than the surface. Under dry conditions the roots are less active in the surface horizons and the plants tend to take up less boron. On the other hand, in quite sandy soils excessive rainfall and leaching may remove some of the available soil boron.

Low soil moisture will also induce or aggravate deficiencies of manganese and molybdenum. Contrasted with this, iron and zinc deficiencies are often associated with high-soil-moisture conditions.

Increased soil moisture has been shown to result in greater amounts of molybdenum and cobalt uptake by alsike clover. There was more of these two elements in the soil solution, and the larger amounts in the plant were explained by greater transpiration and the sweeping of the enriched solution into the plant. Moisture level did not affect the uptake of copper. Manganese becomes more available under moist conditions because of conversion to reduced, more soluble forms.

Placement and Nutrient Absorption. Under drought conditions it is generally accepted that it is best to place the fertilizer in the zone of the soil that retains water for a greater part of the season. As shown in Figure 16-4, deeper placement of fertilizer gave greater absorption of nitrogen under dry conditions. There was no effect under wet conditions. Also in this study, depth of placement had no effect on phosphorus uptake under dry conditions and uptake was quite low. In the case of potassium, placement at 12 in. gave somewhat greater uptake than placement at 2 in. In general, deeper placement of nutrients, so as to be in

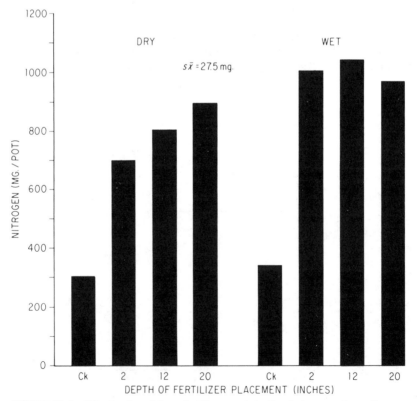

FIGURE 16-4. Effects of moisture, fertilizer, and depth of placement on nitrogen uptake by grain sorghum. [Eck et al., *Agron. J.,* **53**:335 (1961).]

moist soil a greater portion of the season, will result in more efficient utilization. Of course, if a soil is very dry, deeper placement will not be effective. Mention will be made later of the effect of deep banding on water-use efficiency.

Physiological Effect of Plant Nutrients Related to Water Economy

There are numerous nutrient–water interactions in plants, some of which are the following:

1. On a phosphorus-deficient soil added phosphorus speeds up maturity. The plant thus grows for a shorter period and, other things being equal, uses less water. For example, if maturity is advanced 7 days and the plant is using 0.2 in. of water a day, 1.4 in. of water will be saved.
2. A potassium-deficient plant is flaccid. Added potassium increases turgidity and helps to maintain internal water balance and the hydration of the protoplasm. A reduction in turgidity is accompanied by a decrease in stomatal openings to conserve water. In studies conducted in Montana, it was found that barley with adequate potassium had a reduced transpiration rate 5 minutes after exposure to hot, windy conditions. With severe potassium deficiency the transpiration rate was greatly increased and about 45 minutes was required for transpiration to be reduced substantially.
3. Stomatal opening is controlled by active transport of potassium into the guard cells. The guard cells of the closed stomata contain less potassium than those cells of the open stomata. Hence, uptake of CO_2 and photosynthesis by potassium-deficient leaves is less.
4. A high ion concentration in the cell increases osmotic pressure of the cell solute and consequently the plant's ability to withstand high water tension in the soil.
5. Water deficits in the plant affect all processes of cell growth, including division, enlargement, and maturation.
6. Water deficits cause a decreased rate of photosynthesis, which is related to the decreased supply of carbon dioxide caused by the closed stomata. The proportion of starch to sugar is often reduced because of greater hydrolysis of the starch.
7. Quality of succulent vegetables such as celery, lettuce, and cucumbers depends on their state of turgor. Nutrient additions that increase turgor will improve the quality.

How Water Is Lost from the Soil

Water in a soil is lost in three ways:

1. From the soil surface by evaporation.
2. Through the plant by transpiration.
3. By percolation beyond the rooting zone.

The sum of the water used in transpiration and evaporation from soil plus intercepted precipitation is called evapotranspiration. With a more complete cover less water evaporates from the soil directly and more goes through the plant. Adequate fertility and satisfactory stands are among those factors that help to provide more plant cover rapidly and thus get more benefit from the water.

Evapotranspiration can best be described in terms of the net radiation. This is the difference between the incoming radiation and radiation losses from soil and crop surfaces. Net radiation is used (1) to evaporate water from the soil or plant; (2) to heat the air, the soil, or the plant; or (3) in photosynthesis.

With a sparse stand or growth much sunlight will reach the soil, and considerable water may be evaporated directly from a moist soil without passing through the crop. With a heavy crop canopy a blanket of green is presented to the sun and less energy reaches the soil. The soil is kept cooler, the crop provides insulation to maintain a higher humidity just above the soil, and there is less air movement. These three effects are among those helping to reduce evaporation from the soil. It should be kept in mind, however, that even with a heavy canopy a considerable amount of energy still reaches the soil.

Fertilizer affects plant size, total leaf area, and often the color of the foliage. Close rows and adequate stands, along with adequate fertilization, help to provide a heavy crop canopy quickly. For example, in Iowa water use by corn was less in 21-in. rows than in 42-in. rows. Differences in evapotranspiration among crops may be small once a complete cover is developed. Daily use of water with a growing crop on the soil varies greatly from one day to another, depending on soil and atmospheric environmental conditions such as temperature, moisture present, and wind. However, losses of 0.1 to 0.3 in. of water daily per acre are common.

Evaporation from the soil may account for one-third to two-thirds of the total disappearance of water in a crop year in humid areas or where the soil is wet. Experiments in Illinois have shown that when the soil was covered with plastic to reduce evaporation from the soil, corn used 6.1 in. during the growing season in contrast to 13.4 in. when grown normally. Under these conditions a heavy crop canopy would increase water use efficiency.

With local droughts or in arid regions the soil surface is dry, and very little water is lost from the soil. The moisture films between the particles are thin, and little water is transported to the soil surface by capillarity or diffusion of water vapor. Hence on dry soils most of the water use is by transpiration, and low plant spacings are generally employed to reduce the amount of water used by the plants.

Heat advection, in which there is horizontal and vertical movement of air in a turbulent fashion, brings in more heat. In a hot, dry area with a strong wind the heat from the air may contribute to 25 to 50% of the total evapotranspiration. In arid and semiarid areas advection is great, and thus quite variable evapotranspiration may occur.

Effect of Plant Nutrients on Water Requirement

Water requirement is defined as the ratio of the weight of water absorbed by the plant during the growing season to the weight of dry matter produced by the plant. As far back as 1912, workers in Nebraska showed that manure greatly improved effectiveness of water use by corn.

	Pounds of Water per Pound of Dry Ears	
	No Manure	Manure
Infertile	2136	692
Medium fertility	1160	679
Fertile	799	682

Adequate nitrogen decreases the water used per pound of dry matter in grasses (Table 16-3). With the exception of Pangola grass, in which there was a stand problem at the 200 lb rate of nitrogen, the grass receiving the 200-lb rate used only 35 to 40% as much water per pound of dry matter as was used at the 50-lb rate. The wide difference among the grasses is significant, for Common Bermuda used five times as much as Suwanee.

**TABLE 16-3. Effect of Nitrogen on Water
Used per Pound of Dry Matter**

	Water Used per Pound of Dry Matter		
	50 lb N/A	*100 lb N/A*	*200 lb N/A*
Common Bermuda	8275	3962	2941
Coastal Bermuda	2012	1206	722
Suwanee Bermuda	1515	915	572
Pensacola Bahia	2652	1633	1054
Pangola	2546	2049	2625

Source: Burton et al., *Agron. J.,* **48**:498 (1957).

Water Use Efficiency. Water use efficiency is the yield of crop in bushels, pounds, or tons per inch of water—from the soil, rainfall, and irrigation. When cultural practices increase yields, water use efficiency is increased. Yields of crops have increased greatly in the past 20 years, on essentially the same amount of water. Improved soil and crop management practices have brought this about.

Conservation Tillage Saves Water. Tillage which leaves large amounts of surface residues and a rough surface helps to save water—2.8 in. in wheat ground in Nebraska and 20% more available soil water in Indiana as examples.

Increases water infiltration. Vigorous plant growth, heavy crop residues, and extensive root growth help.
Decreases evaporation from the surface.
Gradually deepens the root zone through chiseling or subsoiling, more organic matter.
Increases the reservoir of water in the soil.
An extra inch or two of soil water will help carry a crop through low rainfall periods.

Maximum-Yield Research Shows High Water Use Efficiency. In New Jersey yields of 185 bushels of corn and 71 bushels of soybeans were obtained with 14.2 in. of water during the growing season. State average yields were 102 bushels of corn and 25 bushels of soybeans.

In Indiana, yields of 85 bushels of soybeans and 226 bushels of corn were obtained on 40 in. total rainfall with 16 in. during the growing season. State average yields were 40 bushels of soybeans and 129 bushels of corn.

In Michigan, a 2-year average of 10 tons/A of alfalfa hay was obtained with 33.3 in. of water during the growing season. State average yield was 3.4 tons/A.

In California, added nitrogen increased wheat yield from 84 bu/A to 124 bu/A and water use efficiency from 3.5 bu/in. to 5.2 bu/in. (irrigated). State average was 80 bu/A (Robinson et al., 1979).

Irrigated Soils. Fertility is one of the important controllable factors affecting plant growth and it is important to have adequate amounts. This principle is well illustrated in Figure 16-5. Adequate nitrogen markedly increased the yield of Coastal Bermuda hay. At the same time the amount of water per ton of hay decreased from 18 in./ton with no applied nitrogen to 3 in./ton with 1000 lb/A of nitrogen.

On a low-phosphorus soil, phosphorus banded before seeding increased the amount of alfalfa hay obtained with each inch of water. At each water level the 264-lb/A

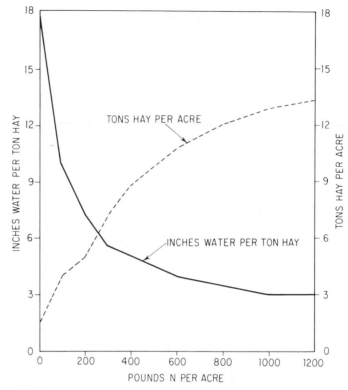

FIGURE 16-5. Effect of nitrogen on water-use efficiency by Coastal Bermuda hay. Phosphorus and potassium were applied in liberal amounts. [*Texas A and M College Prog. Rep. 2193* (1961), given on page 192 of *Fertilizer Salesmen's Handbook,* published and copy-righted by National Plant Food Institute, 1963, by permission of copyright owners.]

of phosphorus treatment increased the tons per inch of water 35 to 40% over the 44-lb/A treatment.

A diagrammatic presentation of these data for alfalfa, as well as the effect of nitrogen on cotton related to soil-water stress, is shown in Figure 16-6. Increased water stress reduced the alfalfa yield about the same proportion under high- and low-phosphorus treatment. With low nitrogen on cotton the yields were already so low that increased water stress had little effect.

Under very high fertility, irrigated conditions with a yield of 299 bu/A, bushels of corn per inch of water reached 11.6 (Table 16-4). With medium fertility the bushels per inch was 8.3. Note the increase in amount of residues.

TABLE 16-4. Effect of Fertility Level on Corn Yield, Amounts of Residues, and Bushels per Inch of Water *

	Corn Yield (bu/A)	Residues Returned (tons/A)	Bushels per Inch of H_2O
Medium fertility, irrigated	214	5.0	8.3
Very high fertility, irrigated	299	6.4	11.6

* High P and K soil, two-yr average.
Source: R. L. Flannery, New Jersey Agricultural Experiment Station, personal communication.

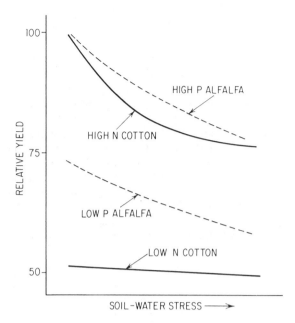

FIGURE 16-6. Fertility–irrigation relationships for cotton and alfalfa on a sandy Arizona soil. [Stanberry et al., *SSSA Proc.,* **19**:303 (1955).]

In many parts of the world large sums of money are being spent for irrigation without proper attention to cultural practices. After the lack of moisture is eliminated by irrigation, a great number of things may limit yields (Figure 16-7). Because of these other factors, there can be many disappointments. It must be kept in mind that if yields of 300 bushels rather than 150 bushels of corn or 14 tons rather than 7 tons of alfalfa are to be obtained, the nutrient removal is at

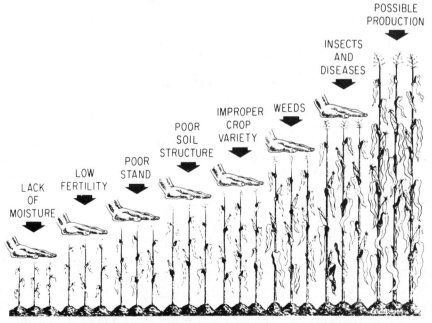

FIGURE 16-7. After the lack of moisture is eliminated by irrigation, a great number of things may limit yield. Careful attention is needed to get the most return out of the water, fertilizer, and other inputs. (*USDA, SCS Bull. 199.*)

least doubled. This means that the crop must obtain more of these elements from some source, whether from the native soil supply, manures, or fertilizers.

Unirrigated Soils. Periods of moisture shortage occur in most cropping areas and this is particularly true in the Great Plains. Getting the most bushels out of every inch of water is important every year and adequate plant nutrients play a key role. Nitrogen and phosphorus are particularly important for highest wheat yields when water is limiting.

In Nebraska studies, 121 field experiments on four crops were evaluated. Optimum fertilizer treatment increased water use efficiency of grain crops an average of 29%. This increase was almost in proportion to yield response to fertilizer. It was emphasized that this represents a major economy in crop water use and is significant in view of shrinking water resources.

In Montana, nitrogen increased wheat yield, evapotranspiration, and bushels per inch of water. Without added nitrogen, water extraction was largely limited to the upper 3 ft. With nitrogen, water was extracted to 6 ft.

	0 N	60 lb N	240 lb N
Yield of wheat (bu/A)	24	46	54
Evapotranspiration (in)	8.7	10.7	12.4
Water remaining in 7-ft profile (in)	7.1	5.2	3.8
Bushels per inch of H_2O	5.1	6.7	6.4

In North Dakota the same trends were observed for wheat, but fertilizer had little effect on the amount of water remaining in the profile. Yield, rainfall, and soil would affect the latter. In one experiment on corn with no nitrogen, the yield was 107 bu/A, and with 200 lb of nitrogen, 177 bu/A. Bushels per inch of water were 3.7 and 6.1, respectively, still relatively low. With barley in Alberta, efficiency of moisture use was more than doubled by fertilizer and fertilizer plus manure treatments.

The responses to deep band or dual placement were discussed in Chapter 13. Some of the positive effects occurred in areas subjected to moisture stress and were due to placement of the nutrients deeper in the soil where the soil was more likely to be moist. A comparison of the effects of two application techniques on the moisture use efficiency of wheat is shown in Figure 16-8. Bushels per inch of water were increased greatly with dual injection application. The water-use efficiency was improved from 3.5 bushels per inch of water to 6.2 bushels with the highest rate of fertilizer, 77-60-0 of N-P_2O_5-K_2O.

Much additional information could be cited, including sugar beets in Colorado, cotton in California, cotton in Alabama, potatoes in Idaho, grain sorghum in Texas, hay in South Dakota, corn in Wisconsin, and wheat in Saskatchewan and Manitoba. However, the data shown serve to illustrate the effect of adequate nutrients on obtaining greater efficiency in the use of water.

Fertilization and Water Extraction by Roots

Most crops use water more slowly from the lower root zone than from the upper soil. The top quarter is the first to be exhausted of available water (Figure 16-9). In periods of stress the plant must draw water from the lower three-fourths of the root depth.

The favorable effects of applied plant nutrients on the mass and distribution of roots when soils are deficient are well known. This subject was discussed in Chapter

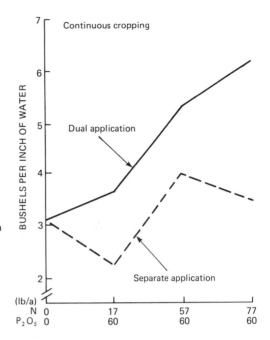

FIGURE 16-8. Effect of fertilizer placement on water use efficiency in wheat. Dual = anhydrous ammonia and ammonium polyphosphate solution injected prior to seeding; separate = anhydrous ammonia injected prior to seeding and ammonium polyphosphate solution applied with seed. [Houlton, *Great Plains Univ.—Ind. Soil Fert. Workshop,* Denver, Colo. (March 6–7, 1980).]

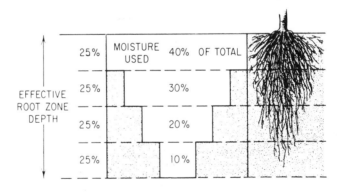

FIGURE 16-9. The top quarter of the root zone is the first to be exhausted of available moisture. Certain management practices, including adequate fertilization, help to develop a deeper root system to use the moisture from the lower root zone. (*USDA, SCS. Bull. 199.*)

13 and illustrated in Figures 13-7 and 13-8. With soils requiring the addition of plant nutrients, the plant may extract water from a depth of only 3 to 4 ft. With fertilization the plant roots may be effective to a depth of 5 to 7 ft or more. Hence the effective depth of the reservoir from which the plant can draw water is increased. If the plant can utilize an extra 4 to 6 in. of water from the lower depths, the crop can endure droughts for a longer period of time without disastrous results.

Studies at the Midwest Claypan Soil Conservation Experiment Farm at McCredie, Missouri, showed that on August 17 there were 1.04 in. of available water in the top 42 in. of soil under well-fertilized corn (Figure 16-10). When no fertilizer was applied there were 4.5 in. On the well-fertilized area the amount of water needed per bushel of corn was 5,600 gal, whereas on the low-fertility area it was 21,000 gal. Growers cannot afford to be this wasteful of water.

It should be emphasized that in areas in which the subsoil is dry, increased fertilization will not help crops to penetrate the soil farther to get more water.

Soils Differ in Their Water-Holding Capacity. Texture, structure, and organic matter content influence the water-holding capacity of soils, which may

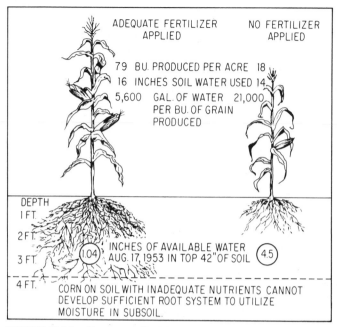

FIGURE 16-10.   Corn on soil with inadequate nutrients cannot develop a sufficient root system to use subsoil moisture effectively. [Smith, *Missouri Farmers Assoc. Bull.* (1963).]

vary from less than 1 in. on sandy soils to more than 2 in./ft of soil on silt and clay loams. The capacities in the surface 5 ft of representative soils in Illinois are as follows:

Oquawka sand	5 in.
Ridgeville fine sandy loam	7 in.
Swygert silt loam	9 in.
Muscatine silt loam	12 in.

However, crops root differently in different soils because of compact soil horizons, zones of unfavorable pH, high concentrations of aluminum and manganese, and so on, or inadequate nutrient supply. The approximate depth of rooting in three soils illustrates this. Note the difference in available water.

	Depth of Roots (ft)	Water Available (in.)
Clarence silt loam	3	6.5
Saybrook silt loam	4.5	10.5
Muscatine silt loam	5+	14

Stored Water and Fertilizer Recommendations

The relation of stored soil water to crop responses to fertilizer has received much attention in semihumid and low-rainfall regions. In such areas the rainfall during the growing season is low and stored soil water is of importance.

TABLE 16-5. Maximum Increase in Yield of Wheat (bu/A) from N-P on Nonfallow Land as Related to Stored Moisture at Seeding and to Rainfall During the Growing Season (66 Trials)

Rainfall from Seeding to 20 Days Before Harvest (in.)	Available Soil Moisture (in./4 ft soil)		
	0–2 (Low)	*2–4 (Medium)*	*4–6 (High)*
> 8	7.1	10.0	15.0
6–8	5.0	9.5	16.4
< 6	2.4	5.9	10.5

Source: Norum, *Better Crops Plant Food,* **47**(1):40 (1963).

An example of responses to N-P in 66 trials with wheat in North Dakota at three levels of stored moisture and three levels of rainfall is shown in Table 16-5. Note that even with the lowest levels of moisture, yield responses were obtained.

On the basis of such information, nitrogen recommendations in some states are made according to stored soil moisture. However, most of these states also base their recommendations on the residual NO_3-N in the soil profile, or fertility level, as related to yield potential. This approach is based on the concept that adequate fertility will help to make better use of the water that is available. With a low amount of water in the profile the probability of a high response to nutrients is less.

In some states a systematic survey of the moisture in the soil profile to rooting depth, 4 to 6 ft, is made in late fall or early spring. This information must then be weighed against the probability of summer rainfall. The approach used for wheat production in eastern Washington was described in Chapter 12.

Figure 16-11 shows that additional water supply (stored + rainfall) will increase yield of wheat to a point, but if fertility is limiting, additional water will not be beneficial. Even with a very low supply of water, fertilizer was beneficial.

Fertilizers have an indirect effect on the amount of stored water in the soil profile. When there is a response to fertilizer, an increased amount of vegetative

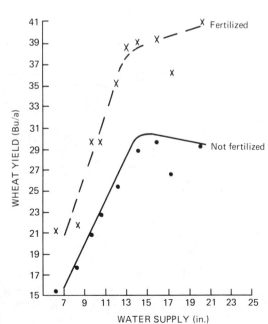

FIGURE 16-11. Wheat yields as affected by soil fertility levels and moisture supply. [Wagner and Vasey, North Dakota State Univ., Coop. Ext. Serv., *Soils Fert. 1* (1971).]

cover is produced in the growing crop. Runoff of intense rains is retarded and infiltration increased. In the nongrowing season the greater amount of residues on the soil surface help retain the water.

In some irrigated regions fall irrigation after crops are harvested is practiced as an insurance policy, particularly with flood irrigation. On deep soils up to 12 in. of water can be stored for the next year's crop. This has several advantages. Water is usually lower in price in the fall. Wetting and drying and freezing and thawing over the winter may improve soil structure. Sometimes, water is short during the growing season.

Fertility Pays in Dry and Wet Seasons

Adequate fertility helps to minimize the dips in production due to a poor season. An example of the effect of lime and P-K on corn grown in a corn–soybean–wheat– hay rotation over a 13-year period is shown in Figure 16-12. The soil was an acid planosol low in fertility in southern Illinois. It is evident that fertility was beneficial in both very dry and very wet seasons.

An indirect effect of fertility was that the limed plots in this experiment could usually be plowed 1 to 2 weeks before the unlimed plots were dry enough to support a tractor. This faster drying is probably the result of more crop residues, improved physical structure, greater root volumes, and deeper root penetration. Beneficial effects of liming on soil structure were reviewed in Chapter 11.

Periodic fluctuations in crop yields related to rainfall have been noted in the Netherlands. Average yields of wheat and rye after some dry years amount to 1.5 times those obtained after a succession of wet years. Pea yields are about three times higher. Water-soluble phosphorus and exchangeable potassium rise in dry periods and fall in wet periods, whereas pH is the reverse.

Although the amount of available nitrogen is affected by alternating rainfall periods, the effect of winter rainfall on nitrogen response is paramount in most cases. The effect of greater winter rainfall on a lower carryover of nitrogen is shown in Figure 16-13. It should be theoretically possible to eliminate some of the yield

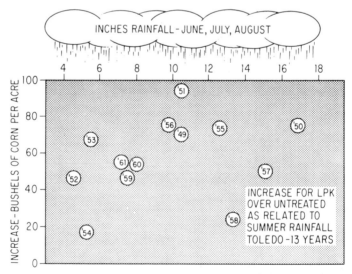

FIGURE 16-12. Increase in corn yield from lime-P-K, as related to June, July, August rainfall—13 years in southern Illinois. Corn grown in rotation with legumes. [Lang et al., *Better Crops Plant Food*, **47**(1):16 (1963).]

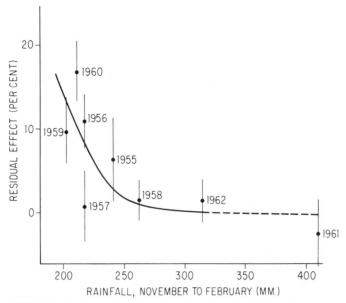

FIGURE 16-13. Relation between amount of rainfall from November to February and average residual effect of nitrogen in different years in the Netherlands. Vertical lines through dots indicate standard errors. [van der Paauw, *Plant Soil,* **19**(3). Copyright 1953 by Martinus Nijhoff, The Hague.]

fluctuations at a high level of production by appropriate fertilization and management.

The effects of weather and of technology on corn yields have been studied in several Corn Belt states. Calculated yields were compared with actual yields and were determined by considering weather factors, together with the technological changes. It is of interest to note the increases from 1930 to 1980 (Figure 16-14). Many factors, including adequate nitrogen and other plant nutrients, have helped

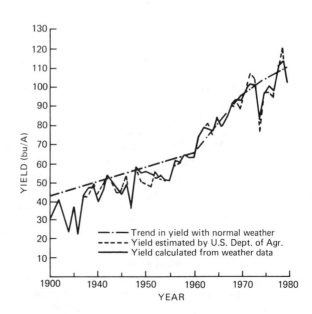

FIGURE 16-14. Actual and calculated yields of corn for Illinois, Missouri, Indiana, Iowa, and Ohio, 1930–1980. [Thompson, *Better Crops Plant Food,* **66**:18 (Spring 1982).]

to bring this about. Similar studies have been made for soybeans, wheat, and grain sorghum.

Take Advantage of the Good Years

Farmers work on probabilities because of uncontrollable weather factors, much of it related to moisture. When they manage all the controllable factors to the best of their ability on a long-term basis, the probability is increased for higher yields and profits.

It is important to have inputs in place to be able to take advantage of an excellent year and to maximize the use of water. Although extreme drought years are infrequent in most areas, even if there is no response to fertilizer in such years, most of the fertilizer, particularly phosphorus and potassium, will be in the soil for the next crop. Progressive farmers obtain much better yields than their neighbors in both good and poor years.

"If you cannot or do not wish to irrigate, start improving your soil through conservation tillage and high-yield management. It will take 4 or 5 years to start showing results," says Herman Warsaw, an Illinois farmer. After a long-term conservation tillage program, in one 5-year period he had a 293-bushel/A corn average on 13 in. of rainfall during the growing season. He is a believer in residue management and tillage to save soil and water. Infiltration of water is increased, he has created a deeper root zone, and there is greater water storage capacity. The corn crop has a larger reservoir of water in the soil on which to draw during periods of moisture shortage.

Summary

1. A good or a poor season is often related to the amount and distribution of rain received during the growing period.

2. It has been known for many years that plant nutrients applied on deficient soils will increase the crop yield per inch of water.

3. Increased moisture tension which reduces the percentage of many elements in plants is related to thinner water films around the soil particles, hence less diffusion to the roots.

4. Response to potassium is greater in dry years and in very wet years.

5. A high ion concentration in the cell increases the osmotic pressure of the cell solute and the ability of the plant to withstand high water tensions in the soil.

6. Water is lost from the soil by evaporation from the soil surface and from the plant by transpiration. The sum of these two losses plus the evaporation of intercepted precipitation held on the plant parts is called evapotranspiration.

7. On moist soils a heavy crop canopy helps to reduce direct evaporation from soils and a higher percentage of the water is used by the plant. Many factors including higher plant stands and fertility bring about a heavier crop canopy.

8. On dry soils little water is evaporated directly from the surface. Low plant spacings are generally employed to reduce the amount of water used by the plants in dry areas.

9. Water use efficiency is the yield of crop in bushels, pounds, or tons per acre-inch of water from the soil, rainfall, and irrigation. Any practice that promotes plant growth and the more efficient use of sunlight in photosynthesis to increase crop yields will increase water use efficiency.

10. Many examples are available to show the effect of fertilizer on increased yield of crops per inch of water.

11. Large sums of money are being spent for irrigation in many parts of the world. Limiting factors, including lack of fertility, are often holding down yields and disappointing results are obtained.

12. Adequate fertility promotes a more extensive and deeper root system. The effective depth of the reservoir from which plants can draw water is thus increased. Compaction and low fertility discourage deep rooting.

13. Adequate fertility helps to avoid the drastic dips in crop yields caused by inadequate rainfall or even excessive water.

14. Growers in the past have worried about having enough moisture to get the most out of the fertilizer. In the future they will worry about having enough fertilizer to get the most out of the moisture.

15. Conservation tillage, combined with excellent management practices over a long-term period, improves water intake, creates a deeper root zone, and provides additional water for the crop to draw on during periods of moisture shortage.

Questions

1. Describe a good or a poor season in an irrigated area and in an unirrigated area.

2. Describe the three mechanisms of nutrient absorption.

3. In a given soil volume, why is absorption of many nutrients by plants decreased as soil moisture tension increases?

4. Explain the greater response to potassium in dry years and in wet years. Why is boron generally less available under dry conditions?

5. How might low soil moisture affect interpretation of the potassium leaf analysis?

6. Why does a higher ion concentration in the cell increase drought resistance?

7. What is evapotranspiration? How does a heavy crop cover affect the losses in the various components of evapotranspiration? Is there more total water loss with a greater yield?

8. Explain the difference in evaporation losses from a moist soil surface and a dry soil surface.

9. Define water use efficiency. Why is it so important in agriculture? List factors that affect water use efficiency.

10. What is the effect of adequate plant nutrients on water-use efficiency? Why?

11. Explain the effect of adequate nutrients in increasing the extent of the root system. Why is this important in drought periods?

12. On what soils in your area will root penetration be limited by lack of fertility and by physical condition in the lower soil horizons?

13. How might placement of nutrients affect uptake in a dry year?

14. What soils in your area have a low water-holding capacity? Why?

15. Why is stored water of importance in dry regions? What advantages are there in irrigation in the fall after crops are harvested?

16. Are there irrigated farms in your region in which full returns are not being obtained from an investment in irrigation? Why?

17. Average yields of many crops are much higher than 20 years ago with about the same amount of rainfall. Why?

18. Explain how conservation tillage increases water available to the crop. Why is the crop better able to withstand periods of moisture shortage?

Selected References

American Society of Agronomy, "Research on water." *ASA Spec. Publ. Ser. 4* (1964).

Barber, S. A., "Rainfall and response," *Better Crops Plant Food*, **47**(1):6 (1963).

Barber, S. A., "Water essential to nutrient uptake." *Plant Food Rev.*, **10**(2):5 (1964).

Barber, S. A., "Soybeans do respond to fertilizer in a rotation." *Better Crops Plant Food*, **55**(2):9 (1971).

Bauer, A., "Fertilized wheat uses water more efficiently." *Farm Res.*, **24**(3):4, North Dakota Agr. Exp. Sta. (1966).

Black, A. L., "Adventitious roots, tillers, and grain yields of spring wheat as influenced by N-P fertilization." *Agron. J.*, **62**:32 (1970).

Black, A. L., "Long-term N-P fertilizer and climate influences on morphology and yield components of spring wheat." *Agron. J.*, **74**:651 (1982).

Briggs, L. J., and H. L. Shantz, *USDA Bur. Plant Ind. Bull. 285* (1913).

Brown, P. L., "Water use and soil water depletion by dryland winter wheat as affected by nitrogen fertilization." *Agron. J.*, **63**:43 (1971).

Burton, G. W., G. M. Prine, and J. E. Jackson, "Studies of drouth tolerance and water use of several southern grasses." *Agron. J.*, **48**:498 (1957).

Eck, H. V., and C. Fanning, "Placement of fertilizer in relation to soil moisture supply." *Agron. J.*, **53**:335 (1961).

Escamilla, E., R. Voss, and J. R. Webb, "Effect of prices and moisture stress on N rates that maximize two economic criteria for corn." *Agron. J.*, **71**:609 (1979).

Garman, W. H., *The Fertilizer Handbook*. Washington, D.C.: National Plant Food Institute, (1963).

Grimme, H., K. Nemeth, and L. C. von Braunschweig, "Some factors controlling potassium availability in soils." *Int. Symp. Soil Fert. Eval. Proc.*, **1**:33. New Delhi: Indian Society of Soil Science, (1971).

Henderson, D. W., R. M. Hagan, and D. S. Mikkelsen, "Water use efficiency in irrigation agriculture." *Better Crops Plant Food*, **47**(1):46 (1963).

Hoyt, P. B., and W. A. Rice, "Effects of high rates of chemical fertilizer and barnyard manure on yield and moisture use of six successive barley crops grown in three gray Luvisolic soils." *Can. J. Soil Sci.*, **57**:425 (1977).

Ignazi, J. C., "Influence of climatic conditions on the response to phosphate in experiments on field crops." *Phosphorus Agr.*, **70**:85 (June 1977).

Johnson, J. and W. Wallingford, "Weather-stress yield loss." *Crops Soils*, **36**(6):15 (1983).

Kubota, J., E. R. Lemon, and W. H. Allaway, "The effect of soil moisture content on the uptake of Mo, Cu, and Co by alsike clover." *SSSA Proc.*, **27**:679 (1963).

Lang, A. L., L. E. Miller, and P. E. Johnson, "Fertility pays in flood or drouth." *Better Crops Plant Food*, **47**(1):16 (1963).

Lutz, J. A., Jr., and G. D. Jones, "Effect of peanut hulls on the performance of corn." *Agron. J.*, **70**:784 (1978).

Mengel, K., and L. C. von Braunschweig, "The effect of soil moisture on the availability of K and its influence on the growth of young maize plants." *Soil Sci.*, **114**(2):142 (1972).

Murick, J. T., D. W. Grimes, and G. M. Herron, "Water management consumptive use, and N fertilization of irrigated winter wheat in western Kansas." *USDA, ARS Prod. Res. Rep. 75* (1963).

Odell, R. T., *Univ. Illinois Agron. Facts SP-16* (1956).

Olsen, S. R., F. S. Watanabe, and R. E. Danielson, "Phosphorus absorption by corn roots as affected by moisture and phosphorus concentrations." *Soil Sci. Soc. Am. Proc.*, **25**:289 (1961).

Olson, R. A., C. A. Thompson, P. H. Grabouski, D. D. Stukenholtz, K. D. Frank, and A. F. Dreier, "Water requirement of grain crops as modified by fertilizer use." *Agron. J.*, **56**:427 (1964).

Potash & Phosphate Institute. *Maximum Yield Systems—How They Improve Water Use Efficiency*. Atlanta, Ga., (1983).

Power, J. F., D. L. Grunes, W. O. Willis, and G. A. Reichman, "Soil temperature and P effects upon barley growth." *Agron. J.*, **55**:389 (1963).

Power, J. F., D. L. Grunes, G. A. Reichman, and W. O. Willis, "Effect of soil temperature on the rate of barley development and nutrition." *Agron. J.*, **62**:567 (1970).

Robinson, F. E., D. W. Cudney, and W. F. Lehman, "Nitrate fertilizer timing, irrigation, protein, and yellow berry in Durum wheat." *Agron. J.*, **71**:304 (1979).

Romig, R. E., and H. F. Rhoades, "Interrelationships of soil moisture level at planting time and nitrogen fertilization on winter wheat production." *Agron. J.*, **55**:123 (1963).

Sawhney, B. L., and I. Zelitch, "Big pump for stomata." *Better Crops Plant Food*, **55**(3):3 (1971).

Skogley, E. O., "Potassium in Montana soils and crop requirements." *Montana Agr. Exp. Sta. Res. Rep. 88* (1976).

Smith, G. E., "Soil fertility—basis for high crop production." *Missouri Farmers Assoc. Bull.* (1953).

Stanberry, C. O., C. D. Converse, H. R. Haise, and O. J. Kelley, "Effect of moisture and P variables on alfalfa hay production on the Yuma Mesa." *SSSA Proc.*, **19**:303 (1955).

Stewart, B. A., "Fertilizers and water," in *The Fertilizer Handbook*. Washington, D.C.: The Fertilizer Institute, 1982, p. 137.

United States Department of the Interior, Bureau of Reclamation, and United States Department of Agriculture, Soil Conservation Service, "Irrigation on Western farms." *Agr. Inf. Bull. 199* (1959).

van der Paauw, F., "Periodic fluctuations of soil fertility, crop yields and of responses to fertilization as affected by alternating periods of low or high rainfall." *Plant Soil*, **17**(2):155 (1962).

van der Paauw, F., "Residual effect of N fertilizer on succeeding crops in a moderate marine climate." *Plant Soil*, **19**(3):324 (1963).

Viets, F. G., Jr., "Fertilizers and the efficient use of water." *Adv. Agron.*, **14**:233 (1962).

Voss, R. D., "Phosphorus most limiting factor for corn in Iowa." *Proc. 22nd Ann. Fert. Agr. Chem. Dealers Conf.*, Iowa State Univ. (1970).

Wagner, D. F., and E. H. Vasey, "Refining fertilizer recommendations by state areas, stored soil moisture conditions and rainfall probabilities." North Dakota State Univ., Coop. Ext. Serv., *Soils Fert. 1* (1971).

Watanabe, F. S., S. R. Olsen, and R. E. Danielson, "P availability as related to soil moisture." *Trans. 7th Int. Congr. Soil Sci.*, 3:450 (1960).

Yao, A. Y. M., and R. H. Shaw, "Effect of plant population and planting pattern of corn on water use and yield." *Agron. J.*, **56**:147 (1964).

Chapter 17

Interactions of Plant Nutrients in a High-Yield Agriculture

GEORGE W. Cooke, former Chief Scientific Officer of Britain's Agricultural Research Council, has said: "In a highly developed agriculture, large increases in yield potential will mostly come from interaction effects. Farmers must be ready to test all new advances that may raise yield potentials of their crops and be prepared to try combinations of two or more practices."

Importance of Interactions

An interaction occurs when the response of one or a series of factors is modified by the effect of one or more other factors. A positive interaction occurs when the response to two or more inputs used together is greater than the sum of their individual responses. There is an add-on effect.

There are many types of interactions, including those between:
Two or more nutrients.
Nutrients and a cultural practice such as planting date, placement, tillage, plant population, or pest control.
Nutrient rate and hybrid or variety.
Hybrid or varieties and row widths or plant population.

Many interactions vital to high yields are not of significance with average yields. This is why there has not been too much concern about them even though balanced fertility and limiting factors have been emphasized. However, with higher and higher yields, increasing pressure or stress is being placed on the plant and on the various practices or factors contributing to those yields. Thus, in the future, interactions and their recognition will be the key to significant progress toward maximum yields in research and maximum profit yields for farmers.

The effect of a plant-nutrient application is often difficult to separate from the other agronomic improvements that make up the package of production practices, which include variety, pest control, tillage, timeliness, water management, plant spacing, and so on, due to the numerous positive interactions between fertility and other inputs.

This is illustrated in Table 17-1. Fertilizer contributed most to the increase in yields of wheat and teff. However, in the case of corn, introduction of an improved variety in the recommended practice gave a marked yield increase.

TABLE 17-1.   Yield Improvement from Fertilization and Management in Ethiopia

Crop	Yield Increase over Check (%)		
	Farmer Practice + Fert.	*Recommended Practice − Fert.*	*Recommended Practice + Fert.*
Wheat	64	15	91
Corn	81	80	156
Teff	84	2	94

Source: Mathieu, *Progress Report. Third Consultation on the FAO Fertilizer Programme.* Rome: FAO, 1978.

The interaction between response to fertilizer and various management levels is shown schematically in Figure 17-1. Additions B, C, and D represent the introduction of improved management practices. As improved practices are implemented, yields are increased and larger quantities of fertilizer can be used more effectively. A fundamental objective of the overall agronomic effort must be to lead farmers from one phase of improved practices to the next.

The potential exists for greatly increasing agricultural output through improvement and integration of the numerous components of yield. Agronomists, through close collaboration with their colleagues in associated and related disciplines, can make tremendous contributions to this global task. The lessons and experiences of the past, particularly during the preceding 25 years, will be invaluable in designing and implementing the exceptional research and development efforts needed.

General Carlos P. Romulo in *Strategy for the Conquest of Hunger* (1968) stated the challenge concisely:

> The test of technology is yield. Unless scientists can produce on their experiment stations yields which are high by world standards, we can be sure that we have not yet mastered the technical problems confronting the producer. If the yields

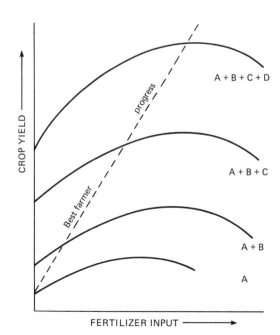

FIGURE 17-1. Relation between adoption of management practices, B, C, and D, and response to applied fertilizer.

are high on the experiment stations but national average yields are low, we can be equally sure that the scientific advances are on the shelf and that for some reason the farmer is unable to make profitable use of them.

The maximum yield concept is directly related to fertilizer efficiency. The higher-yielding crops, with more extensive root systems and with a greater nutrient need, will take up more of the applied nutrients.Various types of interactions will be considered in the following sections.

Interactions Between Nutrients. A common interaction is between potassium and nitrogen. It may be difficult to obtain a response to potassium under low-yield conditions if other nutrients are limiting or if management practices are inadequate. Under such conditions growth is slow, and unless potassium is seriously limiting, some soils will release potassium at a rate adequate to meet the needs of the crop. With adequate nitrogen and phosphorus and improved management practices, there is more rapid growth and the potential for response to potassium, sulfur, and the other nutrients is improved.

This is illustrated for potassium in Figure 17-2. With 30 kg/ha of nitrogen there was little response by rice to potassium and the potassium soil-test value would lead to erroneous conclusions. However, when 90 kgN/ha of nitrogen was applied, the response to potassium was practically linear up to the highest rate applied.

Negative interactions with micronutrients can become severe when aiming for high yields. On a soil in Kansas leveled for irrigation, low in organic matter, phosphorus, and zinc, adding phosphorus or zinc separately decreased corn yields (Table 17-2). When both were applied a substantial positive interaction occurred, increasing the yield by 44 bu/A.

The effect of soil pH on response of corn to P_2O_5 banded beside the row is shown in Figure 17-3. With a pH of 6.1 there was little response to phosphorus, but with a pH of 5.1 there was about a 20-bushel response to 70 lb/A of P_2O_5. Also, liming alone increased yields substantially. It is important that the liming program not be neglected in the total crop production program.

Interactions Between Nutrients and Plant Population. Increasing plant population may not give full return unless there is an adequate quantity of avail-

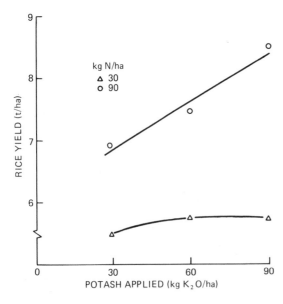

FIGURE 17-2. Nitrogen level affects response of rice to potash. (Malavolta, *Nutrição mineral e adubação de arroz irrigado*. São Paulo, Brazil: Ultrafertil S.A., 1978.)

TABLE 17-2.   Negative Responses on Corn Turned into Positive Interactions

P_2O_5 (lb/A)	Zn (lb/A)	Yield (bu/A)
0	0	131
80	0	119
0	20	109
80	20	175

Source: Ellis, *Kansas Fertilizer Handbook,* Dept. of Agronomy, Kansas State Univ. (1967).

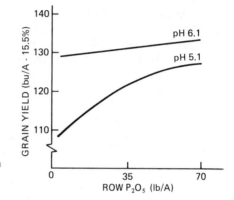

FIGURE 17-3.   Soil pH and row P_2O_5 interact on corn. [Schulte, *Better Crops Plant Food,* **66**:10 (1982).]

able plant nutrients. Similarly, increasing the plant nutrients without a sufficient number of plants will not give a greater return. In Table 17-3 increasing the plant population with 80 lb/A of nitrogen only increased corn yield 46 bu/A. With 240 lb of nitrogen, increasing the plant population increased the yield 76 bushels. At 12,000 plants per acre, increasing the nitrogen to 240 lb/A gave an increase of only 37 bu/A, but with 36,000 plants the increase was 67 bu/A, to a total of 231 bu/A.

In California, with a low plant population (18,000), 200 lb/A of nitrogen increased corn yield 15 bushels. However, with 34,000 the increase was 65 bushels.

Interactions Between Plant Population and Planting Date.   Plant population interacts with planting date. Plants tend to be shorter with earlier planting

TABLE 17-3.   Interaction of Nitrogen and Plant Population to Increase Corn Yields (bu/A)

Corn (plants/A)	Nitrogen (lb/A)		
	80	160	240
12,000	118	138	155
24,000	151	178	202
36,000	164	210	231

Source: Rhoads, Florida Agr. Exp. Sta., *Quincy Res. Rep. 78-1* (1978).

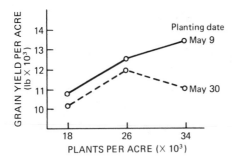

FIGURE 17-4. Planting date and population interact on corn. [Arjal et al., *California Agriculture,* Univ. of California (March 1978).]

and a higher population can be utilized (Figure 17-4). With the later planting date there was a decrease in corn yield. Plants are taller with the May 30 planting date and competition for light at this date would be expected to be higher with higher plant population.

Interactions Between Nutrients and Planting Date. Planting date has a marked effect on response to nutrients. Earlier planting dates for spring-planted crops result in higher yields. Note the greater response of soybeans to increased potassium soil-test level with earlier planting, 13 bushels versus 6 bushels (Table 17-4). Also, potassium level had a greater effect on increasing seed size and on decreasing seed disease.

TABLE 17-4. **Effect of the Planting Date on the Response of Soybeans to Potassium**

Planting Date	Yield (bu/A) at Soil K Level:			Increase (bu/A)
	Low	Medium	High	
May 27	40	47	53	13
June 16	40	44	46	6
July 8	31	36	37	6

Source: Peaslee, Univ. of Kentucky, personal communication.

In Montana, with hard red spring wheat, planting on May 5 gave a 12-bu/A response to phosphorus, while planting June 2 gave only a 3-bu/A response.

In California, planting corn on May 9 gave an 89-bu/A response to 200 lb of nitrogen, while planting on May 30 gave only a 48-bu/A response.

Interactions Between Variety and Row Width. Varieties or hybrids may vary in their response to plant spacing. Note in Table 17-5 that soybean

TABLE 17-5. **Effect of Row Width and Soybean Varieties on Yield**

Variety	Yield (bu/A) at Row Width of:		Increase (bu/A)
	30 in.	7 in.	
A	58	68	10
B	66	79	13
C	63	83	20

Source: R. L. Cooper, Ohio State Univ., personal communication.

variety A gave a 10-bu/A response to 7-in. rows over 30-in. rows, while variety C gave a 20-bu/A response, up to 83 bu/A.

Interactions Between Nutrients and Placement. Sometimes it is not what is done but how it is done that influences response to nutrients. Placement can cause marked yield differences. This was illustrated in Chapter 13 when the effects of dual placement of NH_3-N and phosphorus were shown to have a marked effect on wheat yields.

The effect of placement on response of soybeans to phosphorus is shown in Figure 17-5. At the highest rate, broadcasting the phosphorus gave only a 4-bu/A response, but drilling below the seed gave a 28-bu/A response. This may be due in part to moisture relationships. Under dry conditions broadcast phosphorus would be in the dry surface soil layer and positionally unavailable to the roots. When the phosphorus is drilled below the seed, the probability is greater that the phosphorus will be in moist soil and more available to the roots.

Interactions Between Nutrient Placement and Conservation Tillage. Soil and water conservation practices to reduce soil erosion have become a national priority. Conservation tillage includes any tillage system that reduces the number of trips over a field and increases surface residues. It ranges from a single disking to chiseling to ridge planting to no-till. Building up soil-nutrient levels through proper liming and fertilizer applications before initiating conservation tillage, if possible, is important. Then monitoring these levels through soil tests is essential.

Surface accumulations of phosphorus and potassium, surface accumulations of residues, and cooler temperatures and higher moisture in the spring often dictate a change in fertilizer use. In some situations there appears to be a need for higher levels of nutrients in the soil, particularly for potassium, and an increased need for fertilizer banded at planting.

Note the greater response of corn to 40-40-40 banded beside the row where the soil was not plowed (Figure 17-6). The response was 42 bu/A compared to 16 bu/A where the soil was plowed. Even with the higher amounts broadcast, the responses were 20 and 9 bushels, respectively.

In Alabama, on a soil very high in phosphorus, 10-34-0 banded near the row at planting increased grain sorghum yields 7 bu/A under conventional tillage (moldboard plowing) but 16 bu/A under no-till.

It has been found in some situations that higher rates of nitrogen and perhaps sulfur are required under no-till systems than under conventional tillage. This is

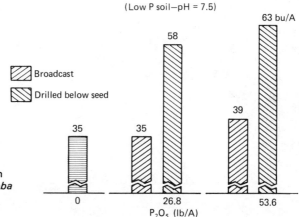

FIGURE 17-5. Phosphorus rates and placement interact to increase soybean yields. [L. D. Bailey, 21st *Annu. Manitoba Soil Sci. Meet.,* pp. 196–200, Univ. of Manitoba (1977).]

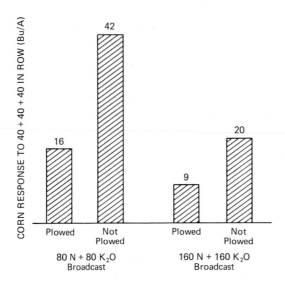

FIGURE 17-6. The response to fertilizer banded beside the row at planting was greater in the unplowed areas. [Schulte, *Better Crops Plant Food,* **63**:25 (1979).]

shown for nitrogen in Figure 17-7. Ammonium nitrate was surface-applied. The surface residue mulch helps to maintain a more moist condition under no-till. This may increase microbial populations and could potentially lead to greater immobilization of surface-applied nitrogen and sulfur and faster nitrification and denitrification.

There are several points to consider:

Subsurface placement of nitrogen could avoid some of these losses.

After a period of years in no-till, the nitrogen–microbial activity–residues system could come to somewhat of an equilibrium and differences on nitrogen needs between no-till and conventional diminish.

Under some conditions the yield potential is greater under no-till and this would require more nitrogen.

On higher-organic-matter soils, increased tillage increases the nitrogen mineralization.

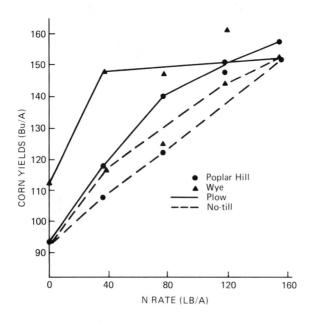

FIGURE 17-7. Tillage and nitrogen rates interact on corn yields. [Bandel et al., *Agron. J.,* **75**:782 (1975).]

Interactions Between Nutrients and Hybrid or Variety. Within a given environment, one hybrid or variety may produce a greater response to applied nutrients than another hybrid or variety.

In Figure 17-8 the Dare soybean variety produced a higher yield and responded more to K_2O than did the Bragg variety on a very low potassium soil.

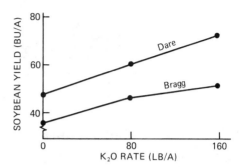

FIGURE 17-8. Soybean varieties and response to potassium interact. [Terman, *Agron. J.*, **69**:234 (1977).]

Some corn hybrids are genetically able to produce much greater yields from higher rates of applied nutrients than others. Table 17-6 shows a striking example of fertility–hybrid interaction. At the lower fertility level the corn hybrids differed by 19 bu/A. At the higher fertility level there was an 85-bu/A difference between the hybrids. Hybrid A did not have the genetic capability to utilize the higher fertility level.

When a farmer or researcher is aiming for higher and higher yields, selection of the hybrid or variety that will respond to a high-yield environment is the key to attaining these yields. Since there is little information on interaction, two or three hybrids or varieties are often used in field trials to compare responses.

Other Interactions of Nutrients. Many other types of nutrient interactions with production factors could be mentioned. Moisture was covered in Chapter 16. Temperature is also important. Low temperatures reduce the amounts of nitrogen, phosphorus, and sulfur mineralized from organic matter. Also, low temperatures reduce diffusion rates and hence nutrient movement to roots. Therefore, response to nutrients may be much greater under cool conditions.

The effect of interactions of nutrients on crop quality can be marked. Increased protein, phosphorus and potassium can occur in grains or forages. Addition of sulfur can narrow the N/S ratio in forages and improve their nutritional value. On low-potassium soils added potassium for soybeans can decrease mold and disease in the seed. Added potassium can increase percent plump kernels in barley.

Nutrient–pesticide interactions occur. This is illustrated in Table 17-7 on a low-potassium soil. A fungicide increased soybean yield 6 bu/A and potassium increased yield 3 bu/A. Both together increased the yield 10 bu/A. Percent germination of the seed produced also improved. Soybean leaf diseases were decreased by the application of both fungicide and potassium.

TABLE 17-6.   Interaction of Fertility and Corn Hybrids to Increase Yields

N (lb/A)	P_2O_5 (lb/A)	K_2O (lb/A)	Yield (bu/A) of: Hybrid A	Hybrid B	Increase (bu/A)
250	125	125	199	218	19
500	300	300	227	312	85

Source: R. L. Flannery, New Jersey Agricultural Experiment Station, personal communication.

TABLE 17-7. Interaction of a Fungicide and Potassium to Increase Soybean Yields and Germination

	Yield (bu/A)	Germination (%)
Control	35	84
Fungicide	41	90
Potassium (K)	38	95
Fungicide + K	45	96

Source: A. Y. Chambers, Department of Agricultural Biology, University of Tennessee, Knoxville, Tenn., personal communication.

Explaining Interactions

Many interactions have been observed in field experiments. However, there has been little attempt to explain the reasons. What happens is known but not why.

M. E. Sumner has made extensive studies on the DRIS system mentioned in earlier chapters. He has found this system to be superior to the critical-level approach. The inability to deal adequately with the variation in nutrient concentration on a dry-matter basis with age is probably the greatest disadvantage of the critical-level approach.

The DRIS system takes into account nutrient balance and establishes the relative order of nutrient requirement by the crop. The recognition of the importance of nutrient balance in crop production is an indirect reflection of the contribution of interactions to yield. The highest yields are obtained where nutrients and other growth factors are in a favorable state of balance.

To help explain interactions there is a real need for multifactorial field experiments, that is, experiments in which several factors are tested in all combinations. Supporting plant and soil analyses data are needed in the interpretations. The DRIS system for interpreting plant analysis has the ability to handle the interactions of many factors simultaneously. Norms in the form of ratios have been established. Values outside the normal range reflect imbalances.

Conservation tillage and increased emphasis on banding of fertilizer increases the problem of obtaining a representative soil sample. Hence plant analysis will assume greater importance in evaluating the nutrient status of the soil and the plant. The DRIS system is a realistic approach to using plant analysis.

Building Maximum-Yield Systems

The importance of exploiting interactions in building maximum-yield systems cannot be overemphasized. When one practice or group of practices increases the yield potential, the nutrient requirement will be increased.

R. L. Flannery in New Jersey started his maximum-yield research in 1980 by setting research yield goals of 300 bushels of corn and 100 bushels of soybeans per acre. His results for 1980–1983, shown in Table 17-8, speak for themselves.

D. W. Dibb interprets these results this way:

> The importance of this accomplishment is not in the fact that any major, new scientific technology was developed; in fact, Dr. Flannery just put together current technology that was available. The significance lies in the fact that the technology was put together in a way that allowed the components to interact

TABLE 17-8.   Building Maximum-Corn-Yield Systems, 1980–1983

	Treatment Yield (bu/A)	
	Soybeans	Corn
1980	94	312
1981	93	285
1982	109	338
1983	118	284
4-year average	103	306

Source: R. L. Flannery, New Jersey Agricultural Experiment Station, personal communication.

TABLE 17-9.   Increased Corn Residues and Water Efficiency with Maximum-Yield Management

	Corn Yield (bu/A)	Residue Returned (T/A)	Water-Use Efficiency (bu/in.)
Medium fertility, irrigated	214	5.0	8.3
Very high fertility, irrigated	299	6.4	11.6

Source: R. L. Flannery, New Jersey Agricultural Experiment Station, personal communication.

positively in consistently high corn and soybean yields. The next challenge in this maximum yield research is to identify, characterize and learn to manage the specific positive interactions, and then further refine the system for even higher yields.

Flannery states: "The biggest lesson I have learned from maximum yield research is this—don't leave anything to chance."

Just as in other significant research, spin-offs develop in maximum-yield research (Table 17-9). The higher yields returned more residues to the soil, which, properly managed, should decrease erosion potential and increase infiltration rates. In other words, this is a soil-building process resulting from higher yields with proper management.

Also, water efficiency is improved as yields increase. In many areas of the world, water is a major concern in production. Positive interactions that improve water use efficiency while increasing yields are of great importance.

This maximum-yield research is not an isolated example. Similar successes in maximum-yield research programs on soybeans, wheat, alfalfa, and other crops are also emerging.

Certainly, Cooke's statement quoted earlier that "large increases in yield potential will mostly come from interaction effects" is being substantiated in these maximum-yield research results. Also, as breakthroughs occur in genetic engineering, rhizosphere technology, plant growth regulators, and related areas, they will only be successful as the technology is integrated in a manner that allows positive interactions to be expressed.

Summary

1. A positive interaction occurs when the response to two or more inputs used together is greater than the sum of their individual responses.

2. Nutrients may interact with each other and with various cultural practices, such as hybrid or varieties, plant spacing, planting date, placement, tillage, and pest control.

3. Interactions are more important at higher yield levels because more stress is placed on the plant and the various growth factors.

4. As improved practices are adopted and yields are increased, larger quantities of nutrients can be used more effectively.

5. The maximum-yield concept influences fertilizer efficiency. The large crop with a more extensive root system and greater need for nutrients takes up more of the applied nutrients.

6. It is impossible to obtain full response to a given nutrient if other nutrients or management practices are limiting and yields are low.

7. Increasing the plant population or narrowing the row width will not give full return if plant nutrients are lacking.

8. Proper planting date helps ensure more adequate returns from increased plant population and plant nutrients. Yield potential is increased and more nutrients are required.

9. Deeper placement of plant nutrients, particularly in areas subject to drought stress, is more likely to give a greater response than placement at or near the surface. Deeper placement improves the probability that the nutrient will be in moist soil and more available to plant roots.

10. With conservation tillage and the accompanying surface accumulations of phosphorus, potassium, and residues, and cooler temperature and higher moisture in the spring, there is often a need for a change in amount and placement of fertilizer. In some situations there may be a need for higher nutrient levels in the soil and an increased need for starter fertilizer.

11. Higher rates of nitrogen may be required in no-till, at least initially, because the surface residue helps maintain a more moist condition. This may lead to increased microbial population, greater immobilization of surface applied nitrogen and faster nitrification and denitrification. Subsurface application of nitrogen could help avoid some of the losses.

12. There may be a distinct difference among hybrids or varieties in response to applied nutrients. Because the extent of these interactions is usually not known, several hybrids or varieties are often compared at various management levels by the researcher and the farmer.

13. Responses to nutrients may be much greater under cool temperatures because mineralization of certain nutrients from organic matter will be reduced and/or diffusion rates of nutrients are decreased.

14. Nutrient–pesticide interactions occur. For example, inadequate plant nutrition favors certain plant or seed diseases. Added nutrients will decrease disease, as will a fungicide or a nematacide. However, both together may have an even greater effect.

15. The significance of Flannery's maximum-yield research is that the technology was put together in a way that allowed the components to interact positively in consistently high corn and soybean yields. The next step is to identify, characterize, and learn to manage the specific positive interactions and then further refine the system for even higher yields.

16. A spin-off from Flannery's research is the greater amount of residue, which when properly managed, should decrease erosion and increase water infiltration rate. Water efficiency was also increased. Positive interactions that improve water use efficiency are of importance in most areas.

17. Efficiency in the use of all inputs is vital. As yields are moved up, inputs such as land, labor, machinery, fuel, seed, and pesticides are used more efficiently. Proper fertilizer use is an important part of an efficient crop production system.

Questions

1. What is a positive interaction? Illustrate.

2. Why are interactions more critical at higher yield levels?

3. Why do higher yields improve fertilizer efficiency?

4. Explain why it is impossible to obtain full response from an applied nutrient if the level of another nutrient is inadequate.

5. How does plant nutrient supply affect response to increased plant population? Why?

6. Explain why deeper placement of plant nutrients is likely to give a greater response than shallow placement in some areas.

7. Why does planting date affect response to fertilizer?

8. Explain why conservation tillage often causes the need for a change in fertilizer use.

9. Why might higher rates of nitrogen be required for no-till as compared to conventional tillage (moldboard plow)?

10. What is the best way to determine how hybrids compare in a high-yield environment?

11. Name two reasons why responses to nutrients are usually greater under cool conditions.

12. Explain why there might be a fertilizer–pesticide interaction.

13. In addition to the very high corn and soybean yields obtained by Flannery, what was the most significant accomplishment?

14. Why would higher yields help decrease erosion and increase water infiltration?

15. How would you go about setting up a maximum-yield-system area on a farmer's field? Whose help would you enlist?

16. Explain the various aspects of efficient fertilizer use as related to Figure 17-1.

17. How do fertilizer and management enter into the efficient use of water?

Selected References

Bailey, L. D., "Nutrient requirements of soybeans." *21st Ann. Manitoba Soil Sci. Meet.*, pp. 196–200, Univ. of Manitoba (1977).

Bandel, V. A., S. Dziema, G. Stanford, and J. O. Legg, "N behavior under no-till vs conventional corn culture." *Agron. J.*, **75**:782 (1975).

Cooke, G. W., *Fertilizing for Maximum Yields*, 3rd ed. New York: Macmillan, 1982.

Dibb, D. W., "Agronomic systems to feed the next generation." *Agron. J.*, **75**:413 (1983).

Ellis, R., *Kansas Fertilizer Handbook*, p. 68. Dept. of Agronomy, Kansas State Univ. (1967).

International Potash Institute, Series of papers on maximizing yields. "Agricultural yield potentials in continental climates." *16th Colloq.*, International Potash Institute, Bern, (1981).

Jones, G. D., J. A. Lutz, and T. J. Smith, "Effects of P and K on number, size, and chemical composition of soybean nodules and seed yield." *Agron. J.*, **69**:1003 (1977).

Malavolta, E., *Nutrição Mineral e Adubação de Arroz Irrigado*. São Paulo, Brazil: Ultrafertil S.A., 1978.

Mathieu, M., *Progress Report. Third Consultation on the FAO Fertilizer Programme*. Rome: Food and Agriculture Organization of the United Nations, 1978.

Munson, R. D., "Efficiency of uptake of P and interactions of P with other elements," in *Phosphorus for Agriculture—A Situation Analysis*. Atlanta, Ga.: Potash & Phosphate Institute, 1978.

Munson, R. D., "Potassium, calcium and magnesium in the tropics and subtropics." *IFDC Tech. Bull. T-23* (1982).

Murphy, L. S., "Potassium interactions with other elements," in *Potassium for Agriculture—A Situation Analysis*. Atlanta, Ga.: Potash & Phosphate Institute, 1980.

Nelson, W. L., "A 'Blueprint' for maximizing yields of soybeans. Agricultural yield potentials in continental climates." *Proc. 16th Colloq.*, International Potash Institute, Bern (1981).

Potash & Phosphate Institute. "Positive interactions and maximum economic yields," in *Maximum Economic Yield Manual—A Guide to Profitable Production*, Chapter 4. Atlanta, Ga., 1982.

Pretty, K. M., "The world-wide efforts of professional agronomists." *J. Agr. Ed.*, **9**:105 (1980).

Rhoads, F. M., "Water and nutrient management for maximum corn yields in north Florida." Florida Agr. Exp. Sta., *Quincy Res. Rep. 78-1* (1978).

Romulo, C. P., "Strategy for conquest of hunger," in *Strategy for the Conquest of Hunger* (Proc. Symp.). New York: Rockefeller Foundation, 1968.

Schulte, E. E., "Buildup soil K levels before shifting to minimum tillage." *Better Crops Plant Food*, **63**:25 (Fall 1979).

Schulte, E. E., "Corn response to row phosphate." *Better Crops Plant Food*, **66**:10 (Summer 1982).

Skogley, E. O., "Potassium in Montana soils." *Montana Agr. Exp. Sta. Res. Rep. 88* (1976).

Sumner, M. E., and P. M. W. Farina, "Phosphorus interactions with other nutrients and lime in field cropping systems," in J. K. Syers, Ed., *Phosphorus in Agricultural Systems*. New York: Elsevier (in press).

Terman, G. L., "Yields and nutrient accumulation by determinant soybeans as affected by applied nutrients." *Agron. J.*, **69**:234 (1977).

Touchton, J. T., and W. L. Hargrove, "Grain sorghum response to starter fertilizers." *Better Crops Plant Food*, **67**:3 (Spring 1983).

Wagner, R. E., "Interactions of plant nutrients in a high yield agriculture." *Spec. Bull. 1*. Atlanta, Ga.: Potash & Phosphate Institute, 1980.

Welch, C. D., C. Gray, and J. N. Pratt, "P and K fertilization for Coastal bermudagrass hay production in E. Texas." *Texas A & M Fact Sheet L-1861* (1981).

Appendix

Common Conversions and Constants

Conversion Factors for English and Metric Units

To Convert Column 1 Into Column 2, Multiply by:	Column 1	Column 2	To Convert Column 2 Into Column 1, Multiply by:
	Length		
0.621	kilometer, km	mile, mi	1.609
1.094	meter, m	yard, yd	0.914
0.394	centimeter, cm	inch, in	2.54
	Area		
0.386	kilometer^2, km^2	mile^2, mi^2	2.590
247.1	kilometer^2, km^2	acre, acre	0.00405
2.471	hectare, ha	acre, acre	0.405
	Volume		
0.00973	meter^3, m^3	acre-inch	102.8
3.532	hectoliter, hl	cubic foot, cu ft^3	0.2832
2.838	hectoliter, hl	bushel, bu	0.352
0.0284	liter, l	bushel, bu	35.24
1.057	liter, l	quart (liquid), qt	0.946
	Mass		
1.102	ton (metric)	ton (English)	0.9072
2.205	quintal, q	hundredweight cwt (short)	0.454
2.205	kilogram, kg	pound, lb	0.454
0.035	gram, gm	ounce (avdp.), oz	28.35
	Pressure		
14.50	bar	lb/inch^2, psi	0.06895
0.9869	bar	atmosphere, atm	1.013
0.9678	kg (weight)/cm^2	atmosphere, atm	1.033
14.22	kg (weight)/cm^2	lb/inch^2, psi	0.07031
14.70	atmosphere, atm	lb/inch^2, psi	0.06805
0.1450	kilopascal	lb/inch^2, psi	6.895
0.009869	kilopascal	atmosphere, atm	101.30
	Yield or Rate		
0.446	ton(metric)/hectare	ton (English)/acre	2.240
0.891	kg/ha	lb/acre	1.12
0.891	quintal/hectare	hundredweight/acre	1.12
1.15	hectoliter/hectare	bu/acre	0.87

Convenient Conversion Factors

Multiply	By	To Get
acres	0.4048	hectare
acres	43,560	square feet
acres	160	square rods
acres	4,840	square yards
bushels	4	pecks
bushels	64	pints
bushels	32	quarts
centimeters	0.3937	inches
centimeters	0.01	meters
cubic feet	0.03382	ounces (liquid)
cubic feet	1,728	cubic inches
cubic feet	0.03704	cubic yards
cubic feet	7.4805	gallons
	29.92	quarts (liquid)
cubic yards	27	cubic feet
cubic yards	46,656	cubic inches
cubic yards	202	gallons
feet	30.48	centimeters
feet	12	inches
feet	0.3048	meters
feet	0.060606	rods
feet	$^1/_3$ or 0.33333	yards
feet	0.01136	miles per hour
gallons	0.1337	cubic feet
gallons	4	quarts (liquid)
gallons of water	8.3453	pounds of water
grams	15.43	grains
grams	0.001	kilograms
grams	1,000	milligrams
grams	0.0353	ounces
grams per liter	1,000	parts per million
hectares	2.471	acres
inches	2.54	centimeters
inches	0.08333	feet
kilograms	1,000	grams
kilograms	2.205	pounds
kilograms per hectare	0.892	pounds per acre
kilometers	3,281	feet
kilometers	0.6214	miles
liters	1,000	cubic centimeters
liters	0.0353	cubic feet
liters	61.02	cubic inches
liters	0.2642	gallons
liters	1.057	quarts (liquid)
meters	100	centimeters
meters	3.2181	feet
meters	39.37	inches
miles	5,280	feet
miles	63,360	inches
miles	320	rods
miles	1,760	yards
miles per hour	88	feet per minute
miles per hour	1.467	feet per second
miles per minute	60	miles per hour
ounces (dry)	0.0625	pounds
ounces (liquid)	0.0625	pints (liquid)
ounces (liquid)	0.03125	quarts (liquid)
parts per million	8.345	pounds per million gallons water
pecks	16	pints (dry)
pecks	8	quarts (dry)
pints (dry)	0.5	quarts (dry)
pints (liquid)	16	ounces (liquid)
pounds	453.5924	grams

Convenient Conversion Factors

Multiply	By	To Get
pounds	16	ounces
pounds of water	0.1198	gallons
quarts (liquid)	0.9463	liters
quarts (liquid)	32	ounces (liquid)
quarts (liquid)	2	pints (liquid)
rods	16.5	feet
rods	5.5	yards
square feet	144	square inches
square feet	0.11111	square yards
square inches	0.00694	square feet
square miles	640	acres
square miles	27,878,400	square feet
square rods	0.00625	acres
square rods	272.25	square feet
square yards	0.0002066	acres
square yards	9	square feet
square yards	1,296	square inches
temperature (°C) + 17.98	⅘ or 1.8	temperature, °F
temperature (°F) − 32	⅘ or 0.5555	temperature, °C
ton	907.1849	kilograms
ton	2,000	pounds
ton, metric	2,240	pounds
yards	3	feet
yards	36	inches
yards	0.9144	meters

Western Fertilizer Handbook, 6th ed. Produced by the Soil Improvement Committee, Cal Fertilizer Association. Danville, Ill.: Interstate Printers and Publishers, 1980.

Weight of Grain per Bushel

Crop	Weight (lb)
Barley	48
Canola/rapeseed	50
Corn	56
Flax	56
Oats	32
Rye	56
Sorghum	56
Soybeans	60
Wheat	60

Index